'Really comprehensive and interesting. A profoundly important book - a genuinely original view of how we might overcome some of the most fundamental problems facing effective climate change policy around the world today.'

Lord Anthony Giddens, *Fellow of Kings College, Cambridge and former Director of the London School of Economics*

'A seminal book that challenges conventional wisdom about growth, innovation and climate policy. After reading it you understand why the tools that mainstream economics has provided policymakers have been unequal to the task. In filling out the picture, *Planetary Economics* contains a crucial practical message: the fact that energy is at the core of the economic machine is both the challenge and the opportunity, and three pillars of policy together are needed to steer it in safer directions.'

Laurence Tubiana *Former Senior Advisor to the French Prime Minister and Director of Global Public Goods at Ministry Of Foreign Affairs; currently Professor at* Sciences Po *and* Columbia University*, and President of the Board of Governors of the French Agency for Development (AFD)*

'The book is compulsory reading for policymakers and academics for understanding the broader challenges of environmental change. What makes the book such an outstanding contribution is the way it brings together the fields of energy, environment, innovation, behavioural economics and macroeconomics. Its key policy message is a timely call for policymakers to act decisively, so that our societies can have the confidence to invest and innovate in solving the great environmental challenges of our time.'

Prof. Marcel Fratzscher, *President of German Institute for Economic Research (DIW Berlin)*

'This important book sets out a clear and comprehensive theory of how transition can be achieved and a convincing and well evidenced argument that it will be, if only we apply the policy tools and approaches available. Proposed policies for building a low carbon economy often reflect very different ways of thinking: economists favouring markets and carbon pricing, scientists and technologists a vision of transformative innovation (fearing the short-term bias of markets and cost benefit analysis), whilst some environmentalists seek to engage people's enthusiasm to achieve changes in personal behaviour. Michael Grubb argues cogently that a successful strategy must combine all three approaches, and illustrates their respective roles and complementary nature.'

Lord Adair Turner, *former Director-General of Confederation of British Industry and Vice-Chairman of Merrill Lynch Europe; from 2008-12, Chair of UK Financial Services Authority and UK Climate Change Committee; currently Senior Fellow, Institute of New Economic Thought*

'The defining features of climate change – long-time horizons, global reach, and uncertain but possibly catastrophic impacts – have stymied policymakers and strained many of the standard tools of economics. In *Planetary Economics*, Grubb and colleagues document and then tackle these characteristics head on. They organize the debate, fill in gaps, and present a unified framework for thinking about the problem, the best policy solutions, and how they need to fit together. This is an important read for anyone looking for a sensible and comprehensive way forward.'

Billy Pizer, *Associate Professor of Public Policy at Duke University and former Deputy Assistant Secretary for Environment and Energy, US Treasury*

'Combines extraordinary breadth with depth, all written with exceptional clarity. In explaining why energy and climate change take us beyond the traditional boundaries of economics, it is a landmark study which should expand the horizons of economics itself – as well as mapping out how, in doing so, we can solve some of the most pressing problems of our time.'

Catherine Mitchell, *Professor of Energy Policy, Exeter University*

'An ambitious book that looks at nothing less than the future of human and planetary well being. It is smart, provocative, and unconventional at every turn. Grubb shows that the world's energy and economic systems are on the wrong track. Profoundly new thinking, including policies, will be needed to transform these systems and put the planet on track to much lower and safer levels of global warming pollution. The elements of that new thinking—such as smarter innovation policies, cleaner energy networks, and better values—are all around us today. Grubb's big contribution is to stitch them together and chart a path for change.'

David Victor, *Professor and Director of Laboratory on International Law and Regulation at University of California San Diego, and author Global Warming Gridlock and The Death of the Kyoto Protocol*

'A book of extraordinary scope and ambition that will challenge readers to think more clearly and carefully about some of the biggest issues this planet and its people face. Grubb has no ideological axe to grind and no particular allegiance to one scholarly school or another – instead he draws on a range of theoretical frames to shed as much light as possible on the dynamics governing our energy systems, all with an eye toward figuring out how to steer them in a safer and more secure direction. Anyone who works in this area will be provoked to think more wisely as a result of reading this book.'

Michael Levi, *Senior Fellow for Energy and the Environment, US Council on Foreign Relations*

Planetary Economics

How well do our assumptions about the global challenges of energy, environment and economic development fit the facts?

Energy prices have varied hugely between countries and over time, yet the share of national income spent on energy has remained surprisingly constant. The foundational theories of economic growth account for only about half the growth observed in practice. Despite escalating warnings for more than two decades about the planetary risks of rising greenhouse gas emissions, most governments have seemed powerless to change course.

Planetary Economics reveals the surprising links between these seemingly unconnected facts. It argues that tackling the energy and environmental problems of the twenty first century requires three different domains of decision-making to be recognised and connected. Each domain involves different theoretical foundations, draws on different areas of evidence, and implies different policies.

The book shows that the transformation of energy systems involves all three domains - and each is equally important. From them flow three pillars of policy – three quite distinct kinds of actions that need to be taken, which rest on fundamentally different principles. Any pillar on its own will fail.

Only by understanding all three, and fitting them together, do we have any hope of changing course. And if we do, the oft-assumed conflict between economy and the environment dissolves – with potential for benefits to both. *Planetary Economics* charts how.

Michael Grubb is Senior Researcher and Chair of Energy and Climate Policy at the Cambridge University Centre for Climate Change Mitigation Research UK, and Senior Advisor on Sustainable Energy Policy to the UK Energy Regulator Ofgem. He is also Editor-in-Chief of the journal *Climate Policy*. His former positions include Chair of the international research organization Climate Strategies, Chief Economist at the Carbon Trust, Professor of Climate Change and Energy Policy at Imperial College London, UK, and Head of Energy and Environment at Chatham House.

Jean-Charles Hourcade is Research Director at the Centre National de la Recherche Scientifique (CNRS) and at the École des Hautes Études en Sciences Sociales, France. He is Professor at the École Nationale des Ponts et Chaussées (E.N.P.C.) and was formerly Director of the Centre International de Recherche sur l'Environnement et le Développement (CIRED). He has advised the UNDP, UNEP, OECD, IEA and World Bank several times.

Karsten Neuhoff is Head of Climate Policy at the German economics research institute Deutsches Institut für Wirtschaftsforschung (DIW) and is Professor at the School of Economics and Management at the Technical University of Berlin, Germany.

Planetary Economics

Energy, climate change and the three domains of sustainable development

Michael Grubb
with Jean-Charles Hourcade and Karsten Neuhoff

LONDON AND NEW YORK

First published 2014
by Routledge
2 Park Square, Milton Park, Abingdon, Oxon OX14 4RN

and by Routledge
711 Third Avenue, New York, NY 10017

Routledge is an imprint of the Taylor & Francis Group, an informa business

British Library Cataloguing in Publication Data
A catalogue record for this book is available from the British Library

Library of Congress Cataloging-in-Publication Data
Grubb, Michael.
Planetary economics : energy, climate change and the three domains of sustainable development / Michael Grubb, with Jean-Charles Hourcade and Karsten Neuhoff.
pages cm
Includes bibliographical references and index.
1. Energy policy–Economic aspects. 2. Energy development–Economic aspects. 3. Energy development–Environmental aspects. 4. Climatic changes–Environmental aspects 5. Sustainable development. I. Hourcade, Jean Charles. II. Neuhoff, Karsten. III. Title.
HD9502.A2G78 2013
338.9'27–dc23
2013033929

ISBN: 978-0-415-51882-6 (hbk)
ISBN: 978-1-315-85768-8 (ebk)

Typeset in Times New Roman
by Deer Park Productions

To:

Jo, Leonia, Arthur and Tara

For their eternal love
and unending patience

and

to all those working to sustain humanity's development
on our finite planet

Contents

List of figures, tables and boxes xiv
Preface xix
Acknowledgements xxiii
List of abbreviations xxv

1 Trapped? 1

1.1 Introduction 1
1.2 The physical challenges – energy, resources and climate 2
1.3 Evidence and theory: energy and economy 14
1.4 Evidence and theory: costing the climate 20
1.5 Trapped? 29
Notes 33
References 39

2 The Three Domains 46

2.1 Three conceptions of risk and opportunity 47
2.2 Three fields of theory 53
2.3 Three economic processes 59
2.4 Three realms of opportunity 62
2.5 Three pillars of response 67
2.6 Domain alignments 68
2.7 Conclusions 72
Notes 74
References 77

PILLAR I
Standards and engagement for smarter choices 79

3 Energy and emissions: technologies and systems 81

3.1 The energy system 81
3.2 Smarter buildings: efficiency and heat supply 84

3.3 Cleaner production: manufacturing and materials 90
3.4 Purer power: electricity without carbon 97
3.5 Tackling transport: transforming vehicles and fuels 105
3.6 Resources 110
3.7 Smarter systems and scenarios 114
3.8 Conclusions 116
Notes 117
References 124

4 Why so wasteful? 129

4.1 Introduction 129
4.2 Evidence: trends and potentials 130
4.3 How much more? 133
4.4 Explanations: barriers to and drivers of change 135
4.5 Behavioural realities: individuals 142
4.6 Behavioural realities: organisations 147
4.7 Better buildings: a dynamic view of long-term potential 150
4.8 Conclusions 152
Notes 152
References 157

5 Tried and tested: four decades of energy-efficiency policy 160

5.1 Introduction 160
5.2 From exhortation to institutions 161
5.3 The end-use toolbox 165
5.4 Integration and application to buildings 178
5.5 A measured (but inadequately measured) success 182
5.6 Bounding rebound 186
5.7 Better buildings: prospects and trade-offs 188
5.8 Missing pieces 190
5.9 Conclusions: beyond the efficiency paradox 192
Notes 193
References 197

PILLAR II
Markets and prices for cleaner products and processes 203

6 Pricing pollution: of truth and taxes 207

6.1 Introduction 207
6.2 Why price matters 208
6.3 First steps first: reforming energy subsidies 211
6.4 Pricing for energy security: gasoline taxes 213
6.5 Pricing for pollution: the principles 215
6.6 Carbon costs, 'co-benefits' and 'double dividend' 217

6.7 Response options in the Second Domain 220
6.8 Tax versus cap-and-trade: which is more efficient? 223
6.9 Pricing pollution in practice 226
6.10 The politics of pricing 229
6.11 Conclusions 231
Notes 232
References 234

7 Cap-and-trade and offsets: from idea to practice 237

7.1 Introduction: slow train coming 237
7.2 Scope and coverage 238
7.3 Evolution of the EU Emissions Trading System: caps and prices 240
7.4 The EU Emissions Trading System: ten lessons 247
7.5 Investment, predictability and confidence 255
7.6 The Kyoto Mechanisms 263
7.7 The emerging global landscape 270
7.8 Conclusions 273
Notes 275
References 280

8 Who's hit? The distributional impacts of carbon pricing and how to handle them 284

8.1 Introduction: who pays? 284
8.2 Profit and loss with emissions trading 284
8.3 Industry impacts – winners, losers and movers 289
8.4 Tackling 'carbon leakage' – an evolutionary approach to production, consumption and trade 292
8.5 Consumer price impacts 296
8.6 Who's hurt (and what to do about it)? 298
8.7 Energy access and carbon pricing in developing countries 301
8.8 The false god of a global carbon price 302
8.9 Conclusion 303
Notes 305
References 308

PILLAR III
Strategic investment for innovation and infrastructure 311

9 Pushing further, pulling deeper: bridging the technology valley of death 315

9.1 Introduction 315
9.2 The glittering prizes 316
9.3 Energy research and development (R&D) – public and private 318

9.4 On learning 321
9.5 The innovation chain 324
9.6 When the chain is broken 326
9.7 The technology valley of death 331
9.8 Pushing further 334
9.9 Pulling deeper 338
9.10 Porter's kick 341
9.11 Industrial strategy revisited – risks, rewards and principles 342
9.12 Conclusion and synthesis: the multiple journeys 344
Notes 346
References 352

10 Transforming systems 356

10.1 Introduction: over the bridge? 356
10.2 No modest task 357
10.3 Systems: evolving, locked and opened 359
10.4 Cool cars in the Americas 364
10.5 Transforming European electricity 372
10.6 Urbanisation and Asia 378
10.7 Integrating transitions 383
10.8 Conclusions: the carbon divide 384
Notes 387
References 391

11 The dark matter of economic growth 396

11.1 Introduction: on transformation and economic prescriptions 396
11.2 Economic growth and the search for the Philosopher's Stone 398
11.3 Probing the past: energy in economic development 405
11.4 Framing the future: insights and blind spots from integrated assessment models 409
11.5 Changer de conversation *416*
11.6 Back to the future: industrial renaissance 419
11.7 Lifting depression: finance and investment 425
11.8 Conclusion 428
Notes 429
References 438

Integration and implications 445

12 Conclusions: changing course 447

12.1 Introduction 447
12.2 Three Domains: a short résumé 448
12.3 On optimality, security and the evolution of energy systems 452

12.4 Order from complexity: the economics of changing course 454
12.5 Not just nails 460
12.6 Only connect: integrating policy across the three pillars 464
12.7 Expanding horizons 468
12.8 Joint benefits 474
12.9 An application: energising European recovery 479
12.10 The Three Domains in economic development 483
12.11 Conclusions 485
Notes 488
References 493

Appendix: The importance of inertia and adaptability – a simple model 495

Index 506

Figures, tables and boxes

Figures

1.1 Energy trends by region and fuel, 1965–2012 3
1.2 Historical crude oil prices with GDP growth and consumption, 1960–2012 5
1.3 Coal and gas end-user prices, 1980–2012 8
1.4 Global trends of (a) atmospheric CO_2 concentration and (b) average surface temperature, 1880–2012 11
1.5 Trends in energy productivity and decarbonisation, historically and potentially required trends to 2030 and 2050 13
1.6 Trends in carbon 'efficiency', or intensity, by region and globally, 1980–2008 15
1.7 Per capita CO_2 emission trends in relation to wealth – trends of major countries, 1990–2008 16
1.8 The risk matrix: an assessment framework for evaluating the social cost of climate change 23
2.1 Three conceptions of risks and their application to climate change 48
2.2 Three fields of theory 54
2.3 Resources and economic outputs in the Three Domains 60
2.4 The three realms of opportunity 65
2.5 The three Pillars of Policy 69
2.6 Alignments within each domain 71
2.7 The fields of theory in the wider context 72
3.1 Energy and emission flows within the fossil fuel system 82
3.2 Final energy use in US buildings 87
3.3 Global energy use and CO_2 emissions in industry 91
3.4 Potentials and technologies for reducing CO_2 from steel blast furnaces 93
3.5 Estimated costs of carbon capture and storage in different applications 95
3.6 Opportunities in energy-intensive supply chains: from primary materials to products 97
3.7 Declining solar PV module costs 102
3.8 Projected cost of low-carbon electricity sources in Northern Europe 104
3.9 Global energy use in transport 106
3.10 Vehicle efficiency in different national standards and models 108
3.11 Projected demand and supply of liquid fuels 113
3.12 Global Energy Assessment Pathways for 2030 and 2050 116

4.1 Efficiency gains from improved insulation and heating in UK housing, 1970–2007 132
4.2 Global potential for apparently 'negative cost' emission reductions by 2030 133
4.3 Energy efficiency barriers and drivers 136
4.4 Proportion of Carbon Trust recommendations to UK business implemented: dependency on payback period 138
4.5 Behavioural dependence on intention, habits and facilitating conditions 146
4.6 Per capita energy consumption in buildings in different regions 151
5.1 (a) EU energy efficiency labels and (b) market share of appliances in Europe over time since introducing labels 168
5.2 Trends of electricity price and energy efficiency improvement (US refrigerators) 170
5.3 (a) Fuel economy trends of US light vehicles, 1975–2005 (b) Evolution of standards in different countries – historical relationship between vehicle ownership and GDP per capita 171
5.4 International adoption of corporate energy-efficiency regulations and incentives 175
5.5 Combined effect of different instruments for energy efficiency 179
5.6 Energy efficiency policies matrix for building sector 180
5.7 Long-term energy savings from efficiency improvements in all sectors 184
5.8 Global scenarios for buildings energy use under different influences 188
5.9 The better buildings dilemma: cheap (now) or deep? 189
5.10 The scope of consumer-driven emissions 190
5.11 Embodied energy in buildings 191
6.1 The most important diagram in energy economics 209
6.2 Energy subsidies by fuel in non-OECD countries, 2010 212
6.3 Automotive gasoline prices and taxation rates 214
6.4 Estimates of local environmental co-benefits in different countries 218
6.5 Abatement options in the Second Domain 221
6.6 Stylised benefits and costs of cutting emissions: implications for 'tax or cap'? 224
7.1 Coverage of the EU ETS 238
7.2 Evolution of EU ETS and CER prices 243
7.3 Factors affecting 2008 emissions ($GtCO_2e$) 245
7.4 Free allocation and auctioning in the EU ETS Phases I, II and III 246
7.5 Source of price instability in an emissions trading system 250
7.6 Allocation methodologies and efficiency 253
7.7 Impact of (a) banking and (b) floor on carbon price in ETS 259
7.8 Steadying mechanisms for emissions trading systems 263
7.9 Evolution of the International Carbon Market 265
7.10 Project volumes – from validation to issuance of Certified Emission Reductions 266
7.11 Existing and emerging carbon pricing systems 272
8.1 Profit and loss depend on the combination of free allocation and cost pass through to final products 286
8.2 Desirable and undesirable effects of carbon prices 287
8.3 Carbon cost impact, local premium and trade cost 288

8.4 Impact of carbon pricing on EU industry sectors and their share of the EU economy 290
8.5 Options for tackling leakage 293
8.6 Breakdown of average domestic gas and electricity bills in the UK, 2013 297
8.7 Household energy expenditure in the UK in different income groups, 2008 299
8.8 The energy rating of the housing stock and the fuel poor in the UK, 2006 300
9.1 Immediate costs and future benefits of low-carbon technology innovation 317
9.2 Total OECD government RD&D budget for different energy technologies, 1974–2011 319
9.3 R&D expenditure by top companies in different sectors as per cent of sales, 2011 321
9.4 Cost reduction associated with a doubling of installed capacity or production volume ('learning rate') for various energy technologies 323
9.5 The innovation chain 325
9.6 Technological diffusion pathways 327
9.7 Innovation intensity and the broken chain 328
9.8 (a) The technology valley of death. (b) The funding gap under R&D plus undifferentiated demand-pull subsidies 332
9.9 The multiple journeys of the innovation chain 345
10.1 Registration of new light duty vehicles in Brazil per fuel type, 2000–2010 367
10.2 Global scenarios for the evolution of vehicle fleets 369
10.3 Transport transitions and the European electricity network 371
10.4 Urban population between 1950 and 2050 378
10.5 A comparison of the built-up areas of Atlanta and Barcelona to the same scale 379
10.6 Two kinds of energy future – the carbon divide 385
11.1 Long-run historical trends in energy consumption per capita for a few major economies 407
11.2 Global mitigation cost trends over the century in three different economic models: the RECIPE comparison 414
11.3 Global mitigation cost trends over the century: dependence on infrastructure investment and other complementary policies in one model (IMACLIM) 415
11.4 Gross world product in 2050 (excluding environmental benefits) from a wide range of models 417
11.5 Waves of innovation 420
12.1 Global trends: default and potential trajectories 456
12.2 Public and private returns from the Three Pillars 463
12.3 An integrated package: interactions between the Three Pillars 466
12.4 Potential joint benefits in energy and climate policy 478
A1.1 Implications of inertia and adaptability for optimal responses 500

Tables

5.1 Direct energy rebound in developed countries for different end-uses 186
6.1 Some existing carbon taxes 228
9.1 OECD typology of industrial policy instruments by policy arena 344
12.1 Different classes of decision-makers, investments and drivers 468
12.2 High-level application of the Three Pillars in the European context 481

Boxes

	The Three Domains	xxi
2.1	On the comparative importance of all three domains	63
3.1	Energy units and conversion factors	85
3.2	The physics of energy efficiency and exergy	86
3.3	Carbon capture storage and utilisation	94
3.4	The solar surprise	101
3.5	Aviation and shipping	107
4.1	A lofty tale of normal housing	142
4.2	Behavioural economics and energy use – 'the largest social science experiment in the world'	145
5.1	Energy agencies: a tale of two cities	162
5.2	The global spread of energy-efficiency policies	174
5.3	Supplier regulation and energy-efficiency obligations	181
5.4	New approaches from analysing gaps and aligning driver-barrier analysis: the UK CRC energy efficiency scheme	183
7.1	Options for 'structural reform' of the EU ETS	261
7.2	Additionality in the CDM	268
9.1	Iterating innovation	327
9.2	Building the missing link – the Carbon Trust's technology accelerators	336
9.3	How not to do market pull – the history of UK renewables policy	340
11.1	Global energy economy modelling with induced innovation: the RECIPE results	412
12.1	Representing the future: UK legislation	473

'Countries are doing political handstands today to open their economies to market forces ... what they haven't yet discovered is that the market can drive development in two ways – sustainable and unsustainable. Whether it does one or the other is not a function of an invisible hand but of man-made policy.'

– *Jim McNeill, 'Sustainable Development – getting through the 21st Century', Address to J. D. Rockefeller 150th Anniversary Conference, Institute for Research on Public Policy, Ottawa, October 1989*

'.. A wide cast of characters shares responsibility ... [including] economists like me, and people like you. Somewhat frighteningly, each one of us did what was sensible given the incentives we faced. Despite mounting evidence that things were going wrong, all of us clung to the hope that things would work out fine, for our interests lay in that outcome. Collectively however, our actions took the world's economy to the brink of disaster, and they could do so again..'

– *Rajan Raghuram, Governor of the Central Bank of India and former Chief Economist at the International Monetary Fund, on the financial crisis* (R. G. Raghuram, *Fault Lines, Princeton University Press, 2010 (p. 4)).*

'The great tragedy of science – the slaying of a beautiful hypothesis by an ugly fact'

– *Thomas Henry Huxley, Collected Essays (1893–4), 'Biogenesis and abiogenesis'*

Preface

When Laurence Tubiana, a former advisor to the French Prime Minister, asked me to give a talk, she posed an unusual question. Has economics actually helped, or has it hindered, our response to the global challenges of energy and environment? It was a provocative challenge – I was Chief Economist at the UK Carbon Trust and a Senior Researcher at Cambridge University's Faculty of Economics – but also an intriguing one.

People usually want answers to a more pragmatic question. We are on an increasingly unsustainable course: the world has discovered what Jim McNeil warned about a quarter of century ago. Why on earth do we – and, in particular, governments – seem to be so incapable of acting to protect our future and what we should be doing instead?

There is a third question, which connects and underlies both of these: how well do the theories which guide our thinking, acknowledged or not, fit the facts? This book is grounded in that question, from which it seeks to answer the other two.

The answers in front of you took four years to develop – or indeed, half a lifetime – for the insights and conclusions are gathered from a career that has spanned tremendous diversity and an enduring fascination with the disconnection between the different parts of the energy conundrum. I started studies in natural sciences and physics, but my interests in energy propelled me into power systems engineering and research on the economics of innovation and integration at the UK's top technical university. From there I moved to study policy as senior researcher and then Head of Energy and Environment at Chatham House (the Royal Institute of International Affairs), where I encountered the chasms between perspectives from different disciplines and different parts of the world.

By the end of the 1990s, it was plain that the seemingly unending disputes between scientists, engineers and economists would never be resolved without learning from the accumulated evidence of efforts to implement solutions. My conjoined economics appointments over the next decade provided the ideal vantage point from which to experience the often disconnected worlds of business and consumers on the one hand, and governmental policy and economic advice on the other.

Even against such a background, assessing what we have learned over the past two decades is a task too large for any individual. In the effort, I have been extraordinarily fortunate to be joined by two of the leading researchers in Europe, each bringing their own wealth of insight and experience: Professor Jean-Charles Hourcade, a top French economist and founder of the French Centre International de Recherche sur l'Environnement et le Développement, and Professor Karsten Neuhoff, Head of the Climate Policy department at the German economics research institute DIW Berlin.

Our central conclusion is that three quite distinct domains of human decision-making need to be recognised, understood and connected. Each corresponds to different theoretical foundations, draws on different areas of evidence and implies different pillars of policy. No single one is either right or wrong, but rather describes different parts of the proverbial elephant; none on its own offers an adequate explanation of all that matters, or in isolation offers a sufficient foundation for policy.

Over the past few decades, the assumptions and theories of the second of these domains – commonly known as neoclassical economics – have tended to dominate economic thought and its influence on policy. Excessive focus on the short-run gains from maximising competition with light regulation led the West to accumulate debt, and ultimately to the credit crunch. The same approach in energy cut costs but weakened investment, including investment in innovation. The solution to energy's environmental impact – and particularly climate change – was held to be pricing carbon, but this has proved both very hard and so far unstable, leaving the atmosphere to accumulate the natural debt of the resulting wastes and our descendants to deal with the consequences.

The ultimate test of theory should be not its elegance but how well it fits the evidence. Global energy and environmental issues take us beyond all the reasonable boundaries of neoclassical assumptions, to include realms where other domains better explain the realities. There is huge uncertainty and limited foresight, and the 'externalities' imposed on others may be bigger than the economic transactions involved. The systems are massive, complex, constantly developing and sporadically unstable. Energy consumption and emissions are driven mainly not by the conscious, considered decisions of major players, but by the individual choices and ingrained patterns of behaviour of almost every person on the planet. Energy systems and technologies have evolved, and will continue to evolve, over periods of decades and centuries; they adapt in ways guided by innovation and infrastructure, but with great inertia. Both the fossil fuel markets and the atmosphere span all nations of the world.

Trying to understand such a system with the tools of traditional economics is like trying to understand the Big Bang with the tools of Newtonian mechanics – that epochal explanation of motion in the world we experience around us. Arguably the most successful and powerful scientific theory in history, it nevertheless only describes the processes that matter within certain scales, not everything. To make sense of the very small and the very big, physics had to develop quantum mechanics and relativity. Similarly, traditional economics offers useful and important insights into major parts of the problem. For understanding the grand challenges, however, its normal tools and assumptions are just hopelessly out of their depth: they need to be flanked by the other domains.

In so doing, our enquiries also shed light on an enduring economic conundrum. It is more than half a century since Nobel Economist Robert Solow demonstrated that classical theory around the accumulation of resources could not adequately explain observed economic growth. His 'residual' – the 'dark matter' of economic growth – was attributed to innovation. We conclude that explaining it in practice means expanding horizons to encompass the other two domains, and we illuminate the underlying processes as they relate to energy and environment.

Charting the map of which domain works where is at the heart of good policy, in both economics and environment. The answer to Laurence's question is that economic advice and modelling have been helpful where they have been applied within corresponding boundaries, and unhelpful where they have trampled unwittingly across them.

The book presents evidence that in relation to the grand challenges of energy and environment, each of the three domains is of roughly comparable importance; moreover, they are interdependent. From the three domains flow three pillars of policy – three quite distinct kinds of actions that need to be taken that rest on fundamentally different principles. Any pillar on its own will fail. Only by understanding all three and fitting them together do we have any hope of changing course and, if we do, the oft-assumed conflict between economy and the environment, and the associated political gridlock of burden-sharing, largely dissolves. In its place is revealed the real structure of the challenge, which is about investment and returns. Smart policy, integrated across the three pillars, can attune to local concerns whilst also delivering the investments for long-term conjoined economic and environmental benefits.

This book is full of facts. To those that want a simple world that accords with elegant theory, the facts may seem ugly. The implications, however, are not: they point to ways in which we can indeed solve some of the biggest challenges of the twenty-first century.

Michael Grubb
Cambridge, June 2013

The Three Domains

The central argument of this book is that sustaining economic progress involves harnessing three distinct domains of economic decision-making and development.

First Domain: *Satisficing* describes the tendency of individuals and organisations to take many of their decisions based on habit, routines and inbuilt assumptions. In understanding this domain, well-established insights from fields as diverse as psychology, management science and transactional analysis have been further extended in recent decades through progress in behavioural and organisational economics. This domain corresponds with a human tendency to *ignore* risks if they are sufficiently remote, intangible or hard for individual decision-makers to influence.

In energy and environment, satisficing behaviour explains notably a huge scope to improve energy efficiency with net economic benefit, and associated potentials for *smarter choices*. This underlies the first Pillar of Policy, *Standards and Engagement*.

Second Domain: *Optimising* describes the domain of considered decisions which attempt to make optimal choices based on economic factors. This reflects traditional assumptions around market behaviour and corresponding theories of neoclassical and welfare economics. Particularly for public authorities, the corresponding approach to risks (or other adverse impacts) is to evaluate them and assume potential to *compensate* for adverse impacts through the wider collective gains of efficient behaviour. This underpins the idea of cost/benefit analysis, which requires aggregating and weighing up overall costs and benefits. The overall framework also yields the principle that market prices should reflect costs and benefits to the extent possible. Research in recent decades has helped to illuminate many complicating factors, including business and commodity cycles and herd behaviour in associated financial markets.

In energy and environment, this domain best approximates the day-to-day operational and expenditure decisions of most energy companies and major energy users. It can also help to explain (though not predict) the periodic and costly instability of fossil fuel markets. The environmental impacts arising from fossil fuels and the practical difficulty of directly compensating the victims of pollution or future generations further increases the benefits of choosing *cleaner products and processes*. All this underlies the second Pillar of Policy, namely the appropriate use of *Markets and Pricing*.

Third Domain: *Transforming* describes the ways in which complex systems can develop over time and be influenced by the strategic choices of large entities, notably governments and big multinational corporations. Understanding of this domain has developed enormously with progress in theories of complexity and chaos reflected in the economic field through the insights of evolutionary and institutional economics. In terms of approaches to risk, this domain corresponds most closely with the overriding need for *security* in our lives and institutions, with the ultimate expression being national security. Since security has historically been considered almost exclusively in terms of defining and defending boundaries around individuals, groups and nations, the concept has hardly yet been extended to the level of collective global concerns, but this may best describe the nature of global energy and climatic risks.

In the evolution of energy and economic systems, the key processes concern *innovation* and the development of *infrastructure*. The energy sector spends an exceptionally low fraction of its turnover on R&D and is wholly dependent upon extensive, complex and long-lived infrastructure. Decisions on these shape the boundaries of feasible future choices, and this underlies the third Pillar of Policy, of *Strategic Investment*.

These Three Domains operate at increasing scales of time, geography and society (i.e. individuals or the institutions of collective decision-making). None are exclusive of the others, but rather describe the processes which may dominate for different types of decisions at the different scales. Their relative importance, in consequence, depends on the nature of the question being addressed. This book presents multiple lines of evidence to suggest that all three domains are equally important to transforming the global energy system over the next few decades. It follows that successful response to global energy and environmental challenges requires using all three Pillars of Policy simultaneously, and understanding and harnessing their interactions.

Each pillar in this book is examined through three chapters, preceded by an overview of the key points. Each outlines the associated theories, but places most emphasis upon the underlying evidence, and the lessons from experience. This serves to demonstrate both the reality of the corresponding domain, and the complexity, potential – and limitations – of policies confined to any single pillar. From this springs our analysis of how we could do better and profit from it: by avoiding the myth of a magic bullet, whether in policy or technology, and instead combining, shaping and strengthening the forces for changing course across all three Pillars of Policy.

The last chapter of Pillar III notes that theories of economic growth have always had to appeal to forces that could not be satisfactorily explained within the confines of Second Domain assumptions. This 'dark matter of economic growth' has increasingly been ascribed to changes in behaviour and regulation, infrastructure, and innovation in technologies and institutions. The First and Third Domains encompass these processes; the Three Domains thus offer a more holistic approach to understanding economic progress and growth, which cannot be adequately accounted for in neoclassical theories. From understanding all three together flows the conclusion that effective policy on energy and climate change, utilising the three pillars, is entirely consistent with (and could ultimately enhance) economic progress. Energy is crucial for development, and good policy, on both economics and the environment, ultimately springs from the same well.

Michael Grubb
Jean-Charles Hourcade
Karsten Neuhoff

Acknowledgements

This book embodies a theme that energy and economic systems evolve; the same might be said of the project itself. For its origins I am indebted to the UK Economic and Social Research Council (ESRC) and the Engineering and Physical Sciences Research Council (EPSRC); their core funding of the University of Cambridge Electricity Policy Research Group (EPRG), under the thematic programme on Towards a Sustainable Energy Economy, financed my increased time at Cambridge to embark on a broad assessment of what we had learned from two decades of research and policy on these topics. I am particularly indebted to Professor David Newbery, the founder and then Director of EPRG, for creating this superb environment for intellectual enquiry into issues of profound practical importance, and for his faith, support, insight and wonderful intellectual acumen.

Cambridge EPRG was also the ground which brought me together with Karsten Neuhoff, who agreed to spread his extraordinary energy even further by joining as a co-author on the book, with a leading role in particular on the Pillar II Chapters (6-8) concerning economic instruments. Jean-Charles Hourcade, one of Europe's most long-standing and creative intellects across the fields of energy, environment, economy and development, completed our trio, bringing along with him the expertise he had built up at CIRED in Paris. He and his team played the leading role in Pillar III, particularly the sweeping Chapters 10 and 11, which link the 'micro' treatise on energy economics and policy design, to the wider economic challenges of our time, including the future of the European economy and Eurozone. I am deeply indebted to both Karsten and Jean-Charles for their faith in the project and their innumerable intellectual contributions across the span; whilst each had a leading area, the final product is an integrated whole that combines all our perspectives.

In turn, the three of us are all indebted to a huge cast of characters that have helped to create *Planetary Economics*. Each pillar of the book was supported by able research assistants. Siobhan McNamara, Tim Laing and Davide Cerruti contributed research support mainly on Pillars I, II and III respectively. Paul Drummond, then at Imperial College, helped to guide Chapter 3 to completion. Aofie Brophy-Haney and Junko Mochizuki deserve particular thanks for invaluable additional inputs on Chapters 4 and 5.

Aurélie Méjean deserves exceptional thanks for carrying the project from Cambridge to CIRED, and playing a key role not only in pulling together with her new colleagues the daunting scope of Chapter 10, but also as our key expert on biofuels. A measure of the breadth involved is our gratitude also at CIRED to the contributions (in alphabetical order) of Paolo Avner, Ruben Bibas, Christophe Cassen, Etienne Espagne, Dominique Finon, Céline Guivarch, Minh Ha Duong, Jun Li, and Adrien Vogt-Schlib. Aurélie and Simone Cooper of Climate Strategies also deserve the major thanks for streamlining and translating the materials for Chapter 11 into an accessible and compelling account of how our overall thesis relates to economic growth theories.

The coordination of the huge effort owes much to Claudia Comberti at Climate Strategies, an international research network established to help connect applied academic research with policy particularly at a European level. Claudia quickly became an indispensable member of the project, contributing research support across the chapters and producing many of the graphics. A number of members contributed valuable comments and assistance particularly on the Pillar II chapters, notably Misato Sato, Anna Korppoo, Axel Michaelowa and Andreas Tuerk, whilst Michel Colombier, Chris Beauman and Hans Juergen Stehr also helped to maintain my conviction that the effort to make sense of such a massive range of work was an important and useful endeavour. Working at the interface of academic research and practical policy is not easy; a share of royalties from the sale of this book will go to support Climate Strategies.

Claudia helped to organise review meetings at Cambridge University, Oxford University (with thanks to Nick Eyre), Imperial College London (with thanks to Rob Gross), CIRED Paris, and DIW Berlin. Claudia also marshalled an extensive list of academic reviewers and we are grateful to the many who contributed useful comments. I would particularly like to thank Michael Spackman and Steve Mandel for comments on the early chapters, and Liz Hooper, Emi Mizuno, Catherine Mitchell, Antoine Dechezleprêtre and Gregory Nemet for comments which provoked a major recasting of Chapter 9.

The book is deliberately grounded in the combination of research and practical experience – which regrettably, is far less common than it should be. I am indebted to Tom Delay, Chief Executive of the Carbon Trust, whose faith that an established academic could also be practically useful in a delivery organisation created a half time post that combined institutional responsibilities with the academic freedoms of my role at Cambridge. I believe and trust that we all benefited. I am also grateful to several former colleagues at the Carbon Trust for their support, tolerance and insights along the way, along with data for the analysis and, long after I left, help with design of some key graphics.

Similarly, I am grateful to Sarah Harrison for making a similar arrangement in my appointment as Senior Advisor at Ofgem, where I have learnt yet more, and had to try and work out some practical implications of all this research for how a regulator can best protect the interests of both present and future consumers.

As the project expanded beyond all reasonable bounds of scale and time, I am also grateful to Doug Crawford-Brown at the Cambridge Centre for Climate Change Mitigation Research, to which I moved after the Research Council funding for EPRG at the Economics Faculty expired. 4CMR hosted the final review meeting and provided the institutional context for employing Sushmita Saha, to whom I am most grateful for her intensive work organising the final stages of the book ready for submission to the publishers, including updating of data, imaginative work improving graphics, as well as tutorials in Mathematica.

Finally, I am of course also indebted to our publishers. Jonathan Sinclair-Wilson takes the credit – if that is the right word – for committing his company Earthscan to the book, however long it took; and I am grateful to Robert Langham and then Andy Humphries for seamlessly continuing the project with enthusiasm when Taylor and Francis acquired Earthscan.

All journeys end – or at least, reach defining points. Many others have travelled parts of this journey, or helped with important elements. It is impossible to name all, but your contributions are gratefully acknowledged. Writing a major book is also a journey of learning from the publications of innumerable academics around the globe, whose efforts have thus also all contributed to the result. Whilst responsibility for the final product and any errors therein rests with the authors, our gratitude to all those who helped knows no bounds. We hope you think the final product has justified the efforts.

Abbreviations

AEEI	autonomous energy efficiency improvement
ADEME	Agence Française pour l'Environnement et la Maîtrise de l'Energie
AEA	Atomic Energy Agency
AGECC	UN Advisory Group on Energy and Climate Change
ARPANET	Advanced Research Projects Agency Network
BECCS	biomass energy with carbon capture and storage
BEV	battery electric vehicle
BOS	balance of system
BRT	Bus Rapid Transport
Btu	British thermal unit
¢	US cent
CAFE	Corporate Average Fuel Economy
CAGE	Competitive Advantage in the Global Economy
CCC	Committee on Climate Change
CCICED	China Council for International Cooperation on Environment and Development
CCL	Climate Change Levy
CCS	carbon capture and storage
CCU	carbon capture and utilisation
CDM	Clean Development Mechanism
CEGB	Central Electricity Generating Board
CEPI	Confederation of European Paper Industry
CEPR	Centre for Economic Policy Research
CERN	European Organisation for Nuclear Research
CERT	Carbon Emission Reduction Target
CFL	compact fluorescent lamp
CHP	combined heat and power
CLCF	Centre for Low Carbon Futures
CRC	Carbon Reduction Commitment
CRIEPI	Central Research Institute of Electric Power Industry
CSIRO	Commonwealth Scientific and Industrial Research Organisation
CSP	concentrating solar power
DECC	Department of Energy and Climate Change
Defra	Department for Environment, Food and Rural Affairs
DGEM	Dynamic General Equilibrium Model
DOE	Designated Operating Entity

DRI	direct reduced iron
DTI	Department of Trade and Industry
EAIC	Executive Agency for Competitiveness and Innovation
EBRD	European Bank for Reconstruction and Development
ECEEE	European Council for an Energy Efficient Economy
EEA	European Environment Agency
EEC	Energy Efficiency Commitment
EIA	Energy Investment Deduction Scheme
ENSO	El Niño/Southern Oscillation
EPBD	EU Performance of Buildings Directive
EPRI	Electric Power Research Institute
ESCO	energy service company
ESMAP	Energy Sector Management Assistance Program
ESRL	Earth System Research Laboratory
EST	Energy Savings Trust
ETP	Energy Technology Perspective
ETS	Emissions Trading System
ETSAP	Energy Technology Systems Analysis Programm
ETSU	Energy Technology Support Unit
EUA	European Union Allowance
EWEA	European Wind Energy Association
FFV	flexible-fuel vehicle
GDP	Gross Domestic Product
GEA	Global Energy Assessment
GHG	greenhouse gas
GtC	gigatonnes of carbon
GVA	gross value-added
GWEC	Global Wind Energy Council
HCEI	Hawaii Clean Energy Initiative
HEES	Home Energy Efficiency Scheme
HVDC	high-voltage direct current
IAEA	International Atomic Energy Authority
ICCT	International Council on Clean Transportation
ICE	Institution of Civil Engineers
IEA	International Energy Agency
IEEA	Intelligent Energy Executive Agency
IETA	International Emissions Trading Association
IIASA	International Institute for Applied Systems Analysis
INSEAD	Institut Européen d'Administration des Affaires (European Institute of Business Administration)
IP	intellectual property
IPCC	Intergovernmental Panel on Climate Change
IPTS	Institute for Prospective Technological Studies
IRC	International Residential Code
ISO	International Standards Organisation
JI	Joint Implementation
JIAG	Joint Implementation Action Group
JRC	Joint Research Centre

LDV	light duty vehicle
LED	light emitting diode
LEED	Leadership in Energy and Environmental Design
MER	market exchange rate
MITI	Ministry of International Trade and Industry (Japan)
Mtoe	million tonnes of oil equivalent
NAMA	Nationally Appropriate Mitigation Action
NAP	National Allocation Plan
NASA	US National Aeronautics and Space Administration
NBER	National Bureau of Economic Research
NCDC	National Climatic Data Center
NDRC	National Development and Reform Commission (China)
NFFO	Non-Fossil Fuel Obligation
NGL	natural gas liquid
NIMBY	'not in my back yard'
NOAA	US National Oceans and Atmosphere Administration
OECD	Organisation for Economic Cooperation and Development
Ofgem	UK Office of Gas and Electricity Markets
OPEC	Organisation of Petroleum Exporting Countries
PAT	'Perform, Achieve and Trade'
PHEV	plug-in hybrid electric vehicle
ppm	parts per million
PPP	purchasing power parity
PV	photo-voltaic
R&D	research and development
RAM	Renewable Auction Mechanism
RD&D	research, development and demonstration
RGGI	Regional Greenhouse Gas Initiative
RIIA	Royal Institute of International Affairs
ROC	Renewable Obligation Certificate
SEEA	System of Environmental and Economic Accounting
SUV	sports utility vehicle
toe	tonnes of oil equivalent
UEDP	underlying energy demand trend
UKCCC	UK Climate Change Committee
UNEP	UN Environment Programme
UNFCCC	UN Framework Convention on Climate Change
UNIDO	UN Industrial Development Organisation
VAT	Value Added Tax
VED	Vehicle Excise Duty
WCI	Western Climate Initiative
WEC	World Energy Council
WTO	World Trade Organization

1 Trapped?

1.1 Introduction

This should be easy – at least in theory. Fossil fuels are finite. So too is the atmosphere. Sooner or later, our energy systems have to change.

Economic theory predicts that the price of a limited resource will rise as it is depleted, until it reaches a point at which alternative technologies compete and take over. Economics recommends that if there is an 'external' impact, like health or environmental damage associated with burning coal, then governments should tax emissions to reflect the cost of damage caused or impose a cap on emissions to the same effect.

In fact of course it is not easy – no one seriously claimed otherwise. Fossil fuels have been at the heart of economic development. Coal-based steam powered the Industrial Revolution in the eighteenth and nineteenth and centuries, and electricity and the internal combustion engine between them did much to shape the twentieth. Energy provides the heat to warm our homes and to produce and transform industrial materials, the electricity that lights our buildings and powers our appliances, communications and entertainment systems, fuel and fertilisers for the agricultural 'green revolution', and the motive power to transport ourselves and our goods. The benefits have been enormous.

The link between energy and development has been clear. The economic boom of industrialised (mostly) nations after the Second World War – roughly the third quarter of the last century – was accompanied by a trebling of global energy demand.[1] That came to an abrupt halt with the oil shocks of the 1970s. The resumed and more global economic growth since the late 1980s – often attributed to economic liberalisation in the 'developing countries' – followed the collapse of oil prices in the mid-1980s and was again accompanied by surging energy demand, which has doubled again over the past quarter century.

Yet all is far from well. Over a third of the global population – 2.5 billion people – still live in grinding poverty and depend on traditional wood and other biomass for cooking and heating; half of them remain unconnected to electricity.[2] The growth potential – and need – is huge, but they have missed out on the interlude of cheap fossil fuels. The gyrations of global oil prices since the late 1990s and steep increases since 2005 have caused real hardship and been implicated in further economic recessions.

Along with relentless growth in global fossil fuel use has come a series of environmental problems, of growing scale and reach. At first, these impacts concerned contaminant pollutants, like smog and sulphur, or the side-effects of extraction. Dealing with these formed major policy battlegrounds in the twentieth century, but with hindsight proved relatively easy to deal with. CO_2, however is not a contaminant but the fundamental product of burning fossil fuels. With concentrations already at levels not seen for millions of years, it continues to accumulate relentlessly in the atmosphere.

This book is about the interplay of theory, evidence and policy implications applied to these challenges. Despite the mind-boggling scale of the issues, the fact remains that the traditional economic theories of such problems are quite simple and date back more than three-quarters of a century.[3] They focus particularly on the role of prices and the trade-off between costs and benefits. We can now also match these theories against the evidence of practical experience of these systems and responses: almost forty years of efforts to tackle oil dependence since the 1970s, and two decades of policies to tackle CO_2 emissions.

The experience is sobering. It points to the fact that the systems involved are far more complex than any single theory assumes. Industrialised countries have struggled to reduce CO_2 emissions and most of the emerging economies – the majority of the world's populations – are still following in the footsteps of the fossil-intensive Western model of development. A view is emerging that the problems are just too big to solve. Vast investments are flowing into new frontier developments in harder-to-reach fossil fuels, and domestic and international discussions on climate change talk increasingly about how to cope with the climate impacts implied by the failure to control emissions – and who should pay for it.

The core argument of this book is that the dominant theories simply do not match the scale of the challenge – they have been extended out of their depth. Different approaches point to different bits of the problem – different 'parts of the elephant' – without an overall picture. This in turn has made it far harder to map coherent responses and gain the political consensus required to implement them. Behind the failure of policies to adequately get to grips with the problems lies a failure of theory to reflect crucial realities – and hence our apparent inability to get on a sustainable course.

This book is addressed to governments and researchers, and to the wider public interested in understanding more deeply the real problems and opportunities. It explores the gap between classical theory and the accumulated evidence and what this says about the options and policy implications. The evidence comes both from extensive research in modern branches of economics and other disciplines and the practical experience of policies. It is about how these systems behave, what research has concluded, what policy-makers have tried and what the combined results imply.

Since this book stresses evidence – empirics – as the foundation for useful theory, the rest of this chapter presents some basic evidence about the nature of energy and related environmental systems, particularly their economic dimensions. The next chapter then sets out a new framework for thinking about the issues – one which emphasises that different theories fit different scales, and that the key need is to understand their assumptions, boundaries and relationships: how in reality they may complement rather than compete as explanations and guides. This then defines the structure for the rest of book, which examines the key ideas, data and experience at each of these levels. The final chapter then draws all this together to offer a wider and more integrated understanding of the economics of energy transformation and the implications for practical policy.

1.2 The physical challenges – energy, resources and climate

Energy trends

Broken down by fuel and region, Figure 1.1 illustrates the extraordinary increase in global energy consumption over the past 45 years – at first dominated by the industrialised countries, giving way more recently to growth particularly in Asia. Along with the march of oil consumption, most striking is the recent explosive growth of coal use there, along with global rising use

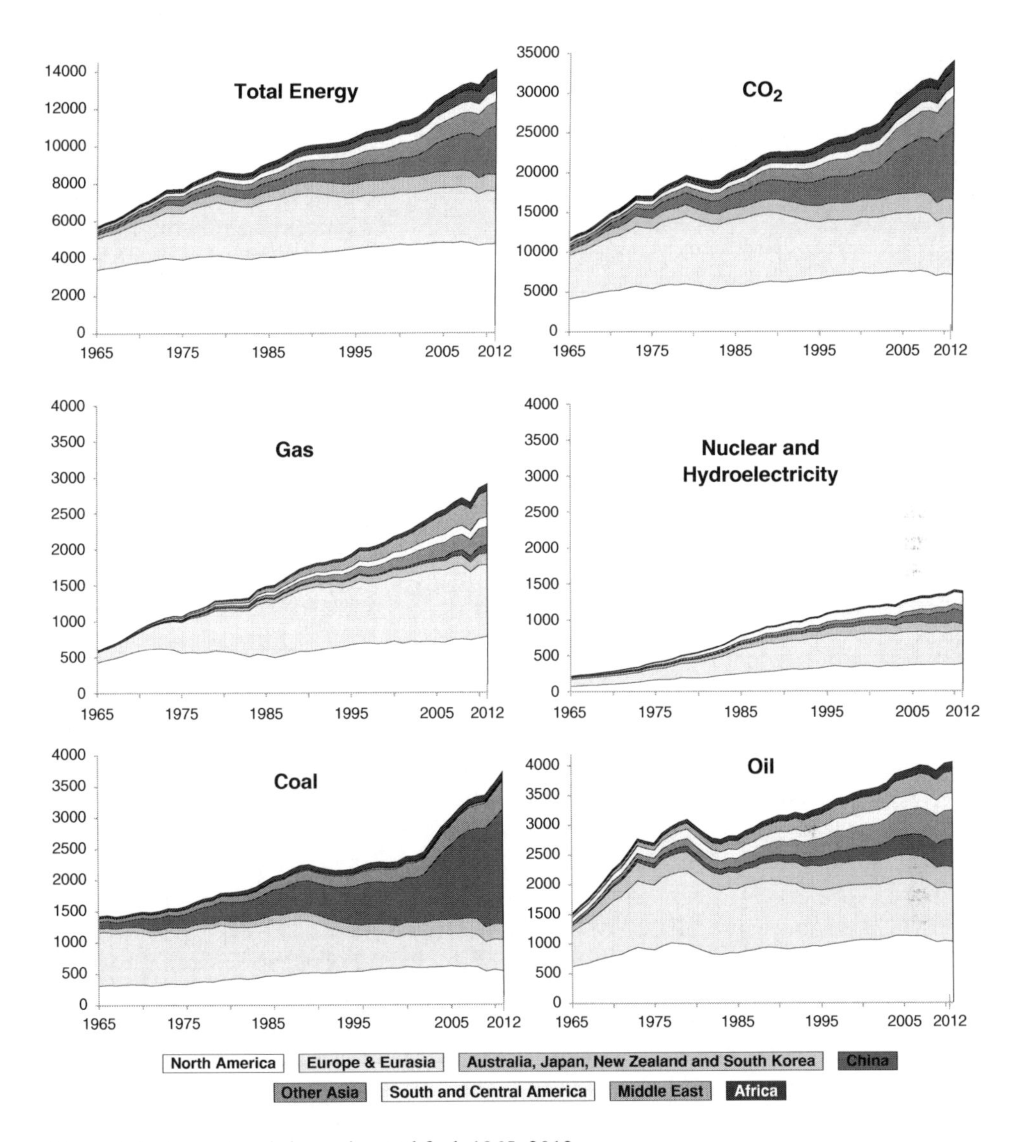

Figure 1.1 Energy trends by region and fuel, 1965–2012

Note: The charts show for each fuel the total energy consumption by region, in million tonnes of oil-equivalent.

Source: BP Statistical Review of World Energy (2013).

of natural gas. After an initial spurt of nuclear power, non-fossil energy sources overall only kept pace over the past quarter century, at just under 15 per cent of the global total; continued growth in hydro and more recently wind energy has compensated for a slow-down in nuclear generation. The overall trend in both energy and CO_2 emissions has been relentless and the most severe global recession in living memory – 2009 – dented it for just one year.

Growing energy consumption is of course a mix of good and bad news. It reflects rising living standards for billions of people – including connection to electricity and the use of fossil fuels displacing potentially more damaging use of local forest and other biomass burning for heat and cooking. Since the 1980s roughly a billion people have been connected per decade.

In fact, the most important single energy contribution is not shown in Figure 1.1, namely energy efficiency. As examined more closely in Chapter 4, in the decades since the oil shock economic growth exceeded that of energy and emissions in almost all regions – reducing their energy (and carbon) intensity, defined as the ratio of energy to GDP. The idea that GDP and energy/emissions are locked together is a myth, but the improvement in most regions has been outstripped by the rate of GDP growth. The decline in global energy intensity has also slowed in the past decade, though the pattern is quite varied between countries, and a few leading countries have since the mid-1970s roughly doubled their GDP without increasing their energy demand (Chapter 5, note 51).

In addition, many countries showed evidence of an 'energy ladder' in their mix of sources, with a shift towards electricity and lower carbon fuels over time (though the boom in Asian, particularly Chinese, coal has been a crucial exception).[4] Yet globally, all this has been at a far slower pace than GDP growth itself. At a global level, nothing in these data suggests that the pressures on fossil fuels will lessen – quite the contrary – and there is little sign of a global turnaround.

There are two basic problems with this trend: what goes in, and what comes out.

What goes in: the fossil fuel roller coaster

If the modern world is built on fossil fuels, then its most important single component – the global oil market – has proved disturbingly unstable. Cheap oil, boosted by mid-century discoveries of massive 'supergiant' fields mainly in the Middle East, helped to fuel a quarter century of rapid economic expansion after the Second World War.

That period came to an abrupt end with oil shocks in the 1970s (Figure 1.2). As the countries that sat on the bulk of global oil reserves asserted their control of it, against an increasingly unstable political backdrop in 1973, the global oil price more than doubled. Economic growth rates plummeted as inflation rose and huge amounts of finance were wrenched away from the industrialised world, slowly returning as recycled petrodollars. Six years later, with the Iran–Iraq war, oil exporters tightened the throttle and prices doubled again, precipitating another global economic downturn.

Global oil demand at the time was dominated by the industrialised countries. Stung by the high prices and loss of control, they turned to other fuels for electricity generation, and consumers moved from oil heaters to coal or gas and bought more efficient cars.[5] The rich country governments also invested hugely in new frontiers of oil development and established strategic stockpiles, while the US entrenched its military presence in the Middle East. The Organisation of Petroleum Exporting Countries (OPEC) had to cut back supply in a bid to maintain the price – which finally crashed in 1986 when Saudi Arabia refused to continue 'carrying the can' for cutbacks. After a turbulent 15 years, prices in real terms had returned almost to pre-1973 levels.

Oil became traded in a big global market. In principle this allowed forces of supply and demand – albeit supported by some subsidies and the military guarantee in the Middle East – to manage global oil. Analysts trusted that economic forces would ensure adequate supplies by maintaining investment in new fields, continuing discoveries (which to date they have, with some help) and smoothing prices (which they have not). The oil confidence that corresponded with, and helped to define, this age was neatly encapsulated by a front cover of *The Economist* just before the Millennium: its issue 'Drowning in Oil' famously predicted an enduring world of US$5 per barrel oil.[6]

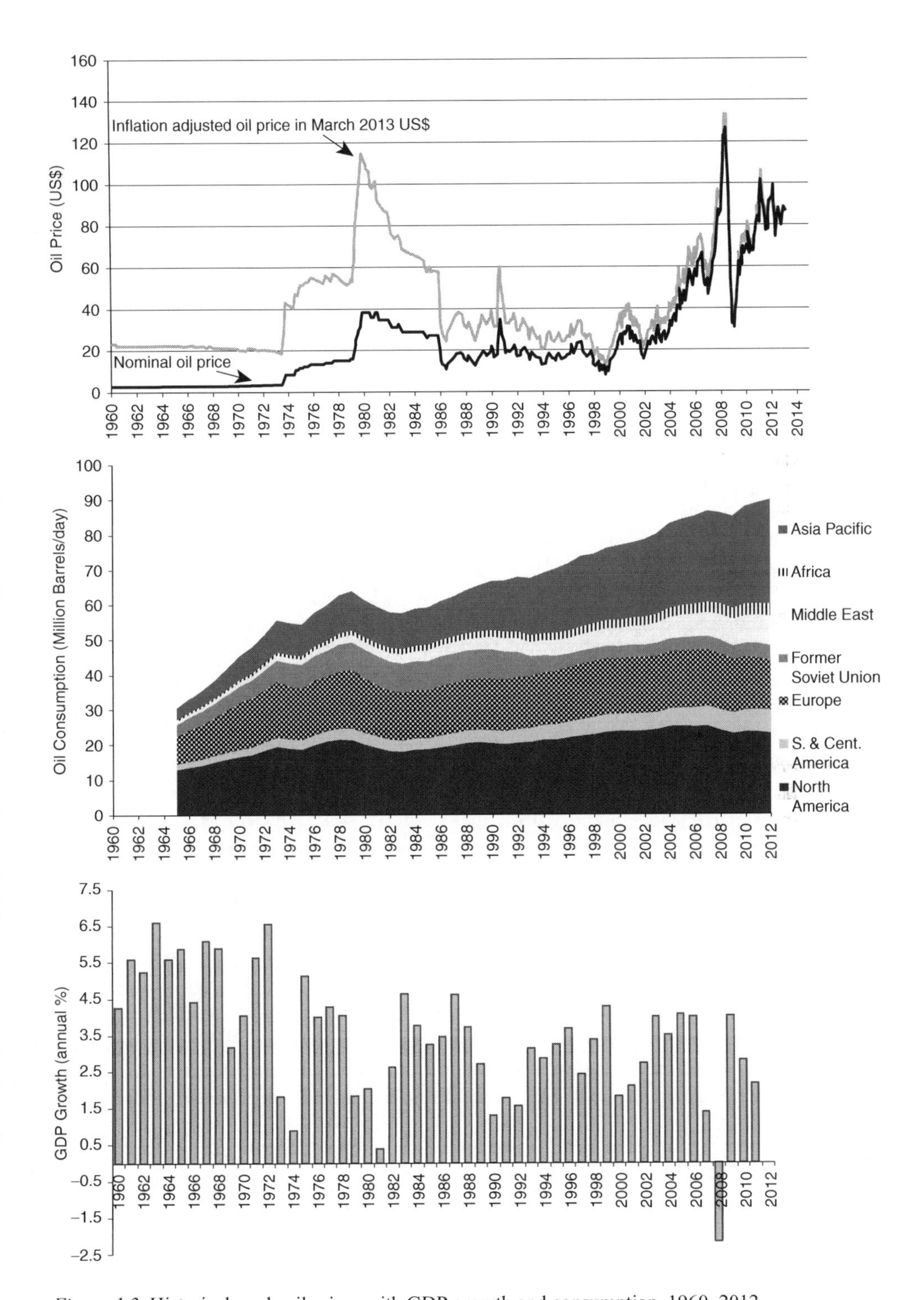

Figure 1.2 Historical crude oil prices with GDP growth and consumption, 1960–2012

Source: Real crude oil monthly average prices of West Texas Intermediate 1945–2008 (from Hamilton 2009a); adjusted to $2008 using CPI ratios; 2008–12 data from IEA (2013).

Its equally famous retraction did not take so long. Within just a few years and without any singular shock to the system, from 2004 the price rose to heights undreamed of, peaking in the summer of 2008 at over $140/barrel, almost ten times the level a decade earlier, higher in real terms even than the peaks of the 1970s. The speculative bubble burst, but the shock – to prices, confidence and to some economies – remains.

Economic consequences

The cost of volatile prices has been huge. Although economic downturns or crises are never caused by energy price increases alone, there is no doubt that the 1970s oil shocks helped to knock the global economy from growth at 4–5 per cent/year down to around 1 per cent (Figure 1.1). In the US, of the eleven postwar recessions, ten were preceded by a sharp increase in oil prices.[7] As well as the 1970s shocks, the most extensive econometric study to date finds significant recessionary effects attributable to the oil price impacts of the First (1990) and Second (2002–3) Gulf Wars; the high international oil prices of 2007–8 were again followed by global economic downturn. This time, the oil price rise was less abrupt and most of the rich economies were more insulated (by many factors including high domestic oil taxation, which provides a buffer, and less dependence on manufacturing) and the recession was driven mainly by the credit crunch of accumulated debt. But the link, however indirect, is significant: the recycling of petrodollars helped to fuel easy credit and the financial bubble, and the inflationary effects of escalating oil prices caused governments to increase interest rates, thereby helping to expose the ugly underbelly of debt. A UK study estimates that fossil fuel price volatility in the coming decades could cost the UK around 0.5 per cent GDP – a similar order of magnitude to the commonly estimated costs of CO_2 controls.[8]

Decades of urban sprawl have left people (particularly the poor) more directly exposed to the impact of higher oil prices on their welfare and purchasing power. Thus oil prices do not need to provoke a recession to wreak havoc for some – especially consumers that have come to depend upon imported oil. In the rich world, US drivers, being used to cheap gasoline, felt the pinch hard. The repercussions for the poor in oil-importing developing countries was far greater. For the 'least developed countries' – a group defined as having a per capita income of less than US$300 p.a. – each increase in oil price of US$10 a barrel can knock around 1.5 per cent off their GDP, and even more for some of the lowest-income countries.[9] One reason for this is that poorer countries tend to spend relatively more of their total imports on energy – meaning that any price fluctuation hurts more. Energy imports cost more than 20 per cent of export earnings in 35 countries with 2.5 billion people, and exceed 10 per cent of GDP in 15 countries with 200 million people; Guyana, for example, spent nearly 19 per cent of its 2009 GDP on importing crude oil.[10] A price shock along with a global recession hits such countries doubly hard, combining rising import costs with declining exports.

For the rural and urban poor who still depend on kerosene for lighting and cooking, and already struggle to find the fuel to get their goods to market or to work, the consequences can be devastating. The oil price rises more than doubled the cost of cooking fuel and transportation in rural areas and remote suburbs of developing countries. In many African countries, the oil price increases wiped out the expected benefits from western aid and debt relief. Energy is intertwined with development.[11]

Ironically, it is not only the oil importers that suffer from the oil roller coaster. A whole literature has grown around the 'resource curse' of countries seemingly blessed, but ultimately trapped, by dependence on the wealth of a single resource: the resulting high exchange rate squeezes out other economic activities, can create huge internal tensions (sometimes leading to civil war) and leaves the economy severely exposed to fluctuating commodity cycles.[12] If 'economic catastrophe' is defined as losing half of national or regional GDP, oil exporters in the 1980s ranked alongside the developing country victims of a few massive natural disasters.

Countries have varied greatly in their ability to manage the paradoxical 'resource curse' and associated high economic rents – the surplus when prices vastly exceed production costs. Such rents are a source of eternal struggle between governments (for tax revenue) and industry (for which use of profits includes further exploration, and lobbying to protect their interests) and interest groups within and outside government, further distorting any ideal of a rational economic system in the national (or global) interest.

Not just oil

Of course, oil is not the only fossil fuel. Coal continues as a mainstay for industrial processes and power generation in many countries. It is scattered throughout the earth's crust in abundance. Gas reserves have steadily expanded with discoveries and the march of new technologies, with particular excitement around the development of shale gas, released from rocks with hydraulic 'fracking' (a technology which is also helping to release more oil reserves). Yet none of these fuels are immune from price volatility – indeed, as Figure 1.3 shows, over the past decade both gas and coal have seen international price fluctuations almost as dramatic as (and following on from) those of oil.

There are enough connections between the different fossil fuels and the common driver of power-hungry economic growth in the emerging economies that volatility in one tends to transmit to all. Moreover, the cycles in such major commodities appear intrinsic, though impossible to predict in detail: low prices encourage more consumption and deter investment, whilst high enough prices induce conservation and a new wave of investment, to explore further and to get at sources previously considered too difficult or costly.

'What goes in' has other impacts too. There are strong connections between energy, water and food; extracting and converting fossil fuels can use a lot of water, and both are important inputs to agriculture. Indeed global food price rises have increasingly been associated with those of fossil fuels, in part because of the rising use of energy for both machinery and fertiliser production.

Of course, much of this is a sign of markets working and reflecting the real and perceived risks of scarcity. Unfortunately volatility also shows how bad we are at predicting and how sensitive these systems are to errors and to the unexpected. Business is prone to cycles and short-term financial markets amplify these.

Much of the literature on fossil fuels focuses on the projected exhaustion, or to be more precise 'peaking', of global oil production. This is countered by pointing to the continual response of new oil finds and new technologies for extraction. To an important degree, this enduring debate misses the point. Many of the frontier developments involve ever bigger scales and timescales of investment. Current enthusiasm for shale gas (and shale oil) echoes the excitement around past major extensions of production. Quite apart from continuing uncertainties about scale and cost, it does not change the fundamental fact that the competition between demand, depletion and discovery is neither predictable nor stable. Much of it involves successively more difficult resources and ever bigger gambles.

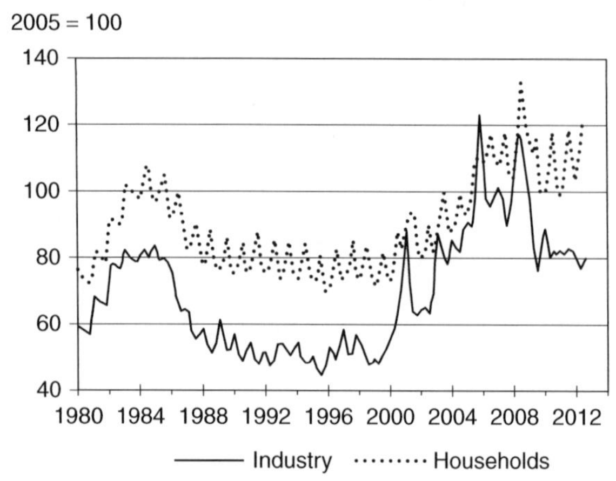

(a) Coal end-user prices

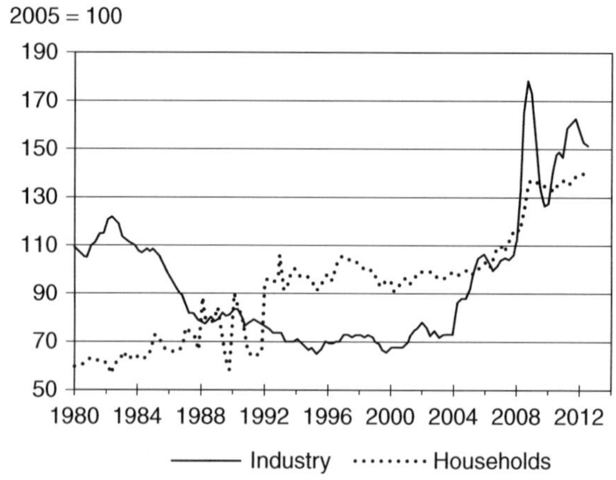

(b) Gas end-user prices

Figure 1.3 Coal and gas end-user prices, 1980–2012

(a) Coal end-user prices

(b) Gas end-user prices

Notes: The Chart shows (a) coal and (b) gas prices to end users from 1980 to 2012, averaged across the Member Countries of the International Energy Agency (IEA).

Source: IEA (2012): Energy Prices and Taxes, Quarterly Statistics, Fourth Quarter 2012.

What comes out: pollution and climate change

These challenges may ultimately be dwarfed by the impact of 'what comes out'. What comes out, of course, is a range of waste products. Along with all its benefits – which include the benefits of displacing the burning of sometimes dirtier traditional fuels – fossil-fuelled industrial development has often brought, or threatened, serious damage to health and the environment.

Indeed history reveals a general and crucial pattern. New resources offer the potential for wealth to those who own them and benefits to those who use them. They become, respectively, sellers and buyers. The environmental side-effects, however, predominantly impact others 'external' to the transaction. The development of environmental policy has been largely about these other groups gaining awareness and doing something about it, eventually forcing action. Economically, it is called 'internalising the externality'. As Chapter 6 explains, that step is essential to the presumed benefits of markets: without it, markets may just amplify activities which damage others.

Half a century ago, pollution-induced smogs across the cities of the industrialised world, from London to Los Angeles via Tokyo, killed tens of thousands of people and made urban air pollution a hot topic. The resulting legislation largely drove coal burning out of cities and forced measures to clean up cars. Many cities in the developing world are grappling with similar problems in similar ways. In the 1970s, the Organisation of Economic Cooperation and Development – the club of rich countries – agreed the 'polluter pays' principle that the cost of clean-up should be paid not by taxpayers (through governments), but by emitters – principally industry, with the consequent struggle over whether the costs are absorbed by shareholders or passed on to consumers.

As the cost of environmental controls has risen, more attention has been devoted to quantifying the benefits (as well as to the design of efficient policy). A huge amount of work has led to some interesting answers. Earth, medical and economic sciences are all complex, and as scientific understanding has improved this has put a greater spotlight on the economic questions, such as how we value morbidity (debilitating illness) and mortality (premature death). For most pollutants, however, analysts have edged towards consensus. The results suggest that regulation has generally lagged behind the scale of impacts by a sizeable margin.

In aggregate, 'local pollution' damage in industrialised countries is equivalent to a few per cent of GDP, exceeding control costs, but declining as environmental regulations have been strengthened and pollution decreases. A US assessment of the cost-effectiveness of its air pollution control programmes under the Clean Air Act (1970–90) concluded that the costs had been around US$0.5 trillion while the benefits were around $22 trillion.[13] In many developing countries, controls are still weaker and damages are bigger.[14]

Traditional fuels – such as wood and dung – are a major source of health problems in poorer countries, partly indoors. They also contribute to the extension of deforestation and degraded land. The main environmental damages associated with fossil fuels are also to do with air pollution, but at larger scales. The two biggest culprits are deaths associated with emissions (mainly sulphur and particulates) from coal combustion, and urban air pollution (particulates, and ozone created from emissions together with methane) mainly from vehicles.

Coal (and oil-based electricity generation) comes out particularly badly from such estimates, however one views it. Two major reviews in the US suggest that the environmental damages from coal exceed its commercial value on almost all measures, the studies being respectively in leading scientific (New York National Academy) and economic (American Economic Review) journals. The former estimates the environmental cost at several times the price; the latter, at one to five times its direct economic 'value-added' in the US economy (with a central estimate of about twice).[15] European studies yield similar conclusions.[16] Overall the US National Academy study estimated the unpriced impacts of energy supply at $120bn, of which electricity accounted for half and transport most of the rest – *excluding climate change or security-related costs.*

That is *despite* extensive pollution controls in the US. Many developing countries still lag behind and are paying the price. In winter 2012–13, millions of people in Beijing and many other Chinese cities had to stay indoors to avoid crippling air pollution. Rigorous 'monetised' estimates of China's environmental degradation span a wide range – they clearly amount to several per cent of Chinese GDP – and in spring 2013 the new Chinese government announced accelerated programmes to clean up its energy system, costed at several hundred billion dollars.[17] Ignoring externalities does not make them go away, it just damages lives and stores up trouble and costs for later.

By far the most voluminous output of course is not a trace pollutant that can be filtered, scrubbed out or otherwise neutralised but is the essential product of burning carbon – namely CO_2, the concentration of which is steadily accumulating in the atmosphere. Together with other greenhouse gases, including water vapour, CO_2 has the basic radiative property of trapping heat in the lower atmosphere and thus helps to keep the surface of the earth much warmer than it would otherwise be. This is undisputed.[18]

Pre-industrial CO_2 levels in the atmosphere can be estimated in various ways at below 280 parts per million (ppm), then rising slowly to a pause associated with the Second World War, and direct measurements since 1957 have tracked a rapid increase following the post Second World War explosion of fossil fuel use and associated CO_2 emissions (Figure 1.4(a)). The earth's natural systems absorb some fraction back, notably into the oceans; the CO_2 accumulated in the atmosphere since the Industrial Revolution represents only about half of everything we have emitted.[19] Even so, the concentration in spring 2013, at the peak of the seasonal cycle, touched 400ppm for the first time since the early Pliocene – 3 to 5 million years ago.[20]

Underlying temperature trends are complicated by significant natural variability and the fact that the oceans absorb a lot of additional heat, so temperature can lag behind concentration increases potentially by some decades. In addition, temperature impacts can be offset by cooling from other things we emit (like sulphur). As the CO_2 levels climbed and the big emitters cleared up some of the other pollution in the 1970s and 1980s (notably sulphate aerosols, which had a cooling effect while producing acid rain), warming emerged more clearly. As illustrated in Fig.1.4(b), which shows the compiled results of three independent research centres, atmospheric warming has been in lurches rather than a smooth trend, as might be expected from such a complex system. Nevertheless in the last half century each decade has been warmer than the one before. On most datasets 2010 was the hottest year seen to date, followed closely by 2005, and 1998 which had an exceptional *El Niño* (Pacific Ocean current) event. The last 15 years contain 13 of the hottest on record.[21]

Other indicators of rising temperatures include measured ocean surface temperatures, and the widespread retreat of both glaciers on land and sea ice. The measurable warming of such a vast amount of water is perhaps the most potent indicator of the scale of additional heat being trapped. Another warming trend has been that of the lower atmosphere – the troposphere – as measured directly by radiosonde and subsequently satellite-based measurements.[22]

There is also a measured cooling of the upper atmosphere (the stratosphere).[23] This helps to pin down possible causes of the observed trends. Challenges to the idea that CO_2 is responsible for warming had mostly focused on the idea that other factors – such as solar variation – could be driving the observed warming. There are, after all, clearly other sources of big global temperature variation (witness the ice ages). However, quite apart from the lack of any evidence to suggest that solar radiation has systematically increased in recent decades, this could not explain the cooling of the stratosphere. As the greenhouse effect strengthens, trapping more heat in the lower atmosphere, less heat escapes to warm the upper atmosphere. Its cooling, combined with surface warming, adds further to the conclusion that the main cause is thickening of the greenhouse blanket.[24]

Global average temperature is of course a crude (and inherently variable) proxy for all the changes involved. Changing weather patterns, ecosystems, ice cover and numerous other

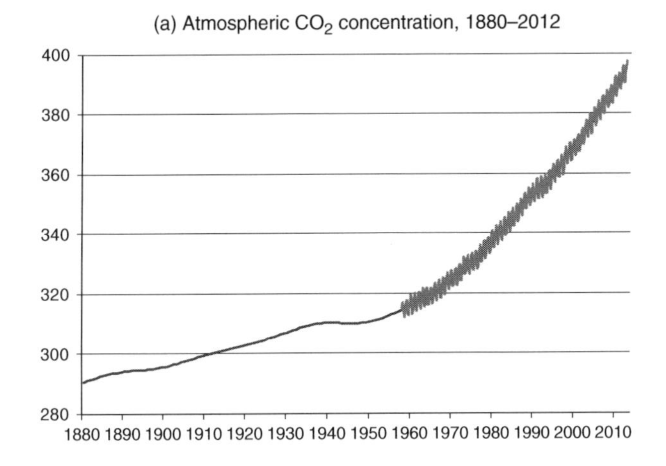

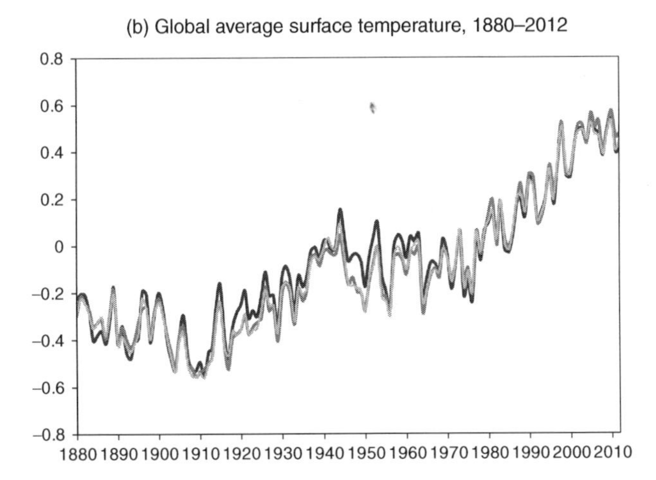

Figure 1.4 Global trends of (a) atmospheric CO_2 concentration and (b) average surface temperature, 1880–2012

(a) Atmospheric CO_2 concentration, 1880–2012

(b) Global average surface temperature, 1880–2012

Notes: (a) Global CO_2 concentrations in parts per million (ppm) by volume, derived from ice core data 1880–1957, direct measurements including seasonal fluctuations since 1958. (b) Global average surface temperatures since 1880 as calculated by three independent research centres.

Sources: From datasets compiled respectively by the US National Oceans and Atmosphere Administration (NOAA), US NASA and UK Hadley Centre. The datasets are all publicly available: CO_2 data from NOAA-ESRL annual data ftp://ftp.cmdl.noaa.gov/ccg/co2/trends/co2_annmean_mlo.txt; HadCRUT4 From http://www.metoffice.gov.uk/hadobs/hadcrut4/; NASA GISTEMP From http://www.metoffice.gov.uk/hadobs/hadcrut4/; NASA GISTEMP From http://data.giss.nasa.gov/gistemp/; NOAA NCDC From http://www.ncdc.noaa.gov/cmb-faq/anomalies.html.

indicators and impacts of warming are now well established by scientists. Their attribution to global warming induced by (mainly) CO_2 emissions remains partly contested by a small but diminishing band of sceptics; the accumulated evidence is overwhelming. The projection of what climate change – and the linked phenomenon of ocean acidification – will mean in practice, however, amplifies uncertainties and raises continuing controversy. The details are beyond the scope of this book, which rests on one basic proposition: that the accelerating accumulation of CO_2 and other heat-trapping gases in the atmosphere, way outside the ranges seen for millions of years, poses risks, the magnitude of which – as discussed below – are extremely hard to quantify or even bound in terms of usual economic or other decision-making metrics.

Ultimately, however, the same basic point remains: the capacity of any natural system to absorb waste without damage is finite. Professor Nicholas Stern (see *A Stern warning about time*, p. 25) observed that high carbon growth will *eventually* kill itself: the rising costs and stresses of escalating climatic instability and sea level rise combined with the ongoing struggle to tap more difficult or remote hydrocarbons make it unsustainable. The question is how long we have, how far it may go – and how much avoidable damage it may do on the way.

Global goals and disconnects

The obvious globally shared goal should be to improve human welfare without exacerbating local or regional environmental damage or risking 'dangerous anthropogenic interference' with the climate system.[25] For the 2.5 billion people without access to modern energy services, gaining that access is the obvious and appropriate priority. The UN has proposed a goal of energy access for all by 2030.[26]

For people and regions that have surmounted this basic step, the choices become more complex. As indicated, excessive exposure to global fossil fuel markets – particularly, but not only, oil – has proved unexpectedly costly. As the globally connected population rises from the current 4.5 billion to 7 or 8 billion over the coming couple of decades, and if the demands of those already connected continue to rise, the pressures on fossil fuel resources will rise much further against the backdrop of declining reserves; no one really knows the likely implications.[27]

Rising fossil fuel consumption will be accompanied by environmental impacts. As noted, the evidence is that environmental policy has consistently lagged behind the desirable level of protection, damaging health and unnecessarily burdening continued development. However, these are choices with mainly national consequences and, as noted, sooner or later self-generate pressures to clean up driven by suffering domestic populations. The global climate challenge is of a different order: the reach is global and the lags intergenerational.

On quantities and ratios

Considering the evidence, many scientists have argued that governments should seek to limit global warming to at most a two-degree rise above pre-industrial temperatures, and to ensure a very low probability of significantly higher global temperature increases. This implies major emission reductions over coming decades. The '2 degree target' has been etched into international agreements, and the most widely agreed goal is to halve global emissions by mid-century; stabilising the atmosphere would ultimately require net emissions to be brought close to zero.[28]

Whether a 2°C limit or halving greenhouse gas emissions by mid-century is the right goal is not the topic of this book. This opening chapter *is* concerned partly with a related but different question: *why* such goals have provoked controversy, particularly between scientists and economists, and why the accumulated scientific confidence has not remotely resolved economic disputes.[29] The book itself is concerned with the central disconnect – between the finite nature of fossil fuel reserves and the atmosphere, and the apparently inexorable increase, despite goals to the contrary, in global fossil fuel consumption and emissions. Why have governments in theory signed up to the goal widely supported by the scientific community, while apparently being unwilling or unable to take adequate steps towards it? In probing the experience, the book explores why progress has been so patchy and what this implies. It concludes that part of the problem has been the incomplete and disconnected nature of underlying assumptions and that a more integrated understanding is key to change.

Economic indicators often use ratios. We noted above that energy intensity – the ratio of energy to GDP – has fallen globally by an average of around 1.5 per cent per year (varying significantly between regions and over time as indicated in Chapter 4). Put the other way round, in 1980, the world generated US$2.8 billion GDP for every million tonnes of oil equivalent of energy supplied; by 2010 that had risen to almost US$5 billion. But that progress was outstripped by the pace of economic growth – so global energy consumption still grew.

Figure 1.5 shows the combined trend of efficiency (moving to the right) and decarbonisation (moving down the chart). Along with the progress in energy efficiency, the solid (historical) line shows that despite big programmes in nuclear and renewable energies and a move to cleaner natural gas, the *carbon intensity* of the world's energy system – carbon per unit energy – has barely budged since 1990. The relentless growth of fossil fuel consumption and associated emissions in many parts of the world has accelerated even further in the last decade with the rapid construction of coal-fired power stations in Asia (and particularly in China).

Energy resources and the atmosphere are physical systems, and it is the total quantity that matters. Geology and physics are not interested in either per capita or per GDP trends but only in the totals. Cutting per capita energy consumption does not cut the total if it is

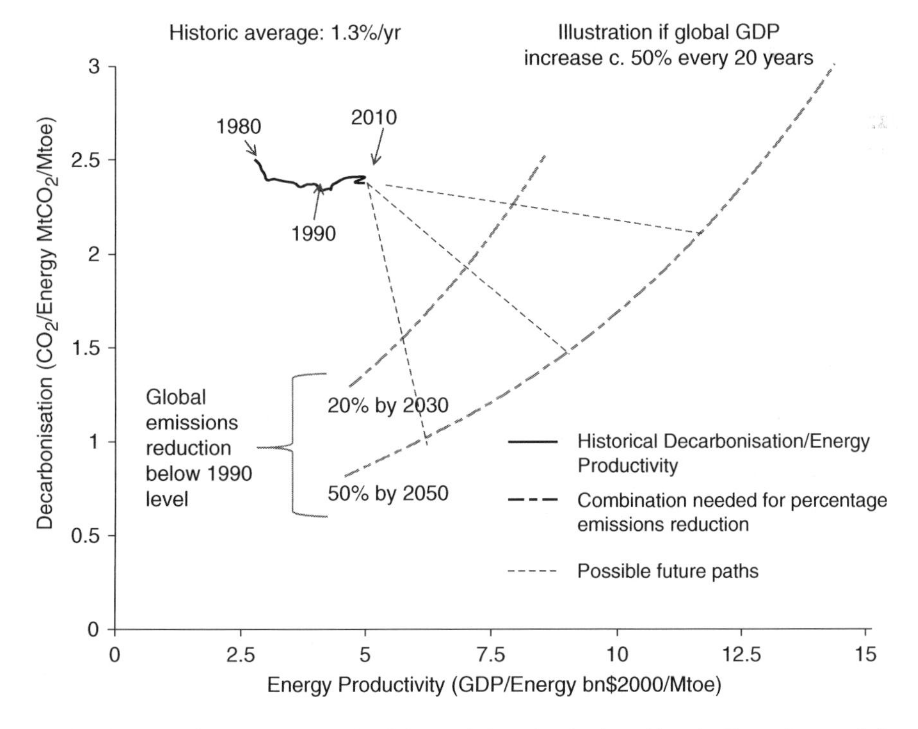

Figure 1.5 Trends in energy productivity and decarbonisation, historically and potentially required trends to 2030 and 2050

outstripped by population growth and reducing emissions intensity is of limited help if it falls so far short of GDP growth rates. The energy system overall seems to be gathering pace in the wrong direction. The meagre historical progress compares starkly against the challenges of reducing fossil fuel dependence or moving towards a stable climate. Figure 1.5 also offers a view of possible implications of the goal to halve mid-century CO_2 emissions. For simplicity, assume a doubling of global GDP to 2050:

- Doing everything by decarbonising energy supplies would involve a dramatic and abrupt turn compared with anything we have seen – it would mean doubling the *level* of decarbonisation in the next twenty years, and doing so again over the subsequent twenty while replacing the vast majority of fossil fuels in a hugely expanding energy system.
- Doing everything by energy efficiency would imply ramping up the pace we have seen – since the oil shocks – several fold: just for the 20 per cent cut, we would need to double the pace of improving energy productivity, and if global GDP doubled out to 2050 it would mean doubling the global rate – sustaining globally for four decades rates of improvement significantly higher even than the circa 3 per cent per year average over the past four decades in leading European economies (Chapter 5, note 51).

In reality, the only sensible approach is a combination of both, which at least sets each to somewhat less mind-boggling levels of ambition. For example, as shown in Figure 1.5, we could still halve global emissions by doubling energy productivity and halving the carbon intensity of energy supply – known as the 'Factor 4' solution.

Thus accelerating efficiency and decarbonisation are the pivotal global challenges of energy and climate change in the twenty-first century. Sustainable energy development will require changing course to become much more efficient and far less dependent on fossil fuels. Despite more than two decades of effort, we do not seem to have been doing well. We need to take a hard look at why.

1.3 Evidence and theory: energy and economy

Component trends

Some common myths in energy can be readily dispatched by a quick look at past trends in different regions. Before the oil shocks, it was often assumed that wealth, energy consumption and emissions inevitably increase together. It was then observed that many pollutants in fact followed what became dubbed the 'environmental Kuznets curve' – initially increasing with industrialisation, then peaking and declining as countries become more efficient, more technologically sophisticated and more sensitive to environmental impacts with corresponding legislation; they reach a point of wealth beyond which emissions start to decline.

CO_2 emissions, however, do not fit either pattern. Growth and emissions uncoupled, but economists have searched in vain for any clear evidence of a Kuznets curve peaking for CO_2. The evidence is in fact far more interesting; the later the countries developed, the lower their peak in energy-intensity. Studies in the 1990s identified 16 countries whose per capita emissions peaked with the oil shocks of the 1970s. Some of those reversed as prices declined, but by no means all – others continued declining. Figure 1.6 and 1.7 summarise the key trends:

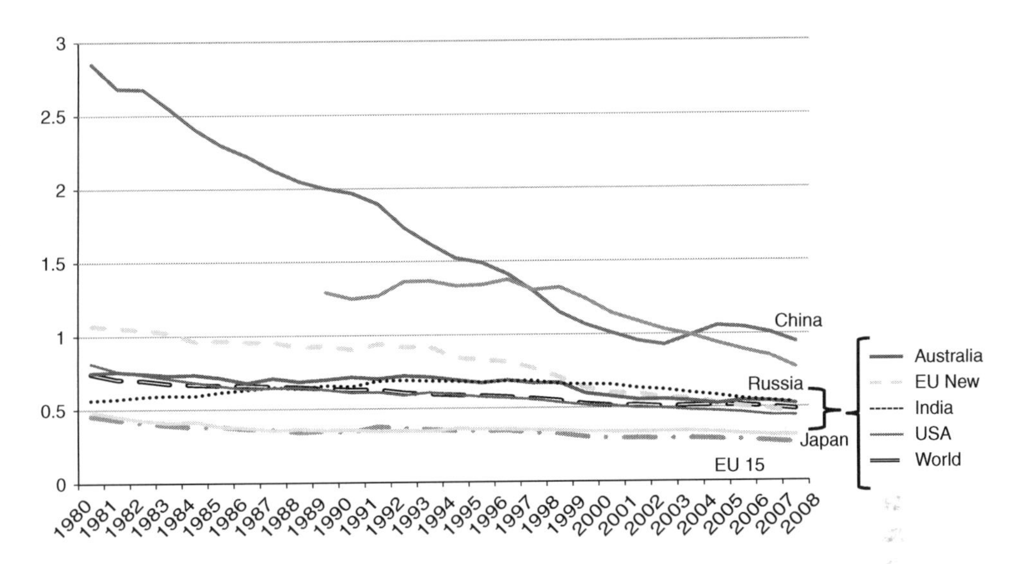

Figure 1.6 Trends in carbon 'efficiency', or intensity, by region and globally, 1980–2008

Source: Authors. Data from IEA (2010) and World Bank (2011).

- Figure 1.6 shows the development of carbon-GDP intensity (CO_2/GDP) in different regions.[30] The inefficiency of the 'former centrally planned' economies (those of the Soviet Union and China) is clear. The rapid efficiency improvement of eastern European countries after joining the European Union and having to adopt more market-oriented policies, including dropping huge energy subsidies, is striking. India initially increased carbon-GDP intensity – reflecting a general pattern of early stage development as countries move out of traditional biomass into fossil fuels – but has since improved pretty much in line with the US and Australia. Per unit of GDP these countries in turn consume 50–100 per cent more energy and emit twice as much CO_2 compared to Japan or the EU (in aggregate).
- Figure 1.7 shows the evolution in per capita emissions plotted against the level of per capita wealth (for key countries). (Figure 11.1 in Chapter 11 shows parallel and much longer-term trends in per capita energy consumption which show that India and China today use far less energy per capita than industrialised countries did at the same level of income.) There is a clear pattern of emissions rising in the early to mid stages of economic development – up to around $10,000 per capita – though even here there is huge divergence between. Notably, energy-related CO_2 emissions in Brazil are less than half those of many other mid-income countries for the same levels of wealth; Russia and east European countries have been much higher still, but economic recovery in these countries has not been accompanied by corresponding emission increases.

Thus economic growth has *not* uniformly increased emissions once countries have reached a basic stage of industrialisation. Above incomes of about $10–15,000 per capita, there is little sign of consistent relationship and indeed emissions per person seem to have roughly stabilised in many industrialised countries for some decades, and more recently for some of the most advanced developing countries of Asia and Latin America. However, there is clear

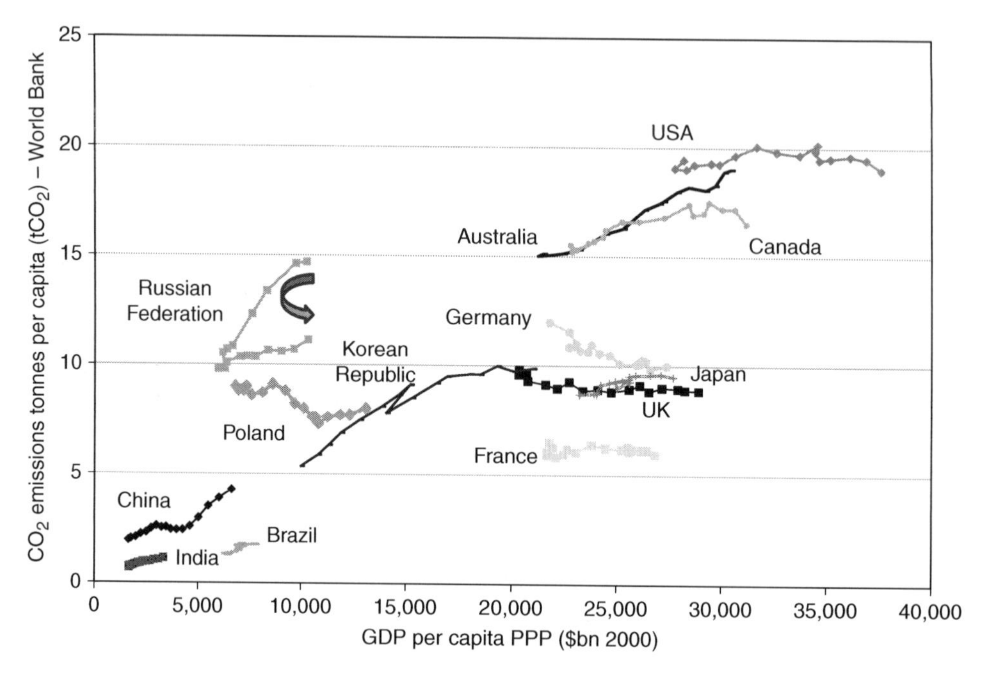

Figure 1.7 Per capita CO_2 emission trends in relation to wealth – trends of major countries, 1990–2008

Source: Authors. Data from World Bank (2011) and IEA (2010).

divergence between North America and Australia on the one hand, and the major industrialised economies of Europe and Asia on the other:[31]

- Japan and the European economies (together with New Zealand), which maintained higher prices on gasoline after the 1970s oil shocks and accepted emission caps under the Kyoto Protocol, have mostly kept emissions stable or declining at below $10tCO_2$ per capita.
- The US stayed close to $20tCO_2$ per capita and has been joined by Australia, with Canada slightly lower due mainly to its large base of hydropower. US emissions have also fallen recently as coal is displaced by domestic gas.

Different geographies, industrialised specialisation and trade patterns somewhat narrow the gap but explain less than half of this divergence.[32]

The pattern reflects the trends in carbon intensity in Figure 1.6. The first group appear to be notably more energy and carbon efficient and there is no sign of convergence between the two groups: if anything, there are some signs of divergence. Stronger policies to improve energy efficiency in Japan and Europe, detailed in Chapter 5 of this book, appear to have had an impact, and Europe has additionally started to reduce its carbon intensity as it has strengthened climate change-related policies.[33]

Among the effects of trade and economic structure, the impact of shifts in most of the developed (OECD) countries to import more carbon-intensive goods has been notable. Despite impressive domestic reductions in energy and emissions in some sectors and countries, detailed further in Chapter 4, it is uncertain whether any region has yet succeeded in reducing its overall 'carbon footprint'. Conversely, a substantial part of the explosive growth in Chinese emissions is due to its growing role as the world's manufacturing centre. Yet there remains no doubt that at national incomes above US$10–15,000 per capita, economic growth and emissions have become increasingly decoupled. The supertanker can change course, albeit only slowly and with difficulty. But the simple assumption that emissions and economy remain locked together as incomes rise beyond basic levels is just plain wrong. The reality is far more complex – and consequential.

Future trends and possibilities thus become all the more interesting. Most of the world's populations reside in the emerging and developing economies. The apparent stabilisation and 'open jaw' of different per capita emission levels at incomes above $10–15,000 per capita is then hugely significant. Given the weight of billions of poor people, the global average income level is around $10,000 per capita. Across Asia overall and much of Latin America, it averages at around $6,000 but is rising rapidly. It makes a massive difference whether their future trajectories emulate those typical in the US and Australia in moving towards 20tCO_2 per capita, the 10tCO_2 typical in Europe and Japan or closer to the 5–6tCO_2 per capita of France.

Yet even a long-run global average at this level, coupled with population still rising towards around 9 billion people by mid-century, would imply a 50 per cent increase in global emissions. To halve them instead would mean emulating the current energy sector emission levels of Brazil, around 2tCO_2 per capita.[34] Additionally critical in this, of course, is whether any of the industrialised groups can sustain and indeed accelerate a trend of absolute reductions in energy intensity and CO_2 emissions, extend this reduction to their overall carbon footprint and thus provide a more generic 'existence proof' of wealthy, low-carbon economies.

Uncertainty and the lack of foresight

Economic theories and policies based on 'rational expectations' require some foresight, at least if the results are to be remotely efficient. The volatility outlined earlier – and the sorry history of energy price forecasts – suggest a big problem which seems to be shared equally between governments and markets.

Examples are just too numerous. Consider how the UK and others drained the vast wealth represented by its North Sea oil (and gas) reserves. The bulk of these oil reserves were extracted and sold at oil prices below US$20 per barrel – one-fifth of the price a decade or two later. The UK, of course, was only playing by the same rules – and expectations – as everyone else. If the industry and government had had more foresight – or patience – it could have been extracted more slowly and most of the reserves sold later at many times that value, doing much to cushion the current impact of recession and debt. Many of the models used to project global energy assume 'perfect foresight' out to at least mid-century.

There has clearly been a bias in energy *price* forecasts, but it has been neither up nor down. It is the power of the present. When energy is cheap, we tend to assume it will stay that way. When the price rockets, we tend to assume that is the future. We attach heavy weight to our present experience and predictions themselves are heavily influenced by the present.[35] At times of low prices, people are reluctant to invest; they are more likely to wait. That same reluctance to look ahead – and scepticism – also often informs attitudes to environmental policy.

Therein lies one source of energy instability which helps to explain the long-run volatility noted above. In more recent years it has been doubtless more amplified by speculation on commodity prices, and the herd behaviour that can induce. All this, and the growing complexity and interdependencies of systems, may make global fossil fuel systems more, not less, vulnerable to 'Black Swans' – seemingly improbable events, not considered (or almost entirely ignored) before they occur, but which nevertheless happen and have huge consequences.[36]

While energy prices have been marked by long-run uncertainty and volatility, projecting other aspects of energy seems to have more systematic bias. The evidence in Chapters 6 and 7 points to a history of inflated forecasts of economic growth and industrial energy demand in the advanced economies and of the costs of environmental controls. With hardly any exceptions, environmental constraints have turned out cheaper to deliver than expected. After all, before we have tried, we cannot know all the ways to cut pollution, and as demonstrated in this book, when we try, we learn. Put these factors together – a tendency towards inflated projections of economic growth and/or emissions and an inability to foresee all the different possible ways of responding to incentives – and you get a pattern now well attested: achieving emission targets usually turns out cheaper than predicted. The history of the European Emissions Trading Scheme, which caps CO_2 emissions in Europe, is just one, big, example: as detailed in Chapter 7, it was opposed on the grounds of being too costly, but its persistent problem has proved exactly the opposite – carbon prices too low and inconsistent to support any low-carbon investment.

In short, the future is uncertain, and our perceptions of it – and projections of the costs of change – tend to discount risks and surprises, and bias us towards the 'status quo'.[37]

The response to energy prices

Markets respond to prices, and prices mediate a balance between supply and demand. Some of the errors in energy forecasting, though, stem from oversimplifying the relationship between energy demand and prices, which economists estimate as a 'demand elasticity' – how much energy demand goes down (or up) for a given percentage rise (or fall) in prices. Analysis is easy if the relationship is constant. It is not – far from it.

In the first place, what goes up does not come down; periods of high energy prices have seen rapid improvements in energy efficiency. These slow but do not reverse when energy prices go down again (Chapter 4). Prices can also induce many other indirect actions (Chapter 5). It remains clear that responses accumulate over time; the long-run elasticity can be much bigger than the short-run response (Chapter 6). New sources of 'price' – such as having to pay for something that was previously free – can also grab attention to a greater degree than price alone would suggest (see, for example, Chapter 7).

Another aspect of price is its tendency to affect everyone, but unequally; so its *distribution matters*, and energy price *rises* are hugely political. Governments have fallen over them, and even though economic theory suggests pricing pollution to be the most efficient way of reducing emissions, political opposition has stymied many attempts to do this. 'Least cost' to society overall is of little interest to those impacted by price rises (Chapter 8). 'Political economy' should thus be a central concern of any practical advice (yet is widely ignored in many economic recommendations).[38]

There is one fascinating additional fact. Instinctively, we would expect energy prices to determine energy bills: that sounds obvious. But a wider view shows this is not true. The amount that different countries (at similar levels of economic development) spend on energy is remarkably constant, despite sizeable differences in energy prices between them (Chapter 6). Moreover, despite even bigger changes in energy prices over time, the proportion of income that countries actually spend on energy is astonishingly constant: data assembled by the Russian energy scientist Bashmakov finds a 'stable' range of energy costs between 8 and 10 per cent of national income across the rich countries for over half a century, and estimates that in fact the relationship has held across more countries for centuries.[39] Deviations outside this range seem to provoke a disproportionate response (in the literal sense) that soon brings the balance back. This is a far cry from the idea of constant price elasticities (Chapter 6). Some reasons for this emerge from the analysis in Chapters 4 (for a lower threshold) and 8 (for a higher threshold) in this book. It has big implications for policy, yet in the West his work has been largely ignored.

Innovation and adaptability

One important factor explaining the above observations is that technologies evolve and systems adapt.

If energy were a fixed and immutable input to economic production, it would be hard to explain how energy and CO_2 could differ so widely between countries at similar levels of economic development. As noted (see note 32) geography alone cannot explain the gap, nor do trade and related structural differences in economies (e.g. specialising in higher or lower emitting economic activities), though all these play a part.

Innovation is one part of the explanation. The development and diffusion of new technologies is neither global nor instant. Chapter 3 illustrates the huge potential for innovation in energy systems, and the process of innovation is a broad theme across the book. Chapters 4 and 5 explain why some better technologies may languish for decades and the impact of policy in promoting their take-up, particularly concerning energy efficiency. The same is true more broadly in innovation (Chapters 9 and 10). Chapter 6 data on price impacts (Figure 6.1) points to apparent discrepancies between national and international indications. All this indicates that any *simple* model of price response misses key factors, and yet most economic analyses retain price as *the prime* indicator of economic cost and responses, if only for the lack of any alternative. This book presents a broader way to think about the relationships of price, cost and change.

A resulting theme that emerges throughout this book is the adaptability of human systems. It occurs in many ways, at many levels. Systems evolve. Infrastructure develops to support different patterns. And, people get used to things. People living in Europe and Japan do not fret every day about the fact that they only consume half as much energy as the New World economies – nor celebrate the fact that they emit half as much. They are, for example, more used to higher gasoline prices and the public transport systems that tend to go along with that. Not only components, but systems become more efficient in ways built in, not an ongoing cost. Adaptation to different circumstances is a demonstrable fact, with increasingly strong theoretical understanding as outlined in the next chapter and explored further in Chapters 10–12.

All this suggests that our energy-economic systems, given time, have a large capacity to adapt to a wide range of possible future requirements and constraints. *Planetary Economics* traces the numerous lines of evidence for this and the associated theoretical foundations in patterns of energy use (Chapters 4 and 5), more detailed comparison between countries' historical response to energy prices (Chapter 6), sectoral trends and possibilities (Chapter 10) and the role of innovation in macroeconomic growth (Chapter 11). This has major implications demonstrated in the concluding chapter – particularly when combined with an understanding of its siblings, inertia and irreversibility, also illustrated quantitively in the Appendix.

Inertia and irreversibility

The flip side of this adaptive capacity of energy systems is inertia. Having become accustomed to one pattern and system, people do not like change. Too much and too rapid change, indeed, can induce huge costs and social stress – as with the oil shocks. There are very real costs to change.

This is of course a very fundamental political phenomenon. Most of the major social and economic reforms in history encountered huge resistance at the time. Bitter opposition to the abolition of slavery, which was claimed to bring economic ruin, is but one example. The same is fundamentally true of many dimensions of our social and economic systems. For all the interest in new fads, fashions and technologies, humans collectively are largely creatures of habit.

Finally many developments are almost irreversible. This of course includes the use of finite resources and (for most practicable purposes) the accumulation of CO_2 in the atmosphere. Urban settlements, transport infrastructure and transmission systems also have enormous inertia, involving both physical infrastructure and land-use decisions that are difficult if not impossible to reverse even over decades (Chapter 10). We cannot simply change our systems overnight if problems arise.

All this makes it impossible to summarise usefully the impact of many energy sector decisions in terms of a single 'net cost'. What fuels will it use (or displace) and what future fuel prices are assumed? How long will the investment last and will it open up (or close off) other options or technologies? Can it be reversed, and will it endure? It is often difficult to quantify benefits in practical 'cost-benefit analysis', with various approaches adopted against the background assumption that some estimate is better than none.[40] However, this book argues that the difficulties take on a distinctive scale and form for energy and environmental systems. The themes of uncertainty, adaptability and inertia in the energy system itself – costs as well as benefits – are important conceptual hooks that thread throughout the book, and we return to illustrate their implications in the final chapter.

At present, the technologies, systems and habits we have, spread and accelerated by global economic trends, are pushing the world towards ever greater consumption of fossil fuels and associated emissions. People and institutions, within and outside government, are struggling to project and understand the likely consequences and to thus take appropriate action. This uncertainty brings us to the other set of complicating issues and associated evidence: trying to put a value on the environmental benefits of changing our course.

1.4 Evidence and theory: costing the climate

There are innumerable attitudes to climate change. Often they are influenced by issues that have little to do with the science itself, but reflect basic attitudes of trust (or not) in scientists or governments, religious beliefs, environmental and ecological values, underlying (or related) assumptions about the stability or fragility of complex systems and one's sense of risk and responsibilities towards others and the future.[41]

Governments face divergent views all the time, and cost-benefit appraisal is one central tool they often seek to navigate a rational, or at least justifiable, course. The appraisal explicitly estimates and balances the costs and benefits of a given action, usually in terms of money-equivalence. Indeed, most economists and many governments regard that as the theoretical heart of good decision-making. In concept, it is almost tautological: the benefits of a decision should exceed the costs. In application, it has been more controversial, not least in environmental policy. Decades of research have slowly narrowed monetised estimates of the local environmental and health damages associated with fossil fuels (see, for example, note 16).

The attempt to put a value on the benefits of cutting CO_2 emissions – and ultimately of stabilising the atmosphere – has, however, generated more heat than anything else, and raised debates among economists that seem surreal to others. Uncertainty of course exists in many areas; that is nothing new. The economic argument suggests that it is still better to rely on a 'best guess' of benefits, not least for consistency of decision-making. The holy grail of climate economics is to estimate the 'social cost of carbon emissions' – a measure of the aggregate damages caused by emitting another tonne of CO_2. In reviewing attempts to do this, however, this chapter concludes that something more fundamental is at stake.[42]

Trying to turn scientific projections into quantified estimates of 'monetised' damages raises four big issues: *What* impacts (and potential adaptations) are costed? *Whose* concerns are represented (and how)? *When* do these costs occur (and how do we weight that)? And how do we deal with *risks*, including possible planetary risks?

The first big-name economist to take a real interest in the climate issue, Professor Bill Nordhaus, initially attempted to quantify the cost of climate change in the 1990s. He estimated the cost of raising sea defences, of damage to crops, etc. His initial conclusion was that it was not worth paying much to avoid climate change – a mere few $ per tonne of CO_2, which might add a few per cent to the cost of gasoline and do little to change emissions.[43]

Nothing testifies better to the gulf between the way scientists, environmentalists and economists tended to view the problem. Professor Rob Stavins, at Harvard, once commented that he had never encountered any issue that so fundamentally divided these communities.[44] The rest of this chapter explores why, and what we have learned.

Costing what counts

Following Professor Nordhaus' initial efforts, many other economists attempted to more fully cost the potential impacts of climate change on additional sectors such as timber, energy, water supply, etc. Numerous such studies have now been conducted, as reviewed by the most prominent authors.[45] One approach compared productivity between countries at different temperatures and overlaid projected warming. This 'comparative static' approach suggests that there are optimum temperatures for most of the sectors, which lie somewhat above the average temperatures typical in mid-latitude regions. The implication drawn was that a warmer world would have modest impact in mid-latitude regions, but be particularly bad for tropical regions. Since mid-latitude countries dominate world GDP, the net impact of climate change was projected to be modest across the century, in part because the escalating costs in the second half of the century were substantially discounted (see discussion below).

Such analyses of course spawned numerous responses. Such 'comparative static' studies assumed smooth adjustment to a warming world, ignoring the likely costs of transitions. Fundamentally, human civilisation in every region has developed around its present climate; it does not follow that regions could seamlessly switch to different patterns represented by different societies in different climates. Proponents retorted that might be true, but nor was it fair to assume that people would just sit there waiting to be hit: people near the sea would retreat, moving rather than losing valuable shorelines; farmers would adjust crops and develop more heat- or drought-resistant crops rather than see their existing crops wilt.

In other words, people would adapt, and it is this adaptation that should be costed. A response in turn was that no one had perfect foresight of the future climate they had to adapt to and also that political systems were not rational or known for their foresight. All this led to entertaining debates about whether models should assume 'dumb farmers' or 'clairvoyant' farmers, King Canutes or retreating coastal resorts.[46] The devastation wrought to New Orleans because of inadequate preparations for known hurricane risks did a lot to boost the case that 'clairvoyant adaptation' was unlikely. The decision to resettle the city rather than abandon it also underlined human reluctance to give way to nature's encroachment in the most 'cost-effective' way that economics would seem to suggest.

Notwithstanding such details, the approach is essentially 'reductionist': reduce the climate problem to its constituent impacts and estimate the cost of each component. Taken in isolation, this can easily be dressed up to make cutting emissions look plain silly. The 'sceptical environmentalist' Bjorn Lomborg argued that if we want to save polar bears, the first thing is to stop shooting them; if we want to reduce heat deaths, try air conditioning; if we want to stop malaria, drain the swamps and provide drugs; and even if we want to save the Maldives, it would be a lot cheaper to fund massive sea defences than to implement the Kyoto Protocol.[47] This makes for great rhetoric but little more – obviously cutting emissions contributes to all these objectives and would reduce innumerable other impacts of climate change around the world. Its costs can only sensibly be compared to all the costs of impacts and adaptation added up across all the climate damages we may worry about together.

That is what the mainstream economists tried to do. Still, despite all these debates about how to calculate numerous potentially quantifiable components, one essential fact remains on the side of the economic sceptics in this reductionist approach: the explicitly 'climate-vulnerable' sectors of modern economies do not account for much GDP. However much it costs to adapt agriculture, forestry, air conditioning and even shorelines, for example, it cannot really get close to the value of the whole economy.

That, however, was by no means the end of the debate: it is actually just a testament to the torch shone on one part of it. The real scale of the analytic challenge is highlighted in Figure 1.8. This sketches out the dimensions of the problem in both the kinds of climate change that could occur (rows), and the scope of what possible impacts are considered (columns), illustrated with some examples.

To explain all the elements fully would take a good many pages, but the main point should be clear. Some things we can take a stab at measuring. Others look almost impossible.

The bottom row is where some of the more nasty possible physical impacts would come in: clearly, the reductionist studies have not costed the possible impact of major instabilities in the earth's systems, as outlined in explaining Weitzman's Dismal Theorem below.[48] The last column speaks to the social equivalent – the distance between theoretical bit-by-bit studies and the potential realities of multiple stresses on real societies. All-seeing planners and strong societies could absorb considerable climate changes, but the capacity to do so faces many real-world constraints. Drawing in part on wider development literature on the economics of natural disasters, a few brave souls have ventured to construct models in which poorer societies are unable to recover from one extreme climate event before the next disaster strikes, leaving such countries trapped in a cycle of under-development – with massive consequences for their wealth and welfare.[49]

The fragility of some small countries and 'failed states', like Haiti and some in sub-Saharan Africa, spring to mind. A more unnerving example is Pakistan's 2009 floods which killed thousands and whose impact on internal stability may yet to be fully played out. A historical study suggests that temperature fluctuations have been correlated with declining agricultural yields, frequency of war and population declines.[50] Consider also, for example, what a drying of the Amazon rainforest would do not only to global CO_2, but to the societies

What kinds of climate changes?	Which kind of impacts?		
	Market	Non-market	Multiple stresses and socially contingent
Projection (trend)	Coastal protection Dryland loss Energy (heating and cooling)	Heat stress Wetland loss Ocean acidification Ecosystem migration/ termination	Displacement from coastal zones Regional systemic impacts
Climate variability and (bounded) extremes	Agriculture Water Storms	Loss of life Biodiversity Environmental services	Cascading social effects Environmental migration
System changes and surprises	'Tipping point' effects on land, resources	Higher order social effects Irreversible losses	Regional collapse Famine War

Figure 1.8 The risk matrix: an assessment framework for evaluating the social cost of climate change

Note: 'Socially contingent' costs may be understood as those that may be amplified by the inability of society to respond to impacts effectively, such as failures of governance, inability to act collectively, or the frictions associated with migration or deeper disturbances.

Source: Developed by the author from Watkiss and Downing (2008), Jones and Yohe (2008) and Downing and Dyszynski (2010).

living there, and what major changes to river systems of the Middle East, or on the Indian subcontinent, might to do those regions' fragile peace. Obviously, sufficient international assistance, well-delivered and implemented, could forestall many problems but, in reality, mechanisms for adaptation, compensation and cost-sharing are inevitably weaker at international levels. This may increase the probability of adverse effects propagating across regions (including through migration), blurring any distinction between 'winners' and 'losers'.

The point is not that such things will happen, but that they have not been evaluated in most of the reductionist studies that claim to put a cost on climate change – because we've no idea how to quantify them. Some have attempted, but the authors of Figure 1.8 noted that over 95 per cent of the studies that seek to put a monetary value on climate impacts have focused on only two out of the nine elements of the matrix, namely the market and non-market costs associated with smooth projected change (and many of those neglected transitional costs of shifting systems from one climate to another, still-changing climate). The estimates of Nordhaus and others try to quantify a wide range of measurable impacts, but the light shines only on those components that are, relatively, easy to quantify. The rest are either ignored or just have crude numbers assumed with little real basis.

The lurking fear of a 'Black Swan' in a perturbed climate system – or in societies' ability to handle climatic change simultaneously with rising pressures on food, energy and other natural resources – cannot be eradicated. This, for better or worse, further opens up the field to almost anyone's perspective – and brings in yet more complexities.

(Not) willing to pay, (not) willing to accept?

Anyone who thinks that putting a cost to climate damages is a dry, technical debate about risks between economists and scientists should go back in time, to the adoption of the Second Assessment Report of the Intergovernmental Panel on Climate Change (IPCC) in 1996: the only time, in fact, when the IPCC was asked by its government sponsors to assess the overall economics of climate change damages. The attempt nearly destroyed the institution, and remains the only occasion in which authors of an IPCC chapter formally dissented from the politically negotiated summary of their work. Most of it boiled down to the question: who?

To estimate 'non-market' impacts that cannot be measured directly in terms of money, economics has developed various approaches, the most classic being to ask people what they would pay for getting something – or accept in return for losing it. Unfortunately these approaches do not give consistent answers. It is not uncommon to find that 'willingness to accept compensation' for an adverse impact differs wildly from 'willingness to pay' to prevent it – but there are no clear grounds for choosing between these measures. Willingness to pay, of course, is limited by one's own wealth – whereas some things we may hold to be dearer than any amount of money.

That is when interests can clash, and societies in practice do not rely primarily on economic measures to decide the outcome. If the government wants to knock down houses to make way for a new road, it does not generally ask the residents how much they would pay to stay nor how much compensation they would like for being evicted. We have planning procedures, courts and appeal procedures to decide whether to go ahead and to determine compensation to victims forced to leave their homes.[51] At the end of this book we return to this theme within national decision-making.

Suffice to say we do not currently have those procedures and institutions across national borders, so how to value and aggregate impacts globally is deeply political.[52] One can only guess at the amount of compensation the small island states or drylands of Africa might seek for suffering climate change. They cannot simply dictate the answers, but nor can their concerns just be ignored – or capped by their poverty.

The most potent case of this problem of international valuation led the IPCC Second Assessment into very hot water in 1995. To cost climate change, economists needed to put a cost on deaths – with a 'value of statistical life', which was carefully defined as 'the value people assigned to a change in the risk of death among a population'. Unfortunately of course they never agreed on which 'people' should 'assign the value'. The standard economic approach yields a 'value of statistical life' that depends on national per capita wealth; it would indeed be absurd for India to try and put the same resources into modern medical services as the US, when its people suffer many more basic threats to life and health.

As long as the relevant decisions of nations remain essentially separate, a different 'value of statistical life' in different places is a simple reality; it is a measure of the resources that a country can wield to try and protect its citizens. However, applied internationally such a 'willingness (and ability) to pay' approach implies that lives in poor countries count for less than in the rich.

The developing country representatives in the IPCC were correspondingly less than amused. Climate change is not an internal issue but concerns impacts by some nations on others, and in aggregate by richer nations on poorer ones. A recent assessment estimates that Africa, the Indian subcontinent, Latin America and smaller island states may experience more than 75 per cent of the climate damages this century, but they account for less than a third of the cumulative emissions even just in the past 50 years.[53] Whose value of statistical life should then be used in evaluating the damages? Are poor victims being expected to pay – and hence valued by their poverty – for damages imposed by the rich? Who should pay for damages imposed by one country on another, and who has a right to put a value on the lives of other people in other societies and on what basis?

Intellectually, it all fell apart on these fundamental ethical questions, leaving bitter suspicions about global cost-benefit analysis that persist. Some time after the dust had settled, the original IPCC authors – quietly modifying their original position that governments were being irrational in objecting so much to their cost-benefit calculations – observed that the lack of any consistent global decision-making authority meant that economics should not be expected to come up with a single globally applicable answer to 'the cost of climate change'.[54]

A Stern warning about time

Following these intemperate rows over the 'value of life', governments largely abandoned efforts to reach any agreement on the global cost of climate change. Economists still needed some unified measure, but different camps of analysts went their own ways, and different governments listened to different camps. Even among rich world analysts, a fundamental schism grew, and one result was the launch of the Stern Review.[55]

Professor Nick Stern was then Chief Economist at the UK Treasury and former Chief Economist at the World Bank. After the 2005 G8 Summit on climate change ended in acrimony and incomprehension, the UK government asked him to lead a major study on the economics of climate change. He arrived to the task with little prior knowledge about the subject, but his background in development made him more inclined than most to take a long-term view, and gave him a keen sense of the ethical underpinnings of the tools that many of his fellow economists had taken for granted. Foremost among these tools was how to weigh future costs and benefits against the present: the value of time.

Standard economics expresses this value of time as the discount rate – an annual percentage by which future money is 'discounted' back to present value equivalents. This normal practice carries an implication that because climate change is a very long-term problem – with impacts that will accumulate over decades and centuries – the discount rate matters a lot. Most previous economists trying to calculate the cost of climate damages had gravitated towards the central rate used by Professor Nordhaus, typically around 5 per cent per year – though he himself later moved to a more sophisticated treatment with rates that declined over time (see below).[56]

Compounding 5 per cent per year weights any impacts just 30 years away at just over one-fifth (0.215) of their actual value. After 60 years, they are discounted by a factor of more than twenty (0.215×0.215). By the end of this century, impacts are weighted at only 1 per cent of their actual value (0.215^3). After two centuries they are discounted by a factor of almost 30,000. The long-term thus becomes effectively irrelevant in the calculation.

There are strong reasons to make some discounting of future costs and benefits. In evaluating business projects it is a standard tool, related in part to risk. Business looks to higher rates of return on investment for taking risks: successful projects help to create wealth; bad ones may risk bankruptcy. Governments have a far higher capacity to absorb such risk and government bonds consequently carry relatively low interest rates (the social cost-benefit analysis of 'projects' that are privately financed but carry public benefits may need to combine private financing costs and the social discount rate).[57] Individuals also clearly value money more now than in the future – but the way in which they do so is not consistent (see Chapters 2 and 4). Both these facts, now undisputed, introduce an important oddity: the 'value of time' used to assess the seriousness of the problem may be very different from that used in costing solutions. This is an additional reason for the approach developed in this book, which avoids the attempt to find a unifying metric by focusing instead on the different processes operating at different scales.

Despite the difficulties, the standard assessments had used a descriptive approach based upon the observed short-termism of economic indicators such as risk-free interest rates – with 5 per cent widely adopted as standard (a rate which after the credit crunch and an era of almost zero rate bonds looks a somewhat quaint expression of the confidence of the era). Several economists had for a long time been disputing this approach, but it took the Stern Review to rewrite the map of the debate.[58]

Stern took a fundamentally different approach. Climate damage is about public policy, not private profits. Discounting combines two factors: one, expectations about rising wealth, which would make a given amount of money less valuable in the future than today; and the other, an ethical judgement about how much weight to place on future human welfare – expressed economically as a 'pure rate of time preference' (the Ramsay Rule).[59] Stern argued that there was no ethical basis for governments to discriminate against future generations (as he expressed it) in this way. Consequently, the expectation that future generations will be richer than us was the only defensible justification for discounting costs and benefits over time. This led Stern to a discount rate of 2–3 per cent per year, consistent with expectations of future global economic growth rates. Suddenly, future damages mattered a whole lot more than in the calculation of most previous economists.

There was more to come. Along with the long timescales, a defining characteristic of climate change is its uncertainty. That does not mean you can pick an average or 'best guess'; it means you have to consider a whole range of possible outcomes. Some of those outcomes – the right-hand column or the bottom row of the 'risk matrix' in Figure 1.8 – could be really bad. In the worst, future generations may not be so much better off – in which case, the remaining rationale for discounting crumbles. To be more precise, economics should not apply a single rate for discounting the future; rather the rate has to reflect how good or bad things get. The net result is that bad outcomes, which were rendered virtually invisible by discounting in the standard approach, really mattered in Stern's calculations. A 'social cost of carbon' was calculated to quantify costs per unit of emissions, and came out at US$85 for every tonne of CO_2 – or just over $300 per tonne of carbon. One of the world's top applied economists, heading a major government review, was saying that climate change is a really huge issue that potentially threatens global development: 'the biggest market failure in history', as he put it.

Failed reconstruction

There were of course criticisms of Stern's approach. Many focused on the 'prescriptive' approach to discounting he used – one based on ethical reasoning of how governments *should* weigh the future. Instead, it was argued that observed risk-free market interest rates (like government bonds) reflect a revealed preference in the population for discounting.

Such a *descriptive* approach would discount future impacts (climate or other) much more heavily.[60]

The Cambridge Professor Partha Dasgupta took a different tack and challenged some of Stern's internal reasoning, in particular the consistency between how Stern weighted the welfare of inequalities within present generations versus future ones. This raised fundamentally complex and long-standing debates about how economics relates wealth to welfare, but was itself subject to charges of inconsistency.[61] Swathes of other debate, from numerous perspectives going way beyond the theoretical considerations of how to aggregate over time and distributional impacts, ensued.[62]

The model that Stern used was designed to represent a wide range of uncertainty (and it did). Its author noted that the range of estimates of the 'social cost of carbon' had in fact got much wider over time, particularly on the upside, with the two biggest factors being assumptions around atmospheric sensitivity and the discount rate. Runs with a wide range of scientific and economic parameters suggested a '90 per cent confidence' that the damages from emitting a tonne of carbon lay in the range $10–222/tC, while using Stern's discount assumptions raised this to $60–1,025/tC – suggesting remarkable foresight in Downing's originally provocative claim that the 'social cost of carbon emissions' could be anywhere between $10 and $1,000/ tCO_2.[63]

The US government subsequently made a determined effort to resolve disagreements about discounting by bringing together many of the key leading protagonists, a roll-call of many of the best economists who had debated the issue. They reached consensus about some fundamentals – the principles of the Ramsey Rule and the fact that discount rates could sensibly decline over time. They could only agree to disagree about the practical implications, however.[64]

In some ways the situation turns out to be even more confusing. Dasgupta followed through his criticism of Stern by probing more deeply how economists represent the welfare of present and future citizens compared to the welfare of rich and poor today ('inequality aversion'). He noted that higher inequality aversions (for which he presented evidence) can restore a significant discount rate, reducing the discrepancy between Stern's 'ethical' and the 'market' discount rates: we discount the future more not because we do not care about our descendants, but because we weight more heavily our expectation that they will be richer than ourselves – so we conclude that it's justified for us to spend, rather than save or invest for their benefit. Unfortunately, Dasgupta found that a higher 'inequality aversion', applied across generations, which also makes the overall valuation *even more* sensitive to uncertainties. He concluded that with sufficiently high uncertainty, 'no optimum policy exists … consumption discount rates cannot be defined and social cost-benefit analysis of projects becomes meaningless.'[65]

Overall, the Stern Review's probing of uncertainty and discounting stimulated a huge set of debates without anything approaching real economic consensus on the costs of climate change. But there was yet another 'elephant in the room'.

A tale of tails: Weitzman's Dismal Theorem

There are not many bigger names in economics than Professor Marty Weitzman, and it was natural that the American Economic Association – in their guise of Editors of the *American Economic Review* – should turn to him to write a critique of the Stern Review. His judgement probably confounded all expectations: in a nutshell, he argued that Stern was right about the seriousness of the climate change problem, but for the wrong reasons. He turned the spotlight onto not just the smooth changes projected, but the possibility of major disruptions of unknown severity. Economists, he warned – even Stern – did not really capture the importance of potential catastrophic outcomes in their cost-benefit analysis.

The next year he built on this with one of those papers that change the terms of the debate. His core question was where do scientists, or indeed anyone, get evidence about the likelihood (or not) of catastrophic impacts. He pointed to some of the scientific literature on potential climatic instabilities, like the collapse of the Amazonian rainforest system or the Indian monsoons, of shifts in the fundamental ocean circulation patterns or runaway feedbacks of Arctic methane release. He developed a mathematical theory of the distribution of possible outcomes, based on the fact that the more extreme the event, the less data we will have about it. When he attempted to apply this to a cost/benefit analysis of climate change, his final equation 'blew up', failing to yield any sensible finite result, for reasons that are masked in common practices of probability calculations.[66]

He emphasised that what matters is what we don't know: the extent to which the more severe climatic disruptions may be less likely, relative to the rising scale of human suffering they might cause. In his analysis, the latter came to dominate the equation and the average expected costs – the costs weighted by their probability – might be 'infinite'. He thus shone the spotlight on the bottom row of Downing's 'risk matrix' (Figure 1.8) and, for some regions, the right-hand column. Such risks might be in the 'tails' of the range of possible outcomes, but this did not mean they could be ignored: it simply meant that we could never assemble enough data on the likelihood (or not) of such events before the welfare loss dominated the 'cost-benefit' calculation.

Weitzman dubbed this conclusion his 'Dismal Theorem'. Economics relied on providing quantified answers to questions and cost/benefit trade-offs are at the heart of an economic approach. Yet no matter how much effort they make trying to estimate the individual components of climate damages, they would always still be 'looking under the lamppost' of a problem for which the potential key – the risk and damage associated with catastrophic disruption – would by definition be unknowable.

Thus he argued that a holistic approach to costing climate change comes down to two guesses: whether and how humanity might cope with severe climatic disruptions (and how to value it); and whether the risk of such events declined in a 'thin tail' (more rapidly than the costs rise) or a 'fat tail' (less rapidly). He argued that being uncertain even about the *nature* of the risks – a 'distribution of possible distributions given the limited data' – implied a 'fat tail' as the most reasonable assumption.

In essence, Weitzman had developed a mathematically formal economic rationale for the scientific precautionary principle: uncertainty is not a reason for inaction, but rather a reason for caution about interfering in crucial, complex systems when we cannot know the full range of consequences. Perhaps the scientific principle was too intangible for government decision-makers trained in the philosophy of cost-benefit analysis. Weitzman had shown that applying this 'rational trade-off' approach consistently to all elements of the risk matrix led to the same basic conclusion: be very cautious about changing the atmospheric heat balance of the only planet we have.

Not surprisingly, having spent the last two decades trying to count the potential economic costs of climate change – filling the literature with papers estimating various 'reductionist' components of adaptation and volumes of economic debate on valuation over time – Weitzman's message was not one that many of his colleagues found appealing. Yet the literature on the 'social cost of carbon' had already generated a disturbingly wide range of answers – from a few to several hundred dollars per tonne of CO_2. Weitzman's Dismal

Theorem, in effect, was just extending this to argue that no one could realistically determine the upper bound.[67]

1.5 Trapped?

The holy grail of an economic cost-benefit calculation applied to climate change is to establish the 'social cost of CO_2 emissions' – the damage inflicted by emitting each tonne. As shown, the calculation turns out to be just about as uncertain as everything else – in fact more so. It does not provide an objective answer in a world of conflicting views, but unavoidably reflects back the assumptions and values that people bring to the table. Climate policy will have to contend with 'incommensurable benefits estimates'.[68] Even the world's top economists have accepted that they cannot reach an agreement about how to tackle the problem, even on the single component of how to weight impacts over time, let alone other dimensions.

If the problem is defined in terms of trying to reach global agreement on the scale and urgency of the problem, there appears to be no prospect of a rational quantified answer that will command consensus. In this book we argue that moving forwards requires far more attention to the structure of the mitigation costs as well as the benefits of action. Based on this reasoning, the concluding chapter explains why the economic case for stronger action on climate change requires neither belief in a risk of catastrophic outcomes nor a Stern-like discount rate.

Ultimately, psychology, behavioural studies and political economy are also required to shed light on the situation – since we are talking about human decision-making. As detailed in Chapter 2, numerous studies have shown that individuals do not have a consistent, 'calculated' approach to risks: perceptions are strongly influenced by social context, past experience, vested interests, associations in people's minds ('associative processes') and events which simply affect people's perceptions ('affective' processes). One might hope that professional experts were immune to such factors, but the evidence suggests otherwise: on reviews of past judgements, only those who receive frequent and timely feedback on the accuracy of their assessments perform well – weather forecasters and bridge players being leading examples.[69]

This, more than anything, probably provides the answer to Stavins' question as to why climate change seemed to divide the economists and scientists at Harvard/MIT more than any other issue. Profound uncertainties provide the scope; the nature of the problem, the reasons. Economics is mainly about marginal changes to stable systems, and how to make the best trade-offs based on relative costs and values. Scientists are more used to dealing with complex systems that can exhibit highly non-linear behaviour, including phase transitions (solid to liquid to gases as the temperature itself changes smoothly, 'butterfly' effects, oscillations between different equilibrium states, and so forth). Managing risk and profiting from it is the essence of economic growth; scientists are more likely to be called to advise on disasters. The disciplines are trained to think differently and their subject matter tends to reinforce this belief. Perhaps the different disciplines also attract people predisposed to think in corresponding ways.

In brief, after two decades of debate, we do not seem to have reached any helpful resolution about the costs of climate change. The effort to provide clear, objective advice to policy-makers – about how much it might be worth spending to combat climate change – has resulted in a series of profound – ethical and other – questions, intemperate disputes and deep imponderables. Why have attempts to cost climate change got into such a tangle and what does it imply?

As outlined above, the debate over what to measure and how to measure it was just the start. The field of economics has found itself embroiled in accusations of seeking to defend the indefensible, in terms of damage inflicted by one region's actions on others. What to some may seem a reasonable weighing of costs and benefits appears to violate others' basic principles of ethics and security, with a disconnect between what some regions seem willing to pay and what others accept as a reasonable effort.

The issues in weighing the future are even more profound. The future generations have no voice – and analytic attempts to give them one, in terms of weighing their interests through discounting, also seem to end up in knots, while pitting present generations against their own descendants.

The challenge is not just technical but involves ethics and judgement. On all three of the 'distributional' aspects noted above – valuation of non-market impacts, their aggregation across countries and weighting of future impacts – paying greater explicit attention to the ethical issues indicates that climate change is a more serious issue than suggested by the simpler earlier calculations of (mostly Western) economists. Not surprisingly, a leading philosopher described climate change as 'the perfect moral storm'.[70]

To cap it all, two top economists have then argued that economics has come face to face with a fundamental inability to answer the question – and from two completely different angles. Dasgupta's conclusion is that uncertainties over how to aggregate diverse but potentially severe impacts on different people over space and time mean that 'there is no optimal policy'. Weitzman's Dismal Theorem suggests that by definition we can never know enough about long-term, planetary risks to provide objective numbers about the things that matter most.

This book – through the framework presented in the next chapter – offers a partial way out of the resulting conundrums. Chapter 6 illustrates one implication – broadly accepted – that the long-term goal is best expressed in terms of 'safe' emission levels based on the scientific estimation of risks, not by largely fictitious attempts at cost-benefit calculations. Much of the rest of this book concerns the fact that we nevertheless seem unable to make serious progress towards the agreed goal, and why a fuller understanding of 'planetary economics' can reduce both the theoretical tensions and ease the policy challenges. But first it is useful to look harder at the conceptual problems that underlie the present state of affairs.

Inside out

The tangled attempts to 'cost the climate' suggest that there is something more profound at work – and there is. Economics classically conceives of climate change as an 'externality' – an external cost that our emissions impose on the world around us and which therefore needs to be appropriately costed, with a 'social cost of CO_2' as outlined earlier.

It is a valiant effort, but is ultimately doomed, because it concatenates two ineradicable problems.

The first problem is trying to find an 'aggregate' answer in the absence of any agreed set of aggregating rules. There is no global decision-maker today, let alone one covering the present and the future. We have neither an objective theory nor agreed institutions that can establish how to add up welfare across different groups, countries and generations – something that governments do all the time within their own jurisdiction, with the check-and-balance of voting if too many people do not like how they do it. Around the world and over time, energy policy and climate change involve winners and losers, and perhaps more viscerally, aggressors and victims. In the absence of agreed political authority, human history suggests we may resolve such situations by negotiation or by conflict, but not by pure abstract analysis.

The other problem is that defining climate change as an externality is inside out. The economic system depends on natural resources and exists within a social framework which in turn depends upon the natural environment. Obviously, modest perturbations to the natural environment – ones with limited damage or which we can clean up – can easily be accommodated. That is how economists are taught to think about environmental costs and policy.

But if the framework itself comes under threat – in this case, the environmental systems that sustain the social fabric whether within individual regions or globally – it becomes meaningless to try and put a single number on it.[71] Countries do not put a single number on their security; nor indeed do they seek to assign a value to the security of others. Each country, rather, reaches a political consensus on how much to invest in the security of its citizens and has a right to defend itself against threats to its security. Beyond a critical level, climate change looks less like an 'externality' and more like a threat to security.[72]

Of course, climate change is not the only possible issue of this nature. Scientists' efforts to probe natural 'planetary boundaries' have identified nine possible global environmental stressors, of which humanity is already estimated to have entered a dangerous zone for three of them – biodiversity loss, interference with the nitrogen cycle and climate change. The sheer scale of these issues over time and geography means that many of the principles of planetary economics explored and developed in this book doubtless apply in some ways to these other big, long-term, cumulative and global issues. It is these factors that distinguish such problems from the 'classic' environmental issues.[73]

Approaching planetary boundary problems in terms of security and risk management does *not* create a carte blanche to put the environment before everything else: how much a country spends on its military, intelligence and diplomatic services, for example, reflects the resources it has and chooses to make available, given its assessment of the potential threats. However, the next chapter shows how it can form *part* of a more useful way of understanding the issues and implications. First though, we need to take a brief look at some of the 'easier answers' proposed – and why they are not.

Driving blind

There have of course been many responses proposed to avoid the implication that big emission reductions will be needed. One was an extension of the traditional argument, namely that it would be cheaper to 'adapt' to climate change itself. Adaptation is of course important to deal with effects that are already occurring or inevitable. However, its limitation is obvious: the more serious the impact the harder it is to adapt, and the more global the change the harder it will be for the international community to help each individual region. All serious adaptation studies emphasise the extent to which adaptation is constrained by the lack of foresight and by likely institutional limitations. The authors of the risk matrix (Figure 1.8) suggested that adaptation might offset 75 per cent of the costs of mild climate impacts, but only 10 per cent of the costs of severe global impacts.[74]

Another argument was that if signs of an impending global climate disaster did start to emerge, humanity could always then cut back on emissions. Unfortunately, there is too much inertia in all the main systems involved – economic as well as natural – to make this a safe or even plausible bet. If the atmosphere were stabilised at a given level, temperature would continue to rise for decades and ice sheets would continue to melt for millennia. Even if humanity stopped emitting overnight, then a century later the atmospheric concentration would still only be a third of the way back to pre-industrial levels.[75] Moreover a central dimension of this book stresses that turning the global energy system around could take almost as long as the natural systems. We are dealing with massive, interrelated systems (atmosphere, ocean, cryosphere and our industrial economies) – each of which has enormous inertia. The 'characteristic timescale' of the systems involved are typically 50–100 years or more; even the economic effects of reducing emissions within the energy system play out over decades.[76]

The final option – if adaptation threatens to be overwhelmed by climatic instability and it is too late for emission cutbacks – has been presented as 'geo-engineering': deliberate but counteracting interference. A Royal Society survey of proposed measures concluded that the 'least bad' option would be to inject sulphur into the stratosphere, to form droplets to reflect solar radiation – an option high on affordability and effectiveness in offsetting global temperature but unfortunately low in safety. While average temperatures might thus well be reduced, it is unclear what the combined impacts would be for individual regions and what other side effects may be anticipated. Not surprisingly, it seems we can only try and tackle our creation of one global risk by creating another. In most respects geo-engineering, at least in the form of trying to engineer the planetary heat balance, appears an option of last resort.[77]

Given all these factors, the 'do nothing' (or very little) approach lacks credibility. In effect it diminishes many of the ethical dimensions (impacts on the poor and future generations of 'central' climate change projections), ignores unquantifiable risks (which equates to assuming they do not exist) and then proposes somewhat unconvincing responses if it all goes horribly wrong. It is like driving into thick fog and keeping one's foot on the accelerator, on the grounds that we do not have a clear view of problems ahead and hope that we can brake or otherwise adjust in time if and when a problem appears. It is hardly a compelling approach.

Yet, effective action remains muted.

Pessimism squared: the problem of collective action

All this underlines an additional and crucial feature of the reality: there is no global decision-maker. There are close to 200 sovereign countries in the world, and those that make significant contributions to global energy consumption and emissions span all continents and cultures.

If the problems are viewed as ones of sharing out the burden of cutting fossil fuel consumption and emissions, then traditional economics has an apparently fatal *coup de grâce* to offer. The cold, hard logic of national cost-benefit assessment implies that a country could not be expected to take on costs bigger than the benefits that would follow. A book by Scott Barrett (2005) most thoroughly laid out the classical argument, and its logical conclusion was stark: a binding agreement on sharing the burden is impossible. Even if a deal was reached, how could it be enforced? It would always face being undermined by 'free-riding' of non-participants. On this view, economic theory predicts that humanity will be incapable of reaching any agreement remotely able to solve the problem.[78]

In fact it was even worse than this. Psychological studies have shown what we all know from common experience: people's beliefs and perceptions are strongly influenced by their interests. Human beings like to believe what suits them. Since climate change denial has in most parts of the world ceased to be plausible, opposing action has increasingly relied upon arguing that someone else should do more instead. Against this background, it is incredible that the world ever managed to reach an agreement about climate-change policies, least of all an architecture as ambitious as the Kyoto Protocol. In fact the subsequent history of the protocol underlined the scale of the global climate problem: to reach a deal, the US Administration had gone out on a limb that its domestic constituencies would not support. Delivering an apparent international success (the Kyoto Protocol) came at the price of domestic failure (inability to deliver at home, thus undermining the whole deal). Twelve years later, at Copenhagen, the US tried a different approach. Instead of arguing for a binding international structure like Kyoto, it reduced international structures to a minimum, on the grounds that this would be more acceptable to the US Congress and to other countries. Yet the US still could not deliver domestically, as explained in Chapters 6 and 7.

Where, then, does that leave the prospects? Traditional economic analysis, taken to its cold logical conclusion, seems to imply that humanity can never solve global problems of such a scale. The patchy progress to date would seem to confirm this. Indeed the unending debate between 'prescriptive' and 'descriptive' discount rates mirrors a gap between how people might think we *should* behave and how we actually *do*. Moreover in much of the 'developed world', people are now burdened by inherited debt and the need to look after ageing parents and plan for their own old age. Whatever the ethical case, or possible risks, this breeds a reluctance to pay to protect the future against uncertain risks.

We seem to be doubly trapped. There is precious little sign that human societies have the capacity to look ahead with an objective and ethical eye that could lead them to invest adequately to avoid unreasonable costs and risks – even for themselves, let alone for the sake of future generations largely in other countries. And even if they could, countries seem trapped by the dilemma of 'collective action' – an inability to agree on what a 'fair' distribution of the effort might be, or to ensure adequate participation. In that sense, the traditional theory is self-defeating and thus an invitation to despair.

This is why it is so important to look afresh at how traditional theories compare to reality. A problem conceived principally in terms of who bears the cost of action, with benefits which are remote, collective and seemingly impossible to quantify, seems almost impossible to solve. Yet this worldview informs the mindset of negotiators and commentators alike in frustrating any attempts to reach an international deal. If it is true, we do indeed seem trapped.

Or, to use Einstein's famous saying, the kind of thinking that has got us into a problem is unlikely to get us out of it.

Notes

1. Global Energy Assessment Report (2012: 113). In the physical units of exajoules, global primary energy consumption rose from about 100EJ in 1950 to 300EJ in 1975. These data include biomass, which is often left out of energy statistics; the relative growth in fossil fuel use was considerably bigger still.
2. IEA World Energy Outlook (2011).
3. See Pigou's work on taxing externalities (Pigou 1920), Coase's development of the idea of tradable permits (Coase 1960) and Hotelling's renowned early resource depletion (Hotelling 1931). The economic theory behind cost-benefit analysis actually dates back to the mid-nineteenth century.
4. For example, Burke (2010).
5. Global oil demand stabilised for some years after the 1970s price shocks – industrialised consumption declined out to 2000 (including the rapid fall with economic transition in the former

Soviet Union), but developing countries' consumption increased and global demand has risen strongly since the millennium.

6. 'Drowning in oil', *The Economist*, 6 March 1999, p. 19.
7. Including the recessions of 1973–4, 1978, 1981–2, August 1990, 2001 and 2007–8, as pointed out by James Hamilton, Professor of Economics at the University of California, San Diego (Hamilton 2009b). See also note 8.
8. Dispute remains about causality of oil–GDP relationships, as the direct cost impacts are insufficient to explain recessionary effects; one view is that the impact of rising oil prices on inflation prompted central banks to raise interest rates, and it is this that slows the economy and amplifies debt problems (see Segal 2011). The most extensive set of papers analysing oil price variations, financial speculation and the historical impacts on GDP are collected in a Special Issue of the *Energy Journal* (M. Manera (ed.), 'Financial speculation in the oil markets and the determinants of the price of oil', *Energy Journal*, 34 (3), 2013). The associated analysis of GDP impacts is by Morana (2013). The UK analysis on the impact of energy price volatility on future GDP was prepared by Oxford Economics for the UK Department of Energy and Climate Change as part of the UK's Energy Market Reforms, available on http://www.decc.gov.uk.
9. World Bank (2005) *World Development Indicators*.
10. Global Energy Assessment Report (GEA, 2012), Chapter 5.
11. World Bank (1992) *Development and the Environment*.
12. For example, for a recent study see Janus (2012). The World Bank has also published studies in the area, e.g. Djankov and Reynal-Querol (2007) on 'The causes of civil war'. The double-edged impacts of oil on African development have been discussed by the World Bank in a series of studies (e.g. *Ghana Report*, PREM4 2009) and subsequent work on Sudan among others.
13. US Environmental Protection Agency 1997 on the Benefits and Costs of the Clean Air Act, Washington, DC, cited in Tietenberg (2006: 67). Tietenberg's book contains an extensive review of cost-benefit studies of US pollution legislation.
14. Bollen *et al.* (2009).
15. The order of magnitude does not depend on assumed climate change-related damages, since in these studies climate damages are assumed to be smaller than those from local pollutants. Epstein *et al.* (2011) focus on the life cycle cost of coal and estimate that the overall 'external' costs of coal in the US amount to 'a "third to over one-half of a trillion dollars annually. Accounting for the damages conservatively doubles to triples the price of electricity from coal per kwh generated"; air pollution is about half the total in their central estimates, which focus on the Appalachian region.' The other study focuses upon the principles of including environmental effects in national economic accounts (Muller *et al.* 2011). In 'Environmental accounting for pollution in the United States economy' – interesting also as being authored by researchers often associated with a more sanguine view on the costs of climate change – Muller *et al.* estimate the environmental cost of air pollution from coal power stations in the US to equate to about 3¢/kWh, and notes that the environmental damage per unit of sulphur emissions is also much higher than the current cost of sulphur emission permits, i.e. the cap set under the US cap-and-trade scheme is much too lenient. The $120bn figure is for 2005, from the full US National Academy of Sciences study, *Hidden Costs of Energy* (Committee on Health, Environmental, and Other External Costs and Benefits of Energy Production and Consumption and the National Research Council 2010).
16. The European Environment Agency summarises the results of a series of large studies of external costs associated with power generation, concluding that the external impacts of hard coal amount to 5–15 €¢/kWh, down to about 2 €¢/kWh for the most modern 'clean coal' technology. Lignite is worse, oil comparable and gas combined cycle has the lowest (at about 1–4 €¢/kWh). The upper estimates in each case are significantly influenced by the assumed cost of carbon (19 (low) – 80 (high) €/tCO_2).
17. Official measurements of fine (<2.5 micrometre) particulates in Beijing's air rose to a record 993 micrograms per cubic metre on 12 January 2013, compared to World Health Organisation guidelines of no higher than 25. The response announced on 5 March was to accelerate programmes on energy efficiency, refinery upgrades, renewable energy and others, estimated to cost 2.57 trillion yuan ($380 billion) out to 2015 (Bloomberg, 5 March 2013).
18. These radiative properties can be measured directly in laboratories – as has been known for almost two centuries – and in the sky. We also know that Mars, which lacks much atmosphere, is icy compared to Earth, while Venus, which has a thick blanket of CO_2, is roasting – each to a degree far greater than can be explained by the difference in distance from the sun. Our greenhouse gas

blanket keeps the earth inhabitable. The Earth is on average 33°C warmer than it would be without a greenhouse gas layer (Le Treut *et al.* 2007).

19. Currently, ecosystems are estimated to absorb about half of anthropogenic CO_2 emissions (oceans about 24% and land about 30% (Munang *et al.* 2013).
20. Recent estimates of CO_2 concentrations during the Pliocene period suggest a range of around 330–420ppmCO_2; see Bartoli *et al.* (2011); Pagani *et al.* (2010); Seki *et al.* (2010).
21. Much has been made of recent years being cooler than the exceptional year of 1998. The global average surface temperature in 2011 was 0.92°F (0.51°C) warmer than the mid-twentieth-century baseline and was the ninth out of the ten warmest years in modern meteorological records to have occurred after 2000 (NASA, January 2012).
22. A radiosonde is a device used in weather balloons that measures certain atmospheric parameters (temperature, humidity, pressure, etc.) and transmits them to a fixed receiver. For many years, scientists were puzzled by apparent discrepancies, particularly from subsequent satellite data, motivating several hundred research papers and two in-depth expert panel assessments. Debates were finally resolved in part by the discovery of errors in correcting for satellite drift. With advancing understanding, recent reviews conclude that there is no evidence for discrepancy between datasets or between modelled and measured temperatures. The US Climate Change Science Program stated: ‘This significant discrepancy no longer exists because errors in the satellite and radiosonde data have been identified and corrected.’ For more detail, see Thorne *et al.* (2011) and McCarthy *et al.* (2008).
23. http://www.metoffice.gov.uk/hadobs/hadat/msu/anomalies/hadat_msu_global_mean.txt, global ASCII series T4. The impact of major volcanic eruptions is also identifiable in this lower stratospheric cooling record.
24. Hegerl and Zwiers (2007); Ramaswamy *et al.* (2001); Karl *et al.* (2006). Richard Lindzen, perhaps the most prominent critic, proposed a theory that water vapour effects in the atmosphere could at least partially offset surface warming rather than amplify it as most scientists expect (Lindzen *et al.* 2001); the IPCC AR4 WGI, section 8.6.3.2.1 gives a brief summary of the surrounding literature, including the numerous objections and a ‘vigorous debate’ over the legitimacy of Lindzen’s theory (Randall *et al.* 2007).
25. This phrasing forms the fundamental Objective of the UN Framework Convention on Climate Change 1992, namely ‘Stabilization of greenhouse gas concentrations in the atmosphere at a level that would prevent dangerous anthropogenic interference with the climate system. Such a level should be achieved within a time-frame sufficient to allow ecosystems to adapt naturally to climate change, to ensure that food production is not threatened, and to enable economic development to proceed in a sustainable manner’ (UNFCCC, Article 2, ‘Objective’).
26. UN Secretary-General Advisory Group on Energy and Climate Change (AGECC 2010).This led on to the ‘Sustainable Energy for All’ initiative of the UN and World Bank (www.sustainableenergyforall.org), and declaration of 2014-2024 as the ‘UN Decade of Sustainable Energy’. Bazilian and Pielke (2013) stress the scale of ambition implied, if more than 2 billion people are to be connected to modern energy at levels commensurate with economic ambitions. For detailed technical assessments see GEA (2012). These analyses underline the point that both higher efficiency and rapid development and diffusion of low carbon sources will be needed if the energy access goals are to be met without resource struggles, or climate change negating the intended development benefits.
27. This book deliberately avoids delving much into the more diverse debates around ‘peak oil’ and projections of global fossil fuel supplies and depletion. Most predictions have in time been proven wrong. A more useful view is to look at the huge scale of investment and innovation required to open up new, previously untapped resources; see the opening to Chapter 10.
28. Gases other than CO_2 do play an important role and reducing them may offer some ‘quick wins’ – with major reductions in some already achieved – but CO_2 is at the heart of the problem, accounting for about 80% of the total greenhouse gas emissions from industrialised countries, for example. CO_2 is shared between atmosphere and other ‘sinks’ (vegetation and surface layers of the oceans) but once emitted is present in these for many centuries. Hence stabilising the atmosphere requires net emissions to be brought to near zero levels.
29. Indeed, just as this book was going to press, the website of the popular economist Tim Harford headlined a story ‘The Science May Be Settled, But the Economics Isn’t’, with comments that included Harvard’s Marty Weitzman describing climate change as ‘a hellish problem that is pushing the bounds of economics’ (http://t.co/Jqk0CrPQil).

30. The ratio of CO_2 to GDP combines measures of national energy intensity (E/GDP) with carbon intensity (C/E). In practice, with carbon intensity relatively static, the graph is a very close approximation of trends in energy intensity, the most significant deviation being due to relative shifts between China (more carbon intensive) and Russia (less so) over time.
31. Asian countries including Singapore, South Korea and the Philippines appear to have reached close to stable per capita CO_2 emission levels. Latin America in aggregate appears to have stayed substantially below $10tCO_2$ per person, aided in large part by the large hydro resources and biomass energy programmes of Brazil, which is among the most carbon efficient countries in the world.
32. Bataille *et al.* (2007). This study sought to statistically test for the impact of 'inflexible' (climate, geography) and 'somewhat flexible' (fossil fuel trade, industrial structure and access to low emitting power sources) factors in the different per capita emissions in 2002. The gap would be significantly lowered by adjusting for these factors; most notably US emissions could be 24% lower. However, more than half of this was due to the assumption that the US does not have the same access to low carbon electricity, which is highly debatable with respect to nuclear, wind or more recently solar and also hydro imports from Canada. The majority of the observed international gap remains reasonably attributable to different prices and policy.
33. Close examination of the data indicate that the EU and Japan have very similar aggregate energy intensities, but Europe has decarbonised more in recent years.
34. Brazil of course faces its own problems of emissions from forestry not included in these fossil fuel statistics. There is also serious debate over hydro-related emissions, not included in the figure (see Chapter 3, note 49).
35. The psychological literature outlined in Chapters 2 and 4 would predict this: in behavioural economics it is called anchoring. It is an effect amplified in markets, where current prices reflect to an important degree the collective expectations about the future.
36. Taleb (2007) presents extensive evidence of the importance of 'Black Swan' events in modern society, and argues that the increasingly sophisticated and interconnected systems make societies more, not less, vulnerable to their occurrence. Of course, shortly after publishing his first edition he had to make a major update to take account of the financial crisis, the biggest Black Swan event since 9/11. See also references to 'X-events' in Chapter 2.
37. The financier George Soros, who made billions speculating on the irrationality of financial systems, has subsequently established an 'Institute for New Economic Thinking' based largely upon the premise that established economics misses these fundamental attributes of real economic behaviour.
38. The most obvious examples of the gulf between recommendations of 'pure' economics and political economy are to be found in pricing, as detailed throughout Pillar II of this book (Chapters 6–8, including the epigraph to Chapter 6). Most striking are persistent recommendations for a global carbon trading system, ignoring the politics of international financial transfers at potentially huge scales, or for global carbon taxes, which would in effect require ceding tax sovereignty in an important area to global decision-making rather than national budgetary processes.
39. Igor Bashmakov (in Barker *et al.* 2007). Bashmakov finds an even tighter relationship for expenditure by final consumers on energy (as a share of gross output) as 4–5% for the US and 4.5–5.5% for the OECD. This is first of his three laws. The second is the law of improving energy quality, and the third is the law of growing energy productivity.
40. For a recent review in relation to US regulation see Sunstein (2013/14).
41. A good overview is to be found in Hulme (2009).
42. Indeed, the UK government is unlikely to be alone in having abandoned the effort to base the 'carbon price' used for cost-benefit appraisal. Instead, it is linked to the carbon prices estimated to be required to achieve the science-derived goal, established in the UK Climate Change Act, of 80% emissions reduction by 2050. Conveniently, the number by 2030 is similar to that roughly estimated by the UK Treasury in some of the earlier efforts. The resort to a 'cost of mitigation' approach in fact has many precedents in environmental policy where the benefits prove almost impossible to quantify.
43. Nordhaus and Boyer (2000). Similar lines of reasoning and calculations include European 'economic sceptics' such as Richard Tol and Bjorn Lomborg, the latter relying heavily on Tol's papers and whose most recent book popularises this way of thinking about the problem (Lomborg 2007). In recent years, Professor Nordhaus has laid more stress on the uncertainties and risks associated with climate change, revising upwards his estimates of 'optimal control costs'; see, for example, 'post-Stern' references cited below.

44. Robert Stavins, personal communication 2005. As noted in the text, one possible explanation for apparently systematically different attitudes between scientists and economists was that whereas natural science often involves studies of highly non-linear and threshold events – like the phase transitions from ice to water to steam – economics is classically about evaluating marginal changes and the trade-offs involved in substituting one thing for another. Economics teaches people to think in the continuum associated with incremental changes, whereas the biggest concerns of most scientists about climate change are about the risks of discontinuities and unknowns in the climate system, and impacts (e.g. on food supply or severe storm damages) that cannot readily be substituted by economic inputs. Economists are also used to relying on market responses to changes – scientists are often wrong about issues like resource depletion because they underestimate the extent to which the rise in price as a resource depletes will provoke a response such as exploration or innovation to find more or a substitute. In climate change, however, any such a feedback loop would be both too slow (to the point of irrelevance) and would still depend on government policy to link impacts to emissions (see section 'Driving blind'). Other factors may also explain the tendency to different attitudes between the disciplines. There is of course risk in caricaturing differences between disciplines, but Stavins' observation was and remains strikingly relevant as a generalisation, notwithstanding important divergences in individual attitudes.
45. For example, Mendelsohn and Williams (2004).
46. King Canute was an English king in the Middle Ages, known partly for the story of him sitting on a chair on the beach commanding the tide not to rise (and, of course, getting wet). Kinder historians insist that he did so to prove to his courtiers that only God himself could control nature.
47. *The Skeptical Environmentalist* (Lomborg 2001) and subsequent *Smart Solutions to Climate Change: Comparing Costs and Benefits* (Lomborg 2010).
48. Potential 'tipping points' in the climate system include possible instabilities in: the Arctic sea ice, the major ice caps (Greenland and West Antarctic), the Hindu-Kush-Himalaya-Tibetan glaciers, the permafrost and carbon stores, the boreal forest, the Atlantic thermohaline circulation, the El Niño/Southern Oscillation (ENSO), the Amazonian rainforest, the West African monsoon and Sahel, the Indian summer monsoon and rainfall patterns in south western North America.
49. Hallegatte *et al.* (2007).
50. Zhang *et al.* (2007).
51. Homes are in fact a relatively easy case, because they are regularly bought and sold, and planning procedures may offer compensation based upon market value – which is itself the intersection between 'willingness to pay' and 'willingness to accept' for similar properties in the aggregate housing market. For such cases, societies develop rules for overriding individual interests and often base this on the average imputed value. Countries, lives and livelihoods are not tradable goods, however.
52. This also relates closely to long-running economic debate about the divergence between 'willingness to pay (to prevent)' and 'willingness to accept compensation (for)' adverse impacts. These often differ – sometimes very substantially, spawning another wider literature.
53. Das and Srinivasan (2010).
54. The political struggle over the 'value of the statistical life' applied to climate change was played out in the IPCC Second Assessment, Working Group III (Bruce *et al.* 2006). An account of the conflicts and some of the follow-up debate is given in Grubb *et al.* (1999, Annex II).
55. Stern (2006).
56. One important generic development in the economics of discounting was the recognition that uncertainty means that the *effective* discount rate falls over time towards the lower of the plausible range, simply because these are the values that dominate long-term costs when assessed across the range of possibilities (Weitzman, in Portney and Weyant 1999). For specific developments in discounting applied to climate analysis see the subsequent notes here.
57. For such projects, a *theoretical* case can be made to try and use different discount rates for different components of a project, based on assessment of their *systemic* risk. In practice this is unlikely to be practicable, given the extreme difficulty of assessing such component contributions to systemic risk. After considerable debate, UK regulators broadly adopted an approach to calculate the major contributions to financial costs at the 'weighted average cost of capital' of the firms, but then to discount the total resulting costs over time at the 'social rate of time preference' (see Joint Regulators Group, 'Statement: Discounting for CBAs involving private investment but public benefit', Ofcom, London, 2012).

58. This includes work by Cline and others as early critiques of Nordhaus' approach to valuation and discounting: the debate was already covered in the IPCC's Second Assessment Report (1996).
59. This reflects the standard Ramsay Rule in economics, that the discount rate should be the sum of the pure rate of time preference, plus the product of expected economic growth and inequality aversion. The UK Treasury official guidance for public policy evaluation is for a 3.5% per year discount rate for the first thirty years of a project, declining thereafter. This attempts to capture the reasoning in the academic literature, and is based on an assumed sustained average per capita economic growth rate of 2%, which some years after the credit crunch looks, to say the least, optimistic.
60. Though note that risk-free, market interest rates have, since the credit crunch, also fallen to very low levels; the instability in this metric must also call into question its appropriateness for weighting impacts over centuries. Nordhaus brought his analysis and critiques of Stern together in his book *A Question of Balance* (Nordhaus 2008). See the previous section for general discussion of the issues in costing climate impacts, and the following notes for more detail on technical but key assumptions around discounting and other aggregation techniques.To this author, it remains hard to see how an accepted formula for deriving a social discount rate can in effect be reverse-engineered by assuming the answer must equal observed market (opportunity cost) rates. This implies a view that the market-based rate gives an accurate and legitimate reflection of a conscious preference to devalue the welfare of future generations, which is surely a contentious assumption, not supported by any of the behavioural and psychological studies covered in Pillar I of this book.
61. DeLong (2006). Analytically, one issue is that the standard economic approach to aggregation assumes that individual utility rises as the logarithm of wealth: money is more important to the poor than the rich in ways such that an X per cent rise in any individual's wealth brings the same additional welfare benefit. This underpins equating a given percentage rise in GDP, if everyone's wealth rises by the same percentage, with an equivalent increase in aggregate welfare. Different assumptions, as Dasgupta argued for, vastly complicate the challenge of equating changes in wealth with welfare, and start to make the debate more about explicitly redistributive policies, nationally and internationally, than about climate change.
62. Reactions included those of Weitzman (2007a), Arrow (2007), Yohe (2006), Lomborg (2006), Baer 2006, Neumayer (2007), Ackerman and Finlayson (2006), Byatt *et al.* (2006). See also Dietz *et al.* (2007).
63. Hope (2006), Watkins and Downing (2007). Note the Downing estimate of '10–1,000' is in $/ tCO_2, an upper value which corresponds roughly to the former range – *c.*$250/$tCO_2$ – rather than the highest levels associated with very low discount rates.
64. 'All of us agree that the Ramsey formula provides a useful conceptual framework for examining intergenerational discounting issues. The key question is whether it can be put to practical use ... one approach is to view the parameters ... as representing policy choices (the "prescriptive" approach to discounting); another is to base estimates ... on market rates of return (the "descriptive" approach) – Arrow *et al.* 1996. Those who favor the prescriptive approach argue that the parameters of the Ramsey formula could be based on ethical principles [... or inferred from ... public policy decisions ... or] from attempts to elicit social preferences using stated preference methods. Others of us who favor the descriptive approach suggest that [the parameters] could be inferred from decisions in financial markets, although behavior in financial markets, even for longer-term assets such as 30-year bonds, is likely to reflect intragenerational rather than intergenerational preferences.' At several other points also, the document refers explicitly to differing views rather than reaching any consensus. The authors were Arrow, Cropper, Gollier, Groom, Heal, Newell, Nordhaus, Pindyck, Pizer, Portney, Sterner, Tol and Weitzman (Arrow *et al.* 2012). Other recent contributions on discounting for long-term public policy include a series of working papers by Michael Spackman for the Grantham Institute, London School of Economics, UK, which are particularly clear and useful for pinpointing different perspectives in the debate; his broad conclusion is that economists have to find some other methodology than discounting to guide policy on intergenerational problems, which matches precisely one important conclusion of this book (Chapters 1, 2 and 6; see also notes 57–63).
65. Dasgupta (2008).
66. The essence of Weitzman's argument was that most probability/ risk studies assumed a normal (Gaussian) distribution of risks for mathematical convenience, which Weitzman argued was logically inappropriate to problems like climate change where the distribution of extreme risks is fundamentally unknowable and unobservable until it is too late (Weitzman 2007b).

67. The considerable literature sparked by Weitzman's Dismal Theory includes Weitzman (2009a, 2009b, 2010, 2011), Pindyck (2011) and Nordhaus (2011).
68. Jacoby (2004).
69. Classic studies of the accuracy of 'expert' judgements include Murphy and Winkler (1984) and Keren (1987). Toxicology experts seem to make better judgements than the general public, as one would hope, but still seem subject to bias (Kraus *et al.* 1992).
70. Gardiner (2011), in IPCC, AR5, Chapter 4.
71. This inability to quantify instability arises in part because money is the metric – and the value of money hinges fundamentally upon social stability based, among other things, on respect of property rights and fulfilling promises to pay.
72. This is not a new observation. Indeed, in recent years there has been an explosion of analysis, from the CIA to the most abstract academic studies, viewing climate change as a security issue. One recent publication (Mabey 2011) summarises '… the growing consensus in the security community that climate change presents significant risks to the delivery of national, regional and global security goals. Through sea level rise, shortages of food and water and severe weather events, climate change will have significant impacts on all countries, which in turn could affect their social stability and economic security. In the coming decades such impacts will increase the likelihood of conflict in fragile countries and regions. Peaceful management of even moderate climatic changes will require investment in increased resilience in national and international security and governance systems …' See also the Chatham House report on *Resources Futures* (Lee *et al.*, 2012), especially section 4.4. For ongoing assessments in this area see the Centre for Climate and Security, http://climateandsecurity.org.
73. The classic analysis of 'planetary boundaries' was published in 2009, authored by 26 of the world's leading environmental scientists, as 'a safe operating space for humanity' (Rockstrom *et al.* 2009). Nitrogen forms part of an estimated 'biogeochemical flow boundary', along with phosphorous for which the article argues that humanity 'may soon be approaching the boundaries'. This warning was also applied to three others – freshwater use, change in land use and ocean acidification. Stratospheric ozone depletion is now stabilised (thanks to an effective international treaty and subsequent global action), and the article did not attempt to quantify the other two listed (chemical interference and atmospheric aerosol loadings). The interesting feature is that the analysis clearly took account of inertia in physical systems, but neglected socio-economic inertia. Thus its placement of climate change was based on an assessment that the current atmospheric concentration already dangerously exceeds the levels seen for millions of years, and will lead to long-term changes in ice, oceans and weather patterns with potentially unsustainable consequences – not on the projection that emissions and concentrations are likely to keep rising.
74. Downing (2012).
75. Solomon *et al.* (2009).
76. Hallegatte (2005) was one of the few authors to consider jointly both the inertial characteristics of the physical systems, the economic systems and the various feedbacks involved. The conclusion is dismal: the problem cannot be effectively managed 'reactively'. Inertia in energy (as well as climate) systems means that, after any decision to act, it can take decades for the full impact of policy changes to unfold.
77. The UK Royal Society (2009). Note that 'geo-engineering' is a term used for two very different forms. One concerns measures to stimulate the carbon cycle, for example by iron fertilisation of ocean plankton to increase carbon uptake. The other involves direct intervention in the planetary heat balance.
78. Barrett (2005), *Environment and Statecraft*. Barrett's book, in its concluding pages, suggested that the answer would have to lie in technological innovation, but as detailed in Pillar III of this book, that does not on its own offer a credible alternative. Barrett himself gave a somewhat less pessimistic overall assessment in his later book *Why Cooperate?* (Barrett 2010).

References

Ackerman, F. (2008) 'The new climate economics: the Stern Review versus its critics, in J. M. Harris and N. R. Goodwin (eds), *Twenty-First Century Macroeconomics: Responding to the Climate Challenge*. Cheltenham, UK and Northampton, MA: Edward Elgar.

Ackerman, F. and Finlayson, I. (2006) 'The economics of inaction on climate change: a sensitivity analysis', *Climate Policy*, 6 (5).

AGECC (2010) *Energy for a Sustainable Future. The UN Secretary-General's Advisory Group on Energy and Climate Change: Summary Report and Recommendations*, 28 April. New York: UN, available from: http://www.un.org.

Anderson, D. (2007) 'The Stern Review and the costs of climate change mitigation – a response to the "dual critique" and the misrepresentations of Tol and Yohe', *World Economics*, 8 (1): 211–19.

Arrow, K. J. (2007) 'Global climate change: a challenge to policy', *Economists' Voice*, June, pp. 1–5.

Arrow, K. J., Bolin, B., Constanza, R., Dasgupta, P., Folke, C., Holling, C. S. *et al.* (2006) 'Economic growth, carrying capacity and the environment, Policy Forum', *Science*, 268: 520–1.

Arrow, K. J., Cropper, M. L., Gollier, C., Groom, B., Heal, G. M., Newell, R. G. *et al.* (2012) *How Should Benefits and Costs Be Discounted in an Intergenerational Context? The Views of an Expert Panel*, RFF DP 12-53. Washington, DC: Resources for the Future.

Arthur, B. (1988) 'Competing technologies, increasing returns and lock-in by historical events', *Economic Journal*, 99: 116–31.

Arthur, B. (1994) *Increasing Returns and Path Dependence in the Economy*. Ann Arbor, MI: University of Michigan Press.

Baer, P. (2006) 'Adaptation: how pays whom?', in W. N. Adger *et al.* (ed.), *Fairness in Adaptation to Climate Change*. Cambridge, MA: MIT Press.

Barker, T., Bashmakov, I., Bernstein, L. and Bogner, J. (2007) *Climate Change 2007: Mitigation of Climate Change Summary for Policymakers*, Cambridge, UK and New York: Cambridge University Press.

Barrett, S. (2005) *Environment and Statecraft: The Strategy of Environmental Treaty-Making*. Oxford: Oxford University Press.

Barrett, S. (2010) *Why Cooperate? The Incentive to Supply Global Public Goods*. Oxford: Oxford University Press.

Barrett, S. and Stavins, R. (2003) 'Increasing participation and compliance in international climate change agreements', *Politics, Law and Economics*, 3: 349–76.

Bartoli, G., Honisch, B. and Zeebe, R. E. (2011) 'Atmospheric CO_2 decline during the Pliocene intensification of Northern Hemisphere glaciations', *Paleooceanography*, 24 (4).

Basalisco, B. (2012) *Discounting for CBAs Involving Private Investment, but Public Benefit*. London: Ofcom, Joint Regulators Group.

Bataille, C., Rivers, R., Mau, P., Joseph, C. and Tu, J.-J. (2007) 'How malleable are the greenhouse gas emission intensities of the G7 nations', *Energy Journal*, 28 (1): 145–70.

Bazilian, M. and Pielke, R. (2013) 'Making energy access meaningful', *Issues in Science and Technology*, Summer 2013.

Beckerman, W. and Hepburn, C. J. (2007) 'Ethics of the discount rate in the Stern Review on the economics of climate change', *World Economics*, 8: 187–208.

Bollen, J., Guay, B., Jamet, S. and Corfee-Merlot, J. (2009) *Co-benefits of Climate Change Mitigation Policies: Literature Review and New Results*. Paris: OECD.

Bruce, J., Lee, H. and Haites, E. (eds) (1996) *Climate Change 1995: Economic and Social Dimensions of Climate Change*, IPCC Report. Cambridge: Cambridge University Press.

Bürer, M. J. and Wüstenhagen, R. (2009) 'Which renewable energy policy is a venture capitalist's best friend? Empirical evidence from a survey of international cleantech investors', *Energy Policy*, 37 (12): 4997–5006.

Burke, P. J. (2010) 'The national level energy-ladder and its carbon implications', *Environment and Development Economics*, 18 (4): 484–503.

Byatt, I., Castles, I., Goklany, I.D., Henderson, D., Lawson, N., McKitrick, R. *et al.* (2006) 'The Stern Review: a dual critique – economic aspects', *World Economics*, 7 (4): 199–229.

Carpenter, R. E. and Petersen, B. C. (2002) 'Capital market imperfections, high-tech investments, and new equity financing', *Economic Journal*, 112: 477.

Carruth, A., Dickerson, A. and Henley, A. (2000) 'What do we know about investment under uncertainty?', *Journal of Economic Surveys*, 14 (2).

Cline, W. R. (2004) 'Meeting the challenge of global warming', in B. Lomborg (ed.), *Global Crises, Global Solutions*. Cambridge: Cambridge University Press.

Coase, R. H. (1960) 'The problem of social cost', *Journal of Law and Economics*, 3: 1–44.

Cohen, L. and Noll, R. (1991) *The Technology Pork Barrel*. Washington, DC: Brookings Institution.

Committee on Health, Environmental, and Other External Costs and Benefits of Energy Production and Consumption and the National Research Council (2010) *Hidden Costs of Energy: Unpriced Consequences of Energy Production and Use*. US National Academy Press.

Crafts, N. (2012) 'Industrial policy: past, present and future', *Competitive Advantage in the Global Economy (CAGE)*. Warwick University.

Das, D. and Srinivasan, R. (2010) *Comparative Study of Carbon Dioxide Emission by Major Fuel Consuming Countries*. International Seminar on 'India Emerging – Opportunities & Challenges', New Delhi.

Dasgupta, P. (2006) 'Comments on the Stern Review', Seminar on the Stern Review's Economics of Climate Change, Royal Society of London, 8 November.

Dasgupta, P. (2007) 'Commentary: the Stern Review's economics of climate change', *National Institute Economic Review*, 199: 4–7.

Dasgupta, P. (2008) 'Discounting climate change', *Journal of Risk Uncertainty*, 37: 141–69.

DeLong, B. (2006) 'Partha Dasgupta makes a mistake in his critique of the Stern Review', 30 November, online at: http://delong.typepad.com/sdj/2006/11/partha_dasgaptu.html.

Dietz, S., Hope, C. W. and Patmore, N. (2007) 'Some economics of "dangerous" climate change: reflections on the Stern Review', *Global Environmental Change*, 17: 311–25.

Dietz, S., Hope, C., Stern, N. and Zenghelis, D. (2007) 'Reflections on the Stern Review (1): a robust case for strong action to reduce the risks of climate change', *World Economics*, 8 (1): 121–68.

Djankov, S. and Reynal-Querol, M. (2007) *The Causes of Civil War*. World Bank, Financial and Private Sector Vice Presidency, Indicators Group.

Downing, T. E. (2012) 'Views of the frontiers in climate change adaptation economics', *WIREs Climate Change*, 3 (2): 115–212.

Downing, T. E. and Dyszynski, J. (2010) *Frontiers in Adaptation Economics: Scaling from the Full Social Cost of Carbon to Adaptation Processes*. Oxford: Global Climate Adaptation Partnership.

Economist, The (2009) 'Drowning in oil', 6 March, p. 19.

Epstein, P. R., Buonocore, J. J., Eckerle, K., Hendryx, M., Stout III, B. M., Heinberg, R. *et al.* (2011) 'Full cost accounting for the life cycle of coal', in R. Costanza, K. Limburg and I. Kubiszewski (eds), *Ecological Economics Reviews*, Annals of the New York Academy of Sciences, 1219: 73–98.

Freeman, C. (1974) *The Economics of Industrial Innovation*. Harmondsworth: Penguin Books.

Gardiner, S. M. (2011) *A Perfect Moral Storm: The Ethical Challenge of Climate Change*. Oxford: Oxford University Press

GEA (2012) *Global Energy Assessment – Toward a Sustainable Future*. Cambridge, UK and New York: Cambridge University Press, and the International Institute for Applied Systems Analysis, Laxenburg, Austria

Gompers, P. and Lerner, J. (2001) 'The venture capital revolution', *Journal of Economic Perspectives*, 15 (2).

Grubb, M. (2004) 'Technology innovation and climate change policy: an overview of issues and options', *Keio Economic Studies*, 41 (2).

Grubb, M., Haj-Hasan, N. and Newbery, D. (2008) 'Accelerating innovation and strategic deployment in UK electricity: applications to renewable energy', in M. Grubb, T. Jamasb and M. Pollitt (eds), *Delivering a Low-Carbon Electricity System: Technologies, Economics and Policy*. Cambridge: Cambridge University Press.

Grubb, M., Vrolijk, C. and Brack, D. (1999) *The Kyoto Protocol: A Guide and Assessment*. London: Earthscan.

Hallegatte, S. (2005) 'The time scales of the climate-economy feedback and the climatic cost of growth', *Environmental Modelling and Assessment*, 10 (4): 277–92.

Hallegatte, S., Hourcade, J.-C. and Dumas, P. (2007) 'Why economic dynamics matter in assessing climate change damages: illustration on extreme events', *Ecological Economics*, 62 (2): 330–40.

Hamilton, J. D. (2009a) 'Oil prices and the economic recession of 2007–08', *VOX*, 16 June.
Hamilton, J. D. (2009b) *Causes and Consequences of the Oil Shock of 2007–08*, NBER Working Paper No. 150002.
Hegerl, C. H. and Zwiers, F. W. (2007) *Understanding and Attributing Climate Change*, Chapter 9, AR4, WG1, International Panel on Climate Change. Cambridge: Cambridge University Press.
Hope, C. (2005) *Exchange Rates and the Social Cost of Carbon*. Cambridge: Cambridge University Press.
Hope, C. (2006) The marginal impact of CO_2 from PAGE2002: an integrated assessment model incorporating the IPCC's five reasons for concern', *Integrated Assessment Journal*, 6 (1): 19–56.
Hotelling, H. C. (1931) 'The economics of exhaustible resources', *Journal of Political Economy*, 39 (2).
Hulme, M. (2009) *Why We Disagree About Climate Change: Understanding Controversy, Inaction, and Opportunity*. Cambridge: Cambridge University Press.
International Energy Agency (2011) *World Energy Outlook*. Paris: IEA
IPCC (1996) *Second Assessment Report*. Cambridge: Cambridge University Press.
Jacoby, H. D. (2004) 'Informing climate policy given incommensurable benefits estimates', *Global Environmental Change: Special Edition on the Benefits of Climate Policy*, 14 (3): 287–97.
Jamasb, T. and Köhler, J. (2008) 'Learning curves for energy technologies: a critical assessment', in M. Grubb, T. Jamasb and M. Pollitt (eds), *Delivering a Low-Carbon Electricity System: Technologies, Economics and Policy*. Cambridge: Cambridge University Press.
Janus, T. (2012) 'Natural resource extraction and civil conflict', *Journal of Development Economics*, 97 (1): 24–31.
Jones, R. and Yohe, G. (2008) 'Applying risk analytic techniques to the integrated assessment of climate policy benefits', *Integrated Assessment Journal*, 8: 123–49.
Karl, T. R., Hassol, S. J., Miller, C. D. and Murray, W. L. (eds) (2006) *Temperature Trends in the Lower Atmosphere: Steps for Understanding and Reconciling Differences*. Asheville, NC: National Oceanic and Atmospheric Administration, National Climatic Data Center.
Keren, G. (1987) 'Facing uncertainty in the game of bridge: a calibration study', *Organization Behavior and Human Decision Processes*, 39: 98–114.
Knight, F. H. (1921) *Risk, Uncertainty and Profit*. New York: Houghton Mifflin.
Kraus, N., Malmfors, T. and Slovic, P. (1992) 'Intuitive toxicology expert and lay judgements of chemical risks', *Risk Analysis*, 12: 215–32.
Le Treut, H., Somerville, R., Cubasch, U., Ding, Y., Mauritzen, C., Mokssit, A. *et al.* (2007) 'Historical overview of climate change science', in S. Solomon, D. Qin, M. Manning, Z. Chen, M. Marquis, K. B. Averyt *et al.* (eds), *Climate Change 2007: The Physical Science Basis*, Contribution of Working Group I to the Fourth Assessment Report of the Intergovernmental Panel on Climate Change. Cambridge, UK and New York: Cambridge University Press.
Lee, B., Preston, F., Kooroshy, J., Bailey, R. and Lahn, G. (2012) *Resources Futures – A Chatham House Report*. London: Chatham House.
Lindzen, R. S., Chou, M.-D. and Hou, A. Y. (2001) 'Does the Earth have an adaptive infrared iris?', *Bulletin of the American Meteorological Society*, 82 (3): 471–32.
Lomborg, B. (2001) *The Skeptical Environmentalist*. Cambridge: Cambridge University Press.
Lomborg, B. (2006) 'Stern Review: the dodgy numbers behind the latest warming scare', *Wall Street Journal*, 2 November.
Lomborg, B. (2007) *Cool It: The Skeptical Environmentalist's Guide to Global Warming*. New York: Knopf.
Lomborg, B. (2010) *Smart Solutions to Climate Change: Comparing Costs and Benefits*. Cambridge: Cambridge University Press.
Mabey, N. (2011) *Degrees of Risk: Defining a Risk Management Framework for Climate Security Current Responses*, E3G, February.
McCarthy, M. A., Thompson, C. J. and Garnett, S. T. (2008) 'Optimal investment in conservation of species', *Journal of Applied Ecology*, 45: 1428–35.

Martinot, E. (2001) 'Renewable energy investment by the World Bank', *Energy Policy*, 29 (9): 689–70.

Mendelsohn, R. and Williams, L. (2004) 'Comparing forecasts of the global impacts of climate change', *Mitigation and Adaptation Strategies for Global Change*, 9: 315–33.

Mendelsohn, R., Ariel, D. and Williams, L. (2006) 'The distributional impact of climate change on rich and poor countries', *Environment and Development Economics*, 11 (2): 159–78.

Morana, C. (2013) 'The oil price-macroeconomy relationship since the mid-1980s: a global perspective', *Energy Journal*, 34 (3): 153–90.

Muller, N. Z., Mendelsohn, R. and Nordhaus, W. (2011) 'Environmental accounting for pollution in the United States economy', *American Economic Review*, 101 (5): 1649–75.

Mumtaz, A. and Amaratunga, G. (2006) 'Solar energy: photovoltaic electricity generation', in T. Jamasb, W. J. Nuttall and M. G. Pollitt (eds), *Future Electricity Technologies and Systems*. Cambridge: Cambridge University Press.

Munang, R., Thiaw, I., Alverson, K., Mumba, M., Liu, J. and Rivington, M. (2013) 'Climate change and ecosystem-based adaptation: a new pragmatic approach to buffering climate change impacts', *Current Opinion in Environmental Sustainability*, 5 (1): 67–71.

Murphy, A. H. and Winkler, R. L. (1984) 'Probability forecasting in meteorology', *Journal of the American Statistical Association*, 79: 489–500.

Murphy, L. M. and Edwards, P. L. (2003) *Bridging the Valley Of Death – Transitioning from Public to Private Sector Financing*, NREL/MP-720-34036. US Department of Energy.

Nemet, G. (2008) 'Demand-pull energy technology policies, diffusion and improvements in California wind power', in T. Foxon, J. Köhler and C. Oughton (eds), *Innovation for a Low Carbon Economy: Economic, Institutional and Management Approaches*. Cheltenham: Edward Elgar.

Neuhoff, K. (2004) *Large Scale Deployment of Renewables for Electricity Generation*, Cambridge Working Papers in Economics, No. 0460, available online at: http://www.econ.cam.ac.uk/electricity/publications/wp/ep59.pdf.

Neumayer, E. (2007) 'A missed opportunity: the Stern Review on climate change fails to tackle the issue of non-substitutable loss of natural capital', *Global Environmental Change*, 17: 297–301.

Nordhaus, W. D. (2006) *The 'Stern Review' on the Economics of Climate Change*, NBER Working Papers 12741. National Bureau of Economic Research.

Nordhaus, W. D. (2007) 'A review of the Stern Review on the economics of climate change', *Journal of Economic Literature*, 45 (3): 686–702.

Nordhaus, W. D. (2008) *A Question of Balance: Economic Modeling of Global Warming*. New Haven, CT: Yale University Press.

Nordhaus, W. D. (2011) *Estimates of the Social Cost of Carbon: Background and Results from the RICE-2011 Model*, NBER Working Papers 17540. National Bureau of Economic Research.

Nordhaus, W. D. and Boyer, J. (2000) *Warming the World: Economic Models of Global Warming*. Boston: MIT Press.

Pagani, M., Liu, Z., LaRiviere, J. and Ravelo, A. C. (2010) 'High Earth-system climate sensitivity determined from Pliocene carbon dioxide concentrations', *Nature Geoscience*, 3: 27–30.

Pielke, R. A. Jr (2007) 'Mistreatment of the economic impacts of extreme events in the Stern Review report on the economics of climate change', *Global Environmental Change*, 17: 302–10.

Pigou, A. C. (1920) *The Economics of Welfare*. London: Macmillan.

Pindyck, R. S. (2011) 'Fat tails, thin tails, and climate change policy', *Review of Environmental Economics and Policy*, 5 (2): 258–74.

Portney, P. and Weyant, J. (1999) *Discounting and Intergenerational Equity*. Washington, DC: Resources for the Future.

Ramaswamy, V. *et al.* (2001) 'Radiative forcing of climate change', in J. T. Houghton *et al.* (eds), *Climate Change 2001: The Scientific Basis*, Contribution of Working Group I to the Third Assessment Report of the Intergovernmental Panel on Climate Change. Cambridge, UK and New York: Cambridge University Press.

Randall, D. A. *et al.* (2007) 'Climate models and their evaluation', in S. Solomon *et al.* (eds), *Climate Change 2007: The Physical Science Basis*, Contribution of Working Group I to the Fourth Assessment

Report of the Intergovernmental Panel on Climate Change. Cambridge, UK and New York: Cambridge University Press.

Raven, R. (2007) 'Niche accumulation and hybridisation strategies in transition processes towards a sustainable energy system: an assessment of differences and pitfalls', *Energy Policy*, 35 (4): 2390–2400.

Read, P. (1994) *Responding to Global Warming: The Technology, Economics and Politics of Sustainable Energy*. London: Zed Books.

Rockström, J. W., Steffen, K., Noone, Å., Persson, F. S., Chapin, III, E. Lambin, T. M. *et al.* (2009) 'Planetary boundaries: exploring the safe operating space for humanity', *Ecology and Society*, 14 (2): 32.

Schumpeter, J. A. (1939) *Business Cycles: A Theoretical, Historical and Statistical Analysis of the Capitalist Process*. New York and London: McGraw-Hill.

Segal, P. (2011) 'Oil price shocks and the macroeconomy', *Oxford Review of Economic Policy*, 27 (1): 169–85.

Seki, O., Foster, G., Schmidt, D., Mackensen, A., Kawamura, K. and Pancost, R. (2010) 'Alkenone and boron based Pliocene pCO_2 records', *Earth Planetary Science Letters*, 292 (1–2): 201–11.

Solomon, S., Plattner, G.-K., Knutti, R. and Friedlingstein, P. (2009) 'Irreversible climate change due to carbon dioxide emissions', *PNAS*, 28 January.

Stern, N. (2006) *The Economics of Climate Change*. Cambridge: Cambridge University Press.

Stiglitz, J. E. and Wallsten, S. J. (2000) 'Public-private technology partnership', in P. V. Rosenau (ed.), *Public-Private Policy Partnerships*. Cambridge, MA: MIT Press.

Sunstein, C. (2013/14) 'Unquantifiable', *California Law Review*, forthcoming, online at: http://papers.ssrn.com/sol3/papers.cfm?abstract_id=2259279.

Taleb, N. N. (2007) *The Black Swan: The Impact of the Highly Improbable*. Random House.

Thorne, P. W., Lanzante, J. R., Peterson, T. C., Seidel, D. J. and Shine, K. P. (2011) 'Tropospheric temperature trends: history of an ongoing controversy', *WIREs Climate Change*, 2: 66–88.

Tietenberg, T. H. (2006) *Emissions Trading: Principles and Practice*. Washington, DC: Resource for the Future Press.

Tol, R. S. J. (2006) 'The *Stern Review* of the economics of climate change: a comment', *Energy and Environment*, 17 (6): 977–81.

Tol, R. S. J. and Yohe, G. W. (2006) 'A review of the *Stern Review*', *World Economics*, 7 (3): 233–50.

Tol, R. S. J. and Yohe, G. W. (2007) 'A Stern reply to the reply of the review of the Stern Review', *World Economics*, 8 (2): 153–9.

UNDP (2005) *International Human Development Indicators*. New York: UN Publications.

US Environmental Protection Agency (1997) *The Benefits and Costs of the Clean Air Act*. Washington, DC.

Watkiss, P. (2011) 'Aggregate economic measures of climate change damages: explaining the differences and implications', *Wiley Interdisciplinary Reviews – Climate Change*, 2 (3).

Watkiss, P. and Downing, T. (2008) 'The social cost of carbon: valuation estimates and their use in UK policy', *Integrated Assessment Journal*, 8 (1): 85–105.

Watkiss, P., Anthoff, D., Downing, T., Hepburn, C., Hope, C., Hunt, A. and Tol, R. (2005) *The Social Costs of Carbon (SCC) Review: Methodological Approaches for Using SCC Estimates in Policy Assessment*, Final Report, Defra, UK.

Weitzman, M. L. (2007a) 'A review of the Stern Review on the economics of climate change', *Journal of Economic Literature*, 45 (3): 703–24.

Weitzman, M. L. (2007b) *Structural Uncertainty and the Value of Statistical Life in the Economics of Catastrophic Climate Change*, NBER Working Papers 13490. National Bureau of Economic Research.

Weitzman, M. L. (2009a) 'On modeling and interpreting the economics of catastrophic climate change', *Review of Economics and Statistics*, 91 (1): 1–19.

Weitzman, M. L. (2009b) *Risk-Adjusted Gamma Discounting*, NBER Working Papers 15588. National Bureau of Economic Research.

Weitzman, M. L. (2010) *GHG Targets as Insurance Against Catastrophic Climate Damages*, NBER Working Papers 16136. National Bureau of Economic Research.

Weitzman, M. L. (2011) 'Fat-tailed uncertainty in the economics of catastrophic climate change', *Review of Environmental Economics and Policy*, 5 (2): 275–92.

Weitzman, M. L. (2012) 'The Ramsey discounting formula for a hidden-state stochastic growth process', *Environmental and Resource Economics*, 53 (3): 309–21.

Wilkins, G. (2002) *Technology Transfer for Renewable Energy: Overcoming Barriers in Developing Countries*. London: Royal Institute of International Affairs.

World Bank (1992) *Development and the Environment*. Washington, DC: World Bank.

World Bank (2005) *World Development Indicators*. Washington, DC: World Bank.

World Bank (2009) *Ghana Report*, PREM4. Washington, DC: World Bank.

World Bank (2012) 'Analysis of Sudan post-oil shutdown economy', *Sudan Tribune*, 9 May, online at: http://www.sudantribune.com/DOCUMENT-World-Bank-Analysis-of,42534.

Yergin, D. (1991) *The Prize: The Epic Quest for Oil, Money and Power*. New York: Pocket Books/ London: Simon & Schuster.

Yohe, G. W. (2006) 'Some thoughts on the damage estimates presented in the Stern Review', *Integrated Assessment Journal*, 6 (3): 65–72.

Yohe, G. W. and Tol, R. S. J. (2007) 'The Stern Review: implications for climate change', *Environment*, 49 (2): 36–42.

Yohe, G. W., Tol, R. S. J. and Murphy, D. (2007) 'On setting near-term climate policy while the dust begins to settle – the legacy of the Stern Review', *Energy and Environment*, 18 (5): 621–33.

Zhang, Y.-C., Rossow, W. B., Stackhouse, P. W., Romanou, A. and Wielicki, B. A. (2007) 'Decadal variations of global energy and ocean heat budget and meridional energy transports inferred from recent global data sets', *Journal of Geophysical Research*, 112.

Zhengelis, D. (2007) *What the Stern Review Really Said*, Tyndall Briefing Note No. 22. London: Beyond Stern: Financing International Investment in Low Carbon Technologies and Projects, 19 February.

2 The Three Domains

'... there is always an easy solution to every human problem – neat, plausible, and wrong.'
– H. L. Mencken (1949: 443)

The enquiring mind yearns for a universal theory – a way to make sense of the world. Most people, indeed, see the world through the lens of some theory, whether knowingly or not. Experience and observation can tell us when these theories are wrong, or at least when they are inadequate for the situation at hand. The most common mistake is to assume that a theory which holds true in one situation will similarly be applicable to all.

The evidence presented in Chapter 1 illustrates that energy and climate change raise questions of extraordinary reach and complexity, with responses that have been limited, contested and as yet unequal to the challenges. Despite decades of effort, progress seems glacial. Modern energy systems have brought huge benefits, yet the way we produce and consume energy is unsustainable. The world remains as dependent on fossil fuels as ever, and CO_2 concentrations in the atmosphere rise relentlessly. Billions of people still do not enjoy the benefits of modern systems. Ensuring universal access to modern technologies is a priority, but seven billion increasing to nine billion carbon-intensive consumers would further exacerbate the global pressures on resources and environment.

Theoretically, the principal 'operating system' of classical economics – which has helped to deliver many of the benefits of modern energy systems, along with their attendant problems – is based on assumptions of rational human behaviour trading off and balancing costs and benefits. Yet the evidence presented in Chapter 1 suggests that there is no simple way of summarising the costs associated with changing energy systems: they seem still to embody large inefficiencies and damaging side effects, and their development comprises a complex mix of uncertainty, innovation, inertia and irreversibility, with multiple objectives to be delivered including energy access and security. Quantifying the benefits of curtailing emissions is even harder – with results being wrapped in the mists of how much climate change societies can handle at what cost, and never-ending debates about how to weight the welfare of people in different places and in future generations. Added to this, Weitzman's Dismal Theorem plays a role somewhat like Heisenberg's Uncertainty Principle – developed into an implication that by the time we know whether there is a real risk of catastrophe it will be too late to avert it.

Our responses – as individuals and our governments – can seize upon such uncertainties as either a serious risk to be managed or as an excuse for inaction, unmoved by apparently remote planetary threats. We face being trapped by how the uncertainties around 'costs and benefits' combine with resistance to action, together with the collective nature of the problem: the 'pessimism squared' with which the first chapter concluded.

This, fundamentally, points to the need for a deeper understanding that better reflects the realities of what we are dealing with. For the world is far more complex than any single theory assumes: these challenges operate at scales which transcend anything that humanity has had to face, or even think about, before. This is a problem, but also an opportunity. The core argument of this book is that a deeper, more integrated understanding that embraces different dimensions is possible and can inform more effective responses.

Specifically, this chapter argues that to make sense of the issues, we need to understand the world in terms of three different domains which are characterised by different time horizons and scales – and different associated processes. These domains apply across a swathe of relevant topics, spanning ways of conceiving the challenges through to the practical policy solutions. Specifically, the chapter delineates:

- *three conceptions of risk*, ranging from too small or remote to bother to being potentially a dominating fear about our collective security – different basic frameworks, which can also describe the attitudes people may take to opportunities;
- *three fields of theory*, that encompass at one end the psychology of individual and organisational behaviour, through to the ways in which complex, interrelated systems of technology, infrastructure and institutions evolve over decades;
- *three economic processes*, from simply making better use of 'win-win' opportunities that involve no trade-offs, through to the development of new technologies and systems that expand the range of what is possible;
- *three realms of opportunity*, which reflect at one end the considerable scope to save energy and emissions in ways that apparently also save money, through to blue skies options that would require extensive innovation.

These in turn underpin the *three pillars of policy*, grounded in the characteristics of the different domains, all of which are necessary to offer the potential of comprehensive long-run solutions. These then provide the core structure for the rest of the book.

2.1 Three conceptions of risk and opportunity

Seven billion people (and rising) make for a lot of decision-makers. Through the choices they make in their daily lives, almost all contribute in some way to the problems at hand, even if in very diverse and unequal ways. Many of them also vote or otherwise contribute to national positions and policies. Their attitudes matter.

Most of the seven billion don't like maths (which is why this book avoids equations, except in the Appendix), and are even less comfortable with the kinds of probabilities and uncertainties that characterise so many of the issues in energy and environment.

Some have already been seriously affected by volatile energy prices or natural disasters. Beyond the impact of energy prices per se, Chapter 1 noted that resource struggles have on occasion, and climate change could, threaten the security of some nations. Ultimately international stability could be at stake. If these issues could threaten our collective security, why has more not been done? Viewing energy and climate change as potential security issues seems just to widen further the gulf between the potential threat and the action, between the warnings and apparent apathy. It also poses awkward questions about how 'security' relates to the desire for cost-effective, proportionate responses. The approach seems to solve one paradox at the expense of creating others.

In his book *Thinking, Fast and Slow*, the Nobel Laureate Daniel Kahnemann illustrates how perceptions of risk are influenced by all kinds of factors that seem logically irrelevant (compared to a formal appraisal, of the kind that statisticians might attempt to use). The root of this (and much else) is that the human mind has two different kinds of systems for decision-making: the immediate and the deliberative. The research he covers shows how this distorts the way we see things, yet also why it is essential – we could not possibly cope if we had to make deliberative calculations about everything we do.[1]

These two mental centres of decision-making are not wholly independent and their relationship is complex. We deal with many things in our daily lives through habit, if we have no particular reason to think about or focus our attention otherwise. Such circumstances may lead us to ignore a risk – usually with no adverse consequence, but occasionally to our peril. However, prior experience or a wide range of other factors may lead us instinctively either to trust or fear something ('affective' processes), or our perceptions may hinge on association with other experiences ('associative' processes) – whether or not the associations are logically relevant. These processes lead us to sometimes ignore, and sometimes hugely exaggerate, different kinds of risks.

Though less studied in the academic literature, similar processes may affect how we perceive or react to opportunities. Some may be steadfastly ignored, unless and until we have experienced the benefits or seen other people doing so. Others may seize our imagination, for example the belief that technology, whose impact seems so transformational around us, will somehow similarly save us from having to do anything difficult.

Making sense of all these imponderables shows that there are at least three different basic conceptions of risks and opportunities. Indeed, human beings (and their societies) tend to react in one of three main ways, as summarised in Figure 2.1.

Risk conception	Basic belief	Typical strategy	Societal process	Timescale of climate change
Indifferent or disempowered	Not proven, or 'What you don't know can't hurt you'	'Ignorance is bliss'	Environmental group campaigns *vs.* resistance lobbying	First few decades of climate change
Tangible and attributed costs	Weigh up costs and benefits	Act at costs up to 'social cost of carbon'	Technocratic valuation and politics of pricing	As impacts rise above the noise – next few decades
Disruption and securitisation	Personal or collective security at risk, climate change as a 'threat multiplier'	'Containment and defence'	Mitigate as much as practical and adapt to the rest	Ultimately, for all (systemic and global risk) Most vulnerable, sooner, with international spillover

Figure 2.1 Three conceptions of risks and their application to climate change

Level 1: below the radar – 'ignorance is bliss'?

The most common conception dominates most of our lives so much that we take it for granted. We do not decide, but we simply carry on doing things in the ways we are used to, ways that seem adequate because nothing has made us think or act otherwise.

Many different situations carry varying degrees of risk. The evidence shows that if a risk is too small – *or is perceived to be too improbable or remote* – we simply ignore it. There is a threshold level for concern, below which people may not actively respond to the risks of even possible catastrophic events. Living near a volcano is an obvious example. Tsunamis, too, were a known hazard in the Pacific – yet governments did not even establish warning systems, and people were happy to live and work on the coast without them.

This first level of risk perception is one of practical *indifference* to risk warnings. The difficulty is further compounded by the fact that most people find it hard to distinguish between huge differences in probabilities if they remain low enough: 1 in 100 is still a hundred times more likely than 1 in 10,000 but is generally not perceived as such. This does not bode well for responding objectively to a risk characterised in part by making highly improbable extremes much less unlikely.[2] All this of course is a problem for climate change, whose main impacts are likely to arise from making various types of extreme events less improbable over time.

The fact that many of the victims of natural disasters may be in other countries or future generations amplifies further the 'psychological distance' of climate change in particular from the experiences and concerns of most people. The combination of these factors is important – people are unlikely to get on a plane if told that '1 in 100' is the risk of it dropping out of the sky due to a known mechanical problem, yet the same probability of planetary disruption may register as too remote to worry about. Risk perception is thus also a matter of distance, in time, geography and other ways. Climate change is far away on most of these indices.[3]

Thus this first level characterises most people's experience and attitudes in daily life and is one dimension of how they perceive energy and environmental problems – and most of all climate change. They cannot clearly identify and relate to the additional risks in question; there is a lot of natural weather variability, and most societies can accommodate that pretty well. Not many have yet been seriously touched by additional climate change. That should be of no surprise. It is standard maths that when two varying things are combined, the variability of the larger one tends to disproportionately swamp the total. Given the inertia in the earth's systems, we are still in the low foothills of climate change. Except for the most extreme situations like marginal glaciers or the Arctic icecap – or for those, like scientists, who are consciously measuring it – it hardly intrudes on our daily experience.[4] As any politician can testify, it is not easy to motivate people about things they do not personally experience.

Moreover, people often feel they can do little about such problems anyway – they feel 'disempowered', whether by the sheer scale of the global problem or the inadequacy of local options (such as poor public transport even for those who would be otherwise motivated to use it). In contrast, people perceive any public policy, taxes or standards as a tangible trouble, as one additional imposition on their lives. Overall, from a psychological and political perspective, there is little paradox between climate change as a distant possible threat to global stability and our inaction. Most people have more immediate and more tangible concerns which they feel they can do something about.

The combination of indifference and disempowerment feeds complacency. The lack of attention towards seemingly *remote risks* is one aspect of what we term 'First Domain' behaviour.

As outlined in the next section and explored more in Pillar I (notably Chapter 4), First Domain characteristics are equally relevant to personal decision-making concerning energy consumption. Lack of attention, force of habit, social norms, constraints and other factors can lead to behaviour that can seem very wasteful. Energy costs may seem too small or disconnected from our daily decisions to bother with. As a result people leave lights on, windows open, engines running – and often live or work in poorly insulated buildings. Again: most of us, much of the time, operate in a 'default' domain of habit, which can merge into carelessness and wastefulness.

In the 1950s, some leading economists developed ideas of 'bounded' rationality and coined the term *satisficing behaviour* to explain conditions in which people are satisfied *enough* with current habits that they can't be bothered to try and do better.[5] Psychologists find it easier to comprehend. Change requires effort, and people don't like making an effort unless they see a need to. Mainstream economics, having acknowledged in principle the phenomenon and given it a name, in practice then largely ignored it so that they could focus on what is generally seen as the core focus of their discipline: optimising approaches to minimise costs and maximise benefits.

Level 2: optimisation, compensation and cost-benefit

If something draws our attention to a cost or risk, we may try to evaluate it more carefully. Clear evidence and unambiguous data can then help us make deliberative choices, to match costs and benefits often with an attempt to quantify each. This second level of risk conception encompasses what is often characterised as 'rational' approaches: evaluating costs or risks that we notice and take account of, and to which we respond in ways calculated (in the broad sense) to maximise our net benefit. This is broadly the domain of market behaviour, conscious investment, actuarial assessment of insurance risks and premiums, and public cost-benefit appraisal. It is the domain in which all manner of things can be traded off and balanced consistently, aiming to *optimise* response particularly through monetary valuation. The underlying assumption is that risks, or damages, can be *compensated* usually in terms of monetary metrics.

This second level of risk perception, combined with an evaluation of costs and benefits, characterises the Second Domain of behaviour and is the dominant domain for many of the larger actors in economic systems. Those taking investment risks are compensated by a higher expected economic rate of return. Energy companies project costs and energy prices to evaluate projects. They may study portfolio choice and new technological options, and evaluate scenarios of future developments. If their actions may impose risks on local populations, they are generally regulated; sometimes they have to compensate for damages.

Individual consumers may also sometimes operate in this Second Domain, but usually only in very discrete ways – for example, when pushed to make explicit choices between different energy suppliers at different costs, or with clear information about the running costs of major purchases like cars, or when sharply rising energy prices kick them out of the attitudes of First Domain behaviour. If there is large uncertainty, we seek insurance to hedge against unpleasant outcomes.

In climate change, this Second Domain characterises the way in which economists have tried to estimate the potential damages and compare these to the potential costs of cutting emissions. Chapter 1, however, has shown how this effort has provoked more questions than answers: questions about how we value risk, time, people in other countries and our successor generations – amid deep uncertainties.

In some societies, extreme events have brought the issue of climate change forcefully onto the public radar, provoking calls for tougher action. Trying to quantify the aggregate global damage in monetary terms, however, is not merely towards the impossible end of difficult, for the reasons detailed in Chapter 1. Nor is it the language of the concerns expressed by most environmentalists and scientists. Concerns about planetary risks and threats to our security draw attention to the third way in which risks are conceived.

Level 3: security and strategy

The third conception of risk takes us into the arena of security and strategy, in the face of an acknowledged (or perceived) threat to fundamental human needs or to the integrity of social systems (which are generally provoked when fundamental needs of their populations may be threatened). This is the arena in which risks, perceived or real, can be transformed into strategic opportunities. It is the domain of national security planning and building political alliances around perceptions of strategic risks and goals. There is usually deep uncertainty, of a kind that cannot be sensibly calculated – economists call it 'Knightian' uncertainty (Knight, 1921).

The difficulties are encapsulated by 'Black Swan' events – risks which may not even be perceived at all before the event, but can actually have a huge impact. Inductive reasoning may lead us in the wrong direction concerning such risks – as with the famous example of a turkey.[6] Our psychological and social processes – and indeed formal appraisal – can often *obscure* the risks. Attitudes are then in the first level, below the radar.

While 'the Black Swan' idea refers to an essentially random example coined by the popular author Nicholas Taleb, the academic literature is increasingly concerned that the growing complexity of human systems may make unexpected extreme events more likely and/or more devastating. Such events, associated with growing complexity and the impossibility of understanding all possible interactions, are acquiring a different name – X-events.[7]

Whatever the terminology, the common characteristic is that such events are not expected and largely ignored – unless and until they happen or something makes them suddenly of concern. Attitudes may then flip, perhaps to wildly exaggerating the risks and taking (or demanding) precipitate action. '9/11' has become the totemic example of a 'Black Swan' event; the financial crisis, the exemplar of possible 'X-events'.

This takes us to the third conception of risk, most relevant to events that combine deep uncertainty with the risk of high consequence – presumed low probabilities but with impacts that would be simply deemed *unacceptable*. The essential approach is then not a cost-benefit calculation but a strategy to avoid or mitigate unacceptable outcomes. To see this, consider how many debates on the defence budget are informed by probabilistic estimates of the likelihood and financial costs of foreign invasion or the economic gains of projecting military power. At an individual level, we may purchase insurance not on a cost-benefit basis, but to make ourselves feel safer.[8] But the more extreme problems are ones that even insurance companies may fear to touch – for example, if they have no idea about the probabilities, if the consequences would be too catastrophic and/or if the premiums would be politically unacceptable. In some countries, flood-zone protection is beginning to fall into this category.[9]

Unacceptable outcomes cannot always be avoided. For some of these outcomes, insurance can play a role in mitigating the effects, though often incompletely. We might buy insurance to protect ourselves against wrecking our car, but governments generally mandate insurance to cover 'third parties' as a legal requirement. Should your driving cause a fatal accident, the insurance is not much use to the victim, but to those who remain it is preferable to nothing. Insurance is an imperfect alternative to avoiding some of the more profound risks we face.

Energy has clear security dimensions in its own right and at several levels in our societies. At a personal level, acknowledged statistics point to the 'excess winter deaths' among poor people who cannot heat their homes. Geopolitically, oil has particular and obvious security dimensions, linked with securing supplies particularly from the Middle East. In fact most energy sources in isolation would carry some risk to 'security of supply'.

In the environment, the difference between the second and third levels of risk conception is not at all about whether environmental issues matter – yet the division is surprisingly sharp. The former corresponds to what is known as *environmental* economics, and it seeks to quantify environmental damages and to devise policy responses that optimise our response. *Ecological economics*, however, is far more concerned with the stability of ecological systems and, to an important degree, the values that underpin our assessment of what can be used and what should be protected.[10] Closely related, the literature has long debated the distinction between 'weak' and 'strong' sustainability – the former applied to environment and resources that can ultimately be substituted (e.g. by greater wealth), the latter concerned with those that cannot.

In accounting terms this logic is reflected in a distinction between monetary-equivalent and physical accounting. Following many years of debate, in February 2012 the UN Statistical Commission agreed to extend the System of National Accounts – used since the 1950s as an international standard for measuring economic activity – to become the System of Environmental and Economic Accounts (SEEA). This defines a framework for both a monetary account (of items for which reasonable valuation techniques exist), to be paralleled with physical asset accounts, to measure and track the quality and quantity of environmental assets in physical terms.[11]

In climate change, this third level of risk conception reflects the deeper worries of the scientists. In terms of the 'risk matrix' (Chapter 1, Figure 1.8), natural scientists worry about the bottom row – the lurking fear that the accumulation of heat-trapping gases will ultimately trigger some global instability in the earth's oceanic or climatic systems. Social scientists, in the broadest sense, also worry about the right-hand column of the risk matrix – the possibility that regional climate changes, particularly (and initially) in weak societies, may stress food supply or invoke other impacts that lead to social upheaval (see Chapter 1, note 72).

Yet as we have noted, in practice most people and political systems have more pressing concerns than climate change. More than a billion poor people have yet to be connected to an electricity grid – clearly a more pressing priority. The impact of the oil shocks on some of the struggling developing country importers is, similarly, a more immediate threat to well-being and security. In such countries this inevitably eclipses the vague, long-term threat of climate change.

When compared to a global, long-term perspective of planetary risk, being more worried about energy access and prices may seem collectively irrational – but it is absolutely human. Basic needs must be met before people can contemplate more systemic ones. For most people and individual decision-makers, prioritising these more immediate concerns is entirely rational, because any individual, or indeed country, on its own can do more to manage its energy poverty (e.g. by connection and insulation programmes) and risks (e.g. by diversification) than it can do to reduce global emissions. Actions to cut CO_2 emissions that exacerbate severe energy problems or worsen security of supply will not be acceptable. It will be infinitely easier to pursue responses to climate change that also address these more tangible concerns. From this flows the 'Third Domain' of decision-making, which lays the stress on the *transformation* of systems over time to meet multiple objectives, including security.

This potential then helps us to understand the need to align approaches across all three levels – seeking synergies between immediate priorities of basic development with personal and national economic enhancement and strategic goals of energy and climate security. Each has a role and involves different types of decisions and decision-makers – a theme which permeates this book.

Risk conceptions: conclusions and implications

Understanding these three levels of risk conception offers a way of attempting to make sense of the reality: how society can comprise scientists, environmentalists and some politicians, concerned that humanity's future may be at stake, while the economic systems carry on regardless. Alongside is a general public that inclines to apathy – but which may in some circumstances also swing to the opposite extreme, of fear.

This stark contrast between the first and third levels points to a real dilemma stemming from the social dynamics of risk perception. These extremes are strongly influenced both by 'associative' and 'affective' processes, and hence are potentially unstable. This is particularly true given the impact of 'availability' effects – the ease of recalling or relating to relevant risks. People may ignore floods or extreme heat risks (including, for example, forest fires) before the event. However, communities which have suffered such events are more likely to believe in climate change – even if, to a scientist, they are just one more notch on the crowded barometer of accumulating evidence.[12] The resulting profile in the media may grab the attention of more people. This makes political attitudes to regulating emissions intrinsically unstable.

The efforts of economics and cost-benefit calculus are one way of trying to strike a balance between the extremes of doing not enough on the one hand and possibly panicking on the other. However, given the fundamental imponderables in quantifying costs and benefits noted in Chapter 1, this is impractical. Instead, a balance between the second and third levels of risk conception generally motivates the setting of thresholds designed to try and maintain safety – such as the 2°C target, or the closely related 50 per cent reduction in mid-century emissions, as outlined in Chapter 1.

The retort from economics is that such a threshold implies a cost anyway, so that the two are logically the same – the key is to consistently express how much we should be willing to spend to mitigate the problem. The implication drawn is that, however wide the uncertainties are, society should debate and negotiate its way to adopt a best guess of the 'social cost of carbon emissions'. That, the argument goes, is the essence of a rational approach, and fits logically with the idea that a carbon price (at that level) would be the most efficient way of responding.

The rest of this chapter probes an implicit assumption in this: namely that all responses can be usefully measured and compared on a common metric of present-cost-equivalent (to be more precise, discounted net present value). It turns out that the 'risk conception' side of the ledger is not the only one which needs to be understood at three qualitatively different levels.[13]

2.2 Three fields of theory

In parallel to the three different basic conceptions of risk, there are qualitatively different processes of decision-making and responses relevant to our energy systems. These involve different actors taking different kinds of decisions on different scales of time and space – which, hence, underpin qualitatively different kinds of response.

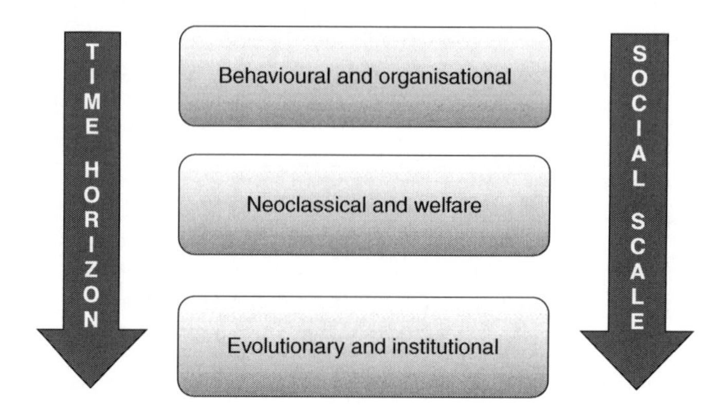

Figure 2.2 Three fields of theory

Note: The three central fields are frequently denoted as different branches of economics and is the terminology used in this book.

The argument rests on intellectual foundations which have developed substantially in recent decades. This section outlines some of the main avenues of enquiry, which are picked up, tested and applied in the rest of this book.

Numerous lines of research and evidence, as illustrated in Figure 2.2, seem to form naturally into three main groups, which can be readily ordered on the basis of the time horizons and scales to which they are most relevant:

- *Behavioural and organisational economics* – which share a fuzzy boundary with basic psychology and management theory – dig into the realities of how the behaviour of individuals and organisations diverges from the traditional assumption of 'rational economic man' (or perfect organisations). In energy, these theories help to explain the apparently wasteful nature of human energy behaviour and the enduring potential to cut both energy consumption and costs, as explored under Pillar I (Chapters 3–5).
- *Neoclassical and welfare economics* is the workhorse framework for most economic analysis, and it underpins the basic rationale for markets and price instruments. It has developed well beyond its simple origins, to help us understand phenomena ranging from business cycles and financial crises through to patterns of trade and investment flows. In energy, these theories can help to illuminate the relationship between energy and carbon pricing, investment and macroeconomic policy, as explored under Pillar II (Chapters 6–8).
- *Evolutionary and institutional economics* explores the long-run development of economic systems and their dependence upon certain institutional fundamentals. It has emerged as a broad discipline uniting studies of technological change, the development of complex systems, the role of institutions in economic growth and development studies. In the energy sector, evolutionary and institutional economics helps to inform a long-run understanding of the different ways in which energy and associated systems have evolved, and might do so over the course of the century. A particular dimension of interest is the relationship between innovation and economic growth, as explored under Pillar III (Chapters 9–11).

In general, the ordering illustrated above follows progressively larger scales of time and space – and the decision-making entity. Behavioural economics is of most obvious relevance to individuals or discrete organisations, at local or sometimes national scales, on timescales of months to a few years. Neoclassical and welfare economics is particularly concerned with how economic systems – and the individuals and organisations that they are comprised of – may respond with time horizons up to a couple of decades. Efforts to push its insights over longer timescales start to run up against the phenomena explored in evolutionary economics and studies of institutional development. These are concerned with phenomena operating over decades, often applying not only across whole societies but relationships between them and the diffusion of ideas and technologies across wider regions.

The main features and associated references, particularly with respect to energy systems and related decision-making, are developed in the relevant chapters of this book. This chapter gives a broad overview, together with a few core general references to introduce the topics.

Behavioural and organisational economics: homo instinctus[14]

Of the three levels of theoretical development indicated above, the first has probably gained the highest public attention in recent years. This is partly because it rests on *experimental* techniques, which have been able to demonstrate just how far individuals and their organisations stray from 'rational' decision-making. Much of this research has focused on perceptions of and responses to risk, as outlined in the previous section. However, it also encompasses far wider aspects of human decision-making, including energy choices.

Most micro-economic theory ultimately rests on the assumption that human beings, to a reasonable approximation, are economically rational, that is they take decisions to maximise their economic welfare, with rational expectations based on a relatively stable set of values and preferences. Of course our own experience tells us this is an approximation, but it took painstaking behavioural research to illuminate systematically how and in what ways people in general, and organisations, stray from that theoretical assumption.

Research into 'behavioural economics' spans some decades. Many of the key findings are brought together in Kahnemann's book *Thinking, Fast and Slow*, which also traces the neurological roots of two systems of decision-making: the rapid and instinctive response, and the calculating, deliberative part of the mind.[15] The insights and their implications have been popularised in several other books, notably *Nudge*.[16]

The findings of behavioural economics have been applied to understanding decision-making in financial markets and a book by Diamond and Vartiainen (2013) charts six other fields where the ideas have already proved useful but have not yet been fully incorporated.[17] This book adds another: energy.

As noted, mainstream economics itself acknowledged the reality of non-optimising behaviours, in the form of 'satisficing theories', more than half a century ago (note 5). However, it was mainly left to psychologists and others to explore more systematically, though economics is reclaiming some of the territory with the tools and terminology of behavioural economics.

With global energy consumption and emissions involving seven billion decision-makers, it is not surprising that the resulting insights turn out to be important. Far from being 'rational optimisers', the way people consume energy in particular is frequently dominated by habit, inertia, aversion to the perceived risks of doing anything differently and a wide range of other factors. The main lines of analysis and their relevance to energy – particularly energy efficiency – are sketched in Chapter 4 of this book, which also considers recent critiques of behavioural economics and notes that in fact they tend to support the underlying approach of this book: the key is to work out when and in what ways the behavioural insights are significant, relative to the influence of mainstream economic forces.[18]

Individual psychology is moreover only a small part of the story. The essential components of behavioural economics also reverberate through organisations. This forms largely different fields of literature, naturally again with different research communities. Economic studies have tended to focus on the problems that arise between 'principals' and 'agents' within and between organisations. Economics has also expanded to include mathematical theories of networks and communications, and how constraints on both compromise classical assumptions of organisational or market efficiency. The management studies literature tends to use different language and has a more action-oriented agenda – how to run better companies – on which the ideas seem limitless.

Yet all these contrasting fields boil down to one basic point: neither theory nor evidence supports the idea of organisations as soulless, efficient machines delivering a single objective (generally assumed least cost and maximum profits, in the case of private sector organisations). They are complex and imperfect structures. As with the modern insights of behavioural economics, this points to a large potential for more efficient behaviour, if energy consumption is seen to be an issue worth managing (see Chapter 4).

Neoclassical and public welfare economics: homo economicus

In the middle of Figure 2.2 lies neoclassical and welfare economics which has been the dominant framework for much policy analysis. Its basic assumptions of rational foresight with stable preferences and technologies make it amenable to mathematical precision and clarity and its use of optimising techniques help to inform policy. In idealised systems there is a theoretical 'optimum equilibrium' state for economic systems, defined by the most efficient use of a fixed set of resources. Price then plays the pivotal role in achieving a state of 'general equilibrium', which also aligns private and social interests, as prices equilibrate to match the cost to producers with the value to consumers.

Let us state once, clearly and firmly, that the insights of neoclassical economics do not all hinge on its core assumption of 'rational representative agents' as a precise description of reality. Its conclusions remain relevant for as long as that is a better assumption than other alternatives (for example, assuming perfect governments). This applies also to policy interpretations: an 'imperfect market' does not on its own justify interventions, or throwing out the baby with the bathwater.[19]

Moreover, over the years, neoclassical economics has encompassed a vast array of developments. These span an improved understanding of the role of *expectations* and *economic and business cycles* – and more generally the interplay of current prices and longer-term expectations in driving economic decisions. For example, research on *business and economic cycles* helps us understand better why economies are not simple, stable entities but are subject to all manner of fluctuations on timescales of a few years to a few decades, and real-world behaviour includes the complex determinants of finance and investment under uncertainty.[20]

'Public welfare economics' is a term used professionally to encompass wider studies of how economic policy can seek to maximise public welfare. Compared to branches of neoclassical economics which seek to describe how an economic system behaves and may respond given its core assumptions, welfare economics is broader and more prescriptive. Broadly interpreted (with the narrower neoclassical framework as a specific formulation that can input to it), it is the dominant theoretical embodiment of Utilitarian philosophy, taking the view that the objective of government should be to seek the 'greatest good for the greatest number'. On this basis, the classical aim is to maximise utility through cost-benefit analysis. The challenges in applying this to energy and climate change have been outlined in the opening chapter of this book, and the conclusions (Chapter 12) revisit the question of 'impact assessment' in the context of the Three Domains.

The core assumptions of welfare economics are less restrictive than the 'rational agents' assumption of neoclassical economics, and concern most fundamentally the ability to *estimate and aggregate* individual welfare, as represented by personal utility, so as to get a measure of the collective good, denominated in terms of money or money-equivalent.

Areas of research particularly relevant to energy and environmental studies concern developments in resource economics and the choice of instruments for influencing prices, expectations and investment. More disputed, though sometimes also under this heading, has been growing attention to the distributional features of economic development, and influences of political economy on policy choice. Some branches of institutional economics, particularly concerned with the governance of firms, also fall within this broad field.

The boundaries of these 'mainstream' economic theories with the other domains considered in this book are interesting, and potentially contestable. For example, one of the biggest intellectual revolutions in economics was Keynes' *General Theory*, which like other major developments was a product of its time. It was developed in the 1930s to explain how a world in which the existing theory assumed that economic agents would soon return to a state of general equilibrium with close to full employment, actually remained mired in devastating recession. Among other insights, Keynes brought attention to many behavioural factors: that consumers could be deterred from spending, and companies from investing, by a lack of confidence; and the *stickiness* of many economic features (such as resistance to working for reduced wages and the way this would further reduce spending power and confidence as consumers). He showed how these could lock the system up in a vicious circle of depression from which government intervention and expenditure was the main way out.

Without using the same terminology, Keynes was of course pointing in part to features that are now well recognised in behavioural economics, with implications widely recognised in mainstream economics – including the fact that business actions tend to reinforce economic cycles through effects related to swings in confidence and herd behaviour. Indeed some of the leading proponents of neoclassic growth theories acknowledged that Keynesian ideas had to be brought into play to explain shorter-term processes of adjustment (Chapter 11).

At the other end of the timescale, many of the axioms of neoclassical economics become increasingly strained in the long term. Welfare economics assumes that people know what is good for them on all timescales – that preferences expressed as personal utility functions are fixed – and tend to assume, for example, that technology costs are given, independently of the economic environment. It also generally assumes that wider institutional 'rules of the game' are reasonably secure and stable. Based on the implied, well-defined characteristics of supply and demand, all this means that an optimum implied point or path of 'least cost' equilibrium can be identified.[21] Uncertainty complicates, but does not in itself undermine, this fundamental framework.

Surprisingly, there seems to have been little explicit analysis of the appropriate scales within which such assumptions are reasonable or *sufficiently good* to be useful. The experiments of behavioural economics help to illuminate the lower levels at which assumptions of market-based rationality and consistency break down. At the higher end, assumptions of foresight, time-consistency, stable preferences and institutional structures, and static or steady development of technologies, are clearly less and less plausible the longer the timescales involved. 'Growth theory' seeks to extend the time horizons of economics to encompass economic growth processes and the accumulation of knowledge. But as we show in Pillar III (particularly Chapter 11), these formal growth theories remain bound by constraining assumptions that clearly clash with the realities of the long timescales characteristic of energy and climate change problems.

Which takes us to the third level.

Evolutionary and institutional economics: homo evolvens

Probing these boundaries points beyond, to the fields of *evolutionary economics* and the economics of the *institutional environment*.[22] The broad range of studies under these headings emphasise the extent to which economic systems are *evolutionary* in character, and depend upon past developments, the direction of innovation, inherited infrastructure and the nature of societal norms and institutions. The literature in these areas in fact spans more widely to encompass technological, systems, economic history and institutional, legal and social disciplines. The evolutionary perspective thus explores the ways in which modern economies are shaped by the interplay of technology, infrastructure, institutions and interests.

Broadly, the technology and economic perspectives underline the role of technological systems and how they evolve in ways that are 'path dependent': previous choices determine future possibilities, and the types of technology developed and utilised respond to these forces. In economic terms, the production possibilities that define a supply curve in a market become *malleable*, depending on the economic environment and investment choices accumulated over previous decades.

Recognising the central role of institutions in economic performance informs the literature widely known as New Institutional Economics, which is particularly concerned with how the institutional environment defines the 'rules of the game' – notably the rules of law (including and particularly in the economics sphere the rule of property rights). The growth of this literature is partly attributable to bitter experience.

More than half a century's efforts of trying to foster global economic development led ineluctably to the conclusion that institutions are central. The idea that the main development need was for rich countries to channel money to the poor failed to produce the scale of results hoped for. This led the World Bank and other international funders to focus more on the policy environment and funds became increasingly contingent on the 'structural adjustment' of policies – accompanied by huge controversy and still patchy success. Conclusions from the post-communist Russian experience were similar, with one evaluation noting ruefully that before recommending mass privatisation (including that of natural resources) as the solution to everything, Western experts should have read the literature on the importance of the institutional environment.[23]

Major crises – like wars and the collapse of communism – offer rare (and usually bungled) opportunities for rapid reform at this level. Otherwise, changes are very slow, as evidenced by the long struggles to establish systems of property rights. The European Union itself can be understood as a development in this Third Domain: it has already been under development for more than half a century and the Eurozone crisis shows that its development is still in the early stages, with the crisis defined by the institutional failure of adopting a single currency without adequate common fiscal mechanisms.

While at first sight there is little connection between the apparently diverse fields of studying long-term phenomena, this is largely false. For example, the Nobel economist

Kenneth Arrow notes that 'Truly among man's innovations, the use of organisation to accomplish his ends is among both his greatest and his earliest',[24] to which a review of institutional economics adds:[25]

> We cannot fail, however, to be awed by the profound importance of technological innovation. Inasmuch as these two work in tandem, we need to find ways to treat technical and organisational innovation in a combined manner.

The scope for, and entwined nature of, technological and institutional innovation has profound implications. As we show later in this book (in Pillar III), it calls into question whether there is a single 'least cost' future path, from which we can calculate the cost of any deviation. Rather, we face choices between different evolutionary paths for our economic and social systems – choices that are continually being made, but which have enduring consequences. Some broad evidence is included in reviewing energy system technologies in Chapter 3, and brought fully to bear on the strategic choices we face in Chapters 10 (for energy systems) and 11 (for their implications for economic growth). The concluding chapter of this book then highlights the way in which the First and Third Domains do, indeed, correlate to issues of organisational and technological innovation.

These three broad fields of theory, collectively, provide the foundations for the exploration of *Planetary Economics* as applied in this book: while mainly about practice and policy, it rests on the structure of theoretical developments of Figure 2.2. Whether you, the reader, consider this to be a critique of economics or an application of its insights at the frontiers of economic research is a personal choice: we think of it as the latter, exploring the appropriate boundaries this implies and their implications.[26]

The core of the book rests on three policy pillars, as outlined at the end of this chapter. Each pillar seeks to draw on a range of developments in the corresponding theories. The essential, underlying argument is that while traditional economics offers tools that may be adequate in considering problems of limited scope ('marginal' changes to the existing system) over bounded time periods and within individual countries, that toolbox is far too limited with respect to problems on the scale and timescale of energy and climate challenges. Neoclassical economics has an important role, but is only part of a much richer understanding of economic affairs: the frontiers of research are far more interesting, rewarding – and relevant to questions of energy and environment.

2.3 Three economic processes

Having sketched out these themes, it is useful for a moment to return to some economic fundamentals, because this can illustrate with surprising simplicity how the three fields relate to each other in terms of economic process.

Economics, at its most basic level, is about allocating resources in ways that seek to meet human needs and desires. Resources – whether of people, capital, land, mineral deposits or energy – are not infinite. Much of economics concerns the inherent trade-offs, whence the term 'the dismal science'.

To many minds, an intense attraction of classical economics stems from its promise to reconcile the individual with the collective interest, through the 'magic of the market' in allocating these resources as efficiently as possible: once property rights have been defined and enforced, the market can achieve this and hence deliver both private and common good. The underlying idea appears to be independent of time and domains but, as illustrated in the previous section, it is not.[27] The other domains matter, and a key intellectual challenge is not only in understanding the component parts but how the three domains fit together.

Figure 2.3 offers a simple conceptual translation of this aspect of the 'three fields of theory' in relation to energy and environment. The top panel shows the simplest possible case of resource trade-offs: a fixed line determines what may be termed the 'best practice frontier' – the best available way to produce economic output or welfare (horizontal axis) for a given use of a key physical resource (vertical axis). Reducing use of that resource comes at a cost of lost economic output, at least in the short term, because other factors (e.g. more capital or labour or some other more expensive resource) must be used – substituted – instead. The frontier defines that trade-off. Obviously, the idea can be generalised to any number of resources and associated dimensions. For the purposes of linking this explanation

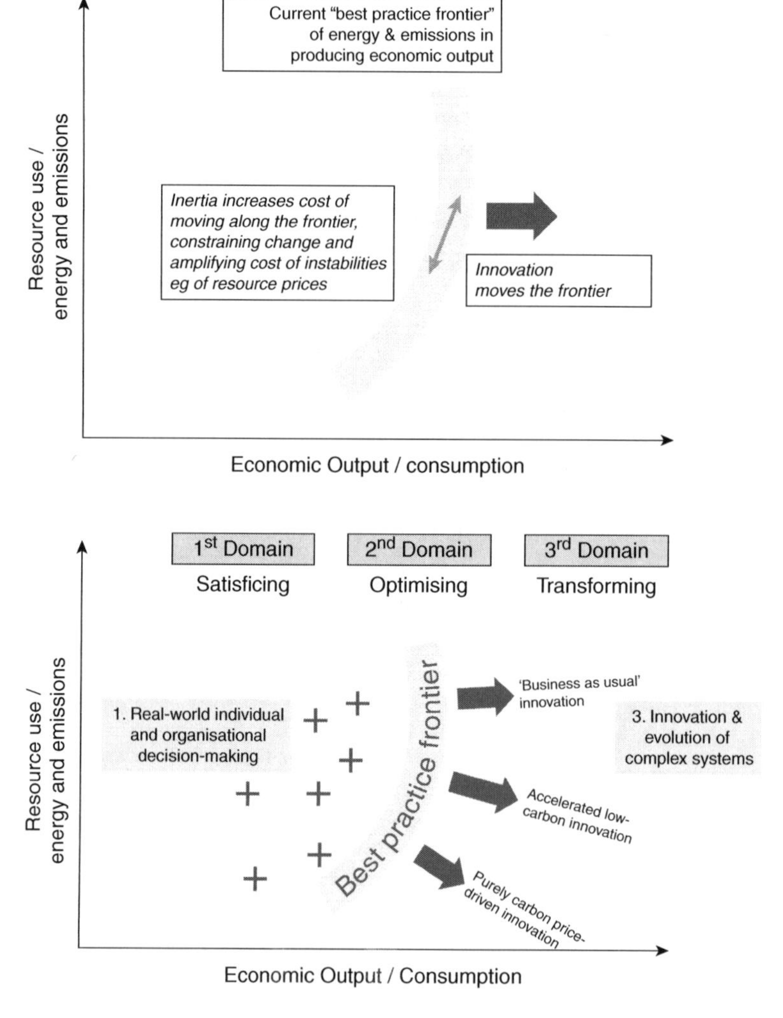

Figure 2.3 Resources and economic outputs in the Three Domains

(a) Resource trade-offs in the neoclassical domain conditions.

(b) Resource trade-offs and opportunities in the three domains.

to the topic of the book, think of the vertical axis as representing energy use and associated emissions, the horizontal axis everything else.[28]

Much of welfare economics is, fundamentally, about how systems can be designed to optimise the trade-offs defined by such curves across the enormous range of resources available – largely as determined through relative prices, the prime mechanism for transmitting information about the scarcity and value of the different resources available.

The simpler levels of analysis are time independent – hence the term 'equilibrium' economics. In practice time enters in many ways. One important way is through the inertia implied by existing investments, which make it costly to move rapidly along the curve from using one mix of resources to another. Data discussed at the beginning of Pillar II (Chapter 6) point to evidence that the energy systems have considerable capacity to adjust to price changes, but only given long time periods; the systems literature reviewed under Pillar III provides extensive additional evidence that inertia in energy and economic systems is very large. If the cost of fossil fuel rises suddenly, for example, we cannot quickly or cheaply switch away from such fuel. The pain is immediate and the shock may lead to some short-term responses, but bigger and less painful adjustments accumulate over years and decades. The debate on the macroeconomic impact of oil shocks (Chapter 1) also points to the costs associated with volatile prices of key economic inputs. Such inertia in itself has important implications.

Time also enters through movement of the curve itself – as a result, for example, of technical change, which increases the output available from the use of a given resource. The classical simplifying assumption is that this process is either *autonomous* – it occurs like manna from heaven – or *optimal* – the product of economic agents maximising their benefits from innovations.

In this illustration, the other two domains enter as shown in Figure 2.3(b). Phenomena of the first domain are represented by points to the left of the line which produce less output than potentially available using 'best practice' for the same resource use. Clearly, this creates the *potential* for improvements apparently *without* any trade-offs – defined in economics as 'Pareto improving' – a gain without anyone having to lose out. The general assumption is that self-interested individuals and organisations, given the freedom to do so, will naturally move to the 'frontier'.

When reality differs from this ideal, neoclassical economics generally attributes this to various forms of market and institutional failures, as well as time lags in catching up with a moving target as innovation moves the best practice frontier. However, as noted in the previous section, First Domain (or 'satisficing') behaviour in individuals implies that, actually, most people fall well short of this. So do real organisations. It is principally the experimental evidence underpinning behavioural and managerial economics which has revealed and explained this as a *systematic* feature of human psychology and organisation, rather than an anomaly that people or organisations will quickly and spontaneously deal with. Note, however, that such behavioural and organisational theories only form part of the First Domain realities considered in this book, which comprise all the factors which lead to societies operating far short of the 'frontier'.

In the field of energy, this is reflected in what is often called the 'efficiency gap'. There is abundant evidence around the way that people use energy and waste it, and the huge potential for greater energy efficiency (Chapter 3). The reasons explored in Chapter 4 show that this is due to a diverse mix of contractual and market failures (such as the split of incentives between tenants and landlords), personal behavioural traits, principal–agent asymmetries and related problems within organisations. The former, in particular, are an example of 'missing contract' problems in economics and obviously reduce efficiency, however defined. Many other factors, again backed by extensive evidence as detailed in Chapter 4, explain the extent to which the real world falls way short of the 'best practice frontier', particularly if this is understood in terms of the social optimum.[29]

On the other side of the line, the 'best practice frontier' defined by available technologies and systems moves over time: technologies and organisational structures improve to allow the same output with less input of resources. Indeed, as discussed further in Chapter 11, innovation is widely acknowledged as a prime driver of economic growth. The Third Domain is defined by the question not of *whether* the curve moves – whether there is innovation – but by *how* it moves – the pace and type of innovation.

The simplest workhorse assumption is innovation that is 'resource neutral' – it occurs equally across all inputs. This is obviously unrealistic. Nothing fundamental changes if the pace of innovation – in this case the efficiency of using resources – is projected to vary for different resources. However, innovation economics gets far more interesting when it looks at how innovation efforts may be channelled, typically to reduce the use of more expensive inputs. As noted, innovation also applies to the development of rules and institutions that help to determine how well human societies make use of the resources they have.

These possibilities help to define the Third Domain, in which technical and institutional innovation can influence the whole direction in which systems evolve. Over time this may fundamentally reshape the 'best practice' frontier. Against this background Figure 2.3 illustrates three schematic choices in relation to energy and climate change:

- 'business as usual' innovation, which may predominantly move the curve to the right and ignore in particular the value of CO_2-reducing innovations (particularly if they are not priced);
- innovation driven purely by rising carbon prices, which will tend to move the curve downwards;
- accelerating innovation in the overall energy system with an emphasis on low carbon technologies.

It turns out there are many reasons why energy-related systems tend not to innovate with the pace that might be expected (Chapter 9) or in the direction that is needed given the reality of carbon constraints (Chapter 10). As measured by the intensity of research and development expenditure, we show that energy is one of the least innovative sectors in our economies, and prone to getting locked into established patterns and technologies – which may ultimately not be good for either the environment or the economy.

2.4 Three realms of opportunity

Presented in this way, it is hard to dispute the existence of the Three Domains. Given the focus of mainstream economic research and much policy on the second domain, however, there remains a legitimate question of scale: how much do the other domains actually matter?

This is an empirical question: it demands insight from data, not just ideological principles. This book presents several different lines of evidence to suggest that, at least in the context of energy and environmental challenges, all three domains matter. The data suggest, indeed, that in the context of mid-century goals around energy and climate change, each domain may be of similar magnitude in terms of its potential significance.

The evidence comes in many different forms, from micro to macro, and from technology to institutional studies. It emerges most comprehensively from a survey of all the ground covered in this book – by a careful look at data and experience in each of the domains. We summarise a few key indicators in Box 2.1.

Box 2.1 On the comparative importance of all three domains

One could acknowledge the existence of the Three Domains and still believe that one of them eclipses the others in importance. In reality their relative significance depends on the topic being considered. Our contention is that for transforming the global energy system over a period of a few decades, all three domains are of comparable importance.

The most simple indication comes from engineering data on the potentials for reducing the use of fossil fuels and CO_2 emissions, explored in Chapter 3 and best brought together in the form of a global cost curve, such as shown in Figure 2.4. This seems to suggest that the first two domains are the most important, with the first dominating energy efficiency potentials (Figure 4.2) while the second dominates possibilities for switching from high to low carbon energy sources (Figure 6.5). However, as the text explains, this is simplistic. For example the relevance of the engineering evidence on energy efficiency is strongly disputed by many economists, as detailed in Chapter 4, and extending the time horizons beyond 2030 obviously increases the importance of Third Domain (innovation-related) potentials (e.g. Chapter 10). Multiple lines of evidence thus inform our contention that all three matter.

Efficiency theory and practice. Attempts to explain away 'negative cost' potentials on the assumptions of the Second Domain – that people and societies cannot really be so wasteful in market economies – prove uncompelling. In addition to the engineering evidence, a huge literature charts numerous barriers to energy efficiency, while behavioural and organisational studies provide further compelling theories; engineering estimates certainly overlook some 'hidden costs', but also 'hidden benefits' (Chapter 4). The impact of efficiency programmes has been charted and improved energy efficiency over the past few decades has decoupled energy use from GDP at 1–2% per year on average, in ways that cannot be explained principally as a direct (Second Domain) impact of price-driven substitution effects. Some countries with sustained and focused energy-efficiency policies have averaged close to 3% per year improvements for decades (Chapter 5).

Price responses. Chapter 6 (e.g. Figure 6.1) highlights the crucial importance of energy prices, but also notes that the relationship of price to energy intensity varies by a factor of 2–4, depending on whether it is measured between or within countries: this points to a large divergence between the pure market responses to price changes compared to more enduring systemic changes across all domains. Moreover, the response of countries to price rises has been much bigger than the converse response to price falls – frequently several times as big – which cannot be explained purely by classic substitution effects. Cross-country measurements are consistent with Bashmakov's constant of energy expenditure (Chapter 1): both their divergence from in-country price responses, and the asymmetric nature of such responses point to the comparable importance of all the domains. Major price shocks have provoked structural, behavioural and policy reform, and innovation, which deliver enduring adjustments consistent with First and Third Domain processes.

Innovation and growth. Chapter 9 demonstrates the weakness of innovation in energy systems left to themselves, in terms of their low R&D intensity and the important role of policy in accelerating it, by forging better links along the chain of innovation from idea to widespread use. Chapter 10 confirms the enormous potential for innovation and infrastructure to deliver radically different systems over time, while also tracing the complexity of energy networks and infrastructure which tend to lock us into current systems. Finally, in the penultimate chapter we show that a long-standing economic debate – the 'Solow residual' in the classical models of economic growth – can only plausibly be explained with reference to the forces at play across all three domains: the simple accumulation of resources does not remotely account for observed growth over the twentieth century. Across whole economies, the other domains are significant; in energy, we conclude, they are disproportionately so.

Perhaps the simplest data to illuminate the potential present and future significance of the three domains in this context lies in engineering assessments of the scale and cost of opportunities to reduce CO_2 emissions. Against the backdrop of energy systems analysis, innumerable studies have now evaluated these; the next chapter briefly reviews many of the technical possibilities. The neatest way of summarising the results is with an 'abatement cost curve'. This plots the potential scale of each option for reducing emissions, against its cost-effectiveness, measured in cost per tonne of CO_2 abated. Figure 2.4 summarises in this way the results of one of the most extensive international efforts to do this, a study by the McKinsey global consulting company which assessed the potential of 161 different measures in each of 21 different country/regions to cut emissions by 2030. Because this book is about energy, the figure focuses on the energy-related components.[30]

Compared to their 'business-as-usual' projection of sharply rising emissions, the analysis suggests the potential to cut emissions in 2030 by around 25 billion tonnes of CO_2 – almost two-thirds of current global energy-related emissions and a scale which, if true and implemented, would turn projected rapid emissions growth into a substantial reduction from present levels.[31]

More relevant to this discussion, the general pattern – reproduced in almost all such efforts – comprises three distinct parts of the curve:

- On the left are options which appear to save a substantial amount of money as well as emissions when assessed on a common economic basis (e.g. discount rate) but which are not currently taken up. A classic example is better building insulation, though there are innumerable others as illustrated throughout Pillar I of this book. These suggest a potential for *smarter choices*, in the broadest sense. The McKinsey curve suggests that global emissions in 2030 could be cut by almost 5,000 million tonnes of CO_2 per year (compared to 'business as usual') by options which also save more than \$20/tCO_2; the full potential of 'negative cost' options is double this. Broadly, such options thus seem to account for around a third of the global potential.
- In the middle part of the cost curve are proven emission-reducing options with costs that are comparable to or more costly than current, carbon-emitting technologies (when put on a comparable basis).This is the zone defined by *substitution* of high carbon by low

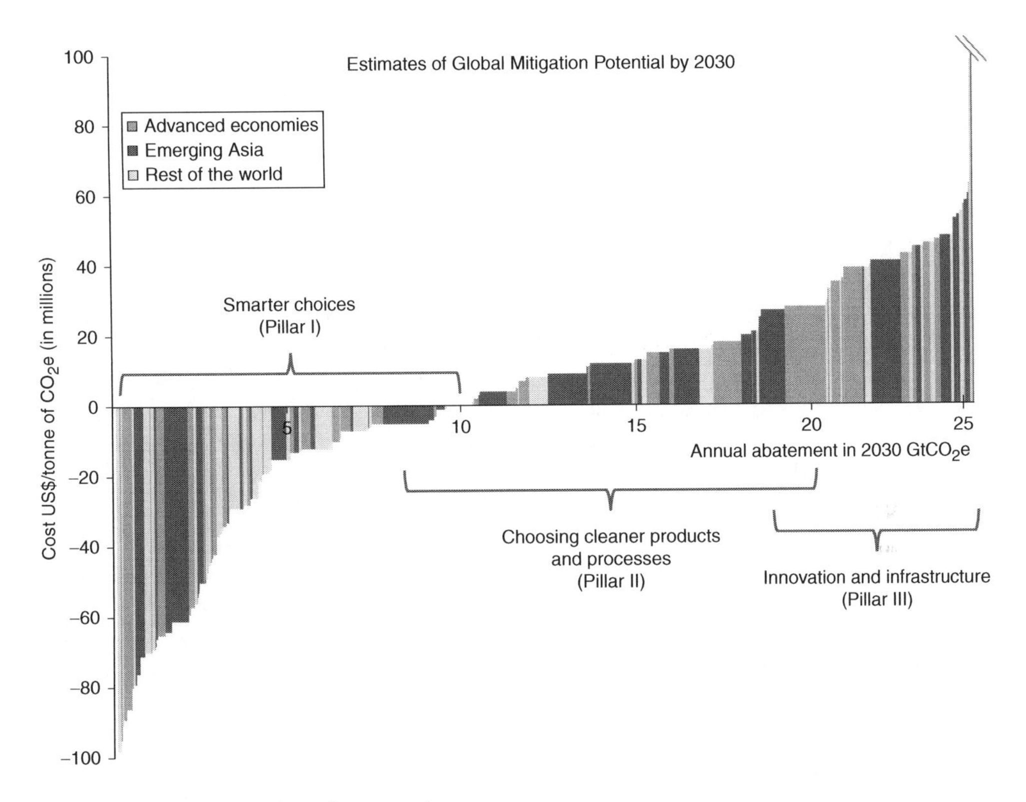

Figure 2.4 The three realms of opportunity

Note: The chart shows the 'McKinsey curve' of estimated potential to cut global CO_2 emissions by 2030, in terms of the cost of measures (vertical axis) and the scale of measures (horizontal axis). Only measures related to the energy sector are included. For details of some of the options represented by the individual bars see Chapter 4 (for the negative cost options) and Chapter 6 (for the central zone).

Source: Generated by the authors using source data from the McKinsey report, *Pathways to a Low Carbon Economy* (2009).

carbon options, and the (carbon) price required to make it worthwhile. These *substitution* options – including some illustrated in the chart as somewhat negative cost, since the boundary is actually far from precise – were assessed to offer the biggest single realm of emission savings, broadly between a third and half the total.[32]

- On the right are options which appear much more costly at present, but which are identified as part of longer-term potential. Carbon capture and storage – for which component technologies are all known and proven but for which there is little experience at a commercial scale – is a classic example. The costs appear high by present standards so innovation is required, and/or they may reflect costs caused in part by technologies that do not 'fit' existing infrastructure. There is obviously a grey area of overlap with the second realm of opportunity (substitution of proven options). In the nature of innovation, it is hard to be confident about the overall scale of options that might emerge; the chart was of course confined to specific known options.

With the backing of a pre-eminent global consulting company, the McKinsey curve is well known in the global energy climate debate. Many energy analysts dislike it, but interestingly sometimes for conflicting reasons. Specialists in energy efficiency have been doing technical cost curves with negative costs for decades and, as we show in Chapter 4, think McKinsey underestimated some of the potentials.[33] Most energy economists in contrast argue that the 'negative costs' on the left-hand side are largely fictitious or irrelevant, for example on the grounds that the engineering studies do not consider 'hidden costs' or the 'rebound effects' of energy efficiency (see, for example, Chapters 4 and 5, and Figure 4.2 for an elaboration on this).

The central region of the curve covers almost every extant technology. Each has its advocates and detractors who dislike the role given to particular technologies (nuclear, large-scale hydro, wind, biomass, CCS, etc.) and point out that differing assumptions could result in different ordering and scales that might lead to their favoured technology being bigger (or render their *bête noire* unnecessary). And on the right-hand side, most engineers can think of technologies that should be there, if only someone spent the money to develop them properly – McKinsey focused only on reasonably well-known technologies, not on the potential for more radical innovation to extend the curve further to the right with new options not yet proven.

Figure 2.4 also codes the options according to three broad regional groups: the potentials in the rich (OECD) countries, in the emerging Asian economies[34] and the rest of the world. The global potential is roughly equally divided between OECD and Asian emerging countries, with a smaller potential in other regions. There is also not much difference in how the different realms of opportunity are divided across these regions: all types are evident in all regions. If anything, there is some indication that the developed economies feature more at both extremes of the cost curve – the highest and the lowest (most negative) cost options. Some possible reasons for this are touched on in Chapters 4 and 6.[35] Most fundamentally, however, the opportunities are not concentrated mainly in rich, emerging or poor countries, but are spread across all regions.

Most critics of the McKinsey curve point out real weaknesses. But they miss an important message conveyed by the three parts of the cost curve. These three realms of opportunity are associated not just with different costs, but they correlate directly with the three economic processes illustrated in the previous section. They are in effect an estimate of the potential scale of these processes in the energy sector (over a couple of decades), which in turn relate to the three fields of theoretical development outlined. The data suggest that all three matter. The details can be argued over, but the point is that such data make it plain, to anyone with an open mind, that *all three domains deserve serious attention*.

The most intensive academic attacks have indeed been on the 'negative cost' conclusions, since these seem to clash so fundamentally with neoclassical assumptions: if these options save so much money, people would do them anyway. Are people (or economic systems) really so wasteful? Explanations are hypothesised about 'hidden' costs, with deeper questions about whether, in practice, governments could (or should) induce smarter choices. And even if governments can do so effectively, shouldn't they do so anyway without claiming it as part of climate policy? Pillar I of this book considers all these objections more closely – drawing also on the data of experience – and finds them wanting.

Though the ire was directed mostly at this 'first realm' of opportunities, some similar objections were also raised regarding the 'third realm'. No one disputes that innovation will be important, but shouldn't markets deliver innovation anyway, as a 'natural' process? Shouldn't, indeed, other policies 'wait' until innovation reduces their costs? Can governments be effective in fostering innovation, and wouldn't that just divert 'innovation resources' away from other ventures? These are among the issues considered under Pillar III of this book (Chapters 9–11).

Such questions are important, but also reflect an ideological search for explanations and arguments that could preserve the pre-eminence of the 'second domain'; they stem in part from reluctance to acknowledge the intellectual underpinnings of the other domains, trying to fit everything into a neoclassical framework. Yet as noted, almost every person and organisation on earth contributes to energy consumption and emissions; it is not surprising if barriers and behavioural quirks add up to a huge potential (the first domain). And since the issues span decades – to the mid-century and beyond – and economic incentives to do something about CO_2 emissions are only just emerging – it is not surprising if there is huge additional potential in the scale and direction of innovation and infrastructural development (the third domain).

Most important, these challenges involve questions on which we can again seek evidence (though some also entwine issues of political philosophy). That is, in part, the quest of this book, drawing upon data and experience. The next chapter outlines the technological options, with the subsequent chapters exploring more deeply the theoretical evidence and experience. The empirical evidence behind these different realms of opportunity is overwhelming and cannot be explained away by, for example, claims that the data ignore 'hidden costs' or the idea that markets will produce 'optimal innovation'.

The data in each case points to a conclusion that each domain is not only real but very significant – to a degree which very broadly suggests each to be of similar magnitude concerning energy system responses over the next few decades. In short, the first and third domains are not minor curiosities, marginal to the 'real' economics. They form an intrinsic and essential part of the landscape.

Thus the data – when combined with all the evidence that explains the underlying processes as covered through this chapter – points to three realms of opportunity, all with huge potential:

- *smarter choices* by people or organisations, particularly regarding the way we use energy;[36]
- *substituting cleaner products and processes*, where the economics of prices, market structure and investment are at the core, and many of the biggest opportunities lie in changing the conversion systems (particularly electricity);
- *innovation and infrastructure* investments which reduce the cost of options that are currently too expensive or are not currently available, including large-scale developments to exploit, convert, deliver and use fundamentally different energy resources.

The policy question then is what can in practice can be done to realise these three realms of opportunity?

2.5 Three pillars of response

This brings us to the final triad: the three pillars of policy responses. Debate on energy and climate policy has veered between different preferred approaches: energy efficiency, carbon pricing and technology policy. Carbon pricing has been the approach most favoured by economists, for perfectly good reasons articulated in Chapter 6, but has tended to be the one least favoured by politicians and the general public.

The tangle of competing policy preferences can be mitigated by revisiting the 'cost curve' of abatement options and recognising that each of the three main parts corresponds to a different type of change that is needed if we are to move towards more secure and less carbon-intensive energy systems. For each of these, there are leading – though not exclusive – policy approaches:

- *Smarter choices* can be fostered through appropriate regulations and engagement. 'Regulation' is typified by standards that ensure adequate insulation in buildings, remove inefficient products from the market or require producers to display clear and unambiguous information to inform consumers. *Engagement* starts with such information and may build upon it in many ways to increase attention to, motivate and facilitate better choices. This forms the first pillar of policy examined in this book (Chapters 3–5).
- Most economic decisions to buy, sell or invest in *cleaner products and processes* depend on *markets and prices*. Measures that affect absolute and relative prices – including carbon pricing – will tend to be the strongest, most effective and efficient levers operating throughout the economic system. In practice their impact will also depend on *market structure* – many big energy sector investments are in sectors (like electricity) which, if not directly conducted by state entities, are nevertheless strongly influenced by the rules and regulations that shape the market structure and related terms of investment. This forms the second pillar of policies examined in this book (Chapters 6–8). It also has particularly strong spillover effects on the other processes, in various ways charted in the concluding part of this book.
- *Innovation and infrastructure* can be accelerated and guided in low-carbon directions partly by price incentives, but the analysis in Pillar III shows why private investment in energy innovation is neither adequate in scale nor likely to lead mainly in a low-carbon direction, unless there is either public involvement or other factors influencing strategic expectations. Infrastructure is also crucial. Hence, the key determinant will be *strategic investment* – investment which, due to public support or other influences, looks beyond the short-term returns to invest in ways that support the evolution of more efficient and lower-carbon energy systems.

These form the three pillars of policy. They are illustrated in Figure 2.5 in a way that seeks to underline that each pillar has a prime focus of impact on the economic processes and opportunities in the respective domain, but also influences the others. These interactions are, in fact, not peripheral but central to the overall argument of this book. The final chapter under each pillar outlines these interactions, while the concluding chapter delves deeper into the relationships between these policy pillars and why stable and effective responses require simultaneous progress on all three fronts.

2.6 Domain alignments

This brings us to the culminating argument of this chapter and a core thesis of this book. In terms of theory, *there is important alignment between the different conceptions of risk and the different pillars of response*. Thus individuals and organisations completely indifferent to energy resource or climate risks may still be motivated to take action in the First Domain, for example due to cost savings, co-benefits or brand appeal. At the opposite end, governments

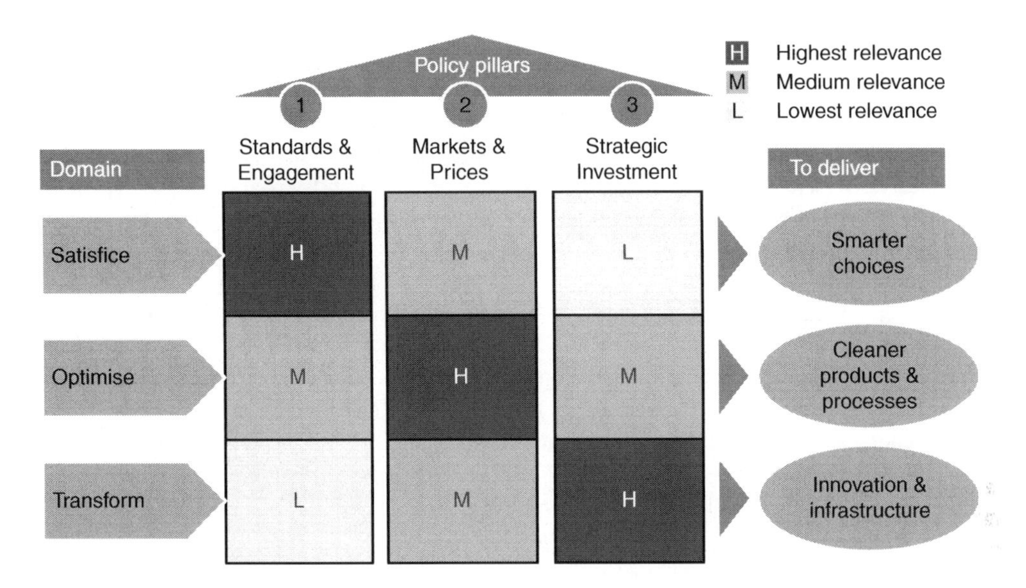

Figure 2.5 The three Pillars of Policy

Note: The row titles refer to the realm of opportunity, while the column headings refer to the policy pillars. The size of the circles in each cell refers to the degree of relevance of the policy to each of the opportunities.

Source: Authors.

and major multinational companies looking ahead to the middle of the century may recognise the need to contain climate risks with deep emission reductions and to act on the huge infrastructural or strategic positioning implications. Neither sets of decisions are fundamentally dependent on a unifying energy or carbon price. Nor do they intrude on one another. They involve different actors, timescales, processes and instruments. Neither the risk nor response can be reduced to a single monetised metric of comparison.

First Domain: ignore/satisfice

In the First Domain, attitudes to risk may be characterised by lack of interest, scepticism or indifference often reinforced by a sense that nothing can be done anyway – a tendency to ignore risks when they are unknown or distant (in geography, time or psychologically), unless something makes them tangible (which can lead to overreaction). This aligns with 'satisficing' behaviour, a dominant role for habits and the inertia of existing behaviours, beliefs and organisational structures: the weight of the 'status quo'. However, people (and organisations) can move closer to the 'best practice' frontier with net benefit, *irrespective of their concerns about energy or beliefs about climate change*. The corresponding opportunity is for policies that lead to smarter choices, either as the default (e.g. through standards) or active choice (through engagement). Of course, expanding awareness of the global challenges could be a useful motivating component for some, and indeed there are reasons explored in the final chapter why an important linkage remains. By improving efficiency, this also helps to prepare the ground for the policies of other domains that are also necessary.

Second Domain: compensate/optimise

The Second Domain, characterised by efforts to optimise choices and associated presumptions of neoclassical economics, includes a presumption that risks can also be compensated and thus incorporated in economic calculus. Both require foresight, planning and calculation. The conception of the challenge is focused upon efforts to quantify. How big is the 'externality' – the cost or risk imposed on others – that should be factored into pricing if markets are to be efficient? For investors in energy projects and large consumers, what is the price of energy, the risk of future price rises – including the prospect of governments acting to 'internalise the externalities'?

While this is relevant across the entire energy system, it is particularly salient for the big producers and consumers of energy, who will indeed make serious efforts to optimise their response based on price expectations. The opportunity is to substitute cleaner production and products, and the obvious mechanisms are around pricing. It may not in practice be possible ever to reach analytic agreement on the 'social cost of carbon', but the carbon price can and should emerge as a reflection of the overall *social* 'willingness to pay' to help reduce dependence on fossil fuels – with debates that of course may also influence the pull of consumer (and voter) demand for cleaner energy.

Third Domain: secure/transform

The Third Domain is characterised by strategic planning and investment to secure the integrity of our energy, economic and environmental systems and to transform them to keep within safe limits. What is the risk of resource depletion leading to major instabilities in global fossil fuel markets? Or of severe climate change events destabilising food, water or other essential systems and requiring emergency responses? How may energy systems over time be transformed within safe limits? The focus tends to be physical, not monetary, because the questions involved are systemic, not marginal – as needed to avoid the 'inside out' mental trap noted in concluding Chapter 1.

The most obvious choices are for government, but do also include the strategic calculations of multinational companies about what kind of world they want to prepare and invest for. The unstable nature of public risk perceptions, outlined in the first part of this chapter, highlight the *strategic risks* of being on the wrong side of the issue, for both major companies and political parties.

These are rational calculations, but they are *not* driven by numerated cost-benefit calculations and price is not the main tool. They are strategic decisions, reflecting the judgement of government departments and ministers and company boards. Given the need for deep emission reductions by mid-century, what kind of energy research should be supported, what kind of electricity networks or urban transport systems should we invest in? The phenomenally complex debates about long-run discount rates or the probability of catastrophic events, reviewed in Chapter 1, are not central in this context: what matters is defining a strategic goal that respects natural limits, is based on the available evidence and is moving towards it.

We show in Chapter 6 (Figure 6.6) that this goal-oriented approach is a perfectly rational consequence of the corresponding risks and uncertainties. In the concluding chapter we also show how the different pillars can correspond to different types of decision-makers – and hence how the overall framework can also help to move analysis to a more practically useful level than the hypothetical – and in practice misleading – universal 'representative agent' of much aggregated economic analysis.

Risk conception/ domain	Dominant scale	Decision framework	Field of theory	Mitigation economic process	Realm of opportunity	Pillar of policy/ response
Ignore/ satisfice	Short term/ local	Indifferent or disempowered	Behavioural and organisational	Move closer to the 'best practice frontier"	'Smarter choices'	Standards and engagement (Pillar I)
Compensate/ optimise	Medium term/ regional	Costs/impacts are tangible and significant	Neoclassical and welfare economics	Make best trade-offs along the frontier	Substitute cleaner production and products	Markets and pricing (Pillar II)
Secure/ transform	Long term/ global	Transformational risks and opportunities	Evolutionary and institutional	Evolve the frontier	Innovation and infrastructure	Strategic investment (Pillar III)

Figure 2.6 Alignments within each domain

Thus there is an important degree of alignment of characteristics between the different elements of each domain. This is summarised in Figure 2.6.

This alignment amplifies the value of understanding the different domains. A powerful reason for the intellectual dominance of cost-benefit analysis is that all decisions carry some implicit or explicit cost and we want to compare this cost against the benefits. Hence, in relation to climate change, we need to estimate or negotiate a 'social cost of carbon'. Indeed, if it is not explicit, our valuation of it can be *revealed* in terms of costs of the policies adopted. This seems to be at the heart of policy *consistency*, and how far different policies diverge from that cost can (and often is) cited as a measure of policy inefficiency. This approach is indeed at the heart of rational Pillar II policy, and it would be central if all decisions involved similar cost-benefit trade-offs. But they do not, because not all decisions are in the same domain. Since Pillar II policies are also those which attract the fiercest political opposition (as discussed across Pillar II), this may be just as well. Pillar III points out examples in which large divergence in the apparent cost of policies may be perfectly rational.

If climate concerns can motivate policies that carry multiple benefits and save costs, for example (Pillar I), they make sense irrespective of the social value of carbon reductions (or beliefs thereof). And scientists' 'best guess' about what constitutes 'safe' long-term concentration levels, and the corresponding cap on feasible global (cumulative) emissions, does a great deal to indicate the kind of innovation and infrastructure that we require (Pillar III). It makes little sense to invest in large-scale coal-to-liquid technology, for example, given the credible risk that using such technology as a way of dealing with oil scarcity would destabilise the global climate and violate everyone's security. This is a matter of strategic judgement, not a cost-benefit calculation dependent upon the discount rate and equity weightings.

In terms of policy, the grand challenges of our energy systems can be tackled, but only if we work with all three pillars of policy, and understand how they link to each other and to the larger domains outlined. Indeed, the key policy challenge is to pursue all three pillars in ways which are mutually reinforcing. Working with the grain of domain alignments can thus do a great deal to defuse the apparently imponderable dilemmas of trying to address everything within a single analytic framework. This is a key focus of our concluding chapter.

Conducting this exploration also sheds light on one of the most contentious claims in the economics of environmental policy by the Harvard Business professor David Porter. Pointing to the success of the Japanese auto industry and other cases in the mid-1990s, his *Porter hypothesis* conjectured that environmental regulation could improve economic performance, by stimulating greater efficiency and innovation responses.[37] It has remained a controversial claim ever since, and never deeply explored in relation to climate change and the subsequent policy experience. There is today a corresponding debate about *Green Growth* – which in its strong form claims that tackling global energy and environmental issues will boost economic growth.

Our penultimate chapter, through delving into the structure of economic growth theories, demonstrates why it is logically impossible ever to resolve the debate about the Porter hypothesis in general, or the strong form of claims that Green Growth will boost economic performance even irrespective of its environmental gains. Nor can the possibilities be ruled out. Something else is possible, however: to exploit the multiple opportunities at hand. In the final chapter, we consider the nature of possible *joint gains*, and the mechanisms by which they could operate within and across the three pillars. From this, we can see better what would constitute good policy in the face of the uncertainties and the real-world features outlined in the first chapter.

2.7 Conclusions

The enquiring mind may search for a universal theory but the quest is likely to be frustrating. Challenges of energy and environment span every person on earth, including complex socio-economic and technological systems, future generations and the planet itself. Organising the problem into the different domains of behaviour and analysis set out in this chapter seems a far more productive line of enquiry. Having done so, we can now step back to view the larger landscape.

Think of the main areas of theoretical development as first sketched in Figure 2.2 as forming the triple-deck of a sandwich as illustrated in Figure 2.7. Having probed the nature of these developments and considered how they manifest in energy systems, we can start to sketch some of the boundaries:

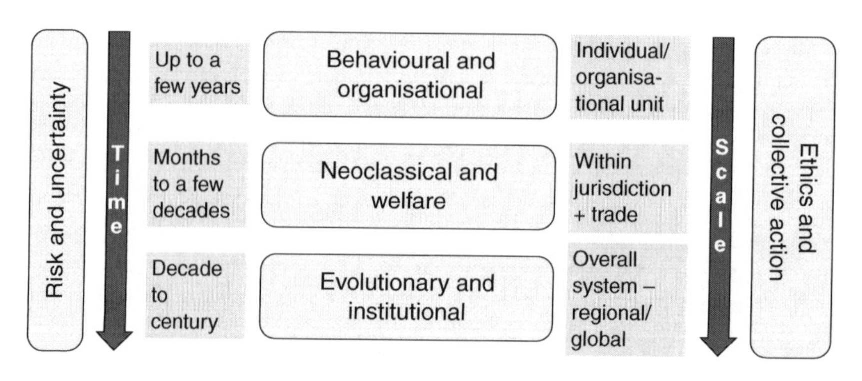

Figure 2.7 The fields of theory in the wider context

- First Domain phenomena seem particularly crucial up to timescales of a few years – as evidenced in Chapter 4, it is, for example, extraordinary how many energy efficiency investments pay back within such timescales. Individuals tend to be the key decision-makers, whether on their own or within the context of most organisations *outside* the energy sector, where energy expenditures tend to be modest and devolved.
- Second Domain phenomena tend to dominate on timescales of a few years to a couple of decades – the timescales within which investors generally seek returns. They often involve decisions by organisations *within* the energy sector and energy-intensive industries, and governments or government agencies (for example, regulators), acting within their well defined jurisdictions.
- Third Domain phenomena tend to dominate on longer timescales – decadal to a century or more. These are timescales over which industrial innovations tend to mature and diffuse, and education and infrastructure contributes to national development. The decision-makers may be governments and organisations with a particular long-term view (particularly multinational companies) and of course innovators.

On either side of the central domains are two cross-cutting areas: our understanding of risk and uncertainty, and of ethics and the problems of 'collective action' – coordinating responses across different groups (particularly countries). Chapter 1 and the first section of this chapter have outlined the challenges of *risk*, which feed into all three of the central elements. Uncertainty is endemic in human affairs and the rational response is neither to ignore an issue because it is uncertain nor to base policy on a central 'best guess'; rather it feeds into different processes at different scales and timescales. The challenges of ethics and international coordination cannot be avoided, but are beyond the scope of this book: a core argument is that we can still make a lot of progress without (yet) resolving the big issues of global coordination, and that *resolution of the global coordination issues also hinges upon a deepened and broadened understanding of the three domains*.

Given the complexity of the issues, there is surprisingly little *overlap* between the different domains: they concern different timescales, processes and decision-makers. It thus offers a framework that can help avoid the internecine warfare between different academic disciplines. Most disciplines have something to offer, and the most useful debate is not about which is right, but what the complementary contributions are and where the boundaries lie. That is something to be charted by evidence and experience, not ideological cat-fights.

Accepting these different spheres of analysis and decision-making also turns out to be practically useful. It offers a pragmatic but systematic approach to understanding energy and environmental challenges and options, and for evaluating responses based on the experience of the past few decades of both research and policy-making.

The rest of this book is thus structured around the Three Domains and the associated pillars of policy. Each pillar comprises three chapters, covering broadly:

- the nature of the challenges, technical data and opportunities arising, and key points of academic debate;
- the empirical evidence and policy experience;
- the scope for better policy, based on the combination of theoretical developments and practical experience, and how linkages between the different domains and pillars may help.

The final chapter brings together all this analysis to look again at the whole. It illustrates a way to get a simplified view of some implications, in the form of distinguishing ongoing costs from transitional efforts. It revisits how the three domains and fields of theory relate to each other, and summarises how the three pillars of policy can complement and reinforce each other. It shows how and why policy based on any one of the pillars alone is unsustainable and consequently why policy based on traditional economic theory has struggled so badly – as have those who sought to reject it. It testifies to the value of experience and of integration. Only by understanding the combined forces of behavioural realities, markets and pricing, and innovation and infrastructure together, can coherent responses be built to change course in ways that ultimately will benefit all.

Notes

1. Kahneman (2011). Kahneman, Tversky and their colleagues use the terms 'System 1' thinking and 'System 2' thinking to describe these respective mental processes. In this book I avoid this terminology because of the risk of confusion with the 'Three Domains'. The two systems of thinking are often called respectively 'experiential' (i.e. based on experience, not to be confused with experimental!) and 'cogitative'. In many cases, experiential (System 1) thinking equates with the First Domain and cogitative (System 2) thinking equates with the Second Domain, while the Third Domain involves processes that unavoidably involve combinations of both. However, the relationship could be stretched too far and I leave it to others to map more precisely the relationship of mental decision-making processes to the architecture of the Three Domains as developed in this book.
2. Evidence on the perception of risks of catastrophic events and related difficulties in understanding relative probabilities is presented in Camerer and Kunreuther (1989) and Kunreuther *et al.* (2001). The literature will be extensively reviewed in the IPCC Fifth Assessment Report, Working Group III, Chapter 2 (forthcoming 2014).
3. Psychologists identify at least four dimensions of psychological distance: temporal, geographical, social and uncertainty. In plainer language, we can more easily connect with, understand and be motivated by issues that are closer to us in time, place and social group, or about which there is little ambiguity. From the standpoint of most rich consumers that drive most of the problem, climate change scores poorly on all four dimensions (Spence *et al.* 2012). I am grateful to Sonia Klinsky for drawing my attention to the contrast with getting onto an aeroplane.
4. Specifically, the variability of two 'random variables' added together is the square root of the sum of each squared. Thus suppose we indicate natural climate variability by 3 in a robust society that could actually cope with a total climate variability of 5. An additional human-induced variability of '1' would only increase the total variability experienced by 0.16, a value of 2 would only increase the total by 0.6. It would then start to rise more rapidly: if climate variability matched the natural level (3) the total experienced variability would be 4.25. Human-induced variability would need to reach '4' before the total reached the societal capability to cope (5); by that point, the human-induced changed would be dominating and, unless it had been brought under control much earlier, would be rapidly escalating further.
5. Simon (1953; reprinted 1997) See also Chapter 4, notes 29–33.
6. Taleb (2007) details the numerous logical errors relating to Black Swan events, of which a common key component is inductive reasoning. The classic example is the empirical accumulation of evidence by a turkey that it is safe and well cared for, with every day strengthening the empirical basis for that belief – until Thanksgiving or Christmas, depending upon where it happens to live.
7. Casti (2012).
8. At the individual level, how many readers of this book have conducted a quantified estimate of the value they get from all their insurances and balanced it against the costs? If they did, most of the world's insurance industries might go bust. Security is worth a sizeable premium, even for individual and non-life-threatening risks.
9. There is a large literature on 'low probability high impact' events and the limitations of insurance approaches, but more detailed treatments are beyond the scope of this book. For practical purposes in considering the challenge posed for approaches to emissions mitigation, the security analogy suffices.
10. For an excellent overview of both the theory and practice, see Shmelev (2012).
11. http://unstats.un.org/unsd/publication/SeriesF/SeriesF_78E.pdf

12. See, for example, Michel-Kerjan and Kunreuther (2011).
13. The same basic proposition has emerged from leading researchers in several places, usually with the common currency that it derives from practical experience working in the energy sector. This author came to the broad view of the 'three pillars of policy' partly as a result of experience working at the UK Carbon Trust during the first half of the 2000s; I first presented it in the course of a lecture tour in Australia in 2005. The Stern Review (Stern 2006) itself carried three chapters on policy instruments, broadly corresponding to each pillar. The International Energy Agency (Hood 2011) published a superb overview of the case for 'Summing up the parts: combining policy instruments for least-cost mitigation strategies'. The German Oeko-Institut (Matthes 2010) makes the same basic case with an interesting variant, presenting 'infrastructure-related potentials' as a separate category spanning right across the cost-curve of options, whereas this book considers infrastructure alongside innovation as part of strategic investment for long-term benefits.

    This book seeks to build upon the common insights in several ways. This chapter suggests an alignment between the three pillars and the three levels of risk conception, and shows that each pillar rests on distinct underlying conceptual and theoretical foundations – forming the 'Three Domains'. The other sources noted tend to illustrate the 'carbon price-related' potentials as biggest, with the other two flanking; the global cost-curve data shown in Figure 2.5 are suggestive of a more even balance between the categories. Correspondingly, the IEA study refers consistently to the first and third pillars as 'complementary to a carbon price'; this book argues that all three are complementary to each other and ultimately inseparable, and hence that it is a mistake to elevate any single pillar above the others. Hence the book's approach, to examine each in depth in equal measure, and extend the analysis to consider their corresponding role in wider macroeconomic phenomena (Chapter 11) and then trace their intimate interactions and cross-cutting implications (Chapter 12).
14. The Latin 'instinctus', drawn from 'instigere', to stimulate.
15. Kahneman (2011). For a note of recent critique of behavioural economics by Levine – which actually turns out to be a discussion of boundaries and relative magnitudes – see Chapter 4, note 34.
16. Thaler and Sunstein (2008).
17. Diamond and Vartiainen (2013) illustrate the ideas of behavioural economics with reference to public economics, development, law and economics, health, wage determination and organisational economics. It does not appear to have been until 2001 that the first behavioural experiments relating to energy use were conducted (Chapter 4, Box 4.2), and this escaped their attention.
18. See Chapter 4, note 34.
19. Obviously, this is based on sets of assumptions that are rarely met in full, but the sensible literature on 'market failures' also recognises a principle of 'remediableness': the existence of market failure is not sufficient to justify intervention, unless 'a superior *feasible* alternative can be described and *implemented* with expected net gains' (Williamson 2000: 601). Distributional considerations are important in this context and form a never-ending source of debate about public policy, but this is not the topic at hand – except insofar as distributional impacts have been one important factor impeding the practicability of using 'economic instruments', as charted in Pillar II.
20. To add to the very extensive pre-existing literature on business cycles and economic cycles, and the role of financial markets dating right back to famous examples of 'economic bubbles' in the eighteenth and nineteenth century, the credit crunch of 2008/9 and its enduring economic impact have stimulated a vast outpouring of economic analysis. It is impossible to do justice to this, but some of the flavour (and references) may be captured from the book titled *The Age of Instability: The Global Financial Crisis and What Comes Next* (Smith 2010). The book *Fault Lines*, by the former Chief Economist of the International Monetary Fund (Rajan 2010), explains some of the mechanics by which financial markets have amplified instabilities and argues that the systemic causes of the credit crunch have not been dealt with.
21. That is, for any market, supply and demand curves – the amount of a good supplied and bought at a given price – can be specified and projected.
22. For general references on evolutionary economics see Pillar III, Chapters 9–11. Note that a broad review of New Institutional Economics classifies different levels of the discipline and defines analysis of 'institutional governance' as 'Level 3', on timescales of 1 to 10 years, distinct from the analysis of 'institutional environment', which he classifies as 'Level 2' on timescales of 10–100 years. In this schema, Level 1 consists of the informal cultural context, the 'Embeddedness' of informal institutions, customs, traditions, norms and religion (Williamson 2000: 595–613).

23. Williamson (2000: 609).
24. Arrow (1970: 224).
25. Williamson (2000: 600).
26. Many of the notes in the chapters that delve deeper into the specifics refer to leading economists and indeed many Nobel Laureates who have probed the realities and advanced the boundaries of understanding.
27. This diverges from the treatment of Williamson (2000), who presents neoclassical economics as time-independent. This seems irreconcilable with the data or logic presented in this book. It *may* be true at a more abstract level, in terms of the evolution of adaptive 'meta-preferences' (see Chapter 12, note 14, on recent research by Professor Dr Christian von Weizsacker).
28. Precision on the topic of trading-off of resources would tend to draw the bottom axis in terms of other resources rather than economic output or welfare. The curve would then be the other way round, and technological progress would involve it moving in towards the origin of the graph. It is drawn in the way illustrated to provide a simple visual consistency with a number of other graphs in the book, in which the public policy objective can be interpreted as moving right (increase welfare), while reducing energy consumption and emissions (moving down). Note that the frontier in economics is often called the 'possibilities frontier', which is potentially confusing as soon as one brings innovation into consideration.

    Also, economics textbooks would typically draw the curve as a straight line and hardly ever show it bending backwards. However, it is obvious that there is such a thing as excessive energy consumption or emission levels which would damage economic output – and, indeed, plenty of evidence that this is not uncommon, in part through subsidies (see Chapter 6).
29. An important – though by no means the only – dimension of all this concerns time. Individual and organisational decisions on energy consumption and end-use investment have short time horizons: Pillar I documented the overwhelming evidence that opportunities to cut energy consumption with paybacks of only a year or two are often still not enacted – and not implementing measures available that could pay back within four or five years seems the norm. This is far more short-sighted than the timescales of most decisions involved in generating or supplying energy, and hence introduces a *systematic* bias towards producing an excessive level of energy and emissions. Again, while the underpinning theories have been sketched in the previous section, it is developed along with the detailed evidence in Chapter 4.
30. In terms of overall greenhouse gas emissions, there are also important opportunities in agriculture and forestry. The McKinsey study estimated a large potential to reduce projected emissions from these sectors, which was contested by a number of land use experts, given the enormous complexities of land tenure and competing land uses. Agriculture and land-use issues are not the topic of this book and the associated estimates are not including in the figure.
31. The McKinsey curve is a map of *potential*. The estimates were criticised for the implications in particular of very rapid global deployment in low carbon electricity sources, cutting emissions by over 10MtCO_2, and also large savings in industry; both of these were much bigger than the estimates of the IPCC Fourth Assessment (see note 33). On the other hand, the McKinsey curve was also criticised as being *pessimistic* in its assessment of the potential emission reductions in buildings, where its estimated potential of 3.5GtCO_2 compared to the IPCC estimate of 5.8GtCO_2. McKinsey attributed part of this to differing assumptions about the 'baseline', stating that they included a higher rate of buildings efficiency improvement in the baseline because it is so cost-effective.
32. The boundary between first and second realms of opportunity – negative and positive cost – depends among other factors on the assumed discount rate. The McKinsey curve assesses options at a 3.5% discount rate. This is a rate appropriate to assessing *public, risk-free* benefits, but gives a misleading impression of the market economics of the different options, which would normally be much higher. This would generally increase the relative costs of most of the options that involve investment, since low-carbon sources tend to be more capital-intensive, compared to fossil fuel investments.
33. Most notably, the IPCC Fourth Assessment. The Intergovernmental Panel on Climate Change, which convenes leading academics to review the state of knowledge, conducted similar analysis and found an even greater 'negative cost' potential, arising particularly from its analysis of possibilities for improved energy efficiency in buildings (IPCC 2007). See Chapter 4 for further details.

34. The raw source data obtained from McKinsey was processed into three main global regions, based on economic progress and growth – the OECD group (depicted in blue), the emerging economies of China, India and the Middle East (colour coded in orange) and the rest of the world (shown in green). Note that the emphasis is on deciphering key trends and patterns of behaviour in energy, industries, buildings and transportation. Data pertaining to land cover and land use has not been used in the analysis.
35. A plausible reason is that OECD economies have far more established capital stock. Where this is inefficient – due to First Domain characteristics such as poor building stocks as noted in Chapter 4 – this provides a large negative cost opportunity. The McKinsey analysis, however, seems to largely assume that emerging economies will build a far more efficient stock in the 'baseline', leaving less scope for 'negative cost' abatement. Conversely however, abatement which involves premature retiring stock – such as dealing with CO_2 from existing coal plant by either premature retirement or by retrofitting CCS – may be more costly than installing clean rather than dirty generating stock in the course of rapid economic development.
36. These 'Pareto improving' choices – which improve one thing without any enduring trade-offs against loss elsewhere – are not confined to energy efficiency, but energy efficiency, particularly in buildings, is, for reasons explained in Chapter 4, the largest and totemic example.
37. Key references on the Porter hypothesis include Ambec *et al.* (2013), Wagner (2004) and Xepapadeas and de Zeeuw (1999). For discussion in relation to energy and environment, see Chapter 9, section 9.10 ('Porter's kick').

References

Note: For more references on each of the Three Domains see the corresponding sections of this book.

Ambec, S., Cohen, M. A., Elgie, S. and Lanoie, P. (2013) 'The Porter hypothesis at 20: can environmental regulation enhance innovation and competitiveness?', *Review of Environmental Economics and Policy*, 7 (1): 2–22.

Arrow, K. J. (1970) *Essays in the Theory of Risk-Bearing*. Amsterdam: North Holland.

Camerer, C. F. and Kunreuther, H. (1989) 'Decision processes for low probability events: policy implications', *Journal of Policy Analysis and Management*, 8: 565–92.

Casti, J. (2012) *X-Events: The Collapse of Everything*. William Morrow; reprint edition: HarperCollins.

Diamond, P. and Vartiainen, H. (2007) *Behavioral Economics and Its Applications*. Princeton, NJ: Princeton University Press.

Hood, C. (2011) *Summing Up the Parts: Combining Policy Instruments for Least-Cost Mitigation Strategies*. Paris: International Energy Agency.

IPCC (2007) *Fourth Assessment Report – WG3*. Cambridge: Cambridge University Press.

Kahneman, D. (2011) *Thinking, Fast and Slow*. New York: Farrar, Straus, & Giroux.

Kunreuther, H. C., Novemsky, D. and Kahneman, D. (2001) 'Making low probabilities useful', *Journal of Risk Uncertainty*, 23 (2): 103–20.

Levittt, S. D. and Dubner, S. J. (2005) *Freakonomics*. New York: William Morrow.

McKinsey & Co. (2009) *Pathways to Low Carbon Economy*. New York: McKinsey.

Matthes, F. C. (2010) *Greenhouse Gas Emissions Trading and Complementary Policies: Developing a Smart Mix for Ambitious Climate Policies*. Frieberg: Öko-Institut e.v. (http://www.oeko.de).

Mencken, H. L. (1949) *A Mencken Chrestomathy: His Own Selection of his Choicest Writings*. New York: Random House.

Michel-Kerjan, E. and Kunreuther, H. C. (2011) 'Redesigning flood insurance', *Science*, 333 (6041): 408–9.

Rajan, R. G. (2010) *Fault Lines*. Princeton, NJ: Princeton University Press.

Shmelev, S. E. (2012) *Ecological Economics: Sustainability in Practice*. London and New York: Springer.

Simon, H. (1953) *Models of Bounded Rationality*. MIT Press Classic (reprinted 1997).

Smith, D. (2010) *The Age of Instability: The Global Financial Crisis and What Comes Next*. London: Profile Books.

Spence, A., Poortinga, W. and Pidgeon, N. (2012) 'The psychological distance of climate change', *Risk Analysis*, 32: 957–72.

Stern, N. (2006) *The Economics of Climate Change*. Cambridge: Cambridge University Press.

Taleb, N. N. (2007) *The Black Swan: The Impact of the Highly Improbable*. London and New York: Random House.

Thaler, R. H. and Sunstein, C.R. (2008) *Nudge*. New Haven, CT: Yale University Press.

von Weizsacker, C. C. (2011) 'Homo oeconomicus adaptivus', *Journal of Comparative Research in Anthropology and Sociology*, 2 (2): 147–53.

Wagner, M. (2004) *The Porter Hypothesis Revisited: A Literature Review of Theoretical Models and Empirical Tests*, Public Economics 0407014. EconWPA.

Williamson, O. E. (2000) 'The New Institutional Economics: taking stock, looking ahead', *Journal of Economic Literature*, 38 (3): 595–613.

Xepapadeas, A. and de Zeeuw, A. (1999) 'Environmental policy and competitiveness: the Porter hypothesis and the composition of capital', *Journal of Environmental Economics and Management*, 37 (2): 165–82.

Pillar I

Standards and engagement for smarter choices

Overview

Global energy demand and emissions are driven mainly by our need for warmth and comfort in buildings, for goods and services provided by industry, and for transport. Options abound for meeting these needs in better ways, but the enormous diversity of energy-related activities precludes any single technology providing a 'magic bullet' solution.

The cheapest, most compelling opportunities lie in increasing the efficiency of energy use. Despite major improvements in recent decades, overall we still consume about ten times as much energy as is physically necessary for the activities we enjoy. Numerous technologies and other options for improving efficiency are demonstrably cost-effective today: many measures would pay back, in terms of energy savings, within a few years (where they require investment, with 'rates of return' often far exceeding 15 per cent per year). There are also many options for low carbon supply technologies; most of these are currently more expensive than fossil fuels, though the gap is narrowing rapidly for some of these technologies. However, the scope for 'smarter choices' is not confined to end-use energy efficiency: expanding horizons reveal important opportunities for 'smarter systems' overall (Chapter 3).

Developments in behavioural and organisational studies help to explain why the way we use energy remains so wasteful. Patterns of energy use reflect a tension between habits, intentions and the options practically available to individuals as constrained by their infrastructure and their organisational and technological environment. There are many structural barriers which impede more efficient choices. The most pervasive occur when energy characteristics are embodied within far larger choices (such as property, where occupiers typically pay the bills but have limited – or for rented property almost no – scope to improve thermal performance). However, such barriers are hugely amplified by the intrinsic First Domain characteristics of decision-making in relation to recurring, incidental costs: habits, inertia, myopia and rules-of-thumb tend to dominate.

Modes of decision-making in organisations can mirror those of individuals to a surprising degree, yielding similar patterns of inefficient energy use. Organisations often treat energy costs as unavoidable or incidental – i.e. not managed optimally. Public bodies often restrict the ability to invest; many corporations still leave opportunities for energy efficiency untapped and/or demand rates of return similar to those of individual consumers, which is far higher than for their core business (Chapter 4).

Almost four decades of policies to enhance energy efficiency have yielded big benefits. These efforts have helped to largely stabilise per capita energy demand and/or emissions in richer countries, and in several cases have reduced it. The global spread of such policies has also helped to reduce the exposure of energy importing countries to international energy price fluctuations. Evaluation confirms that energy-efficiency policies have generally curtailed energy demand more cheaply than equivalent investment in new supplies (also avoiding the external impacts associated with most supply options) – thus creating both economic as well as environmental benefits (Chapter 5).

Though this book focuses on principles and lessons from experience in industrialised countries, the global spread of energy efficiency policies is notable and testifies to their multiple benefits. The process of economic development itself involves moving closer to the 'best practice frontier'. In addition to strengthening the institutions and infrastructure required to develop markets, setting appropriate standards and engaging citizens on issues of common concern (like energy) is also an important part of development. In terms of enhancing human welfare, the biggest 'energy efficiency gap' is being disconnected from modern energy services. The manner in which 20 per cent plus of the planet's population achieve 'energy access' will have enduring consequences.

Two major global studies published in 2012 concurred on three core points about the future of energy efficiency. First, there is a huge remaining practical potential for improvement. Second, tapping this potential is central to sustainable energy futures. Third, buildings, which account for over a third of global energy consumption (and more, through their indirect use of energy in building materials like concrete and steel), represent the biggest single sector of opportunity. Huge increases in incomes and building floor areas over the next few decades mean that default trajectories could add 50 per cent to global buildings-related energy consumption by mid-century; conversely, strong policies could reduce thermal-related energy consumption to well below current levels globally, and substantially curtail electricity growth associated with the rising use of appliances and IT (Chapters 3, 4, 5).

Pillar I policies could be expanded further with net benefits, most notably by going beyond the historic focus mainly on the technical energy performance of products. Tackling habits of energy use aided by IT (such as smart meters) and extending the lessons of energy-efficiency policies to address the broader energy and carbon footprint of construction and consumption are important areas for future attention (Chapter 5).

However, historically the overall scale of the impact of improved energy efficiency has also been less than hoped for, often with slow uptake even of highly cost-effective options, and with a significant part of the gains 'taken back' by a combination of rebound effects and the growing overall scale of consuming activities. Pillar I policies still offer the most immediately attractive potential, but to achieve their full potential – sustaining global energy efficiency improvements at 2 per cent per year or more over the coming decades – they ultimately have to be combined with policies on the other two pillars (Chapter 5).

3 Energy and emissions

Technologies and systems

To solve a complex problem, first understand its components. That is a basic principle, yet much public debate on energy and the environment seems to ignore it. Sweeping generalisations about clean technologies to save us make little sense without understanding what it is that needs to change, what is possible and what can be significant.

This chapter sets out to explain the essential components that make up our energy systems, how they relate and the possibilities for doing better. The energy system can be understood in terms of flows between three levels: (a) end-uses that draw their energy from (b) different channels of conversion and distribution, which are powered from (c) primary fuels and the systems for extracting them. To some degree, these three different levels of the energy system – use, channel and fuel – also map onto the three domains described in the previous chapter. Energy demand depends heavily upon the choices of individuals and energy-consuming organisations, with all the First Domain characteristics that go along with this. The channels that supply these are dominated by large investments, weighing up financial costs and benefits (Second Domain). The development of the primary resources and associated connections across the energy system involve the exceptionally long timescales, innovation and strategic and security choices typical of the Third Domain.

This mapping is far from exact, and this chapter starts by outlining the main components at each level and the main flows. It then explores the key technologies at each stage, organised around three main blocks of energy supply and demand. The final section, which touches on energy resources and systems overall, also illustrates how the wider concept of 'smarter choices' will involve looking ahead and integrating not just across the pillars, but across different components of our energy systems.

3.1 The energy system

Uses, channels, fuels

The global energy system is surprisingly easy to break down into a few main components. Three main categories of commercial energy-consuming activities each account for close to 30 per cent of global energy consumption; they are supplied through three main types of channels; and most of the energy going in comes from burning the three fossil fuels of coal, oil and gas, supplemented by nuclear and renewable energy sources (Figure 3.1):[1]

- *Buildings* account for over 30 per cent of global energy demand: heating for space, water and cooking is increasingly supplied by natural gas, alongside ever-growing

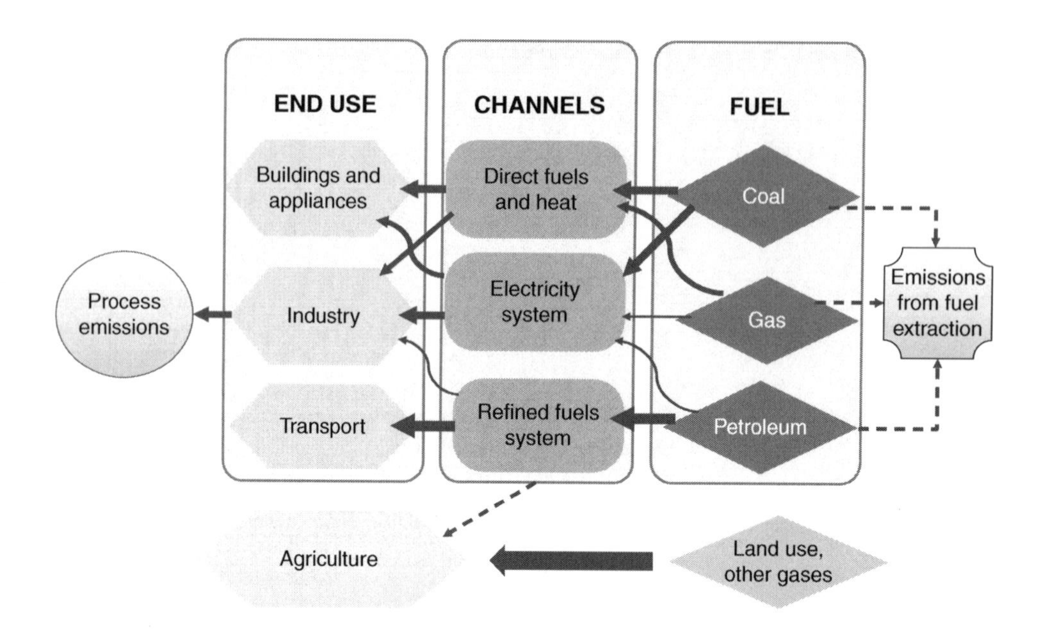

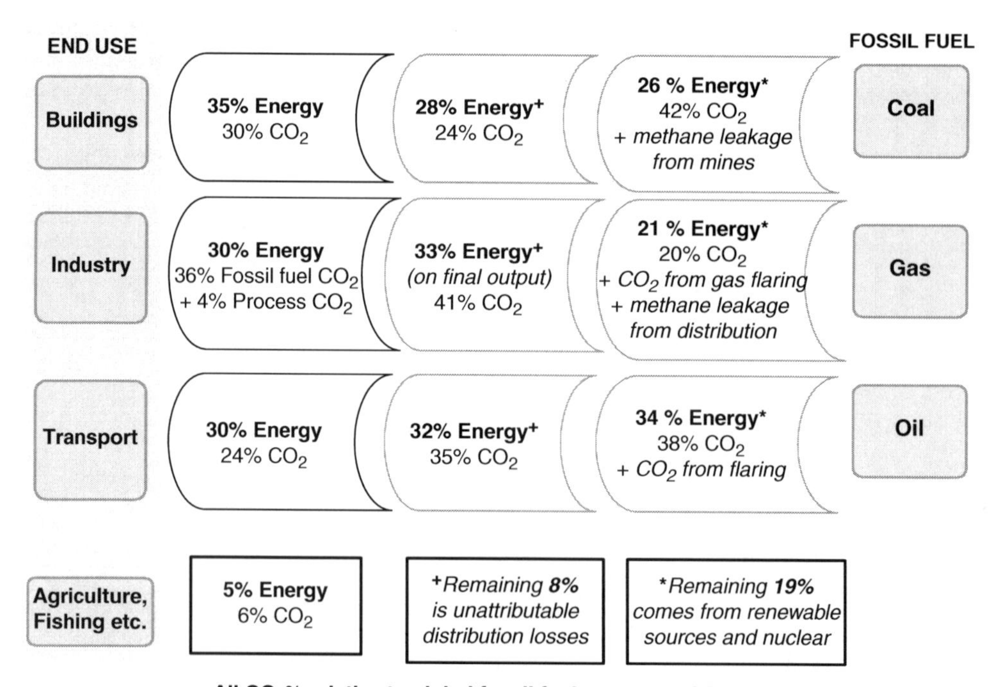

Figure 3.1 Energy and emission flows within the fossil fuel system

Note: The lower panel gives the numeric breakdown at each stage illustrated in the upper panel. Numbers at each step in the chart (fuel, channel, end-use) independently add to 100%.

Source: Author, with all data for the lower panel from the IEA, accessed through ESDS, at http://esds.ac.uk.

consumption of electricity for home appliances, lighting, electronics and other gadgets, as well as for air conditioning in warmer climates. For buildings not connected to energy networks, oil or biomass remain the main fuels.

- *Industry* consumes similar amounts of energy in total. As we will see there is a sharp distinction between heavy industrial processes fuelled mainly by coal (the most carbon-intensive fuel) together with some petrochemicals, and the more ubiquitous use of gas and electricity across industry. Additional greenhouse gas emissions from industrial processes bring industry's share of global CO_2 emissions to around 40 per cent.
- *Transport* accounts for over a quarter of global energy and emissions, fuelled almost entirely from oil refined into gasoline or diesel. Adding other smaller uses – mainly agriculture which also relies heavily on petroleum products – gives a total emissions impact comparable with the other two main blocks.

This is just the 'commercial' energy system. A measure of the world's inequality is that an estimated 2.5 billion people hardly participate in this; they rely mainly on traditional biomass (e.g. locally collected wood) for heating and cooking, and half of them have no connection at all to electricity networks (GEA 2012, Chapter 5. IEA (2012b) puts the number at 2.6 billion). Helping them get a better life will mean connecting them with commercial energy – expanding the global system further. How to do so while making the whole thing sustainable is an intrinsic part of the global energy challenge.

Most people would draw the chart in Figure 3.1 the other way round – to follow the physical flow of energy. Instead, Figure 3.1 is drawn to emphasise the three main types of end-use that drive the entire system – with all its benefits and problems. The fuels and channels just feed these activities.

In exploring the energy system and potential to change, the nexus of buildings and the direct supply of energy for low-grade heating forms an 'exemplar' for exploring First Domain processes and the associated policies of Pillar 1. Industry and electricity generation, in contrast, are far more dominated by big private sector investors and other entities, more likely to carefully evaluate and trade-off the costs and benefits of specific investments, in which the inefficiencies appear less and the scope for substitution – for example, the choice of different fuels for powering electricity systems – is far greater. These are classic Second Domain characteristics.

These blocks have in common that that they are *stationary* energy systems, supplying energy demands often through fixed physical infrastructure (grids and pipelines). This means that they are closely entwined with the channels of both direct fuel delivery (notably gas) and electricity. Coal comes into the picture both as the key fuel for heavy industrial processes and as a major source of power generation. Per unit of electricity generated, coal emits almost twice as much CO_2 as gas, and its role in these systems accounts for around 40 per cent of greenhouse gas emissions. Natural gas – for heating buildings and increasingly fuelling our ever-expanding appetite for electricity – accounts for close to a fifth of global emissions. An additional 5 per cent of end-use emissions relating to energy come from oil and gas flaring, and methane leaks from coal beds and gas distribution.

In sharp contrast, energy for *transport* is dominated by oil, almost all of which goes through refineries, which also produce the smaller and more specialised products for industrial and agricultural uses. Through these and other routes, oil accounts for about another 40 per cent of global CO_2 emissions.[2] The technologies and structures for transport, and particularly alternative modes, vehicles and fuels, turn out to be particularly extensive, complex and long-lived, with a strong component of Third Domain characteristics.

Changing these systems is not for the fainthearted. The physicists' favoured unit of energy is the joule – and our economies now consume about 500 million million million, or 500 exajoules, annually. That's equivalent to over ten thousand million tonnes of oil equivalent (Mtoe). Almost four decades after the first oil shock rocked the world economy, and after two decades of efforts on climate change, we remain as dependent on fossil fuels as ever, which still account for 80 per cent of global primary energy supply.

This chapter's brief exploration of technologies and options for change is broadly structured along these lines. Of course there are numerous linkages. Industry provides the materials required for both buildings and transport, as well, indeed, as energy extraction and conversion. The location of buildings and industries affects the demand for transportation. Some oil is still used for power generation as well as for chemicals and heat. Nevertheless Figure 3.1 still provides the essential structure. After skimming through the main components, this and subsequent chapters will look at the options and policies for change.

3.2 Smarter buildings: efficiency and heat supply

Despite the improvements noted in Chapter 1, our technologically advanced societies remain astonishingly inefficient in their use of energy. Just *how* inefficient is hard to define, but it is interesting to estimate the *physical* limits. This is subtle because energy in fact is never really 'consumed', just converted to other forms, and some types of energy can do more useful work than others. This can be measured in terms of 'exergy' efficiency: comparing the primary energy in fuels to the useful work extracted (e.g. the chemical transformations in industry, the warmth indoors relative to outside, the photons emitted from a light bulb, the movement of vehicles and their contents).

By that measure, the world is about 10 per cent 'exergy' efficient, with big sectoral variations (see Box 3.1). And that is before taking account of the 'service efficiency' – whether the services supplied are actually targeted to what we want (e.g. heating or lighting a whole building when we actually just want to be personally warm and to see clearly our desk or immediate surroundings). A reasonable estimate is that *in theory*, the world only really needs around 5 per cent of the energy that it actually uses to provide the comforts we enjoy.

There is little dispute that the scale of apparently 'unnecessary' use is biggest in buildings and in the supply of the 'low-grade' heat required for heating, cooling and cooking. A key question then is the extent to which practical solutions exist to help close the theoretical gap.

Building energy use

Buildings and appliances account for around a third of *fossil fuel* energy and CO_2 emissions globally. In developed countries, this dwarfs use of non-commercial fuels (e.g. collecting firewood) and is mostly in the form of gas or electricity. In developing countries there is a far wider use of both coal and kerosene, and use of wood, dung or other fuels collected locally. Almost all these other fuels have adverse impacts on health in the household and local and regional environments, and moving to cleaner fuels – through connection to grids and urbanisation – tends to be a major feature of economic development anyway. Our focus here is on managing the overall energy demand.

Keeping warm dominates the need in the northern hemisphere while keeping cool does so in more southern climes. Space heating and cooling together often account for at least 50 per cent of energy consumption in both commercial and domestic buildings but homes use a lot more hot water. The share of electricity-consuming activities is greater in commercial

Box 3.1 Energy units and conversion factors

Within the framework of the Système International (SI) adopted by physical scientists globally, the fundamental unit of energy is the joule (J) and the unit of power – the rate of energy conversion – is the watt (W), equal to one joule per second. The units joules and watts are generally too small to be useful in energy accounting. Multiples are kilojoules (kJ, × 1,000), megajoules (MJ, × million), gigajoules (GJ, × 1,000 million), terajoules (TJ, × million million), petajoules (PJ, × 1,000 million million) and exajoules (EJ, × million million million). Global primary energy consumption is around 500EJ annually.

The rich history of energy has created a plethora of other units which are widely used in energy industries and analysis, where SI units are rarely used, the most common being million tonnes of oil equivalent (Mtoe). (Confusingly, the oil industry itself prefers barrels: there are 7.33 barrels in one tonne of oil.) For the main accounting of primary energy, this book uses Mtoe. The standardised 'tonne of oil equivalent' equates to 41.868 GJ; therefore 1Mtoe = 0.0419 EJ.

Electricity, however, is typically measured in units of kWh or multiples thereof, and 1 GWh (a million kWh) is physically equivalent to 3,600 GJ (since there are 3,600 seconds in an hour), or 3.6 GJ. National electricity generating capacity is typically measured in GW and output in TWh (i.e. a million million watt-hours per year): a 1GW power plant operating flat out with no interruption would generate 8.76 TWh in a year (since there are 8,760 hours in a year).

However, no electricity generator is a hundred per cent efficient and the thermal stations which dominate most systems lose up to two-thirds of the energy as waste heat (the most advanced gas turbine combined cycles bring this close to 50%), so the energy input to electricity is typically much bigger than the output. There are various conventions for converting non-thermal electricity sources into primary energy equivalent, which can make a big difference to the apparent significance of sources like nuclear power and renewable electricity. Where such data are presented in this book, the convention of the data source is used.

Other energy units include calories, British thermal units (therms, often used in the gas industry: 1 million Btu = 1,055 MJ), kW-years, tonnes of coal equivalent and quads (= quadrillion Btu).

CO_2 emissions are now generally measured in (million) tonnes of CO_2/yr, and other radiatively active gases are converted to CO_2-equivalents generally on the basis of their equivalent radiative impact over a century. Each tonne of carbon converts to 3.7 tonnes of CO_2 when oxidised. For a sense of scale, 1.4 tonnes of coal has an energy content of around 1 toe and contains somewhat over 1 tonne of carbon and hence emits close to 4 tonnes of CO_2 when fully burnt (coal is quite variable, so the numbers are not exact). Roughly, the energy-to-CO_2 ratio is 20% better for oil, and 20% better again for gas. The higher efficiency of gas power plants compared to coal explains the oft-cited numeric that gas is twice as clean as coal in terms of CO_2 emissions.

buildings than in residential – in the US these account for almost equal shares of overall energy, while, for example, lighting accounts for 10 per cent of domestic energy use compared to over 20 per cent in commercial premises (see Figure 3.2).

Box 3.2 The physics of energy efficiency and exergy[a]

According to the first law of thermodynamics, energy is never destroyed but converted into other energy forms. However, there is a cascade of *types* of energy from 'high-grade' (like electricity or mechanical energy, which can be converted into any other energy type with little or no loss) to 'low grade' (like heat, for which there are physical limits on how much can be converted into electricity or motion).

Energy efficiency (i.e. the ratio of useful output and energy inputs) does not take account of this 'quality' of energy, and its varied ability to perform useful 'work'. Scientists measure this quality of energy, and therefore potential for useful work, in terms of *exergy*. Exergy efficiency, then, helps to highlight how well (or not) we harness the 'work' from inputs into the energy system. Numerous studies have traced energy flows through the stages, typically known as: primary, secondary, final, useful and ultimate services. The biggest losses tend to occur between the 'final' and the 'useful' – for example, the energy used in a light compared to the lumens coming out of it.

Globally, only about 11% of the total work available from our primary energy is *used*, in this sense, and it varies by sector:

- Heat, mainly for buildings, typically utilises only 10% of the theoretical input available.
- Industry as a whole is a lot better at utilising energy, often reaching 25–30%.
- Lighting, cooling and many electronic services use only about 2% of the potential (e.g. much of the energy in lighting comes out as heat, not light).
- Transportation is better than average, harnessing about 17% of total available work from its energy inputs.

Capturing some of this wasted work through improving devices or using waste heat productively, we could theoretically save 80–90% of our current global energy consumption and enjoy the same services we receive today.

We can also probe the efficiency of the *services* themselves. Could we use less energy-intensive materials to do the same job? Do we need to light the whole building, or just some rooms or surfaces? What defines how much heating we really need if a building is well insulated, since we don't actually *consume* any – and mostly, we just want to keep our own selves warm, not all the air in the building? And what if the energy physically needed to accelerate a vehicle can be recaptured when it brakes, so the only real need is to overcome air and a bit of rolling resistance?

Estimating the potential for *service* efficiency can be more of an art than a science. An estimate by the National Academies (2009) is that the overall global efficiency of the energy system is around 5%. Not surprisingly therefore, the scope for improving energy efficiency overall is huge.

Note

a Data drawn from a recent systematic review of energy and exergy efficiency in Cullen and Allwood (2010); Ertesvåg (2001) provides detail on industrial energy. Cullen and Allwood estimate a theoretical savings potential of 420EJ, of a total global consumption of 500EJ. See also the US National Academies report (2009).

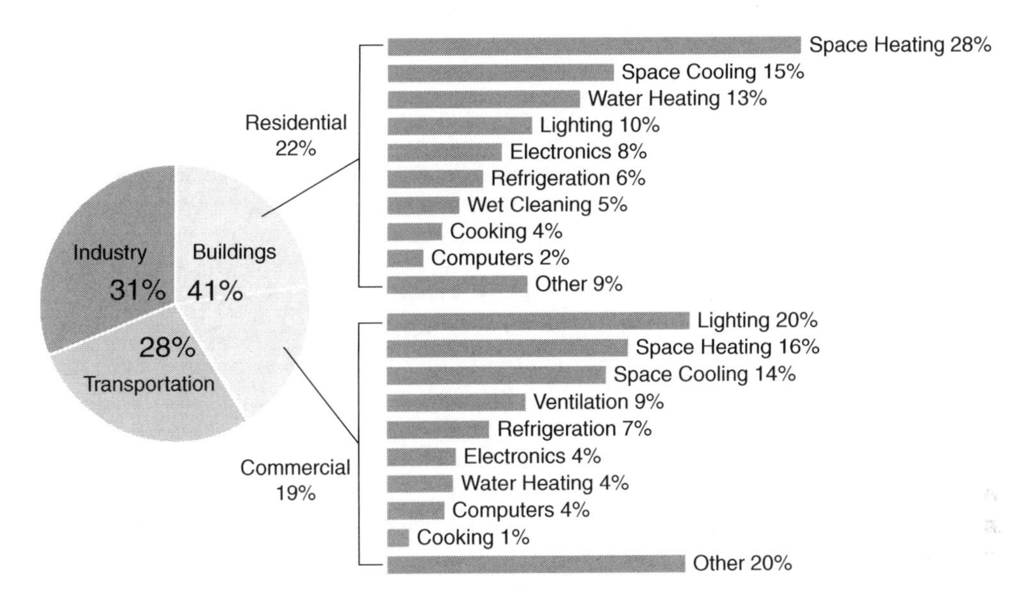

Figure 3.2 Final energy use in US buildings

Source: Adapted from Figure 10.6 in the Global Energy Assessment (2012). The data are derived from the US Energy Information Administration. The chart shown is from the draft GEA; the final version is in a less convenient form and shows a greater proportion of energy use allocated to 'other' in both residential and commercial sectors.

Among energy analysts the buildings sector is notorious for its staggering levels of inefficiency – and for variations between countries. Many countries are cursed by a buildings stock that leaks energy like a sieve.[3] Reasons for the appalling energy performance of buildings in many countries are explained further in Chapter 4. Unfortunately things are only improving slowly, and many of the mistakes are being repeated in the scramble for urbanisation in developing countries – though in many of these cooling, not heating, may become the biggest source of waste from uninsulated buildings and poor equipment.

Fortunately, over and above the basic measures of 'good practice' there are lots of additional ways to curb our energy wastage in buildings. Where it hasn't already been done, installing insulation in cavity walls and lofts (attics) can dramatically cut the energy required for heating across northern latitudes. Modern control systems on the flow of air can cut heating and cooling requirements. The use of double glazing can have the same effect, and 'smart glass' can adjust the amount of light and heat as appropriate. Many of these options can be retrofitted to existing buildings. Simple measures such as limiting the waste heat from appliances in conditioned spaces can also help to reduce cooling loads.

For new buildings, modern architecture – often borrowing from the old – can design efficient building envelopes and make more use of 'passive' heating and cooling. Taking into account solar reflexivity, ventilation throughout buildings and using evaporative coolers or radiant chilled ceiling cooling can all help to reduce the energy required for heating and cooling.

We can also do a lot about lighting. Much less than 10 per cent of the power consumed by traditional bulbs comes out as light. Compact fluorescent lamps (CFLs) and new light emitting diode (LED) bulbs consume a tiny fraction of this and can pay back their higher purchase cost in a few years. As discussed in Chapter 5 their use is growing rapidly, with some countries phasing out traditional bulbs. Smart building design can also maximise the use of natural light.

The integrated impact can be staggering: combining efficient building envelopes with such measures has reduced energy for heat by factors of 5 to 30.[4] So a great deal can be done about the big blocks of energy and emissions associated with heating and lighting the buildings we live and work in – good practice and modern technologies dangle the prospect of the biggest energy 'free lunch' on the planet.

The biggest technical challenge in buildings comes from the explosion in electrical appliances. Adding to the basic fridges, cookers and cleaning equipment, use of telecommunication, computer and entertainment equipment has ballooned, and so has their demand for electricity. In many there remains a lot of scope for improvement. The most efficient appliances require two to five times less energy than the least. Reducing power demand in both active and standby modes offers big opportunities for efficiency savings. The scope is more modest than for heating or lighting but the 'cost-effective potential' to improve the efficiency of appliances is still around 30 per cent by 2020.[5] In the longer term the installation of IT into standard appliances such as washing machines and fridges could allow them to draw power mostly when it is cheap and clean – a part of the 'smart grid' discussed later.

A world away, it is technically even easier to improve biomass stoves widely used for cooking in developing countries; electric and even solar cooker alternatives bring not only greater efficiency but also a myriad of health co-benefits from cutting the associated air pollution that kills millions worldwide.[6]

And as the efficiency of energy use in a building improves, it becomes easier to use local renewable energy, discussed below, to supply what's left.

Additional improvements are possible if horizons expand to thinking of urban areas as a system. Urban areas act as 'heat islands' which can be cooled with trees and the whitewashing of buildings; integrated development can also more easily allow district heating and cooling, as well as potentially reducing transport needs (outlined below).

Key challenges

However, a few clouds hang over this apparently happy outlook. Why are we so wasteful? Unless we understand why apparently modern, sophisticated societies waste so much energy, we are unlikely to change things. This is the focus of Chapter 4. What we have learned about how to change things is then the topic of Chapter 5. These two chapters show why it is not *easy*, but also show how much can be realistically achieved if we abandon the idea that delivering cost-effective improvements should be simple. It isn't.

Another feature is the sheer timescales involved. Buildings last decades or even centuries.[7] Not all measures can be retrofitted – and many, only at some cost and inconvenience. This also means that there is a large danger of 'lock-in' to inefficient buildings. China is currently constructing approximately 2 billion square metres of new buildings each year – equivalent to 25 million average UK homes and far outpacing even the US. India is projected to construct more buildings between 2008 and 2020 than it had in total in 2007. Much may thus hinge on whether Asia in particular does it a lot smarter than most of the developed countries managed, a transformation challenge outlined in Chapter 10.[8]

Finally, the buildings sector is a major consumer of some of the most carbon-intensive manufactured products, namely steel and cement; construction utilised almost 50 per cent of all steel in 2007.[9] This is consistent with estimates that the carbon 'embodied' in an average UK dwelling exceeds 30t CO_2 – which could amount to several decade's emissions from energy consumption in the building, and implies that just constructing the UK's housing stock emitted close to one billion tonnes of CO_2.[10] As buildings become more efficient in their *use* of energy, these 'indirect' emissions from constructing houses and offices will become more important, as outlined in concluding Chapter 5. Truly 'zero-carbon' buildings are indeed a major challenge.

Energy for low-grade heating and cooling

Heating and cooling form the biggest part of that challenge, and along with cooking and industrial heating account for the quarter of global energy and emissions consumption that comes from burning fossil fuels directly, i.e. not using them to generate electricity or turning them into refined products (e.g. transport fuels). Beyond efficient buildings and boilers (which can also deliver sizeable savings), we need to diversify heat sources.[11]

In fact much heat also comes directly from non-fossil fuels, but we do not usually count it. The sun provides some heat for every building on earth, and the wind can help cool them. As noted above, using 'passive' heating and cooling can help cut energy use and emissions in buildings, using the natural resources around us more efficiently. Despite the best natural lighting and ventilation, however, most regions still require extra inputs.[12]

Moving from coal to gas provides heat more efficiently and with lower emissions where there is a gas distribution infrastructure. Moving from burning coal to more efficient wood (biomass) fired boilers can provide efficient low-carbon heating where mains gas or electricity is not available. Throughout most of history, of course, 'biomass' – energy from wood and other plants – was the main way of supplying heat and it remains so in many developing countries. Its use is declining with development, yet in the industrialised countries, use of modern biomass boilers and processed biomass (e.g. pellets) is small but growing.[13] Advanced biomass conversion technologies such as anaerobic digestion, gasification and pyrolysis can also be used as part of a heating strategy.[14]

Heat can also be supplied directly, through 'district heating'. Heat pipes allow waste heat sources (and/or bigger combustion facilities) to be used – including, for example, the use of spare heat from underground transport systems and combined heat and power (CHP) stations; they also make it easier to use biomass for heating.[15] CHP plants vary in scale from city-level installations, common in countries like Sweden and much of Eastern Europe, down to household-level gas-fired cogeneration and even fuel cells (micro-CHP[16]). The former require not just the plants, but a whole infrastructure to move the heat from source to use, including the measurement and marketing systems required to bill for the heat and to fund the system. Though heat distribution systems have improved enormously, a key determinant (of both efficiency and cost) remains the density of heat demand: more dispersed dwellings *or* improved building insulation thus lessens the practical scope for district heating.

Beyond improving buildings and piping (often waste) heat, the sun can heat water for hot water and/or for space heating (usually through a heat exchanger), providing up to 70 per cent of domestic hot water needs. In Sweden and Denmark, solar thermal energy is used in conjunction with district heating systems at low cost, demonstrating its potential even at high latitudes – at lower latitudes, it is increasingly used in space cooling.[17] Thus solar thermal technologies may be used to provide both heating and cooling at different latitudes and different times of the year, displacing much of the fossil fuel demand required for the same job.

Because the heat required in homes is low grade, it is also amenable to 'upgrading' with heat pumps. These move heat from one place (the outside air or ground) to another (the inside of a building to be heated), like a refrigerator or air conditioner in reverse (or in normal operation if used for cooling). Already deployed widely in Scandinavia and increasingly the United States, heat pumps can easily deliver three or four times as much heat as the energy they consume. Moreover, they run off electricity – so they could pretty much complete the job of decarbonising buildings if they can use low-carbon power. There are potential capacity issues with widespread use in dense urban areas.[18]

Obviously there are constraints on each option. The amount of energy available from wastes is obviously limited by the waste stream, generally to a few per cent of urban heating demands. Pressures and tensions around biomass supplies more broadly are outlined later in this chapter, but it still plays a key role in countries with low population density and/or fertile soils or large forests. Some countries are well placed to tap geothermal resources. But large global use is likely to be hampered by limits on resources and other constraints.[19]

Low-carbon heating is thus possible by drawing on a range of different technologies which seek to improve efficiency and match mostly local non-fossil resources to the remaining needs. Unlike the simpler measures associated with buildings insulation, the more ambitious task of decarbonising heating starts to involve more clearly questions about the relative costs of different options, and the complexities of systems – infrastructural, contractual and other dimensions – for delivering them. In the frame of this book, energy for buildings has a strong component of First Domain characteristics, but the more ambitious goals for the sector also clearly and powerfully invoke the need to get to grips with the other domains of relative prices, innovation and infrastructure.

3.3 Cleaner production: manufacturing and materials

Globally, industry dominates coal use (including generation for industrial electricity consumption) and associated CO_2 emissions. Moreover, in stark contrast to buildings, there are some good reasons to be pessimistic about our ability to do much about industrial energy use.

Half of industrial energy consumption (and a bigger share of emissions) come from just four massive sectors, as shown in Figure 3.3: iron and steel, cement, non-ferrous metals (mainly aluminium) and chemicals (including petrochemicals). If you want a measure of the scale, consider all the concerns raised about the environmental impact of flying and note that just cement and steel production together in total emit almost five times as much CO_2 as international aviation. Growth is surging; some tentative decline in developed country industry emissions is eclipsed by – and indeed partly reflects imports from – expansion in the emerging economies such as China, India and Brazil (and partly reflects a shift towards those regions).[20] Many of the new facilities use the latest technologies – the scope for abatement through better technology could lie just as much through improving existing plants in the rich economies, though there is no shortage of dirty old steel mills in Asia too.

In these and other energy-intensive primary industries, energy is already a big cost – so if much could be done, cheaply, to cut down on energy use, one would expect this to have already happened. Also a surprising share of their emissions is associated directly with the chemistry of the conversion processes rather than combustion or electricity consumption: largely overlooked in energy statistics, these add almost 30 per cent overall to the energy-related emissions of these sectors, one eighth of the overall industry total.[21] Many companies say there is not much physically that can be done about such process emissions, and that there is not a lot more they can do about their energy consumption – a view partly supported by exergy analysis (see Box 3.2). Also many of these commodities are quite widely traded – leading to fears that if one region imposes a cost of carbon, industries might simply migrate to escape the controls (Chapter 8). In stark contrast to the huge potential in buildings – which have the added bonus of the inability to migrate – it makes for a bleak picture.

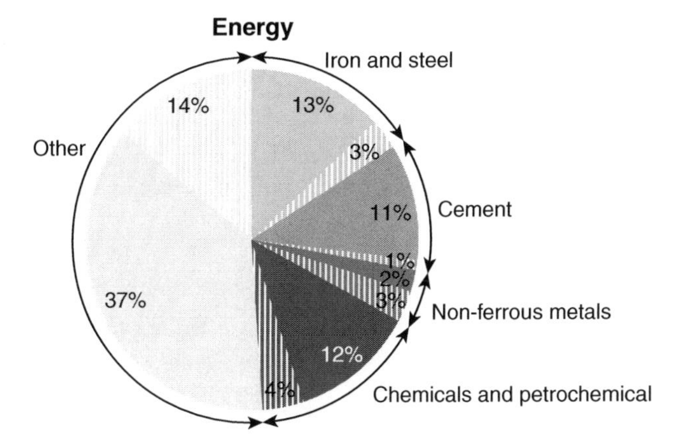

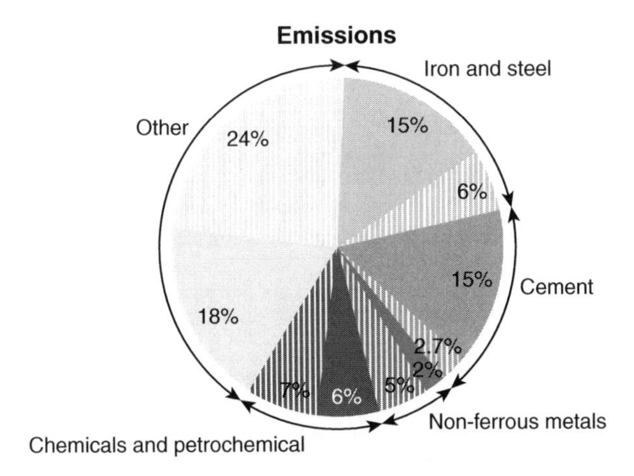

Figure 3.3 Global energy use and CO_2 emissions in industry

Note: The hatched areas represent indirect emissions through electricity consumption by the sectors.

Source: Data from IEA (2010).

Not quite. We probably *can* do a lot about industrial energy consumption and emissions, given time and intelligent policy. This chapter outlines technology options; the real challenge lies in policy across all the three pillars of this book.

First, though industries pay attention to their most *energy-intensive* processes, cross-cutting options offer an often untapped potential. Bear in mind that almost half of industrial energy and emissions are *not* from the big energy-intensive sectors. These less intensive activities are prone to many of the behavioural and other characteristics discussed in the next chapter. This is backed up by the technological evidence. Motor systems account for around 60 per cent of industrial *electricity* use, but typically more than half of this is lost before reaching the end-use work. The technical savings potential is estimated at 15–25 per cent – suggesting, for example, that improved electric motor systems across swathes of manufacturing could save close to 100MtCO_2/year in *each* of the EU and US, and much more globally.[22]

Major industrial facilities tend to be better at utilising waste energy than in the buildings sector, at least within individual plants and factories; industry-specific CHP can provide the high-grade heat required along with helping to meet electricity demand.[23] Overall exergy efficiency, as noted, is higher. Nevertheless big savings are still possible from recovering heat from some industrial processes and feeding it into others (often in the form of steam), and through more advanced heat storage and transfer technologies. A key barrier remains the difficulty in creating markets for waste heat.

The broader availability of natural gas could help sustain a move from coal to gas which might cut industrial CO_2 emissions by 10–20 per cent.[24] Tapping more methane from landfill, waste material or biomass could amplify the resulting savings. This is well established in some industries – for example, wood waste in the Scandinavian pulp and paper industries. Generating both electricity and heat using bagasse – the remains of sugarcane after the juice is extracted in the sugar industry – is becoming the norm in the Brazilian ethanol industry.

But what about some of the biggest, energy-intensive sectors and processes? The rest of this section considers just the two biggest emitters – steel and cement manufacturing – from which it draws some broader points of principle.

Iron and steel

Iron and steel production is the biggest energy user and emitting industrial sector, with steel produced through three main processes:

- *Blast furnaces* reduce iron ore to pig-iron, using coke or coal, which is then processed into steel. This process produces over two-thirds of global steel output – and the lion's share of steel sector emissions.[25]
- *Electric-arc furnaces* melt scrap to produce crude steel which can then be reformed into various different products. This uses only 30–40 per cent of the energy of blast furnaces (the associated emissions depend on the source of the electricity) but of course depends upon availability of scrap from recycling.[26]
- *Direct reduced iron* (DRI) technology which, if fuelled from natural gas, can halve emissions compared to blast furnaces, but the raw iron must then be fed into electric arc plants for processing – with variable overall impact on emissions.[27]

Differing levels of sophistication, scrap availability and fuels help to explain large international differences in overall emissions per tonne of steel. Numerous potential improvements remain possible.[28] Even within blast furnaces, the potential for improvement has been far from exhausted; current best practice is 20 per cent better than the global average and can still be significantly improved. Figure 3.4 shows this, together with a range of options for further improvement.

The big savings in blast furnaces would come from radical innovations. Installing carbon capture (Box 3.3) onto blast furnaces or DRI plants could make these processes low-carbon, with estimated emission savings potential of 100MtCO_2 by 2030. Use of coke derived from biomass rather than fossil fuels or electrolysis (for which emissions again would depend on the source of electricity) offer other routes to deep emission reductions from steel-making.[29]

However, savings do not necessarily depend on massive investment in new ways of making steel or stripping out and burying the CO_2. For example, *reusing* steel, rather than recycling it, can cut the need for heat for recasting the steel. If steel can be extracted from buildings without it being damaged then it can be reused for new buildings. But perhaps the most basic issue is that

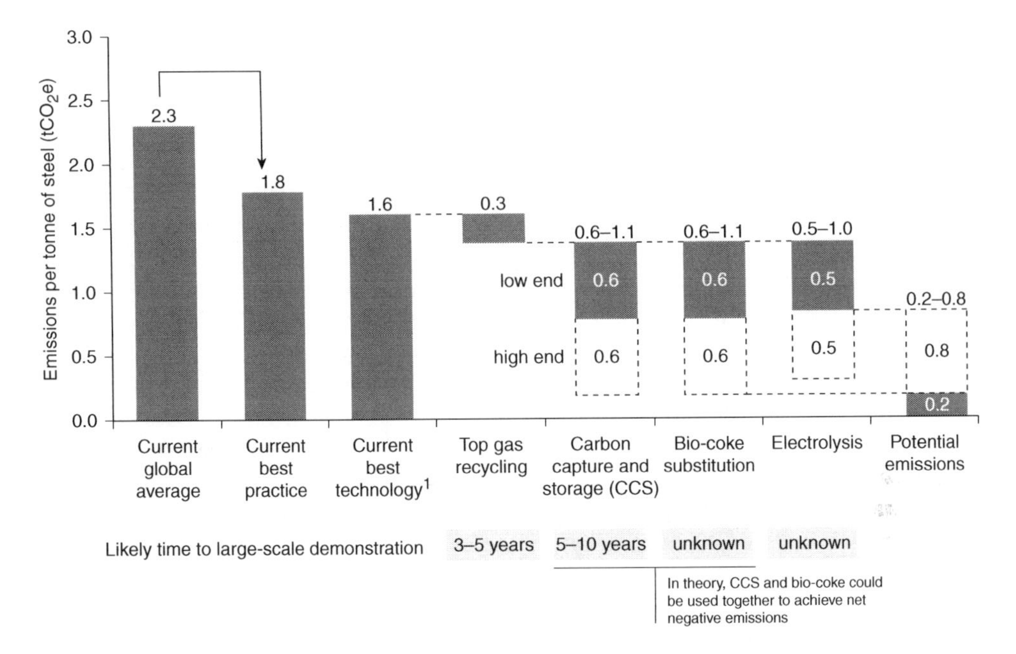

Figure 3.4 Potentials and technologies for reducing CO_2 from steel blast furnaces

Notes: Approximate data only.[1] Current best technically available includes smelt reduction-based technologies (currently at demonstration stage only). The emissions associated with 'electrolysis' assume decarbonised source of electricity.

Source: Carbon Trust (2011): International Carbon Flows – Steel; see: http://www.carbontrust.co.uk/Publications.

energy analysts have got used to thinking about efficiency purely in terms of energy. Bigger potentials may lie simply in considering the efficiency with which we use the materials. It is not uncommon, for example, for 50 per cent of the steel or aluminium supplied to be lost as offcuts.[30]

And, as noted later, homes can be made with wood – and still are, in many parts of the world. Even in this most traditional of technologies, innovation may have an important role to play: pre-stressed laminate wood beams offer a serious potential option for many construction applications, and indeed this is likely to be the most cost-effective use of biomass for reducing industrial emissions.[31]

Cement

Cement production is the most carbon-intensive of all major global industries – the third biggest energy user and second highest CO_2 emitter. In the standard procedure, heating limestone in a kiln (energy emissions) drives off an even greater volume of CO_2 (process emissions) to leave clinker – nodules which are subsequently ground with gypsum to make Portland cement for blending with other materials into concrete.

Moving from outdated wet-process clinker kilns to dry-rotary kilns cuts energy use, and additional efficiency improvements are possible from, for example, efficient grinding technology, low-temperature waste heat recovery, and the use of precalciner and preheater systems. With such measures, modern rotary kilns can approach the physical limits – though most remain far from that.[32]

Then there are many measures 'beyond efficiency'. In most regions, coal provides almost all the heat and the clinker is rarely considered separately from the cement that it is then

Box 3.3 Carbon capture storage and utilisation

Industrial and power plants using coal are the most obvious candidates for carbon capture technology to strip out the CO_2, which in principle can then be stored (for example, in depleted gas fields). There is also some scope for carbon capture and *utilisation* of the CO_2 stream (CCU), since some chemical and industrial processes (including enhanced oil recovery, where it is injected to push more oil out of wells) utilise CO_2 as inputs.

Most efforts on CCS have focused on power generation, but the principles are equally applicable to any concentrated CO_2 source, such as chemicals, steel and cement plants. Extracting CO_2 from simply burning fossil fuels requires a lot of separation. Some industrial processes and 'oxy-fuel' combustion technology in new power generation plants create high-purity CO_2-rich steams that facilitate capture.

Though most components of CCS are proven, applying the integrated technology at scale faces many uncertainties around costs and indeed political acceptability, given the need for extensive pipelines to carry the CO_2 and perceived risks around transport and storage. The most optimistic studies suggest the potential to capture up to 7 billion tonnes CO_2 by 2030, a quarter of global emissions, though this would require fitting CCS to almost every major power and industrial facility over that time, which appears increasingly implausible. In practice, progress to date has been slow – the costs thus remain somewhat speculative – and there are significant uncertainties on the trends and acceptability as well as costs.

Unlike most other options, CCS is an additional discrete investment specifically for the purposes of CO_2 removal, with little in the way of 'co-benefits' or other complicating factors, and as such its costs can be well expressed in terms of the cost per tonne of CO_2 removed. As indicated in Figure 3.5, which shows the most recent estimates, this is likely to vary between different applications. Separating CO_2 from some gas processing and chemicals processes may cost less than \$30/t CO_2, but such opportunities are limited. Stripping CO_2 from coal plants is estimated to cost around \$60/tCO_2, which would more than double the cost of power generation (see Chapter 6). Its application to steel and cement plants is likely to cost as much per tonne/CO_2 or more (though some sources indicate potential for novel industrial approaches which could reduce these costs) and in most other applications costs are likely to be even higher.

Utilising captured carbon for valuable applications – like enhanced oil recovery or various industrial processes – avoids the complications of storage and has the potential economic bonus of getting value from the CO_2 waste. The scope for practical uses of CO_2 remains unclear, as it depends on other kinds of industrial innovation.

Overall these factors make carbon capture at scale probably a long-term rather than a short-term option, but it remains an important part of the 'global portfolio'.

Source: For an extensive survey see the Global Energy Assessment on integrated fossil fuel and CCS (Chapters 12 and 13); and IEA Energy Technology Perspective (2012a), Chapter 10.

ground into. But all sorts of fuels can be used, and under the incentives in Europe (Chapter 7) cement plants have begun burning tyres and other wastes or biomass to cut their CO_2 emissions. Moreover, some companies have developed cements that use far less clinker, including using other waste products such as slag from steel mills and/or fly-ash from coal power stations. As discussed in Chapter 7, such measures have already significantly cut cement emissions and could do so by much more.[33]

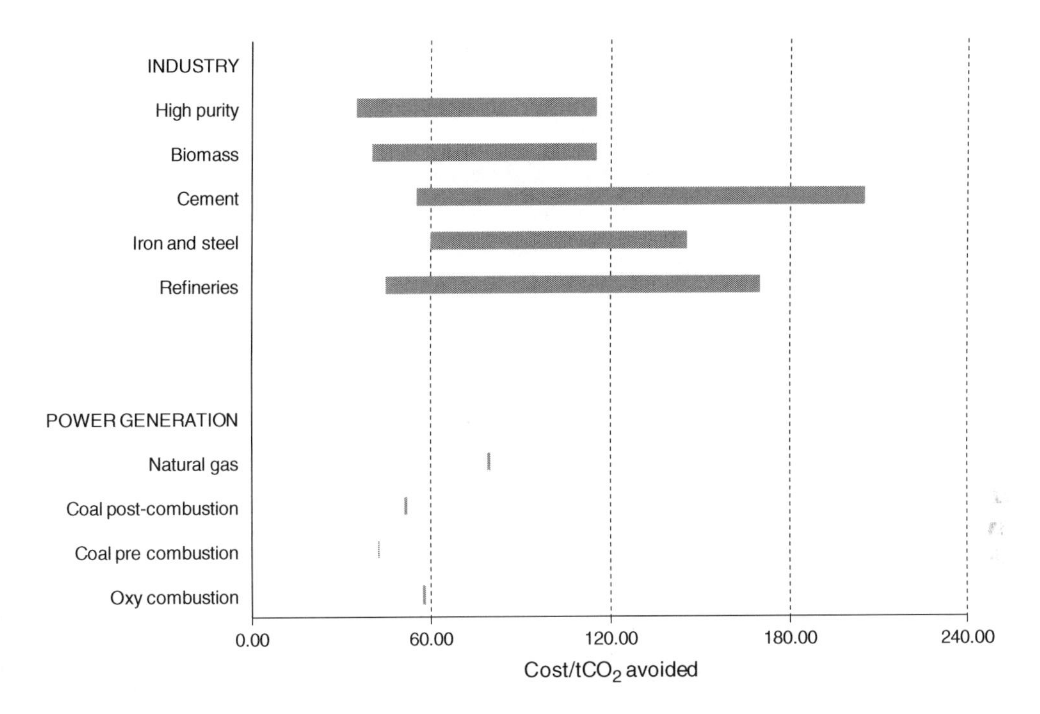

Figure 3.5 Estimated costs of carbon capture and storage in different applications

Notes:

- Source data for the industries adapted from IEA (2012) ETP Figure 10.8, p. 350, which gives the typical ranges of regional average cost of applying CCS in each sector including the cost of capture and transport, but exclusive of utilisation options that may generate revenues.
- The source data for power generation has been adapted from IEA (2012) ETP Table 10.2, p. 345, which gives point estimates of power generation and no ranges, though it is not clear why the cost of CCS for power generation should not also carry comparable uncertainties. For reference, a price of \$50/tCO_2 would add about 4¢/kWh to the cost of power from coal generation and a little over half that for gas generation.

More radical developments are also possible:[34]

- Carbon capture and storage (Box 3.3) could be applied to cement kilns as for steel.
- Geopolymer cement, based on key waste products from the production of steel (slag), powder (ash) or indeed concrete, reduces or even avoids the emissions associated with clinker altogether and is starting to emerge in niche markets. With current technologies the resulting cement is not as strong as traditional cement which would limit the range of applications.
- Cements based on magnesium oxide are also possible, which actually absorb carbon dioxide from the environment in the first year of its use.

A recent report summarises 19 different technologies for lower carbon cement at various stages of development.[35] The most radical could offer the enticing vision of turning one of the biggest emitting sectors into a source that sucks CO_2 from the atmosphere and locks it away in our buildings. This would give us two ways – the use of advanced wood technologies or advanced cement technologies – of turning buildings into 'carbon sinks' instead of sources. When we have a problem with CO_2 in the atmosphere, along with massive urban expansion, what better place to lock some of that carbon up?

Implications

Looking at these two apparently 'hopeless cases' in terms of industrial emissions illustrates the risk of taking too narrow a view. Some of the patterns evident in iron and steel and cement may be common to other industrial sectors. We can still improve energy efficiency. We can use some different inputs of energy (biomass or gas instead of coal) or other materials (e.g. slag or fly-ash instead of clinker in cement). We may be able to use carbon capture or electrification with zero carbon sources.

Most of that is within the production process. In addition we can use the product more efficiently, increase recycling and/or substitute it with something that involves lower emissions (such as the use of pre-stressed laminate wood instead of concrete in buildings). Finally, there may be truly radical options which transform an entire sector from being one of the biggest sources of CO_2 into something which could help to solve the problem.

Such positive possibilities are unlikely to be confined to these two sectors. Indeed, pulp and paper is another major heavy industrial sector, yet the Confederation of the European Paper Industry published a decarbonisation 'roadmap' whose title, *Unfold the Future – 2050 Roadmap to a Low-Carbon Bioeconomy*, underlines the potential for an industry which has renewable resources (i.e. trees) as its raw resource to play a bigger role in a low-carbon economy.[36] For raw aluminium, we cannot rule out radical innovations like inert anodes in aluminium that eliminate process emissions, or the transformations in cement indicated above.[37] Hydrogen is an important input in oil refineries and ammonia and chemical production – currently this comes from fossil fuels, but it could come from biomass or electrolysis of water using renewables, adding to the flexibility and implicit storage capacity of the electricity system. The list is long.

Moreover, greater savings potential in industry may become available through increasing *materials* efficiency. The mantra of 'Reduce, Reuse and Recycle' has been prominent in waste management for years. It is possible to reduce the amount of raw carbon-intensive commodities we use in constructing buildings, cars and the materials around us – we just don't know by how much. Reusing existing products, as in steel, could be important. Increasing recycling helps to reduce energy and emissions – with an extreme example being recycled aluminium, which uses just 5 per cent of the energy needed to produce new aluminium. Many people have got used to recycling cans along with paper and glass, which all contribute. And there may be surprising linkages such as the ability to replace clinker with blast furnace slag.

To tackle industry emissions, therefore, we need to start thinking about the overall supply chain efficiency of the systems and the relationship between different sectors, as suggested by Figure 3.6. In the academic literature it is called understanding the 'industrial ecology' and making full use of it.

The key to unlocking such potential is good policies to foster materials efficiency and innovation, and to ensure that the full supply chain reflects the full costs and is open to new and innovative entrants, reuse and recycling. Indeed, much of the emerging literature on materials efficiency echoes that of energy efficiency, pointing to the importance of First Domain effects that help to explain the degree of present wastage and the theoretical scope for big improvements. Though price is crucially important, this does point to the relevance of the other domains as well.

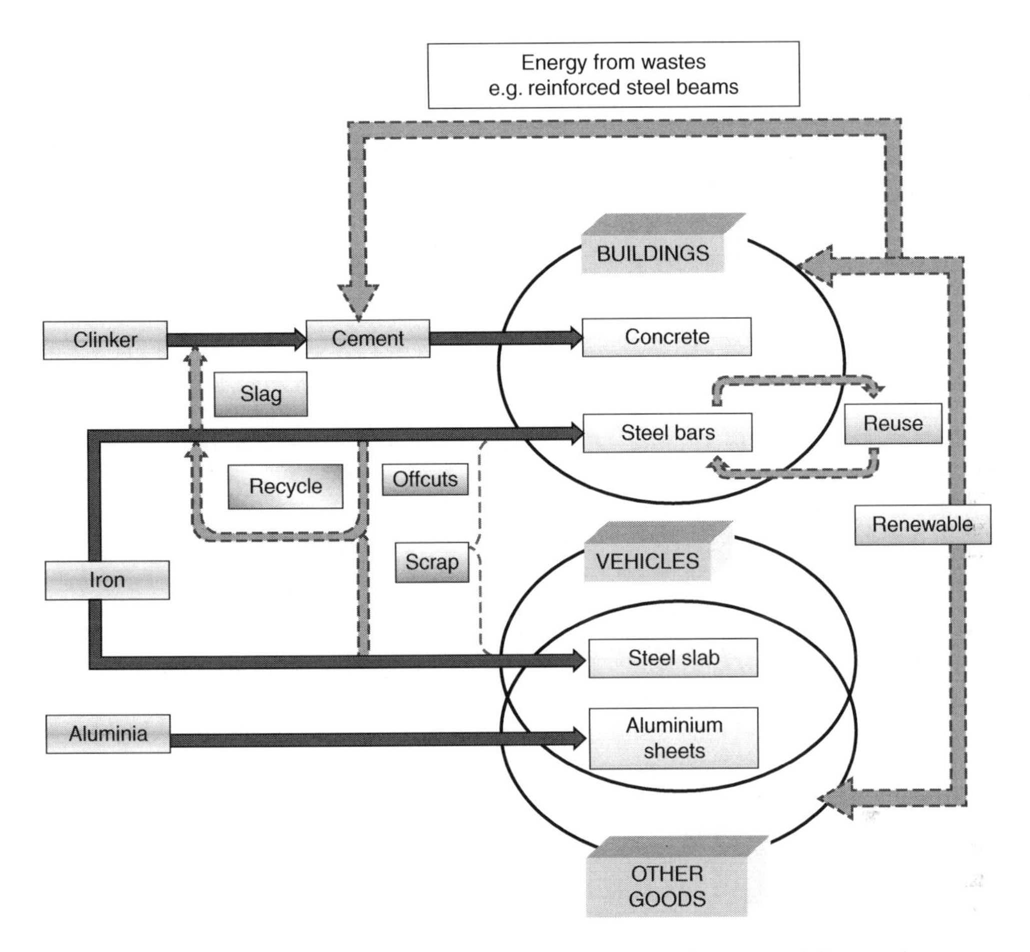

Figure 3.6 Opportunities in energy-intensive supply chains: from primary materials to products

Source: Authors.

3.4 Purer power: electricity without carbon

Electricity offers options. The growth in the vast array of activities which it powers has driven an inexorable rise in global electricity demand. Beyond these, there are major additional possible applications, for example powering heat pumps or electric vehicles.

There are also many ways of generating electricity. One rapidly rising way is from natural gas. Carbon emissions from natural gas-fired generation are about half those of coal-fired generators. Recent excitement over shale gas has led to talk of a 'golden age of gas', in which greater use of gas for power generation also contributes to reducing emissions by displacing coal in the short term. However, this comes with some risk of longer-term 'lock-in' to gas for several decades, though the emission consequences are moderate compared to those of the current dash for coal in Asia.

Clearly the world needs other, zero carbon, options – and it has them. They fall broadly into three categories: carbon capture and storage fitted to either coal or gas, nuclear power and various renewable sources.

Carbon capture and storage

The potentials and problems of *carbon capture and storage* have been noted above (Box 3.3). The broadest summary is that for power generation, as in industry, it appears to be proving slower, harder and more expensive than initially hoped.

Nevertheless, full-scale demonstration plants are emerging and the characteristics of potential CO_2 storage reservoirs have been well mapped out, increasing confidence in the scale and integrity of potential storage. The cost estimates for CCS noted above (Figure 3.5) suggest that in most applications for power generation, this would raise the cost of electricity generation to over 10¢/kWh.

Apart from this cost, as the technology starts to mature, there are two fundamental factors that point to constraints on the pace and scale of CCS. One is the technical and political complexity of the systems, including the need for both long-term investment and planning consent for CO_2 pipelines and disposal facilities. Public unease about CO_2 transport disposal – whether or not justified – has been growing, at least in Europe. This has led Germany, once expected to be a leader in the field, to halt all major demonstration projects.[38]

The other factor concerns investment risks. Because CCS involves fitting substantial additional technology to clean up emissions – a classic 'end-of-pipe' fix, albeit at an unusual scale – it *inevitably* adds costs. Consequently, it can only ever be an economic proposition if there is a carbon price (or other policy mechanism) sufficient to recoup those costs. The uncertainty around the European carbon price (Chapter 7) and the political difficulties facing other schemes have, at least, stretched out the time horizon before it seems plausible for CCS to offer an economic investment. Alternative financing mechanisms (analogous to renewable energy feed-in tariffs) to support larger-scale deployment have yet to be established. If other mechanisms are not established and carbon pricing is too weak, there is no strategic upside to justify private CCS investments. The current anaemic progress on carbon pricing thus risks leaving CCS as a technology which may be demonstrated but not deployed at scale.

Nuclear power

Nuclear power provides about 14 per cent of generation worldwide and provides a large share of total generation in some countries, most notably France (around 80 per cent). The history of nuclear power construction has been something of large-scale boom and bust, highly influenced by political factors and safety concerns – the so-called 'Chernobyl Syndrome', reignited in the wake of the Fukushima accident in Japan in March 2011. Among the casualties of this, Japan retreated from nuclear and Germany decided it would close all its reactors by 2022, with some other European countries following suit.[39] The long-term disposal of radioactive waste material remains a subject of intense debate.

Despite these misgivings, nuclear is a mature technology that can provide low-carbon, near continuous power and is seen by many as a crucial option, both to diversify power supplies and increase the share of low-carbon power generation. Rapid expansion of nuclear capacity in Asia (excluding Japan), especially in China, shows that many emerging economies do not share concerns like those of Japan and many European countries – at least not enough to risk slowing the rapid expansion of power generation to fuel economic growth.[40]

Aside from safety fears and the potential legacy of radioactive waste, the other barrier is cost. Expectations in the 1960s of electricity from nuclear 'too cheap to meter' were for decades afterwards derided. Indeed the world has inherited not only a controversial option but also a paradox and lessons. In Europe, French nuclear power now generates some of the cheapest electricity, yet the cost of new plants looks daunting and nothing will happen without major government backing.

The lessons concern capital, complexity and commitment. Nuclear plants involve large upfront costs on each plant and the need for a complex web of industries. Scale and coordination can cut down production costs, but is worth undertaking only for large programmes. Early warnings that nuclear power could not pass the 'market test' are confirmed by the French example.[41] The programme launched in 1973 benefited from an enduring national political consensus that yielded well over €100bn of capital investment and transformed the entire French energy system.[42] It eventually delivered cheap electricity – thanks to some combination of early public subsidies/hidden costs, and the fact it was planned and sustained over a 15-year period with such a level of confidence that it could attract large volumes of capital (including foreign) at low financing costs.

Such conditions, institutional in nature, are not easy to recover. The belief that oil would be 'too expensive to buy' helped to cement the consensus behind the French programme in the oil-shocked 1970s. But the sharp drop in oil price in the 1980s reversed this sentiment and much of the existing fleet built during this period is approaching retirement. If nuclear is to even maintain its role, let alone expand its share, new plants would need to be widely built. However, the only two nuclear plants constructed in Europe in the past decade have suffered huge overruns of both time and cost, and this, together with the post-Fukushima backlash, has severely dented the prospects for a 'nuclear renaissance' in the rich world and raised the expected costs. Construction plans in Korea and China, however, appear much more significant.

There are some radical alternative approaches. Around a dozen new reactor designs ('third generation') are at various stages of development.[43] Thorium reactor technology has raised particular academic excitement, tackling (at least partially) several areas of concern.[44] Currently at the theoretical stage, 'fourth generation' reactors, if constructed would build on these benefits and more – but such reactors are likely to be many years from commercial deployment. Nuclear fusion (with its zero-waste, unlimited energy promises) still looks a long time away and is unlikely to become an option even by 2050.[45]

With continued technological development and rising energy prices, nuclear may once again become an attractive option, at least for some regions. When and by how much remains unclear and contested; the capital intensity and timescales of construction (>5 years) and programmes (>15 years) cannot tolerate uncertainty. A rising value of low-carbon generation would help, but not suffice, in the absence of central direction, financial support and long-term confidence.[46]

As discussed in Chapter 7 a credible long-term price for carbon is hard to provide in the forthcoming years. If these various obstacles were overcome, the most optimistic scenarios suggest that nuclear could account for up to 25 per cent of total generation capacity by 2050. The known constraints, however, make higher levels implausible.[47]

Primary renewable energy sources

The *theoretical* potential of the renewable energy resource is, like nuclear, huge. The amount of solar radiation that reaches the Earth's surface is several thousand times the rate of global energy demand. The tiny fraction of this that is fixed by plants still amounts to several times global energy demand. The winds that circle our planet dissipate every week as much energy as we consume globally in a year. The energy from rainfall on mountains, water in ocean currents and tides, and the heat under the surface of the earth provides additional, varied resources. At issue is our ability to harness these resources in reliable, affordable ways.[48]

Hydro-electricity represents a mature and relatively cheap technology, contributing about 15 per cent of global electricity production. Large untapped potential remains in some regions (such as central Asia), and use of small scale 'run-of-the-river' plants is increasing worldwide.[49] Yet in many regions the best sites are already being used and other big sites would face huge opposition. Projections suggest that hydro expansion will struggle to do any more than keep pace with rising electricity demand, so that the fraction is likely to remain close to today's levels.[50]

Production from non-hydro renewables has trebled since 1990, although still only approaching 3 per cent of total generation so far.[51] Many scenarios project extremely rapid growth, particularly from wind and solar energy.

Wind power has expanded dramatically over the past 20 years and is now also quite well established with wide deployment in Europe, the USA and Asia. Projections are of course uncertain, but suggest that wind energy could contribute around 10 per cent of global electricity by 2030. This would far from exhaust the global potential, and wind's contribution could double (or more) by mid-century.[52]

Since large-scale deployment began 20 years ago, the cost of wind energy has more than halved, despite a period of rising costs in the mid-2000s when demand for turbines outstripped supply and combined with rising steel and other commodity prices. Upfront capital accounts for most (usually over 80 per cent) of the costs, and the energy in the wind is very sensitive to windspeed, so that the cost of energy can vary a lot between different sites. In reasonable locations it is now one of the cheapest forms of renewable energy, below 10¢/kWh.[53] Further cost reductions are identifiable and expected.[54]

Going offshore can access stronger and more steady winds, but also enormously complicates the technology and its maintenance and adds to transmission costs. This can double the capital costs compared to onshore, and operation and maintenance costs also rise.[55] Offshore wind is still a much younger industry, with wide variations in both reported costs and projections. It is also more dependent upon the development of transmission networks and advanced transmission technologies to reduce costs. Offshore wind generation costs are currently in the range 15–25¢/kWh. Experience and scale economies, the use of HVDC (high-voltage direct current) cables for grid connection, innovative 'floating foundations' and new methods of construction such as 'all-in-one' installations all offer opportunities for significant cost savings.[56] However, the scale and complexity of offshore wind is such that realising the full potential could require efforts on a par with the French nuclear programme of the 1970s and 1980s, though it has the advantage of building from the base of the now-established onshore industry and existing oil and gas offshore expertise.

Solar electricity has long been touted for its huge potential. There are two main routes: photo-voltaic (PV) cells which directly convert sunlight to electric current; and concentrating solar power (CSP) which uses mirrors to focus sunlight to heat water or other fluids into steam to drive a traditional turbine.

PV has the big advantage of being modular – available at almost any scale, it has evolved through small specialist applications. Until quite recently, its costs seemed prohibitive for bulk power production and its contribution remained tiny. Its remarkable development in the past few years (see Box 3.4) has confounded all expectations. The rate of solar PV uptake by individuals and retail companies in response to incentives has also been a surprise to policy-makers, pointing to the personal appeal of being able to have your own clean energy from your roof.

In temperate climates, the total contribution of such PV is limited by the seasonal and daily mismatch between the sun and energy demand, but the opposite is true in hot regions, particularly where demand is driven by air conditioning needs. All this makes rapid expansion

Box 3.4 The solar surprise

Solar cells first came to public attention with their use for powering spacecraft, and they slowly expanded to specialist applications like boats, remote communications and other mobile applications – valuable, but tiny on the scale of global energy use. Crystalline silicon or (c-Si) PV cells were (and remain) by far the most common.

As remote applications grew over the decades, the costs fell as global installed capacity surpassed 1,000MW – still trivial in terms of global energy. Since about 2000, governments began to provide 'feed-in tariffs' for grid-connected applications. The resulting explosion of PV (along with IT) use initially strained global supplies of key input materials. 'Thin film' modules, which require far less material than c-Si modules and can be more easily mass produced in sheets, also began to enter at scale.[a]

For much of the 2000s costs remained stubbornly around $5/pW (peak watt). As supply chains caught up and China invested heavily in large-scale production, from 2009 the cost of both types started to tumble. At the start of 2012, c-Si modules had approached the long-held 'magic' cost barrier of $2/pW (Figure 3.7(a)).

Over the past few years, the global market has expanded at around 40%/yr – an unprecedented rate of growth for a new energy technology – to reach over 60,000MW in 2011. Governments have systematically underestimated the rate of take-up by the commercial sector and individuals' response to feed-in tariffs. On average over the past few decades, expansion has been accompanied by continuing cost reductions at a long-term average rate of about 22% price drop for every doubling of modules produced (Figure 3.7(b)).[b]

The remaining capital costs are 'balance of system' (BOS) costs which include all other items needed to make a PV system functional.[c] The average cost of energy across the lifetime of a PV system, including all capital and operating costs, was an average of 16¢/kWh in early 2012 for c-Si systems and a fraction less for thin-film – a drop from over 30¢/kWh and 24¢/kWh respectively in 2009 – highlighting the blinding speed at which costs have been dropping (although the range remains high).[d]

The combination of rapid expansion, cost reduction and the financial crisis have led to many of the feed-in tariffs being hastily revised. The long-run trend and implications of the 'solar surprise' have yet to be seen. Both c-Si and thin-film module efficiency are projected to improve by around a third – increasing energy production and reducing energy costs. The continued development of new, more advanced 'third generation' PV technology (such as concentrated PV), could build upon this further. Innovation in the technological aspects of BOS (such as the development of micro-inverters integrated into the PV module itself), standardisation of components across systems and manufacturers (bringing economies of scale) and competition would reduce installation costs. If such innovations occur, the PV energy costs should meet or even beat the lower values predicted for northern European conditions in 2020 and 2040 in Figure 3.8.[e]

Notes

a The most common (and economical) thin-film technology is cadmium telluride (CdTe). Thin-film technologies – often known as 'second generation' – cost less per peak watt (pW) but this advantage is offset by lower conversion efficiency.

b IRENA (2012c).

c The main balance-of-system components include inverters (to convert the DC energy produced into usable AC) and mounting structures (typically forming about 5% each of the total cost of a PV system), and 'soft costs' such as management, administration and planning – contributing up to 50% (IRENA 2012). Other significant costs include site preparation, wiring, data monitoring systems and physical labour to install the system. In off-grid systems, energy storage (batteries) adds significant additional cost (GTM Research 2011).

d Higher-end estimates remain over 30¢/kWh (e.g. Bazilian et al. 2012).

e For example, IRENA (2012).

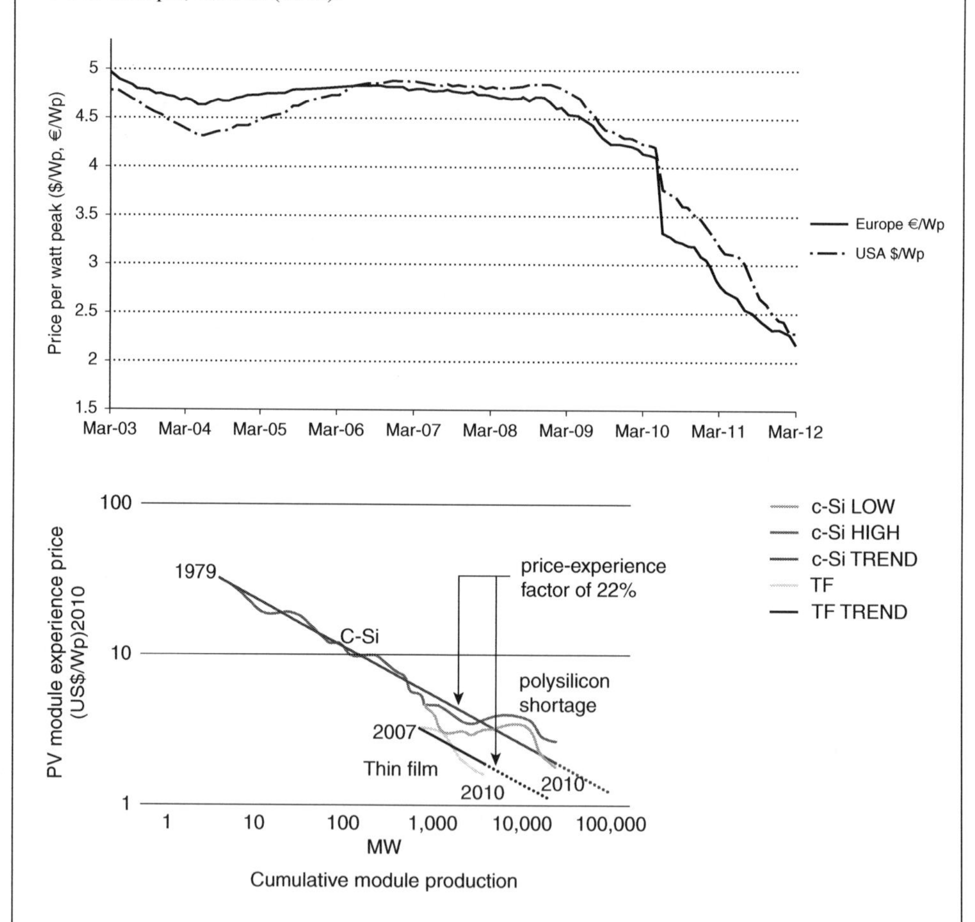

Figure 3.7 Declining solar PV module costs

Source: Adapted from EPIA (2011) *Solar Photovoltaics: On the Road to Competitiveness*, September, Figure 4.

seem far more plausible, especially in solar-rich regions, though the costs still remain higher than conventional energy.

Concentrating solar power requires abundant and direct solar radiation to operate effectively.[57] This, combined with the land use, makes it most appropriate to near-desert conditions. The potential advantage relative to PV is that modern systems include underground heat storage, which means it can provide continuous ('baseload') power. CSP plants – with parabolic troughs emerging as the most favoured approach – are being deployed in Spain and southern USA, with discussion of CSP plants in solar-rich North Africa potentially transmitting power back to Europe.

CSP costs are dominated by the capital investment – at least $4,000/kW – and estimates for the total cost of CSP range from about 10¢/kWh upwards, largely depending on location.[58] There are important economies of scale – larger plants can collect more energy for moderate small additional cost. Standardisation, incremental improvements and competition may also reduce costs and there are more specific opportunities. Most studies project average costs for large-scale CSP use may be around today's lower boundary of about 10¢/kWh.[59]

Another rapidly growing source of renewable electricity is biomass. Wood can easily be pelletised and traded, and hence, unlike liquid biofuels, there is far less concern about indirect emissions or competition with food crops. The cheapest application is through the mixing ('co-firing') or conversion of existing coal power stations. Dedicated stations can focus on electricity or a range of more sophisticated applications, from co-generation to full-scale 'biorefineries' that can generate a flexible mix of heat, power and liquid fuels.[60] The main debates are about the extent to which forests (or forested land) should be harvested for energy.

Rain, sun, wind and biomass are the main 'global' renewable resources, but they are not the only ones or indeed in some regions the most important. Some countries on the Pacific 'ring of fire' (like Indonesia) have huge potential to tap geothermal energy. Coastal geography may also concentrate tidal energy, creating opportunities for tidal barrages using largely conventional dam technologies.[61] Globally wave and tidal resources are limited, but islands and archipelagos in particular may be well placed to tap locally or nationally important concentrations, and innovations in design and turbine technology may reduce costs and expand the scope.[62]

Hydro, biomass and geothermal energies come with in-built energy storage; the other renewables fluctuate with the natural energy flows underpinning them. This limits their contribution, according to how well matched they are to demand and the ability of the electricity system to absorb them and provide other backup. This is outlined in the final section of this chapter, but does not preclude the potential for renewables to make large contributions.

Other concerns such as the 'indirect' emissions are largely unfounded.[63] For some technologies there may be constraints arising from the material available for construction at the scale envisaged. Notably, many advanced technologies use 'rare earth' materials, and the supply of at least five such materials used in wind and solar technologies could be constrained (though the history of technology is that generally such constraints can be circumnavigated, given time and innovation).[64]

The dominant constraint at present tends to be the fact that renewable energies have an impact on the landscape, and thus symbolise a tension between 'conservation' and 'sustainability'. Gaining planning permission can face opposition from NIMBYism ('not in my back yard'), which has been prominent in relation to wind energy in the UK, for example, slowing down renewable energy development and driving up the costs.

This complicates assessment of their likely contribution, but plenty of scenarios now suggest that renewable energies *could* become the dominant sources of electricity during the second quarter of the century, first regionally and then globally by mid-century.[65]

Conclusions

Each of the three main categories of low-carbon electricity – CCS, nuclear and renewables – has their strengths and drawbacks, their supporters and detractors. National priorities and political preferences play an important role.

One key determinant will be costs. In practice, the costs of all options vary regionally.[66] The relative contribution of renewables in hot climates may be particularly dependent on the further development of solar energy. In temperate climates, a wider range of renewables and other options may compete. Cost projections for 2020 and 2040 in northern Europe are shown in Figure 3.8. The broad message is as follows:

- In the shorter term (to 2020), onshore wind energy is the most obvious, cheap and large-scale low-carbon source. Recognising in particular that the CCS options remain undemonstrated and cannot be available at scale on this timescale, most others are likely to cost more than 10p/kWh, clearly above the range of new fossil fuel-based costs.

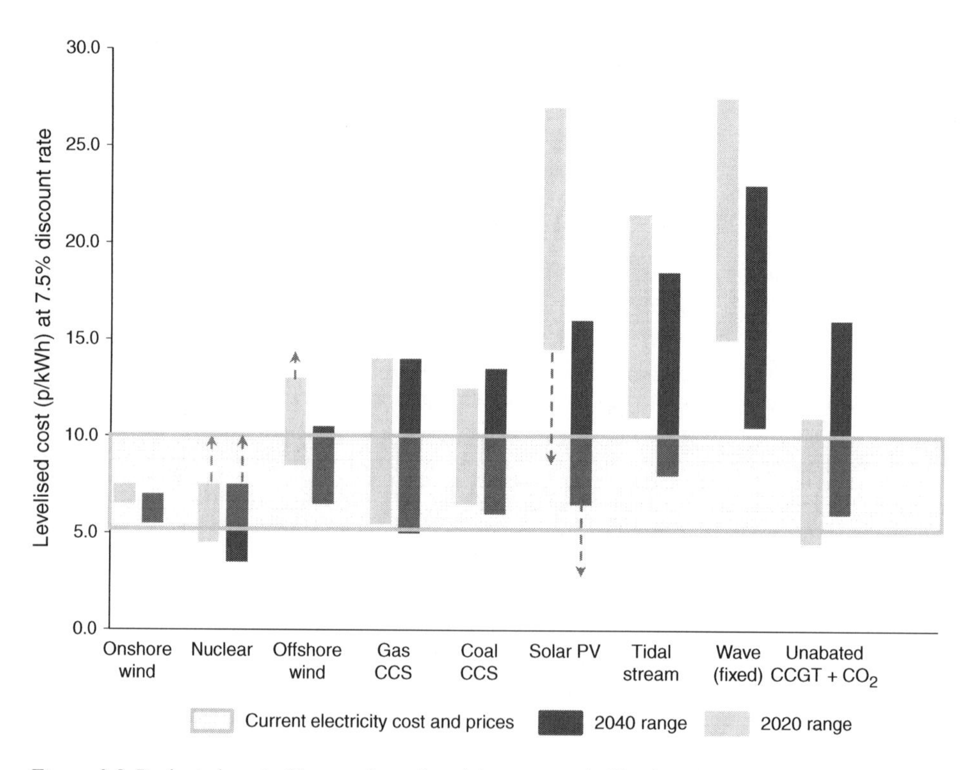

Figure 3.8 Projected cost of low-carbon electricity sources in Northern Europe

Note: The bars show ranges of cost estimates for different low-carbon electricity sources in 2020 (lighter) and 2040 (darker). The band across (5–10p/kWh) shows a range of estimated present-generation costs from new conventional plant investments (gas and coal, with uncertainty about future fuel prices). Units are UK p/kWh: 1 p = 1.2 €¢/kwh = 1.5 US¢/kWh. As indicated by the red dotted lines, since this graph was produced, estimated nuclear costs in the UK have risen dramatically (more than doubling, in the context of negotiations about actually financing a new nuclear station); also there is some evidence that offshore wind costs are proving higher than estimated earlier – with the UK draft 'strike price' for 2018/19 proposed at 13.5p/kWh over 15 years (note 67). The most recent estimates of CCS costs also imply costs towards the upper end of the bars illustrated. In contrast, PV costs have tumbled (see Box 3.4) so that the previous PV cost range for 2020 (15–25p/kWh) is higher than some prices already available in 2012.

Source: Climate Change Committee calculations, based on Mott Macdonald (2012).

- Beyond this, the cost of low-carbon electricity is clearly capped below 15p/kWh and is likely to be much lower, with the central range of most low carbon options projected at 10p/kWh or less. This – combined with the wide array of options available – caps the plausible cost of deep decarbonisation in power generation.

Compared to the right-hand side of Figure 3.8 – which shows the cost of fossil (gas) fired generation – the challenge is thus upfront investment to secure both future cost reductions and a power system that is more stable and economically secure against possible future crises in fossil fuel markets and climatic risks. Decarbonising power generation is affordable, but does come at a short-run premium cost.

Also, while most established power generation choices appear to be very much 'Second Domain' processes – rational investment planning and optimisation based on generating costs – the low-carbon options are more complex. The design of electricity markets typically favours low capital cost plants for multiple reasons, introducing some inherent bias against low-carbon sources. Shortly before going to press, 'strike prices', announced as part of the UK Energy Market Reform, clarified the prices for renewable energy on 15-year contracts required to support rapid growth in the UK.[67] The centrality of both innovation and systems development (grid networks, systemic backup capacity, planning and pipelines) for both CCS and most of the intermittent renewables, points to a strong role of Third Domain processes – which are also important for nuclear, given its long lead times, capital intensity, fuel and disposal concerns, and planning/public acceptance issues.[68] The biggest exception to this is PV – for which, because of its potential household scale, the evidence suggests also has strong, and positive, First Domain characteristics that help to account for the 'solar surprise'.

3.5 Tackling transport: transforming vehicles and fuels

Alongside keeping warm, using gadgets and consuming materials, our desire to move people and goods around has continued to expand. Energy demand for transport globally and associated emissions have roughly doubled alongside other uses since the 1970s; transport accounts for over two-thirds of global oil consumption and a quarter of global CO_2.[69] The oil price shocks of the 1970s which drove oil out of many other uses had little sustained impact on transport; oil's sheer convenience as a transport fuel has largely withstood all attempts to find alternatives.

Road transport accounts for around three-quarters of all transport energy use, with aviation and shipping at about 10 per cent each. Rail and other modes consume far less (Figure 3.9). About two-thirds of road transport is for passengers, the rest for goods.[70]

One thing missing from this is all the other transport, e.g. the journeys we make by walking or cycling. In much of the developing world animal power remains important. As Asia in particular aspires to western levels of mobility, transport use via commercial energy sources is expected to grow rapidly, potentially swamping any increase in energy efficiency or alternative fuels.

Although vehicles may only last a decade or two (though many last longer), the infrastructures last many decades – or more likely centuries. As outlined previously, the interstate highways construction in the middle of the twentieth century, as part of US recovery after the Great Depression, left the US unusually dependent on oil. The current choices of the emerging Asian economies are likely to have similarly enduring consequences.

Like homes, transport can become a very personal issue, but economics and infrastructure do strongly influence choices. One difference is that outside North America, most consumers already pay a high price for transport, including high taxes, to try and curb national oil

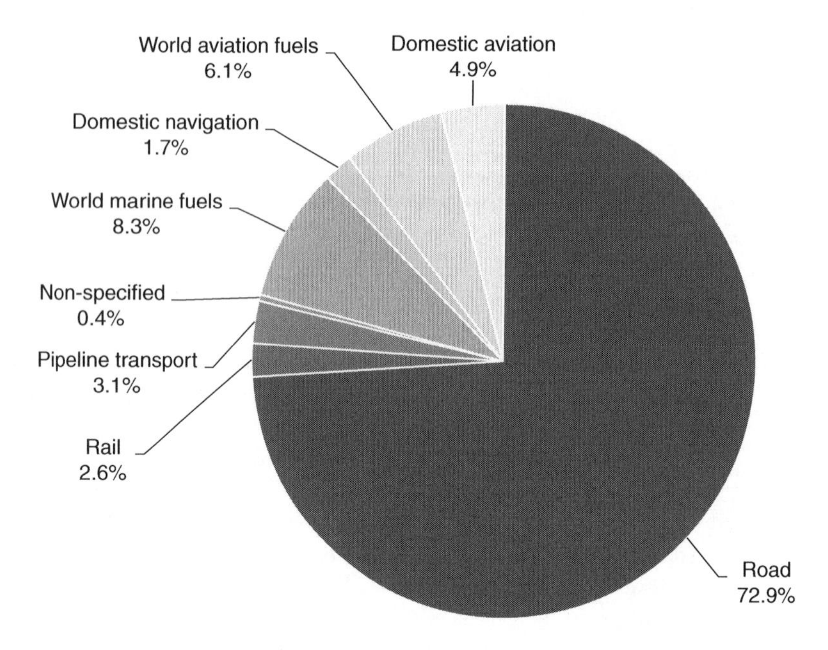

Figure 3.9 Global energy use in transport

Data source: IEA (2012).

dependence (Chapter 6). This, together with other fiscal measures like road tolls or car purchase taxes, has cumulatively done a lot to lower oil demand (and emissions).

Going further along this road depends on how (and how much) we travel, the efficiency of the vehicles we use and the fuels they consume.

How we travel: modal choice

Travel choices depend on our wealth, infrastructure, options and preferences. For short distances we could walk or cycle more – with potential health and other benefits from the reduced local emissions and congestion, and of course more exercise in an age of growing obesity. Journeys under two miles typically account for around a quarter of trips in the UK, but urban design has a strong long-run influence (see also Chapter 10).

Electric bikes extend the range and make 'cycling' more attractive for many – especially in crowded cities. This has become a striking phenomenon in China, with tens of millions of such bikes now being sold annually. In addition to the obvious factors of improving (mostly battery) technology and increasingly choked roads, affordability is important – growing incomes of the poor combine with rising gasoline prices and electric bikes are the cheapest of all motorised vehicles. Local pollution is another factor, and some Chinese cities (such as Suzhou) have now banned the use of fossil-fuel based motorcycles or three-wheelers.[71]

At the other end of the surface transport spectrum, for journeys of tens to hundreds of kilometres, the long trend shift from railways to cars has reversed in Europe with the advent of high-speed trains. The incentive is both environmental and commercial, cutting travel times between major centres and indeed helping to establish the political hub of Europe as

Box 3.5 Aviation and shipping

Emissions from aviation attract a lot of attention, though they account for just a tenth of CO_2 emissions from transport. However, aviation also releases large quantities of nitrogen oxides – a group of potent greenhouse gases. Its water vapour emissions can also be a potential greenhouse gas when emitted at altitude. The climate impact of aviation is typically assessed to be around twice that of the CO_2 emitted, though there is a lot of variation.[a] The real problem is the relative lack of mitigation options, combined with projected growth rates.

Reducing aviation's energy use and emissions is hard. Aviation is closer to physical efficiency limits than most other sectors. Planes have to use around half of their energy in just keeping in the air, with the other half being used for propulsion. Even cutting speeds wouldn't reduce emissions much. Despite further efficiency improvements, given expected growth in flying, CO_2 emissions could increase fourfold between 2000 and 2050.[b]

Some switch to lower (net) carbon liquid fuels may be possible; planes can run on biofuels and hydrogen has been proposed for the longer term though this poses big problems for storage and/or safety and increases water vapour emissions. Any changes would be slow due to long design and plane lifetimes.

Growth in the demand for short-haul flights might be lessened by the spread of high-speed rail networks (see text). The final and more contentious options would involve trying to curtail overall demand for long-distance travel: for example, video-conferencing for business and concerted efforts to make nearby holiday destinations more attractive. In some industrialised countries, it might conceivably be possible to keep aviation emissions around present levels; anything more ambitious seems hard to imagine.[c]

Shipping represents a similar percentage of global emissions but generally receives far less attention than aviation. There is greater scope for emission reduction through operational measures such as speed reduction and load optimisation, and still some scope for design improvements to reduce energy use; large sails or high-altitude kites offer additional theoretical options.[d] Because the areas are bigger and weight matters far less, it also much easier to utilise alternative fuels ranging from natural gas to biofuels.

Notes

a Intergovernmental Panel on Climate Change (1999) *Special Report: Aviation and the Global Atmosphere*; main points updated in subsequent Third Assessment (2001) and Fourth Assessment (2007).

b IEA (2008).

c Ceron and Dubois (2005) and Wilbanks *et al.* (2007: 380). In the UK, the construction of a third runway at Heathrow was initially made conditional upon establishing whether UK aviation CO_2 emissions could plausibly be held to 1990 levels by 2050. A report by the UK Climate Change Committee (UK CCC 2009) found that it could be possible, but only by combining optimistic assumptions. The planned runway remains on the table as a policy option.

d Marintek (2000). The Global Energy Assessment (2012: 51) and detail in Chapter 9 state that energy (and emissions) can be reduced by 4–20% in older ships and 5–30% in new ships through a combination of technical measures, while cutting a ship's speed from 26 to 23 knots can yield a 30% fuel saving.

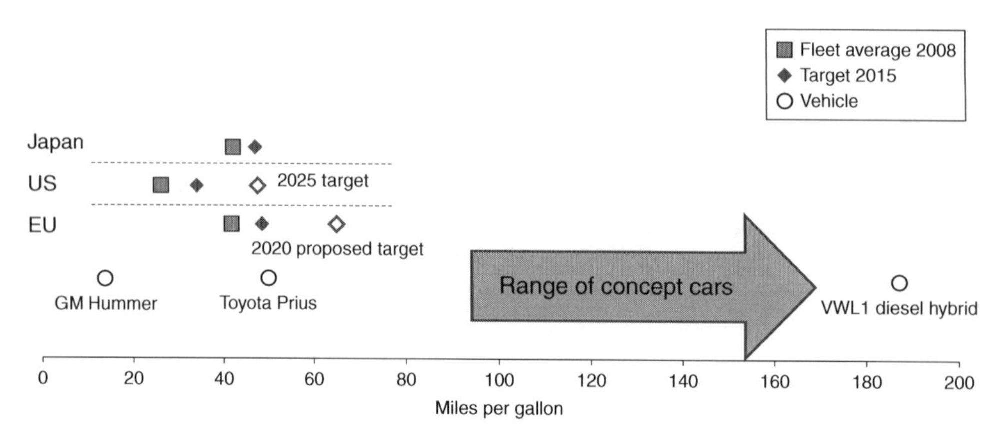

Figure 3.10 Vehicle efficiency in different national standards and models

Source: Derived by the authors from data in a range of sources including US DOE (2012) and ICCT (2011); see also Chapter 5 for future trajectories of vehicle efficiency standards.

the network of cities London–Paris–Brussels–Amsterdam–Cologne, now all linked with around two hours travel from the Brussels hub. This is the only development so far to make a real dent in short-haul aviation, with high-speed trains becoming the preferred mode of travel for distances up to several hundred kilometres.[72]

On the boundary between behaviour and efficient technology, driving habits can make a surprising difference. Cutting speed reduces energy consumption, as does smoother acceleration and braking, maintaining correct tyre pressures or switching off engines while idling. Efficient driving – backed up by better enforcement of speed limits – can cut 20 per cent off vehicle fuel consumption.[73]

Efficient vehicles

Big gains can come from better vehicles. It really shouldn't take so much energy just to move weight from one place to another. Despite improvements, most cars remain very inefficient – as any comparison between countries and models shows (see Figure 3.10).

Existing fleet differences reflect culture, fuel prices and the politics of the auto-industry. However, the pressures of both oil prices and environment have created a lot of innovation. The iconic Toyota Prius hybrid has become joined by many competitors; small efficient diesel cars such as the VW BlueMotion series can match it, and the EU's proposed 2020 target for vehicle efficiency would go further. Indeed as outlined in Chapter 5, vehicle efficiency standards have increased sporadically; only recently has the US adopted standards which will by 2025 lead it to converge towards the levels adopted in many other countries (including developing countries). All these in turn are still a lot less efficient than existing electric vehicles such as the Nissan Leaf, or others with low weight and frontal area which can double or treble efficiency. At the time of writing, the TESLA is widely promoted as the leading such vehicle commercially available – attractive also in its power and performance characteristics – and several other 'concept cars' are moving towards production.[74]

Overall the savings made from improving car efficiencies can be substantial.[75] One factor that may limit the long-run savings from vehicle efficiency measures are 'rebound' effects, as efficiency lowers the cost of driving. This is discussed in Chapter 5, but forms another reason to look beyond pure efficiency to the possibility of other fuels and systems.

Fuelling transport

Given the sheer convenience of personal vehicles and the limitations on efficiency alone, any transition away from oil must ultimately imply new fuels.

Some options require little change to vehicles. Plant matter doesn't have to accumulate for millions of years to turn into a liquid fuel, and ethanol is the most established way of getting a good fuel directly from sugars or grains. After the oil shocks, Brazil pursued the development of ethanol from sugarcane, with steadily increasing yields and declining costs in production, while the march of technology and policy has culminated in 'flexifuel' vehicles that can run on almost any mix of gasoline and ethanol (Chapter 10). Brazilian ethanol is now cost-competitive with fossil fuels without any subsidy or government support.[76]

US support for ethanol production from corn has more to do with supporting farmers than any environmental improvement which is distinctly ambiguous. On the same pretence, Europe has pushed for the production of ethanol from wheat and sugar beet, and biodiesel from rapeseed. The production costs of biofuels in the US and in Europe have been well above the costs of ethanol in Brazil, and they cannot compete without government subsidies or protection.[77]

Intrinsically many of these 'first-generation' biofuels – loosely defined as production from conventional crops – have the potential limitation of using crops developed for other purposes, so they may compete with those applications (such as food production). Efficiencies are intrinsically limited, and yield improvements may require extensive fertiliser use, irrigation or other inputs.[78]

'Second-generation' biofuels involve more advanced technologies to convert other sorts of biomass. These include improved technologies to convert waste biomass, such as stalks or husks, from current food production. Energy-only crops such as miscanthus or other fast-growing grasses can greatly extend the resource for heat applications and minimise competition with the human food chain; the big opportunities seem to lie in lignocellulosic biofuels which are made from such sources, from wood (such as poplar and willow), and from urban, agricultural and forestry biomass wastes.[79]

For traditional vehicles and infrastructure – reinforced by legislation in some countries – ethanol use is constrained by a 'blend wall'; cellulosic butanol, however, can be more easily used in existing infrastructure and is increasingly favoured.[80]

Still more exploratory, 'third-generation' liquid fuels can be produced directly from processing algae, which could remove the constraints of land and water competition altogether. Some algal fuels have a high energy density and could start to replace diesel and aviation fuel.[81]

The common currency is using natural photosynthesis to produce liquid hydrocarbons from carbon in the atmosphere – renewable oil. Hydrocarbons have the virtue of very high energy density and potential for extraordinary rates of power output; they thus have a natural advantage for heavier duty roles where power density is paramount, like short freight journeys or for aviation.[82]

There are, however, a far wider range of options for fuelling transport more generally. Natural gas can fuel vehicles in various ways, depending on the extent to which either the fuel or the vehicle is transformed – compressed for onboard tanks, or converted into synthetic fuels for use in current fleets. Either route cuts oil consumption and local emissions, but the CO_2 savings are marginal.

There are two options for more a radical transformation which abandons hydrocarbons entirely, by switching to electric motors:

- electric vehicles which use batteries on board; or
- fuel cell vehicles which store energy as hydrogen and convert to electricity on board, producing only water vapour as waste.

Periodically, some enthusiasts have predicted a future in which hydrogen would be the main energy carrier. It was hoped that fuel cell vehicles would overcome problems of slow charging and limited battery range. Car producers built some prototypes of fuel cell cars but technical difficulties relating to storage (hydrogen needs to be kept under very high pressures) and refuelling times have proved hard to solve.[83] More attention to the system also raised the question whether, if electricity is required to produce hydrogen through electrolysis, it would be simpler just to use the electricity directly.

For electric vehicles, the combined obstacles of limited range and the lack of refuelling infrastructure have been 'worked around' with the emergence of hybrid electric cars. Vehicles such as the Toyota Prius and the Honda Insight utilise both a traditional internal combustion engine and also a battery-powered electric motor. More advanced plug-in hybrids are starting to emerge as essentially plug-in electric vehicles with the combustion engine merely as a back-up. Hybrids remove the 'range-anxiety' associated with electric cars. Costs have been an obstacle, with the first decade of hybrid cars costing typically $30–40,000, but are falling. Very roughly, the new generation of such advanced vehicles offers around 50 per cent greater efficiency at an additional cost of a few thousand dollars. How quickly that pays back for the buyer, of course, depends on fuel prices and how much the vehicle is driven.[84]

Despite all the hype and struggle between advocates of fuel cell *vs* battery-electric vehicles, they have a great deal in common – including the potential advantages of an electric-drive train, which increases efficiency and allows regenerative braking (storing and reusing the energy of braking), and opening up transport to a wide range of low-carbon energy sources. Except for variants in which hydrogen might be produced directly from biomass sources, clean electricity would be needed for either – whether to produce hydrogen using electrolysis or to charge the battery. The main determinants are trade-offs between the greater storage capacity of hydrogen set against the potential safety concerns and infrastructure requirements of the 'hydrogen economy'. All sorts of variants could be considered, including storing hydrogen at depots for fuel cell vehicle fleets.[85]

There is no reason why the contest between hydrogen and electric vehicles should result in only one winner: there are many different types of transport needs and many different configurations, so the best trade-off may well vary with circumstance. An evolving mix of electric and fuel cells appropriate to different transport applications is entirely plausible. Paths towards 'transport transitions' are becoming clearer, as sketched in Chapter 10, whether through hybrids to electric-only – or indeed approached from the opposite end, scaling up the electric bikes currently sweeping Asia to look more and more like motor cars.

3.6 Resources

As illustrated in Figure 3.1, this chapter has reviewed energy technologies and systems 'back to front' – starting with how we use energy and then looking at the major conversion technologies and systems. It is now time to explain how this links to the physical resource base – the more natural starting point.

The world is not short of energy resources overall, clean or dirty, but there *are* important constraints on many options. These arise from the distribution of accessible reserves and the costs, environmental impacts or the timing of extracting, converting, and/or matching these reserves to demand. Many such constraints depend ultimately on technologies and systems.

Thus the amount of 'coal' in the ground is essentially unlimited if it includes all lumps of solid fossil carbon throughout the earth's crust. The same is true of natural gas if we include all sources of methane. Oil is more complex and problematic – cheap accessible reserves are steadily depleting – but there is a lot more in deep and less accessible deposits, or squeezed between rocks. Of the non-fossil deposits, there are vast amounts of uranium and thorium, though again concentrated reserves are far more limited. Radioactive decay in the earth also underpins the heat accumulated as essentially unlimited quantities of geothermal heat – but is only practically exploitable from the far more limited 'hot spots' nearer the surface.[86]

The true renewable resources are ultimately derived from the sun, which radiates the earth with energy at a rate around eight thousand times greater than we use it. The fraction converted into winds is, physically, still around 200 times greater. Wind and biomass are constrained mainly by conversion technologies and the amount of surface area we can devote to convert these resources – not by the absolute physical flows. The most constrained renewable resources are those of water – hydro, wave and tidal – and yet hydro dominates renewable energy contributions where concentration through river flows means it can be tapped at low cost, albeit with ever growing controversy.[87]

The major constraints on global energy contributions thus arise from limitations of technologies, systems and acceptability. For millennia, human development hinged on the use of renewables with traditional techniques, including wind energy with sails and old windmills; this was very constraining. These were largely superseded with the arrival of steam energy as a means of exploiting concentrated coal resources, facilitating the explosion of energy demand with the Industrial Revolution. As late as 1977 a tract on 'Energy or Extinction' proclaimed that renewables were far too limited to make relevant contributions to global energy, but this was based on assuming both inexorable growth in energy demand and the technologies of the previous century.[88] If there is one thing energy history should have taught us, it is humility.

In practice, per capita energy demand has roughly stabilised in the rich world as illustrated: the aerofoil dynamics of modern wind turbines are an order of magnitude improvement on the old wind-drag-based old mills, the conversion efficiency of solar devices has soared while costs have tumbled, and the growing interconnection of power systems enables us to tap into more remote resources.

Similarly if the standard for measuring global biomass resources is the ancient technology of fermenting traditional crops to make alcohol, the global potential is hopelessly limited. In stark contrast, successful development of algae as a fuel, honed for that purpose, could open a huge resource either from the oceans, or in sequestering CO_2 from power stations. To a large degree it is developments in technologies and systems that extend the practical potentials.

This helps to redefine the nature of the challenges. Thus the UK was once known as an 'island of coal in a sea of oil and gas' – an unlikely place to try and lead low-carbon developments. But the easy deposits of all of these have been substantially depleted, and the low-carbon agenda has accelerated efforts to exploit relatively abundant resources of wind and water – which face other kinds of constraints, to do with acceptability, cost and integration. Similarly it is often said that China and India are bound to use their huge reserves of coal, but they have equally huge potential resources of sun and wind; it is technologies, systems, costs and policy choices which will determine which will be used and which left untapped.

Of course many resources have limits, as beautifully mapped out in a book by David McKay, who matches UK energy needs against the potential resources available.[89] At local and regional levels, the density of urban energy demand typically vastly exceeds that of renewables except for direct solar radiation; networks to bring energy from outside are essential. At the national level, countries tend to deplete their own fossil fuel deposits first, so remaining reserves tend to become increasingly concentrated, at least until major price shocks lead to new waves of exploration and investment. Systems for international trade are already well established.

At the *global* level, the key constraints are on liquid fuels. Despite all the developments in vehicles noted, moving away from liquid fuels will be hard for two reasons: they offer the most convenient, dense way of storing energy on board; and current transport systems, through global systems of trade, refining and distribution, have evolved for them. Even if there is a transition towards electric or fuel cell vehicles, the constraints on the pace and scale of it suggest that large quantities of liquid fuels will still be required to power transportation over coming decades.

But conventional oil is limited, as indicated above. Most of the evidence points to global production peaking this decade; the optimists push 'peak conventional oil' a decade or two further back. Existing production provinces are declining; most of the new ones are even more difficult and remote, and geopolitical risks loom large in the calculus of many governments.

The default trajectory will be for firms to extend their existing business model and interests, and pursue massive investments in unconventional oils – such as the Canadian oil sands and heavy oil in Venezuela, along with increasing amounts of offshore deep-sea drilling. The costs are much higher than for producing conventional oil, though they are declining with investment, and resources are limited and geographically concentrated, just like their conventional counterparts. Extracting shale oil through the 'fracking' of shale rocks is rapidly increasing, but again at higher cost and energy inputs. Even if the demand for liquid fuels were constant, such a path would increase emissions of CO_2 and indeed many other pollutants – converting unconventional oil sources to useable gasoline can add 10–25 per cent to emissions compared to a standard barrel of oil, though like many things this should also decline with technological improvements.[90] Still, a recent study argues that the central challenge is neither resources nor costs, but the rapidly rising energy inputs required to extract a barrel of frontier or shale oil.[91]

The lower carbon route for liquid fuels would be to massively increase our use of biofuels, but this also faces major social, environmental and systemic challenges. The current stock of 'first-generation' biofuels is limited in efficiency and benefits (particularly where pursued as an agricultural support policy). More advanced biofuels may help to alleviate a number of these concerns but their development has been slow.

In practice, we are seeing – and will see – a lot more of all these activities: heavy and unconventional oils, and mainstream biofuels based on conventional, land-based crops. Relative to conventional oil, both remain small, but spurred on by oil prices that are high, uncertain and periodically unstable, both are rapidly growing: the 'breakout' in Figure 3.11 shows the growth of unconventional oil and biofuels from negligible contributions in 2000, to around 5 million barrels/day by 2012, with biofuels about a quarter of the unconventionals but growing even faster. It is perhaps ironically symbolic that competition seems to be emerging between dirty fuels in the industrialised countries and cleaner fuels from the developing world, symbolised by Canadian oil sands and Brazilian ethanol respectively. But the fact is that neither offers a global, enduring solution to fuelling global transport demands.

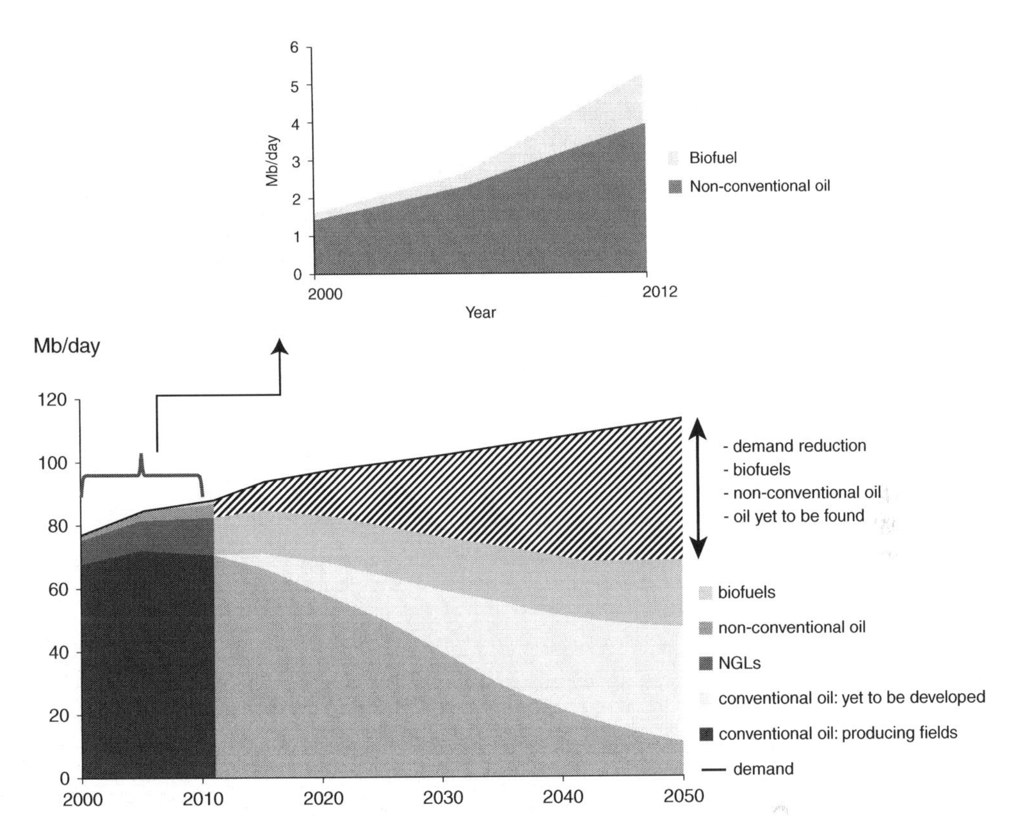

Figure 3.11 Projected demand and supply of liquid fuels

Source: Adapted by authors with data from IEA (2012).

The strategic setting is illustrated in the main part of Figure 3.11, which reflects the most recent (2012) projections from the International Energy Agency.[92] This illustrates the decline in production from existing oil fields, which will only partly be offset by production from known reserves that have yet to be developed. The growing gap between that and rising demand for liquid fuels has to be filled by unconventional deposits, new oil finds and biofuels. Based on current technologies and resource estimates, if biofuels lead to 'fill the gap' their potential limit is reached within little over 20 years; the world then has to move heavily into unconventional fossil fuels but even this fails to meet demand by mid-century. This combination would probably imply a transport sector that is as carbon-intensive as today, split between fuels that are on average around 50 per cent worse and 50 per cent (or more) better.

Further down the path towards more and more carbon-intensive fuels, we could move beyond the constraints of unconventional oils to embrace coal-to-liquid technology, thus opening up global coal resources for the transport sector. The technology is broadly known, but the scale of investment and the damage to both local and global environments would be staggering. The carbon damage could be moderated by CCS, or instead by gas-to-liquid technologies, but the conversion also adds to costs and emissions compared to other uses for gas and risks a lock-in to gas-based infrastructure.

Some combination of advanced biofuels (second and third generation) at scale, and a move to alternatively fuelled vehicles powered ultimately from zero carbon electricity sources, will thus be essential eventually. If demand grows as projected, decarbonising transport fuels in aggregate (by avoiding heavy dependence on *more* carbon-intensive resources) will require these to start playing a major role globally during the 2030s. Accelerating energy efficiency to curtail the growth in the residual fuel demand could give another decade or so.

3.7 Smarter systems and scenarios

Liquid fuels come with inherent storage, as do vehicles. For applications other than transport, however, demand can be just as varied and location-specific as renewable resources. In cold climates, the peak need is for space heating on cold winter evenings; solar energy is not much use then. In sharp contrast, a major rising demand is for air conditioning in hot climates; solar PV may be well matched there, which is one reason why the cost reductions in PV are so important. In subtropical areas the storage that can come with solar thermal generation may give it a decisive edge. In temperate climates, the seasonal patterns of wind energy may be well matched to energy demand, but shorter-term fluctuations remain. The practical resources thus also depend on an energy system that can help to mix and match sources of supply and demand, and has storage or other capacity for backup if needed. In short, making the best of physical resources hinges upon the combination of improved technologies and smarter systems.

The challenge of matching energy sources to demands comes to the fore with energy sources that fluctuate (wind and solar) in producing a carrier (electricity) that is hard to store. On small systems – such as islands or remote villages – this can be a major constraint. Without the advantage of geographical dispersion, a sophisticated set of storage and load management is often necessary to balance intermittent resources.[93]

For feeding renewables into larger electricity systems, in practice it doesn't matter much while the contribution from the variable source is relatively small – up to 20 per cent of electricity typically being taken as a benchmark. For contributions much above that, the system costs associated with 'backup' (e.g. for when there is little or no wind) and 'balancing' (to cope with fluctuations) start to escalate. The main value becomes increasingly confined to displacing fuel (and emissions), not generating capacity.

However, numerous factors can help to alleviate these constraints. The wider the geographical spread of installations and/or diversity of renewable sources, the less the problem. Wider transmission networks, including interconnection with neighbouring systems, generally facilitate both. This may also enhance access to storage, most obviously if interconnection gives access to hydro systems, and indeed typically carries a number of other potential advantages. Both transmission and distribution technologies have improved greatly. Eradicating huge distribution losses in India through investment in high voltage distribution systems is a realistic short-term option, for example.[94] Continent-wide grids are under discussion or development in most continents.

Storage does not have to be big and remote: indeed it is already implicit in much demand, for example the thermal storage capacity implicit where electricity is used for heating and cooling buildings, and for refrigeration and washing machines. This offers the potential for 'net demands' to be responsive over limited periods.

The technology for enabling this – 'smart meters' – also gives consumers far better, real-time information about their electricity demand and can transmit this information back to the supplier – saving everyone the time and hassle of physical meter readings. Two-way smart meters also enable decentralised generators – solar PV, small wind, energy from waste facilities at small factories or large stores, for example – to feed power back to the main system. Many countries are now engaged in rolling out smart meters. The combination of smart meters with advanced transmission and power electronics on the distribution system underpins the 'smart grid' of the future.[95]

The biggest gains may emerge as transport starts to electrify. Cars could be plugged in overnight – or at the company car park – to be charged when power is the cheapest, or potentially return some power to the grid if the price is right. There are concerns about impact on the lifetime of current batteries, but the links between electricity and transport could also be simpler: the high battery performance required in vehicles means that batteries are typically replaced when they still have many years potential service as storage in less onerous applications – whether in buildings or, for example, mediating power and conversion flows at local transformer stations. The possibilities seem endless.

'Smarter systems' can also apply at the level of individual facilities. The Global Energy Assessment devotes much of its chapter on fossil fuel conversion to 'coal-biomass co-processing systems with CCS that provide both liquid fuels and electricity', mapping the theoretical advantages of such integrated systems – including of course flexibility, and the ability to retrofit onto and repower existing coal plants.[96] A radically different approach to the benefits of integrated thinking is the 'Sahara forest' project in Qatar. This starts from the simple proposition that the deserts have too much sun and not enough water or fertile soils, and has developed an integrated combination of desalination, solar cooling and agriculture with production of both power and experimental algal fuels.[97]

In short, we do not know all the ways in which 'smarter systems' may develop, or the opportunities they may open up if and as energy systems become more interconnected. We just know there is huge potential, which can further extend the scope for efficient use of energy and access to clean energy sources.

All these observations and developments basically open up further the range of future possibilities. This is illustrated by the most recent clutch of global energy scenarios, worked through in unprecedented levels of detail. The International Energy Agency's Energy Technology Perspectives (2012) has produced global scenarios consistent with global temperature increases of 2, 4 and 6°C; all are possible. The Global Energy Assessment explored three main groups of scenarios consistent with the dual goals of Energy Access for All by 2030, and the 2°C target, which varied mainly in the degree of energy efficiency. These are summarised in Figure 3.12. Notable in this is the range in global energy demand, which in 2050 spans from a 30 per cent increase to a doubling, compared to current levels. The difference represents more than half of today's global energy stock, and obviously carries radically different implications for the amount of new energy infrastructure needed.

Moreover, all the GEA scenarios illustrated in Figure 3.12 project massive growth in renewable energy. It is hard to construct any vision that connects almost 9 billion people to modern energy systems while cutting global CO_2 emissions by 50 per cent without this – but as charted in Chapter 9, scenarios for renewables also vary enormously.

There is only one way to interpret such wide ranges. Our future energy systems will be constrained not by the technologies or resources available, but by the choices we make. The huge range of scenarios reflect, in large measure, policy choices. The radically different outcomes for global energy demand will depend to an important degree on the Pillar I policies considered in the next two chapters. Pricing and market structures (Pillar II) will also be crucial to this, and also to the expansion of low-carbon supplies. Yet the ultimate response will also hinge on the scale and nature of innovation and infrastructure, and associated industrial development fostered by Pillar III policies.

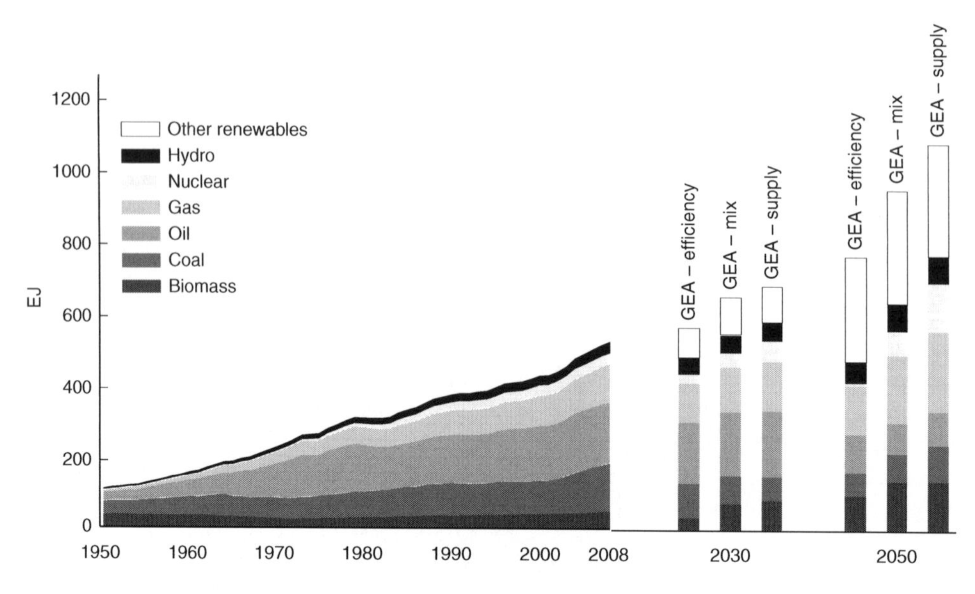

Figure 3.12 Global Energy Assessment Pathways for 2030 and 2050

Source: GEA (2012).

3.8 Conclusions

Energy systems are complex, but this *tour d'horizon* of energy uses and technologies has outlined its basic structures and underlined a few big themes.

There is no 'magic bullet': energy uses, conversion systems and prime resources are far too many and varied for that. Innovation is crucial, but hanging policy on the hope for a 'breakthrough' technology betrays a fundamental misunderstanding of the role and nature of energy systems – and the sheer diversity and dynamics of innovation.

There is, rather, a huge potential for improvement on innumerable fronts. A striking statistic is that we consume about ten times as much energy as is physically necessary for the services we demand – and we could furthermore meet underlying aims with lower service demands (see Box 4.2). Numerous practical technologies exist for more efficient use, many of them plainly cost-effective, above all in the buildings sector and other heating applications. Why they are not more widely adopted and how this might change is the major focus of the rest of this First Pillar of responses.

This does not supplant the need to change energy supplies. In this, resources are not the defining problem: the world has plenty of both fossil and non-fossil resources. Technical constraints arise from the limitations and costs of technologies to access, convert and connect these resources to energy demands; the other constraints concern the environmental impacts and social acceptance of the different options.

These factors play out through two main mega-systems:

- Technologies and systems for low carbon electricity, particularly a range of renewables, have developed rapidly but most remain more expensive than cheap coal-based power generation (depending on the nature of the coal resources and environmental constraints).

For decarbonising power supplies, the balance of choices between these renewables, nuclear and carbon capture and storage on coal and ultimately gas plants may play out quite differently in different regions, reflecting both physical factors and social and political choices. Across these three generic options, deep decarbonisation appears possible at power generation costs of around 10p/kWh (15¢/kWh) or less.[98]
- The era of transport dominated by cheap oil for internal combustion engines is approaching its twilight. The alternatives involve a contest between higher and lower carbon options in both the mid-term (first-generation biofuels *vs* unconventional oil deposits) and the longer term (heavier carbon deposits and coal-to-liquid technologies *vs* advanced biofuels and/or alternative transport systems based around electricity and fuel cell vehicles.

To make the best use of available technologies and resources, energy systems are becoming more integrated. Systemic technologies including advanced transmission and smart grids will over time transform possibilities.

Given the vast array of options outlined in this chapter, the future is open. The potential for 'smarter choices' is not confined to energy efficiency but ultimately is a possibility for the entire energy system, providing that policies help to support the adoption of cleaner products and processes and the development of appropriate infrastructure, and drive the enormous potential for continued innovation towards lower rather than higher dependence on fossil fuels. The question is how.

Notes

1. Non-commercial sources of energy include fuel wood collected mainly in developing countries for home heating and cooking, and passive solar heating gains. These relate primarily to building energy use and are much harder to account for, in terms of both energy and climate change impact (see Buildings section of this chapter), than commercial energy. The UN Secretary-General Advisory Group on Energy and Climate Change (AGECC 2010) proposed a goal of energy access for all by 2030; the achievement of connecting about 2 billion people to electricity globally from 1990–2008 suggests this to be feasible. A number of developing country governments have defined basic or lifeline 'energy entitlements' within the range 20–50kWh electricity, and the equivalent of 6–15kg of liquid petroleum gas for cooking per household per month. For more data on the overall scale of the challenge see Chapter 19 of the Global Energy Assessment (GEA 2012).
2. Because oil is the easiest to store and transport, it also supplies energy for remote applications, such as kerosene for light and heat and diesel for power generation.
3. This is visible both in comparative data (e.g. energy per square metre, corrected for temperature differences) but also through common observations. Visitors to the UK from Scandinavia, where they adopted strong building efficiency standards decades ago – or from other cold regions where good insulation makes winters liveable – are easily shocked. A friend who moved from Wisconsin in Canada to London couldn't believe his flat had its hot water pipes running uninsulated outside the building – to heat the street, perhaps?
4. IPCC (2007).
5. For surveys see IPCC (2007), Chapter 6, and the Global Energy Assessment (2012), Chapter 10.
6. For detailed analysis see the Global Energy Assessment (2012), chapters on Energy Poverty/Access and Environment/Health.
7. For example, less than 1% of UK building stock is replaced each year (Roberts 2008).
8. IEA (2009). Though not directly comparable, the US constructed around 800,000 homes in 2009 (US Census Bureau 2010).
9. World Steel Association (2008).
10. Hammond and Jones (2008) cite an estimate of 800MtCO_2. Vaze states 'these indirect emissions from building a UK house are about 25% of the total for building, heating and providing

hot water to it' (2008: 155). The Global Energy Assessment (Chapter 10, p. 664) concurs that '... the construction of a building contributes as much as 25% to total indirect energy use, with higher values for very high efficiency buildings.'

11. The UK Energy Savings Trust estimates that installing the most efficient boiler compared to the worst could save households a quarter of their energy bills and cut emissions per house by around a tonne of CO_2 a year.
12. For the most extensive survey of building efficiency potentials, costs and limits, and also the potential of wider urban-based systems, see the Global Energy Assessment (2012), Chapters 10 (Buildings) and 18 (Urbanisation).
13. In the UK, the 'Bioenergy Capital Grant Scheme' and 'Renewable Heat Incentive' promote the use of bioenergy for heating (DECC 2012a).
14. Anaerobic digestion uses the natural process by which micro-organisms break down organic matter in the absence of oxygen to produce biogas (around 60% methane, 40% carbon dioxide), which may be used in place of natural gas to fire boilers or CHP plants. Gasification is a thermochemical process by which material is heated to high temperatures in the presence of a controlled volume of oxygen, producing combustible synthetic gas ('syngas') which may be used similarly to biogas. Pyrolysis is a similar process but takes place in the total absence of oxygen (DECC 2012b).
15. CHP with district heating is more common today than household-level installations, which cost more per unit of energy. In a UK study CHP with district heating was projected as likely to yield net cost savings, while costs from a micro-CHP field trial were at least £270/tCO_2, highlighting the cost reductions needed for micro-CHP to become cost-effective (AEA 2008).
16. For micro-level CHP, Stirling engines and fuel cells can provide both electricity and heat within buildings (including homes), but the technology is expensive, although prices may reduce with technical development and deployment – of which field trials in many countries, including the UK and Germany, are a start along the road.
17. Otherwise known as 'thermally driven chillers', in which the solar energy heats a fluid that carries heat to a generator of an absorption chiller (IEA 2011).
18. Heat pumps work by moving heat from one place to another, one place gets warmer and one cooler. On cold days densely populated areas will move a lot of heat from either the air or the ground outside into homes, and there are question marks over the capacity of the ground or air to supply enough heat (or absorb enough when cooling is required) in such areas.
19. IPCC (2011). Biomass is only likely to generate 5% of total electricity in the OECD by 2030, with a slightly higher 10% in non-OECD regions where access to suitable resources is greater. Geothermal is only predicted to reach 2% of total generation by 2030. See also notes 86 and 87 on resources.
20. IEA (2009). In some projections developed country industry emissions could fall by 10% up to 2020, but some of this would be due to 'offshoring' production to developing country producers and importing instead.
21. The biggest categories of process emissions are CO_2 associated with the reduction of iron ore to pig iron for steel, and the conversion of calcium carbonate into clinker for cement: both involve using the carbon from fossil fuels to extract oxygen (then released as CO_2) to leave the primary mineral. Other categories include process emissions from aluminium smelting and various chemical conversion processes.
22. The IEA (2008) estimate savings of 20–25% from improved motor drives. The proportion of industrial *electricity* (not overall electricity) consumed by motor drives has been estimated at 65% for the EU-25 (de Keulanaer *et al.* 2004) and 63% for the US (Xenergy 1998). All references from IPCC (2007), Chapter 7: Industry. For more extensive and recent references see the Global Energy Assessment (2012), Chapter 8.
23. The concept of exergy (see Box 3.2) is crucial here – the usefulness or 'quality' of energy, putting more of the energy that we produce to the uses it's most suitable for and capturing anything we waste and putting it to productive uses means we need to generate less in the first place, getting more from less. The real challenge is in implementation and in getting all of us to be smarter about using energy – the topic of Pillar III.
24. IPCC (2007).
25. *Iron and Steel Industry – Analysis and Data* (OECD 2009).
26. Ibid.

27. IEA (2006). India's steel sector has a large share of small, inefficient coal-based DRI units, coupled with an electricity sector dominated by coal generation (Sreenivasamurthy 2009).
28. See, for example, Worrell *et al.* (2010).
29. IEA (2006). The combination of bio-coke together with CCS has been claimed as a route to 'negative emissions', but this is less to do with steel production per se than the fact that growing biomass and then burying the CO_2 from its combustion can be considered as a negative emission – whether the biomass is used in a steel plant or power plant, for example.
30. This would require developing a supply chain for such steel and also the creation of deconstruction plans for buildings so steel can be extracted in future. For an overview see reports from the WellMet 2050 programme: Allwood *et al.* (2011). For more information see Allwood and Cullen (2009).
31. UK Committee on Climate Change (2011). Online at: http://www.theccc.org.uk/reports/bioenergy-review.
32. For an extensive appraisal of more efficient cement technologies see, for example, Hasanbeigi *et al.* (2012).
33. The IPCC Fourth Assessment (2007) estimated the potential for cement emission reductions at 30%.
34. Good overviews of cement technologies are in IEA (2009b), Kitson and Wooders (2012) and the IEA Energy Technology Systems Analysis Programme (ETSAP) Technology Brief 103 (IEA 2010b). Each contains interesting different degrees of emphasis and optimism (or not) about the potential for the more radical 'low or negative' cement technologies. The IEA ETSAP study summarises these 'new low-carbon or carbon-negative cements' as follows:

    > The mechanical properties of these novel cements are similar to those of regular Portland cement. The necessary geological resources for the raw material feedstock are available on all continents. *NOVACEM* cement is based on magnesium oxide (MgO) and special mineral additives. It offers the prospect of lower-carbon cement through the use of an innovative production process which can use a variety of non-carbonate-based feedstocks and a novel cement composition that accelerates the absorption of CO_2 from the environment by the manufactured construction products. *CALERA* cement is a mixture of calcium and magnesium carbonates and calcium and magnesium hydroxides. Its production process involves bringing sea-water, brackish water or brine into contact with the waste heat in the flue gas of power stations. *CALIX'S* cement is produced by the rapid calcination of dolomite ... CO_2 emissions from the flash calciner can be captured using a separate CO_2 scrubbing system ... the carbonated sorbents are then recycled back to the first step where they are decomposed again and the CO_2 released is captured and potentially sequestered.

35. Hasanbeigi *et al.* (2012).
36. CEPI (2011).
37. Although this is still at a research stage and not expected to be commercially viable before 2020; see the Aluminium Association (2003).
38. An academic policy update (Reichardt *et al.* 2012, available from http://www.responsesproject.eu) reports:

    > However, several barriers within the policy mix have recently slowed down CCS innovation activity, calling into question the planned contribution of CCS to a low-carbon power system in the EU. At EU level, the major barriers are the low price of EUAs and uncertainty about future prices, as well as uncertainty about emission targets for the EU within a future international climate policy agreement. In addition, because of national administrative procedures, some CCS projects were not able to meet the application and implementation deadlines set by the EU's EEPR and NER 300 programmes. In response to these factors, several firms have substantially decreased their CCS technology demonstration activities.
    >
    > Within Germany, the failure to implement the EU CCS Directive has been identified as the main barrier to CCS. Since Germany has not yet passed a law governing the storage of CO_2, there is no legal basis for CO_2 storage. This regulatory gap together with a lack of social

acceptance currently prevents larger scale CCS innovation activities in Germany, including demonstration projects. As a result, the first full-scale German CO_2 capture project planned by Vattenfall in Jaenschwalde has been terminated.

39. The plan was to immediately close eight reactors (which were already offline), and to follow with the remaining nine by the end of 2022 (Heinen-Esser 2011).
40. In China 26 reactors were under construction with an estimated 51 additional reactors planned when assessed in World Nuclear Association (2012).
41. Flavin (1983).
42. France constructed 50GW of nuclear power stations in the decade 1980–90, and nuclear provides around 80% of French electricity. After many years of speculation, the costs of the programme were finally revealed in 2000, since when data have been more transparent. The number cited is estimated from data in Grubler (2010).
43. Including accelerator-driven subcritical reactors, which produce far less radioactive waste material, and thorium reactors (World Nuclear Association 2012).
44. The claimed benefits of thorium are typically that it is three times more abundant in the earth's crust and is more 'fertile' than mined uranium – it does not require enriching. Reactions produce far less waste material, which is far less radioactive (with half-lives measured in decades rather than millennia) and is not directly compatible with weaponised use. Thorium also does not oxidise like uranium, and so storage of waste is also safer. Numerous other benefits surround this option (IAEA, 2005).
45. Again, the Global Energy Assessment (2012), Chapter 11, gives a useful overview of nuclear technologies and issues.
46. For French assessments see Finon and Staropoli (2001) and Finon (2013). Brinkerhoff (2011) estimated the levelised cost of nuclear capital construction *alone* to be at least 5p/kWh in a UK context; the capital is estimated to be 70% of the total cost of producing electricity over the nuclear plant lifetime, and appears to have expanded even more as the UK has moved towards contractual negotiations for new nuclear provision. Helm (2011) among others emphasises that current cap-and-trade/carbon pricing systems are not adequate to support nuclear. Just before this book went to press, the UK government completed negotiations for a new nuclear power station, at a 'strike' price of 9.25p/kWh guaranteed for 35 years.
47. IEA (2010).
48. IPCC (2007). The kinetic energy embodied in winds is estimated at 870TJ a second by Hermann (2006). Compared to annual energy consumption of around 500 exajoules, the Global Energy Assessment (2012: 47) summarises:

    > Solar radiation reaching the earth's surface amounts to 3.9 million EJ/yr; accounting for cloud coverage and empirical irradiance ... local availability is 633,000. The energy carried by wind flows is around 110,000EJ, and energy in the global water cycle is more than 500,000 EJ of which 200EJ/yr could theoretically be harnessed ... Net primary biomass production is *c.*2,400EJ/yr, which after deducting the needs for food and animal feed leaves in theory 1330-EU/yr for potential use in satisfying energy service demand ... geothermal head flow to the surface is about 1,500EJ/yr ... oceans absorb on average some one million EJ/yr.

49. Although Chapter 1 acknowledges that Brazil is one of the most carbon-efficient economies in the world, partly due to its significant hydroelectric generation, there is evidence to suggest that hydroelectricity is less low-carbon in the tropics than in temperate climes. When reservoirs are created behind hydroelectric dams, the biomass in the flooded soil and vegetation decays and releases methane which rises through the water and into the atmosphere. Tropical regions typically hold a much higher density of biomass, and so the effect is exacerbated. Some studies suggest that a typical hydroelectric plant in the tropics produces around 20 times the emissions than in temperate regions – up to 3,000kgCO_2e/MWh in the worst cases (Steinhurst *et al.* 2012).
50. IPCC (2007), IEA/OECD (2009).
51. IEA (2010).
52. As of mid-2012 there were 1,503 turbines across 56 wind farms in 10 countries in Europe (EWEA 2012). The IEA/OECD (2009) projected wind at *c.*17% of global power generation by

mid-century; some wind industry scenarios are even more optimistic with forecasts up to almost 30% of generation in 2050 (GWEC 2008).

53. IRENA (2012c). Most of this is the cost of the turbines themselves, with the remainder coming from foundations, grid connection, electrical equipment and construction costs.
54. Options for further cost reduction include the development of lightweight carbon fibre blades, integration of lightweight materials into the turbine tower (reducing dependence on steel and the price volatility it brings), and increasing uptake of gearless drive generators. Increasing standardisation, scale and market competition also plays a significant role. See, for example, IRENA (2012c).
55. Additional cost factors include foundations in the sea bed and the connection distance from land and the electrical grid, requiring long underwater connection cables. Also, offshore maintenance costs are typically around 25% of total costs for direct maintenance of the turbines themselves, spare parts, labour, overheads and insurance (EWEA 2009).
56. Floating foundations are currently more expensive than fixed foundations, but a range of innovations are currently under investigation to dramatically reduce their cost (IRENA 2012c). These include integrated processes where the turbine is fully assembled onshore then transported out to sea and simply secured into pre-built foundations.
57. Over 2,000kWh/m^2/year, at present.
58. IRENA (2012b) cites capital cost estimates of between \$4,200/kW and \$8,400/kW with lifetimes of 30 years plus. The main contributor to the initial investment is the 'solar field' – the array of solar collectors that make up the system – which is in turn dominated by the metal support structure, mirrors and heat transfer systems, as well as land and labour installation costs. Systems with energy storage may cost around 10% more, but the improved use of the investment can reduce lifetime costs. The remaining 20% of costs are operation and maintenance, a large part of which is the replacement of broken mirrors and receivers, and insurance. Lazard (2008) estimates costs from \$0.11/kWh to \$0.36/kWh for mature systems, similar to the estimates from IRENA.
59. Mirrors could be made from lightweight, front-surface reflectors rather than heavy silver-backed glass. Using heat transfer fluids with a higher operating temperature would allow more energy storage. Automation and incremental technological improvements have also reduced operational costs. For a review of cost projections see IRENA (2012b).
60. See Global Energy Assessment (2012) for a detailed exposition of biorefining.
61. The first large-scale tidal energy plant became operational in France in 1966. For a recent account see, for example, Calamia (2011).
62. As a crude illustration, the global wave resource of over 4,000TWh/yr is still ten times UK electricity consumption, for example – and the Atlantic waves carry a disproportionate amount of wave energy to UK shores. Around 200 different wave energy devices are at various stages of development (CSIRO 2011; IEA 2011).
63. No energy source is entirely zero carbon when its full 'footprint' is considered. While there may be no direct emissions from converting renewable resources into electricity, the conversion equipment (the wind turbines, the solar panels, etc.) require energy (and emissions) to construct and operate. This 'footprint' includes energy and emissions from manufacturing, the mining of raw materials, the construction of foundations and indirectly the plant's operation – such as workers driving to and from the site of a wind farm. These emissions, while not zero, still do not touch the emissions associated with fossil fuels. Coal (with 'scrubbing' technology fitted) has a lifecycle emissions estimate of 960gCO_2/kWh and natural gas around half this. Wind (both onshore and offshore) and solar thermal are all under 15gCO_2/kWh. Solar PV is slightly higher at around 32gCO_2/kWh, and nuclear is at around 66gCO_2/kWh. See Sovacool (2008). Hydroelectricity can be more varied as the flooding of vegetation in the creation of the reservoirs can lead to generation of methane, a greenhouse gas which is 25 times as potent as CO_2.
64. Moss *et al.* (2011).
65. A combination of hydro, wind, solar, biomass and geothermal energy can play a huge role in providing a decarbonised power sector by 2030 and more thereafter, supplemented in the longer term by tidal, wave and others such as ocean current generation. The IEA estimates that renewables could provide 37% of global electricity generation by 2030, which is roughly the level expected in the EU by around 2020.
66. CCS costs will vary according to whether power is from coal or gas, the fuel costs, the complexity of planning, the length of CO_2 pipelines required to reach disposal reservoirs and

the scale and nature of reservoirs. Nuclear costs seem particularly sensitive to differences in the planning process, labour costs and other 'socially contingent' factors; renewables costs depend on the local resource and (for larger contributions) the absorptive capacity of the power system.

67. https://www.gov.uk/government/news/new-energy-infrastructure-investment-to-fuel-recovery. The draft strike prices were published as:

Strike prices with Over 1 Gigawatt of Potential Deployment

Renewable Technology	Draft Strike prices (£/MWh) (2012 prices)					Potential 2020 Deployment Sensitivities (subject to VfM and cost reduction) (GW)[1]
	2014/15	**2015/16**	**2016/17**	**2017/18**	**2018/19**	
Biomass Conversion[2]	105	105	105	105	105	**1.2 – 4**
Hydro[3]	95	95	95	95	95	***c.* 1.7**
Offshore Wind	155	155	150	140	135	**8 – 16**
Onshore Wind	100	100	100	95	95	**9 – 12**
Large Solar Photo-Voltaic	125	125	120	115	110	**2.4 – 3.2**

After consultation, the final decision reduced the price for both onshore wind and PV by £5/kWh, and added strike prices for a number of other smaller technology categories (DECC 2013), whilst confirming the 15-year duration of these contracts. See also Chapter 10, note 52 on Brazilian electric power auctions, which on fixed 20-year contracts appear to have achieved significantly lower prices for wind energy in that context.

68. Onshore wind energy (and in some locations biomass and geothermal) appear to be the only low-carbon sources yet in a position to 'compete' largely on Second Domain economics, and even for these the competition is skewed by the fact that in competitive electricity markets, fossil fuels always set the price of power while the low-carbon sources, having very low running costs, are the 'price takers' and thus intrinsically face higher economic risks just from the structure of electricity markets.
69. IEA (2010a).
70. Department of Trade and Industry (2008).
71. Weinert *et al.* (2007). For a popular account see Ramzy (2009) 'On the streets of China, electric bikes are swarming'. One of China's major cities, Guangzhou, has progressively banned traditional motorbikes with infrastucture that does allow pedal cycles through; the implications for electric bikes remain unclear. Online at: http://www.itdp.org/documents/Guangzhou%20Case%20Studies%20-%20Motorcycles%2015-Sep-08.pdf
72. Park and Ha (2006); UK CCC (2009).
73. UK CCC (2009).
74. For example, the VW L1 uses carbon fibre to reduce its weight and seats two people in tandem to reduce the frontal area. Now moving to production, it can achieve over 200 miles per gallon – about 20 times the performance of a Hummer.
75. A standard of 95gCO_2 per km, only slightly tighter than that of the EU's current target of 120g/km, could cut 110Mt of CO_2 by 2020. See Chapter 5 for more discussion.
76. Goldemberg *et al.* (2004) demonstrated that economies of scale and technological advances led to increased competitiveness of Brazilian ethanol against conventional fossil fuels, while Van den Wall Bake *et al.* (2009) later identified increasing sugar cane yields as the main driver of the decrease in ethanol production costs.
77. In particular, significant CO_2 emissions may occur from indirect land-use change: forest or grassland may be converted into cropland to replace land used for biofuel production (Searchinger *et al.* 2008). The largest percentage of biofuel produced in Europe is biodiesel (UNEP 2011),

a biofuel suitable for diesel engines and produced from oil-based crops. For recent economic assessments see IEA (2012 – Energy Technology Perspectives).

78. Bioethanol is produced by the fermentation of sugars found in a variety of feedstock. All plants contain sugars and could in principle be used for bioethanol production. First- and second-generation ethanol are distinguished according to the processes required to extract the sugars out of the crops. In the case of first-generation ethanol, sugar crops (e.g. sugar cane) are processed to remove the sugar (e.g. through crushing, soaking or chemical treatment) before fermentation can start, while grain crops (e.g. corn) are milled and then converted to sugars using a high-temperature enzyme process (IEA 2008)
79. Second-generation ethanol is produced from cellulosic material (e.g. wood) which is broken down into sugars through acid or enzyme hydrolysis (IEA 2004).
80. The blend wall refers to the upper limit to the amount of ethanol that can be blended with petrol. In the US, a 10% blend wall was set by the Clean Air Act. The technical rationale behind this upper limit is twofold. First, car manufacturers do not certify that their vehicles can be operated reliably on higher blends (with the exception of flexible fuel vehicles). Second, it is unclear whether the existing distribution infrastructure could tolerate higher ethanol blends. Butanol, however, can be used in gasoline vehicles in blends up to 85% (IEA 2011). This fuel is similar to gasoline and can be distributed using existing infrastructure.
81. Algae can be used to produce a wide variety of fuels, including ethanol, butanol, diesel and hydrocarbons. So far, algae have been mainly used for biodiesel production (Darzins *et al.* 2010).
82. Algal fuels could offer a suitable alternative for both these applications (heavy freight, and air travel) (IEA 2011).
83. Edwards, Kuznetsov *et al.* (2006); see also Jamasb and Pollitt (2011).
84. For full data see Global Energy Assessment (2012), Chapter 9; combining numerous sources, their key Chart 9.32 (GEA 2012: 609) suggests that increasing vehicle efficiency by 50% would add around $4,000 or more to the upfront cost of a car, based on current and prototype technologies. With typical driving at US fuel prices, this would take 12 years to pay back; at Japanese fuel prices and average driving, around six years.
85. Bossel (2006); see also *The Economist* (2008). The removal of funding for research into hydrogen in the US 2010 budget underlines the doubts and diminishes the prospects for breakthroughs. Transporting large amounts of hydrogen is not easy, neither is creating an infrastructure to produce, pump and distribute it.
86. Extensive data on energy resources are in the Global Energy Assessment (2012), Chapter 7. Another recent assessment is in Mercure and Salas (2012). Their estimates are tens of thousands of EJ for overall oil and gas *conventional + unconventional resources*, and more for coal. They caution that uranium resources may be significantly more limited, potentially constraining nuclear power this century. For geothermal they report a physical energy recharge rate of about 315EJ/yr and a technical potential of around one tenth of that (Pollack *et al.* 1993).
87. Apart from solar (>3,000EJ/y), other renewables are relatively constrained compared to the current total energy consumption of 500EJ/yr (estimated resources are a few hundred EJ/yr for wind and biomass), though they compare better against electricity consumption. Mercure and Salas (2012) also underline the limitations on hydro, wave and tidal resources at 66, 19 and 4 EJ/yr. A mostly renewable long-term scenario in which wind, solar and biomass produce 500–1,000EJ/y could take up to about 12% of global land area, with different renewables competing with each other for land if ocean energy (e.g. offshore wind or algae) is not used. An extensive data table on renewable resources is also given in the Global Energy Assessment (2012: 47).
88. Hoyle (1977).
89. For an exemplary account of the comparative scales of energy demand and supply sources, see David McKay's superb book, *Sustainable Energy – Without the Hot Air* (2008), available free online from http://www.withouthotair.com, which does indeed spell out some of the practical limits on renewable energy and their dependence upon social acceptance.
90. The Global Energy Assessment cites the US National Energy Technology Lab as estimating that *producing* unconventional oils emits 9.5–14 additional kg CO_2/GJ (NETL 2009), compared to 72.6 combustion and 91 overall for conventional oils. Mejean and Hope (2010).
91. Morgan (2012); see also previous note.
92. World Energy Outlook (2012).

93. In fact, because of their unique constraints – e.g. fossil fuel dependency, higher generation cost, smaller capacity and the need for advanced load management – small islands and remote areas often become a test-bed for advanced renewable and grid management (IEA-REDT 2012). In Hawaii, one of the most remote chains of islands in the Pacific, for example, the local government is currently working with US federal and regional partners to demonstrate state-of-the art integration of renewable and energy efficiency technologies, aimed at achieving 70% clean energy use by 2030 (Hawaii Clean Energy Initiative (HCEI) website: http://www.hawaiicleanenergyinitiative.org/).
94. For a discussion see Singh (2008).
95. For an explanation of the concept of Smart Grids see Ipakchi and Albuyeh (2009).
96. GEA (2012), Chapter 12.

> Co-processing biomass with coal in these systems requires half or less biomass to provide low carbon transport fuels as required for advanced fuels made only from biomass, such as cellulosic ethanol. Co-production also presents a promising approach for gaining early market experience with CCS, but CO_2 capture is easier in co-production than for stand-alone power plants … the economics of such systems depend on the GHG emissions price and the oil price … (GEA 2012: 904–5)

97. 'The Sahara Forest Project – Qatar Pilot Plant', opened and presented at the UNFCCC COP18 conference, Doha; see: http//www.saharaforestproject.com
98. In 2007, for example, the Global Energy Technology Strategy Program (a collaboration between several global research institutes) released its report as the culmination of nine years of research into the technological transition to a low-carbon world. Concluding that there is no magic bullet and with the inherent uncertainty of the future, they identified six key energy technology systems on which we should place our focus – CCS, biotechnology and biomass, hydrogen systems, nuclear energy, wind and solar, and end-use energy technologies (Edmonds, Wise *et al.* 2007).

References

AEA (2008) *Review and Update of UK Abatement Cost Curves for the Industrial, Domestic and Non-Domestic Sectors*, Final Report to the Committee on Climate Change, UK.

AGECC (2010) *Energy for a Sustainable Future – Report and Recommendations*. New York: UN Secretary-General's Advisory Group on Energy and Climate Change.

Allwood, J. and Cullen, J. (2009) *Steel, Aluminium and Carbon: Alternative Strategies for Meeting the 2050 Carbon Emission Targets*, R'09 Conference, Davos.

Allwood, J. *et al.* (2011) *Going on a Metal Diet – Using Less Liquid Metal to Deliver the Same Services* and *Prolonging Our Metal Life*. University of Cambridge, online at: http://www.wellmet2050.com

Aluminum Association (2003) *Aluminum Industry Technology Roadmap*. Washington, DC.

Bazilian, M. et al. (2012) 'Energy access scenarios to 2030 for the power sector in sub-Saharan Africa', *Utilities Policy*, 20: 1–16.

Bossel, U. (2006) 'Does a hydrogen economy make sense?', *Proceedings of the IEEE*, 94 (10): 1826–37.

Brinckerhoff, P. (2011) *Electricity Generation Cost Model – 2011 Update*, Report prepared for the Department of Energy and Climate Change, UK.

Calamia, J. (2011) 'Tide turns for turbines', *IEEE Spectrum*, 28 February; online at: http://spectrum.ieee.org/energy/renewables/tide-turns-for-turbines

CEPI (2011) *Unfold the Future – 2050 Roadmap to a Low-Carbon Bioeconomy*. Confederation of European Paper Industry.

Ceron, J. P. and Dubois, G. (2005) 'More mobility means more impact on climate change: prospects for household leisure mobility in France', *Belgeo*, 1–2: 103–20.

Cullen, J. and Allwood, J. (2010) 'Theoretical efficiency limits for energy conversion devices', *Energy*, 35 (5): 2059–69.

Darzins, A., Pienkos, P. and Edye, L. (2010) *Current Status and Potential for Algal Biofuels Production*, Report to IEA Bioenergy Task 39, pp. 43–4.

de Keulenaer, H. (2009) 'Energy efficient motor driven system', *Energy and Environment*, 15 (5): 873–905.

DECC (2012a) *Bioenergy Capital Grant Scheme and Renewable Heat Incentive Scheme – 1st Phase*, Department of Energy and Climate Change, UK.

DECC (2012b) 'Advanced Conversion Technologies – What Is Pyrolysis', Department of Energy and Climate Change, UK; online at: http://chp.decc.gov.uk/cms/advanced-conversion-technologies/

DECC (2013), 'UK Energy Market Reform – Delivery Plan', Department of Energy and Climate Change, UK; online at https://www.gov.uk/government/publications/electricity-market-reform-delivery-plan

DTI (2008) *Energy Consumption in the UK*, Department of Trade and Industry, UK.

Economist, The (2008) 'The car of the perpetual future', September, p. 81.

Edmonds, J. A., Calvin, K. V., Clarke, L. E., Kyle, G. P. and Wise, M. A. (2012) 'Energy and technology lessons since Rio', *Energy Economics*, 34, Suppl. 1: S7–S14.

Edmonds, J. A., Wise, M. A., Dooley, J. J., Kim, S. H., Smith, S. J., Runci, P. J. *et al.* (2007) *Global Energy Technology Strategy – Addressing Climate Change: Phase 2 Findings from an International Public-Private Sponsored Research Program*. College Park, MD: Joint Global Change Research Institute.

Edwards, P. P., Kuznetsov, V. L. *et al.* (2006) 'Sustainable hydrogen energy', in T. Jamasb, W. J. Nuttall and M. G. Pollitt (eds), *Future Electricity Technologies and Systems*. Cambridge: Cambridge University Press.

Ertesvåg, I. (2001) 'Society exergy analysis: a comparison of different societies', *Energy*, 26: 253–70.

EWEA (2009) *Wind in Power*. European Wind Energy Association.

Finon, D. (2013) 'The transition of the electricity system towards decarbonization: the need for change in the market regime', *Climate Policy*, 13 (Suppl. 01): 130–45.

Finon, D. and Staropoli C. (2001) 'Institutional and technological co-evolution in the French electronuclear industry', *Industry and Innovation*, 8 (2): 179–99.

Flavin, M. (1983) 'Excess volatility in the financial markets: a reassessment of the empirical evidence', *Journal of Political Economy*, 91 (6): 929–56.

GEA (2012) *Global Energy Assessment – Towards a Sustainable Future*. Cambridge and New York: Cambridge University Press and Laxenburg, Austria: IIASA.

Goldemberg, J. *et al.* (2004) 'Ethanol learning curve – the Brazilian experience', *Biomass and Bioenergy*, 26: 301–4.

Grubler, A. (2010) 'The cost of the French nuclear scale up: a case of negative learning-by-doing', *Energy Policy*, 38 (9): 5174–88.

GTM (2011) *Technologies, Market Forecast and Leading Players*, Smart Grid HAN Strategy Report 2011, GreenTech Media, 25 January.

GWEC (2008) *Global Wind Energy Outlook*. Brussels and Amsterdam: Global Wind Energy Council and Greenpeace.

Hammond, G. and Jones, C. (2008) 'Embodied energy and carbon in construction materials', *Proceedings of the ICE – Energy*, 161 (2): 87–98.

Hasanbeigi, A., Price, L. and Lin, E. (2012) *Emerging Energy-Efficiency and CO_2 Emission-Reduction Technologies for Cement and Concrete Production*, Berkeley National Laboratory, LBNL-5434E.

Hayward, J. A. and Osman, P. (2011) *The Potential of Wave Energy*, CSIRO, March.

Hayward, J. A., Graham, P. W. and Campbell, P. K. (2011) *Projections for the Future Cost of Electricity Generation Technologies*, CSIRO EP104982, February.

Heinen-Esser, U. (2011) *Germany's Statement in the Plenary Session*, IAEA Ministerial Conference on Nuclear Safety, Vienna.

Helm, D. (2011) Presentation to the Nuclear Industry Forum, UK.

Hermann, W. (2006) 'Quantifying global exergy resources', *Energy*, 31 (12): 1685–702.

Hoyle, F. (1977) *Energy or Extinction? The Case for Nuclear Energy*. London: Heinemann Educational (Open University set book).

IAEA (2005) *Thorium Fuel Cycle – Potential Benefits and Challenges*. Vienna: IAEA.

ICCT (2011) *Global Comparison of Light-Duty Vehicle Fuel Economy/GHG Emissions Standards*, Update: August. Washington, DC: International Council on Clean Transportation. Online at: http://www.theicct.org/sites/default/files/ICCT_PVStd_Aug2011_web.pdf

IEA (2006) *Energy Technology Perspectives 2006: Scenarios and Strategies to 2050*. Paris: OECD/IEA.

IEA (2008) *Energy Technology Perspectives 2008: Scenarios and Strategies to 2050*. Paris: OECD/IEA.

IEA (2009a) *World Energy Outlook*. Paris: OECD/IEA.

IEA (2009b) *Cement Technology Roadmap 2009: Carbon Emissions Reductions up to 2050*. Paris: OECD/IEA.

IEA (2010a) *World Energy Balances*. Manchester: ESDS International, University of Manchester.

IEA (2010b) *Cement Production, Energy Technology Systems Analysis Programme (ETSAP)*, Technology Brief 103. Paris: OECD/IEA.

IEA (2011) *Energy Efficient Buildings: Heating and Cooling Equipment*. Paris: OECD/IEA.

IEA (2012a) *Energy Technology Perspectives 2012: Pathways to a Clean Energy System*. Paris: IEA/OECD.

IEA (2012b) *World Energy Outlook*. Paris: OECD/IEA.

IEA (2012c) *Technology Roadmaps for Bioenergy for Heat and Power, Solar Heating and Cooling, and Hydro-Power*. Paris: OECD/IEA Renewable Energy Division.

Ipakchi, A. and Albuyeh, F. (2009) 'Grid of the future', *Power and Energy Markets* (IEEE), 7 (2): 52–62.

IPCC (1999) *Special Report: Aviation and the Global Atmosphere*. Cambridge: Cambridge University Press.

IPCC (2004) *Climate Change 2007: Mitigation of Climate Change: Working Group III Contribution to the Third Assessment Report*. Cambridge: Cambridge University Press.

IPCC (2007) *Climate Change 2007: Mitigation of Climate Change: Working Group III Contribution to the Fourth Assessment Report*. Cambridge: Cambridge University Press.

IPCC (2011) *Special Report on Renewable Energy*. Cambridge: Cambridge University Press, Chapter 2 – 'Biomass'.

IRENA (2012a) *Solar Photovoltaics*, International Renewable Energy Agency Working Paper, Renewable Energy: Cost Analysis Series.

IRENA (2012b) *Renewable Power Generation Costs – Summary for Policymakers*. Abu Dhabi: IRENA.

IRENA (2012c) *Renewable Power Generation Costs in 2012: An Overview*. Abu Dhabi: IRENA.

Jamasb, T. and Pollitt, M.G (2011) *The Future of Electricity Demand: Customers, Citizens and Loads*. Cambridge: Cambridge University Press.

Jamasb, T. W. J., Nuttall and Pollitt, M. G. (2006) *Future Electricity Technologies and Systems*. Cambridge: Cambridge University Press.

Kitson, L. and Wooders, P. (2012) *Energy Intensive Industries: Decision-Making for a Low-Carbon Future: The Case of Cement*. Winnipeg: International Institute for Sustainable Development.

Lazard (2008) *Levelized Cost of Energy Analysis – Version 2.0*. New York: Lazard Ltd.

McKay, D. (2008) *Sustainable Energy – Without the Hot Air*. Cambridge: UIT Cambridge.

Marintek (2000) *Study of Greenhouse Gas Emissions from Ships*, Final Report to IMO. Trondheim: Marintek.

Mejean, A. and Hope, C. (2010) *Supplying Synthetic Crude Oil from Canadian Oil Sands: A Comparative Study of the Costs and CO_2 Emissions of Mining and In-Situ Recovery*, Electricity Policy Research Group Working Papers, No. EPRG 1005. Cambridge: University of Cambridge.

Mercure, J.-F. and Salas, P. (2012) 'On the global economic potentials and marginal costs of non-renewable resources and the price dynamics of energy commodities', *Papers* 1209.0708, arXiv.org

Morgan, T. (2012) *The Perfect Storm: Energy, Finance and the End of Growth*. London: Tullett Prebon.
Moss, T., Guy, S., Marvin, S. and Medd, W. (2011) *Shaping Urban Infrastructures: Intermediaries and the Governance of Socio-technical Networks*. London: Earthscan.
National Academies (2009) *America's Energy Future*, US National Academies' Report.
National Energy Technology Laboratory (2009) *Consideration of crude oil source in evaluating transportation fuel GHG emissions*, Report No: DOE/NETL 2009/1360.
OECD (2009) *Iron and Steel Industry 2009 – Analysis and Data*. Paris: IEA/OECD.
Park, Y. and Ha, H.-K. (2006) 'Analysis of the impact of high-speed railroad service on air transport demand', *Transportation Research Part E: Logistics and Transportation Review*, 42 (2): 95–104.
Pollack, H. N., Hurter, S. J. and Johnson, J. R. (1993) 'Heat flow from the Earth's interior: analysis of the global data set', *Reviews of Geophysics*, 31 (3): 267–80.
Ramzy, A. (2009) 'On the streets of China electric bikes are swarming', *Time Magazine*, 14 June.
Reichardt, K., Pfluger, B., Schleich, J. and Marth, H. (2012) 'With or without CCS? Decarbonising the EU power sector', *Responses Policy Update No.3*, available from: http://www.responsesproject.eu
Roberts, S. (2008) 'Altering existing buildings in the UK', *Energy Policy*, 36: 4482–6.
Searchinger, T. *et al.* (2008) 'Use of U.S. croplands for biofuels increases greenhouse gases through emissions from land use change', *Science*, 319 (5867): 1238–40.
Singh, A. (2008) *Climate Co-Benefit Policies in India: Domestic Drivers and North–South Cooperation*. London: Climate Strategies.
Sovacool, B. (2008) Valuing the Greenhouse Gas Emissions from Nuclear Power, *Energy Policy*, 36: 2940–53.
Sreenivasamurthy, U. (2009) 'Domestic climate policy for the steel sector, India', *Climate Policy*, 9 (5): 517–28.
Steinhurst, W., Knight, P. and Schultz, M. (2012) *Hydropower Greenhouse Gas Emissions*. Cambridge, MA: MIT Press.
Transportation, T. I. C. o. C. (2007) *Passenger Vehicle Greenhouse Gas and Fuel Economy Standards: A Global Update*.
UK CCC (2009) *Meeting the UK Aviation Target – Options for Reducing Emissions to 2050*. UK Committee on Climate Change.
UK CCC (2011) *Bioenergy Review*. UK Committee on Climate Change, 7 December.
UNEP (2011) *The Green Economy Report*. UNEP.
UNFCC COP18 (2012) *The Sahara Forest Project – Qatar Pilot Plant*. Doha: UNFCCC COP18 conference, online at: http://www.saharaforestproject.com
US Census Bureau (2010) *New Residential Construction*. Washington, DC.
US Department of Energy (DoE) (2011, 2012) Online at: http://www.fueleconomy.gov
Van den Wall Bake, J. D., Junginger, M., Faaij, A., Poot, T. and Walter, A. (2009) 'Explaining the experience curve: cost reductions of Brazilian ethanol from sugarcane', *Biomass and Bioenergy*, 33 (4): 644–58.
Vaze, P. (2008) *The Economical Environmentalist: My Attempt to Live a Low-Carbon Life and What It Cost*. London: Earthscan.
Weinert, J. *et al.* (2007) 'The transition to electric bikes in China: history and key reasons for rapid growth', *Transportation*, 34 (3): 301–18.
Wilbanks, T. J. *et al.* (eds) (2007) 'Industry, settlement and society', in M. L. Parry et al. (eds), *Climate Change 2007: Impacts, Adaptation and Vulnerability. Contribution of Working Group II to the Fourth Assessment Report of the Intergovernmental Panel on Climate Change*. Cambridge and New York: Cambridge University Press, pp. 357–90.
World Nuclear Association (2012) *The World Nuclear Industry Status Report*. World Nuclear Association.
World Steel Association (2008) *Sustainability Report of the World Steel Industry*. World Steel Association.

Worrell, E. *et al.* (2010) *Energy Efficiency Improvement and Cost Saving Opportunities for the U.S. Iron and Steel Industry – An ENERGY STAR® Guide for Energy and Plant Managers*, Lawrence Berkeley Labs, LBNL-4779E.

Xenergy Inc. (1998) *United States Industrial Electric Motor Systems Market Opportunities Assessment*, Final Report prepared for Oak Ridge National Laboratory, TN, pp. E9–E11.

4 Why so wasteful?

'Our worst property is 13 times less energy efficient than the best.'
– Major UK retailer

'You should be pleased if you have the measurement systems in place to know that.'
– Response from another retail company
Both comments at a workshop hosted by the Confederation of British Industry in 2005, as cited in Carbon Trust (2005)

4.1 Introduction

Energy efficiency is boring.

Of course, that is not true for many working in the field. Engineers can delight in inventing more energy-efficient technologies. Corporate energy managers can become fascinated with showing how their organisations could cut energy bills along with emissions. Environmental analysts and activists enthuse about the potential of energy efficiency to save energy, emissions and money all at the same time – as illustrated in this chapter. Economists in turn frequently question the basic belief in a huge potential for cost-effective efficiency, and caution that improving efficiency might just lead people to use more energy.[1] For the vast majority of people and organisations that consume energy the world over, though, it is just not interesting.

All those things put together is what makes it fascinating.

For the way we use – and waste – energy offers among the most striking examples of the First Domain of human behaviour and is a foundation of the First Pillar of policy. If traditional assumptions around cost-benefit trade-offs comprise the Newtonian mechanics of economic theory, and evolutionary economics offers an equivalent of relativity to explain the long-term evolution of large systems, then energy efficiency illuminates the equivalent of quantum physics: the importance and value of properly understanding the decision-making realities of individual human beings and organisations, with all their quirks. After all, it is this which adds up to the global influence of seven billion decision-makers.

The potential for energy efficiency to contribute directly to both economic and environmental goals reflects its roots in First Domain potentials. It is of course not the only such example. The biggest potential gains in global welfare and environmental conditions would come from 'cleaner cooking', for the 2.5 billion people who still rely on traditional cooking stoves and fuels and access to electricity for the 1.3 billion who are not connected to grids. Their plight reflects extreme poverty, lagging development and inadequate institutions, which results in their being stranded far from the 'best practice frontier' (Chapter 2). The solutions, however, lie more in the realm of development policy and finance, which is not the topic of this book. Energy Access for All by 2030 (see Chapter 1, note 26 and Chapter 3, note 1) is widely shared as a crucial development goal, and the Global Energy Assessment (2012) demonstrated that achieving this need not at all compromise the achievement of global environmental goals – the dominant cause of global emissions being high consumption, not the meeting of basic needs.

Using energy more efficiently – the focus of this chapter – is, however, a key component of reconciling national development and global environment, and indeed can address multiple problems. The oft-cited 'rebound' effect does not change this, though it does influence and reflect where the main benefits may accrue.[2] Chapter 3 pointed out that the entire energy system, with all its attendant problems, is driven by a few major categories of demand – in homes and offices, in industry and in transport – and that in each of these, there are abundant technological options to do more with less. Reducing energy demand can address energy security and energy poverty while also promoting economic growth and carbon abatement with little associated costs or risk. Energy efficiency looks like a 'win-win-win'. Who, after all, could oppose being more efficient?

No one – and everyone. Consumer decisions are the ultimate driver of energy and emissions, in multiple ways. As sketched in Chapter 3, the potential for energy efficiency is huge, but at the heart of this observation lies a paradox about our own behaviour and deep-rooted assumptions about our economic systems. Indeed, energy efficiency is but the broadest range of an even larger set of options that seem to offer cost-effective opportunities to reduce energy, emissions and costs. Yet, while engineers point to long lists of potentially better technologies, economists tend to argue that if they were really 'better', people would buy them and use them. Policy-makers struggle to understand and close the gap.

The classical argument has been that if these options are so good, then they will succeed regardless. The engineers singing the multiple praises of efficient technologies have partly, from a policy perspective, been their own worst enemies – when others take this to imply efficiency will come forth to save governments or ourselves from having to do anything difficult.

In practice, we observe a big gap between what is available – let alone possible – and what people and organisations actually buy, use and do. It turns out that energy efficiency is anything but 'easy'. This chapter outlines the overall progress and what we've learned about the nature of the problem. Chapter 5 then looks more closely at the specific policy experience and options.

4.2 Evidence: trends and potentials

As noted in Chapter 1, one of the most striking responses to the oil shocks of the 1970s was a gush of improved energy efficiency as pretty much a worldwide phenomenon. Some of the broad policy responses that contributed to this are sketched in the next chapter. Their motivation was simple: saving energy when its price has shot up. While this is hardly surprising, more interesting is the trends since then and the variations between countries.

Contrary to some expectations, progress did *not* reverse after the collapse in energy prices, except for the sole major (and temporary) exception of Russia.[3] This in itself is important evidence. If technology were static and energy just a 'factor of production', we would expect to see less energy used when energy prices rise, reversing when energy prices fall. Surprisingly, this didn't happen: the trend of improvement didn't reverse. The world in 2006 – before most of the recent price rises – used about 25 per cent less energy per unit of GDP than it did twenty years earlier.

A lot of global energy modelling, based on traditional economic approaches, coped with this by introducing an 'autonomous energy efficiency improvement' – a ubiquitous 'AEEI'.

This was typically set at around 1 per cent/yr and attributed to a kind of technology manna from heaven – the 'Solow residual' (Chapter 11) of energy consumption. But this was also a very poor approximation of the evidence. Closer analysis pointed clearly to an asymmetric response: gains accelerated in response to high prices but were not reversed, and they instead became embodied in the system whatever then happened to prices. In terms of energy efficiency, what goes up rarely comes down. In economic language, the 'price elasticity' (Chapter 6) is asymmetric. Much of this reflects the core facts of the first and third domain processes: technology development and diffusion is strongly influenced by prices, and then becomes embedded in long-lived capital, infrastructure and systems.[4] Some important implications of this are brought together in our concluding chapter.

The oil price shock did prove to be something of an equaliser: the pattern of energy intensity changes since then has been very similar to that of carbon intensity (Chapter 1, Figure 1.5), namely in showing rapid improvement and some convergence. In the 1970s, the US consumed about twice as much energy per unit GDP as western Europe. While all regions accelerated energy efficiency in response to the energy price shocks, change in the US was faster: its energy intensity fell by 20 per cent just in the six years 1980–6, outpacing Japan and more than twice as fast as the EU. However, after this it still consumed a third more energy per unit GDP – a gap that has remained ever since, as both the US and western Europe resumed a more leisurely progress of less than 1 per cent/yr.

Globally, improved energy efficiency since 1990 has met more than half of the new demands for energy-related goods and services; in the US, it has accounted for an estimated 75 per cent since 1970.[5] The savings have spanned all sectors, including manufacturing. Structural change, i.e. a shift to less energy-intensive and higher-value activities, partially explains this fall.[6] From an efficiency standpoint, this is a mixed blessing: as economies moved up the 'value chain', heavy manufacturing industries moved abroad, and imports of energy-intensive goods increased. In the UK, for example, domestic CO_2 emissions have declined by 12 per cent since 1990, but total emissions including embedded emissions of imported goods rose by at least as much.[7]

The story around emerging and the former centrally planned economies is equally striking, though harder to be precise because of data problems (see note 3). There is little doubt that Russia and China were both staggeringly inefficient in their use of energy. As Russia tried to maintain central planning and shield its citizens from energy price rises while its economy stagnated and then collapsed, its energy intensity actually increased. China, in contrast, made dramatic progress, more than halving its energy intensity over the past quarter century. Perhaps the most striking evidence comes from the ten countries of central and eastern Europe which, after leaving the Soviet bloc, had to start converging with the economic policies of western Europe prior to joining the EU. After a decade of having stagnant national energy intensity to the early 1990s, when the USSR collapsed, their energy consumption then declined while their economies recovered, and national energy intensity halved over the subsequent 15 years.

These gains did not come for free: indeed, they were extremely painful. The fury of US consumers at the sharp rise in their oil prices in the 1970s did much to destroy the Presidency of Jimmy Carter. Eastern European citizens – their pain deferred by Soviet energy subsidies – later suffered terribly: accustomed to paying almost nothing for energy (and consequently consuming a lot mostly in leaky, inefficient buildings) people with incomes at a tiny fraction of their western European counterparts had to start paying the same amount for energy, and it really hurt.

Nor, indeed, is the story a simple one of consumers responding to higher prices by using less energy. In fact, as detailed in the next chapter, the energy price shocks of the 1970s

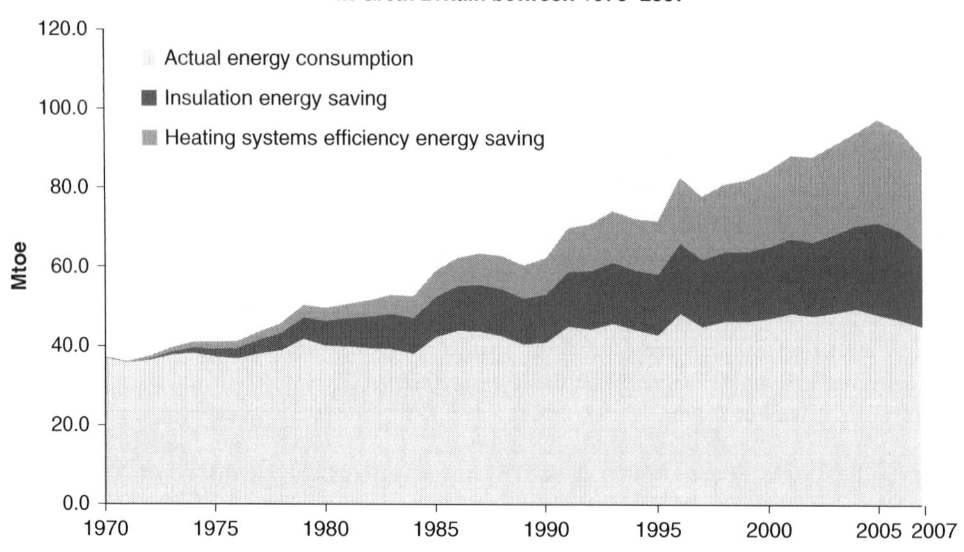

Figure 4.1 Efficiency gains from improved insulation and heating in UK housing, 1970–2007

Note: The savings are based on how much additional energy would be required if the insulation and heating efficiency measures had not taken place. The graph indicates that UK energy demand for heating homes has been roughly constant despite more, larger (and warmer) homes, and the savings have been roughly equal between insulation and more efficient heating systems.

Source: DECC (2009).

mobilised national efforts to conserve energy, some of which endured and others of which relaxed as energy prices declined again.

Such complexity is also seen at a more micro-level, too. Britain today requires almost 50 per cent less energy to heat the average home than it did in 1970, primarily due to more efficient heating systems and insulation (see Figure 4.1). As indicated in the next chapter, this was substantially driven by policy. Yet this remains far below the potential, with only around half of homes having such basic measures.[8] Germany, the Netherlands and the Scandinavian countries have generally led energy efficiency, including the adoption of the more energy-efficient condensing boilers; the UK and others followed sometimes a decade or more later, partly in response to EU legislation.[9]

Properly understood, the data are in fact one of several ways in which energy efficiency provides evidence for the First Domain of human decision-making. The postwar glut of cheap energy induced a thoughtless attitude. We – as individuals and societies – slipped into 'satisficing' our use of energy, i.e. pretty much forgetting about it, and using as much as seemed convenient without thinking about the costs and benefits. The oil shocks gave everyone a nasty jolt. Because change requires effort, societies resisted it – some longer than others. And, as energy prices declined again from the mid-1980s, many of us – individuals, businesses and countries – again lost interest. We inherited improved infrastructure and technology but slipped back into the old habits, as if energy was – emotionally in terms of our time and attention – 'for free'. And those that maintained lower prices, for example by not introducing gasoline taxes, did so more, and remain substantially less energy efficient – by far more than can be explained by classical views on price response, as explained in Chapter 6.

4.3 How much more?

It is striking how inefficient we remain, and this forms both an opportunity for and further evidence of First Domain behaviour. As outlined in Chapter 3, technologies that could provide more using less energy remain abundant. Estimates of the global potential of these technologies vary, but they all come up with big numbers at attractive costs.

Chapter 2 showed the results of one of the most extensive assessments of global potentials for cutting CO_2 emissions relative to the costs of doing so – the 'McKinsey Cost Curve' (see Chapter 2, Figure 2.4). This assessed 150 technology options across 21 countries and regions. On the left-hand side of the figure, as noted, is a portion in which the costs are 'negative': options that appear to cut emissions at a net economic gain, if assessed at the long-run discount rates in the McKinsey analysis. This portion is expanded with some basic sectoral detail in Figure 4.2. It suggests that by 2030 these options could save huge amounts of energy, along with 10 billion tonnes of CO_2, relative to the projected baseline, which already assumes a considerable 'business as usual' gain in energy efficiency.

This represents roughly a third of current global fossil fuel consumption and CO_2 emissions. Set against trend projections, implementing these options alone could come close to turning global increase into declining consumption and emissions over the next two decades. These are just options which already appear, in this analysis, as cost-effective. It is not surprising that McKinsey's analysis – no different in essence from what many energy researchers had been saying for decades, but coming from one of the world's most powerful global consulting companies – helped to provoke huge global interest in what looks like the biggest 'free lunch' on the planet.

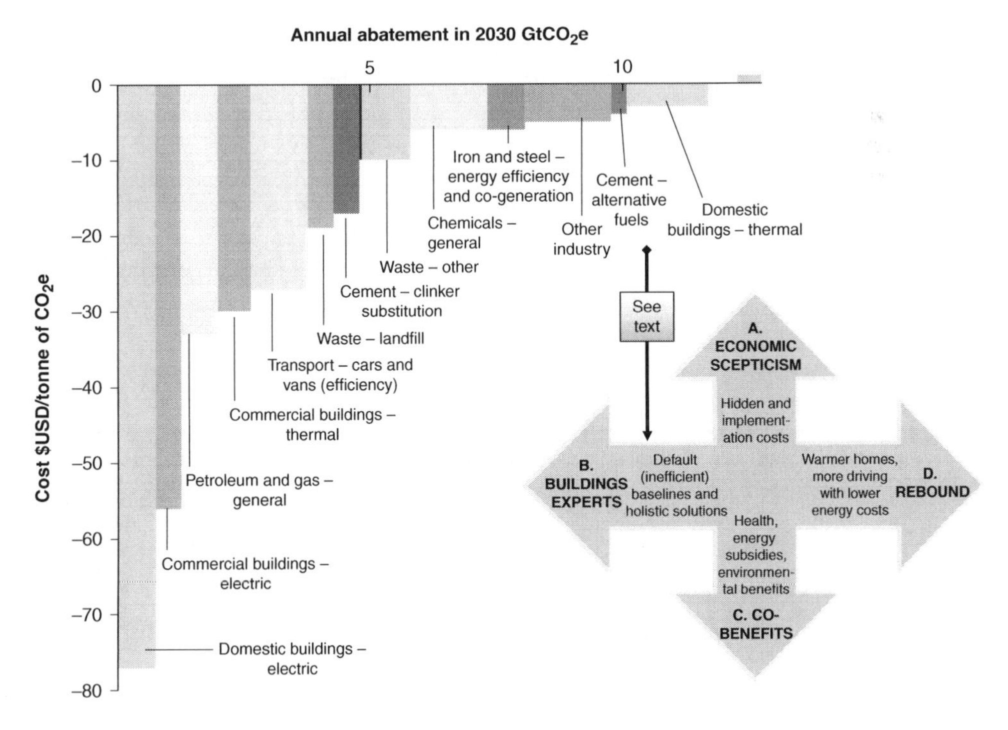

Figure 4.2 Global potential for apparently 'negative cost' emission reductions by 2030

Source: Authors, with data from McKinsey & Co. (2009).

Not surprisingly the 'McKinsey Curve' has been criticised, and in fact from several quite different – and somewhat opposite – directions. First, it provoked a backlash from many economists who refused to believe that modern market economies could be so inefficient – as indicated in Chapter 2 (Figure 2.3). The data moreover undermined a popular view that inefficiencies resided mostly in developing countries.[10] Insofar as they engaged with the data, economists tended to explain it away by saying that McKinsey must have overlooked various 'hidden costs' or other factors that would raise the actual costs ('A' in Figure 4.2). There certainly is some validity in such concerns, but efforts to reconcile the evidence with Second Domain assumptions seemed stretched.[11] The underlying attitude seems to resist the very idea of a 'free lunch' arising from efficiency gaps in the economy. Part of the difficulty seems to lie in equating 'negative costs' with 'no pain', whereas tapping negative costs can imply substantial policy or institutional change, including capacity building and reform of market structures.

From the opposite perspective, a number of leading specialists in energy efficiency, particularly those working on the thermal efficiency of buildings (to reduce heating and cooling demand), argued that McKinsey estimates of technical potential are too conservative. McKinsey estimated that building-related measures over the next twenty years could save energy and money (average cost savings exceeded $30 per ton of CO_2) while cutting global CO_2 emissions by 3.5GtCO_2 annually. In the context of the IPCC Fourth Assessment, this academic community also conducted a major assessment and estimated an even bigger 'negative cost' potential from buildings, totalling about 5GtCO_2 by 2030 ('B' in Figure 2.4). The difference was particularly stark regarding thermal performance in housing. McKinsey assumed in particular that Asia would ensure all new buildings were made energy-efficient anyway (so that higher cost measures were required to go further), and included high costs in rich countries associated with retrofitting their existing stock to high standards. Specialists in eastern Europe also rejected McKinsey's assessment of the potential in that region as implausibly modest.[12] Academic experts also highlight a potential for 'whole buildings' efficiency that could not be captured in the bit-by-bit assessment of McKinsey. Numerous complications arise in assessing the overall potential in energy systems.[13]

Other lines of criticism could pull the estimates in other directions. McKinsey assessed the economics based on local energy prices (hence, for example, the assumed modest potential in Russia). However, these prices are often subsidised and/or do not include the cost of other environmental impacts (externalities) – as flagged in Chapter 1 (section 1.3) and discussed further in Chapter 6. A full economic assessment of the overall social benefits of energy efficiency should include these costs – further enhancing the economic gains ('C' in Figure 2.4). Buildings experts also tend to identify additional 'co-benefits' of improved efficiency, such as improved working and living environments.

Finally, a more generic criticism of energy efficiency is that gains may be taken back by 'rebound' effects – because energy efficiency reduces the cost of doing something, people will do more of it – reducing the energy/environmental benefits ('D' in Figure 2.4), which are 'taken back' as economic benefits (considered in concluding this Pillar, Chapter 5).

There are some merits in all these points; none invalidate the importance of understanding better what share of the apparently huge potential can be tapped, at what pace and under what conditions.

The sectoral pattern of the raw technical potential is interesting. Efficient use of electricity in buildings stands out as offering the biggest economic gains (more than US$50/t$CO_2$), associated with saving a billion tonnes of CO_2. Enhanced efficiency in commercial buildings and light duty vehicles were assessed to offer savings of 2 billion tonnes with economic gains of around US$30/t$CO_2$. Interestingly, general efficiency improvements in the petroleum and gas industries were also assessed as offering similar cost savings. Savings of US$20/t$CO_2$ were identified across a wide range of industrial processes and waste handling. Domestic building insulation offered the largest single 'negative cost' CO_2 savings but McKinsey assessed the economic benefits as marginal – a conclusion disputed strongly by buildings experts (note 12).

Thus although the analysis in Pillar I focuses particularly on buildings, the phenomenon exists in other sectors too. There is sizeable potential to increase the efficiency of light duty vehicles and industrial energy use – options not significantly quantified in the IPCC Fourth Assessment but highlighted in analysis by the International Energy Agency. Energy efficiency makes up the bulk, though not all, of the potential for 'negative cost' savings – 6.5–8GtCO_2 reductions at an average rate of return of 17 per cent on capital invested by 2030, on the McKinsey analysis.[14]

The potential in industry in particular seems puzzling and, after presenting additional evidence from UK programmes, reasons are considered later in this chapter in the context of organisational decision-making.

All this high-level assessment suggests that the financial benefits from investing in energy efficiency should be tremendous, and indeed they can be. Some improvements pay back in months in terms of saved energy costs, and major options like improved loft and cavity wall insulation can pay back in a few years. Others may take longer but still offer much higher returns than typical consumer interest rates. In commercial premises, investment in combined heat and power (CHP) or smart building management systems add to the potential, though with longer payment.[15]

National implementation agencies, as well as international assessments, are now replete with such evidence, which confirms that the potential is not confined either to buildings or to personal energy use. In the UK business sector, the Carbon Trust has estimated through targeted energy audits that there is an opportunity for annual cost and carbon savings of £3.6 billion and 29MtCO_2 respectively, if businesses and organisations were to operate at their most efficient level. The Energy Savings Trust, providing a similar service for the residential sector, estimated a potential for the average household to save £300 annually on energy costs, equivalent to 1.5 tonnes of CO_2 – which across the UK's 26 million households would match the CO_2 savings potential from business, with double the cost savings per year.[16]

There are of course uncertainties about how quickly and how far such opportunities can be taken up, due to the interaction of technological, economic, regulatory and cultural factors. Nevertheless, take-up clearly lags far behind the potential. Inferior technologies continue to prevail with wholly unnecessary levels of cost and energy use – a pattern that appears to be consistent across investments in buildings, transport and industry. The discrepancy between the most economical energy options and consumers' real choices is known as the 'energy efficiency gap'.

To anyone who thinks that people and economic systems are rational, and that modern market economies are efficient, this energy-efficiency gap demands a fuller explanation. So too for policy: unless we can identify what causes the gap, we are unlikely to close it.

4.4 Explanations: barriers to and drivers of change

Faced with the 'energy efficiency gap', the first instinct of many economists and policymakers used to be denial. One of my more thoughtful colleagues in the Cambridge University

Economics Faculty said he couldn't believe cost-effective opportunities existed on that scale because it violated the basic principle of rational economic behaviour.

Beyond denial, the next response is to seek out what impedes a 'rational' response to the apparent economic (and other) benefits of being more efficient. A massive literature now exists citing barriers to energy efficiency.[17] Probably more than a hundred have been identified or proposed, but they can be usefully grouped into four main classes, as illustrated (with key examples) on the left side of Figure 4.3 and explained below:

- financial costs – the barrier of spending money on an energy-efficient investment;
- hidden costs – costs which are real but not captured directly in financial flows;
- market failures – notably 'split incentives' when the person who has to pay for an improvement (e.g. a landlord) may not get the benefit (e.g. the tenant);
- behavioural factors.

Less studied are the specific drivers that might lead people and organisations towards energy efficiency – probably because it is widely assumed to be just the financial savings associated with using less energy: all we need to do is to remove barriers. In fact this is grossly simplistic – as any marketing, sociological or management expert could point out. Alongside the barriers, therefore, the right-hand side of Figure 4.3 shows varied potential drivers, similarly grouped. As illustrated, these seem to be generally outweighed by the barriers.

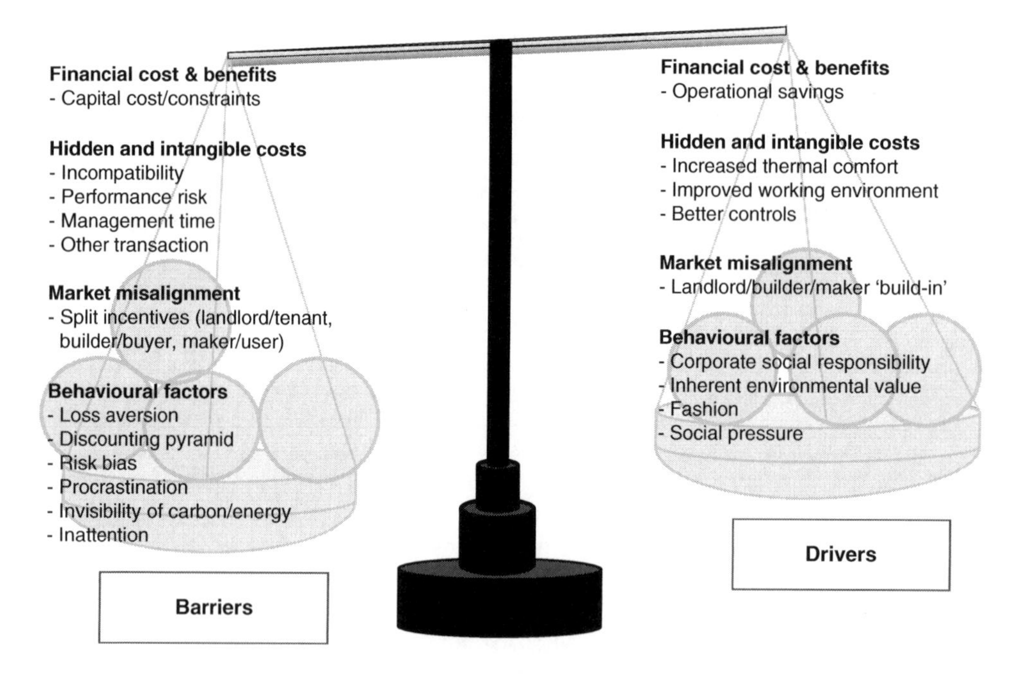

Figure 4.3 Energy efficiency barriers and drivers

Note: The figure shows the four classes of barriers that impede energy efficiency, set alongside the drivers that can offset these, with some of the most common examples from the literature. At present, there is a clear imbalance between barriers and drivers that explains the persistent observed gap between the potential for and take-up of cost-effective energy efficiency options. Things that increase motivation to address these barriers or otherwise increase efficiency – to balance the diagram – have a good chance of being economically beneficial as well.

Source: Adapted from Carbon Trust (2005) and UNEP (2009), Buildings and climate change, UNEP SBCI, Paris (Table 1). See also Ryan (2011) and note 17.

Financing, capital constraints and the discounting pyramid

The first category is the straightforward financial calculation, in which additional investment required (if any) is the main barrier and the financial value of reduced energy consumption is the main driver. Simple enough – it hardly sounds like a barrier: it is the heart of normal economic behaviour.

Unfortunately it is not that simple, particularly if we are interested in whether we are getting the 'optimum' level of energy efficiency. That would first and foremost require the prices to be right, including external costs, which as shown in Pillar II (Chapters 6–8) is frequently not the case: energy is often subsidised, and external costs are rarely fully included, diminishing the weight of energy cost-savings.

More fundamentally, most energy efficiency investments are small decisions made by individuals, whether at home or at their workplace, and they often face real or perceived constraints on expenditure that just don't balance with the way bigger decisions – including those on energy supply – are made. There is a lot of angst in the energy literature at the observation that most consumers take a very short-term view: their 'discount rate' is much higher than for a large corporate investment, for example. They don't have the same access to capital, they often face higher interest rates or they just feel deterred by upfront costs that increase a consumer's perception of the risk involved.

The same turns out to be true at the lower levels of many organisations where energy management and equipment purchase decisions are made: energy managers, for example, are frequently constrained by budgets that only allow them to spend on projects with one or two year payments. Again this may be accompanied by a sense of risk of spending capital with the uncertain returns of saving energy, compared to purchasing a product.

This is empirically shown in the slow implementation and very short payback thresholds demanded within many organisations. Measures with a one-year payback – which save the additional upfront cost within a year and then continue to generate net benefits of lower energy bills – would seem obviously attractive. Yet a survey by the Carbon Trust found that only about 50 per cent of such recommendations from energy audits were implemented in the following years, and only about a quarter of those with 2–3 year paybacks were implemented (Figure 4.4).

In essence, a broad 'pyramid' of discounting exists, in which the smaller the entity – or the lower down in an organisational hierarchy – the more future gains tend to be discounted. And most decisions relating to energy efficiency are small.[18]

A rapid technological turnover enhances a sense of risk by effectively putting a shorter shelf-life on technologies: today's best investments may not remain so for all that long. We can always wait. All this can lead to heavy discounting of the less tangible gains that flow from future energy savings.

Hidden costs

A second class of barrier is from costs that are real but hidden. For example, if new energy-efficient appliances are harder to get serviced or don't fit existing infrastructures – such as a new efficient light bulb not fitting in the lampshade – then there is an added, real cost that may be missed in financial assessments. Getting information and evaluating new decisions also takes time and attention, a real cost: working out which product may be best, and understanding the installation, application and upkeep of the new technology. There is another whole literature and field of economics relating to transaction costs.[19]

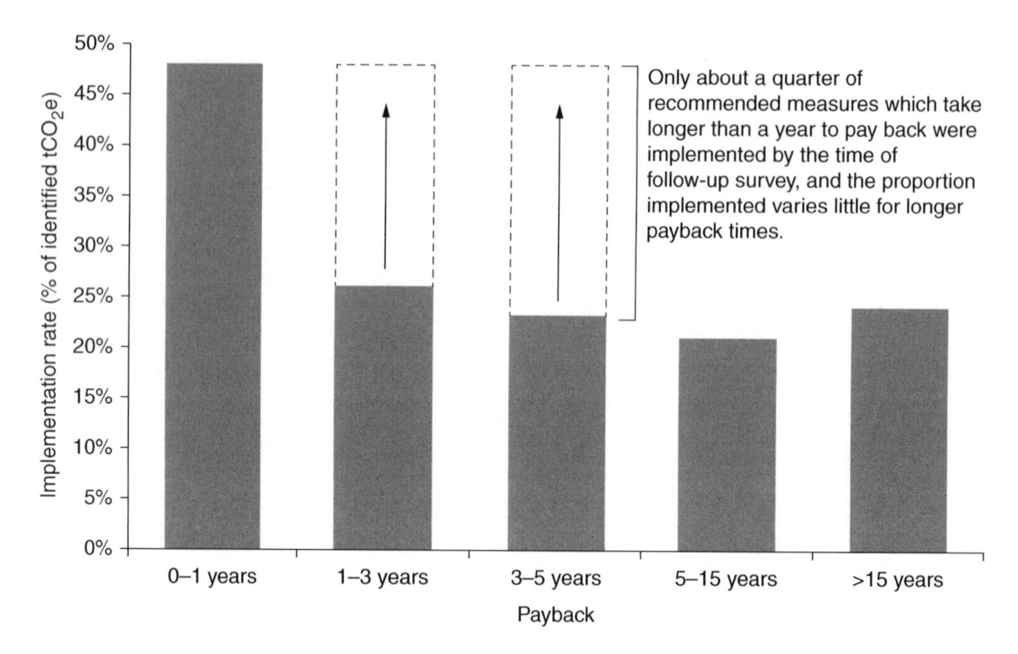

Figure 4.4 Proportion of Carbon Trust recommendations to UK business implemented: dependency on payback period

Note: The graph shows combined responses of public, services, retail and chemical sector regarding recommendations identified in 2006–7.

Source: Carbon Trust, based on Carbon Management/Energy Efficiency Advice Close-out database (personal communication).

Hidden costs can be estimated. As the UK government sought to ramp up its efforts to improve energy efficiency in the early 2000s, it commissioned a study to estimate the scale of hidden costs. For a number of domestic and commercial measures, the consultants estimated that hidden costs might account for a sixth to a third of the apparent 'efficiency gap'. The reactions to this were amusing. The advocates of energy efficiency were concerned at the implication that the cost-effective potential might be up to one-third less than they had claimed. Many government economists, however, were baffled, having persuaded themselves that hidden costs must be the main explanation of the 'efficiency gap'.[20] They aren't.

Making any new purchase involves some search costs – whether or not focused on energy efficiency – and some presumed 'hidden costs' may actually disappear. Some may be fictitious – if new technologies are unfamiliar, consumers may assume a risk, deterring change. Further, many of these costs will decrease over time as the new information is learned and becomes embedded within an organisation or household, and as infrastructures and technologies (such as lampshades and bulb fittings) converge. A large part of hidden costs thus appear as barriers to *change*: they are not enduring. In many cases, underlying barriers may masquerade as financial or hidden costs, but in reality may be more behavioural in nature.

Hidden costs can also be reduced by policy: by providing services and technical support from trusted organisations, and some through policy measures – such as stipulating minimum energy performance standards on products and banning inefficient products.

This bypasses the costs of millions of consumers having to hunt for information on which products to avoid; clear labelling can bring many of the same benefits, as detailed in Chapter 5.

On the 'drivers' side of the ledger, costs aren't the only unseen outcomes. Hidden benefits, too, can arise – and normally do in energy efficiency investment. In addition to decreasing energy bills, a well-designed, energy-efficient building often also has better thermal, acoustic and visual properties. At home, these changes can increase comfort, and at work can increase productivity – one study estimating an increase in labour productivity of up to 16 per cent which would clearly eclipse any upfront cost.[21]

Making a trade-off of investment costs against energy savings assumes that the energy component can be separated and independently costed. One key insight from experience with manufacturing industry (as indicated in Chapter 5) is that this is rarely the case: frequently, more energy-efficient equipment is more modern all round, with better design and better control systems, which often provide a number of hidden benefits that cannot be separately costed. Improved efficiency may then come as a package, as part of an overall process of modernisation – a far cry from the classical assumption that enhancing energy efficiency involves economic trade-offs.

Split incentives: tenant-landlord and other divides

The third class of barrier relates to explicit market failures, particularly relating to 'split incentives': who pays doesn't get the benefit. The biggest example is the split between landlord and tenants in rented property. Most commercial property is rented. Patterns of housing vary the world over, but even in Britain, after three decades of policy intended to create a nation of home owners, a third of domestic housing is still rented.

Owners or landlords are nearly always the ones who would have to organise and pay for major improvements in the energy efficiency of their buildings. They very rarely pay the energy bills. Tenants often are not allowed to make capital investments, but would have little incentive to do so anyway; the next tenants would get the lower bills and the landlord any increase in value. Further, if a building is inhabited by several parties, the energy costs are usually split evenly between them, meaning that any savings that may be generated by one tenant will be shared.

In *theory*, a more efficient building might attract other benefits. Some evidence has begun to emerge that 'eco-labelling' in US commercial offices has both increased occupancy and added a premium to both rental and sales value; preliminary European studies are divided.[22] Whether any effects are big enough, or sufficiently known, accepted and broad enough to overcome the 'tenant-landlord' split incentive, seems very dubious.

Though the tenant-landlord split is the biggest and most ubiquitous example, there are many more examples of split incentives. Common types arise from *embodied hardware, misaligned supply chain incentives* and *visual and non-visual properties effects*.

Motor drives are a good example of embodied hardware. Very few people actually buy motor drives for their own sake: they are embodied in washing machines and a host of other home appliances, and in a wide range of manufacturing equipment. The huge scale of potential savings has been noted in Chapter 3 (see note 22). However, the purchaser may have no choice over the energy efficiency of the drive, and the seller does not incur the energy cost of installing an inefficient one. Not surprisingly, the potential from more efficient motor drives is one of the biggest areas of potential cost-effective energy improvements identified in manufacturing industry.

The contractual structures of industrial supply chains can also easily fail to align incentives on energy, and sometimes introduce the exact opposite. One of the more amusing examples was in the supply chain for making potato chips, where full carbon accounting enabled the manufacturer to identify and make substantial savings in both CO_2 and costs.[23]

Residential retrofitting could also affect visual and non-visual properties, thereby reducing its attractiveness. In the UK, for example, implementing double glazing is hard in conservation areas or listed buildings of historical significance. It often requires additional permission and tailored engineering. External wall retrofitting can also be cumbersome or damaging, as well as expensive. These hard-to-quantify effects serve as persistent barriers against insulating existing properties.

Information and institutional failures and the role of energy service companies

These are the three most widely cited kinds of barriers to energy efficiency. Before turning to the fourth class – behavioural – consider first the role of information. An entertaining experiment, practised by a colleague in the field, was to ask his audience how much they thought it would cost to have cavity wall insulation in their homes and what the value of the energy saved would be. The audience generally didn't have a clue, and when pressed, tended to over-estimate the cost of insulation and underestimate the value of the savings. The entertainment was when this came from economists in his audience who had been arguing that the 'efficiency gap' couldn't exist because people take rational, informed decisions.[24]

Lack of information is often cited as a major barrier in its own right. In fact, as suggested by Figure 4.5, it is bigger than that: inadequate information really permeates and can amplify any or all of the barriers. It impedes and can easily bias 'cost-benefit' decisions, as the anecdote illustrates. The cost of obtaining information is a classical 'hidden cost'. And split incentives can easily be amplified by parties on either side not having a clue what the real opportunities for the other might be.

This is particularly the case when two sides in a transaction – 'principals' and 'agents' – have different information. A whole field of economic theory points to the problems that can arise, including 'moral hazard' and 'adverse selection', resulting in the triumph of worse products – a set of insights which earned three economists the field's equivalent of the Nobel Prize in 2001.[25] Though it was not widely appreciated at the time, some of the empirical data on appliances pointed clearly to this phenomenon. This information helped to underpin the adoption of mandatory energy efficiency labelling in the EU, with the dramatic results shown in Chapter 5.[26]

Of course, information is central to any market and it doesn't come for free. It is most often paid for by the manufacturers who want to sell you a product and, to do that, want to tell you how good it is. It is then part of marketing. Sometimes neutral parties arise as well, both governmental and private bodies established to protect consumers.

In energy, the natural analogy is to expect that energy service companies (ESCOs) should arise, to deliver and market energy efficiency. There has indeed been periodic hype about the potential for ESCOs. They can provide information to clients who might lack the know-how to be more efficient – in principle, a valuable assessment of costs and opportunities. ESCOs can offer a promise of energy efficiency improvements, tailored to the client, with a guarantee of savings. They may also design and develop energy efficiency projects, and in some cases take on the installation and maintenance of the equipment, hence eliminating a burden of transaction costs.

ESCOs initially became prominent in the US during the first oil crisis in the 1970s. The idea spread to many European countries, often receiving significant government support. Given the apparent scale of the potential, one might have expected them to prosper. However, even in the German case, with a strong national drive for energy efficiency and high energy costs, the potential for efficiency gains has been constrained where ESCOs are expected to be the dominant mechanisms. This has been put down to large transaction costs which limit their action to large-scale projects and a lack of consumer demand.[27]

One core problem is that initial financing, wherever it comes from, must be backed by a reasonably assured flow of payments derived from the resulting energy savings. This is known as 'energy performance contracting'. This requires one of the parties to measure, monitor and verify the energy savings over a contracted period of time. It requires a reasonable projection of the savings, and for all to trust the result and to agree up front how the savings will be shared.

The universal lesson is that it 'ain't so easy'. The process needs to start with an energy audit, which an ESCO could conduct to establish the potential. Armed with a good energy audit, however, a company might decide to improve things itself or go to another ESCO to do the implementation. No amount of information will overcome some of the barriers, such as heavy discounting for small-scale savings and the landlord-tenant split. The harsh reality is that it is much easier to sell a piece of kit than a future stream of energy cost reductions (generally estimated by the ESCO), with all the attendant scepticism and uncertainties. Trust is paramount in such a transaction, but is generally lacking.

The reality is that except at times of sharply rising energy costs, consumer demand is not strong enough to overcome these substantial transactional barriers. To try and force the pace, mandatory audits across industries, buildings or transport have been introduced in some countries, but this is relatively recent and hence little is known about their impact.[28] Government-funded organisations offering such a service have made better progress, as discussed in the next chapter.

One could choose to stop there. Acknowledging a role for governments to provide some collective information is one thing, but if the market for energy efficiency is not self-sustaining, does that not prove: that energy efficiency is not as cheap as it looks? That the real explanation for the 'energy efficiency gap' lies in the hidden costs associated with actually trying to get it to happen, and that government has no business trying to subsidise this or force it, or to presume that governments can fix what markets don't? That acknowledging 'market failures' gives no reason to assume that 'government failures' of intervening would be any less?

This question is one reason why it is important to look at the evidence of experience, which is the purpose of Chapter 5. The anomalies already demonstrated and the explanations given create a clear potential for government action targeted at overcoming the barriers to reduce both energy/emissions and costs. The experience accumulated, surveyed in the next chapter, lends strong evidence that policies have cut both energy consumption and costs. Yet as explained there, problems remain, and in some countries energy efficiency remains at risk of being treated as a political football of state *vs* the individual – or, to the extent that programmes require government funding, just becoming a victim of public expenditure cutbacks.

Which is why it is important to dig deeper, in the fourth and final class of 'barrier-driver'.

4.5 Behavioural realities: individuals

The fourth class of barrier-driver offers largely new terrain and new opportunities for both energy efficiency and indeed wider policy. The three 'classical' types of barrier so far considered still assume that individuals and organisations are basically economically rational: that they are motivated by money and that what impedes change is a rational approach to future costs and benefits under uncertainty, hidden costs or specific market failures like split incentives. It is now clear that is a gross simplification – and one that really matters.

Strange as it sounds, our understanding of 'real behaviour' has improved hugely in recent decades, and the messages are stark. Weighing purchase cost against longer-term savings, the presumed mode of *homo economicus*, is rarely the most important factor in consumer choice. Behavioural 'anomalies', far less straightforward and simple, are often much more important. These may include influences from 'social norms', 'rules of thumb', procrastination and a wide range of other considerations from fashion to underlying values.

Box 4.1 A lofty tale of normal housing

To help bring the abstract ideas about energy efficiency costs, benefits, barriers and drivers down to earth, consider a small house in London: my house, comprising the upper two floors of a former collective housing block, sold under Mrs Thatcher's property reforms. It had old wood-framed windows, filament lights and central heating powered by a gas boiler past its prime that fed water to a hot tank.

I checked there was some insulation in the loft and on the tank. I soon fitted compact fluorescent lights – where they would fit (a 'hidden cost' eliminated by new designs a few years later) – and eventually (inertia – a 'behavioural' feature) got around to buying a new 'A-rated' fridge and separate freezer. By that time, they were clearly labelled (reducing the 'hidden cost' of trying to find out what was most efficient – and also tweaking my 'hidden benefit' driver, the label displaying to everyone I had gone for the best). Anyway, by that time, so many people were buying A efficiency fridges that they were almost as cheap as others ('economies of scale').

For a long time, I didn't get round to the rest (inertia again – not a reasoned weighing of costs and benefits). However, one day the local council announced that it was going to do a major windows replacement. Workmen came in – a disruption ('hidden cost') – and soon I had a new set of gleaming, double-glazed and high-efficiency windows, together with a hefty bill to pay back over several years. Left to myself, I would never have found the capital (the upfront cost barrier) or organised it (transaction cost), and the savings probably haven't yet paid back the cost. But the place is less prone to draughts, more secure against break-ins and has less noise from the street (all 'hidden benefits'). By then, however, I had rented out the house and the tenants get many of these benefits ('split incentives').

Two years later I received a letter: on 16 September, unless I declined, workers would arrive to install much deeper insulation in the loft. It turned out I only had two inches – at least six was recommended. The council would pay, not me; I just had to clear the loft. As an exercise in behavioural nudging, combined with a bit of subsidy, it was masterful. To get out of it, I'd have to refuse. (How could I possibly do that?) Rising energy prices also helped the tenants consent to the disruption. It took me two days to clear the loft properly and put things back afterwards (a sizeable 'hidden cost'), but I found some valuable things – including an unused stereo system (!) – and at long last had everything there ordered and sorted, easy to get at again, and with some nice old stuff also given away to charity ('hidden benefits', widely shared), followed by noticeably lower energy bills – the classical benefit, but almost entirely irrelevant to the actual decision.

Some years later, I have no hesitation in pronouncing the cost-benefit balance to be positive, but if the local council hadn't butted in, I would probably never have got around to it. I did eventually replace the ancient boiler with a smaller, neater combi condensing one (aided by a small subsidy), which also meant I could remove the old tank – some chunky metal for recycling and a whole new cupboard space for me (more hidden benefits, this time from technical progress).

Energy consumption (and costs and emissions) are now way down. Due to my own rational, cost-benefit approach to decision-making? You must be joking – and I was Chief Economist at the Carbon Trust. A final, bemusing thought is that some of the simplistic economic models don't only assume free market economies to be 'optimally efficient', they also assume that changes are reversible: that an efficiency measure that may be cost-effective at high energy prices will reverse if the energy price falls. More sophisticated economic models allow for capital lifetimes, typically a few decades. I fully expect my new windows to be there for decades, and they won't be replaced by worse ones. As for the loft, I expect the insulation to be there for as long as the house (probably a century or more). Even if energy were free, you wouldn't find me removing it.

Some of this had been known for a long time – or just seemed common sense. Economics itself had already in the 1950s developed some elements of 'satisficing theory' – signifying the observation that people often seem 'satisfied' with decisions that are clearly not as good as they could be. Since the idea of 'economically rational' choice is at the heart of most economic theory, the conundrum was solved by coining the term of 'bounded rationality', postulating that consumers' decisions were rational 'within the bounds' of their attention and information.[29]

However, credit for establishing 'behavioural economics' as a major discipline is usually traced back to work in the 1970s, notably by Daniel Kahneman and Vernon Smith, who in 2002 received the Economics Nobel Prize for their groundbreaking work. Smith received his for developing laboratory methods for testing assumptions about economic behaviour, Kahneman for applying the basic idea to systematic study of human judgement and decision-making under uncertainty. The field has since exploded with fascinating insights, now amplified by an emerging field of neuro-behavioural studies that trace human behaviour to how the mind works.[30]

There are probably far more relevant insights than this book can cover, but a few seem particularly important for the field of energy efficiency.[31] The intersection of just two features could on their own explain a lot of the energy efficiency gap:

- *We weigh losses more than equivalent gains*: a loss evokes a level of emotional reaction that is not matched by the pleasure of a commensurate gain; we are naturally *loss-averse* instead of weighing up potential losses and gains together.

- *We are inconsistent in our treatment of time*: individuals attach a high weight to things 'now', and discount quite rapidly the value of things in the future in ways that are easily shown to be entirely inconsistent – known as 'hyperbolic discounting'.[32]

Thus if the 'loss' is spending capital and the 'gain' is saving money over subsequent years, the balance is already heavily slanted. This is critical in terms of energy efficiency investments, meaning that the present loss associated with laying out capital will disproportionately eclipse the long-term savings in actual decision-making. Moreover, this can be greatly amplified by any risks and uncertainties we attach to whether we will really get all the money back: a third crucial finding of behavioural economics is that humans have an intrinsically biased approach to risk.

Procrastination is another common – and related – human characteristic: putting off a decision for another day.

We all have some practical experience, if we but reflect on it. For my part, the small house – 'maisonette' – I bought in London 15 years ago has, in its own little way, become a personal symbol of much that we have learned about energy and behaviour (Box 4.1). The house wasted energy; it still does now, but less so. Some improvements I did myself quickly; others waited till I was galvanised by higher energy prices. Some were forced upon me by the local council that operates the estate; for others I was subsidised and/or nudged by smart policy. After I rented the house out, in turn I found myself, to some degree, trying to incentivise and nudge the tenants. Mostly it has been very beneficial. Yet we could all still do better.

Some of these behavioural factors are driven by uncertainty. In the field of energy and environment this is amplified by the unstable nature of fluctuating energy prices and evolving energy and climate legislation. Combined with the other behavioural characteristics noted, uncertainty tends to encourage inertia and discourage change.

There are many reasons for this. The current status quo is what we know and is thus the least risky path to choose. There is an 'endowment effect' – we tend to put disproportionate value on things we already have as opposed to things we could have. This amplifies a tendency to wait for others to take the plunge first – 'vicarious learning' by letting others test the claimed benefits of doing something different.[33] Collective responses are even more resistant to change and information alone often has little impact, but informing people about better practices by their neighbours can exert a much stronger impact – an insight that has formed the basis of a successful energy efficiency company (Box 4.2).

Faced with uncertainty, people tend towards social norms to inform their decisions. With the introduction of new, unfamiliar technologies, we look to those around us for advice on how to respond, and the most common collective response to uncertainty is to resist change.

Although the benefits are clear, we may still fail because our actions don't always follow our intentions. The simple act of turning off lights when you leave a room, for example, is easy to forget; you may have all the right intentions but old habits die hard. So-called 'facilitating conditions' can be important for linking intention, habits and our behaviour. For instance, modern sensors which can turn lights off automatically may greatly reduce the gap between intentions and the outcome. Likewise, where urban planning is amenable to pedestrians and cyclists, more people walk or use a bicycle. As Figure 4.5 implies, many Pillar II and III types of intervention could facilitate better choices.

By thus mapping out the bias against change, behavioural economics adds further to explanations for the slow uptake of more efficient practices or 'new' technologies – new, that is, to people who do not yet have them.

Box 4.2 Behavioural economics and energy use – 'the largest social science experiment in the world'

Lack of consumer engagement contributes to energy wastage and often means that energy efficiency policies aimed at households deliver less than hoped. Most of the research on 'behavioural economics' had not looked at energy, but in 2001 a social science experiment in California explored different ways of motivating consumers to use fans instead of air conditioning as the state struggled with power shortages.

On their own, messages about cost reductions, environmental benefits or power savings to avoid blackouts proved relatively ineffective. However, people who received a message that neighbours were taking such action, showed a 6% decrease in energy consumption. If something is inconvenient, neither modest financial gains nor moral suasion may spur us into action, even if we believe that would be the right thing to do. But social pressure and perceived norms, harnessed correctly, are powerful drivers.[a]

In 2007 two Harvard graduates set out to build a business based on the use of behavioural insights to save energy. Opower partners with utilities who want (or have an obligation to) help their customers save energy. It delivers personalised information on how a household's consumption compares to that of neighbouring homes and includes tips to help consumers take action. Communications expanded from paper to web, email, mobile phones and thermostats. Opower's platform relies heavily on behavioural science insights as well as data analytics and consumer-friendly design, such as:

- neighbour comparisons to utilise not only social norms but implied 'vicarious learning' (people incline to do what they believe others do);
- messages phrased in terms of avoidable losses not savings (harnessing loss aversion – 'You used X% MORE electricity than your efficient neighbours, costing you $Y MORE');
- a commitment to setting an energy reduction goal (commitment devices can focus attention, measure progress and give a sense of achievement – as widely used by WeightWatchers).

In barely six years, Opower has grown to partner with 85 utilities and delivers paper reports to 15 million households, mostly in the US but expanding internationally. The reporting programmes (over ninety to date) use randomised control trials and have been subject to at least 20 independent evaluations. MIT's Hunt Allcott described it as 'one of the largest randomized field experiments in history' and concluded that this kind of behavioural energy efficiency programme saved an average of 2% of electricity consumption, 'an effect equivalent to that of a short-run electricity price increase of 11 to 20%, and the cost effectiveness compares favourably to that of traditional energy conservation programs.'[b]

Opower's programmes overall have saved energy equivalent to the annual consumption of a medium-sized city. Key lessons include that behavioural changes can dissipate over time without proactive, personalised and relevant communication; that consumers need insights, not data; and that consumers must be able to 'see their savings' (to track progress and compare it to others). Given this, savings persist, and tend to increase slightly over time as more users take up more ambitious measures, and this also tends to induce wider participation.[c]

Notes

a Cialdini and Schultz (2004) 'Understanding and motivating energy conservation via social norms', final report. Available at: http://opower.com/uploads/library/file/2/understanding_and_motivating_energy_conservation_via_social_norms.pdf. The fourth message simply said: 'When surveyed, 77% of your neighbours said that they turn off the air-conditioning and turn on their fans. Please join them, turn off the air-conditioning and turn on your fans.'

b Allcott (2011) evaluated nearly 22 million utility bills from over 600,000 households across 17 of the longest running deployments of energy usage statement programmes.

c See Laskey and Kavazovic (2011) and Opower (2013). The four key Opower principles to platform design derived from its experience are that communications must be Simple, Relevant, Actionable and Motivating.

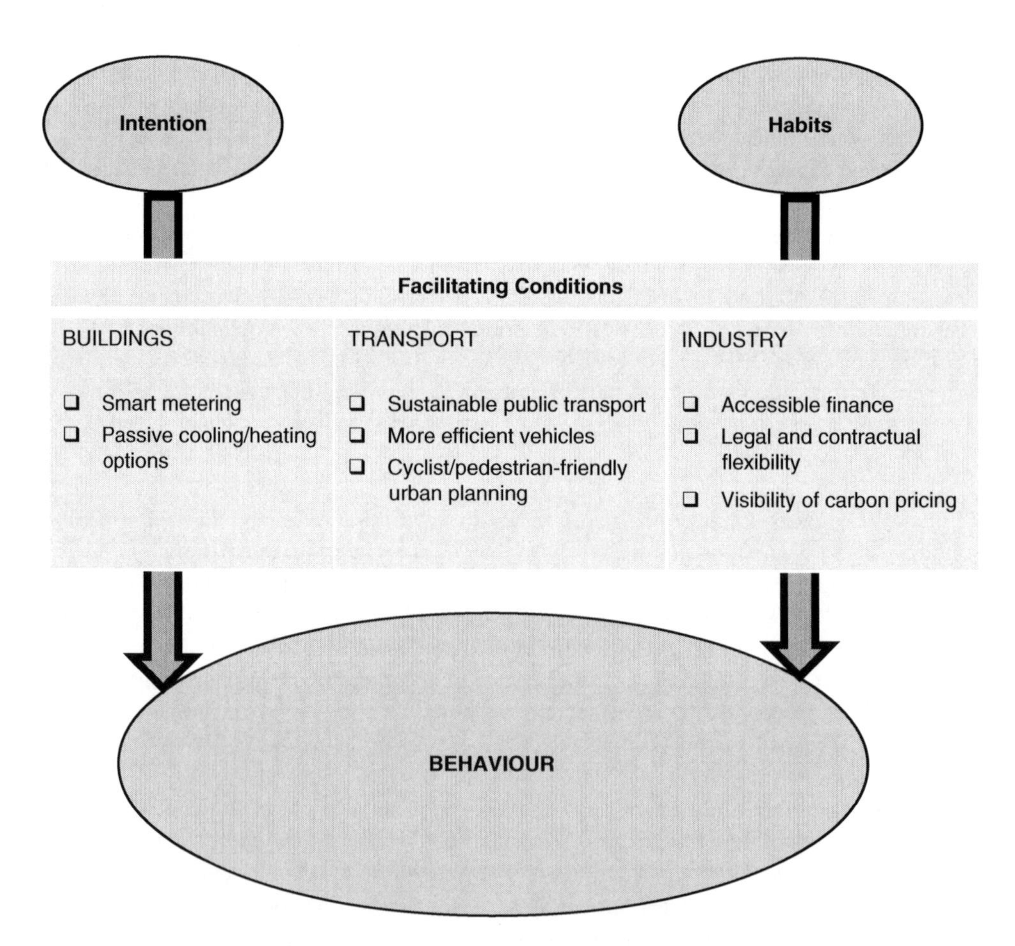

Figure 4.5 Behavioural dependence on intention, habits and facilitating conditions

Source: Adapted from DECC (2011).

The uncertainties cannot be countered simply by offering more information. People may be unaware of information already on offer, distrust it and/or have trouble distinguishing between information offered from a variety of sources. Bombardment might only create withdrawal and a retreat to the existing norms.

With seven billion people on the earth, the behavioural dimension is paramount, and three fundamental points reach to the heart of the First Domain:

- Behavioural realities further explain and amplify the bias against 'economically optimal' levels of energy efficiency and impede the uptake of better technologies.
- The human desire for improvement is in constant tension with the collective resistance to change.
- Human motivations are complex and depend also on social context: money is only one among many motivating factors.

Behavioural economics has its limitations, and in detail there is plenty of scope for debate about the exact boundaries with neoclassical assumptions and expectations.[34] Nor does it suggest 'free' gains – indeed, some features of 'behavioural economics' amplify the challenges. But they also point to ways in which the psychological insights can be used also to motivate improvements. Collectively, this underlines that there may be considerable inertia, but that once improvements are made, the progress is most unlikely to be reversed. The concluding parts of this book (notably, Chapter 12, section 12.4) demonstrate that this is an important, neglected dimension of the global challenge.

4.6 Behavioural realities: organisations

We might all accept that individual decision-making can be 'economically irrational', but we commonly assume that organisations are a lot better. After all, private companies live or die by being the 'best' – where 'best' usually has a pretty clear definition in terms of money-making. Individuals might have their quirks, but mainstream economics usually assumes companies to be rationally 'cost optimising'.[35]

Hardly. The boom in management science reflects the complexities of organisational decision-making and alongside the growth of behavioural economics has emerged an equally formidable literature on organisational theory. DeCanio *et al.* identify four key assumptions common to Second Domain representation of firms, all of which are disputed in the specialist literature that unpicks the actual structure and behaviour of firms and other organisations.[36] Indeed they illustrate the broad categories of firms' inefficiencies with curves that echo uncannily those used in Chapter 2 (Figure 2.3(b)) to illustrate the fundamental definitions of the Three Domains, and many of the specific findings relevant to energy use mirror those identified for individuals in behavioural economics:[37]

- Like individuals, firms often use simple 'rules of thumb' and norms of behaviour derived from default or traditional assumptions in the firm's habits and culture – referred to as 'cognitive bias' in the literature. Network theories of organisations explain why firms adopt simple rules to cope with complexity, which may lead them to miss opportunities not core to their business.
- The 'loss aversion' bias of individuals is matched by conservative bias when they are in an organisational hierarchy: you are unlikely to get blamed for doing things in the

traditional way, but doing something new may carry a high personal risk of being blamed if it goes wrong.

- Motivation of employees to address issues (like energy) is also critical and can depend very strongly on the examples and directions from the top, but also the internal culture that may emerge from the 'bottom up' – largely 'affective' processes.
- The flow of information between a firm and its customers on the one hand, and its investors on the other, is vital but obviously never perfect; the presentation, comparability and reliability of this information is crucial. Hence also the central importance of marketing and 'branding' to provide a simple image for external (and internal) purposes.

Although not much of this literature has focused on energy, it offers a good illustration of the key points. There is overwhelming evidence that organisations, like individuals, waste energy – see, for example, the quotes at the opening of this chapter. In the US, firms were observed to neglect energy-efficient lighting projects with an average annual rate of return on investment of 45 per cent, until targeted intervention prodded them to exploit such opportunities.[38] The accumulated experience of the UK Carbon Trust in working with business found similar potentials and helped to identify huge energy- and money-saving potentials (see, for example, Figure 4.4).

Such findings are reflected in the McKinsey global curve, which interestingly finds that although the potential for thermal insulation is *largest* in homes, it is most *cost-effective* in commercial buildings (Figure 4.1). A plausible reason – in addition to the greater transaction costs of dealing with small buildings – is that in most countries, most organisations rent their property rather than own it (whereas the degree of home ownership varies a lot across different countries). Hence, the 'tenant-landlord' split incentive is even more pervasive, and may be amplified further with frequent multiple occupancy of office space, in which it may be almost impossible to allocate responsibilities for energy use.

Understanding energy decision-making in organisations

For most organisations, energy is not a big cost. Just as for individuals, it can be 'below the radar screen' of senior decision-making. In some cases the situation can even be worse, because it is no individual's money that is being wasted. It is typical in the organisation of many large service sector firms, for example, that energy costs are written off as an unavoidable cost – not part of the controllable cost base against which managers are judged. Moreover, because of their 'status quo' bias, inefficient behaviour is hard to get rid of – as the classic example of a brewer's pump illustrates.[39]

Therein lies a key feature of organisational decision-making: it is all about management, incentives and motivation. One person may know about energy opportunities; that does not mean the organisation will act on them. Organisational economics (and management science, in different language) has long recognised the centrality of 'principal-agent' problems: the many failures that can occur in the relationships and incentives between owner and manager, between boss and employee, etc. throughout a hierarchy. High-level objectives may not permeate; bottom-up information may never be factored in. The result may be a range of 'internal split incentives'.

In theory, energy efficiency investments should compete against alternative capital projects on a comparable economic basis. In practice, energy-related choices may be relegated to simple payback times, or even just lowest first cost.[40] Indeed in most organisations, capital is less available to those lower down the system; it is not uncommon to hear of energy managers restricted to projects which pay back in a year or two, while the board is looking at 20-year investments.[41] 'Peripheral' costs – including energy – are often neglected, even when managing them may well have a much higher rate of return than large projects, which receive more management attention and dedicated budgets.[42]

Of course it might be perfectly rational for senior management not to spend time on peripheral costs, but then bias can easily enter. As noted, sticking with the way things have always been done, or not spending money, is unlikely to attract the boss's criticism; spending money on something new is risky (in economic language, principal-agent theory stresses that 'agents' are risk-averse). Distrust of the energy savings from unfamiliar equipment – or inability to monitor it due to inadequate metering – can amplify the effect. Moreover, the person making the relevant decisions – for example, the engineer responsible for upgrading or replacing motor drives or an IT equipment manager – may have no interest in the energy consumption of the product. Equipment may be specified, procured and/or maintained by individuals who lack the knowledge, information and incentives to minimise operating costs. The same applies to office workers, operators of process equipment and designers. If the costs are not big enough, there is little push to sort out such organisational inefficiencies. Of course, *in theory* good management can organise solutions; in practice it is hard, and rare.[43]

Consequently, organisations can look surprisingly like individuals in the way they waste energy – even the big energy companies. A striking example of the size of opportunities being routinely overlooked was BP's experience in the late 1990s, after deciding to address its own CO_2 emissions by reducing internal energy wastage – a programme which BP estimated ended up saving around £600m per year while reducing the company's CO_2 emissions by around 10 per cent.[44] The key was the combination of measurement, management and motivation. The McKinsey data (Figure 4.2) suggest a big continuing and highly cost-effective potential in the oil and gas industries. Everything is relative; in the context of a massive oil and gas development worth billions, a few million dollars' worth of unnecessary venting may hardly register. Few companies operate at such a scale, but the principle is the same.

Contractual constraints and the public sector

Such energy waste doesn't only occur within companies; the public sector can be notorious for inefficiencies, and contracts between organisations can also amplify problems. One manager complained that UK structures to bring private finance into public projects 'stop you dead in your tracks' in terms of any scope to take initiatives like spending money on energy efficiency.[45] An organisation's management, subcontractors (and potential energy services companies) all have distinct attitudes, incentives and approaches – and the behaviour and intent of each is subject to complex corporate workings and prone to misalignment, miscommunication and mistrust among the parties involved.[46]

Also common, especially in the public sector, is the lack of a budget line for energy efficiency; without such arrangements, offices may not be able to install even simple modifications which pay back in months. UK experience identified many such examples, along with four specific barriers that impede energy efficiency in public sector organisations.[47]

Some final pieces of the picture, though, are quite different for organisations. Good managers are on a constant search for how to motivate their employees and how to appeal to their customers. They, more than anyone, know that both of these go beyond simple money; they involve building a brand which appeals to a range of values. Alongside all the obstacles, this can offer opportunities. The implication of all the above is that if concern about environment can motivate attention to energy use, cost reductions may also follow.

4.7 Better buildings: a dynamic view of long-term potential

Of the numerous examples of wastage in energy systems, the biggest saving potentials seem to be associated with buildings, as indicated in section 4.3. Given the understanding of barriers and behaviour presented in this chapter, this is now readily understandable. Energy costs are a far greater proportion of total costs in energy-intensive industries, and even in transport, than they are in buildings. In these sectors too competitive pressures may come for those who use energy more efficiently or market more efficient products – indeed, at times of high oil prices, vehicle manufacturers may compete on efficiency and driving costs.

In sharp contrast, energy costs in a building are nearly always 'incidental' to the main activity, and we cannot import or even easily switch to another building on the grounds of energy costs: competitive pressures are almost entirely absent. All these factors suggest that energy use in buildings is likely to be *dominated* by First Domain decision-making processes, for both individual and corporate occupants. This is a reasonable hypothesis and is overwhelmingly supported by both the evidence of energy consumption in buildings and the more detailed theories of both individual and corporate decision-making. There is no other sector with such consistently large scope for cost-effective energy efficiency improvements and with such tangible multiple benefits. The debate between McKinsey and the IPCC experts was 'merely' about whether this potential is a big and cost-effective reduction in future trends, or a huge and massively cost-saving additional opportunity.

Looking ahead on the bigger global picture, the scope for – and potential pattern of – changing demand is readily apparent (Figure 4.6). The building sector currently accounts for around 40 per cent of global energy use and a third of CO_2 emissions – and future prospects seem wide open. Average per capita consumption in North American households is over 8,000kWh/yr – an average of almost 1kw continuous energy consumption per person – and it is not much lower in Europe or Russia. Obviously, cold regions have a case to consume more, but this is partly offset by higher building standards and also the growing use of air conditioning in hotter climates. In many developing countries, consumption is still around a fifth of these levels – and could rise sharply.

Another striking story is the huge difference in the commercial contribution. In Japan, this actually exceeds the energy use of the (relatively small) Japanese homes. It is a large and growing contribution in most rich countries, but its tiny role in most developing countries betrays one of the areas where massive growth in energy demand is likely unless policy can counteract this. Changes will need to span both domestic buildings and business: understanding energy efficiency in both homes and commercial operations is thus paramount.

Yet technological options are readily available. As the GEA (2012) concludes:

> Recent advances in materials and know-how make buildings that use 10–40 per cent of the final heating and cooling energy of conventional new buildings cost-effective in all world regions and climate zones. Holistic retrofits can also achieve 50–90 per cent final energy savings in thermal energy use in existing buildings, typically representing profitable investments.

The overall scope is biggest from improving thermal performance and major savings also arise in electricity use for lights and a limited set of appliances.

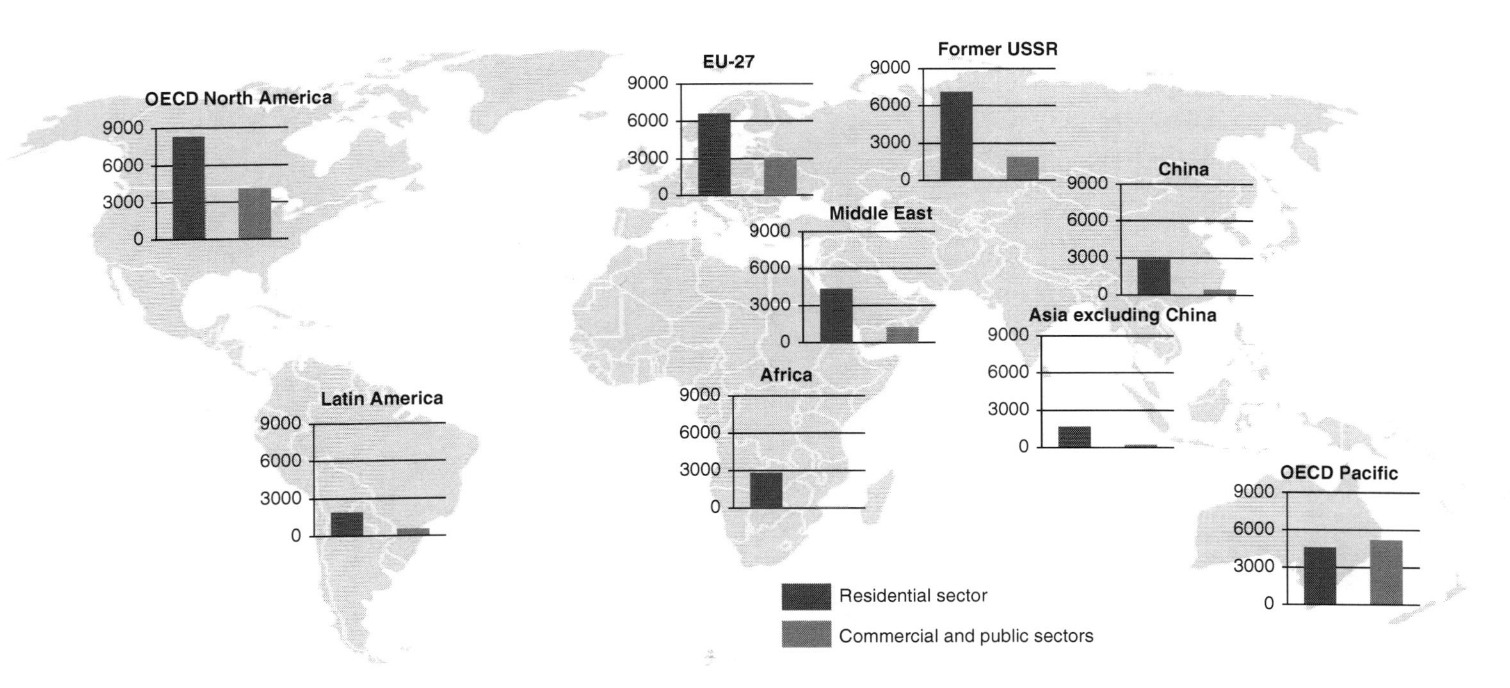

Figure 4.6 Per capita energy consumption in buildings in different regions

Source: GEA (2012), Chapter 10, Figure 10.4.

As a result, the Global Energy Assessment Scenarios show that the buildings sector has the widest 'open jaw' of possibilities, ranging from a 50 per cent increase in global energy and emissions by mid-century, to a 43 per cent reduction if best practice could be adopted globally. After reviewing policy experience, Chapter 5 (Figure 5.7) illustrates this and the influences on what may actually happen. It underlines that we face a dynamic challenge which is highly diffuse: almost every building in the world – maybe a billion individual dwellings and hundreds of millions of offices or other commercial buildings – consumes energy and emits CO_2. Buildings have the largest potential, but the main principles and insights apply to other sectors.

Looking more broadly, an expert's report to the 'G8+5' group of major industrialised and emerging economies analysed the potential and policies, and recommended that the global rate of energy efficiency improvement could and should be doubled to 2.5 per cent a year. This was to be spearheaded by improved performance in the major industrialised countries.[48] Doubling the rate of improvement in this way, they estimated, might almost stabilise global energy consumption, and by 2030 avoid $3 trillion of new energy generation investment, save consumers $500 billion per year and help get the world on a path below 550ppmCO_2 in the atmosphere. The IPCC, surveying this and the wider literature, came up with similar numbers, though the IEA was more cautious.[49] If we are to accelerate average efficiency gains to 2 per cent per year – or more where feasible – we are going to have to get a lot smarter still. We need to understand better which factors get in the way – and how these can be changed.

4.8 Conclusions

The First Domain of human behaviour is endemic and particularly relevant to understanding energy efficiency. To most people and organisations, most of the time, energy, emissions and climate change are invisible, someone else's problem: so too are the potential savings that might come from better investment or changing behaviour. In these cases the inhibiting factors are neither cost nor the inadequacy of technology: the blockages arise from the mis-aligned incentives in buildings markets and ingrained patterns of human and corporate behaviour. Similar features underpin wastage in other sectors, though not to the same scale.

The long-standing literature on energy efficiency is now more fully supported by the developments in behavioural sciences which provide a deeper understanding of why much human decision-making is not 'optimising' but is rather reactive, short-term and risk-averse; it is largely driven by habits, along with affective and associative processes and endowment effects. Responses which neglect these realities will be unnecessarily costly and may ultimately founder.

These underlying characteristics create a clear potential to cut energy consumption and emissions with net economic benefit. That is easy to say. In itself, the broad claim has been around at least since the oil shocks of the 1970s. Scepticism has also abounded. Analytically, the key is to systematically shift the balance of barriers and drivers set out in Figure 4.3 from the former to the latter. Delivering on this will require building on the additional insights and cumulative experience of policies to date. We thus now turn to consider what has been achieved and what we can learn from it.

Notes

1. One clear expression of these two fundamental doubts about the real potential of energy efficiency is by Helm (2011) who warns that 'the energy efficiency role in climate change policy is based on

two flawed arguments: that energy efficiency will substantially reduce general energy demand; and that there are lots of projects with positive returns. The former is hard to substantiate, the latter … is open to serious doubts and depends in any event on the price of energy …' While these doubts echo common sentiments among economists, detailed academic critiques which examine actual data from an economic perspective are less common; for discussion of one of the most detailed and widely cited, see note 11. The purpose of this chapter and the next are to examine such assertions based on accumulated evidence and the theories underpinning the First Domain.

2. Rebound occurs when making some energy-using activity more efficient leads to its increased use because it lowers the energy-related costs of the activity. A high level of rebound offsets the energy-environmental gains, but indicates that the activity was suppressed by high energy costs, so the major benefits are being realised though economic and welfare gains. A low rebound indicates either that the energy costs were trivial, or that the use was largely saturated (e.g. with a building already heated to a satisfactory level). See Chapter 5 (section 5.6) for estimates and discussion of the implications.
3. National energy intensity is the precise term for total energy/GDP. It is, broadly, inversely related to energy efficiency, but since energy efficiency has so many different aspects at so many different levels, the term national energy intensity is used for clarity in relation to national statistics. Data for Russia and other former Soviet countries during this period are not particularly reliable and are also complicated by the uncertainties in appropriate 'purchasing power parity' exchange rates before and during their economic transition. Exchange rate effects also complicate assessment for many developing countries over this period, and in some of the poorer developing countries (including large ones like India) data are further complicated because much of their energy consumption was of traditional, non-commercial fuels and not generally included in energy statistics.
4. For an early summary of evidence and review see Grubb *et al.* (1995).
5. Ehrhardt-Martinez and Laitner (2008).
6. Roughly a third of the decline in manufacturing energy intensity in industrialised countries in the decades after the oil shocks was due to structural change in major IEA economies (IEA 2004; GEA 2012).
7. Willis and Eyre (2011).
8. DECC (2009) reports that in Britain only about 40% of suitable houses had cavity wall insulation, 50% had double glazed windows throughout the house and 57% had the maximum depth of loft insulation. The strengthened targets under the CERT and CESP programmes since then increased these numbers significantly (Chapter 5, Box 5.3 and note 43).
9. Condensing boilers extract the energy in the steam emitted from combustion, increasing efficiency by 10–15%. For many years, the UK government (and engineers) maintained that the technology was not reliable or would be naturally taken up by the market. In a pattern repeated in a number of areas, leading countries in Europe were a decade ahead of others, but the case for regulatory and supply chain intervention (including training of engineers) gradually become overwhelming, to the point at which the leading countries were able to promote legislation at the EU level that eventually got rid of inefficient practices.
10. Many economists had suggested for some time that there might be important inefficiencies in the 'developing' countries, because they had not yet established competitive energy markets. However, the potential for energy efficiency in the McKinsey curve was split roughly equally between the OECD and other regions (Chapter 2, Figure 2.4).
11. A recent articulation of the 'classical' economic scepticism, focused on critiquing the McKinsey curve, is Allcott and Greenstone (2012). The paper introduces some basic mathematics which in practice is only relevant to their analysis insofar as it defines 'investment efficiency' in terms of a parameter g, such that g = 1 is 'optimal' investment. The analysis derives from the assumption that g = 1 unless proven otherwise, placing all the burden of proof on 'disproving' the assumption of rational optimising behaviour in individuals, equated also to the social optimum. The article then argues that the case is *not proven*. For example, it acknowledges real implied consumer discount rates of 15 to 25%, 'not much different from real credit card interest rates', which they take as suggesting that consumer investments in energy efficiency are sufficiently close to 'optimal'. In practice, this confirms one of the points made in this book concerning the potentially high returns to First Domain opportunities. Reluctantly, 'it seems possible that some consumers might be inattentive to energy efficiency when purchasing energy-using durable goods.'

See also Chapter 5, note 43, on their assessment of utility programme costs, despite their earlier acknowledgement that these have both hidden (and unquantified) costs and benefits to customers. Their policy conclusions focused on relatively short-term paybacks, ignoring both the longevity of many energy efficiency investments in buildings and the entirety of Third Domain effects as, for example, with their discussion of vehicle efficiency standards (see Chapter 9), precluding consideration of the impact of regulation on innovation, or the interactions between the different domains (Chapter 12). Constructing the maths to separate investment inefficiencies from the failure to price externalities is useful for analytic clarity, but precludes the possibility that energy efficiency investments would not only reduce unpriced externalities, but also make it politically more feasible to introduce proper pricing (or that nationally lower energy demand might help to curtail high energy prices, though that is a wholly different issue). The underlying challenge remains: many critiques of the McKinsey curve and energy efficiency potentials are possible, and it is entirely reasonable (as this book does) to argue the potential may be smaller, but one cannot reasonably claim the cost-effective potential to be negligible without presenting any more plausible and credible estimates consistent with the evidence.

12. As Figure 2.4, shows, McKinsey estimated an aggregate potential for emission reductions from domestic buildings at just over 2GtCO_2, but with only marginal cost savings – a result which building experts condemned as unreasonably conservative in both scale and cost. The IPCC Working Group III (2007) had assessed a 'negative cost' potential from buildings of 5.5GtCO_2 with net savings exceeding \$20/tCO_2 in their central case. This included much bigger projections of the potential savings from domestic buildings in Asia and the former Soviet and emerging economies. This arose partly because McKinsey had assumed that these countries would ensure that new buildings were more efficient 'in the baseline'. Of the McKinsey total thermal abatement potential in domestic buildings, almost a quarter was associated with 'high cost' measures – more than 250MtCO_2 associated with strong new-build standards in Asia (cost +\$30–40t/CO_2), and 100MtCO_2 associated with retrofits in OECD countries at a cost almost twice this (around +\$60/tCO_2). This of course skews the total; other measures average cost savings around (–) \$25–30/tCO_2. Russian experts also argue that McKinsey vastly underestimated the cost-effective potential for improving the Russian building stock (Bashmakov 2011) and cost curves for the region regularly identify exceptionally large negative cost.
13. An additional critique of the McKinsey curve is that it sees efficiency potential as a sum of individual measures. This fails to capture systematic opportunities available in the sector. A performance-based approach, on the other hand, takes a holistic view and incorporates gains that can be made when building designs are optimised to achieve high efficiency performance (Ürge-Vorsatz *et al.* 2012). Partially offsetting this, global energy efficiency potentials need to be assessed with models that capture the internal consistency of the energy systems to avoid double-counting and capture the fact that efficiency gains are not additive in the energy transformation chain; thus very efficient buildings reduce the net benefit of increased efficiency of the electrical system or of the heat distribution networks. Both the McKinsey and the IPCC assessments claim to do this; it is unclear how completely potential overlaps are excluded.
14. McKinsey & Co. (2008). The IPCC WGIII (2007) assessment of the 'negative cost' potential in other sectors was lower than McKinsey's, but its treatment of transport energy efficiency was more limited and it did not separate out 'negative cost' potentials in industry or power generation at all, on the classical economic grounds that if these existed then industry could be expected to take up the opportunities anyway.
15. For broad surveys see IPCC (2007) and numerous other references in this chapter on building energy efficiency opportunities.
16. Data respectively from Carbon Trust (2011) and Energy Savings Trust (2011).
17. A good overview of literature on energy efficiency – potentials, barriers and policies – is included in several sectoral chapters of the Global Energy Assessment (2012), which is more extensive in its coverage even than the IPCC Fourth Assessment (2007). The intellectual challenge is to get a coherent overall understanding of the implications from the myriad of detailed analyses now available. The International Energy Agency (Ryan 2011; also summarised by Hood 2011) gives an excellent introduction to the wider literature. The IEA classification quite closely reflects that in this book, in highlighting 'principal-agent' problems (asymmetric information and split incentives) and 'behavioural failures' as key; other aspects they group as 'information' problems. Whether 'hidden costs' should be counted as a 'barrier' is semantic (observed behaviour

is optimal to the extent it can be explained by hidden costs); their distinct economic significance lies in the facts that (a) some hidden costs may be readily removed by policy (e.g. the search costs reduced by standards and labels) and (b) they are often transitional, rather than enduring, with the implications highlighted in the concluding chapter.

18. Given the vast chambers of literature that have been filled with discussion of discounting issues, this feature of the 'discounting pyramid' seems to have received little attention – and virtually none as it applies to organisational behaviour. The economic literature that has considered the high discount rates of individual consumers has largely explained it in terms of their higher risk characteristics – capital constraints, aversion to risk and associated cost of capital – with some of the literature arguing that this is a perfectly rational response and that policies which lead consumers to make purchases with longer time horizons therefore reduce welfare. This is only true under a narrow set of modelling assumptions. Research for this book did not uncover any literature on the 'discounting pyramid' in corporate behaviour. The most plausible explanations in both personal and corporate contexts are the behavioural characteristics as discussed in the next section, which provide clear reasons why policies that give more weight to future benefits (less discounting) are likely to be beneficial all round. See Chapter 12 for some discussion of the implications.
19. For extensive discussion of transaction costs in energy efficiency see Sorrell *et al.* (2004).
20. The report was from ENTEC to the Department of Environment, Food and Rural Affairs, I believe around 2004–5, but like innumerable consultants' reports the world over, may now be lost in the mists of time. However, I well recall the polarised reactions to the study, as indicated. Keen readers could always try their own estimates.
21. Lovins and Lovins (1997). Since in an average workplace labour costs tend to be 25 times more than spent on energy, these hidden benefits might vastly outweigh the more obvious benefits of increased energy efficiency. See also GEA (2012), Chapter 8.
22. See Chapter 5, note 46.
23. This example refers to Walkers crisps – a popular brand in the UK, which worked with the Carbon Trust to understand the carbon footprint of making potato crisps. It turned out that the 'life cycle' emissions were dominated by the energy used in frying the potatoes. In turn, this energy requirement is largely driven by the water content of the potatoes. Walkers purchases potatoes per kilogram, and it turned out that some suppliers were actually putting potatoes in humidifiers to increase the sales weight – wasting energy and carbon on both sides. By simply changing their contracts to denominate potato purchases in dry weight equivalent, Walkers cut both costs and emissions – and advertised the fact on their crisp packets through a Carbon Footprint label.
24. Nick Eyre, Oxford University, personal communication.
25. 'Moral hazard' arises when the 'principal' cannot verify the actual performance of an 'agent'. As the principal is unaware what level of service they are receiving, they will not pay more for a better service. This results in adverse selection. For example, when a purchaser cannot observe the quality of a good, then inefficient and efficient goods can sell for the same price and buyers are likely to stick with what they know. Thus the diffusion of new technologies is stalled – indeed worse ones may prosper. Nobel himself did not establish a prize for economics, but its substitute – the Sveriges Riksbank Prize in Economics Sciencies in Memory of Alfred Nobel – has established a pattern of awarding for key breakthroughs in a selected area of research each year. In recognition of the importance of research on asymmetric information, in 2001 the Prize was awarded to George Akerlof for his analyses of adverse selection; to Michael Spence for showing the high value of signalling quality in such conditions; and to Josephe Stiglitz for showing that asymmetric information can provide the key to understanding many observed phenomena, including unemployment and credit rationing. For application to energy efficiency, see Bleischwitz and Andersen (2009).
26. Research in the EU around 1990 clearly demonstrated that the cost of refrigerators had no discernible relationship to their efficiency; see, for example, Herring (1992). I have reason to recall this. At about the time that book was published, I had a conversation with an economist who argued that energy efficiency standards were intrinsically inefficient and would raise costs because they interfered with the market. When I explained the evidence found by the research on the costs of refrigerators, he responded 'Oh, empirics! I'm talking about the *theory*' – clearly expecting that to be the trump card of his argument. Twenty years later, the results displayed in Chapter 5 (Figure 4.5) show just how wrong the theory of perfect markets was in this case.
27. Bleyl-Androschin *et al.* (2009).

28. WEC (2008). An EU Directive on energy end-use efficiency and energy services requires EU Member States to provide 'adequate incentives' for energy consultants and advisors to develop and distribute information on energy efficiency, energy audits and financing recommendations to consumers (Uihlein and Eder, 2009).
29. The classic reference is Simon (1955).
30. The roots of this were developed in the famous article that introduced 'Prospect Theory' by Kahneman and Tversky (1979). The simplest way into the huge literature that has developed since then is through Khaneman's recent book, *Thinking, Fast and Slow* (2011).
31. For a useful overview see Pollitt and Shaorshadze (2011). They classify the four main areas in which behaviour departs from classical economics as (1) time varying discount rates (including procrastination); (2) prospect theory and reference points (including loss aversion, endowment effects and status quo bias); (3) bounded rationality (including choice overload, heuristics and saliency effects); and (4) pro-social behaviour and perceptions of fairness.
32. A classical example is that given a choice between one good in 10 days' time or two in 11 days, nearly everyone will choose the latter; faced with the same choice today or tomorrow, many will choose today.
33. Studies on endowment effects include Brown and Hagen (2010) and Knetsch (2010). For a discussion of how these effects relate to other dimensions of behavioural economics applied to environment see also Baddeley (2012), Chapters 6 and 7.
34. One recent such challenge is the short book by David Levine (2012). The difficulty is that his defence of neoclassical economics and critique of behavioural economics often boils down in part to terminological debates about what is really 'rational'. What is 'rational' is obviously a moveable thing that depends on the boundaries drawn and the levels of information and layers of possible implications considered. The fact is that many of the implications are not what would be associated with traditional assumptions about what neoclassical economic behaviour would imply – or the numerous models based upon this. Moreover, Levine clearly acknowledges the fact that the core issue is about *relative significance*: 'understanding the psychological elements that predominate when economics incentives are weak would be of great value to economics' – he refers to these as *epsilons* (p. 77). The essence of the Three Domains can be understood as concluding that in relation to a problem of the scale and reach of energy and climate challenge, the role of the non-classical (as usually understood) domains are crucial, including the role of 'epsilons' within the First Domain of energy use behaviour. Moreover, Levine does note the analogy with physics: 'Do you condemn quantum mechanics as useless because it cannot predict simultaneously the location and velocity of subatomic particles?' No – and Levine himself goes on to argue that people have unrealistic expectations of what neoclassical economics can predict. This is entirely consistent with a core conclusion of this book and in particular the implications of First Domain realities for the Third Domain phenomena shown in Chapters 10 and 11: no theory can predict the prospect for major transitions or long-run economic growth, nor (in consequence) the long-run economic cost implications of embracing and responding to environmental constraints.
35. Sutherland (1994, 1996).
36. As surveyed in DeCanio *et al.* (2000). These are the assumptions that: (1) the firms have a single unitary objective of profit maximisation, whereas it is actually known that managers frequently seek to maximise size, not profits; (2) that managers focus on the 'factors of production' whereas actually these occupy only a small portion of their time and attention; (3) that technology is exogenous and the firm cannot shape it; and (4) that firms always make optimal decisions whereas actually firms operate in environments too complex to optimise and managers may prioritise avoiding blame.
37. Obviously the overall management literature is huge. The authors are grateful to Aoife Brophy-Haney at the Judge Business School for suggesting a few key examples: for a review of evidence and literature on cognitive bias in firm decision-making see Baddeley (2006); on the relevance of information flows between firms, consumers and investors see Hamilton (1995, 2005; and on the significance of motivation see, for example, Howarth *et al.* (2000) and Siero *et al.* (1996). For early work on network model explanations applied to energy management, see, for example, DeCanio (1993).
38. See, for example, DeCanio (1994); the industry lighting example is from Howarth *et al.* (2000: 483).

39. Fawkes and Jacques (1987) found workers at a brewery adamantly insisted on using a conventional pump simply because they considered it easier to clean; the (large) energy cost-savings of a modern pump (or the fact that it might need cleaning less often) were not seriously considered. See also UNIDO (2011). The literature on the diffusion of innovations is littered with similar examples.
40. Sorrell *et al.* (2004).
41. This was a comment also from workshops hosted by the Confederation of British Industry for the Carbon Trust. Broad theories of how budgets get allocated include Stiglitz and Weiss (1981).
42. Ross (1986).
43. If individual departments were accountable for their own energy costs, they could directly benefit from any savings from investment projects or housekeeping measures. If cost-savings are recouped elsewhere, this incentive is diluted. To introduce such accountability, it would be necessary to sub-meter and bill individual cost centres for their energy use – which would be associated with investment, staff and operational costs. The resulting incentives will be proportional to the importance of energy costs to the individual department and would only be effective if the department had the capacity to identify and initiate energy efficiency improvements. An alternative and preferable approach in many instances would be to place accountability for energy costs with the energy management staff, perhaps with individual posts made self-funding from the savings from energy efficiency improvements.
44. Lord Browne, BP Group Chief Executive, speaking at DTI/DEFRA Conference, 2005; online at http://www.wbcsd.org/.
45. Private finance initiative (PFI) contracts, for example, last over 15–25 years and specifically prescribe how energy use should be managed. Once signed, these contracts make it extremely difficult to make any change, due to legal risks of any damage or reduced performance. With such rigid restrictions, any energy service providers would struggle to convince facilities managers to implement changes (Newey 2011: 52).
46. Ibid.
47. These were characterised as (1) lack of capital restricting any significant investment that was not specifically budgeted; and (2) accounting barriers, organisational fragmentation both within and between government departments, and the lack of skills and resource base. The UK Carbon Trust and government established a specific fund to support public sector energy efficiency programmes, and found that the non-economic barriers and lack of motivations could still impede progress even when funding was available (see Memorandum from Peter Mallaburn, 2006).
48. Expert Group on Energy Efficiency (2007).
49. The IPCC forecast that doubling the rate of energy efficiency would allow the world to hold CO_2 concentrations below 550ppmv, avoid $3.0 trillion worth of new generation and save consumers $500 billion per year by 2030 (IPCC, 2007). However, such aspirations are more ambitious than even the strongest scenario considered by the International Energy Agency, which averages 1.9% per year global efficiency improvement out to 2035 (IEA 2010).

References

Allcott, H. (2011) 'Social norms and energy conservation', *Journal of Public Economics*, 95 (9–10): 1082–95.

Allcott, H. and Greenstone, M. (2012) 'Is there an energy efficiency gap?', *Journal of Economic Perspectives*, 26 (1): 3–28.

Baddeley, M. (2006) 'Behind the black box: a survey of real-world investment appraisal approaches', *Empirica*, 33 (5): 329–50.

Baddeley, M. (2012) *Behavioural Economics and Finance*. Abbingdon: Routledge.

Bleischwitz, R. and Andersen, L.-M. (2009) *Informational Barriers to Energy Efficiency – Theory and European Policies*, MPRA Paper 19937, University Library of Munich, Germany.

Bleyl-Androschin, J. W., Seefeldt, F. and Eikmeier, B. (2009) *Energy Contracting: How Much Can It Contribute to Energy Efficiency in the Residential Sector? Transaction and Life Cycle Cost Analyses, Market Survey and Statistical Potential.* Tagungsbeitrag zur 10th IAEE European Conference 'Energy, Policies and Technologies for Sustainable Economies', 7–10 September.

Brown, G. and Hagen, D. A. (2010) 'Behavioral economics and the environment', *Environmental Resource Economics*, 46 (2): 139–46.

Carbon Trust (2005) *The UK Climate Change Programme: Potential Evolution for Business and the Public Sector*. London: Carbon Trust. Online at: http://www.carbontrust.com/media/84912/ctc518-uk-climate-change-programme-potential-evolution.pdf

Carbon Trust (2011) *International Carbon Flows*. London: Carbon Trust.

Cialdini, R. and Schultz, W. (2004) *Understanding and Motivating Energy Conservation via Social Norms*, Final Report. Available online at: http://opower.com/uploads/library/file/2/understanding_and_motivating_energy_conservation_via_social_norms.pdf; also at: http://smartenergyefficiency.eu/how-behavioural-science-can-lower-your-energy-bill

DeCanio, S. (1993) 'Barriers within firms to energy-efficient investments', *Energy Policy*, 21 (9): 906–14.

DeCanio, S. (1994) 'Agency and control problems in US corporations: the case of energy-efficient investment projects', *International Journal of the Economics of Business*, 1 (1): 105–24.

DeCanio, S. *et al.* (2000) *New Directions in the Economics and Integrated Assessment of Global Climate Change*. Washington, DC: Pew Center on Global Climate Change.

DECC (2009) *Energy Consumption in the United Kingdom*. London: Department of Energy and Climate Change.

DECC (2011) *An Introduction to Thinking About 'Energy Behaviour': A Multi-Model Approach*. London: UK Department of Energy and Climate Change. Online at: https://www.gov.uk/government/uploads/system/uploads/attachment_data/file/48256/3887-intro-thinking-energy-behaviours.pdf

DEFRA (2005) Lord Browne, BP Group Chief Executive, speaking at DTI/DEFRA Conference, online at: http://www.wbcsd.org/

Ehrhardt-Martinez, K. and Laitner, J. A. 'Skip' (2008) *The Size of the US Energy Efficiency Market: Generating a More Complete Picture*, Report No. E083. Washington, DC: American Council for an Energy-Efficient Economy.

Energy Savings Trust (2011) Energy Statistics, online at: http://www.energysavingtrust.org.uk/Easy-ways-to-stop-wasting-energy/Stop-wasting-energy-and-cut-your-bills/Tips-to-help-you-stop-wasting-energy/Energy-saving-tips

Expert Group on Energy Efficiency (2007) *Realising the Potential for Energy Efficiency: Targets, Policies and Measures for G8 Countries*. UN Foundation.

Fawkes, S. D. and Jacques, J. K. (1987) 'Approaches to energy conservation management in beverage related industries and their effectiveness', *Energy Policy*, December.

GEA (2012) *Global Energy Assessment – Towards a Sustainable Future*. Cambridge and New York: Cambridge University Press and Laxenburg, Austria: IIASA.

Grubb, M., Ha Duong, M. and Chapuis, T. (1995) 'The economics of changing course: implications of adaptability and inertia for optimal climate policy', *Energy Policy*, 23 (4): 1–14.

Hamilton, J. (1995) 'Pollution as news: media and stock market reactions to the toxics release inventory data', *Journal of Environmental Economics and Management*, 28: 98–113.

Hamilton, J. (2005) *Regulation Through Revelation: The Origin, Politics, and Impacts of the Toxics Release Inventory Program*. New York: Cambridge University Press.

Herring, H. (1992) 'Energy savings in domestic electrical appliances', in M. Grubb *Emerging Energy Technologies: Impacts and Policy Implications*. Aldershot: Dartmouth, pp. 69–85.

Howarth, R. B., Haddad, B. M. and Paton, B. (2000) 'The economics of energy efficiency: insights from voluntary participation programs', *Energy Policy*, 28: 477–86.

IEA (2004) *World Energy Outlook*. Paris: OECD/IEA.

IEA (2010) *World Energy Outlook*. Paris: OECD/IEA.

IPCC (2007) *Fourth Assessment Report – Working Group III*. Cambridge: Cambridge University Press.

Kahneman, D. (2011) *Thinking, Fast and Slow*. New York: Farrar, Straus & Giroux.

Kahneman, D. and Tversky, R. (1979) 'Prospect theory: an analysis of decision under risk', *Econometrica*, 47: 263–91.

Knetsch, J. L. (2010) 'Values of gains and losses: reference states and choice of measure', *Environmental Resource Economics*, 46 (2): 179–88.

Laskey, A. and Kavazovic, O. (2011) 'Energy efficiency through behavioral science and technology', *XRDS*, 17 (4), online at: http://www.opower.com/uploads/library/file/15/xrds_opower.pdf

Levine, D. (2012) *Is Behavioral Economics Doomed? The Ordinary versus the Extraordinary*. Cambridge: Open Book Publishers.

Lovins, A. and Lovins, L. H. (1997) *Climate: Making Sense and Making Money*, Report C97-13. Boulder, CO: Rocky Mountain Institute.

McKinsey & Co. (2009) *Pathways to Low Carbon Economy*. New York: McKinsey & Co.

Mallaburn, P. (2006) Memorandum from Peter Mallaburn to 'Greening Government' – Report by UK Environmental Audit Committee, House of Commons, UK.

Newey, G. (2011) *Boosting Energy IQ: UK Energy Efficiency Policy for the Workplace*, ed. S. Less. London: Policy Exchange.

Opower (2013) *Fitting a Square Peg into a Round Hole: Behavioral Energy Efficiency at Scale*, White Paper 04. Arlington, VA: Opower.

Ozkan, F. G. and Sutherland, A. (1994) *A Model of the ERM Crisis*, CEPR Discussion Papers 879. London: CEPR.

Pollitt, M. G. and Shaorshadze, I. (2011) *The Role of Behavioural Economics in Energy and Climate Policy*, Energy Policy Research Group Working Paper WP1130, University of Cambridge; also forthcoming as a chapter in R. Fouquet (ed.) (2013) *Handbook on Energy and Climate Change*. Cheltenham: Edward Elgar.

Ross, M. (1986) 'The capital budgeting practices of 12 large manufacturing firms', *Financial Management*, 15 (4): 15–22.

Siero, F. W., Bakker, A. B., Dekker, G. B. and van den Burg, M. T. C. (1996) 'Changing organizational energy consumption behaviour through comparative feedback', *Journal of Environmental Psychology*, 16: 235–46.

Simon, H. A. (1955) 'A behavioural model of rational choice', *Quarterly Journal of Economics*, 69 (1): 99–118.

Sorrell, E., O'Malley, E., Schleich, J. and Scott, S. (2004) *The Economics of Energy Efficiency: Barriers to Cost-Effective Investment*. Cheltenham: Edward Elgar.

Stiglitz, J. E. and Weiss, A. (1981) 'Credit rationing in markets with imperfect information', *American Economic Review*, 71 (3): 393–410.

Sutherland, R. J. (1994) 'Energy efficiency or the efficient use of energy resources?', *Energy Sources*, 16: 257–68.

Sutherland, R. J. (1996) 'The economics of energy conservation policy', *Energy Policy*, 24: 361–70.

Uihlein, A. and Eder, P. (2009) *Towards Additional Policies to Improve the Environmental Performance of Buildings Part II : Quantitative Assessment*, JRC Scientific and Technical Reports. EC Joint Research Centre, IPTS.

UNEP (2009) *Buildings and Climate Change: Summary for Decision-Makers*. Paris: UNEP DTIE Sustainable Consumption & Production Branch. Online at: http://www.unep.org/sbci/pdfs/SBCI-BCCSummary.pdf

UNIDO (2011) *Industrial Development Report*. UNIDO.

Ürge-Vorsatz, D. *et al*. (2012) 'Towards sustainable energy end-use: buildings', in GEA, *Global Energy Assessment – Towards a Sustainable Future*. Cambridge and New York: Cambridge University Press and Laxenburg, Austria: IIASA, Chapter 10.

WEC (2008) *Energy Efficiency Policies Around the World: Review and Evaluation*. London: World Energy Council.

Willis, R. and Eyre, N. (2011) *Demanding Less: Why We Need a New Politics of Energy?* London: Green Alliance.

5 Tried and tested

Four decades of energy-efficiency policy

'Oh, Empirics!'[1]

5.1 Introduction

The huge potential for improving energy efficiency has long been recognised – and acted upon. In the wake of the 1970s oil crisis, most government agencies continued to project that energy demand would rise almost in lock-step with economic growth. In Japan, they decided as a matter of national security to do something about it, and 'Japan, Incorporated' was soon to emerge with the most energy-efficient industry sector in the world. In the US, energy efficiency 'guru' Amory Lovins contrasted the official, nuclear-and-coal-to-the-rescue view with his 'soft energy paths', predicting that a gush of energy efficiency would emerge instead, initiating a long struggle over whether and how government should be involved in fostering this.[2]

Europe offered a microcosm for the various approaches. Strong policies for energy efficiency were already embedded in the Scandinavian psyche, for example strong standards on building energy efficiency to cope with severe winters. Denmark added particular efforts to develop CHP (combined heat and power), which eventually linked nearly all of its thermal power stations with district heating networks.[3] Germany and France strengthened their centralised policies to promote efficiency. The response in southern Europe was more patchy.

The UK found itself deeply confused over the issue. At the time that officials were promoting massive new electricity investments, the Department of Trade and Industry bought 100 copies of a book that reached the heretical conclusion that energy demand might not grow, because energy efficiency carried a larger and more cost-effective potential.[4] Both these views were then largely swept aside in the 'Thatcher revolution' which privatised the energy system and made the UK one of the first countries in the world to have competitive electricity and gas markets. This, it was argued, would deliver the best outcome and would enable the government to wash its hands of any responsibility for the balance between supply and demand.

Yet around the world, analysts continued to point to a huge potential for energy efficiency, regardless of how the system was regulated. Increasingly free-market-oriented governments were at first reluctant to do much about it. But as understanding deepened – and as climate change was added to the list of concerns – many governments embarked on efforts to exploit the potential. Through the ebbs and flows of various political approaches, the UK itself has seen a host of initiatives bolted on to patch up or complement the limitations of the privatised energy model with respect to energy efficiency.

Now, forty years on, what has been done, and what have we learned? On the surface, energy efficiency measures appear to be a powerful way to address energy security and energy poverty while promoting economic growth and carbon abatement with little associated costs or risk. Governments have pursued various ways to speed up the deployment of efficient technologies. Much good has come of these efforts, but they have also underlined one central fact: energy efficiency is not as simple as it seems. What appeared to be a shallow trough impeding optimal energy efficiency has revealed itself as a complex, staggered gorge, still incompletely navigated. As inefficient technologies and practices continue, policies have evolved, with each seeking to improve on the former or plug identified gaps.

This chapter maps out the course that policy has taken. Faith that energy markets would deliver optimal outcomes has given way to increasingly sophisticated regulatory measures. Agencies have been established overseeing a growing scale of programmes to reach every corner of the market, and address each and every barrier – monetary, behavioural, organisational or market based. Legislative priorities have shifted and spread broadly across multiple objectives. Policies have evolved to impose regulations and standards, followed by fiscal incentives, obligations on energy suppliers and other measures.

Experience is the best and most trusted teacher. This chapter draws on a huge range of literature, but it also draws significantly on the lead author's experience as Chief Economist at the UK Carbon Trust.

5.2 From exhortation to institutions

Most governments' initial responses to the energy price shocks of the 1970s reflected US President Jimmy Carter's exhortation at the height of the 1979 oil crisis that energy conservation had become 'the moral equivalent of war'. This was good for headlines but has gone down in history with some derision. If the history of energy conservation was to be judged as a thirty-years war we could hardly claim resounding victory.

Against a diffuse, invisible and uncertain 'enemy', such a level of rhetorical intensity is unsustainable. In some quarters it bred public cynicism rather than private action. 'Policies that did things to things have reduced energy use, while those that did things to people haven't' was one assessment of whether the oil-shock campaigns of the 1970s and early 1980s endured as oil prices declined again.[5] It was a sobering yet possibly premature judgement, as societies developed more measured responses. Most energy efficiency is a 'bottom-up' business. A vast proliferation of local initiatives, often at the town or district level, have developed and complemented, supported or otherwise been woven into broader patterns of national and international efforts.

At the national level, most countries today have established energy agencies of one sort or another or strengthened ministry responsibilities, to develop programmes to foster energy savings and technology innovation.[6] The agencies specialise in the technical skills required to promote energy efficiency and sometimes broader environmental policy. The objectives and the institutional arrangements vary enormously: an interesting and contrasting sample is offered between London and Paris (see Box 5.1).

In Europe, a wide range of national efforts were increasingly complemented, stimulated or reinforced by efforts at the European level. The 'SAVE' programme in the 1990s led to some of the concrete measures sketched in this chapter. The Directive on energy end-use efficiency and energy services outlines an annual target of 1 per cent improvement from 2008 to 2016, towards 20 per cent by 2020.[7] Along with this, the focus and branding has itself varied and developed; what started as energy conservation developed into an emphasis on

Box 5.1 Energy agencies: a tale of two cities

The two cities of Dickens' famous *Tale of Two Cities* – Paris and London – host an interesting sample of the diverse institutions developed to improve energy performance. Each has a range of city-level programmes but the focus here is on national (and international) institutions which feature in this book.

After the first oil shock rocked the global economy in 1973, the countries of the Organisation for Economic Cooperation and Development (OECD) created the International Energy Agency (IEA) – a constitutionally separate but affiliated body – to help manage the rich world's dependence on oil (such was this focus that initially everything else was classified as 'alternative energy'). As the IEA developed into the most authoritative source of data and analysis on international energy issues, it embraced environmental concerns and increasingly stressed the importance of energy efficiency for meeting the wide range of energy-related problems.

Copying the model of the OECD itself, the IEA also conducts country reviews of its members' energy policies and produces the key annual *World Energy Outlook*, as well as more recently the biannual Energy Technology Perspectives which map out a longer-term view of how the world can meet its energy challenges. Its role has always been limited to analysis and meetings: the national sensitivity of energy issues means that governments would never entrust it with the powers to regulate or even run programmes. Perhaps the closest it came to an international programme was its Climate Technologies Initiative.

Elsewhere in Paris, the French government launched its own efforts to drive energy efficiency which evolved over the subsequent two decades into probably the largest and most inflential national agency devoted to clean energy. The initial Agency for Energy Conservation created in 1974 was transformed in 1982 to include broader energy management (e.g. renewable energy programmes), and ten years later, in the run-up to the Rio Earth Summit in 1992, was expanded again to encompass environmental goals (waste management, air quality) in the form of ADEME (Agence de l'Environnement et de la Maîtrise de l'Énergie).

Thanks to its decentralised structure at the regional level ADEME offers expertise for energy efficiency projects and programmes adapted to local conditions, and concentrates its financial support on upstream and capacity building. It also funds R&D programmes and pilot projects in partnership with enterprises and local authorities. In addition, it exerts an important influence through the funding of research grants, its capacity to capitalise experiences, inform businesses and heighten public awareness of the social benefits and economic opportunities that can be provided by energy efficiency. From this basis it has often helped to create or lead many European-level initiatives, spanning the energy-efficiency programmes noted in this chapter and various energy technology development and promotion programmes.[a]

Typically ADEME works within the constraints of a public agency. It is primarily funded by a tax on polluting activities, government subsidies and the revenues of its contractual activities. Though it was weakened in periods of low energy prices it has survived several political changes and – despite cohabiting in a country with a surplus of nuclear generation – its importance is no longer seriously contested.

Across 'La Manche' – the English Channel – the UK followed a very different track. Its main energy R&D capacities had developed around two institutions: the Atomic Energy Agency (AEA) and the Central Electricity Generating Board (CEGB). After the first oil shock, the government spun a broader 'Energy Technology Support Unit' out of the AEA – ETSU being repeatedly (if unfairly) attacked for its pro-nuclear origins and location – and it supplemented this with sporadic government campaigns to promote energy efficiency. Mrs Thatcher's privatisation revolution swept away the CEGB in favour of a competitive electricity market and privatised ETSU into what basically became a consulting company.

The CEGB's culture had developed around building bigger and better coal and nuclear power stations. The government assumed that privatisation would correct this bias and establish a market-based optimum level of energy efficiency, and that competition would drive a new paradigm of energy innovation. Instead, competing companies had a clear interest in maximising sales (hardly a spur to improved end-use efficiency) and to cut costs. A notable early victim was the closure of the former CEGB's research and development facilities.

For a government which claimed to be the guardian of consumer interest and had recently proclaimed its green credentials, this was all potentially embarrassing. In 1992, also spurred in part by the wave of public attention around the Rio Earth Summit process, it established the Energy Savings Trust (EST), with a mandate 'to help people save energy and reduce carbon emissions'. Over the subsequent fifteen years, EST estimates that its programmes saved people £100m on fuel bills and cut their total CO_2 emissions by 140 million tonnes.[b]

Five years after the EST was established, the new Labour government under Tony Blair faced a different problem. It was committed to introduce a carbon price (which ended up as the Climate Change Levy – see Chapter 6). This angered industry, which argued that if environmental concerns drove up energy prices it needed help to deal with the costs and to manage the low carbon transition. Industry trusted neither the EST – a body devoted to consumer needs – nor the government itself to run the programmes required, and so in 2001 the Carbon Trust was established as an independent not-for-profit company. It rapidly grew to be a unique business-oriented entity designed to help diverse parts of UK businesses and the public sector to cut their energy consumption, and expanded programmes on technology development and commercialisation, market development and investor engagements. After five years, independent verification of Carbon Trust calculations showed that its operations had cut 2006/7 fuel bills by £114–171m/year, and had saved over 10MtCO_2 to date.[c] By 2009/10 the energy and carbon savings had roughly tripled: Carbon Trust programmes had helped to save British business £1m/day in energy costs – over £1bn cumulatively – while also helping to build some of the low-carbon industries of the future.

The EST and the Carbon Trust both illustrated the combined benefits and risks of being quasi-independent yet government-funded bodies. Officials from ADEME marvelled at the degree of independence enjoyed by the Carbon Trust which was freer to develop programmes matched to market needs and to innovate, for example, in taking shareholdings in low-carbon companies it created. This is impossible for a government agency like ADEME. Yet EST and the Carbon Trust both faced a basic paradox:

they were expected to complement their government funding with growing levels of private sector funding, and yet state aid constraints barred them from the most potentially profitable activities on the (understandable) grounds that state-funded companies should not compete in private markets. The risks associated with quasi-independence combined with being government-funded became even more apparent amid the credit crunch, the recession and the ensuing transformations, including a new government, which led EST to become a 'social enterprise' while the Carbon Trust navigated the tricky transition to depend mainly on private finance.[d]

Notes

a L'Agence pour les Économies d'Énergie (AEE) was set up in 1974, and in 1982 became 'Agence Française pour la Maîtrise de l'Énergie (AFME), and finally in 1992 l'Agence de l'Environnement et de la Maîtrise de l'Énergie (ADEME).

b From the EST website: '… by providing expert insight and knowledge about energy saving, supporting people to take action, helping local authorities and communities to save energy and providing quality assurance for goods, services and installers'. Originally, the EST was to be funded by a levy on gas and electricity bills, but the idea of a tax on the consumer was strongly opposed by the regulator so that EST was scaled back dramatically with a budget of £20m instead of the originally intended £200m. The 140 $MtCO_2$ is about right (Eyre et al., 2010). The confusion is what can be called the 'double-sum problem'. The 140 $MtCO_2$ seems a lot because it is not only the sum over all the years of EST operation, but also the lifetime projected savings, mainly insulation with a lifetime of 40 years. The money savings on the same basis would be £7–14bn (based on a cost-effectiveness of £50–100/tCO_2). The £100m figure possibly denotes the annual/annual number (Eyre, 2012, personal correspondence).

c An enquiry by the House of Commons Environmental Audit Committee in 2007 examined the Carbon Trust's operations and broadly endorsed its estimates of programme delivery and cost-effectiveness, while also recommending a number of improvements.

d In 2008 the Carbon Trust was suddenly asked to quadruple its programmes for small industry as part of the 'economic stimulus package', and then along with the EST faced the chill winds of recession-induced cutbacks by a new govenrment that in particular saw the Carbon Trust as a creation of the previous (Labour) government and an unacceptable 'outsourcing' of policy which they wanted to keep within central government. In 2011, the EST registered as a charitable 'social enterprise', while the Carbon Trust split into a series of self-funding companies and a much smaller rump of publicly funded activities. So, in effect, from April 2012, the UK has had no publicly funded body to implement energy efficiency, making it probably almost unique in the EU.

efficiency with its implied economic benefit. As the philosophy of wider sustainability and 'green growth' became more established, the EU set up the Intelligent Energy Executive Agency (IEEA) in 2003.[8] A growing awareness that Europe was not on track to achieve its '20 per cent' goal led to stronger measures in the Energy Efficiency Directive in 2012.

The role of quantitative targets for energy efficiency, established by governments and/or their institutions, are complex. They may be in the form of long-term goals, but generally focus on shorter or mid-term targets and/or monitoring requirements. Negotiating such targets can take a lot of political capital, and their status and import can vary enormously. They provide an overall goal and framework, and energy laws may be shaped around the conception of these targets; alternatively, they may soon be forgotten, revised or rescheduled, depending both upon actual progress and shifting political winds. Objectives, institutions and agreements are all very well, but they do not guarantee delivery.

5.3 The end-use toolbox

In practice, governments soon realised they needed to develop a far wider range of tools to try and deliver on the promise of energy efficiency. This chapter looks at three main categories:

- *Information-based tools* are designed to increase the visibility of opportunities for energy saving and their net benefits by lowering transaction costs, building capacity and ensuring an easier access to reliable information.
- *Regulatory standards* specify minimum efficiency levels at which producers must supply new products. Standards do not directly influence consumer values and behaviour but they rather 'bypass' common barriers to nullify potentially sub-optimal choices.
- *Financial incentives* address the issues of capital access and related barriers through mechanisms such as preferential loans, tax credits and investment subsidies.

This section looks briefly at some of the policies, and the lessons, in each of these areas – first as they apply to consumers and products, and then to business and public sector organisations. The chapter then looks more closely at the additional measures undertaken for the buildings sector, including obligations on energy suppliers, before finally stepping back to look at what has been achieved and learned from the experience.

The end-use toolbox: consumers and products

To overcome pervasive barriers, efforts to supply information to consumers have included: information campaigns to provide basic knowledge and create awareness, energy efficiency labels to provide information about the performance of specific goods, and certification and related services to offer more tailored information and increase confidence.

Information campaigns

Many early efforts placed a lot of emphasis on providing information, with the belief that knowledge would help rational consumers make the 'best choices'. Consumer education or conservation programmes generally targeted at behavioural change have been spread thinly across a variety of environmental concerns: from littering, to money-saving efficiency measures, to helping their country through an economically challenging time by enduring a little discomfort. Energy efficiency knowledge is partly a 'public' good and lack of it accounts for numerous problems. This forms the bedrock of why most governments have sought to ensure that consumers are adequately informed – and why markets themselves don't do this.[9]

These programmes certainly had some success during periods of crisis such as the fuel price hikes or electricity shortages.[10] As noted, the enduring impact of transient information is more debatable. Information can be easily forgotten or may be too complex or too hard to relate to. Furthermore, even when consumers might benefit from lower energy costs, it is often not seen as relevant: few people consider kilowatt hours when buying or watching their TV (still less when it is on 'standby' mode). Prospective home-buyers or renters viewing a new house rarely ask about the energy bill and, unless retrofitting is mandatory – as in the case of Sweden – owners are unlikely to perform efficiency improvements before selling or renting their properties.

The barriers outlined in the previous chapter are about far more than just information, but lack of information makes it hard to achieve much else. Information and awareness campaigns are, of course, familiar in the private sector – where it is called marketing. It is decades since Galbraith dispatched the quaint idea that marketing in rich countries is simply about informing consumers how to get what they want; it is just as much about creating wants.[11] Having reviewed the limited impact of many early energy conservation advertising programmes, government-sponsored campaigns eventually caught up.

For example, in Ireland a 'Power of One' campaign was set up with three main goals: (1) to enhance awareness of the sources, associated costs and impacts of energy; (2) to provide consumers with adequate information on the cost and environmental implications of their decisions; and (3) to create a sense of individual responsibility among the public – an attempt to initiate behavioural change. The Power of One programme is broken down into multiple targeted campaigns. The 'Power of One Street' part of the campaign was aimed at households, taking participants from a variety of geographical locations and setting them efficiency targets on a monthly basis, and the results were then reported to the media. The 'Power of One Schools' programme targets transition year students to identify energy inefficiency and ways to promote efficiency locally. The 'Power of One at Work' programme targets owners, managers and employees. Indeed, energy efficiency and marketing are just as relevant for businesses as consumers – an important point which this chapter elaborates in the 'corporate energy efficiency' section.

The institutions set up in the UK (see Box 5.1) similarly learned the importance of marketing, particularly when combined with regional outreach. The Energy Saving Trust established a network of advice centres across the country to provide information ranging from local grants to advice on local installers. The Carbon Trust, building upon the Energy Efficiency Best Practice Programme that it inherited from the government, worked through Regional Development Agencies to reach businesses across the country. The impact of marketing was consistently evident in the surge of enquiries to these centres and to helplines. The Carbon Trust also found that climate change could be an important lever for interest from larger companies.[12]

Labels

If changing 'things' is easier than changing behaviour, labelling them is an obvious way to provide information about their energy use, and it can also help to raise awareness.[13] Energy efficiency labels inform everyone about relative energy performance; aside from the direct benefits of information to the market, the systems and awareness involved are also a valuable precursor to standards and financial incentives.

While providing information sounds relatively innocuous, it proved otherwise. Initial proposals proved controversial; it was, for example, only in 1990 that the UK dropped its staunch opposition to energy efficiency labels.[14] Manufacturers of such goods argued that government had no business 'interfering' and that the costs of measuring and monitoring would be too high. Their next line of defence was to argue that labels should be purely voluntary. Against this backdrop, a number of countries have implemented voluntary programmes while others have taken firmer steps and introduced mandatory schemes.

For example, in the US the voluntary Energy Star label was developed initially as a status label for the top 25 per cent most energy-efficient consumer electronics. The US Environmental

Protection Agency manages the programme, complemented with media campaigns raising public awareness of the ES brand, rewarding firms which perform well and publicising cooperative efforts among industry players. As a product label, Energy Star has been adopted as a de facto international standard for consumer electronics. Following this success, the programme has now been expanded to rate the efficiency of home appliances and commercial and residential buildings. Since 2000, public interest in this efficiency 'brand' has helped to support a tenfold increase in the number of Energy Star products purchased annually in the US.[15]

The programme has, however, also pointed to a limitation. The branding of Energy Star homes tended to appeal to a more affluent clientele, which coincides with bigger homes where the standard is also easier to reach. Despite a study indicating that Energy Star homes are at least 15 per cent more efficient than those built according to the International Residential Code (IRC) of 2004, it seems this was more than offset by the larger size of Energy Star homes, which a study in Arizona found to use 12 per cent more electricity than non-labelled homes. The definition of efficiency, its use and its scope matter.[16]

While voluntary labelling is politically easier to initiate, its impact may be limited because no one *has* to change. Their impact may hinge on an advertising campaign to popularise the label as a new brand because they must rely on market demand and interest to gain success.

Mandatory labels have the advantage of covering all manufacturers equally, thus enabling comparison across the range. They may not have the same drive for 'brand' appeal, but because of this the information may be more robust and trusted.

The first target for mandatory labelling was refrigeration, in which the scale of wastage was most striking. In Europe, after a long struggle, the European Energy Labelling Directive of 1992 established the legal basis for harmonised labelling and product information schemes for household appliances across Europe.[17] The refrigeration industry argued that it needed more time to adapt 'an argument they have been using for the last fifteen years', as the *Financial Times* noted drily at the time, but could not then forestall the mandatory EU scheme which applies a ranking order, A–G, on products' efficiency levels (Figure 5.1).[18] In Europe and elsewhere, the list of goods was steadily expanded to a wide range of 'white goods' – washing machines, dryers, dishwashers – and others, including air conditioners, lights and water heaters, have followed.

The European programme led to a rapid increase in the market share of the most efficient appliances; in the decade after 1995 (when labels were first applied to refrigerators), sales of the A ranking models rose more than tenfold, from 5 per cent to almost 60 per cent. Results have followed a similar pattern across other appliances. In many cases, the combined response of consumers and producers (introducing more efficient models) was complemented by accompanying rebate and information programmes, designed to secure an overall 'market transformation'. Similar approaches have now been adopted in many parts of the world, particularly spanning many countries in Asia, South America and Africa (see Box 5.3). The Australian label, using up to six stars as the symbol of efficiency performance, has experienced similar success and the classification system has also been replicated in Thailand and Korea.

The biggest limitation of the European scheme soon emerged as a conservative set of starting assumptions. With A ranked fridges forming only about 2 per cent of sales at the start of the scheme, it was hard to foresee that within ten years they would dominate the market; the EU had to introduce A+, A++ and even A+++ categories. This 'problem of success' created unnecessary confusion for consumers: consumers may purchase an 'A' refrigerator assuming it to be the best, but with class B and worse fading out entirely, 'A' is now pretty average.

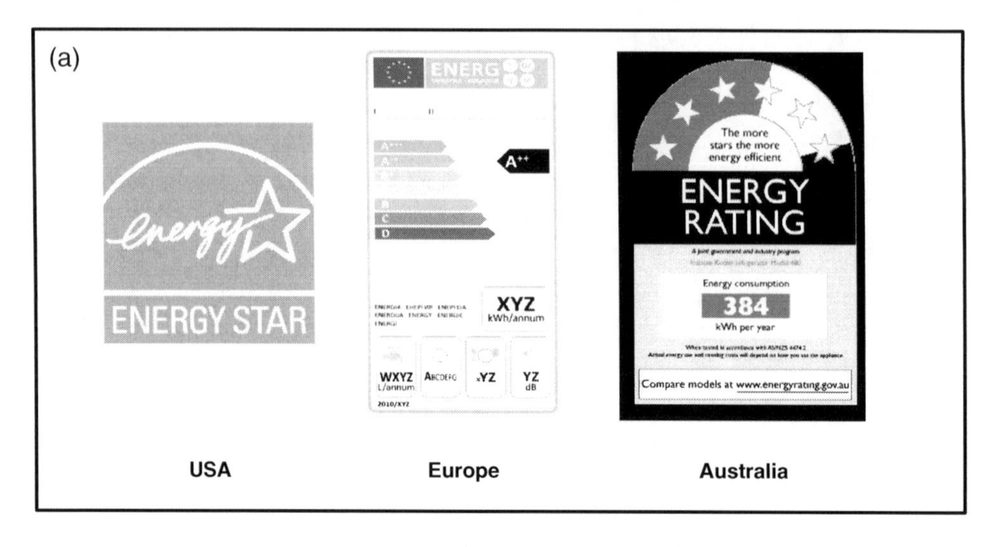

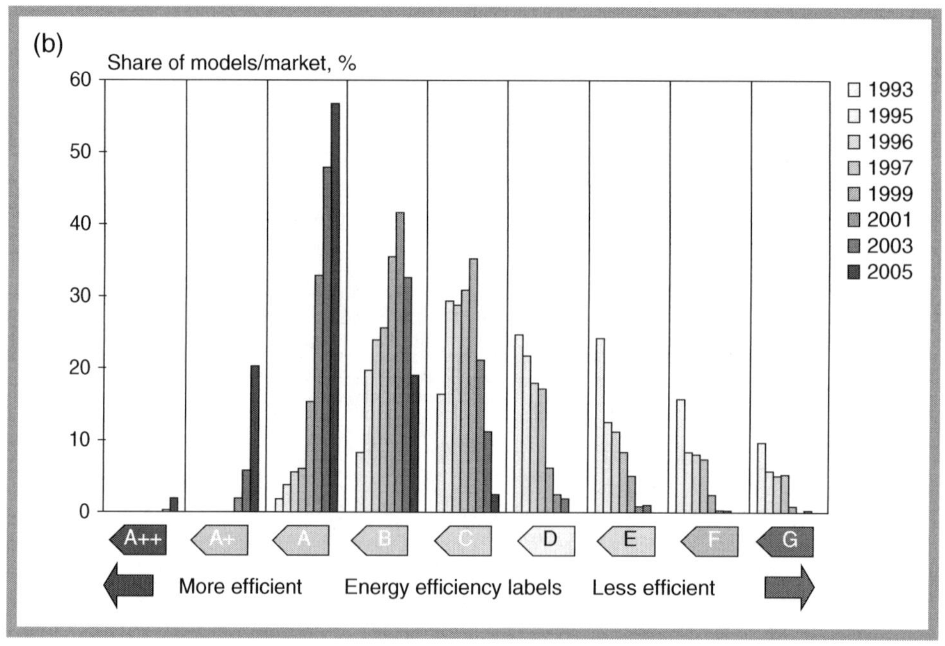

Figure 5.1 (a) EU energy efficiency labels and (b) market share of appliances in Europe over time since introducing labels

Source: IPCC (2007) and Energy Commission (2010).

The evidence is that mandatory labelling has done more than voluntary programmes to increase the market share/number sold of the most efficient goods: by ensuring that all products are labelled and not just the efficient ones, there is greater scope for product comparison, greater awareness among consumers, more risk in the continuation of old product lines and a stronger push to improve. However, politically it can be slow and arduous to impose obligatory labelling, whereas voluntary programmes can be initiated more quickly and more easily.

Building on the undisputed success of its programmes for mandatory labelling of 'white goods', EU policy moved on to vehicles and from there on to the tougher challenge of buildings (see section 5.6).

Labelling schemes vastly reduce the cost of consumers acquiring information (one of the hidden costs in Figure 4.5), and may marginally address some of the behavioural dimensions by drawing attention to the opportunities. This may lift appliances out of the zones of 'adverse selection' and other perverse behavioural anomalies noted in Chapter 4. Yet mapping information against the specific driver-barriers of Figure 4.5 underlines that it does not help households with capital constraints to bear the upfront capital cost of more efficient choices and does little or nothing to address 'split incentives'. Clearly additional medicine is needed.

Standards

Rather than millions of private consumers having to obtain, understand and decide on the best energy buys, governments can cut through the swathe of these transaction costs by mandating a minimum level of performance. An efficiency standard in effect bans less efficient products from the market.

Ironically perhaps, given its free market reputation, it was the US that became the first major battleground for energy efficiency standards. Individual states argued that they had a right to impose standards, and California – facing a looming power supply and demand gap – led the charge in the 1980s with strong standards on refrigerators.[19] Faced with a potential proliferation of different state standards, the industry caved in and started lobbying for a federal standard. The impact is illustrated in Figure 5.2: aided by a rising electricity price, the energy intensity of refrigerators halved from 1970 to 1985 and continued to decline substantially even as electricity prices fell. The EU also built upon its base of product labelling and introduced complementary standards in 1999, for example, banning fridges below 'C' rating.

Efficient lighting offers a textbook study of the issues and interactions that occur between information and standards. Technologies for the fluorescent lighting used in most commercial premises can vary hugely in their efficiency. Inefficient 'ballasts' continued to be used for many years after better technologies became available across the US and Europe. In the US, the Lighting Trade Association testified that they could not understand why the more efficient product was not chosen by consumers, and use of the more efficient technologies tended to track state regulations that outlawed the less efficient version.

After a national standard was enforced in 1988 eliminating transaction costs to customers, a cost-benefit analysis of 5,000 commercial buildings suggested a typical '100 per cent' payback rate – recouping the additional cost within a year.[20] As the technology for 'compact fluorescent' bulbs suitable for household use improved, governments eventually – two decades later – gathered the courage to apply similarly stark but powerful legislation to domestic lighting. Spurred by an Australian ban on filament bulbs, the EU followed suit in 2010. Experience with standards and labels in appliances, lighting and indeed homes globally has proven that despite fears of raised costs, manufacturers or builders have not suffered profit loss or had to reduce the service provided to meet regulation.[21]

For more complex cases like cars, standards may be imposed as a minimum energy efficiency for a product line or as a corporate average. After the first oil shock, the US Corporate Average Fuel Economy (CAFE) standard in 1975 required the average fleet fuel efficiency to double, reaching 27.5 miles per (US) gallon for passenger cars, with lower levels for light

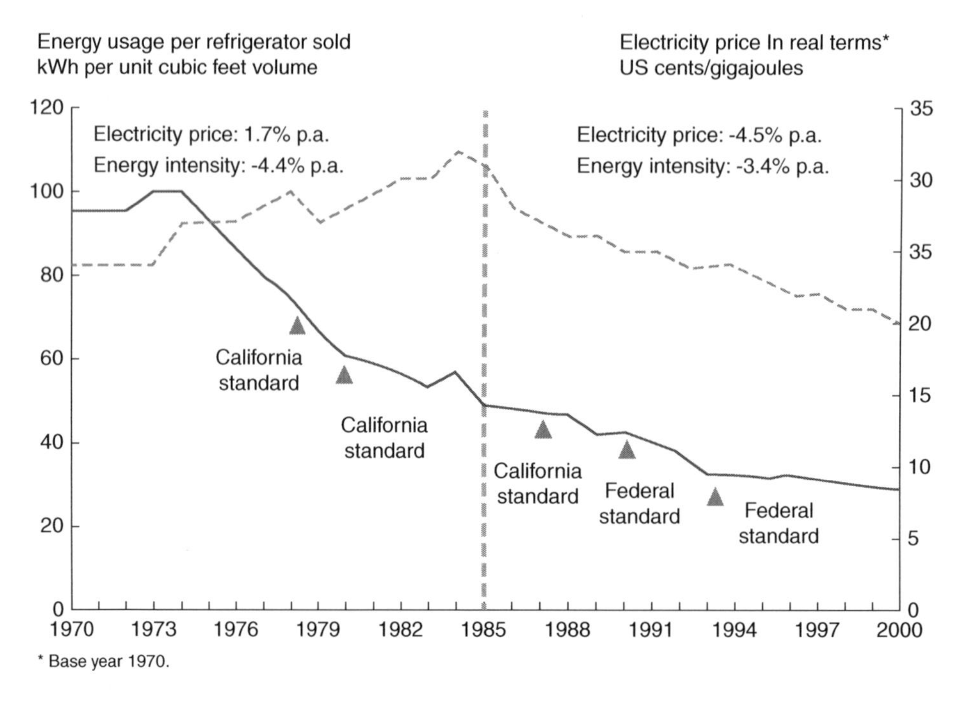

Figure 5.2 Trends of electricity price and energy efficiency improvement (US refrigerators)

Note: A solid line refers to the trend of energy usage per refrigerator sold while a dotted line refers to the trend of electricity price in real terms.

Source: McKinsey & Co. (2007).

trucks in 1985.[22] This clearly reduced fuel use compared to what might have been – on one estimate by 18 per cent – but the standards then became politically locked in, with industry resistance preventing them being tightened for almost 30 years (Figure 5.3(a)).[23]

Vehicle efficiency standards were introduced in many regions in subsequent years, and the US standards over time were surpassed by many others (Figure 5.3(b)). Nowadays, more than 70 per cent of light-duty vehicles sold globally are subject to mandatory fuel or emissions standards.[24] By 2005 North American standards were conspicuously weaker than most other regions, including major emerging economies. Manufacturers also exploited a loophole in the form of Sports Utility Vehicles (SUVs) – a new class of gas guzzlers that qualified as 'light trucks' and enjoyed less stringent fuel economy standards.

It was not until the US stimulus package of 2009, which bailed out key US car companies after the credit crunch, that the US administration was finally able to impose a dramatic increase in US auto standards.[25] The bailout plan explicitly stipulated that General Motors and Chrysler must aggressively invest in fuel-efficient technologies and revive their international competitiveness.[26] With large and inefficient vehicles increasingly growing out of fashion, these industrial leaders had no other choice than to acquiesce. The Obama administration soon announced its first round of the new CAFE standard in May 2010, to achieve a fuel economy of 35.5 miles per gallons by 2016.[27] The second round released in 2012 set a further goal of 54.5 miles per gallon by 2025. These new regulations also include incentives to produce and sell 'game-changing' technologies – electric, plug-in hybrid and

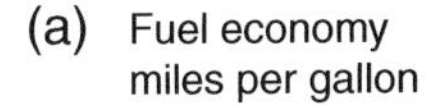

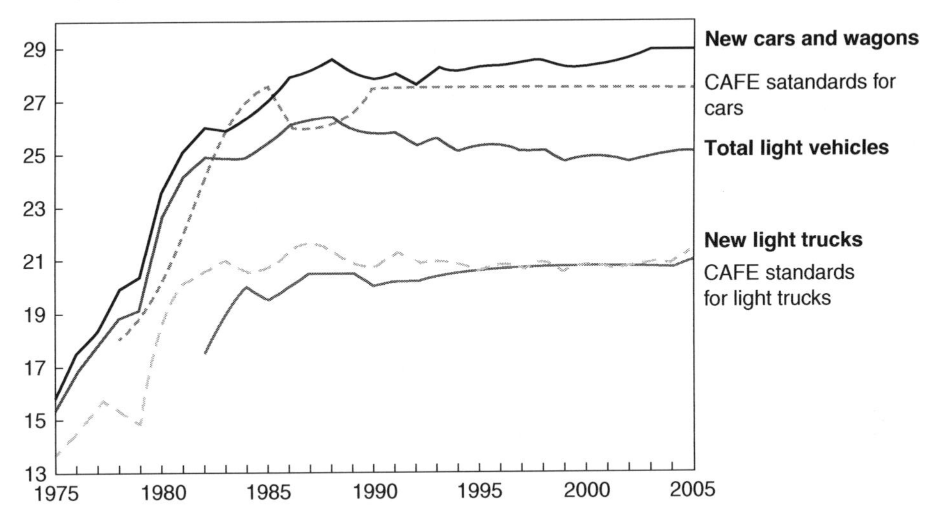

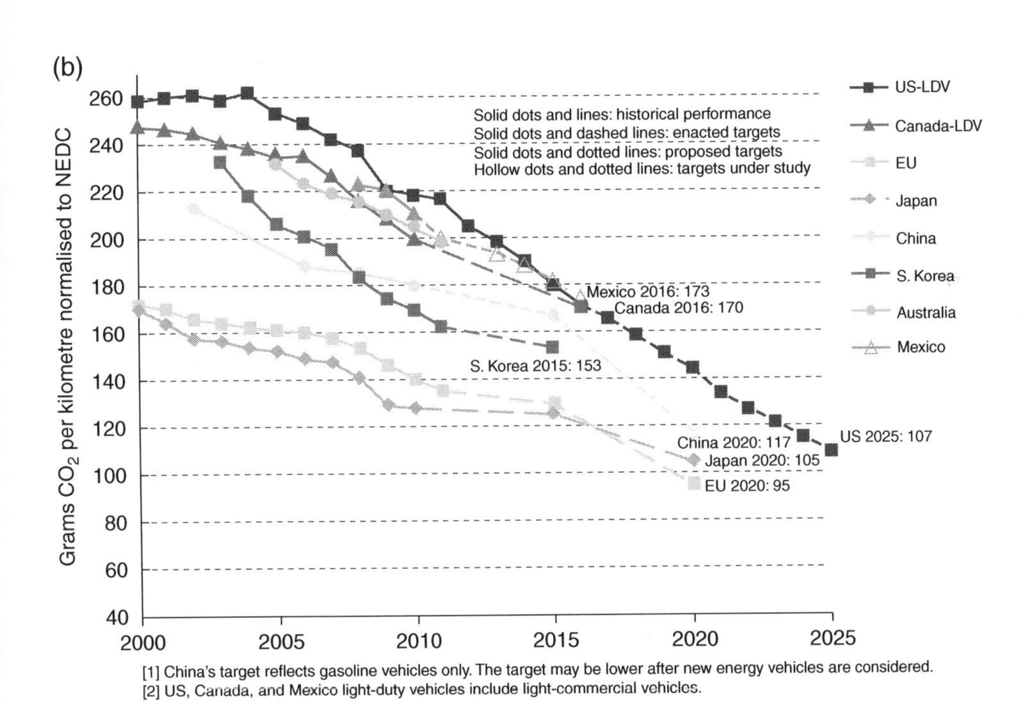

Figure 5.3 (a) Fuel economy trends of US light vehicles, 1975–2005 (b) Evolution of standards in different countries – historical relationship between vehicle ownership and GDP per capita

Sources: (a) McKinsey & Co. (2007); (b) GEA (2012).

fuel cell vehicles – in the form of credit multipliers, which counted these next-generation vehicles as more than one vehicle in manufacturers' compliance calculation. With an expected rebound rate of 10 per cent (section 5.6), the cumulative US savings over the lifetime of new vehicles are estimated to be around 4 billion barrels in fuel and 2 billion metric tonnes of CO_2 equivalent in emissions for the 2017 and 2025 period. The net cost-benefit of the standards is assessed by the government to be hugely positive.[28]

The history of the US CAFE standards illustrates both the strengths and weaknesses of standards. While forcing products to reach a minimum level of efficiency, they do not necessarily raise the standard of the entire sector. They can also be prone to political lock-in and in some cases deter the industry from improvements in leading edge technologies which might encourage regulators to raise the standard. In the US case it took a new US administration, with power over the industry from the bailout, and the combined motivation of concerns about US oil dependence and environment to unblock thirty years of stasis.

The Japanese 'Top Runner' programme developed a smart solution to such problems: the standard was designed to evolve in ways predictably linked to the most efficient products on the market. This alleviates the ponderous process of negotiating and enforcing new targets. Ministerial guidance and industry collaboration ensures that advances in energy efficiency do not plateau. The mandatory standards, based on the market's best performer, encourage product designers to adopt the latest technologies that may be available but have not been widely adopted for commercial use. This accelerates the efficiency improvements.[29] The accompanying labelling schemes also include both mandatory and voluntary components: all appliances must display their absolute energy efficiency, while labels indicating efficiency relative to that of the Top Runner remain optional.

The Top Runner programme originally covered eleven energy-intensive products and has continued to expand. Eighteen products were covered in 2002, 21 in 2005 and 23 in 2009, ranging from household appliances to passenger cars. Auxiliary policy instruments such as tax credits and public green procurement schemes also favour the Top Runner compliant products, while 'name-and-shame' penalties deter non-compliance. The scheme proved highly effective; in the space of six years after its introduction (1997–2003), air conditioning efficiency increased by 68 per cent, fridges by 55 per cent and televisions by 25 per cent, all surpassing their original targets.[30]

The experience illustrated one other important feature of policy: it was not driven purely by energy cost considerations. Concerns about energy security combined with high-profile environmental concerns – 1997 was the year that Japan hosted the Kyoto conference. These concerns coalesced with a sense of industrial opportunity to support ambitious policy, which turned out to be highly effective and economically beneficial.

Fiscal measures

'Money talks': unfortunately, it often says no – particularly to the upfront costs involved in domestic energy efficiency measures. Information and standards can help with some products and can be used for new homes – where energy performance is embodied in a purchase and higher standards add few costs. But they do not overcome the financial and related behavioural barriers noted in Chapter 4. When considering better insulation, a new boiler or heating controls, for example, the upfront costs can be a powerful deterrent.

Fiscal measures such as direct subsidies, low-interest loans and tax credits can help people overcome just that. Such measures have become income-dependent (targeted at lower income groups), technology-specific (focusing on new and proven technologies) or cost-effectiveness based (must achieve specified levels of savings per year), each of which offers targeted incentives for effective policy impact.[31]

Investment subsidies encourage energy efficient purchase and investment by reducing upfront costs and payback periods.[32] Subsidies may be fixed, set as a percentage of the investment or proportional to the energy saved, and were popular during the 1970s and 1980s to fund building retrofits. Their general application is of course more costly to the government and more prone to 'free-riding' by those who might have paid anyway for efficiency improvements, one obvious factor leading to more targeted supports.

Fiscal incentives for lower-income households have promoted household energy efficiency while simultaneously tackling fuel poverty. Poor families often live in inefficient dwellings, so even the simplest options such as loft and cavity wall insulation can go a long way towards improving comfort and ultimately saving energy. In the UK, the Home Energy Efficiency Scheme (HEES) introduced in 1991 offered grants for vulnerable groups of households and the programme insulated 2.5 million such homes – 10 per cent of UK households – during the 1990s.[33] Home insulation incentives fared less well among middle-class or affluent households.

Any consumer subsidy programme faces a basic distributional choice. If the programme is not targeted, there will be 'free-riding' by those who would have bought the more efficient option anyway, and it gets expensive. If it is targeted, poorer groups are more likely to benefit but it becomes far more bureaucratic and narrow, raising transaction costs and reducing aggregate impact.

Incentives through tax deductions have also been used. The US Energy Star programme, for example, offered up to US$500 in federal tax credit or 10 per cent of costs for families purchasing an efficient product or renewable energy generators. They can work well when taxable income and collection rates are sufficiently high but have also been criticised as too inflexible, narrow ranging and lacking clarity.[34] They are more naturally suited to encourage business energy efficiency, with sometimes unexpected results, as detailed in the next section.

The tax system is more easily adapted to raise the cost of inefficient products and this has been particularly important in vehicles. The gas guzzler tax in the US imposes a progressive tax on passenger vehicles which fail to meet fuel economy standards.[35] The Vehicle Excise Duty (VED) in the UK was scaled to add almost £1,000 to the initial cost of the least efficient cars, about half this level annually thereafter and zero for the most efficient (A rated) cars; not surprisingly, this began to exert strong leverage towards more efficient vehicles.[36] Other EU countries have adopted CO_2 emissions into their vehicle tax formulas, including Germany, France, Italy and Spain, while a growing number of countries implement tax reduction schemes for alternative-fuelled vehicles. Explicit taxes on environmental 'bads' also promote cleaner vehicles: in central London, for example, congestion charge exemption is a key driver of electric vehicles and hybrids.

The end-use toolbox: organisational energy efficiency

Inefficient energy use is not confined to the quirks of individuals. As explained in Chapter 4, businesses and other organisations often seem little better, owing to the combined obstacles that can arise from a wide range of structural barriers. The advantages that in principle arise from the greater capacity for rational evaluation of options in an organisation often seem offset by other problems. If no individual is made clearly responsible, it is 'someone else's money' (the organisation's) that is wasted by inefficient energy use. Even if someone is

Box 5.2 The global spread of energy-efficiency policies

The World Energy Council, an international forum of government and energy industry experts, conducts periodic surveys to assess the state of energy efficiency policy implementation. A major survey in 2009 covered 88 countries across Europe, America, Asia and the Pacific, Africa and the Middle East, accounting for 90% of global energy consumption. Key findings included the following:

- *Energy efficiency is increasingly seen as a national priority.* Of all countries surveyed, around two-thirds have established a national agency responsible for energy efficiency policy implementation. Over 90% of such agencies operate as a national ministry or part of one, illustrating the importance of energy efficiency as a national priority. These agencies are typically funded through state budgets and increasingly funded from fuel taxes (e.g. Spain, Norway, Switzerland, Thailand). Also, overseas development funds often support these agencies in developing countries.
- *Quantitative targets are increasingly adopted as key milestones for policy implementation.* Quantitative targets, in the form of savings by volume (e.g. Mtoe), intensity improvements and rate of efficiency improvements are adopted widely as part of efficiency programme goals. The survey shows that around 70% of countries now have quantitative targets, a figure that is almost twice that of 2006. Around two-thirds of targets are based on total energy consumption (either final or primary) while the rest adopt sectoral targets.
- *Efficiency labels are becoming a universal feature of consumer appliances, with standards less so, especially outside of non-OECD countries.* Around 60 of the 88 countries surveyed implement labelling schemes, while ten expect the introduction of such schemes in the near future. Overwhelmingly, the majority (90%) of these schemes are mandatory. Minimum efficiency standards are more popular among OECD countries, but less so in developing countries.
- *Efficiency standards for new buildings are adopted widely in OECD countries, but less so in non-OECD countries.* The majority of OECD countries now have building standards for both commercial and residential buildings. While a number of non-OECD countries are also adopting similar measures (e.g. Singapore, the Philippines, Algeria, Tunisia, Egypt, etc.), there is still a large gap in implementation (see Figure 5.4). In recent years, energy certificate schemes for existing buildings are also emerging, as seen in the EU 2006/2007 Directive on Buildings.
- *CO_2 labelling schemes for new cars are emerging.* CO_2 and fuel efficiency labelling for new vehicles have been implemented in all EU countries. Around ten additional countries including Australia, Brazil, China, Japan, India, New Zealand, South Korea and the United States also have similar schemes. CO_2 labelling can be more effective when used in combination with greener taxation, as seen in some EU countries.
- *Energy audit, reporting and management are less common, but are used to encourage energy efficiency.* Around a quarter of the countries have mandatory energy consumption reporting schemes, and one-fifth have the mandatory appointment of energy managers. These measures are more common among the OECD countries

and typically apply to large consumers in industry and increasingly in the service sector.

- *Financial incentives are widely used in OECD countries and increasingly so in non-OECD countries.* Financial and fiscal incentives are common among the OECD countries where more than 80% implement such measures. In general, financial incentives are more widely used and more popular than fiscal measures; investment and audit subsidies each account for around one-third of all measures. Currently, about one-third of subsidies globally are used for solar water heaters and a quarter for compact fluorescent lighting.

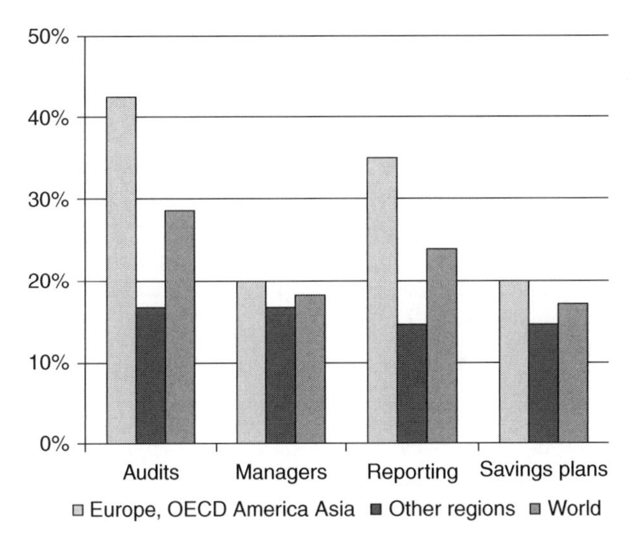

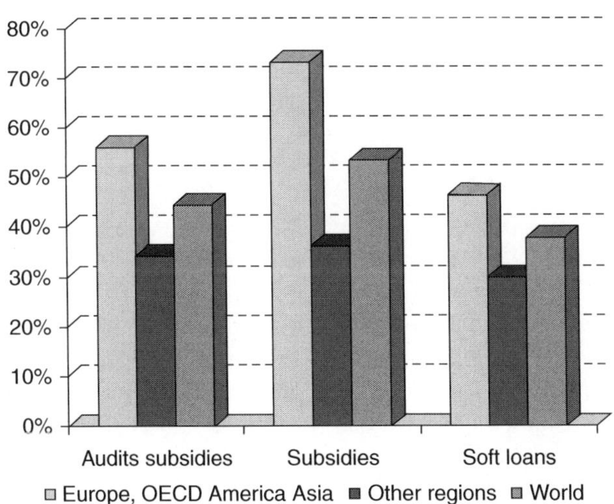

Figure 5.4 International adoption of corporate energy-efficiency regulations and incentives

Note: The figures show adoption by region of energy-related legislation on companies (top) and financial incentives (bottom).

made responsible, they may face externally imposed budget lines or split incentives when energy savings cannot be quantified or attributed to the people empowered to spend. Organisations are replete with the potential for failures in part arising from, in economic language, the problem of aligning the aims of 'principals' with their 'agents' lower down in an organisation – the discounting pyramid being one manifestation of this (Chapter 4). As evidence of such problems has accumulated, growing efforts have sought to improve business energy efficiency as well.

Information

As with individuals, campaigns have helped to galvanise corporate interest. The UK Carbon Trust, charged with transforming the energy efficiency of UK business, soon concluded that trying to reach out to tens of thousands of companies in a piecemeal manner would be woefully ineffective in the absence of wider motivation. Hard-hitting advertising campaigns proved to be effective, yet controversial. The raised business awareness of both risks and opportunities brought companies knocking on the door for the services the Carbon Trust was set up to provide; enquiries increased substantially after each campaign.

Capitalising on interest requires targeted services. The trend in information measures, which originally focused on providing general information (e.g. about energy-efficient machinery and office equipment), has gradually shifted its emphasis towards tailored information (e.g. energy auditing) and related managerial change based around the old principal that 'you can't manage what you don't measure'.

Energy management as a focused activity emerged first and foremost for large energy consuming companies, given their energy costs. As the easy gains were realised, and with the growth of light manufacturing and the service sector, attention has expanded to less intensive consumers – most notably in the commercial sector.

As with domestic consumers, general energy efficiency information is limited in its impact. It has proved most powerful when combined with other tools: case-specific information, subsidies for recommended improvements, other fiscal policies (e.g. tax deduction, low or no interest loans, grants or credits for energy efficiency improvements) and/or corporate emphasis on greener identities can indeed be powerful drivers for energy efficiency.

Energy reporting, disclosure and management

The business principle of 'what is measured gets improved' has proved to be correct, especially with the right set of incentives. Energy reporting schemes require large consumers to monitor energy use with increasing emphasis on public reporting. Mandatory schemes have spread across more sectors and countries on all continents (see Figure 5.4).[37] A next step is to conduct energy audits, which examine a specific organisation's energy use and evaluate saving opportunities. The resulting awareness often results in organisations appointing an energy manager to reduce energy costs on a more ongoing basis.[38] In a number of countries – Denmark, Japan, Romania, Portugal and, more recently, in Singapore and India – energy managers are compulsory for companies above a certain size.

Because of the benefits, such schemes can also be voluntary. The Carbon Trust Standard in the UK requires its participants (of which there are over 900 of all business sizes) to measure their corporate carbon footprint, demonstrate active carbon management and produce an absolute reduction in emissions year on year (or relative efficiency improvement).[39]

Accreditations are provided by an independent body and enable the recipient to display the Standard's logo in marketing and communication materials, taking advantage of the increasing public awareness to reap reputational gain in the eyes of consumers, employees and shareholders. An additional motivation was the insight that the effort gave them into their energy costs and supply chains.

Moving beyond simple behavioural measures to lessen energy wastage requires the purchase of better equipment. Chapter 4 outlined how individuals and organisations often make decisions based on quick rules of thumb. Public sector organisations are often subject to externally-imposed constraints on capital expenditure, targeted at their core function, which hinders their ability to 'invest' in energy savings. Small companies tend to be strapped for cash overall, pointing to a key role for fiscal support (as outlined below).

Corporate energy management has proved to be about much more than just energy consumption and costs. Neither the appointment of an energy manager nor audits alone guarantee that firms will follow up – even big ones for which cash is not a problem. This is particularly true when energy is seen as mere unavoidable cost rather than as part of the core business to be 'proactively managed'.[40] For larger organisations in particular, energy performance improves most when linked with corporate branding, stakeholder engagement and new business opportunities – and the general evaluation of 'good management'.

Emphasis has thus shifted towards leveraging knowledge to drive corporate cultural change in a resource and carbon-constrained world. For example, a recent Australian initiative (Energy Efficiency Opportunities Programme) promotes the energy audit as a 'cultural process', in which a firm strengthens its organisational leadership, management process, communication skills, data accuracy and analysis. While mandatory, it also tries to accommodate as much individuality as possible, to form part of the day-to-day business decision-making processes.

The steady growth of programmes and experiences over the years has, finally, led to a new global 'energy management standard', ISO50001, of the International Standards Organisation. ISO standards are entirely voluntary but have frequently had a considerable impact as a badge of good practice, and ISO50001 sets out a standard set of procedures for good energy management.[41]

Fiscal measures

To help organisations address financial barriers, myriad financial supports have been developed spanning R&D and demonstration project grants, investment grants, energy audit subsidies, preferential loans and tax and other incentives to encourage energy-saving investments.

Grants and subsidies were among the first measures adopted during the 1970s. As illustrated in Chapter 11, energy efficiency R&D has tended to be modest. This may reflect the relative weakness of producer interests and the challenging nature of technology transfer and take-up. Investment grants help firms to upgrade energy efficiency but many governments have limited them to a few promising technologies, given the cost of more broad-ranging grants. Incidents of free-riding, insufficient knowledge as well as cumbersome procedures are often cited as drawbacks of such investment subsidies.

Audit subsidies, preferential loans and tax incentives have also been popular in many countries. Audit subsides, which have frequently been tied to voluntary industrial agreements (e.g. in Denmark, the Netherlands and Sweden), are often less expensive than other financial instruments – they support the analysis, not the implementation. This has the limitations noted above.

Their effectiveness depends largely on take-up, which in turn depends on ancillary policies including low-cost loans and investment subsidies and, more importantly, on energy prices.

Preferential loans, which offer a zero- or low-interest rate loan for specific investments, are also popular, as they can be implemented easily by lending institutions. Some governments have also established credit guarantee schemes as a way to lower risks and encourage lending. However, preferential loans alone are unlikely to offer sufficient incentive in times of low interest rates. Finally, tax incentives such as accelerated depreciation, tax rebates and exemption are also commonly used to stimulate efficiency investments. As with domestic schemes, there is a tension between the complexity of targeting and the risks of 'free-riding'.[42]

When costs may be too prohibitive but potential societal benefits will abound, another option is direct state funding, presenting the combined challenges of agreeing the funding and developing a good way to deliver it. The Carbon Trust, as explained in Box 5.1, was unusual in being state-funded but operated like a private company to deliver energy efficiency across UK business. Apart from the (often tedious) requirements limiting state aid, it had full freedom to adapt its programmes based on market feedback. The experience underlined the importance both of developing a brand that businesses can trust, and of understanding who does, and does not, need fiscal support.

Over time it became clear that financing per se was *not* a barrier for large companies – the key challenge was motivating their interest and, once the value of working with the Carbon Trust had been established over years, the Trust was able to charge for its services to large companies. However, funding is a dominant concern for small and medium-sized enterprises. A zero-interest loans facility, offering four-year revolving loans for energy efficiency investments in this sector proved very popular. It was subsequently seen as so relevant to the perceived wider national challenge of fostering the growth of small enterprises that as part of the UK Stimulus Package in 2009 the government awarded an additional £100m to massively expand the loans programme.

The experience demonstrates that money is only one component, and some apparently 'fiscal' measures can have various unexpected impacts. Notably, the UK operated an Enhanced Capital Allowances scheme, under which companies could claim 100 per cent tax depreciation on their purchases from a government-approved list of efficient technologies. It turned out that while many companies bought from the list, some never even claimed the credit. Instead, the market simply focused on the technologies list as an indication of advanced, efficient technologies. For purchasers, it served as a reference point for good purchases; for manufacturers, it was a benchmark for which they made sure that new products would qualify. In this case the tax allowance itself was probably of secondary importance.

Another approach to funding energy efficiency measures is through obligations placed on energy supplies (Box 5.3). Also known as 'energy efficiency resource standards', such policies have become popular across some US States and EU member countries (Bertoldi, Rezessy et al. 2010). In Italy and France, obligations have been imposed upon power distributors and suppliers respectively, accompanied with tradable energy certificates e.g. White Certificates. In Belgium (Flanders region), Denmark, France, Ireland, Italy, the Netherlands, Poland, Portugal and the UK, these have been imposed on energy companies.[43]

5.4 Integration and application to buildings

Motivations are complex. Money talks, but its message needs to be aligned with other drivers of interest and action. A key conclusion from experience with energy efficiency is the need for a combination of approaches.

This is partly because, as any marketing manager knows, populations are diverse and markets are dynamic. Figure 5.5 illustrates how the three different basic tools of information, standards and financial support can combine over time to 'transform' a market of energy efficient goods. Information can move the curve to the right, particularly engaging the 'leaders'. The bulk of the population might respond much more strongly to financial incentives of various forms. Laggards – the old product lines selling to 'disengaged' consumers – may pay attention to neither. But once the bulk of products have moved and the better alternatives are well established, it becomes more feasible to set an effective standard – in some cases banning an entire class of energy-wasting technology (as with incandescent bulbs). The jump to more efficient fridges from 2000 onwards in Europe (see Figure 5.1) is also attributable to the previous year's ban on those below a 'C' ranking, which indirectly expanded the market for the more efficient ranges. The EU's growing problem since then, as noted, has been that of success, with 'A' class now in reality being the norm. Both the technology and the market have been transformed.

The need for 'policy packages' also arises because different instruments tackle different barriers. Figure 5.6 illustrates how various policies for building energy efficiency in the three classes considered relate to the four types of barrier-drivers. The matrix also includes energy efficiency obligations on suppliers, which have become a major specific source of funding for financial instruments but which take the emphasis away from the end-user (see Box 5.3). No individual instrument addresses more than two of the four classes. Combinations are thus essential, even without the dynamics of a diverse population spanning leaders to laggards.

The need for building and other safety codes is universally accepted, and the most obvious approach to new buildings is to incorporate energy efficiency standards as part of these codes. Almost every country has a different structure of codes with differing attention to energy efficiency. Energy regulation may range from thermal codes for individual components to more sophisticated all-inclusive 'performance standards', and different countries move at different times. It was not until 2005, for example, that the UK dramatically raised energy efficiency standards in its building codes ('Part L'), and then raised the stakes further by stipulating that all new UK homes should be zero carbon by 2016 (Defra, 2007). Such more encompassing

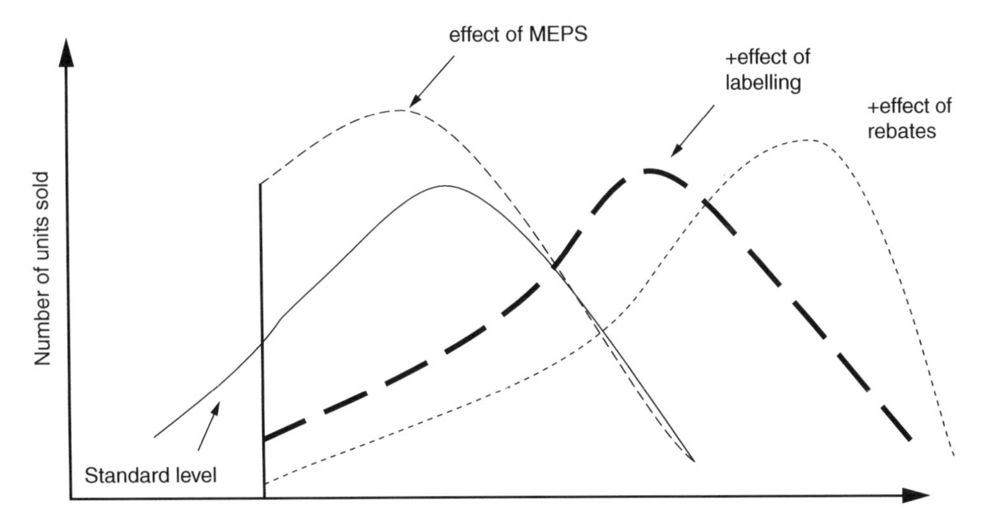

Figure 5.5 Combined effect of different instruments for energy efficiency

Source: Adapted from Jamasb and Pollitt (2011), Chapter 14, Figure 14.4.

	Information Tools			Regulatory Mechanisms			Fiscal Instruments		
	Mnd. Label	Voln. Label	Mnd. Audit	App. Stand.	Bld. Codes	EE. Olig.	CO_2/ E Tax	Tax Exp.	Cap. Sup.
Financial Cost & Benefit	X	X	X	X	X	√	√	√	√
Hidden Cost	X	X	X	√	X	X	X	X	X
Split Incentives	√	√	√	√	√	√	X	X	X
Behavioural Factors	√	√	√	X	X	X	X	X	X
Overall Effectiveness	High	M/H	High	High	High	Med*	Low	High	M/H
Cost Effectiveness	High	Med	Med	High	Med	High	Med	High	M/H

Figure 5.6 Energy efficiency policies matrix for building sector

Note: Mnd. Label (mandatory labelling and certification); Voln. Label (voluntary labelling and certification); Mnd. Audit (mandatory audit and energy management); App. Stand. (appliance standards); Bld. codes (building codes); EE. Oblg. (energy efficiency obligations and quotas); CO_2/E Tax (CO_2 or fuel taxation); Tax Exp (tax exemptions/ reductions); Cap. Sub. (capital subsidies and grants).

* Overall effectiveness measured against transformational requirements.

Source: Adapted from Ürge-Vorsatz *et al.* (2007) and Ryan *et al.* (2011).

approaches have been favoured in California, and in Germany and France helped to lay the basis for performance standards set in the EU's 2005 'eco-design' Directive, which is projected to save an equivalent of 12 per cent of the EU-wide power consumption by 2020.[44]

The sheer diversity of buildings complicates both standard-setting and evaluation, and enforcement can be difficult. The challenges may be exacerbated for commercial premises, which account for around one-third of building energy use in industrialised countries (see Chapter 4, Figure 4.6), given the diversity of uses and complexity of ownership and letting structures.

Existing stock

In industrialised countries, the bigger headache is how to tackle the existing stock of buildings. Technically the biggest opportunities arise during major refurbishment, which is more common in commercial than domestic buildings. Some countries including Germany and Sweden have applied energy standards in refurbishment, and the EU Performance of Buildings Directive (EPBD) establishes the same principle. Of course, both defining and policing this is complex.

Tackling this end of the challenge is easiest for home owners, where there is no 'tenant–landlord' divide – though incentives for long-run measures are still limited by the prospect

Box 5.3 Supplier regulation and energy-efficiency obligations

One route to energy efficiency in existing buildings is to focus upon the companies that supply energy to them. The network elements (for supplying gas or electricity) mean that these are regulated industries. Where companies are subject to direct price regulation, the regulators have to decide what costs can be counted and incorporated in the prices charged. By including energy efficiency measures in the 'rate base', these companies have some incentive to implement such measures on behalf of their consumers – particularly when, as is often the case, it is demonstrably cheaper than constructing new generation or supply infrastructure. This has been a major plank in US energy efficiency policy. In addition, many states in the US introduced a System Benefit Charge to fund energy efficiency, renewable energy and fuel-poverty-related programmes. (Rader and Wiser 1999) Revenues from the US RGGI emissions trading scheme are also largely directed to such programmes (Chapter 7).

Where the electricity market is 'liberalised', with supply companies competing for customers without any direct regulation of price, governments can still impose charges to fund related programmes or obligations on electricity and gas utilities or suppliers which require them to deliver a target level of energy savings. The target is calculated from their energy market share, their percentage of annual shares or the number of customers, and generates what are sometimes called 'white certificates' – accredited energy savings, matching the terminology of 'green certificates' often associated with renewable energy production.

The use of both approaches has spread rapidly, across US states and European countries and beyond. These policies come in many different forms, tailored for the specific market.

In practice much also depends on the implementation details. Initial targets have often been exceeded, but the scale of impact on energy efficiency is more debated. Inevitably, private suppliers seek out the cheapest way of meeting their obligations, and that means focusing on simpler measures, for example associated with lighting, rather than deeper and more costly challenges of long-lived thermal improvement. After several rounds, the targets and scope can iterate towards higher levels of ambition and deal with some of the problems experienced; under the UK CERT programme completed in 2012, over 3.9 million lofts and 2.5 million cavity walls were insulated, with a number of other measures including those targeted at vulnerable customers. The associated CO_2 savings were estimated at over 300 million tonnes of CO_2 over the lifetime of the measures – equivalent to taking all private cars off UK roads for five years.

A brief review of the Danish Energy Efficiency Obligation (which 'is the strongest in relation to industry' and does not contain specific targets on fuel poverty) usefully sets this in the context of European experience, citing a number of positive evaluations of their economic impact, and emphasises the diversity of potential designs. (Bundgaard *et al.* 2013)

Evaluations of US programmes suggest varied but generally positive gross benefits (contested with respect to possible recipient hidden costs (and benefits) – for example, see Chapter 4, note 54 and Box 4.2 and note 11). A number of issues remain in live debate without clear resolution.

of moving house. Several countries now require Energy Performance Certificates of some sort, including the UK, where these have to be included as part of the information supplied whenever a house is sold or rented. Overcoming the tenant-landlord divide has demanded stronger medicine. The EPBD mandates Display Energy Certificates for all public buildings – including some forms of commercial premises – with floor areas greater than 1,000m^2. Walk into any such building in Europe and the chances are you can now see instantly whether it is a good performer.[45]

The principle is that owners, renters and employees do not like a bad rating. This can generate pressure to do something about it across the tenant-landlord divide. Building owners can see a visible benefit if they improve the building. This appears to be reflected in property values, with both rents and selling prices reported to be higher for 'green buildings' in the US and Netherlands, but not (so far) in the UK.[46]

The biggest opportunities globally lie in enhancing energy efficiency in development. While most policy options are in the hands of the respective governments, international finance has been deployed to help developing countries both to develop policy and to finance key elements of energy efficiency programmes. Albeit with a far lower profile than the financing of big controversial projects (like dams and power stations), World Bank support helped to establish energy efficiency financing windows for the development of energy efficiency programmes and service companies in some of the big emerging economies.[47] The typical dilemma and argument is around the case for international public support of measures that appear so 'cost-effective' that countries should be doing them in their own interests anyway; the practical experience is that international finance and expertise can do a lot to raise awareness and accelerate the practical development of domestic energy efficiency efforts.[48]

5.5 A measured (but inadequately measured) success

Has energy efficiency policy been a success?

Forty years after the oil shocks first made energy efficiency a policy issue, there is clear evidence. In terms of either gross energy savings or economic value, the answer is a resounding yes. Yet on neither measure has it delivered overall the scale of benefits that proponents hoped for, or as much as seems to be needed when projected forward. Despite all the efforts, assessments show a continuing gap between the apparent market potential for energy saving and the actual uptake.[49] Despite the presence of clear labels on products, some consumers continue to buy less efficient models. Some inefficiencies and energy wastages are locked in by infrastructure, so that regardless of preference, consumers have limited power in choosing the performance of their energy stock (the 'enabling' box in Figure 4.4).

Measuring energy efficiency gains is not simple. As shown in Chapter 1, most countries have steadily reduced energy use per unit of GDP – the OECD countries cut it by more than a third over 1973–2005. This 'energy intensity' itself is affected by too many other factors (e.g. industrialisation and manufacturing structure, average home size, etc.) to be a good proxy for efficiency in itself. However, a major study across 11 OECD countries which sought to strip out these other effects concluded that energy efficiency accounts for about two-thirds of the decline in energy intensity. As illustrated in Figure 5.7, however, after energy prices fell in the 1980s, the improvement was insufficient to stabilise over energy use.[50]

But these aggregate results hide some telling national differences. Germany, Sweden and Denmark, for example, have improved national energy intensity averaging close to 3 per cent per year over 40 years – typically roughly doubling GDP without a significant increase in energy consumption (along with significant overall decarbonisation).[51]

Box 5.4 New approaches from analysing gaps and aligning driver-barrier analysis: the UK CRC energy efficiency scheme

The energy efficiency toolbox continues to expand as experience grows, particularly with instruments that may combine features and /or target particular additional barriers and drivers. A five-year review of the UK's Climate Change Programme in 2005 found compelling evidence to suggest that the main instrument covering the large but less energy-intensive organisations (mostly in the commercial and public sectors) – the Climate Change Levy (CCL) – had limited impact. The CCL had some initial 'attention' effect, but frequently no one in these companies was responsible for energy management, since it was regarded as an incidental cost spread across stores or offices or other facilities around the country; for many such companies the CCL cost was too small to drive organisational reform.

The resulting CRC scheme (initially the Carbon Reduction Commitment) was designed to tackle this by requiring the *parent organisation* overall to monitor and report its energy and CO_2 emissions across all sites, and to purchase corresponding emission allowances. The results are published – raising reputational issues for the organisations concerned. The combination of factors was designed to help align the 'brand' and 'financial' interests of the organisation (and the corresponding directors at corporate level) and this had a substantial impact. Excessive regulatory enthusiasm and shifting political circumstances led to several changes, but participant emissions appear to have declined by almost 10% in its first year of operation.[a]

Note

a Emission results as reported in UK Environment Agency (2012). The CRC covers several thousand large organisations in the UK. The original proposal was that the instrument would be 'revenue-neutral', with money raised from purchasing these 'CRC' allowances to be offset by reductions in the Climate Change Levy. However, this was vetoed by the UK Treasury, and a complex mechanism was introduced to recycle revenues back to the companies according to a 'league table' of performance. The league table inevitably became complex and contentious, and while motivated by a desire to amplify reputational incentives, it became an Achilles' heel of the CRC since there is no logical way to make fair comparisons across sectors (how can one compare a retail store against a school or a garage?). While proving that such companies care hugely about any reputational comparison, it thus also raised strong opposition and swallowed up vast amounts of effort. Some of this was inevitable resistance induced by an instrument that was designed to drive organisational change; the complexities (and resistance) were hugely amplified by the league table and associated payments (see Grubb et al. 2007 and Grubb and Brophy-Haney 2009). Then shortly before its introduction, the new government, facing fiscal desperation, cancelled the whole idea of revenue recycling (or neutrality), infuriating the regulated sectors. Some league tables remained, but without any direct financial consequence. Following all this, the Treasury then almost cancelled the entire CRC, but it survived and entered into force in 2011. Anyone associated with it knows the scale of managerial impact it had in forcing some of the organisations covered to monitor their overall energy and emissions and assign appropriate responsibilities.

Bottom-up evidence points to the impact of energy efficiency policies. Figure 5.1 shows the remarkable impact of labelling in the EU, which has transformed the market for considerably more efficient appliances and electronics. Standards have clearly increased the average efficiency of product lines with the Japanese Top Runner programme having the most

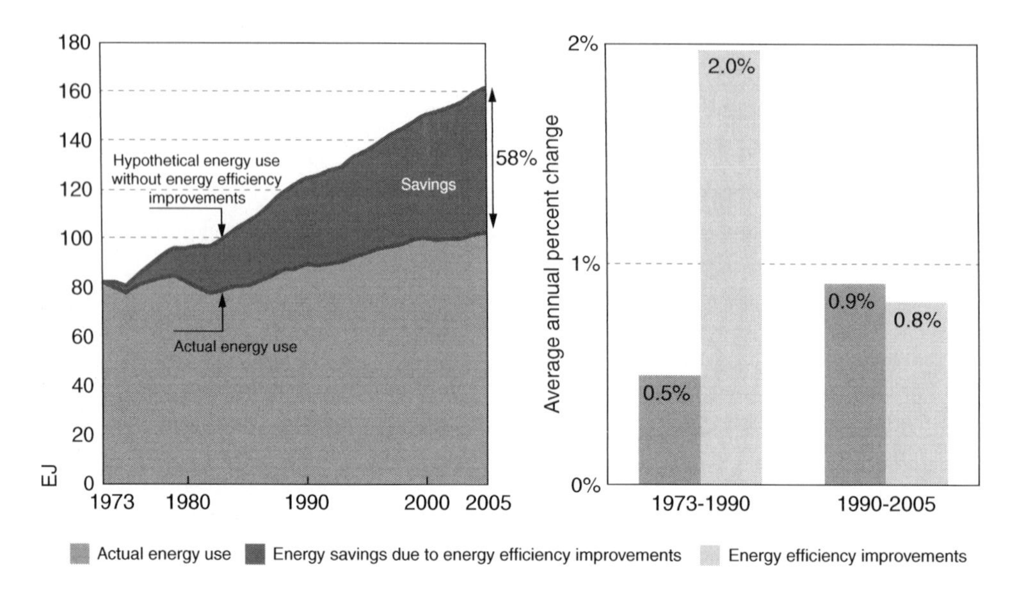

Figure 5.7 Long-term energy savings from efficiency improvements in all sectors

Note: The estimation is based on data from 11 IEA countries (Australia, Denmark, Finland, France, Germany, Italy, Japan, Norway, Sweden, the UK and the US).

Sources: IEA (2008) © OECD/IEA, 2008, Figure 2.7, p. 73.

sustained impact. Fiscal measures (e.g. investment-specific subsidies or tax deductions) have provided funds to stimulate investments in energy efficiency and promote further innovations. Many of these gains in specific product and programme areas were delivered in the era of low energy prices. Energy efficiency improvements slowed after energy prices collapsed, but they did not reverse (as traditional economic theory might predict) or even halt (if all savings were attributable to 'lock-in' effects).

Taken overall, the differences in national trends combined with the micro-data from policies offer powerful evidence for the combined importance of prices with energy efficiency policies.

The International Energy Agency's analysis of mandatory labelling and standards not only quantifies the energy gains for many household products including refrigerators, washing machines and air conditioners, but finds that most of these gains were made without compromising services or product prices. For example, the price of most appliances continued to fall: over 1980–2001, the average real price of refrigerators and freezers declined by more than 40 per cent while average electricity consumption fell by 60 per cent. The policies that drove these improvements have proved cost-effective.

This is particularly true for end-users. Most consumers now have far more efficient products with lower running costs and little attributable increase in capital costs. For most products, the manufacturers learned how to make more efficient equipment with no significant extra cost.[52]

Governments of course incurred the cost of running these policies. An assessment across eight programmes in different countries at very varied scales was also positive.[53]

Data assessing government programmes remains patchy. There have been more assessments in the US than elsewhere, and results vary widely according to both the programme and sometimes the ideological inclination of those carrying out the assessment. However, most do find substantial net benefits, a typical result being the finding that the overall cost of the US residential appliance standards programmes, per unit energy saved, has been about half the average price of electricity during the same period.[54]

In the UK, as noted, programmes by the Energy Savings Trust and the Carbon Trust both clearly have positive benefit/cost ratios. The Carbon Trust estimates that capital expenditure by customers of under £500 million in 2010/11 as a result of the Trust's activities is likely to yield energy savings worth £1.6 billion over the lifetimes of the projects.[55]

Measuring and attributing energy savings and costs is extremely complex. Undoubtedly some programmes have failed to deliver or have been unreasonably expensive. The institutions involved would of course be less keen to publicise these failures, though most countries have audit offices – or critics – to uncover them. But the uncertainties and exceptions do not obscure the basic fact that energy efficiency policies have in aggregate reduced our energy consumption, emissions and costs.[56]

Just because the cynics have been proved wrong, however, does not mean that the success of energy efficiency policies is unqualified. In the first place, it has not been *easy*. The idea that energy efficiency would just 'gush forth' because it was so cheap has given way to the reality of myriad policies and a labyrinth of efforts honed over time to try and tackle the wide and complex range of barriers that account for our extraordinary waste of energy.

From the late 1980s, as energy prices declined, these policies were increasingly motivated by environmental concerns – an observation with wider implications that we return to in the concluding chapter. For countries that sustained the effort, the energy price rises since 2005 have rewarded these efforts further. Yet complexity has continued to grow.

One critique of energy efficiency policy is that it has in fact failed to close the gap; a considerable cost-effective potential for improved efficiency appears to remain. This holds true in both developed and developing countries, where the outstanding potential is even greater. Profitability is especially high for system-level improvement typically with less than two years payback, but uptake has been slow due to many factors.[57] No single instrument can overcome all the barriers, and policies have often been designed in a piecemeal manner, as reflected in the analysis of building policies in the previous section. It also illustrates the continuing march of technological innovation, stimulated in part by the increased markets for energy-efficient goods, combined with the inevitable inertia and lags in the relevant markets. Energy efficiency has been a bit like fossil fuel reserves: the potential keeps moving forward as the identified options are tapped.

The idea that energy efficiency is the answer to our energy problems, however, faces the troubling fact that policies haven't delivered *enough*. The Carbon Trust, despite its undoubted successes and sizeable marketing expenditures, still found carbon management implementation rates below 50 per cent. In the era that got used to cheap energy again, from *c.*1990–2005, the improvements in industrialised countries averaged less than 1 per cent per year, and their total energy consumption rose. That is less than proponents hoped for, and contrasts with the projections in previous chapters of both the potential and the need as being at least 2 per cent per year. Only a few countries have exceeded that projection. The aggregate results have been positive – but not good enough.

One response is that the results in Figure 5.7 do not capture either the strengthened programmes in recent years or the impact of energy price rises since 2005. It is true that a number of stronger measures have only come in quite recently. Examples include the broader building measures noted in the previous section and some wholly new instruments deployed like the UK's CRC. To an extent their impact remains to be seen, but as yet there is little sign of radical acceleration. This poses a problem, and requires digging a little deeper to consider what limits the gains in energy efficiency and what else might be done.

5.6 Bounding rebound

Another challenge to the conclusion that energy efficiency has been a major success is the idea that it doesn't actually save energy – because of what economists call the 'rebound' effect. The most widely cited example is lighting, where new technologies are thousands of times more efficient than candles. Yet as candles were replaced by gas lights, then more efficient oil lamps and then electric bulbs, overall energy demand for lighting increased. People spend as much on lighting today as they did centuries ago, because as the technology has improved they have enjoyed brighter lighting in more places.[58] Energy efficiency is not synonymous with the use of less energy.

Rebound is important and complex, but the field is muddled.[59] Some mistakes stem from applying neoclassical economic perspectives which assume that markets work well, when the main motivation for most efficiency policy is that they do not. It is easy to confuse correlation with cause and effect. People moved from candles to more and better lighting technologies over the ages largely *because* they wanted more and better lighting, and energy costs dominated the overall cost: improving lighting efficiency in such circumstances inevitably led to greater use. But which car you buy is not mainly determined by the fuel cost, and people do not buy more efficient cars *because* they want to drive more. Nor do governments strengthen building energy standards because people want to live in a sauna but cannot afford the heating costs. In that sense, the lighting example is doubly misleading.

At the same time, neither is derisive rejection of possible rebound effects (once cast as the 'boil the rich' theory) helpful; it is important.[60] Fortunately, research in the last few years has greatly improved our understanding. These studies have sought to quantify more closely how much efficiency gains in various areas might have been offset by rebound effects. Quantifying the rebound effect, just as measuring energy saving, is difficult. Some ranges and best estimates of the 'direct' rebound effects in industrialised countries are shown in Table 5.1.[61]

As a possible crude summary, approximately one-fifth (20 per cent) of the energy savings expected from improved efficiency of vehicles and homes (appliances and thermal) has been offset by *direct* rebound – using more because it is cheaper to do so. As a general statistic, however, this is very incomplete.

First, the impact in practice can vary a lot, as indicated in the table. People who cannot afford to heat their homes adequately are likely to take back most of the gains from efficiency by being warmer. Beyond a certain level, however, the demand for heat saturates. Similarly, the

Table 5.1 Direct energy rebound in developed countries for different end-uses

Energy service	*Full range of estimates in literature*	*Best guess*
Vehicles	5–87%	10–30%
Space heating	1.4–60%	10–30%
Space cooling	0–50%	1–26%
Water heating	10–40%	–
Lighting	–	5–20%
Other energy-consuming services	0–49%	<20%

Source: Derived from Sorrell *et al.* (2009); *Greening et al.* (2000).

poor are more likely to cut down on travel to reduce fuel bills and hence drive more if an efficient vehicle helps them to afford it. The rich wouldn't even think about it. This points to a general feature: if and as more people approach such levels of wealth and comfort, saturation effects will tend to reduce rebound. By the same token, however, rebound effects will be greater in developing countries. Developed countries may be close to the saturation of basic energy services, while some 1.6 billion live still without electricity in developing countries.

Rebound also depends on the type of technology. It will tend to be larger for 'general purpose technologies' (e.g. steam engines and personal computers), but less so for 'dedicated energy-saving technologies' which a typical energy efficiency policy promotes.[62] However, there are also two indirect effects:

- To the extent that efficiency saves people money, they might spend that money on other energy-consuming activities. This *re-spending* effect should have an impact akin to the general impact of people becoming wealthier (the 'income elasticity of energy consumption'); the net effect would generally be modest, but not trivial (unless people focus all the money saved on driving or flying!).
- A global increase in energy efficiency will mean less use of fossil fuels, which may lower energy price, thus also tending to partly offset the initial savings.

The overall *scale* of the combined effects is hard to estimate. Research has increased concerns that the *combined* effects of rebound – direct and indirect – from efficiency policies alone could be substantial; one estimate is that the total rebound effects from current suites of energy efficiency policies could offset 30 per cent of the savings by 2020 and offset 50 per cent by 2030 if they are not accompanied by any other policies over time.[63]

Rebound is often viewed negatively and from an environmental standpoint it does lessen the impact of efficiency policies. Viewed more holistically though, it looks very different. In the first place, it is evidence of First Domain effects: if an efficiency standard leads to the purchase or increased use of a product, it proves that the standard has indeed reduced costs to the consumer. Rebound is *an environmental loss but a welfare gain* – making the service cheaper (direct), saving money (re-spending) and potentially reducing the pressure on global energy supplies (fuel markets). It does not negate the benefits of most energy efficiency policies, but it does have the potential to make the impact on energy and emissions less than might be expected at first sight, because some may be taken back as other benefits.

The fact that rebound is bigger in developing countries underlines the fact that they are likely to benefit even more from energy efficiency policies – it reflects areas where poverty was suppressing wants, and helps to alleviate those constraints. Fundamentally, 'rebound' is evidence for the observation in Chapter 11 that effects from the First Domain – and associated Pillar I policies – can contribute to economic growth.

This had led some authors to claim the opposite extreme, namely that efficiency is counterproductive and will make things worse. One recent book lambasts efficiency for this very reason, and concludes that we are essentially doomed to over-exploitation of the planet and the eventual collapse of civilisation.[64] There is no reason why rebound effects need fuel such pessimism. As physical wants (like a warm room) are satisfied, efficiency does contribute more and more to real savings. Rebound probably helps to explain why efficiency in rich countries has not improved as rapidly in the last couple of decades as had been hoped. But as noted, countries that have had strong policies on energy efficiency combined with other pillars stand out for delivering real and sustained improvements (see, for example, note 51). Policies do make a sizeable difference. The notable feature is that these countries have had consistent action across all three domains – not just energy efficiency – underlining why *Planetary Economics* emphasises the central importance of combining all three pillars of policy.

Particularly in an era of low or declining energy prices, the cost of most specific applications declined substantially. While the efficiency of many products may have improved dramatically, people overall became more profligate. Energy efficiency policies will have the most sustained impact when combined with rising prices, as Figure 5.6 suggests and as explored further in Pillar II of this book.

5.7 Better buildings: prospects and trade-offs

Buildings, as shown in Chapters 3 and 4, offer a huge potential for improvement in thermal energy efficiency and other related characteristics. Chapter 4 (Figure 4.6) outlined data on the current situation and the huge prospects for growth by 2050 as the global floor area of buildings is expected to more than double by 2050, but also the huge potential for reductions. Figure 5.8 summarises the prospects for global buildings-related energy and emissions from the detailed studies of the Global Energy Assessment (2012). The results were stark. With 'business as usual', global energy consumptions and emissions associated with the buildings stock could increase by at least 50 per cent. Their 'high efficiency' scenarios modelled a global reduction to 46 per cent below 2005 levels.

For a sector which currently accounts for close to a third of global energy and emissions, this wedge represents a staggering difference in its contribution to either exacerbating or ameliorating global stresses – and policy will have a very strong bearing on which comes to pass.

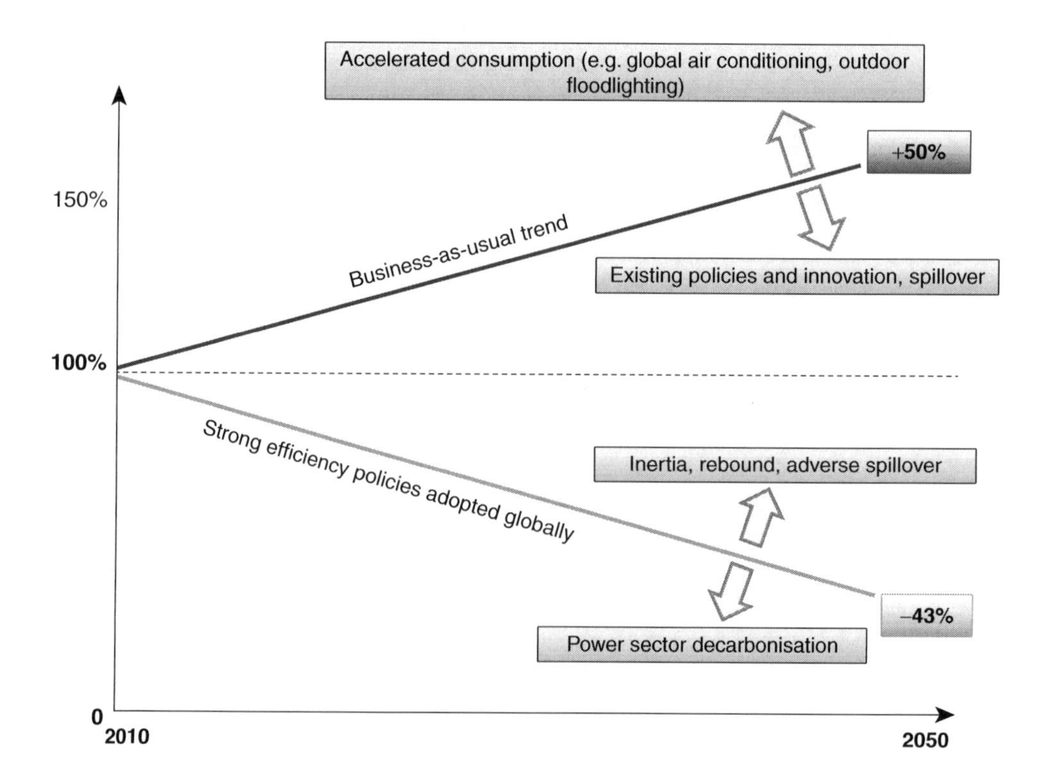

Figure 5.8 Global scenarios for buildings energy use under different influences

For developing countries with rapidly expanding building stock, the single most important factor may be the extent, depth and enforcement of energy efficiency standards for new build. As noted through the debate on the McKinsey Curve (see Chapter 4), the 'baseline' performance of buildings in developing countries is a crucial factor. Even getting to the most 'cost-effective' level in terms of direct energy costs and benefits – broadly the underlying McKinsey assumption – poses a logistical challenge. Implementation is complex in a rapidly developing economy with a rush to expand infrastructure.

There is also a genuine economic dilemma. A common misperception is that everything related to building energy efficiency is currently cost-effective. This is wrong: some measures are really quite expensive. The dominant historical attitude of developing countries has been that climate change is not (yet) their responsibility. However, all the data show that strong thermal performance is far easier and cheaper to incorporate at the time of construction rather than with later retrofits. The McKinsey data suggested that in terms of 'abatement costs', achieving top-range thermal performance might cost €30/tCO_2 for new build but would cost twice that for retrofitted measures. Failing to adopt the strongest standards now could thus have a strong element of lock-in, and would double or more the cost of abatement for developing countries in subsequent decades.

The developed countries mostly do not have the luxury of that choice. Their dominant problem is with existing buildings, where the natural rate of major retrofits is typically little over 1 per cent per year for domestic buildings (somewhat more for commercial). They do, however, have a parallel problem for retrofits can be either 'shallow' or 'deep'. Again: deep retrofits cost more in the short term. Piecemeal improvements that pick off the presently most cost-effective options would be favoured under many of the policy instruments considered earlier in this chapter (especially those that place the onus on supply or other private companies: see Box 5.4). Yet, if and as the screws tighten on global energy and climate concerns, this would necessitate revisiting such properties again, with all the disruption and additional costs this implies. For retrofits in particular there is thus a sharper 'better buildings dilemma' – deep or cheap – as illustrated schematically in Figure 5.9. In addition,

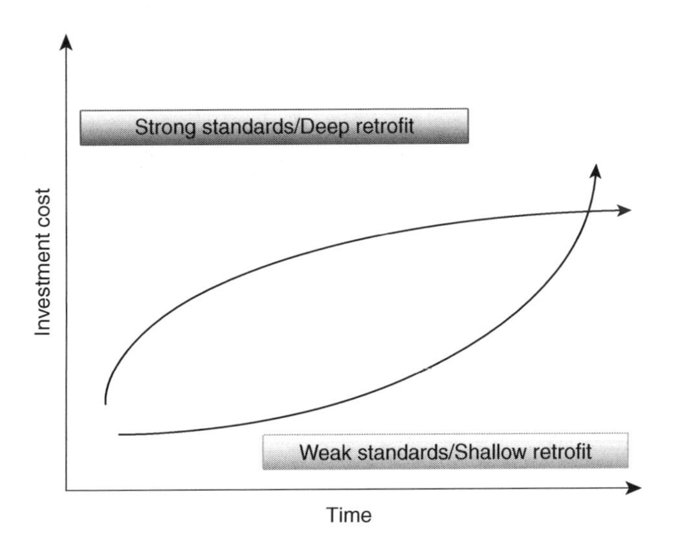

Figure 5.9 The better buildings dilemma: cheap (now) or deep?

developed countries need to accelerate the rate of refurbishment to something more like 2 per cent per year if goals such as the lower trend in Figure 5.8 are to be met.

Finally, as also indicated in Figure 5.8, other factors also bear upon the outcome. The 'baseline' could be even higher if people find major new ways to waste energy, for example if 'deep chill' air conditioning becomes rampant in developing countries and/or people take to garden floodlighting. If the culture is to waste energy, even strong standards and deep retrofits will have limited impact. As noted, there may be important rebound effects that also offset the apparent savings of ambitious scenarios, some of them by spilling over into other sectors. Conversely though, as the sector potentially moves toward high efficiency electric heat pumps, deep decarbonisation of power generation might bring down the carbon footprint for the buildings sector even beyond the trajectory of the Global Energy Assessment scenarios. There is, in short, everything to play for.

5.8 Missing pieces

To dig deeper requires first looking afresh at the *scope* of energy efficiency policies from the standpoint of the ultimate driver: consumption. Energy and emissions depend on the *efficiency* of goods and services, how they are *used* and the energy and emissions *embodied from their production*, as illustrated in Figure 5.10:

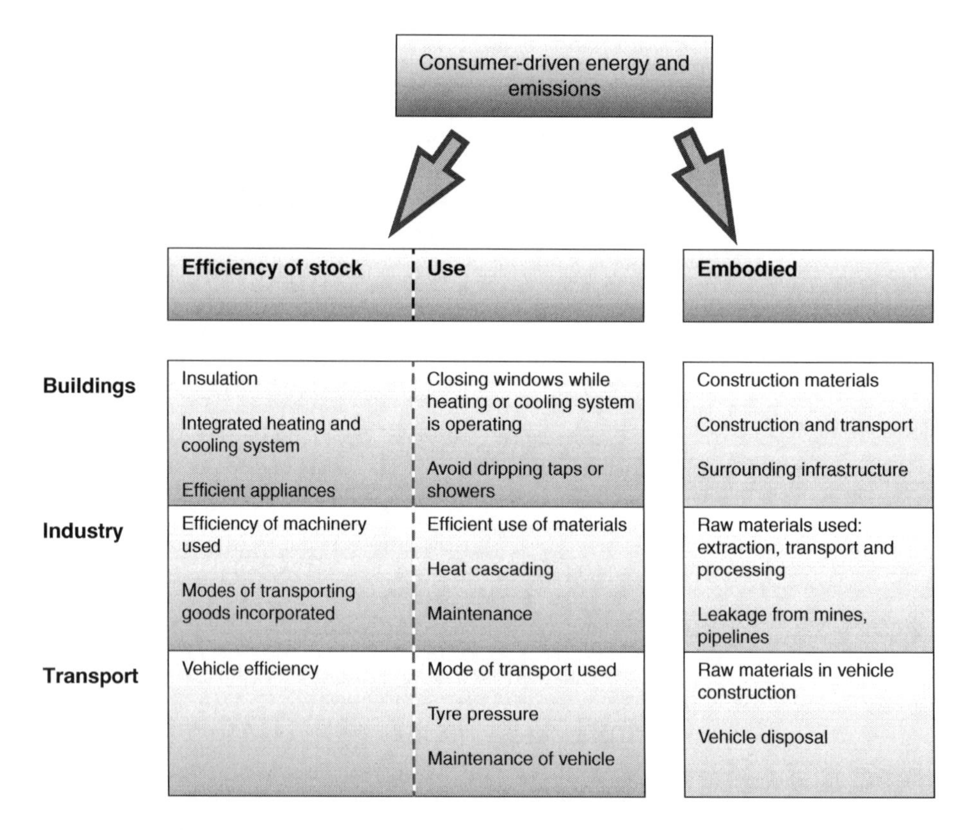

Figure 5.10 The scope of consumer-driven emissions

Note: These categorisations of the consumers' part are not entirely independent of one another. For example, the emissions from the industrial process make up the embodied emissions of consumer goods and services.

- The energy impact of *buildings* depends not just on their efficiency, but on the behaviour of the occupants (e.g. leaving windows open above a radiator) and the energy used in construction, which can amount to many years of occupation. Some of the key issues for commercial buildings are a lack of demand to implement efficiency measures, fragmented supply chain and the disparity between modelled and operational energy use.[65]
- *Industrial* energy use depends not only on the equipment, but on the efficiency of operations and maintenance, the choice of inputs and logistics, and the energy consumption of all the activities that go into the supply chain.
- *Transport* consumption is not just about buying more efficient vehicles: it is about how and how much people drive them, how well vehicles are maintained, whether they choose to go by car, train, air, bike or walk (the mode), and how much they want to travel overall.

As described throughout this chapter, most policy has been targeted at the energy efficiency of products or stock – 'things'. Attention to patterns of *use* has largely focused on information campaigns, with the limited impact on 'people' as noted. The final column, embodied emissions, has largely escaped policy attention. To go further and faster, that will have to change.

As indicated, one reason why improved energy efficiency has not cut energy consumption more is because of rebound. This is unsurprising if people are not engaged with the objective: improved efficiency, by reducing costs, makes it natural that people will use more, and may extend 'satisficing' (First Domain) behaviour if and as energy costs become too small to bother. Yet many of the policies reviewed – notably standards and supplier obligations – largely disempower consumers, removing them further from feeling any responsibility or engagement with their consumption of energy or its consequences.

Moreover, as product efficiency improves, the relative importance of 'embodied' energy grows. Data show the increasing profile of such embodied energy, including as imports, and projections show industrial emissions becoming more not less important as economies decarbonise.[66] Efficient products sometimes take more energy to create – electric cars provide a pertinent example, with the addition of a battery and motor leading to an increase of 50 per cent in embodied emissions.[67] The same principle holds for buildings themselves (Figure 5.11).

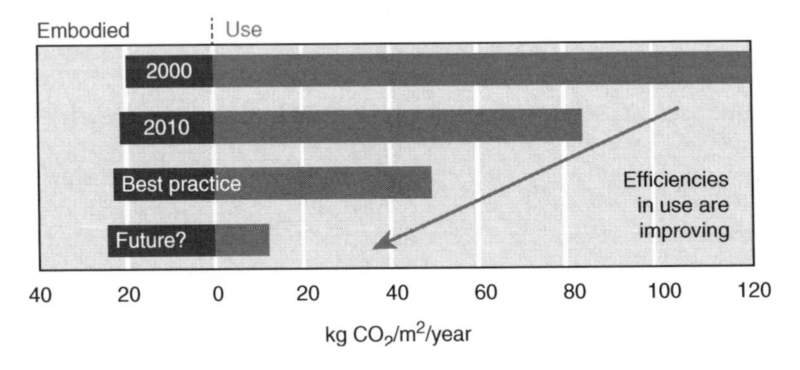

Figure 5.11 Embodied energy in buildings

Source: Estimates drawn by Allwood and Cullen (2012) from a paper by Ramesh *et al.* (2010).

So if there is no attention to energy-in-use or embodied energy, the efficiency effort will get increasingly out of balance and could even in some cases start to be counterproductive. The new horizons for energy efficiency will thus need to engage consumers more and pay more attention to embodied energy.

5.9 Conclusions: beyond the efficiency paradox

To complement the technology survey in Chapter 3 and the theoretical analysis of Chapter 4, the experience of energy efficiency policies surveyed in this chapter confirms the practical reality of the opportunities identified. It underlines the feasibility of actions that result in multiple gains from reduced energy use, emissions and costs. The fears raised by initial industrial or ideological opposition – such as those around appliance and vehicle standards, for example – have largely proved groundless. Energy efficiency policies have spread globally and rarely been reversed. All this confirms the reality and relevance of First Domain phenomena played out in the arena of energy use.

Yet serious questions remain. As noted, despite the strengthening of policies in recent decades, the *rate* of energy efficiency improvements in many (though not all) developed economies has actually slowed. Maintaining and indeed accelerating progress will require us to go beyond the traditional boundaries of such policies, potentially in several ways. In fact, the combined evidence points to two quite distinct challenges. One is practical: the extension of Pillar 1 policies. The other is deeper: understanding their limits.

For extending Pillar I policies, the energy efficiency literature itself points increasingly to the importance of *policy integration*. The review by Brophy-Haney *et al.* refers to the 'increasing awareness about the interactions between single policies and the recognition that energy efficient targets require the coordination of myriad small actions across society', leading naturally to a growing focus on 'integrated policy strategies' required to deliver even the potential in traditional residential energy efficiency.[68]

An important prize lies in expanding the *scope*. While it is natural to think of policies to improve energy efficiency as focusing upon the performance of assets (homes, appliances, cars), the three chapters of Pillar I have all illustrated that the potential scope is much broader. The previous section noted the focus of most policy on the efficiency of stock, not on how it is used or on the energy and emissions embodied in making products. Waste of energy-intensive materials is one obvious, but indirect, way of wasting energy. As noted in Chapter 3, the *systems* can be inefficient. Though energy efficiency is a good *illustration* of the principles and a treasure-trove of practical experience, phenomena of the First Domain – inefficient behaviour and the potential to improve on this – apply far more broadly. As yet, policies have barely scratched the surface of these less direct avenues.

In addition, increased *engagement* remains in its infancy. To date it has mostly been cast in terms of helping people to know more about using energy efficiently. Social scientists argue a potential – and need – to extend well beyond this. Increasing awareness and promoting conscious decision-making is another way to mitigate rebound effects, and aids the development of values that can contribute to resource and environmental management – including attention to the energy and emissions embodied in purchased goods. Energy efficiency policy, to be more effective in the long run, may need to better engage people as both active consumers and citizens, as part of a broader process of social development implied in the search for sustainability.

In the language of Chapter 2, these outstanding possibilities suggest that our societies remain further from the 'best practice frontier' than we thought, and there are more ways to narrow the gap. This echoes wider literature on economic growth, which highlights persistent structural inefficiencies as an important factor inhibiting growth (Chapter 11). The potential to continue expanding both the depth and scope of efficiency policies underlines our contention that Pillar I policies are of comparable importance to the other two pillars – *particularly when combined with continuing progress across all three pillars,* so that the evolution of prices and market structures, and infrastructure and innovation, continue to generate new cost-effective ways to improve efficiency.

The fact remains that the combined impact of all the efficiency policy efforts to date have been less than the apparent potential. Unless the reasons are understood, the same may be true of future policies. The experience points to an apparent paradox, namely the difference between *micro* and *macro* evidence.

The *micro* evidence seems unassailable. Physics, technology and engineering all point to a huge potential for cost-effective energy efficiency improvements across all swathes of the energy system (see Chapter 3). First Domain theories, which span the details of the four barrier-driver classes outlined in Chapter 4 including (but not restricted to) the multiple and overpowering lines of evidence that have converged in the ennoblement of 'behavioural economics', provide a strong theoretical basis to explain the efficiency gap. The empirical evidence from energy efficiency policies surveyed in this chapter confirm that this is not just a theoretical imperfection in our economic systems which then fails the test of 'remediableness' (Chapter 2, note 19): the gains of innumerable policies have been measured (albeit imperfectly), tracked and proven to be positive. Energy efficiency policies have moved individuals, organisations and whole countries closer to the 'best practice frontier'.

Yet looking at the *macro* perspective, there remains an unanswered challenge from classical economics. It can point to the combination of rebound effects, the continuing growth of global energy consumption and the dramatic way in which energy efficiency accelerated after the oil price shocks and then slumped – the challenge of Figure 5.8 – to argue that all the hype around energy efficiency is, at best, overblown. Viewed from the 'top-down' world of aggregated economic systems and macro-trends, it seems less impressive. Viewed from the broad assumptions of the Second Domain, much of the rest may be discarded as trivial compared to the impact of price. In the extreme, Second Domain theorists could point to the global trends and, to misapply Shakespeare, dismiss the whole agenda of 'micro' energy-efficiency policies as 'full of sound and fury, signifying nothing' (see, for example, Chapter 4, note 1).

Who then is right? This challenge cannot, in fact, be tackled without expanding our horizons to the other Domains. For the paradoxical answer, as we will explain more fully in the subsequent chapters, is both.

Notes

1. See the challenge laid down in Chapter 4, note 26.
2. Lovins (1977). Lovins' arguments focused on the technical potential for greater energy efficiency, and its cheapness compared to the mainstream 'hard' energy paths that emphasised energy supply.
3. Tindale (2010).
4. Personal communication with officials regarding Leach (1979). I reviewed the options for UK climate change policy a decade later and had the pleasure of reflecting the ideological confusion of the times with the title, 'Will Britannia waive the rules?' (in Grubb 1991). Eventually, supposedly 'free-market' Britannia did waive the rules and steadily strengthened its energy efficiency policies.
5. Schipper (1987). See also Chapter 6, section 6.1 on asymmetric price responses.
6. WEC (2008).
7. European Parliament (2006).

8. Attractive names to capture the multiple hopes for energy efficiency have become the norm: the Agency became the Executive Agency for Competitiveness and Innovation (EACI) to implement the Intelligent Energy Europe programme (2007–13).
9. Information is most readily available and directly relevant when it serves a private need – usually a producer persuading a consumer to buy something. But much of the information needed for energy efficiency also has the characteristics of a 'public good'. In economic terms, its benefits are non-exclusive, and a person may use it without reducing its availability for others. Of course, information on efficiency also has the characteristics of being a 'private' good. As a result, producers may be willing to tell you how efficient the specific model of car they want to sell is, but clear, reliable, comparative information about all products generally requires some entity acting in the public interest. For discussion see, for example, Ryan *et al.* (2011).
10. Calwell (2010).
11. Galbraith (1958).
12. Between 2005 and 2008, some hard-hitting advertising campaigns around climate change increased measured awareness from around 35% to 65% among UK businesses (a rise in awareness of the Carbon Trust's role in helping businesses reduce carbon emissions). The financial savings available from energy efficiency were frequently not large enough to get the attention of senior management in large companies. When coupled with messages about the risks arising from climate change – the risks to brand arising from higher emissions, from emerging regulations, and the likelihood that energy costs would rise to reflect environmental impacts – the fact that these risks could be mitigated while also saving money had a powerful impact. For smaller companies, however, financial considerations dominated. Despite generating initial interest, turning intent into action remained a challenge (see Chapter 4, Figure 4.4).
13. Bleischwitz and Andersen (2009).
14. For a brief history of UK policy on appliance labelling see Herring (1992). As late as 1989 the Minister responsible had described efficiency labelling as 'unnecessary bureaucracy which would prove tedious and turn people against energy efficiency'. The Environment White Paper the following year finally reversed the policy and expressed support for a European common labelling scheme and energy efficiency standards.
15. Energy Star Overview of 2011 Achievements, online at: http://www.energystar.gov/ia/partners/publications/pubdocs/2011_4-Pager_508c_060812.pdf?e785-c5fc
16. Ryan *et al.* (2011).
17. Uihlein and Eder (2009).
18. See Herring (1992: note 19).
19. The US federal administration had no jurisdiction to prevent state-level standards, and efforts by the Reagan administration to impose a Federal 'zero level standard' were rebuffed.
20. Koomey *et al.* (1996).
21. WEC (2008).
22. Calwell (2010). These CAFE standards respectively equate to 53.16km/gCO_2 for passenger cars and 45.42km/gCO_2 for light trucks.
23. Pew Environment Group (2011).
24. International Council on Clean Transportation (ICCT) (2012).
25. The European car industry successfully resisted fuel standards for many years, but came under increasing pressure. This led to a series of increasingly forceful 'voluntary agreements' and, with the combination of environmental concerns and rising oil prices, eventually to mandatory standards. These are denominated in CO_2. Fuel efficiency standards for new cars and vans in the EU require that the average for every fleet reach 130 grams per kilometre by 2015; this is likely to be tightened towards 100 grams per kilometre by 2020.
26. Weiss and Weidman (2012).
27. US Department of Transportation (2010).
28. On the fuel and CO_2 savings, see US Department of Transportation (2012). An interesting discussion on the US vehicle rebound assumption is to be found at: http://blogs.wsj.com/numbersguy/kicking-the-tires-on-the-rebound-effect-698/. Their estimate initially used higher figures (15–20%) but this was contested, and the DoT now uses 10%. Cost-benefit analysis of the CAFE standards is discussed in Yacobucci *et al.* (2012) who cite 'EPA estimates the total costs of the program to automakers and vehicle buyers at roughly $148 billion to $156 billion, while the benefits are roughly $510 billion to $639 billion over the life of the vehicles covered by the rule, depending on various factors, especially the discount rate.'

29. Swedish Environmental Protection Agency (2004).
30. METI (2010).
31. Fenna (2006).
32. WEC (2008).
33. Vulnerable households were defined as those with people who are pregnant or with children under 16, lone-parent households with children under 16, those over 60, or disabled or chronically ill households.
34. IEA/OECD and AFD (2008).
35. The tax rate increases according to the degree to which vehicles miss the fuel economy standards.
36. VED is an annual charge on vehicle ownership. For Category M vehicles – the least efficient – the duty was set at £950 for the first year of new car purchase and £430 annually thereafter (Ryan *et al.* 2011).
37. At a macro level, energy management and reporting also enables proactive planning of carbon reduction strategies. Within the EU, monitoring and reporting has been extended to CO_2 emissions under the Emission Trading Directive. Also, in October 2009 the US EPA released the Mandatory Reporting of Greenhouse Gas Rule, requiring large emitters to submit annual reporting to inform the development of future mitigation strategies. In some cases, energy consumption reporting is augmented by voluntary energy saving targets (e.g. in the form of government and industry agreements) to bring about economy-wide improvements. This has been popular in many countries such as France, the Netherlands, Denmark and Japan, and more recently in countries like China where the Top 1000 Energy-Consuming Enterprise Program launched in April 2006 stipulates that large industrial consumers, accounting for 33% of the country's total demand, report its energy use and implement energy-saving measures. Energy consumption reporting schemes exist in many countries including Australia, Canada, India, Japan, Mexico and Turkey.
38. Maintenance is an important area of energy management, which ensures that the initial efficient level is kept and that no long-term loss occurs due to wearing of equipment. As such, a maintenance clause has been incorporated into the European directive on buildings for heaters and boilers. Similar regulation exists in Italy and Romania for the transport sector. The upkeep of technologies, so that they are in fact performing at their maximum efficiency level, is vital for energy conservation and transparency of performance. Education and training of operation and maintenance personnel is therefore important, and the benefits of it may indeed outweigh such investment.
39. Carbon Trust (2012).
40. Jollands *et al.* (2010).
41. For details see ISO50001 and the Global Energy Assessment (GEA 2012), Chapter 8.
42. While being widely adopted, technology-based financial supports may be prone to free-riding. For example, an ex-post evaluation of the Netherlands' Energy Investment Deduction Scheme (EIA), where up to 40% of costs may be deducted, revealed that 33% of private firms would have invested regardless of the scheme, while 65% of non-profit organisations would have done the same (Price *et al.* 2005).
43. The UK experience is instructive. The UK's Energy Efficiency Commitment (EEC) placed obligations on suppliers in two rounds. The energy-saving targets were surpassed in both rounds, but there was concern about the effectiveness of some of the measures. Distribution of fluorescent light bulbs accounted for 31% of the qualifying savings in the first year with little evidence on how many were actually used (CCC 2009), while some utilities had double counted insulations. In response to such criticisms, the EEC was replaced by the Carbon Emission Reduction Targets (CERT), which mandated all energy suppliers with more than 50,000 customers to deliver energy savings; individual targets are calculated based on the number of customers they serve, with an overall reduction target of 185MtCO_2 by 2012. Light bulbs were no longer included as of January 2010 and for other appliances, energy savings only count if they exceed the market average for that product. There are no formal traded certificates but energy suppliers have an option to trade certified savings should they wish. (So far, this option has not been popular.) (See also WEC 2008.) The targets under CERT also included specific obligations towards 'vulnerable customers'. The compliance of the system, enforced by the UK regulator Ofgem, was greater than 99%; companies delivered the broad CERT goals but some struggled to implement the measures targeted at vulnerable customers in the most deprived areas across Great Britain, though over 150,000 households had at least one energy-efficiency measure installed, including insulating over 75,000 external solid walls and replacing nearly 43,000 old, inefficient boilers (Ofgem 2013). CERT is in turn being succeeded by a system of 'Energy Company Obligations', intended to complement

the finance-oriented 'Green Deal'. Policy evolves, with each step seeking to fix the problems of the earlier systems; the underlying tensions in getting profit-making companies to deliver mainly public benefits remains.

The literature on cost-benefit evaluation of demand-side programmes is strongest in the US. There seems clear evidence that the cost to utilities is cheaper per kWh than equivalent new generation. The sceptical review of energy efficiency by Allcott and Greenstone (2012) (see Chapter 4, note 10) cites the 'most advanced estimate in this literature' as that of Arimura et al. (2011) who calculate programme costs of 5 or 6.1¢/kWh at discount rates of 5 and 7% respectively. Allcott and Greenstone try to square this with Second Domain assumptions by adding 70% attributed additional costs to consumers (for inconvenience etc.), although they note earlier that demand-side programmes involve intangibilities in both costs and benefits to recipients.

Supplier obligations have raised a number of outstanding design issues (WEC 2008). First, whether it is more effective to impose an obligation on a supplier or a distributor: because the supplier is closer to the consumer, they have greater marketing power, but a distributor might be a more stable organisation to deliver an obligation since they are often regional, regulated monopolies. Second, accounting of savings in the context of multiple policies (e.g. claiming credit for measures required under standards or otherwise incentivised). Third, how to stimulate innovation, since supplier obligations do not increase information or demand in the market. Fourth, how to foster 'comprehensive' approaches, since supplier obligations so far have been focused on individual retrofit measures that may also overlap. Finally, supplier obligations may be undermined by consumer mistrust of energy companies; in such cases, local government or other public organisations may be better placed to pursue energy efficiency. Underlying all this, placing the obligations on private companies inevitably leads to a focus on least cost measures that qualify, with all the pros and cons this brings in the context of seeking more comprehensive transformation, particularly concerning buildings.

44. Tindale (2011).
45. In the UK the Energy Performance Certificates give a classical 'A to G' ranking, along with information packs on cost implications and steps that can be taken to improve a particular building (Defra 2007). These help to drive demand for improvements, such as with the services of the Energy Saving Trust.

    As often in European legislation, the EPBD sets the high-level requirements while leaving many details of interpretation to member states. The UK interprets the requirement to display energy certificates as applying to all buildings with a floor area exceeding 1,000m^2 'that are occupied in whole or part by public authorities and by institutions providing public services to a large number of persons and therefore frequently visited by those persons.' This conceals a major row over whether commercial premises such as supermarkets would also be covered. The UK also chose to make CO_2 the common metric as the way to aggregate heating and electricity consumption, based on a standard grid average. Asset ratings of building energy quality – such as required for the UK's Energy Performance Certificates – form part of the display.
46. US Energy Star and LEED-rated commercial and public buildings were found to attract 3% higher rents per unit area for certified 'green buildings', while selling prices were 16% higher (Eichholtz *et al.* 2009; for other studies of the commercial value of green and energy efficient buildings see also Fuerst and McAllister 2011). Mallaburn and Eyre (2013) also cite references finding market premiums for both domestic and commercial labelled buildings in the Netherlands.
47. Painuly (2009).
48. For a fuller review of international experience see the Global Energy Assessment (2012), particularly Chapters 23–25.
49. Bleischwitz and Andersen (2009).
50. IEA (2009b).
51. PWC (2013). The analysis covered 41 years 1970–2011, with energy use per unit GDP declining at annual percentage rates over this period of 2.7% (Sweden), 3.1% (Germany) and 3.4% (Denmark). On this measure the UK outperformed all (close to 4% per year), but this also relied more on deindustrialisation, with the growth of the financial services sector and import of energy-intensive goods, while Germany and Sweden in particular both maintained strong manufacturing sectors. Progress in the Netherlands (also covered in the study) was much slower at 1.9% per year average. Note that France has had the greatest decarbonisation over the same period due to its nuclear programme.

52. IEA (2009a).
53. IEA (2009a). The programmes on average delivered US$0.6 of net benefits per GJ of energy saved. In general, the larger programmes were found to be less cost-effective than the smaller programmes, possibly as a result of their less targeted nature. The programmes included the Netherlands' appliance labelling scheme, the German KFW soft loans, the British Energy Efficiency Commitment (AID-EE), New York's Energy Star market support programme, New York's EmPower Program, Thailand's Thin Tube CFL programme, the Danish Kitchen Appliances Efficiency Programme and California's multi-family rebate programme. The scale ranged from US$5 million (Danish kitchen appliances programme) to US$4,255 million (the KFW soft loans programme).
54. Palmer *et al.* (2004) estimated an all-in cost of 3.8¢/kWh per unit of electricity saved. A number of other assessments, of varied results and persuasions but all subject to robust review, have been published in the *Energy Journal* of the International Association of Energy Economists, as well as numerous publications of the American Council for an Energy-Efficient Economy.
55. Carbon Trust (2011).
56. It requires careful selection of appropriate indicators and the use of creative approaches including a comparison of 'before-and-after' consumption, a survey of market trends, personal interviews and a statistical analysis of utility bills. While an objective baseline is hard enough to estimate, the measured savings must then be differentiated from 'autonomous gains' and 'free-riding' that would have occurred regardless, as well as other factors that may influence ex-post energy demand (e.g. a change in building occupancy and climate conditions etc.) to properly account for the pure impact of policy intervention. For some discussion see Geller and Attali (2005).
57. Alcorta *et al.* (2012).
58. Tierney (2011). For a full review of lighting and rebound effects in different phases of economic development see Fouquet and Pearson (2012).
59. Turner (2013) gives a detail deconstruction of the many confusions in analysing and quantifying 'rebound' effects.
60. The 'boil the rich' caricature of the rebound debate comes from the absurdity of assuming that people will want ever-increasing home temperatures as their insulation improves and/or they get richer: it points to the importance of understanding saturation effects but, on its own, cannot be generalised any more than the lighting example. For more considered major studies and reviews see Sorrel (2007, 2009), Herring and Sorrel (2008) and Breakthrough Institute (2011).
61. For an empirical estimation to be reliable and informative, an appropriate timescale, system boundary (e.g. sectoral *vs* national energy use) and indicator of energy consumption must be adopted; however, a lack of available data has often hindered reliable measurement (Sorrell 2009).
62. Sorrell (2009).
63. Barker *et al.* (2009).
64. Hallett (2013).
65. Carbon Trust (2009).
66. See, for example, Climate Change Committee (2010).
67. Carbon Trust (2011).
68. Brophy-Haney *et al.* (2011).

References

Alcorta, L., Bazilian, M., De Simone, G. and Pedersen, A. (2012) *Return on Investment from Industrial Energy Efficiency: Evidence from Developing Countries*, Nota di Lavoro 35.2012, Findazione Eni Enrico Mattei. Vienna: UNIDO.

Allcott, H. and Greenstone, M. (2012) 'Is there an energy efficiency gap?', *Journal of Economic Perspectives*, 26 (1): 3–28.

Allwood, J. M. and Cullen, J. M. (2012) *Sustainable Materials: With Both Eyes Open*. Cambridge: UIT Cambridge.

Arimura, T. H., Li, S., Newell, R. G. and Palmer, K. (2011) *Cost Effectiveness of Electricity Efficiency Programmes*, RFF DP 09-48-REV. Washington, DC: Resources for the Future.

Barker, T., Dagoumass, A. and Rubin, J. (2009) 'The macroeconomic rebound effect and the world economy', *Energy Efficiency*, 2: 411–27.

Bertoldi, P., Rezessy, S., Lees, E., Baudry, P., Jeandel, A. and Labanca, N. (2010) 'Energy supplier obligations and white certificate schemes: comparative analysis of experiences in the European Union', *Energy Policy*, 38 (3): 1455–69.

Bleischwitz, R. and Andersen, L.-M. (2009) *Informational Barriers to Energy Efficiency – Theory and European Policies*, MPRA Paper 19937, University Library of Munich, Germany.

Breakthrough Institute (2011) *Energy Emergence: Rebound and Backfire as Emergent Phenomena*. Oakland, CA: Breakthrough Institute. Online at: http://thebreakthrough.org/blog/Energy_Emergence.pdf

Brophy-Haney, A., Jamasb, T., Platchkov, L. M. and Pollitt, M. G. (2011) 'Demand-side management strategies and the residential sector: lessons from the international experience', in T. Jamasb and M. G. Pollitt, *The Future of Electricity Demand: Customers, Citizens and Loads*. Cambridge: Cambridge University Press.

Calwell, C. (2010) *Is Efficient Sufficient: The Case for Shifting Our Emphasis in Energy Specifications to Progressive Efficiency and Sufficiency*. Stockholm: European Council for an Energy Efficient Economy.

Carbon Trust (2009) *Global Carbon Mechanisms: Emerging Lessons and Implications*. London: Carbon Trust.

Carbon Trust (2011) *International Carbon Flows: Automotive*, Report CTC-792. London: Carbon Trust.

Carbon Trust (2012) *Brazil: The $200 Billion Low Carbon Opportunity*. London: Carbon Trust.

CCC (2009) Letter from Lord Krebs to Hillary Benn, 'Advice on evolving methodology for the Climate Change Risk Assessment and Adaptation Economic Assessment'.

CCC (2010) *The Fourth Carbon Budget Report*, Committee on Climate Change, UK Government.

Commonwealth Bank Group (2010) *Energy Efficiency Opportunities Report 2010*. Australia.

DECC (2011) *Carbon Emissions Reduction Target (CERT) – Paving the Way for the Green Deal*. Department of Energy and Climate Change, UK Government.

Defra (2007) *The Building Regulations and the Building (Approved Inspectors etc.) Regulations 2000*, Circular 06/2007. UK Government.

Eicholtz, P., Kok, N. and Quigley, J. M. (2009) *Doing Well by Doing Good? Green Office Buildings*, Center for the Study of Energy Markets (CSEM) Working Paper No. 192, University of California Energy Institute.

Energy Commission (2010) 'Energy Labelling: Commission Sets Up New Energy Labels'. Online at: http://ec.europa.eu/energy/efficiency/labelling/energy_labelling_en.htm

Energy Star (2011) *Energy Star Overview of 2011 Achievements*, Illinois: Energy Star; online at: http://www.energystar.gov/ia/partners/publications/pubdocs/2011_4-Pager_508c_060812.pdf?e785-c5fc

European Environment Agency (2005) Climate Change and a European Low-Carbon Energy System, EEA Report No 1/2005.

European Environment Agency (2012) *2011/12 Compliance in the CRC Energy Efficiency Scheme and Publication of the 2012 Performance League Table*. EEA.

European Parliament (2006) *Directive 2006/32/EC of the European Parliament and of the Council of 5 April 2006 on energy end-use efficiency and energy services and repealing Council Directive 93/76/EEC*. Brussels: EU.

Eyre, N., Flanagan, B. and Double, K. (2010) 'Engaging people in saving energy on a large scale: lessons from the programmes of the Energy Saving Trust in the UK', in L. Whitmarsh, S. O'Neill and I. Lorenzoni (eds), *Engaging the Public with Climate Change: Communication and Behaviour Change*. London: Earthscan.

Fenna, A. (2006) 'Tax Policy', in A. Parkin, J. Summers and D. Woodward (eds), *Government, Politics, Power and Policy in Australia*, 8th edn. Frenchs Forest, NSW: Pearson Education Australia, pp. 448–72.

Fouquet, R. and Pearson, P. (2012) 'The long run demand for lighting: elasticities and rebound effects in different phases of economic development', *Economics of Energy and Environmental Policy*, 1 (1): 83–100.

Fuerst, F. and McAllister, P. (2011) 'Green noise or green value? Measuring the effects of environmental certification on office values', *Real Estate Economics*, 39 (1): 45–69.

Galbraith, J. K. (1958) *The Affluent Society*, US: Penguin.

GEA (2012) *Global Energy Assessment – Towards a Sustainable Future*. Cambridge and New York: Cambridge University Press and Laxenburg, Austria: IIASA.

Geller, H. and Attali, S. (2005) *The Experience with Energy Efficiency Policies and Programmes in IEA Countries: Learning from the Critics*, IEA Information Paper. Paris: IEA.

Greening, L., Greene, D. and Difiglio, C. (2000) 'Energy efficiency and consumption – the rebound effect – a survey', *Energy Policy*, 28 (6–7): 389–401.

Grubb, M. (1991) 'Will Britannia waive the rules?', in M. Grubb *et al.* (eds), *Energy Policies and the Greenhouse Effect*. Aldershot: RIIA/Dartmouth.

Grubb, M. and Brophy-Haney, A. (2009) 'Plugging the gap in energy efficiency policies: the emergence of the UK carbon reduction commitment', *European Review of Energy Markets*, 3 (2): 33–62.

Grubb, M., Wilde, J. and Sorrell, S. (2007) 'Enhancing efficient use of electricity in the business and public sectors', in M. Grubb, T. Jamasb, and M. Pollitt (eds), *A Low Carbon Electricity System for the UK: Technology, Economics, and Policy*. Cambridge: Cambridge University Press.

Hallet, S. (2013) *The Efficiency Trap: Finding a Better Way to Achieve a Sustainable Energy Future*. New York: Prometheus Books.

Herring, H. (1992) 'Energy savings in domestic electrical appliances', in M. Grubb and J. Walker (eds), *Emerging Energy Technologies: Impacts and Policy Implications*. London: Royal Institute of International Affairs.

Herring, H. (1996) 'Is energy efficiency good for the environment? Some conflicts and confusions!', in G. MacKerron and P. Pearson (eds), *The UK Energy Experience: A Model or a Warning*. London: Imperial College Press, pp. 327–38.

Herring, H. and Sorrell, S. (eds) (2008) *Energy Efficiency and Sustainable Consumption: Dealing with the Rebound Effect*. Basingstoke: Palgrave Macmillan.

ICCT (2012) *Global Passenger Vehicle Standards*. International Council on Clean Transportation; online at: http://www.theicct.org/info-tools/global-passenger-vehicle-standards#.Uf7Zhqw9VE0

IEA (2004) *Energy Policies of IEA Countries – 2004 Review*. Paris: OECD/IEA.

IEA (2008) *Energy Technology Perspectives 2008: Scenarios and Strategies to 2050*. Paris: OECD/IEA.

IEA (2009a) *World Energy Outlook*. Paris: OECD/IEA.

IEA (2009b) *Cement Technology Roadmap 2009: Carbon Emissions Reductions up to 2050*. Paris: OECD/IEA.

IEA/OECD and AFD (2008) *Promoting Energy Efficiency Investments – Case Studies in the Residential Sector*. Paris: IEA/OECD and AFD.

IPPC (2007) *Climate Change 2007: Mitigation of Climate Change*, Working Group III Contribution to the Fourth Assessment Report, Intergovernmental Panel on Climate Change. Cambridge: Cambridge University Press.

Jamasb, T. and Pollitt, M. G. (2011) *The Future of Electricity Demand: Customers, Citizens, and Loads*. Cambridge: Cambridge University Press.

Jenkins, J., Nordhaus, T. and Shellenberger, M. (2011) *Energy Emergence: Rebound and Backfire as Emergent Phenomena*. Oakland, CA: Breakthrough Institute.

Jollands, N., Waide, P., Ellis, M., Onoda, T., Laustsen, J., Tanaka, K. *et al.* (2010) 'The 25 IEA energy efficiency policy recommendations to the G8 Gleneagles plan of action', *Energy Policy*, 38 (11): 6409–18.

Kimura, O. (2010) *Japanese Top-Runner Approach for Energy Efficiency Standards*, SERC Discussion Paper: SERC09035. CRIEPI.

Koomey, J. G., Sanstad, A. H. and Shown, L. J. (1996) 'Energy-efficient lighting: market data, market imperfections, and policy success', *Contemporary Economic Policy*, 14 (3): 98–111.

Leach, G. (1979) *A Low Energy Strategy for the UK*. London: Science Reviews Ltd.

Lovins, A. (1977) *Soft Energy Paths*. San Francisco: Friends of the Earth International.
McKinsey & Co. (2007) Curbing global energy-demand growth: the energy productivity opportunity. Online at: http://www.mckinsey.com/insights/energy_resources_materials/curbing_global_energy_demand_growth
Mallaburn, P. and Eyre, N. (2013) 'Lessons from energy efficiency policy and programmes in the UK from 1973 to 2013', in *Energy Efficiency*. Springer.
METI (2010) *Top Runner Program: Developing the World's Best Energy-Efficient Appliances*, Japanese Ministry of Economy, Trade and Industry. Online at: http://www.enecho.meti.go.jp/policy/saveenergy/toprunner2010.03en.pdf
OECD (2008) *Annual Report*. Paris: OECD/IEA.
Ofgem (2013) 'Enforcement Announcement and Press Release, UK Office of Gas and Electricity Markets (Ofgem), 1 May.
Painuly, J. P. (2009) 'Financing energy efficiency: lessons from experiences in India and China', *International Journal of Energy Sector Management*, 3 (3): 293–307.
Palmer, K., Newell, R. and Gillingham, K. (2004) *Retrospective Examination of Demand-side Energy-efficiency Policies*, Discussion Papers dp-04-19. Resources for the Future.
Pew Environment Group (2011) *Who's Winning the Clean Energy Race?* (2011 edn). Washington, DC: PEI.
Price, L. (2005) *Voluntary Agreements for Energy Efficiency or Greenhouse Gas Emissions Reduction in Industry: An Assessment of Programs Around the World*, Proceedings of the 2005 ACEEE Summer Study on Energy Efficiency in Industry. Washington, DC: American Council for an Energy-Efficient Economy, online at: http://ies.lbl.gov/iespubs/58138.pdf
Price, L., Galitsky, C. and Worrell, E. (2005) *xEnd – Use Technologies, Main Drivers, and Patterns of Future Demand: Industry, x Future Technologies for a Sustainable Electricity System*. Cambridge: Cambridge University Press.
PWC (2013) *Decarbonisation and the Economy: An Empirical Analysis of the Economic Impact of Energy and Climate Change Policies in Denmark, Sweden, Germany, UK and The Netherlands*. Available at: http://www.pwc.nl/nl/publicaties/decarbonisation-and-the-economy.jhtml
Rader, N. and Ryan, W. (1999) *Strategies for Supporting Wind Energy: A Review and Analysis of State Policy Options*. National Wind Coordinating Committee, July.
Ramesh, T., Prakash, R. and Shukla, K. K. (2010) 'Life cycle energy analysis of buildings: an overview', *Energy and Buildings*, 42 (10): 1592–600.
Ryan, L., Moarif, S., Levina, E. and Baron, R. (2011) *Energy Efficiency Policy and Carbon Pricing*, IEA Information Paper. Paris: OECD/IEA.
Schipper, L. (1987) 'Energy conservation policies in the OECD – did they make a difference?', *Energy Policy*, 15 (6): 538–48.
Sorrell, S. (2007) *The Rebound Effect: An Assessment of the Evidence for Economy-wide Energy Savings from Improved Energy Efficiency*. London: UK Energy Research Centre.
Sorrell, S. (2009) 'Jevons' paradox revisited: the evidence for backfire from improved energy efficiency', *Energy Policy*, 37 (4): 1456–69.
Sorrel, S., Dimitropoulos, J. and Sommerville, M. (2009) 'Empirical estimates of the direct rebound effect: a review', *Energy Policy*, 37 (4): 1356–71.
Sorrell, S., Harrison, D., Radov, D., Kelvnasand, P. and Foss, A. (2009) 'White certificate schemes: economic analysis and interactions with the EU ETS', *Energy Policy*, 37 (1): 29–42.
Swedish Environmental Protection Agency (2004) *The Top Runner Program in Japan – Its Effectiveness and Implications for the EU*. Stockholm: Swedish Environmental Protection Agency.
Tierney, J. (2011) 'When energy efficiency sullies the environment', *New York Times*, 7 March.
Tindale, S. (2010) *Re-powering Global Communities: Local Solutions to a Global Problem*, online at: http://climateanswers.info/2010/11/repowering-communities-local-solutions-to-a-global-problem/
Tindale, S. (2011) EU Energy Summit, 9 February, online at: http://climateanswers.info/2011/02/9-february-2011-eu-energy-summit/

Turner, K. (2013) '"Rebound" effects from increased energy efficiency: a time to pause and reflect', *Energy Journal*, 34 (4): 25–42.

Uihlein, A. and Eder, P. (2009) *Towards Additional Policies to Improve the Environmental Performance of Buildings Part II: Quantitative Assessment*, JRC Scientific and Technical Report. Luxembourg: Office for Official Publications of the European Communities.

UK Environment Agency (2012) *2011/12 Compliance in the CRC Energy Efficiency Scheme and Publication of the 2012 Performance League Table*. Online at: http://crc.environment-agency.gov.uk/pplt/web/plt/public/2011-12/CRCPerformanceLeagueTable20112012

Ürge-Vorsatz, D., Koeppel, S. and Mirasgedis, S. (2007) 'Appraisal of policy instruments for reducing buildings' CO_2 emissions', *Building Research and Information*, 35 (4): 458–77.

US Department of Transportation (2010) *Lead Agency Annual Report*. US Government.

WEC (2008) *Energy Efficiency Policies Around the World: Review and Evaluation*. London: World Energy Council.

Weiss, D. J. and Weidman, J. (2012) 'How big oil spent part of its $90 billion in profits so far in 2012', *Issue: Energy and Environment*, Center for American Progress, 5 November.

Yacobucci, B. D., Canis, B. and Lattanzio, R. K. (2012) *Automobile and Truck Fuel Economy (CAFE) and Greenhouse Gas Standards*, CRS Report for Congress 7-5700. Washington, DC: Congressional Research Service.

Pillar II

Markets and prices for cleaner products and processes

Overview

Markets connect; prices inform. These are the principles that underpin the role of markets in modern economies. Energy markets are inherently 'imperfect', but they still provide the main way of providing energy to consumers and financing energy producers. Energy prices should in principle convey the real costs of both producing and consuming energy, and of the resulting environmental damages. There is powerful evidence that higher energy prices are ultimately offset by greater energy efficiency (lower energy intensity of the economy); as a result, countries with higher energy prices are not spending more on energy. However, the process of adjustment to higher prices can be slow, painful and complex, and involves all three domains.

In practice, there are numerous energy subsidies. Many developing countries subsidise energy to consumers but at a high cost and with benefits that often flow to more middle-class consumers. Many industrialised countries subsidise fossil fuel production to try and stem their dependence on energy imports. Removing such subsidies is usually advantageous for both economies and the environment but politically can be very challenging (Chapter 6).

Energy prices are political; so too, therefore, are attempts to include in them the costs associated with energy security and environmental impacts. Including such costs in energy prices encourages efficiency, deters damaging activities, and rewards clean energy investment and innovation. The oil shocks of the 1970s provided an impetus for taxing gasoline in many countries, reducing their dependence on oil. Yet because environmental problems tend to be cumulative, the benefits of environmental pricing may be less immediately tangible, making it politically harder and slower.

Pricing is political because prices do not just signal real costs, they change flows of real money. Raising revenue is also an opportunity, however; the net economic impact depends on how the revenues are dispersed, and the appropriate use of revenues can do much to offset any adverse economic impact. Pricing CO_2 is an obvious way to reduce emissions efficiently, and there can be significant co-benefits associated not only with the revenues but with reducing other environmental impacts as well as dependence on international fuel markets (Chapter 6).

CO_2 emissions can be priced either through taxation or by capping emissions with allowances that can then be traded. Each approach has benefits and drawbacks. Proposals in the 1990s for a broad-based energy tax in the US and for a carbon tax across the EU collapsed, though some EU countries did proceed. Cap-and-trade has made more headway, because it directly implements the environmental goal (the cap) and is perceived as offering an easier way of separating the price from the revenue implications; the associated targets can also help to set strategic expectations. The US succeeded in implementing cap-and-trade on sulphur emissions (and some other pollutants), which successfully demonstrated the principle.

The EU subsequently implemented a major carbon emissions trading system (EU ETS), which caps emissions from power generation and heavy industry across 30 countries. Since its inception in 2005 it has cut European CO_2 emissions by 200–500 million tonnes in total – by far the biggest contribution of any climate policy so far globally – with additional reductions elsewhere through 'offset' mechanisms under the Kyoto Protocol. It also focused corporations on a consideration of climate change-related risks. However, the CO_2 price that has emerged from the system has proved to be sensitive to unanticipated developments, and the caps have consistently proven with hindsight to be too lenient, negating thier value as an incentive to low-carbon investment. Many important lessons were incorporated in a much improved design for its third phase which started in 2013, and includes a large-scale shift to the auctioning of emission allowances. However, attempts to improve price stability by linking the second phase (2008–12) to an eight-year Phase III (to 2020) have proved unsuccessful. Multiple factors, including the severe economic recession, led to a huge surplus of emission allowances that have now rendered Phase III itself largely impotent.

Renewed efforts to implement a US Federal cap-and-trade system were more ambitious but collapsed in 2010. However, trading schemes of various kinds (with some carbon taxation) are steadily expanding, particularly in Asia, along with resurgent interest in carbon taxation. The 2013 World Bank survey reported that ten other nations (outside the EU ETS) and 20 sub-national government jurisdictions have implemented or are seriously considering CO_2 pricing, in varied forms (Chapters 6, 7).

The 1997 Kyoto Protocol established the Clean Development Mechanism (CDM) as a regulated system for generating 'emission credits' from emission-reducing projects in developing countries. Despite many criticisms of specific project types and decisions, the CDM successfully implemented key aims and stimulated a rapid expansion of clean-energy industries, particularly in Asia. The use of CDM credits to offset against EU ETS emission caps became its major driver, but the CDM's unexpected strong growth thereby also became an additional major cause of oversupply and collapse in the European CO_2 price. Wider experience suggests that such 'offset' mechanisms can be important in helping to establish systems, contain costs and expand participation, but are unlikely to be a central element of enduring solutions.

Outside the EU, direct emissions trading under the Kyoto Protocol has included links to use revenues for 'Green Investment Schemes', particularly for energy efficiency in eastern Europe, but these have also been hampered by general oversupply. There is a great irony in the EU ETS – and indeed in the Kyoto Protocol targets: whereas almost everyone assumed the big challenges would arise from the difficulty of meeting emission caps, the dominant problems have been about the precise opposite – 'overachievement' which has led to a collapse of the incentives. This underlines inherent uncertainties and biases in fixing near-term caps, which consequently are unlikely to drive deeper levels of innovation and transformation. Designing carbon markets to be more robust as a support to investment requires a mix of quantity with price-based elements in ways yet to be implemented (Chapter 7).

The fact that CO_2 pricing moves money means it affects everyone, but in different ways. The most widespread impacts are those on electricity prices, to which the EU ETS has added a few per cent in Europe. Energy price impacts can disproportionately affect the poor; many countries implement rebates or other targeted supports, but the most enduring response is to strengthen energy-efficiency policies that can help to contain the impact of price rises on actual bills. The huge domestic political complexities of pricing CO_2 and its unequal impacts mean that the world will have to cope with CO_2 prices that vary widely and are implemented through a wide diversity of systems (Chapters 7, 8).

Half a dozen heavy industrial commodities account for a large share of industrial emissions (Chapter 3) and have the potential to make either large profits or large losses from CO_2 pricing, depending on how it is applied. The potential costs induce the fear and risk of 'carbon leakage' – production moving abroad to escape paying for the pollution. The only three options to address this leakage would involve: 'levelling down' the price, by exemption or free allocation; 'levelling up' by establishing the same price on all such production globally; or levelling at the border (i.e. on imports and exports). The first is clearly not an enduring solution and the second is a mirage. As for the third option, no region has yet been willing to consider levelling carbon costs at the border, fearing the technical complexities and political consequences. It thus currently remains an unresolved and growing problem which undermines the entire effort to price carbon.

Ultimately, both CO_2 taxes and cap-and-trade must get to grips with four basic challenges:

- making the gains to people as citizens more tangible than the costs to them as consumers – including by funding programmes to help the 'fuel poor';
- reconciling the political and strategic advantages of cap-and-trade with the realities of uncertain energy projections and the needs of investors;
- maintaining incentives for emission reductions on the most carbon-intensive emitters without inducing 'carbon leakage'; and
- forging links between CO_2 pricing and the other policy pillars in ways that are mutually reinforcing.

Politically, proper pricing is the most difficult of the three pillars of policy yet it must play a central role, especially in market-based economies: pricing 'bads' is good. In reality, like economic development itself, this is likely to be an evolutionary process. There will be many differences among regions, tensions and occasional breakthroughs. The common key will be to establish links both internationally and with the other pillars of policy, to entrench both their effectiveness and political feasibility.

6 Pricing pollution

Of truth and taxes

> 'There appears to be a nearly inverse relationship between those policies that policy analysts tend to endorse as holding the greatest promise … and political feasibility …'
>
> – Rabe (2008: 106)

6.1 Introduction

Prices matter. If there is one lesson from the entire history of economic development (and thought), that pretty much sums it up. If all 'economic agents' – the technical term for you, me, corporations and institutions like government when they buy and sell things – had good foresight and were economically rational (together with a few other key assumptions), then indeed prices would eclipse all other factors as the determinant of our energy consumption and its environmental impact. At least on the surface, policy-making would be easy.

This would be a pure Second Domain world, with little need for the complications arising from the First and Third Domains. The Second Pillar of policy is indeed concerned with prices, their implications and the many complications that have arisen in translating ideas that seem obvious (to most economists) into practical policies.

This also assumes that markets, in some form, play a key role in producing, delivering and consuming energy. In practice, the *form* of markets can vary radically. Crude oil and petroleum products are mostly delivered through competitive markets, but with governments playing big roles not least in pricing – through subsidies and taxes – at both the production and consumption ends. Gas is complicated by the need for pipelines in distribution and often for transmission – monopoly assets that have to be regulated. Electricity generation – the biggest consumer of coal – similarly can only reach markets through wires. The terms on which generators may compete and supply electricity through wires to customers thus inevitably has to be defined and governed by regulators.

This chapter looks mainly at the theory of Second Domain economics applied to energy and environmental issues, together with a basic introduction to the historical experience of trying to use pricing for environmental policy. The next chapter looks at the detailed experience with CO_2 pricing through emissions cap-and-trade. The third and final chapter in the Pillar delves much deeper into the distributional impacts – the winners and losers, the problems these create and the options for addressing them.

6.2 Why price matters

No one likes paying higher prices. Why then are they often advocated by economists?

The basic answer is information: prices should convey the real cost of supplying products and services. If it is worth enough to consumers, they will pay for it. If it is not worth the cost, consumers will consume less. If they are willing to buy more (or pay more), more will be produced. If the product is subsidised in any way, the consumers will not be paying the real cost and their additional consumption will shift costs elsewhere.

Though markets are largely taken for granted in many areas, letting markets (and market prices) rule in such basic and essential commodities as energy has been highly controversial. Instead energy has often been seen as a basic need, necessitating subsidies to ensure that it can be afforded by low-income households.

Moreover, the fact that energy is so integral to so much of society feeds scepticism about whether prices can (or should) really influence energy use. After all, households need to heat and light their homes and power their TVs and cars. Firms need to power their machines and industrial processes. Fossil fuels historically have been endemic to both. Would raising energy prices change much? Curiously, the answer appears to be in part: 'it depends how long they have to adjust' – and therein lies one of the great paradoxes of energy economics.

Figure 6.1 shows that countries with higher average energy prices produce wealth with less energy consumption. Using such a cross-country comparison, the striking result is that across a range of countries, each 10 per cent higher energy price is associated with using energy about 10 per cent more efficiently. The implied 'price elasticity' – a measure of the flexibility of countries' responses to price difference – is around –1. Japan, with prices twice those in the US, uses less than half as much energy per unit of output. At the opposite extreme, until the mid-1990s many eastern European countries kept energy very cheap and used more than twice as much per unit of output. Russia and other former Soviet countries would be almost off the scale; (we have left these and developing countries off the chart because so many other factors complicate meaningful comparisons).[1]

Note that as price doubles and use halves, expenditure on energy remains the same: one striking implication of such a 'unit' (–1) price elasticity is thus that the higher energy prices are offset by reduced energy consumption. Figure 6.1 shows this line of constant energy expenditures. Many countries with higher energy prices do not ultimately seem to end up paying more for energy overall: they use it more efficiently. This is entirely consistent with Bashmakov's Constant of Energy Expenditure noted in Chapter 1 (section 1.3, note 39). Those to the left of the line are spending less; those to the right are spending more. Despite more than double the prices, Japan spent less on energy per unit GDP than the US; so did Germany and France despite their much higher energy prices over the period, measured per unit GDP (or per person).

The puzzle in this is that it seems counter-intuitive to our own experience – it is not that easy to cut energy consumption, a view which seems confirmed by *in-country* measurements of responses to price changes. Estimates vary, but measured this way, elasticities seem much smaller, typically more like –0.2 to –0.5 (depending on the time horizon), a reduction of only 2–5 per cent in energy use for each 10 per cent increase in price. In other words, national level responses to price changes measured within countries seem much less than the comparisons between countries would suggest.

This difference probably reflects several things. One is that the ability to change in the short term is more limited. When one is constrained by existing habits and capital stock (cars, houses, machinery), for example, it takes time to adjust. Responding to higher prices with new investment in cleaner goods and facilities also, inevitably, takes time – and a belief that

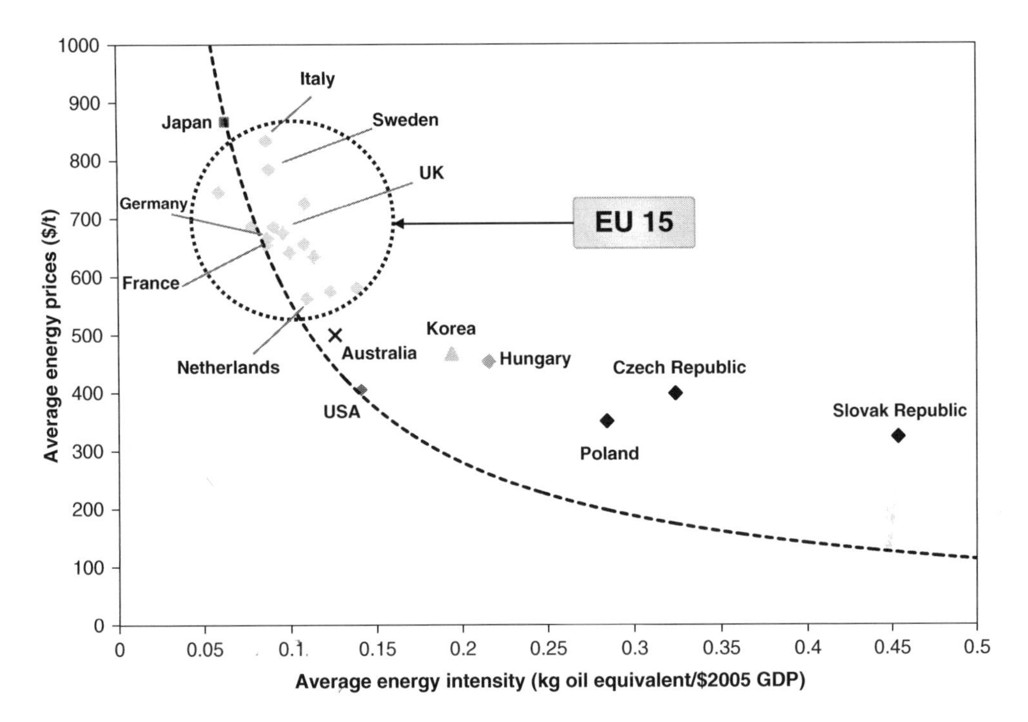

Figure 6.1 The most important diagram in energy economics

Note: Average energy intensity against average energy prices (1990–2005). The dotted line illustrates expenditure of 6% of GDP, which was the level across the US, Japan and the major EU economies of France and Germany, despite wide differences in their domestic energy prices. The data cover a period of rapid GDP growth in these countries and exceptionally low international energy prices, neither proving sustainable. The long-run norm of national energy expenditure has been in the range 8-10% (Chapter 1, p.19).

Source: After Newbery (2003), with updated data from IEA and EU KLEMS.

higher prices will persist. Energy prices can only play a limited role in the short term.[2] It also takes time for prices to feed through in intermediate and final goods, provoking responses through the supply chain as consumers and producers choose less energy-intensive products.[3] Innovators will also respond, bringing more efficient products to market. Thus higher energy prices can change the long-term direction of economies and their production.[4]

In the longer term both households and firms have more options to be more efficient and to buy cleaner goods. Economists have measured how the responses to price changes through these various channels increase, given more time, raising the overall energy efficiency of the economy.[5]

After the oil price roller coaster of the 1970s and 1980s, economists measured something else too: a ratchet effect. They found that the response to prices was highly asymmetric. One (Gately 1993) concluded that 'the response to energy price cuts in the 1980s was perhaps only one-fifth of that for price increases in the 1970s'; another (Dargay 1993) suggested the reason: 'High energy prices induced the development and application of considerably more energy-efficient technologies in all sectors of the economy, many of which will remain economically optimal despite falling prices.'

Economies that had insulated themselves from the effects of higher prices, however – above all the communist countries – continued to have much higher energy intensity, and ended up spending far more of their income on energy as Figure 6.1 illustrates.

Despite the finding that 'energy demand since 1986 (when global oil prices collapsed) seems inconsistent with the notion of constant income and price elasticities', subsequent economic analysis has been limited. Research has shed some light upon ways of representing the enduring improvements in energy efficiency, illustrating the complex respective roles of prices and technology and their interactions.[6] However, most economic models continue to assume that responses to price variations are symmetric and that technology is unaffected by prices.

Within a single country, over a decade or two, the difference may not matter much. For planetary economics – effects of global scope over many decades – it has huge implications, as developed in the concluding chapter. The facts that elasticities measured across countries appear much larger than when measured within a country and that the response of energy demand to price over time has been asymmetric and involved additional external factors are conjoined. They both point to processes impacting on energy intensity that are wide, enduring and involve a mix of direct price-related and other policy-related effects. This is entirely consistent with the evidence summarised in Chapter 5.

Prices do other things too. If the price of a *specific fuel* rises relative to others, people (businesses and consumers) who use that fuel will try to switch to some other energy source. This effect might be far larger than the impact on overall energy consumption. For example, electricity can in principle be generated from any fuel, and in the 1960s oil was most commonly used. The oil price shocks in the 1970s effectively drove oil out of power production in many countries, initially to a mix of coal and nuclear power, and as the restrictions on the use of gas for power generation were removed, to combined-cycle gas plants.

Increasing the price of a conventional or polluting technology will thus tend to drive people to newer or cleaner ones, using market mechanisms that allow the best alternatives to come forward without government direction. Not surprisingly, therefore, economists have long favoured using 'environmental pricing', as elaborated below, and, faced with the challenge of climate change, recommend pricing carbon emissions.

Note the implication that prices may also have an impact on innovation, and in multiple ways. Higher prices for fossil fuels:

- reduce the *size of the economic hurdle* that a new low-carbon technology has to bridge (and thus the scale of private or public innovation support required);
- allow a new technology to *compete earlier* against incumbent technologies – thus also reducing policy uncertainty where support is publicly provided or financing costs for the innovating firms;
- increase both the *public benefit* and the *scale of private rewards* associated with a new low-carbon technology – thus increasing the overall political attractiveness of investing in innovation and the willingness of the private sector to contribute.

These effects are illustrated in Pillar III (Chapter 9, Figure 9.1), but it is worth stressing the importance of environmental pricing for innovation here, because innovation is often presented as an *alternative* to CO_2 pricing. It is hard to think of any more wrong assumption. Proper pricing *spurs* innovation. Some economists indeed assume it is almost the only thing required, maybe alongside public R&D. Pillar III (Chapter 9) explains why this is also wrong. Despite this, the systemic value of pricing, for both rewarding innovation in cleaner products and processes and for encouraging consumers to buy them, remains.

Finally, the pattern observed in Figure 6.1 reflects some even deeper changes. Faced with higher prices, governments themselves are more likely to pursue policies to cut energy consumption (like stronger building standards, high-speed rail networks or other infrastructure) and consumers are more likely to support them when prices transmit the fact that energy is a valuable thing.

So the difference between the 'short-run' elasticities and 'very long-run' elasticities based on cross-country comparisons is a measure of the cumulative response of individuals, the economy and society. Therein lies the challenge. If energy prices rise, then in the short run individuals do end up paying more, and they don't like it. In the long run, society overall becomes more energy-efficient. This is another – and major – facet of the 'Economics of Changing Course' developed fully in the final chapter.

6.3 First steps first: reforming energy subsidies

Many government policies shield consumers from the true costs of energy (though some exacerbate them). Subsidies to fossil-fuel production, power generation and energy use remain widespread. Economists are convinced that most of these subsidies are, ultimately, economically damaging to the countries concerned. This does not, however, make it any easier to tackle the subsidies.

Fossil fuel subsidies

Subsidies can take a wide variety of forms. They can be direct and transparent but are more frequently indirect and hidden. This makes the task of defining and quantifying subsidies difficult. They range from price controls on petroleum fuels in oil-producing countries like Iran and Venezuela, to state aid for coal mining in some European countries and tax breaks for US oil and natural gas production.[7]

Removing fossil-fuel subsidies has been a mantra of international economic discussions for many years (including in statements from G20 meetings of the heads of the largest 20 economies).[8] To move beyond the mantra, however, requires an understanding of the underlying structures and causes. One notable feature is that most subsidies in developing countries are to *consumers*, helping to shield people or industries from the true costs of their energy consumption – or at least hold domestic prices below international levels (a common definition of subsidy). In industrialised countries, by contrast, most of the subsidies tend to go to *producers* – sustaining production in the face of cheaper options like imports or other substitutes.

Consumer energy subsidies were estimated at US$400 billion for the year 2010 across the developing and transitional countries, if measured simply by the difference between domestic and international prices.[9] The largest subsidies are provided by fossil-fuel-producing countries such as Iran, Saudi Arabia and Russia. They tend to give their populations access to cheap energy, well below world prices (Figure 6.2). China and India also feature prominently.

Production subsidies for fossil fuels in the energy sector are estimated at around $100 billion a year.[10] In addition, support provided for renewable energy sources has increased in recent years, following the increase of the deployment volume of new technologies.

Indirect subsidies are harder to track. Governments can subsidise the use of fossil-fuel-intensive technologies through the provision of cheap infrastructure that facilitates them. Building roads or airports with public money represents a subsidy that encourages the use of cars or planes and hides some of the associated costs, for example, and thus form one of the rationales for taxing transport fuels to offset this effect.

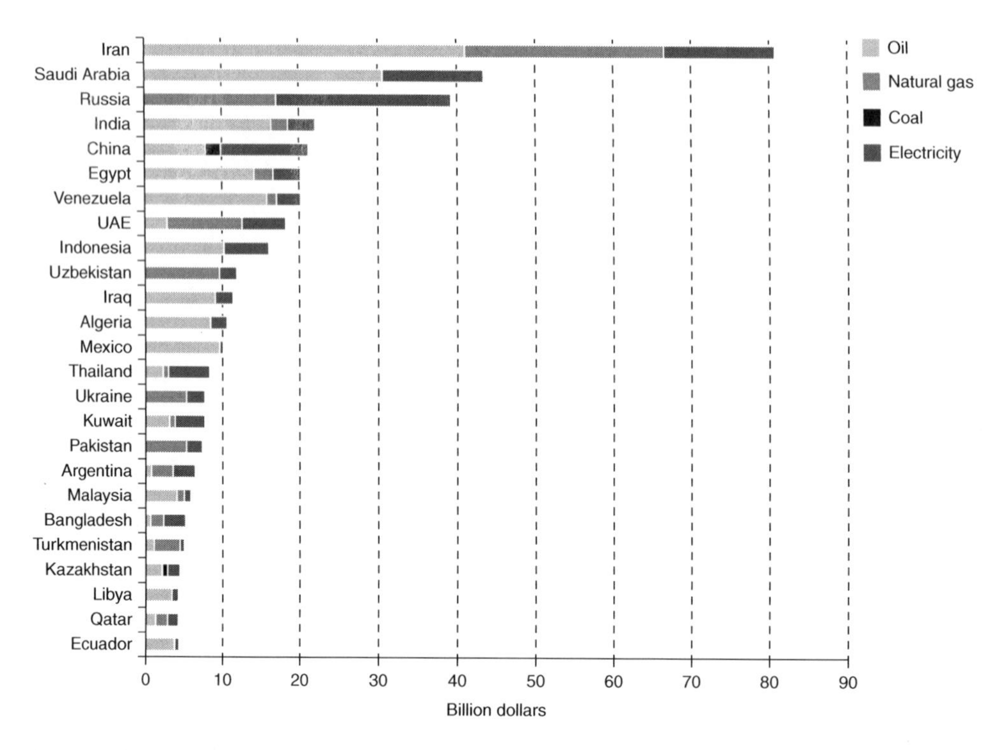

Figure 6.2 Energy subsidies by fuel in non-OECD countries, 2010

Source: IEA (2011) © OECD/IEA, 2011, Figure 14.3, p. 515.

The good and bad of energy subsidies

There is of course a difference between subsidising 'goods' and 'bads'. Subsidies to research and development are almost universal, and in some sectors – including energy – support for new industries can be essential to nurturing their growth due to the nature of the sectors and their incumbent interests (Chapter 9). Renewable energy is an obvious example, and the scale of subsidies has grown as the industries have expanded. By contrast, fossil-fuel subsidies contradict efforts to move away from fossil fuels and cut CO_2 emissions.

Subsidies need to be placed in a wider political and social context. Consumer subsidies are most prevalent in developing countries, where they help to give even the poorest people access to basic needs such as heating, lighting and cooking. In countries such as India they are used in lieu of social security nets, and are a way for the government to provide support to vulnerable members of society who they would otherwise be unable to reach. Such a rationale explains why, for example, many states in India until recently provided free or reduced-price electricity to small farmers. However, as richer consumers use more energy, they also benefit disproportionately from energy subsidies. An estimated 80 per cent of the total benefits of such subsidies actually accrue to the richest 40 per cent of population.[11]

Subsidies that lower final energy prices make life easier for consumers – except that in aggregate they pay in other ways, such as through general taxation or in poor services. The adverse impact on efficiency is now abundantly clear. Under communism, the Soviet Union and its Eastern European satellites heavily subsidised both residential and industrial energy prices in the desire to meet the basic needs of the population and advance industrial activity; cheap Russian energy for Eastern Europe also offered an instrument of political control. All this led the Soviet Union to become among the most profligate energy consumers in the world – off the scale of Figure 6.1. As part of the post-Soviet economic transition, most Eastern European states have managed to phase out energy subsidies, particularly under the pressure of joining the EU. In Russia, large energy subsidies – and associated inefficiencies – remain.[12]

Subsidies to incumbent producers can undermine incentives to invest in new, more efficient equipment or to innovate in alternative technologies. However, they may also be targeted to foster innovation and the adoption of low-carbon technologies – as explained in Chapter 9.

There may be more overt political reasons for subsidies. Producers can be a powerful political force, and indeed a major source of employment in underdeveloped rural regions. Subsidies may help to capture votes. The fossil fuel industry is one of the richest industries on earth; it will try to stay that way, as witnessed by the long-standing struggles over both energy subsidies and efforts to price carbon, as outlined later.

Removing subsidies is neither typically easy, nor is it a panacea – but it nonetheless remains important. Continuing to subsidise fossil fuel production and its use undermines energy efficiency and makes it harder for new energy sources to compete. Removing fossil fuel consumption subsidies alone could reduce global greenhouse gas emissions by as much as 5.8 per cent in 2020.[13] Subsidising fossil fuel production inevitably counteracts attempts to include all costs in the price of energy. Reform will likely be a gradual process and is more likely to succeed when linked to other drivers, like fiscal difficulties, and where combined with measures to address distributional implications. Ensuring that subsidies are in the first instance clearly targeted, soundly based, practical and transparent at least removes some of their most damaging features and makes it possible to map out a clear path to their removal.[14]

To many economists, the principle of removing subsidies needs to be extended further – to the notion that failing to price in environmental and other external costs is itself a subsidy. In practice, there begins another – and politically still harder – story.

6.4 Pricing for energy security: gasoline taxes

The indirect costs of energy span a huge range of potential issues, including the costs of maintaining energy security, the impact of volatile prices and environmental impacts. The practice of using gasoline taxes as a buffer against oil price shocks had already been adopted in the 1950s after the Suez crisis in many European countries like France, Germany and Italy.

After the relatively calm period of the 1960s, the oil price shock of 1973 was a huge wake-up call. The industrialised countries, increasingly dependent on oil imports, had lost control of the world's oil markets; they were hurt and vulnerable. The economists' remedy was seemingly perverse: to increase taxes on oil. This, they argued, was the only way of weaning the rich economies off a dangerous level of oil dependence. It would also start to seize the 'economic rents' – the difference between the cost of getting oil out of the ground and its final price for consumers. There were other motivations too – including the need to raise revenue to fund road construction.[15]

The rates of energy taxation vary dramatically among countries, helping to explain the varying end-user energy prices in Figure 6.1. While many developing countries, as we have seen, still subsidise energy, increasing numbers tax oil fuels. Among industrialised countries, as shown in Figure 6.3, Western Europe and Japan have systematically much higher energy taxation rates than North American governments.

Energy prices are politically sensitive and a common cause of social protest, sometimes leading to the fall of governments. The European experience suggests some of the limits. In the late 1980s, cushioned in part by its own North Sea oil, the UK had a relatively low rate of gasoline taxation. The Conservative government introduced a 'price escalator' – a predictable increase in gasoline tax at 3 per cent per year, on the grounds of both security and environment, which was later increased to 5 per cent per year. UK prices moved from the lower to higher end of levels typical in Europe. Then, triggered by the high crude oil prices in 1999 (which seem puny compared with those a decade later), drivers – commercial and private – rebelled at the combination of high oil prices and fuel taxes, and blockaded refineries. The price escalator was frozen.[16] It was an example of a government using environmental arguments to support taxes that had other benefits, but over-ambition took the policy beyond politically acceptable limits.

Given both the political sensitivities of price impacts, and the government's desire for revenue, governments still fiercely guard their right to tax (or not). In the EU, efforts to harmonise rates of energy taxation across Member States finally culminated in 2003 with a directive on energy taxation. This aimed to remove distortions due to differences in rates between fuels in different Member States, and to encourage greater energy efficiency. After more than a decade of struggle, it was in fact the rump left from Europe's first collective efforts to price carbon.

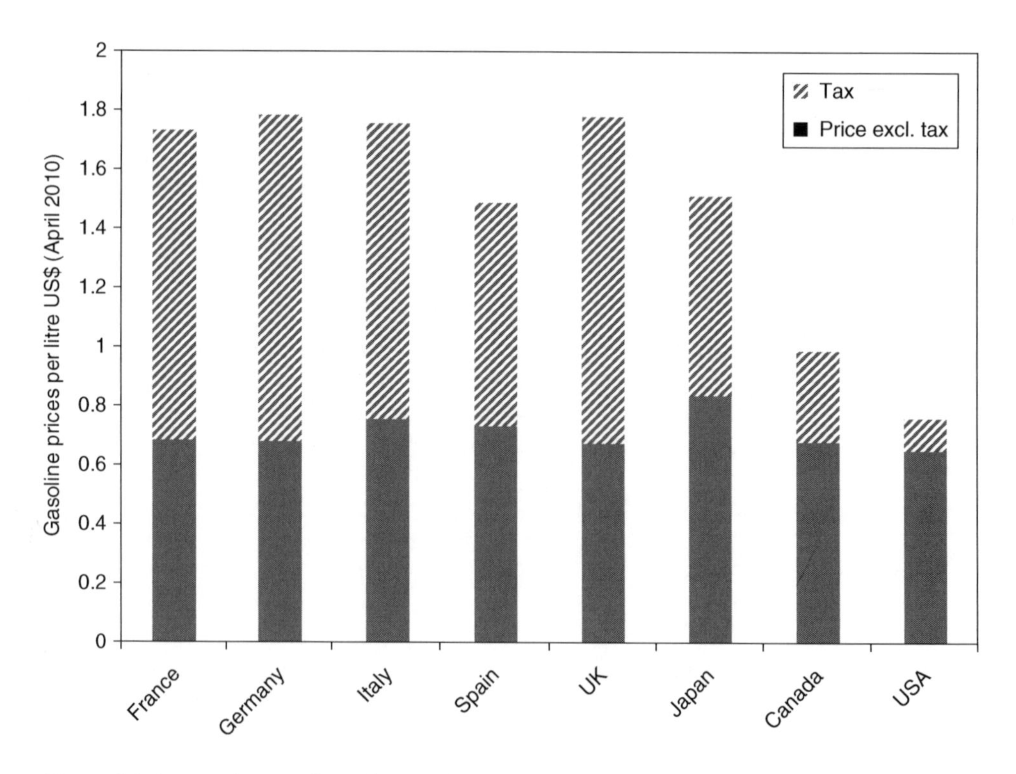

Figure 6.3 Automotive gasoline prices and taxation rates

Source: Authors, using data from IEA (2011).

6.5 Pricing for pollution: the principles

The fact that higher energy prices increase the efficiency of energy use and oil taxation helped to reduce import dependency, signals their possible role in tackling related challenges, including environmental impacts. Of course, price is not the only lever. Environmental issues have traditionally been more often tackled with direct regulation – the banning of dangerous pollutants, requirements to fit clean-up equipment, minimum emission standards, etc. These – which became known in the economics community as 'command and control' measures – were often highly effective. But they also had obvious potential drawbacks, mostly stemming from the dependence on bureaucratic processes to define standards and technologies, which hence easily became inefficient or ossified with little incentive to innovate.

As the scale and reach of problematic emissions increased, the cost and limitations of such approaches got more profile and more attention was given to 'economic instruments' – to price pollution out of the market. In both the United States and Europe, scholars underlined the potential benefits of this approach and – particularly in the aftermath of the oil shocks and the rising profile of acid rain and ozone depletion in the 1980s – governments began to take notice.[17]

This intellectual and policy transition drew on a much longer tradition of economic thought. Arthur Pigou in 1920 was one of the first to develop the idea of 'externalities' – impacts from economic activity which are not naturally priced, such as environmental damage; (risks from energy insecurity are another form of externality).[18] Externalities are now recognised as a major form of market failure – where the market system does not deliver the welfare benefits that it should. Worse, if activities are not exposed to their external costs, then market dynamics can contribute to the growth of these activities, thus amplifying the damage.

Essentially, Pigou was pointing to the fact that unpriced environmental damage amounts to a hidden subsidy. One natural response, therefore, was to try and correct that – to 'internalise the externality' – by raising the price. William Baumol was one of those to develop proposals for 'Pigouvian taxes', to ensure that the cost of damage is factored into the decisions of individuals and firms.[19]

Just a few years after Pigou's seminal work, another approach to the same goal was proposed. Governments could cap the quantity, but instead of directly regulating individual firms, they could define the quantity through tradable quotas, or 'allowances'. With a fixed total quantity, firms could then trade these quotas, until an equilibrium price was reached that reflected the cheapest way of meeting the constraint. In effect, imposing an allowed quantity would create value from cutting back on something that had previously been unlimited (and hence free). Allowing trade would then establish a market, with all the usual theoretical efficiencies compared with simple command-and-control approaches.

Controlling externalities, such as pollution, through such pricing mechanisms does indeed have clear potential advantages. It gives all those in the market the freedom to choose the most appropriate response and the flexibility to hunt out the cheapest options. Imposing a price allows firms to choose how to cut emissions. At a given price, those with low-cost opportunities can reduce emissions by more, while those for which it is far more difficult may prefer to pay the imposed price. The same signal transmits through the economy. The classic book on emissions trading cites 14 studies of the cost-saving impact of implemented economic instruments which frequently exceed 50 per cent.[20]

Pricing environmental externalities in this way potentially allows three types of flexibility: 'what', 'where' and 'when':

- '*What* flexibility' allows the goal (of reducing pollution) to be met by whatever technical measures are available and effective at least cost. The price creates the space, value and market opportunity for cleaner processes, products and services.
- '*Where* flexibility' allows those with the lowest costs of abatement to cut emissions while those facing higher costs may choose to pay the price associated with their emissions. By contrast, regulatory standards have either to impose standard reductions on all firms, despite disproportional impacts, or set individual firm targets, which would become incredibly complex and contentious. 'Where flexibility' allows the decisions to be taken at the firm level, where costs are known, while ensuring that polluters are paying a cost.
- '*When* flexibility' allows the effort to be scheduled over time. As noted, the cost of cutting emissions is likely to change with timescales, as capital stock turns over and different technologies develop in different and unpredictable ways. The 'when flexibility' allows some balancing of when to cut versus when to pay for continuing to emit emissions, and may also reward earlier action.

While economists espoused these virtues, others did raise extensive ethical as well as practical concerns. Taxation would allow emitters to continue polluting, if they paid a price. The cap-and-trade approach was portrayed as even worse, because it seemed to confer a 'right to pollute' up to the specific level. 'Pollution rights' smacked of Middle Ages indulgences (payments to the Church to salve one's conscience – or to reduce one's time in Purgatory – after committing a sin). In practice, US cap-and-trade legislation was generally accompanied by clear legal statements that emission allowances did not constitute a right to emit but merely defined a level deemed currently acceptable. After all, minor levels of pollution may be insignificant (and unavoidable) and there is no prospect of phasing out CO_2 overnight. If the issue is the degree of control rather than an outright ban, ethical argument against price or trading is far more problematic – an economic instrument which imposes a cost or quantified constraint can have clear advantages.

There *are* important ethical dimensions, but these relate more to the distributional consequences and the ways in which different levels of power and wealth may enable economic instruments to be abused or allow 'the rich' to escape constraints in ways that seem unjust. There can also be valid *economic* reasons to constrain the scope of market instruments, as we illustrate in subsequent chapters. For most practical purposes though, by the 1990s the case for pricing pollution had been won in principle.

The case seemed exceptionally compelling for a problem like carbon emissions, which originate from activities throughout the economy and whose impacts are widely spread over place and time unrelated to the point of emission. It seems intuitively impractical and undesirable for governments to micro-manage all the opportunities for emission reductions without the use of pricing mechanisms. Broad-ranging sources, innumerable options for control and far-reaching impacts point to the value of incentives that apply widely across key sectors and indeed the wider economy. Pricing carbon enhances opportunities and incentives to use less carbon-intensive carbon products, processes and technologies. Increasing the profitability of low-carbon choices should boost investment and innovation in them. In fact all these features suggest that carbon exemplifies a pollutant where pricing seems not only appropriate and efficient, but essential.

This is not to claim that CO_2 pricing can do the job on its own – and the Three Domains is a useful framework for understanding why. Amid the huge range of lower carbon options, many factors other than price alone determine prospects and use (Chapter 3), and all those which characterise the First Domain (Chapters 4 and 5) are not *primarily* determined by price, so *pricing on its own* is unlikely to be an effective or efficient way of delivering them. Similarly, the effective use of many technologies depends on adequate infrastructure, and innovation and transformational changes too are a complex process – crucial Third Domain aspects that again cannot be delivered by price alone (Chapters 9 and 10). Even within Second Domain processes, the rest of this Pillar charts the various practical limitations of pricing instruments. Yet economic instruments have a compelling logic and market-based economic systems cannot conceivably tackle dependence on fossil fuels and associated carbon emissions without reflecting their real costs in prices – and hence making them more expensive.

6.6 Carbon costs, 'co-benefits' and 'double dividend'

However, classical economic arguments that pricing carbon is the most efficient approach to reduce emissions initially did little to head off concerns about its perceived costs to the economy. The role of cheap fossil fuels in historic economic development has been noted in Chapter 1. The impact of oil price shocks on economic growth has also been cited as an illustration of the risks of pricing carbon – though in reality a deeply misleading one.[21]

Evaluating the overall economic impacts associated with taxing fossil fuels or otherwise pricing carbon is a complex undertaking. At the most basic level, Chapter 1 (section 1.2) noted that failing to tackle external impacts is a false economy: the costs are borne by others or ultimately come back to bite in very real ways, almost always at higher costs than if they had been tackled in the first place. This chapter has made the basic observation, in the context of economic theory, that failing to internalise an externality advantages polluters and that making them pay for their pollution is ultimately good for economies. The aim of economic instruments is precisely that.

However, the global challenges associated with resource depletion and climate change are unusual in their global reach and huge timescales, which largely disconnect present generations in individual countries from the direct benefits of tackling the problem. This is one among other reasons why climate change in particular has been described as 'the perfect moral storm': the intrinsic inequalities are amplified by the fact that most of the present perpetrators can get away with emitting for free.[22] Hence the debate and concern about the economic costs to individual countries of doing something today to avoid harm that will only appear in the aggregate and long term.

Fortunately, even simple back-of-the envelope reasoning shows many fears to be overblown. Energy typically accounts for around 5 per cent of GDP. CO_2 prices at the highest level seen in the EU ETS add only a modest fraction to final energy costs. The most basic economic modelling confirms that the costs of CO_2 prices at these levels are well under 1 per cent of GDP. As explored in Chapter 8, some individual energy-intensive *sectors* may be much more exposed, but again these only account for a very small fraction of economic output. Nor is there any substance to the idea that the taxation of energy or carbon would kill jobs in aggregate: to a first-order approximation, it would shift some jobs around, from carbon-intensive to lower carbon sectors. The idea that CO_2 pricing at the current levels of ambition could do anything more than set back economic growth by a few months (at most) over a period of some decades, is fanciful and wholly unsupported.

Of course there are complications, and they can cut in both ways. One important issue is the 'environmental co-benefits' of cutting fossil-fuel use. Chapter 1 (note 15) cited a strong consensus that coal burning, in particular, inflicts local environmental damages that are at least comparable to its value in most economies, and often more so. CO_2 pricing, which

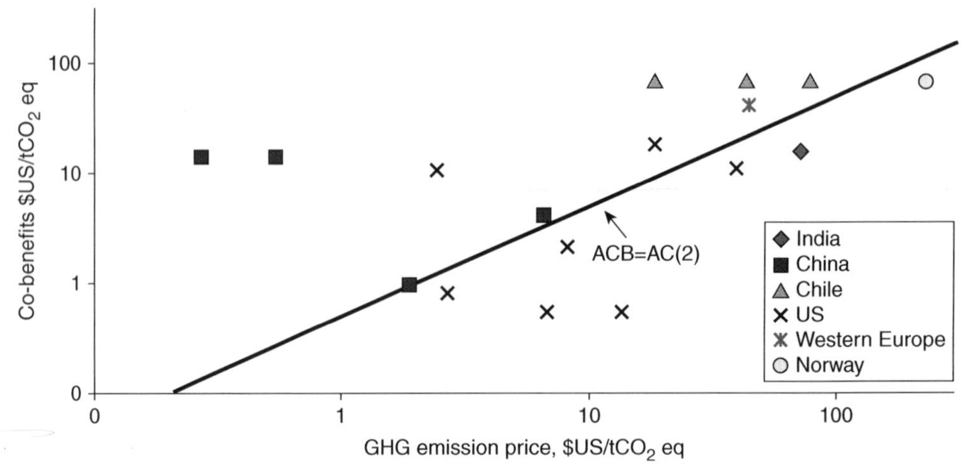

1. For each country, observations represent estimates from various studies and/or for various carbon prices. The base year for estimates is 1996 or the latest available year.
2. The line "ACB=AC" indicates a situation where the average co-benefit is equal to the average cost of abatement. It assumes that abatement costs are a square function of emission reductions; average costs can then be computed as one half of marginal costs (i.e. the carbon price). Points above this line indicate situations where the average co-benefit is higher than the average cost.

Figure 6.4 Estimates of local environmental co-benefits in different countries

reduces coal use, thereby reduces these other impacts too. Indeed, many studies have now charted the environmental co-benefits that might go along with controlling CO_2 emissions. Figure 6.4 shows results from an OECD study along these lines – plotting estimates of air pollution benefits associated with the reductions caused by a given price on CO_2 emissions. More than half the estimates lie above the line – signifying that the air pollution benefits would *on their own* exceed the costs. These co-benefits are exceptionally big in two of the Chinese studies – ten times the CO_2 price! – reflecting the scale of its air pollution problems.

A separate study of Mexico estimates that major reductions in CO_2 would, through the by-product of reduced air pollution, save close to 3,000 lives a year and would prevent almost half a million non-fatal diseases, with associated cost-of-illness savings at around half a billion dollars.[23] A wide range of other studies have now produced similar findings about the environmental co-benefits of controlling CO_2 emissions.[24]

These are not trivial benefits. However, such results do not *in themselves* point to policies aimed directly at curtailing fossil-fuel use or associated CO_2 emissions. The first implication is that countries should do more to control local pollution. Yet it does illustrate additional and very real co-benefits that could come from pricing carbon too. It also raises a question considered again in the concluding Chapter 12 – whether it actually makes more sense to address the various challenges associated with fossil fuels in more integrated, and far-sighted, ways.

Other important potential co-benefits are those associated with energy security – or, more specifically, reducing dependence on international fossil-fuel markets that may be subject to high volatility, over which the buyer usually has little or no control. Usually trade enhances welfare: it gives us access to cheaper resources than countries have domestically. However, as noted (in Chapter 1, section 1.2), volatility itself incurs a cost which is poorly represented in trade theory; domestic energy taxation can help to insulate the economy from the associated price shocks, both directly and by reducing dependence on the related imports.

Perhaps the biggest complication in estimating the costs of using economic instruments is the fact that CO_2 prices do not only correct the price information on fossil fuels to reflect external costs. They also increase the cost of producing energy-intensive products and services, which will (usually) transmit throughout the economy. They also raise revenue that might go to other parts of the economy. Such *general equilibrium mechanisms* point to the need to understand the impact of sectoral change on the wider economy.

Increasing the cost of energy-intensive products increases the total cost to consumers above the direct impact on their energy bills; the cumulative effect thus reduces the purchasing power of a household's income. In the absence of other factors, this has a depressing effect on the economy and raises the *macro*economic costs of CO_2 prices above the technical costs of abating carbon: an amplifier effect. The amplifier effect is also enhanced because of the way that higher relative energy prices will distort economic choices throughout the production chain.[25]

The second and converse factor stems from the money raised through a CO_2 price. For simplicity we will focus on CO_2 pricing and refer to carbon revenues. Paying a CO_2 price at a level sufficient to make a difference – or remotely equivalent to most estimates of climate damages – means a lot of money changes hands. Before the recent economic recession, the value of emission allowances in the European carbon trading system was around €50 billion per year (Chapter 7). If the allowances are given out for free to industry, the effects are complicated – it can generate windfall profits as noted in Chapter 7, which might support investment and innovation but are more likely to go to shareholders with few wider benefits. However, if emitters pay for all or most of their emissions – a tax, or equivalently auctioning emission allowances – the money goes to the government. The societal costs then depend on how the revenue is used.

Because we need to fund public goods, such as defence, education, health and research, there is no blank sheet. Governments could use the money to reduce other taxes to restore consumer's purchasing power (income tax), to increase the incentive to invest (corporation tax) and lower production costs (payroll taxes or any tax falling on labour to fund social security or insurances). Carbon revenues can reduce the need for other such taxes. The net effect depends on whether those other taxes are more or less distorting to efficient economic activity, which depends on the existing tax landscape in a country. One use of revenue recycling could be to minimise the multiplier effect of its propagation throughout the economic sectors by lowering taxes which ultimately fall back on the production costs. This is why, in the European case, most of the general equilibrium studies use revenues to reduce labour taxes, so that the net cost is small or even nil.[26]

The potential 'double dividend' from environmental pricing (the first dividend being the environmental benefit) has fuelled debate among economists over decades.[27] The debate was not over whether any such double dividend exists: it is indisputable that using carbon revenues will reduce the economic impact compared with throwing away the carbon revenues (or giving direct rebates to consumers).[28] Rather, the debate became polarised between weak and strong forms – whether use of revenues merely *reduces* impacts or could result in net economic *gain to the country*.[29]

Fundamentally, however, the debate points to an underlying issue almost ideological in nature regarding assumptions about the current state of the world. The basic argument against a 'strong' double dividend is that economists should relate their analysis of carbon costs to a world in which the economy is assumed to be on its 'best practice production frontier' (Chapter 2, Figure 2.3) – including with the most efficient possible implementation of taxes and other polices. Or at least, the potential for carbon revenues should not be linked to improvement in existing tax structures: they should be *separable*. From this perspective, almost by definition there is no way to price environmental costs *in* to the economy without decreasing the *economic* output – because it is assumed that any options to improve economic performance could and should be implemented *separately*. We return to this point in the conclusion of this chapter for the underlying assumptions relate to the heart of our argument.

Fundamentally, the distinction between the strong or weak forms of double dividend is a second-order issue – an example of becoming obsessed with levels of detail in subtle economic controversies while neglecting more important common messages. The fact is that returning carbon revenues into the production system can do a lot to offset the direct economic impact of CO_2 prices – a conclusion of broad economic consensus.[30] In principle, if we set aside the questions of political acceptability, one can levy a carbon tax of, say, \$200/ tCO_2 without any increase in net average production costs if the revenues are recycled into a corresponding decrease of corporation or labour taxes.[31]

This in fact is a crucial point of contrast between regulation and pricing. Regulation is more popular because the costs are less visible. But regulation can still distort and increase costs (depending on how well aligned they are to tackling contractual or behavioural barriers, as discussed under Pillar I), without generating public revenues that can be used to offset these costs. The weak double dividend may be weaker, but it is still a vital part of the economics of CO_2 pricing.

The net result makes it infernally difficult to be *precise* about the actual macroeconomic impacts of CO_2 pricing – one really needs to assess the whole package and its innumerable interactions. But the *order of magnitude* seems clear. The major empirical assessment of Phase 1 of the European trading scheme, led by the Massachusetts Institute of Technology, estimated the macroeconomic costs to be 'imperceptible' – of the order of one hundredth of one per cent of EU GDP.[32] Most studies project the impact of *future* assumed carbon constraints and associated prices. Either way, most (though not all) estimates of total costs are small, below 1 per cent of GDP or welfare over the next couple of decades.[33] Lost GDP is not the big issue.

Nor, in fact, is overall employment. Emitting less carbon requires shifting jobs from carbon-intensive activities to jobs in cleaner production and improving energy efficiency through activities such as retrofitting existing buildings. There are eternal debates about the net effects of this shift in jobs and some grounds for thinking that in the right conditions, resource- or pollution-based taxes may lead to a modest increase in jobs as money previously spent on fossil-fuel imports is invested domestically in energy efficiency and renewable energy.

6.7 Response options in the Second Domain

As indicated by Figure 6.1, given time, there appears to be large scope for societies to adjust to higher energy prices by increasing efficiency, so as to keep their total domestic energy bill roughly constant. This, however, is a long-term adjustment process spanning all three domains, and has focused mainly on improving efficiency. As underlined by Figure 1.5, tackling climate change and reducing dependence on fossil-fuel markets more substantially critically depends also on decarbonising the energy mix. A price on carbon is the natural instrument to drive substitution towards low-carbon products and processes – options which dominate the middle part of the McKinsey curve. This is shown in more detail in Figure 6.5.

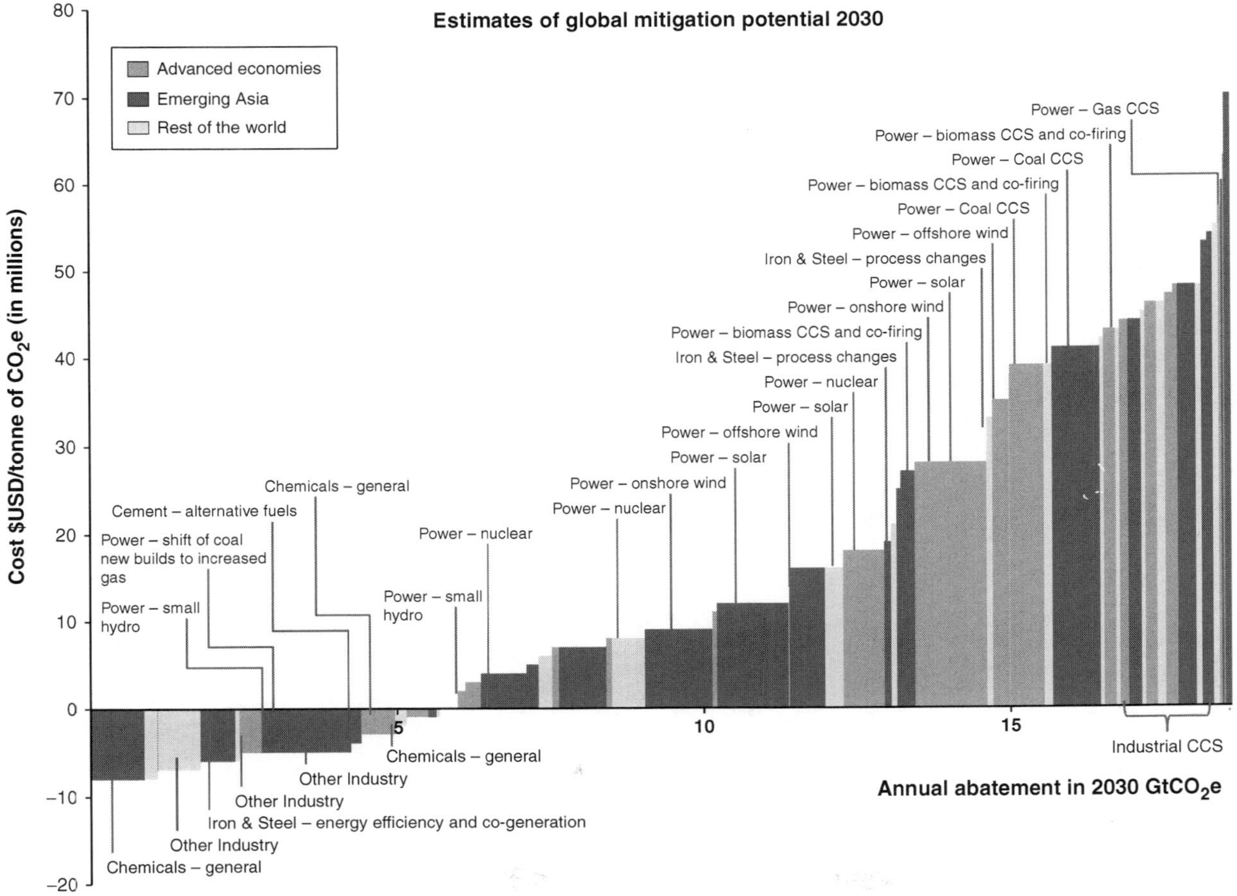

Figure 6.5 Abatement options in the Second Domain

Source: Raw data from McKinsey & Co. (2009).

The usual caveats apply. The curve compares all options on an equal footing at a low annual discount rate of 3.5 per cent – high in terms of the theoretical valuation of climate damages and current returns to government-backed bonds, but far lower than almost anything in a private market. Thus Figure 6.5 includes a portion of 'negative costs' (down to about –\$10/tCO_2), suggesting that measures to reduce emissions also reduce costs. However, in the eyes of private decision-makers many of these may be ignored or not considered to be 'cost-effective' (or only marginally so).[34] Interestingly, this 'slightly cost-saving' portion of the McKinsey curve is heavily populated by options to improve industrial energy efficiency. This may help to explain the impact of the EU emissions trading scheme on industry responses in this area (Chapter 7) – it focused attention on these opportunities, improved their economic attractiveness, and introduced a fear in industry that CO_2 prices might go higher, amplifying the benefits of taking action.

The bulk of the curve, however, is populated by options to cut emissions by up to 10GtCO_2 – almost a third of current global emissions – at costs apparently in the range €5–40/tCO_2. Again, caveats apply. As illustrated in Chapter 3, some costs have already changed significantly, most notably with the sharp cost reductions of 'the solar surprise' (see Box 3.4), while estimated costs have to varying degrees risen for some others (nuclear power, offshore wind and CCS). Moreover, the data compare (of course) monetised costs and benefits, whereas many other factors determine the perceived attractiveness of options, as noted in the technical review of Chapter 3, which in part reflects views of 'hidden costs'.[35]

The options have different timescales and – especially towards the right-hand side – there may be scope for learning, cost reduction and economies of scale as new industries develop. Building up industries takes time and investment, including in infrastructure and innovation, and options that currently appear quite expensive may nevertheless be crucial in 2050 – hence a World Bank study titled *When Starting with the Most Expensive Option Makes Sense* (Vogt-Schilb and Hallegatte 2011).

For all these reasons, the cost curve is *not* necessarily an ordering according to desirability or even priority. It does, however, give an ordering of the potential scale, nature and costs of options, and in that sense a ranking of possible responses to a pure CO_2 price. In this, it confirms that the dominant Second Domain options lie in industry and power generation, as hypothesised in Chapter 3: rational industries making informed, deliberative choices based on calculated financial costs and benefits.

For the power-generation options, however, there is another important caveat. Electricity markets are unusual constructs: all the different options produce an identical product (electrons) that flow through transmission and distribution systems – natural monopolies that have to be regulated. Competition is possible in power generation (and sometimes in transmission assets), but this has to be established and governed by a conscious 'market design', which defines the terms on which producers can gain access to the network and on which suppliers can deliver and (if competitive) market the electrons to customers. Moreover, *most of the options involve substantial and long-lived investment*. Whether that investment is forthcoming depends on future expectations as well as on current CO_2 prices and confidence. The realised costs will also depend on the cost of finance as affected by perception of risks. Thus, much devil lies in the detail, as illustrated in subsequent chapters.

The McKinsey Curve is thus a rough guide to the nature and scale of opportunities and an indication of contingent responses to a CO_2 price – contingent upon a number of other factors. It is suggestive of the level of CO_2 price that might be required to get a given level of emission savings over the next couple of decades, if everything else 'works as it should' – which of course is a simplification. Still, it highlights the huge scale of potential responses in the Second Domain, particularly in industry and power generation, for which putting an adequate price on carbon is a central step. The question then is how best to do it.

6.8 Tax versus cap-and-trade: which is more efficient?

There are two basic ways to price carbon: impose a tax, or set a cap by issuing a fixed quantity of emission allowances that participants can trade. Carbon taxes levy a price directly on either the carbon content of a fuel or the emissions from a production process or the carbon embodied in a final good; the government sets the price and collects the revenue. A cap-and-trade scheme involves the government issuing allowances for a target level of emissions, either through free allocation or through auctions. Firms then need to make sure that they get enough allowances to cover their emissions in a defined period. If a firm requires more allowances to cover their emissions, they can be purchased through the market from firms who can make cuts beyond their allowances; through this trading, a market price for carbon emerges.

If there were perfect markets, no uncertainty and complete information, the two approaches would yield the same resulting level of pollution. Of course the real world is not like that.

Some of the difference comes down to how to set the level. In classical economic theory, the price should reflect the damages caused – or more precisely the *marginal* damage associated with emitting more. The price, or the cap, drives down emissions and avoids the damage costs characterised by the uncertain but rising slope in Figure 6.6.

The costs of emission control generally rise with the degree of emission cutback – as characterised by the declining slope in the figure. In equilibrium theory, the addition of the two curves gives the total costs of damages and control, and policy should aim to minimise the sum of these two. This defines the nirvana of optimal control, whether through taxes (vertical axis) or quantity (horizontal axis). Which instrument is chosen is then almost irrelevant. We can fix either.[36]

The reality of energy and environment of course involves imperfect markets, high uncertainty and incomplete information. Chapter 1 underlined the huge scale of complexity around climate damages, and we attempt to illustrate this with the broad swathe of possible damage estimates in Figure 6.6. Uncertainties also exist about the real cost of emission cutbacks, due to a mix of market and technological uncertainties (think back to Chapter 3) and the speed at which they are undertaken. Many other factors are relevant but these uncertainties are key to the debate.[37]

This means that we face a very different challenge from the one with nice known lines that neatly sum or intersect at the point of optimal control. It is more like the sketch in Figure 6.6, which itself changes over time. The lines are uncertain and non-linear. Small-scale emission cutbacks may be quite cheap, while costs may rise a lot for the more substantial cutbacks, particularly if they are imposed quickly. It is very difficult, if not impossible, to calculate the perfect levels at which taxes or caps should be set.

In these conditions, taxes and cap-and-trade mechanisms are not the same. A tax generates a certain price, but its impact on emissions is uncertain. A cap ensures an emission level is not exceeded, but the price generated by trading quotas may be highly uncertain. Whether a tax or a cap is more efficient and effective will then depend on many factors, including how we rank the various multiple objectives of CO_2 pricing set out above. Economists have been wrestling between the choice of these two instruments in conditions of uncertainty and asymmetric information since a seminal paper by Harvard economist Martin Weitzman

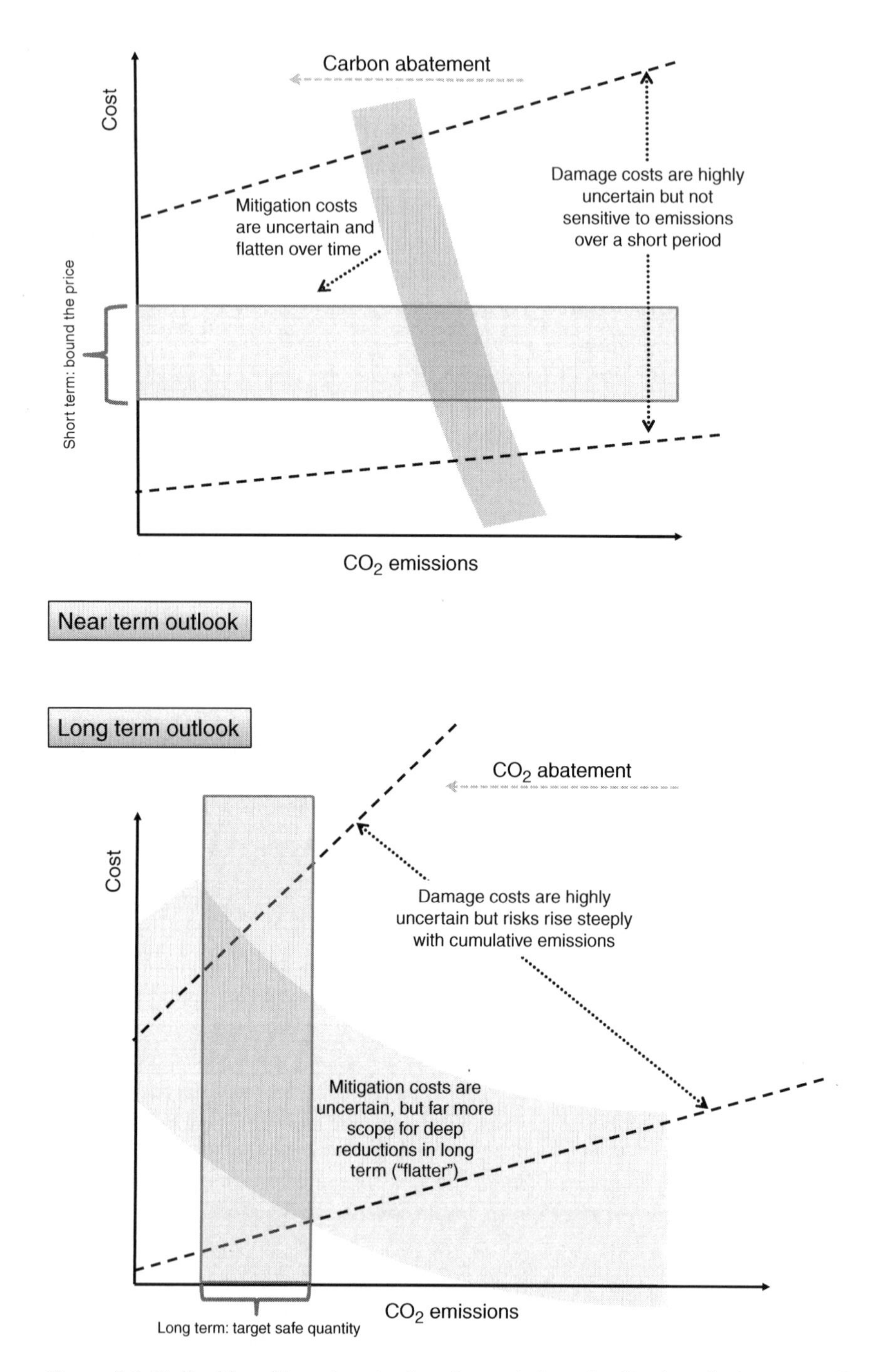

Figure 6.6 Stylised benefits and costs of cutting emissions: implications for 'tax or cap'?

Source: Authors.

in 1974.[38] He showed that it matters whether the abatement cost curve or the damage cost curve is steeper. For a greenhouse gas like CO_2:

- The limits on short-term flexibility mean that costs may rise sharply with emission cutbacks if the cap is set too tight. At the same time, the direct climatic benefits of reduced emissions over a short period are minimal. In the short term therefore a tax would be preferential (which sets a horizontal line on Figure 6.6(a)).
- More time gives more flexibility to innovate additional solutions and to cut emissions, hence the mitigation costs curve is flatter, while the cumulative damages of sustaining higher emissions increases more steeply. In the longer-term it is therefore preferable to define the quantity of allowed long-term emissions (Figure 6.6(b)).

This suggests that a carbon tax is more efficient in the short term, but points more towards a cap- or target-based control for the long-term objective. The essential economic structure thus suggests a price which needs to rise over time until the long-term quantity objective is achieved.

A variety of further dimensions can – and should – be considered. The impact of CO_2 prices – and concomitant revenues – flowing through the economy has been outlined in the previous section. Other important dimensions include how a given instrument, whether tax or trade, will affect industry expectations in the future, particularly relating to investment – a crucial dimension discussed in the context of the European system in Chapter 7.

Amendments to 'pure' tax or cap-and-trade schemes are possible. In Europe the cap was set initially for a three-year first (trial) period, then for five years (2008–12) and then for eight years (2013–20), with the latter set already in 2008. Setting a precise cap 12 years ahead seemed logical and sought to meet industry demands for longer-term certainty but has already proved to be too long given the scale of the uncertainties (Chapter 7). If caps are not set long term, a tax approach could be ramped up if it turned out that emissions were not cut back as much as hoped or if consensus grew on the severity of the problem. The challenge of such announcements of future tax increases is their credibility to investors – recall the UK 'fuel price escalator'. Investors tend to base decisions heavily on existing policies and prices, knowing the political resistance to substantial change.

There are also more hybrid schemes that combine elements of a tax and cap-and-trade schemes, such as price floors and ceilings on trading schemes discussed in the next chapter. The UK even explicitly formulated a carbon tax to complement the emission trading price for coal and gas power stations if the CO_2 price emerging from the trading scheme is too low. 'Hybrid instruments' may have the potential to offer the best mix of the various objectives indicated earlier.[39]

To foster long-term change, an important element is the long-term investment signals that the two instruments may offer. As we have seen, decarbonising the energy system requires large investments over significant timescales. Providing clear signals that low-carbon investments will be rewarded is important. But, again, the implications for choice of instrument are ambiguous.

Another dimension is how consumers may perceive the different instruments – if they do so at all. As a cap-and-trade scheme effectively sets a limit on emissions, this can create the perception that individual behaviour aimed at saving emissions is pointless as it is taken care of by the cap. Any additional emission reduction would only give someone somewhere else in the system scope to emit more. In practice, the success of accelerated emission reductions can encourage tighter future targets domestically or in other regions, and thus consumer initiatives are likely to have a similar impact on emission reductions under a cap-and-trade scheme as under a tax scheme. But the impact is complex and thus difficult to explain. In some countries this has become a fierce point of discussion, notably in the development of the Australian Carbon Pollution Reduction Scheme. Mechanisms to reflect additional reductions more explicitly in the current or future caps are outlined in Chapter 7.

Finally, taxes and cap-and-trade differ in how they respond to macroeconomic conditions. During an economic boom, production increases can boost carbon emissions, thus pushing up the CO_2 price when the economy is doing well. By contrast, when the economy is faltering, reduced manufacturing and transport activities will lower CO_2 prices. This counter-cyclical feature of cap-and-trade may thus aid macroeconomic stability, which has obvious potential attractions – within limits – and might also enhance political acceptance.

6.9 Pricing pollution in practice

Although 'Pigouvian' taxes have been discussed for a number of decades, their use has been limited, at least in their strictest form, although some types of environmental taxation, albeit at rather low tax rates, have started to creep into budgets around the world. Cap-and-trade schemes have been used most notably in the US for tackling 'acid rain' and in the EU for tackling carbon mitigation.

In the mid-1990s the US introduced emissions trading, a unique policy instrument, in an attempt to reduce sulphur dioxide emissions.[40] Sulphur dioxide forms from the combustion of coal, and causes acid rain when it combines with water vapour. Acid rain created major environmental problems across much of the United States and neighbouring Canada in the 1980s and 1990s (as well as across Europe). The amendments to the US Clean Air Act in 1990 aimed to reduce annual SO_2 emissions by 10 million tonnes below 1980 levels.[41] Following lobbying from various sides, including from the Environmental Defense Fund (an American environmental advocacy group), the Clean Air Act amendments introduced a cap-and-trade system.

The US SO_2 system covered almost 500 installations in its first phase, which began in 1995, and rose to over 2,000 in its second phase, which began in 2000. Utilities were allocated freely tradable allowances based on fuel-specific emission rate and company-specific historic fuel consumption. Further annual auctions were available whereby additional allowances could be purchased. The programme has been a clear success. There has been 100 per cent compliance. Indeed, in the first phase, emissions fell to 22 per cent below mandated levels as firms found emissions reductions cheaper than expected and saved surplus allowances for use in subsequent years. Crucially, the emission reductions have been achieved at much lower costs than those anticipated. Market prices were mostly in the range of $100–200 per tonne in contrast to expected prices in the range of $650–850 per tonne at the outset of the programme. Thus, in contrast to initial industry cost estimates of approximately $10 billion a year, the goals have been achieved at around a tenth of that.

Several explanations have been proposed for these much lower-than-expected costs. The competition across abatement options possible with emissions trading encouraged rapid cost reductions. For example, by as early as 1997 costs of flue gas desulphurisation, a major SO_2 abatement technology, had fallen by 40 per cent compared with 1990 levels.[42] Costs of coal transportation fell, allowing wider use of low-sulphur coal. Crucially, the market allowed the true cost of compliance to be revealed.

Indeed, the SO_2 system is another example of the finding that the unit costs of meeting environmental regulations have usually been overestimated prior to programme implementation – irrespective of the policy instrument used to cut emissions.[43] Incentives for industry to overestimate the initial costs of action may combine with the ability of the market to search out hidden cheap options and reward innovative solutions. 'Baseline' emission projections also often turn out to have been unrealistically high. These were all patterns similarly revealed for carbon in the EU ETS the following decade, as discussed in Chapter 7.

The success of the US SO_2 trading scheme led to interest in the use of cap-and-trade for the control of greenhouse gas emissions, and the US led its introduction into the Kyoto Protocol in 1997.

At the time, the EU preferred a policy mix of harmonised carbon taxes and coordinated government measures to promote low-carbon technologies. The European Community, as it then was, first committed to the stabilisation of greenhouse gases in 1990, just before the Second World Climate Conference in Geneva. A series of proposals for Directives followed in 1992, including for a Directive on a Community-wide combined carbon and energy tax. This was proposed to start in 1993 at a relatively low level, but would ramp up towards 2000. The proposal was met with considerable opposition from industry and also from individual Member States. To deal with this, substantial exemptions for energy-intensive industries were proposed, along with a move to tax electricity as per output rather than as per input fuels. The tax was also made conditional on the introduction of similar measures in the other OECD countries. Despite these concessions, which watered down the emission reduction potential of the tax considerably, it still failed to gain the Council support that was necessary for its enactment.

A further attempt was made in 1994 with even greater concessions, including greater exemptions for both energy-intensive industries and also for countries with relatively low levels of emissions, and maintaining a low level of taxation for three years until a review could be conducted. These additional concessions weakened the emission reduction impact of the tax even further, but were still not sufficient to gain the level of support required, and at the end of 1994 attempts at an EU-wide carbon tax were abandoned in favour of attempts to harmonise energy taxes across the Community.

This failure to enact carbon taxation, combined with the success of the US SO_2 system and the US-led incorporation of emissions trading in the Kyoto Protocol, heralded a move toward the introduction of the EU Emissions Trading Scheme (discussed in Chapter 7). An additional advantage of the move from a tax proposal to an emissions trading scheme became apparent during the EU decision-making process. The implementation of an EU-wide tax requires unanimity between all Member States, effectively giving each state a veto. Given the political difficulties the tax encountered, a unanimous decision was implausible. An emissions trading scheme, on the other hand, fell within the realm of environmental policy rather than taxation. As such, it only required qualified majority voting, implying that only a bloc of relatively populous countries could stop its enactment, thus smoothing the political process.

Despite the political failures to enact an EU-wide carbon tax there are, however, some individual countries which have successfully done so. Beginning with Finland in 1990, which introduced a carbon tax levied on all energy products except transport fuels, the four Scandinavian countries (along with the Netherlands) have introduced carbon taxation. Even here, where there are significant environmental constituencies, the taxes adopted have run into significant difficulties. In all four countries the taxes are beset by exemptions and differentiated rates that mean that they are very different on paper from the uniform economy-wide tax that theory dictates. The levels of carbon tax vary between and within the countries, with the level of taxes varying by up to a factor of five in both cases (see Table 6.1). Energy-intensive industries and electricity producers have secured significant exemptions in most of them, invoking competitiveness concerns.

Table 6.1 Some existing carbon taxes

Country	*Year begun*	*Level*	*Exemptions*
Finland	1990	$30/t$CO_2$	Commercial shipping and aviation, fuels for electricity
Norway	1991	$15.93 to $61.76/t$CO_2$	Majority of industry pay reduced rates; fishing and external shipping and aviation exempt
Sweden	1991	Standard rate $104.83/t$CO_2$ Industry rate approx. $23.04/t$CO_2$	Industry taxed at quarter of household rate
Denmark	1992	$16.41/t$CO_2$	Reduced rates for energy-intensive industry in combination with energy-saving agreements
British Colombia (Canada)	2008	Can$10 /tCO2 rising by $5/t$CO_2$/yr to $30/t$CO_2$ in 2012	No exemptions from combustion emissions; does not include process emissions; significant corporate and personal income tax rebates

Note: tCO2 = tonnes carbon dioxide

Source: Sumner *et al.* (2009), with addition for BC by authors derived from Rhodes and Jaccard (2013).

Norway, for example, levies the highest rate of tax on offshore oil and gas, with exemptions for the majority of industry, meaning that the tax only covers about 60 per cent of Norway's CO_2 emissions.[44] Sweden initially levied the same rate upon industry and households as part of a comprehensive tax reform, but after outcries over competitiveness concerns the industry rate was reduced to 25 per cent of the household rate in 1993.[45] It was subsequently raised again to 50 per cent of the level, before being reduced to 21 per cent of the rate in 2004. Denmark has a three-tier approach to levying its tax on industry, comprising one standard rate and three bands of refunds for energy-intensive processes. Industry processes qualify for the refunds if the CO_2 tax burden as a percentage of the difference between sales and purchases is sufficiently high. These exemptions have significantly weakened the emission-reducing impacts of the taxes, even where they have been set at high levels, such as in Sweden. Across Europe of course, taxes on gasoline (or the road duty escalator in the UK) remain one major contributor to controlling GHG emissions.

The challenges for the implementation of sufficient and robust carbon taxes are not limited to Europe. Most notable was the failure of the Clinton-proposed 'BTU' (British Thermal Unit) tax of the mid-1990s. Against the backdrop of fiscal stimulus packages and deficit reduction measures, President Clinton in his 1993 State of the Union address had proposed a:

> broad-based energy tax as the best way to provide [us with] revenue to cover the deficit, because [the tax] also combats pollution, promotes energy efficiency, and promotes the independence economically of the country, as well as helping to reduce the debt.[46]

The devil emerged in debating what form a 'broad-based energy tax' should take. Early fallers in the race included a pure carbon tax, as a result of objections from Senators from coal-rich states. A gasoline tax had already been denounced by the President during his campaign as too 'back-breaking'.[47] This left two options: one a BTU tax based upon the heat content of fuels, or an *ad valorem* tax based upon the sale price of fuels. Resistance to the ad-valorem tax by leading environmental campaigners led to the proposal of a BTU tax. As the tax passed from committee to committee, the lobbying pressure on it grew and grew. Industry and state pressure led to a wide range of exemptions both for industries and specific fuels. The exemptions led to a Bill that had been watered down to a point that weakened its own *raison d'être*: deficit reduction and environmental protection. This undermined support from even its own initial proponents and the proposal died under its own weight. No significant attempts at a broad-based energy tax have emerged over the subsequent 15 years.[48] Many of these messages in Clinton's statement were, however, similar to the objectives of President Obama and plans for US cap-and-trade that emerged a decade and a half later.

In the EU the issue of carbon taxation has recently raised its head again, albeit for activities not covered by the EU emission trading scheme. President Nicolas Sarkozy proposed a French carbon tax in 2007. The proposal was to start at €34/tCO_2, but by the time of the final text it had been reduced to €17 a tonne, which covered oil, gas and coal consumption by business and households, with exemptions for industry and electricity that fell under the EU ETS. The proposed tax featured many tax recycling elements in its early phases, including tax breaks for households, but these were poorly explained and the perception arose that it was merely a revenue-raising tax, hindering wider public support for the instrument. In the end the exemptions for industry fell foul of the highest French court of law, the Conseil d'État, on the grounds that they discriminated against the general public and the tax had to be withdrawn in early 2010.[49]

Across the Atlantic, the governor of British Colombia on Canada's west coast drove through a state carbon tax, starting at $10/t$CO_2$ in 2008 and rising to $30/t$CO_2$. A number of features, including highly visible recycling of the revenues, were designed to make it more palatable to voters. Its co-evolution with proposals for a regional cap-and-trade system for energy-intensive industry (the 'Western Climate Initiative') helped to offset some of the fears of BC industry. Though it was not popular in the state elections of 2010, the party was re-elected on a platform of broad economic competence, with the carbon tax presented as part of the necessary medicine. The world witnessed the first new carbon tax for almost 20 years.

A similar effort failed spectacularly, however, when the donnish former Canadian environment minister Stephan Dion, as Party leader, tried to follow the model at federal level. In the 2010 election campaign his platform of 'green shift' was promptly renamed 'green shaft' by the opposition and became a prime focus of attack. At a time of economic recession the message of 'shafting the taxpayer' proved dynamite; Dion lost heavily, subsequently describing his carbon tax proposals as 'political suicide'. The Australian scheme proved similarly unpopular once it got labelled a 'tax'. Politically, the case for CO_2 taxes cannot be won purely on environmental grounds: it needs clear economic rationale as having something to offer an informed public.

6.10 The politics of pricing

The political problems of enacting either a tax or cap-and-trade scheme are likely to be, and indeed history teaches us that they are, huge. There may be contexts, however, where one or the other is more politically palatable, beyond any pure economic advantages that either instrument yields. In many (and some would argue all) circumstances, political economy factors may dominate over pure economic rationale. In these cases it is important to understand not just what instrument may be best on paper, but what it is possible to enact and what is less likely to be rendered ineffective and indeed obsolete by the political process. Indeed the political connotation of the word 'tax' is very different in North America from that in Europe. The perception of tax as a dirty word in the United States may be enough of a justification to seek other instruments – Robert Stavins, a leading US economist, actively used cap-and-trade non-tax status as a key argument in favour of emissions trading.[50] As described in the next chapter, the Obama administration tried it, but despite coming very close to getting it through Congress, it fell. Despite the Administration's attempts to move the debate away from carbon taxation, one commentator said after that it was dead from the moment the opposition adopted the phrase 'cap-and-tax' – though distrust of trading in the aftermath of the credit crunch undoubtedly played a factor.[51]

For many countries, one additional factor in favour of cap-and-trade is the ability to scale it up internationally, by linking different schemes in different regions. Indeed, the Kyoto Protocol was founded on the idea of emissions trading and related instruments linking globally (as outlined in Chapter 8). A world of unequal and/or missing CO_2 prices raises further challenges. As explained, the EU struggled to implement a Community-wide carbon-energy tax in 1992, which was vetoed by some Member States. The international community lacks even the partial supra-national authority of the EU.

Either taxes or cap-and-trade systems with auctions raise revenues. Although much sought after by governments, there are many political difficulties in trying to simultaneously change relative prices and redistribute income from one group of society to another. Emissions trading with free allocation can change relative prices while largely shielding industry from revenue transfers, thus alleviating industry objections. Indeed, as described in Chapter 7, such initial free allocation gained the EU ETS the support it required to be adopted. In theory the same disconnection can be achieved by setting 'tax thresholds', though this appears to have scarcely been tried in practice. The systems which have been adopted have more simply exempted or reduced tax rates for any sectors for which the revenue transfers (and competitiveness fears) prove infeasible.[52]

The cap-and-trade bills discussed in the US legislation pursued a different avenue. The European experience had pointed to distortions and windfall profits associated with free allowance allocation to emitters. Hence the US Bill envisaged large-scale free allowance allocation to a variety of actors so as to address distributional implications among domestic consumers and to provide resources for energy efficiency improvements. Assuming that these services would otherwise have been financed through public budgets, some level of 'double-dividend' benefit was thus sought. In the end, it proved insufficient.

Businesses may have further preferences between the two instruments. Taxes may offer greater certainty over prices (although governments of course have a history of changing tax levels). Cap-and-trade offers different possibilities. Some firms have seen the EU ETS as an opportunity to profit through trading. The perception that you can make money by being better than the average may lead firms to prefer cap-and-trade regimes over taxes. In more unorthodox trading regimes such as the UK's Carbon Reduction Commitment that covers the commercial and public sectors, firms not only trade permits, but are ranked according to their environmental performance, leveraging sensitivities of these less energy-intensive companies about image and 'corporate social responsibility' (see Chapter 5, Box 5.4).

Ultimately, a crucial factor in choosing between 'tax' and 'cap' is to establish which can be more plausibly implemented, and then amended, to solve any initial problems – and therefore which instrument can survive the political process most effectively and efficiently. A well-designed tax is likely to be better than any cap-and-trade scheme which we can actually implement and vice versa. The question, therefore, is really between a cap-and-trade system with all the exemptions and design features that are necessary to make it politically acceptable, and a carbon tax set at a politically acceptable level, with exemptions and differentiated rates.[53] In CO_2 pricing, it seems, the perfect is the enemy of the good.

6.11 Conclusions

The simple principles that point to the importance of energy prices and the value of economic instruments in reflecting 'external' costs are supported by strong evidence. International data point to the high impact of energy prices in improving energy efficiency, helping to maintain surprisingly constant overall energy bills – the Bashmakov constant (Chapter 1). The economic costs of subsidising energy, including through the profligacy this encourages, are well mapped out. The use of economic incentives to reduce pollution is increasingly well established.

In principle, the efficiency benefits of economic instruments would seem to be particularly large for diffuse, global and long-term problems like climate change; price would thus seem a natural instrument for tackling the larger problems associated with fossil fuels and their emissions. CO_2 prices can support energy efficiency and create incentives to use the least-cost mix of low carbon technologies. This suggests also that their use and intensity (i.e. prices) should rise over time.

These factors suggest an enticing possibility. If the CO_2 price rises at a pace broadly aligned with the wider adjustment effects (such as outlined at the beginning of the chapter), it should be possible to keep energy *bills* broadly constant, at least as a proportion of income. Integrated approaches towards such an outcome emerge as an important theme in subsequent chapters.

Alongside signalling the real costs associated with fossil fuels and climate change, economic instruments inevitably change the pattern of financial flows in an economy. This has the potential to amplify costs or generate benefits, depending upon starting conditions and how the revenues are used. One key factor in assessing net economic results is the assumed starting point of the economy. A CO_2 price is not added to an 'optimal' economy with no fiscal needs, for the pure purpose of reducing emissions. Rather, it will contribute to pre-existing fiscal structures. Integrating energy and CO_2 pricing as part of a wider economic strategy is not just politically important, as noted; it is also a simple expression of reality.

Nevertheless, the impacts on financial flows raise major political issues. The case of gasoline taxation, with widely varying tax rates across countries in the aftermath of the 1970s oil shocks, does illustrate both the feasibility of using such measures – given enough shock to the system – and their multiple roles: they serve to enhance energy security, insulate economies from such shocks in the future, fund roads and public transport and plug fiscal gaps. However, pricing CO_2 on industrial emissions is harder, in part because of competitiveness concerns and more focused points of opposition. Carbon taxes have proved hard to implement, and those that survive are usually watered down and softened with exemptions. Cap-and-trade has proved somewhat more feasible for these sectors, with some concerns examined more closely in Chapter 7. Cap-and-trade cannot forever hide the need for public understanding and acceptance of why pricing 'bads' is a good thing.

Pricing policy is difficult, but has to be part of any serious policy to tackle the big challenges of energy and environment. By the same token, durable CO_2 pricing must inevitably become embedded in the realities of economic and political systems. The question explored in the next chapter, then, is how the major real-world CO_2 pricing schemes adopted to date have fared.

Notes

1. The price elasticity of energy measures how much the energy intensity changes in response to a change in prices. Previous work such as Newbery (2003) have estimated an elasticity of –1 for the time period 1993–9, excluding transition economies. Including these economies and repeating the analysis with all economies gives an elasticity of –1.7 for the time period 1990–2005, while excluding the transition economies gives an elasticity of –1.2.
2. This effect has been studied using both theoretical and empirical methods, for example Pindyck (1979).
3. A survey of the literature associated with this channel can be found in Ang and Zhang (2000).
4. For an investigation into this effect see Newell *et al.* (1999).
5. Atkeson and Kehoe (1994); Steinbuks and Neuhoff (2010).
6. Quote from Walker and Wirl (1993). It appears to have been almost a decade until the next major new contribution to the literature on asymmetric elasticities, by Gately and Huntington (2002). Subsequent analysis sought to preserve symmetric price responses of substitution through introducing a time-varying underlying energy demand trend (UEDT), which captures 'not only exogenous technical progress (or energy-saving technical change) but other important socioeconomic effects'. The most recent analysis concludes that in many cases, both UEDT and asymmetric price elasticities are required to give the fullest explanation of the observed trends in energy demand, and that 'when estimating energy or fuel demand functions a general model allowing for both asymmetric price response and UEDT should initially be estimated, and only if accepted by the data should a more restrictive specification be considered as the preferred specification' (Adeyemi *et al.* 2010). Note that a significant additional element in the analyses is testing for a 'threshold effect' associated with prices rising above the previously highest level in the data.
7. The political difficulties of repealing subsidies can be seen in the US where a recent effort to repeal tax breaks to the oil and gas industry of $35 billion failed in the Senate (Smith 2010).
8. Pittsburgh G20 Leaders Statement, September 2009: 'phase out and rationalize over the medium term inefficient fossil fuel subsidies while providing targeted support for the poorest.'
9. IEA (2010).
10. GSI (2009).
11. IMF (2010).
12. For a more detailed discussion see Ürge-Vorsatz and Miladinova (2006).
13. IEA (2010).
14. UNEP (2008).
15. Newbery and Santos (1999).
16. Pearce (2006).
17. In the UK, this transition was epitomised by the work of David Pearce and colleagues, in 'Blueprint for a green economy' and subsequent volumes in the series (Pearce 2006), and Professor Pearce's appointment as advisor to the Department of Environment. In the US, the iconic event was the conversation of a key environmental lobby group, the Environmental Defense Fund, to support the use of emissions trading as the key instrument for tackling the SO_2 emissions behind acid rain and other accumulating, widespread impacts.
18. Pigou (1920).
19. Baumol (1972).
20. Tietenberg (2005), Chapter 3.
21. See Chapter 1, p.6 for examples. However, two main factors accounted for the economic impact of the oil shocks: the impact of volatility (hence shock – the rapid change in prices relative to existing economic patterns), and the fact that it was heavily traded so that higher oil prices took money out of importing countries – notably the OECD. Neither bears any relation to a well-managed carbon price. The analogy is thus fundamentally false and misleading. Indeed, as noted, gasoline taxation has been an important factor helping to *protect* OECD countries against subsequent oil shocks.
22. Gardiner (2011).
23. Crawford-Brown *et al.* (2012).

24. See, for example, many other papers in the journal *Environmental Science and Policy*, including a Special Issue on Climate Policy and Air Pollution in Europe.
25. This is a 'propagation effect' through the production system: higher costs of energy increase the production costs of energy-intensive industries (steel, cement, glass, non-ferrous materials, chemicals as well as transportation) and these additional costs are passed through all industries (construction, agriculture, infrastructure, manufacturing). This effect is conveyed through what economists call an input-output matrix and on the pass-through of cost increases into price levels at every stage. Thus a €10 increase of the energy costs can ultimately turn into a €15–20 bill for consumers. Note that a *sudden* introduction of a substantial carbon price, or high carbon price volatility, also has the potential to generate dynamic costs associated with the resulting dislocations of capital.
26. Obviously this is true only on average. Energy intensive sectors are losers in this substitution and the labour-intensive ones are winners. For a synthesis of European studies in a period when carbon taxes were hotly debated see IPCC (2002), Chapter 8.
27. The idea was first formed by Tullock (1967) but has been much discussed and criticised, starting with Sandmo (1975). For a review of the literature see Schöb (2003).
28. Giving revenues directly back to consumers may address the impacts on consumer purchasing power but do nothing to offset the distortionary effects on production.
29. The *strong* form of the double dividend – that carbon pricing could yield net economic benefit *to the country itself* – was far more controversial and contested. A key influential paper argued that this was in fact not possible, for quite subtle theoretical reasons (Bovenberg and de Mooij 1994). In practice, this stimulated yet more debate on other mechanisms that might restore a 'strong' double dividend – mechanisms included lower energy imports, the diversification of the fiscal base, transferring more of the tax burden to economic rents and labour market effects (Goulder 1995). Naturally, this did not resolve the issue either, which rumbles on in the academic literature.
30. Goulder (1995).
31. This has been demonstrated in an exercise for the Commission Rocard in France, where a carbon tax levied in 1985 to reach a level of $400/t$CO_2$ in 2010 would have resulted in variations of average production costs spanning from –1% to +2%, depending upon the assumptions made about technical change and overall economic policies, leading to GDP variations of –1% to +0.8%, cumulated over a 25-year period. This corresponds to negligible variation of economic growth rates. (See Rocard 2009; see also Combet *et al.* 2010.)
32. Ellerman *et al.* (2010: 258). Note that in their estimate, the macroeconomic cost is around one-fortieth of the overall value of the EU ETS allowances. For a full review of models see Dannenberg *et al.* (2007). Under an EU-wide '–30%' (from 1990 levels) target, the ETS cap would have to be adjusted from –21% to 36% below 2005 levels. In European Commission (2010) carbon prices were estimated to reach €30 per tonne of CO_2 in order to reach this reduction.
33. For brief reference to global modelling studies see Chapter 11. An example of more detailed and region-specific analysis is the official estimate of the EU's current goal to cut emissions by 20% by 2020, published in 2010 after the 2008/2009 economic crisis (but before the Eurozone crisis), that the overall policy package would cost 0.32% of GDP. This one-off impact on GDP implied that Europe would have to wait until April 2020 to grow to the GDP level that is otherwise projected to be reached in January 2020.
34. That is at the (higher) private sector discount rates, the value of the energy savings from investments in energy efficiency would be discounted more heavily, taking them closer to or into the 'positive cost' characteristics of the Second Domain.
35. Thus, for example, nuclear power involves radioactive waste which risks rendering it unacceptable in some countries; others object to the visual impact of large-scale wind turbines, challenge whether biomass produces net environmental benefits or oppose CCS storage. Energy technologies involve social choice, and options for power generation can to some extent be substitutes for each other.
36. Economic textbooks often show the *marginal* (incremental) costs and benefits. Where they cross – where the marginal benefits intersect with the marginal costs – defines the optimum point. These marginal costs are often depicted as straight lines, which is broadly consistent with the rising curves of total costs in Figure 6.6.
37. See Harrington *et al.* (2000). To complicate life further there are also asymmetric damages – who will be hurt and how to value that – and asymmetric information, including between firms and governments, over what it will actually cost to cut emissions. The clear history from other environmental regulations is one of overestimating the costs of pollution control.

38. Weitzman (1974). Additional work has extended Weitzman's analysis, for example Stavins (1996), Hoel and Karp (2001) and Newell and Pizer (2003).
39. Pizer (2002); Newbery (2013).
40. For a full analysis of the experience of the US Sulphur Trading Program see Ellerman *et al.* (2000).
41. MacKenzie (2007). The 10 million tonnes was not the result of a controversial cost-benefit analysis but was instead chosen for its scientific merit (and the roundness of the number in all probability).
42. McLean (2003).
43. Harrington *et al.* (2000). See also the review of 14 studies in Tietenberg (2005).
44. Andersen (2005).
45. Hammar *et al.* (2013).
46. Clinton (1993).
47. Quoted in Erlandson (1994).
48. For a more detailed discussion of US energy tax policy see Lazzari (2007).
49. For a discussion of the impact of tax recycling in the French context see Combet *et al.* (2010).
50. Stavins (2008).
51. Opponents of cap-and-trade jumped on the phrase cap-and-tax from journalists such as Robert J. Samuelson in the *Washington Post*, politicians such as former Vice-Presidential candidate Sarah Palin, to right-wing think-tanks such as Americans for Prosperity.
52. Pezzey and Jotzo (2013). Presumably one factor deterring the use of tax thresholds is that it may be considered to set a dangerous precedent for taxation and would lead to many of the same complexities that have bedevilled the negotiation of free allowances in emission trading schemes.
53. Pezzey and Jotzo (2012).

References

Adeyemi, O. I., Broadstock, D. C., Chitnis, M., Hunt, L. C. and Judge, G. (2010) Asymmetric price responses and the underlying energy demand trend: are they substitutes or complements? Evidence from modelling OECD aggregate energy demand', *Energy Economics*, 32: 1157–64.

Andersen, M. S. (2005) 'Vikings and virtue: a decade of CO_2 taxation', *Climate Policy*, 4 (1): 13–24.

Ang, B. W. and Zhang, F. (2000) 'A survey of index decomposition analysis in energy and environmental studies', *Energy*, 25 (12): 1149–76.

Atkeson, A. and Kehoe, P. J. (1994) *Models of Energy Use: Putty-Putty versus Putty-Clay, NBER Working Papers 4833*. National Bureau of Economic Research.

Baumol, W. (1972) 'On taxation and the control of externalities', *American Economic Review*, 62 (3): 307–22.

Bovenberg, L. and de Mooij, R. A. (1994) 'Environmental taxes and labor-market distortions', *European Journal of Political Economy*, 10 (4): 655–83.

Clinton, W. J. (1993) *State of the Union Presidential Address*. USA.

Combet, E., Ghersi, F., Hourcade, J.-C. and Thery, D. (2010) Carbon tax and equity – the importance of policy design', *Critical Issues in Environmental Taxation*, VIII: 277–95.

Crawford-Brown, D., Barker, T., Anger, A. and Dessens, O. (2012) 'Ozone and PM related health co-benefits of climate change policies in Mexico', *Environmental Science and Policy*, 17: 33–40.

Dannenberg, A., Mennel, T., Osberghaus, D. and Sturm, B. (2007) *The Economics of Adaptation to Climate Change – The Case of Germany*, Discussion Paper No. 09-057. Centre for European Economic Research.

Dargay, J. (1993) *Are Price and Income Elasticities of Demand Constant? The UK Experience*. Oxford: Oxford Institute for Energy Studies.

de Mooij, R. and Bovenberg, A. (1998) 'Environmental taxes, international capital mobility and inefficient tax systems: tax burden vs. tax shifting', *International Tax and Public Finance*, 5 (1): 7–39.

DECC (2011) *Fossil Fuel Price Shocks and a Low Carbon Economy*. Department of Energy and Climate Change.

Ellerman, A. D., Convery, F. J. and de Perthius, C. (2010) *Pricing Carbon: The European Emissions Trading Scheme*. Cambridge: Cambridge University Press.
Ellerman, A. D., Joskow, P., Schmalensee, R., Montero, J. and Bailey, E. (2000) *Markets for Clean Air: The U.S. Acid Rain Program*. New York: Cambridge University Press.
Erlandson, D. (1994) 'The BTU tax experience: what happened and why it happened', *Pace Environmental Law Review*, 12 (1): 173–84.
European Commission (2009) *Leaders' Statement: The G-20 Pittsburgh Summit*. September.
Gardiner, S. M. (2011) *A Perfect Moral Storm: The Ethical Tragedy of Climate Change*. Oxford: Oxford University Press.
Gately, D. and Huntington, H. G. (2002) 'The asymmetric effects of changes in price and income on energy and oil demand', *Energy Journal*, 23: 19–55.
Goulder, L. (1995) 'Environmental taxation and the double dividend: a reader's guide', *International Tax and Public Finance*, 2 (2): 157–83.
GSI (2009) *The Politics of Fossil-Fuel Subsidies*, Global Subsidies Initiative Report. International Institute for Sustainable Development.
Hallegatte, S. (2007) 'Do current assessments underestimate future damages from climate change?', *World Economics*, 8: 131–46.
Hammar, H., Sterner, T. and Åkerfeldt, S. (2013) 'Sweden's CO_2 tax and taxation reform experiences', in R. Genevey, R. Pachauri and L. Tubiana (eds), *Reducing Inequalities: A Sustainable Development Challenge* New Dehli: TERI Press.
Harrington, W., Morgenstern, R. and Nelson, P. (2000) 'On the accuracy of regulatory cost estimates', *Journal of Policy Analysis and Management*, 19 (2): 297–322.
Hoel, M. and Karp, L. (2001) 'Taxes and quotas for a stock pollutant with multiplicative uncertainty', *Journal of Public Economics*, 82: 91–114.
IEA (2010) *World Energy Outlook*. Paris: OECD/IEA.
IEA (2011) *World Energy Outlook*. Paris: OECD/IEA.
IMF (2010) *World Economic Outlook*. Washington, DC: International Monetary Fund.
IPCC (2002) *Climate Change 2001: Mitigation*, Contribution of Working Group III to the Third Assessment Report of the Intergovernmental Panel on Climate Change, UNEP, WMO. Cambridge: Cambridge University Press.
Lazzari, S. (2007) *Energy Tax Policy: History and Current Issues*, Order code RL33578. Congressional Research Service.
MacKenzie, D. (2007) 'Making things the same: gases, emission rights and the politics of carbon markets', *Accounting, Organizations and Society*, 34 (3–4): 440–55.
McKinsey & Co. (2009) *Pathways to Low Carbon Economy*. New York: McKinsey.
McLean, B. (2003) *Ex-post Evaluation in the United States*. Intervention at the OECD Workshop on the 'Ex-post Evaluation of Tradable Permits', OECD, 21–22 January.
Maten, L. (1999) 'Energy taxation – a historical overview', *International Journal of Global Energy Issues*, 12 (7/8): 304–14.
Newbery, D. M. (2003) 'Sectoral dimensions of sustainable development: energy and transport', *Economic Survey of Europe*, 2: 73–93.
Newbery, D. M. (2013) 'Evolution of the British electricity market and the role of policy for the low carbon future', in Fereidoon P. Sioshansi (ed.), *Evolution of Global Electricity Markets: New Paradigms, New Challenges, New Approaches*. London: Elsevier.
Newbery, D. M. and Santos, G. (1999) 'Road taxes, road user charges and earmarking', *Fiscal Studies*, 20 (2): 103–32.
Newell, R. G. and Pizer, W. A. (2003) 'Regulating stock externalities under uncertainty', *Journal of Environmental Economics and Management*, 45: 416–32.
Newell, R., Jaffe, A. B. and Stavins, R. (1999) *Energy-Efficient Technologies and Climate Change Policies: Issues and Evidence*, Resources for the Future Climate Issue Brief No. 19.
OECD (2013) *Climate and carban: aligning prices and policies,* Environment Policy Paper no.1. Paris: Organisation for Economic Cooperation and Development.

Pearce, D. (2006) 'The political economy of an energy tax: the United Kingdom's Climate Change Levy', *Energy Economics*, 28 (2): 149–58.

Pezzey, J. C. V. and Jotzo, F. (2012) 'Tax-versus-trading and efficient revenue recycling as issues for greenhouse gas abatement', *Journal of Environmental Economics and Management*, 64 (2): 230–6.

Pezzey, J. C. V. and Jotzo, F. (2013) 'Carbon tax needs thresholds to reach its full potential', *Nature Climate Change*, December.

Pigou, A. C. (1920) *The Economics of Welfare*. London: Macmillan.

Pindyck, R. S. (1979) *The Structure of World Energy Demand*, MIT Press Books, Vol. 1. Cambridge, MA: MIT Press.

Pizer, W. (2002) 'Combining price and quantity controls to mitigate global climate change', *Journal of Public Economics*, 85 (3): 409–34.

Rabe, B. (2008) 'States on steroids: the intergovernmental odyssey of American climate policy', *Review of Policy Research*, 25 (2): 105–28.

Rhodes, E. and Jaccard, M. (2013) 'A tale of two climate policies : political economy of British Columbia's Carbon Tax and Clean Electricity Standard', *Canadian Public Policy – Analyse de politiques*, XXXIX, supplement 2.

Rocard, M. (2009) *Rapport de la conférence des experts et de la table ronde sur la contribution Climat et Energie*. Paris: Ministre de Developpement Durable. Available at: http://www.developpement-durable.gouv.fr/IMG/pdf/01-18.pdf

Royal Society (2009) *Geoengineering the Climate: Science, Governance and Uncertainty*. London: Royal Society.

Sandmo, A. (1975) 'Optimal taxation in the presence of externalities', *Swedish Journal of Economics*, 77: 86–98.

Schöb, R. (2003) *The Double Dividend Hypothesis of Environmental Taxes: A Survey*, Working Papers 2003.60, Fondazione Eni Enrico Mattei. Vienna: UNIDO.

Smith, D. (2010) 'Effort to repeal oil tax breaks fails in Senate', reported by Reuters, 15 June.

Stavins, R. N. (1996) 'Correlated uncertainty and policy instrument choice', *Journal of Environmental Economics and Management*, 30 (2): 218–32.

Stavins, R. N. (2008) 'A meaningful U.S. cap-and-trade system to address climate change', *Harvard Environmental Law Review*, 32: 293–371.

Steinbuks, J. and Neuhoff, K. (2010) *Operational and Investment Response to Energy Prices in the OECD Manufacturing Sector*, Cambridge EPRG Working Paper 03/10. University of Cambridge.

Sumner, J., Bird, L. and Smith, H. (2009) *Carbon Taxes: A Review of Experience and Policy Design Considerations*, Technical Report NREL/TP-6A2-47312. National Renewable Energy Laboratory.

Tietenberg, T. H. (2005) *Emissions Trading – Principles and Practice*. Washington, DC: Resources for the Future Second Edition.

Tullock, G. (1967) 'The welfare costs of tariffs, monopolies, and theft', *Western Economic Journal*, 5 (3): 224–33.

UNEP (2008) *UNEP Climate Change Strategy*. United Nations Environment Programme.

Ürge-Vorsatz, D. and Miladinova, G. (2006) 'Energy efficiency policy in an enlarged European Union: the Eastern perspective', in S. Attali and K. Tillerson (eds) *ECEEE 2005 Summer Study. Energy Savings: What Works and Who Delivers?* France: Mandelieu la Napoule, Vol. 1, pp. 253–67.

Vogt-Schilb, A. and Hallegatte, S. (2011) *When Starting with the Most Expensive Option Makes Sense: Use and Misuse of Marginal Abatement Cost Curves*, Policy Research Working Paper Series 5803. World Bank.

Walker, I. O. and Wirl, F. (1993) 'Irreversible price induced efficiency improvements: theory and empirical application to road transportation', *Energy Journal*, 14 (4): 183–205.

Weitzman M. (1974) 'Prices vs. quantities', *Review of Economic Studies*, 41 (4): 477–91.

7 Cap-and-trade and offsets

From idea to practice

> 'Until recently … resorting to emissions trading usually occurred only after other, more familiar approaches had been tried and failed.'
>
> – Tietenberg (2006: 60)

> 'Energy forecasting was invented to make economic forecasting look good.'
>
> – Anon

7.1 Introduction: slow train coming

Proper pricing pays. That, in essence, is the main message of a century of experience with economic systems, including their environmental consequences. The previous chapter summarised the reasons, including why CO_2 seems an almost ideal case for applying the principle. CO_2 pricing can create an efficient and consistent incentive, across a huge range of activities, to cut emissions. In market-based economies, it seems an essential element of policy. Yet Chapter 6 also observed the struggles to implement the principle, particularly over carbon taxation. In most cases, cap-and-trade seems to have emerged as the least intractable way of getting CO_2 emitters to pay their dues. As so often, however, there is much devil in the detail. This chapter charts the many lessons learned in trying to translate the idea into reality.

Despite being the home of cap-and-trade, as outlined in Chapter 6, the US effort to adopt carbon trading federally in 2010 failed: the flagship Waxman-Markey Bill passed the House of Representatives but could not muster the 60 per cent majority required to pass the US Senate. The EU emissions trading system, the first applied to carbon and Europe's 'flagship' policy, is in trouble, and some consider it discredited – along with a range of mechanisms associated with it.

Not surprisingly, there is renewed talk of carbon taxation as a better alternative – a convenient display of the risk of political debates going in circles, given that most carbon trading proposals were born out of the wreckage of efforts to tax carbon in the 1990s. The real lesson is that pricing carbon is inherently difficult, however one goes about it.

Despite all the problems and opposition, carbon pricing is spreading. The emerging systems have the opportunity to learn from experience, both good and bad – and they need to.

The dominant concerns are no longer about the ethics of allowing companies to trade emission quotas for reasons explained in Chapter 6, but do involve issues of trust and confidence in our ability to design and manage complex systems. The US system for capping sulphur emissions, as outlined in Chapter 6, seemed relatively straightforward – with the benefit of hindsight. Carbon has proved a lot tougher.

While new schemes are operational (California, some city schemes around the world, Chinese regional pilots) and in law (Korea), our main operational experience comes from two main and interlinked systems: the EU Emissions Trading System, and the international mechanisms built into the Kyoto Protocol, particularly for project-based 'offsets'. This chapter charts what has been done, what we have learned, what is emerging – and, most importantly, where we might go from here.

7.2 Scope and coverage

The various cap-and-trade schemes designed and debated over the past decade have differed significantly in their scope and coverage. The focus of the EU ETS is on capping carbon dioxide emissions from power generation and heavy industry, covering over 11,500 energy-intensive installations across Europe. As indicated in Figure 7.1, the majority of emissions covered are from power generation and other combustion, with most of the rest from cement, steel, refineries and chemicals production. As the focus is on industrial producers, emissions from domestic consumers, transport and agriculture are all excluded.

This exclusion of these other sectors was a source of criticism from some quarters, particularly as proposals for more ambitious schemes developed in the US, Australia and New Zealand. The narrow focus of the EU ETS was, however, one of the key reasons for its implementation. By excluding potentially difficult sectors such as transport and agriculture, the system was simplified. The same was even more true for the Regional Greenhouse Gas Initiative, which only covered power generation in seven North-East US states.

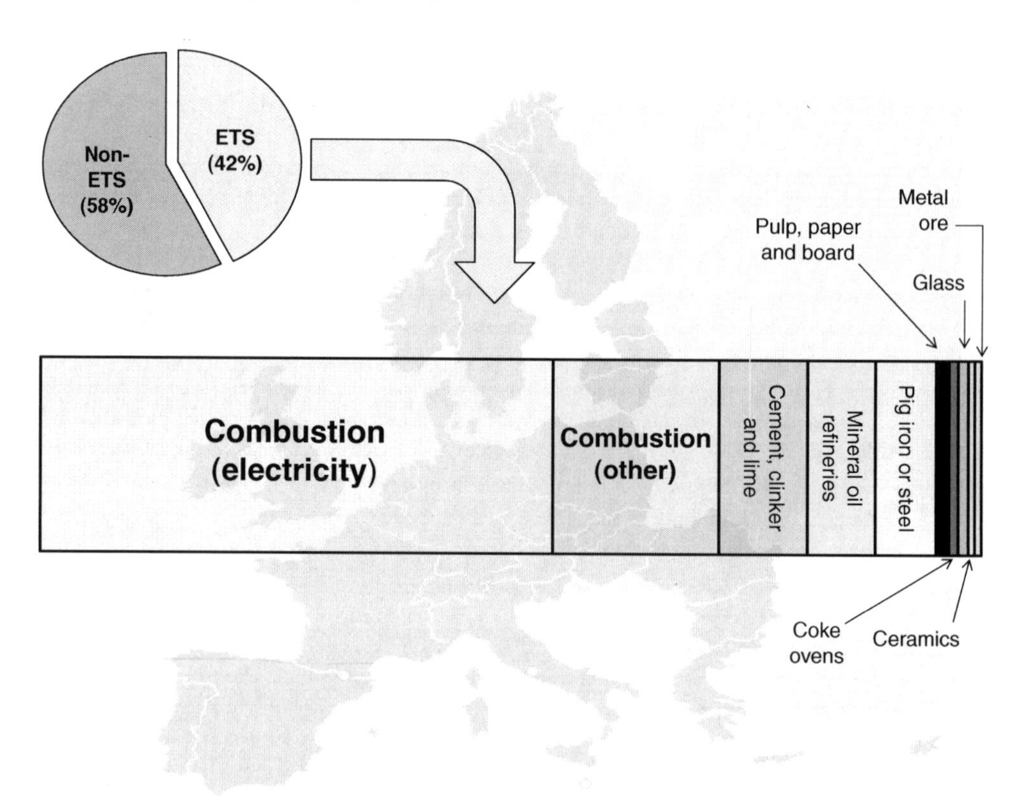

Figure 7.1 Coverage of the EU ETS

Note: The darker shaded countries are those participating in the EU ETS. The bar shows the distribution of verified emissions over 2005–8 between the different sectors covered under the EU ETS.

Raw data source: European Environment Agency (2013) and Ecofys *et al.* (2009).

Going upstream

The US Waxman-Markey Bill for a federal US system was far more ambitious; it aimed to cover the great majority of US emissions, including transport, by holding oil producers accountable for the carbon in the fuel they sold – 'upstream' application of the carbon cap. The final version of the bill that passed the House of Representatives expanded to almost 1,500 pages to accommodate the various sectors and provide for compensation to several constituencies. The original Australian Carbon Pollution Reduction System (first discussed in 2007 but delayed in 2010) had similar levels of ambition and proposed to bring in agricultural emissions from 2015.

Upstream systems have the clear attraction of involving fewer entities and consequently relatively low implementation and administration costs, along with wider coverage of emissions. Some academics went further and argued that the whole climate change problem should be tackled entirely upstream, by capping fossil fuel production and imports – essentially rationing fossil fuels. This contrasts with the EU ETS which applies the cap at point-of-emissions, requiring installations to surrender allowances for the carbon dioxide emissions that they actually emit.

Upstream systems would rely heavily on well-functioning markets for the price signal to filter through the economy. The price perceived by all actors downstream from fossil fuel producers is the same as with an equivalent tax – assuming no manipulation. An upstream pricing system – whether cap-and-trade, or a tax on the carbon content of fuels – is the ultimate in its dependence on Second Domain assumptions. Its claim to efficiency lies in assuming that a single carbon price across all fuels is efficient, irrespective of existing tax and other structures; that the economic system transmits the prices efficiently, without manipulation, speculation or rent-seeking; and that the consumers downstream are pure market actors responding principally to price.

In practice, of course, there are drawbacks. Their wide scope amplifies the political problems, making them more complex and creating larger constituencies to lobby against them. Moving the point of control upstream, e.g. accounting for automotive emissions already at the refinery level, amplified the sense that the cap-and-trade systems were essentially taxes in disguise – in effect, the consumers of these fuels were not engaged at all, and the only impact on them would be higher prices. Even worse for critics and voters was the fear that they would generate industrial profits at the expense of consumers.[1] So far, the only successful system taking this route is doing so in an incremental way, with the Californian system due to expand in 2015 to include CO_2 from transport upstream.

… or going downstream

Alternatively, hypothetical 'downstream' systems would hold final consumers accountable not only for their own emissions, but also the CO_2 emitted in the electricity (and in some proposals emitted during the production of other goods) they consume. At least half a dozen different variants on this theme of 'personal carbon trading' have been proposed.[2]

Some elements of downstream proposals have also been enacted, most notably in the Tokyo municipal emissions trading scheme, and a number of countries have made some progress with 'downstream' taxation, including a UK 'climate change levy' on business energy consumption. We touch on these points later.

However, the main scope and coverage of cap-and-trade has been on the point of emission, and focused on entities where prices are realistically the *primary* influence on decision-making – and consequently where the gain of using a carbon price is most likely to dominate over the pain.

For example, Pillar I articulated the innumerable ways in which energy use in buildings – particularly heating – is in fact determined by innumerable other factors, with overwhelming evidence that people in their homes are frequently unaware and uninformed about efficiency improvement opportunities and may have little control over their energy consumption. The inclusion of gas pipeline or distribution companies in a trading scheme may thus not yield much in the way of emission reductions, but could inflict a lot of pain on those in poor dwellings – a problem considered further in Chapter 8.

Additionally, since gasoline is already heavily taxed for transport use in Europe and Japan (see Chapter 6, Figure 6.3), carbon pricing would mostly be a modest addition to pre-existing taxes with little real impact. The scope for consumers to respond is also limited without new technology and infrastructure – Third Domain factors for which price on its own is a blunt tool. All this and the administrative complexity explain why road transport is not included in the EU ETS.

The sectors likely to be most responsive to carbon pricing and most able to manage cap-and-trade are thus those dominated by commercial decision-making for whom energy is a major cost. Heavy industry across the world generally fits into these categories, while competitive electricity markets – where generators can in principle avoid exposure by using lower carbon generation – fit the bill particularly well, given the scope to generate power from a range of sources including lower carbon fuels. Focusing on these sectors also enabled the EU ETS to limit its administrative complexity. Monitoring and verifying emissions from 12,000 sources is much easier than doing the same for several hundred million households (or drivers). In this respect, the scope and coverage of the EU ETS seems entirely appropriate. Yet its problems have lain elsewhere.

7.3 Evolution of the EU Emissions Trading System: caps and prices

Background

The EU ETS is no minor beast. Designed initially mainly among the 15 West European Member States, it automatically extended to the 12 countries of Eastern Europe which joined the EU over 2005–7; three neighbouring states (Norway, Liechtenstein and Iceland) also subsequently joined, so that it now spans 30 countries. In capping more than 40 per cent of emissions across most of the European continent, it is by far the world's most ambitious attempt to tackle CO_2 emissions.

It took some time for people to realise the economic implication: by charging for a major activity that was previously free, namely emitting CO_2, it created assets worth many tens of billions of euros a year (more than ten times the scale of the much-vaunted US sulphur system). It has been seen as the flagship of European efforts to tackle climate change, and if you live in any of those 30 countries then it has most probably affected your electricity prices. Globally, it emerged as the fulcrum of the international system under the Kyoto Protocol. It deserves serious attention for it is replete with lessons.

As outlined in Chapter 6, the EU ETS grew out of the wreckage of attempts to enact an EU-wide carbon tax, combined with the inclusion of trading mechanisms in the Kyoto Protocol in 1997.

After some years of deliberation, a proposal emerged in 2001 and the resulting Emissions Trading Directive was formally issued in October 2003, with the system to begin with a three-year Phase I from 2005 to 2007.[3]

Installations covered by the EU ETS receive European Union Allowances (or EUAs), which they are then required to surrender commensurate with their emissions annually. The allowances are freely tradable, not only with other installations, but also with financial intermediaries and indeed with individuals who are able to register and trade them.[4] Initially, almost all the allowances were given out for free (as discussed later). So if a company could cut its emissions below its initial allocation cheaply, it could thus make a profit by selling its surplus allowances to another company which finds emissions reductions more expensive. Thus the economic ideal of cap-and-trade is achieved, with a carbon price emerging from the trading market which settles at the lowest possible cost of achieving the overall emissions cap.

The EU ETS was designed to fit with the Kyoto Protocol. Its first trial period (2005–7) was followed by a second period timed to coincide with the first Commitment Period of the Kyoto Protocol (2008–12). Indeed, the EU ETS was legally coupled with the Kyoto Protocol: exchanges of EU allowances across borders were matched by transfers of national Assigned Amount Units under the Kyoto Protocol. The EU ETS is thus the embodiment of the original US ideal of international emissions trading, with the private sector incentivised to hunt out emission reductions wherever they would be cheapest, with a disregard to national borders.

This linkage was extended further by coupling the EU ETS to the international 'emissions offset' mechanisms established under the Kyoto Protocol, most notably the Clean Development Mechanism discussed later in this chapter. The ETS Linking Directive[5] allowed credits from these mechanisms to be purchased by EU installations to offset against their own emission caps. This involved a controversial debate about how much the EU should rely on such international credits. The conclusion that they should be 'supplementary to' Member States' obligations to take action to reduce domestic emissions has in practice not remotely resolved the issue, as discussed later.

Beyond all this, the Directive also established the ETS as an enduring system, designed for successive phases to secure ongoing emission reductions, building upon the experience of earlier phases. This also facilitates a 'futures' market, reflecting expectations about the future cost and value of emission allowances.

Phase I 2005–7

The first phase of the EU ETS, from 2005 to 2007, was always perceived as a learning step to build experience for the later Phase II (2008–12) which would coordinate with the first Kyoto period. It was used to establish principles and capacity within the EU Commission, the industry and Member State governments, along with the first carbon price.

The cap and the distribution of allowances to emitting industries were decided by each Member State through a process of National Allocation Plans (NAPs). Allocating allowances was in effect distributing the value, which European Member States were adamant must remain their preserve. Thus was born a model in which the *trading* system was harmonised across Europe – the definitions, coverage, monitoring, enforcement and trading mechanisms required to enable allowances to be freely traded across the continent – while the *initial allocation* of those allowances remained national. The dominant economic theory anyway was that allocation did not matter to the operation of the system – which would deliver the cap set efficiently irrespective of how allowances were distributed. In practice, countries allocated almost all allowances freely to firms based upon historical emission levels.

The Phase 1 NAPs by Member States were dogged by the lack of accurate historical data on which to base allocations. There had been little previous attempts to collect (or indeed had been a demand for) such installation level emission data. This meant that many NAPs were produced on the basis of very little information and were subject to intense lobbying, especially from large firms.[6] This was combined with tight deadlines and revisions of some of the plans under scrutiny by the European Commission.[7]

The first phase rules allowed a small role for auctioning – capped at 5 per cent of the total. In practice, only Denmark proposed to auction that amount, with a small number of other countries proposing smaller percentages.[8]

With these details settled, the system began operating on 1 January 2005. Europe had a cap on industrial CO_2 emissions – and, as was hoped, a price emerged as intended (see the left-hand line in Figure 7.2). Indeed, the first year of the system saw the carbon price rise as the system swung fully into operation, and as natural gas prices rose.[9] The carbon price stayed between €20 and €30t/CO_2 for a year, engendering a strong sense that the EU ETS was operating as planned, though some industries expressed alarm at the potential for very high carbon prices as gas prices continued to rise. In practice, utilities found other ways of cutting emissions, including switching operation from less efficient brown coal to highly efficient hard coal plants. The first academic study, published by the US MIT, subsequently credited the EU ETS with cutting Europe's CO_2 emissions that year by about 90 million tonnes – making it far and away the biggest contribution to emission reductions in Europe, or indeed anywhere on the planet.[10]

Cracks started to show, however, as information about the level of emissions in the first year became apparent.[11] Many industries and countries knew that *their own* emissions were significantly below their allocation; they soon discovered that others' were too. The data revealed an overall surplus in the system. This had a dramatic impact on prices which tumbled, and then in late 2006 fell towards zero as it became plain that Phase 1 overall had more allowances than plausible emissions. By the end of Phase I, in 2007, there was an overall surplus of about 200MtCO_2 or just under 3 per cent of total emissions.

The Commission came in for strong criticism for having sought to 'ban banking' of allowances from Phase I forward for use in the next phase.[12] This would have avoided the price crash, as allowances would still have had value. The ban was intended to protect the integrity of the next phase in case anything went wrong in Phase I, and would also ensure its consistency with Kyoto Protocol targets. The lack of banking made Phase I a self-contained system, which ended after three years – after which any surplus allowances would be worthless.

In practice, 'the market' itself had no problem: the participants' attention simply turned to the next phase, in which from 2005 there was already 'futures' trading based on expectations of the value of Phase II allowances. The developments during Phase II, however, turned out to defy all expectations.

Phase II 2008–12

Phase II of the EU ETS ran from 2008 to 2012 in parallel with the Kyoto Protocol's First Commitment period. It was intended to incorporate lessons learnt from Phase I, and as a consequence of monitoring and reporting pursued under Phase I, the NAPs could now be based upon real installation-level data.[13]

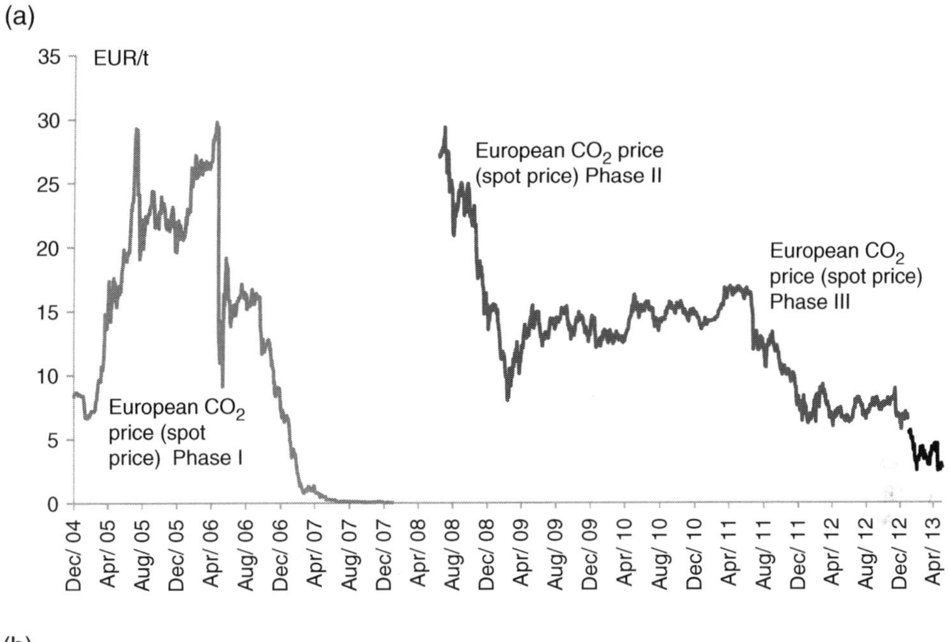

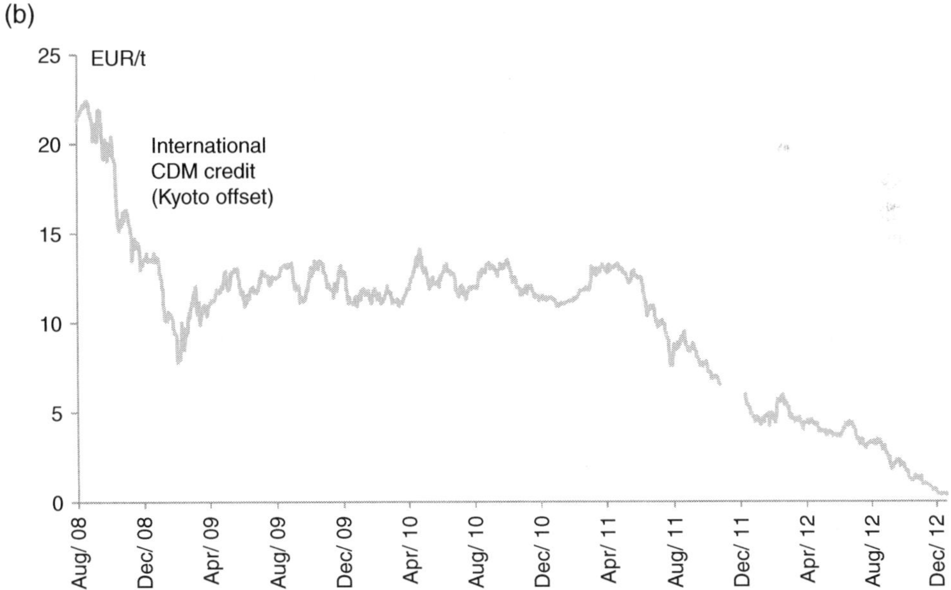

Figure 7.2 Evolution of EU ETS and CER prices

Data source: European Climate Exchange, Bluenext.

The allocation process was similar to Phase I, with Member States producing NAPs that had to be approved by the European Commission. However, countries and their industries had become more conscious of the value of the allowances, and there was far stronger lobbying. Yet, as Phase II was linked to Europe's commitments under the Kyoto Protocol, there was clearer guidance over appropriate targets – and this gave more ammunition and legal powers to the Commission in its policing of the allocations. The stage was set for an epic confrontation.

In the absence of any CO_2 controls, all projections suggested a slight increase in emissions, which was sharply at odds with the EU-15 Kyoto targets. The NAPs which countries proposed initially for Phase II mostly offered modest cutbacks relative to projections of sharply rising emissions – but in aggregate, would have resulted in an increase of around 5 per cent relative to 2005 levels. This was not only inconsistent with the EU's aggregate Kyoto target, but it would also have left a precariously thin margin below 'business-as-usual' emission projections – creating an obvious risk of another price collapse. The European Commission, in its regulatory role, rejected most of the allocation plans as inadequate, which heralded a furious political battle with most of the Member States. Ultimately, Germany, followed by other west European countries, withdrew their threats of legal challenge, and the Commission established common principles for total allocations which cut back the total cap by about 10 per cent, based on the legal nature of the EU's Kyoto commitments.[14] This resulted in an overall emissions cap for Phase II of almost 6 per cent below the (now verified) 2005 level (though even that was to prove inadequate).[15]

However, the battle revealed an additional and ominous division. The dozen 'New Member States' of eastern Europe had been in the process of joining the EU during the same period as the development of the ETS; they were thus presented with a new regulatory system in which they had scarcely been involved, nor was it fully developed as part of the EU Accession package. They felt it was thrust upon them by the dominant western European countries without adequately considering their views or interests, creating a political rift that has endured. In Phase II the east European Member States challenged the Commission's cuts on different legal grounds and the dispute dragged on, though the volumes involved were modest compared to the overall Phase II battle.[16]

As the Commission's victory in cutting the cap became established, the forward trading carbon price for Phase II rose to a high of around €25/tCO_2. As Phase II started in 2008 it rose higher still, tracking rising fuel prices and expectations about increasing public support for, and stringency in, the implementation of climate policy. By mid-2008, carbon prices again touched on €30 per tonne. Despite the growing economic crisis, subsequent analyses strongly suggested that the EU ETS was the biggest single influence ensuring that emissions in that year stayed below the (annual) cap set for Phase II (see Figure 7.3).

The basic Phase II framework had allowed countries to auction up to 10 per cent of their allowances, but initial proposals were much lower. Initially the only significant proposals for auctioning came from the UK which planned to auction 7 per cent of its allowances. In Germany the Parliament then weighed into renegotiating the German plans to increase the auction volume up to 9 per cent. The EU ETS was also becoming recognised as a potential source of government revenue.

The outcome also required Member States to set out a nationally agreed cap on CDM/JI credit purchases. Across all Member States, the CDM/JI cap totalled about 1,680MtCO_2 across Phase II. Both credits and EU allowances could be banked forward to the next phase if they were unused during Phase II. This time, it seemed safe to do so.

Designing Phase III, 2013–20

As Phase II moved into operation, Europe began to look ahead to the next phase. It was agreed that the EU could not wait for the outcome of global negotiations on a successor to

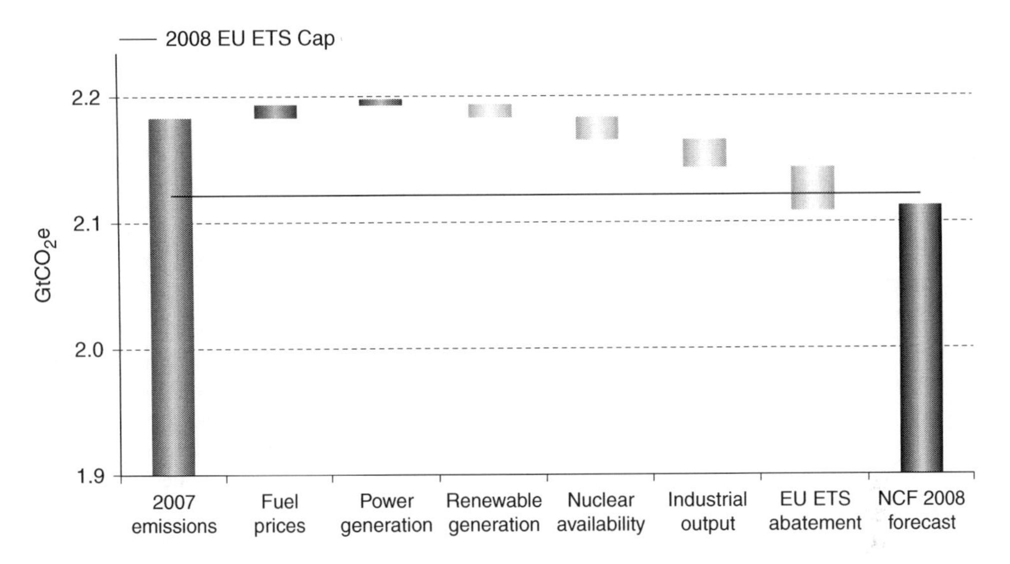

Figure 7.3 Factors affecting 2008 emissions ($GtCO_2e$)

Source: New Carbon Finance (2009).

Kyoto's first-period commitments; industry wanted some sight of the future. The EU ETS was designed to continue as an enduring feature of European climate policy irrespective of international developments and reflections began in earnest on the design of Phase III. This time, with a five-year Phase II, there was at least more time to learn.

Politically the Phase II allocation battle had marked a seismic shift: Member States were exhausted by the allocation process and the Commission had been forced to step in with common rules that largely dictated the totals of each national allocation. The Commission boldly followed this through to the logical conclusion and proposed that Phase III allocation rules should be set centrally and be applied across all Member States. Thus 'National Allocation Plans' were consigned to history.

The Member States, who had so forcefully protected their prerogative over allocation only five years earlier, acquiesced with barely a murmur. Indeed, each Member State had faced arguments and accusations from its industries, claiming that other EU countries were being more lenient, offering more free allowances to their competitors. Central allocation finally overcame an exhausting and unproductive competition between participating countries – a pressure to 'race to the bottom'. However, it risked inflaming further the resentments in Eastern Europe, whose countries adopted a tough negotiating stance and gained a number of special derogations.

The overall level of ambition was set when the European Council of Ministers in March 2008 agreed that Europe should aim to cut its emissions by 20 per cent (from 1990 levels) by 2020 unilaterally, with the promise of a 30 per cent cut in the event of an effective global deal.

Moreover, as also outlined in the next section, the warnings of economists that power generators would accumulate 'windfall profits' due to free allocation had proved unambiguously correct, establishing a political context which enabled its next phase to finally move (in that sector) to the economic ideal of auctioning allowances in full (excepting special treatment for eastern Europe).[17] Heavy industry continued to receive free allowances, but the sheer impossibility of doing facility-by-facility negotiations across 27 countries helped to establish the principle that allocation would be 'benchmarked' to best practice – further removing distortions and tightening the incentives, as outlined below. Thus Phase III finally brought a logical, coherent design that seemed to respect the best economic principles.

Full of renewed confidence in the EU ETS, it was agreed that Phase III would span a full eight years, to 2020, with allocations appropriate to the overall goal agreed by the EU Council. The resulting caps, along with the move to auctioning, are shown in Figure 7.4.

The move to an eight-year period for Phase III reflected the confidence in the system and the continuing demands of industry for more certainty about the future. Unfortunately, the long time period, along with the relatively weak 'unconditional' 20 per cent target, has proved to be disastrous: internal and global developments combined to prove 'Murphy's Law' – if anything can go wrong, it will.

The Clean Development Mechanism, after a slow start, had gathered pace remarkably. Emission-reducing projects in developing countries got off the ground more quickly and at a larger scale than anticipated, motivated in part by the promise of good prices for emission credits sold into Europe. By the end of 2012, more than 2 billion Certified Emission Reductions had been generated under the CDM. This by far exceeded earlier expectations and meant that the maximum volume granted for imports to Europe could be utilised.[18]

As Phase II began, Europe's economy was slowing and its industrial emissions declined more sharply than expected – partly through the success in cutting emissions, but increasingly due to the impact of the gathering economic crisis. With energy prices escalating globally and the financial credit crunch erupting in autumn 2008, and as the subsequent Eurozone crisis unfolded, emissions under the EU ETS fell dramatically.[19] Correspondingly, carbon prices that had been projected to rise steadily from €20–30/tCO_2 towards €40/tCO_2, began a

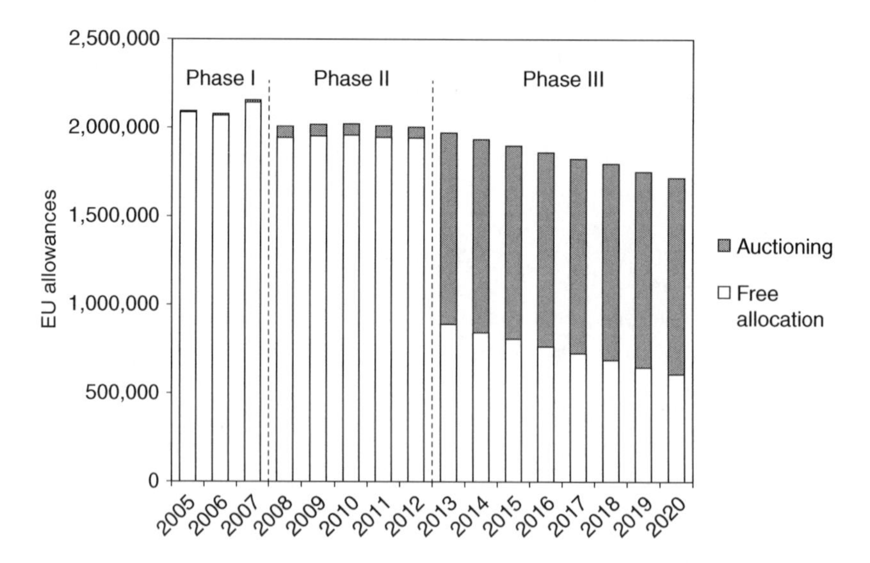

Figure 7.4 Free allocation and auctioning in the EU ETS Phases I, II and III

Note: The rules for EU ETS Phase III allow some flexibility for the new Member States of the EU (mainly Eastern Europe) in allocating free allowances to their power sector. The figure assumes the maximum allocation of free allowances under the Directive, and illustrates an upper bound to the overall level of free allocation in the EU ETS for Phase III.

Source: Aggregated from Chart 7 in Grubb *et al.* (2009).

precipitous decline. In the midst of this, the Copenhagen climate conference (December 2009) did not deliver the desired global deal, so the EU targets stayed at a 20 per cent reduction – nowhere near enough to absorb the growing surplus.

Despite this apparent 'perfect storm', for the next two years these forces seemed roughly balanced against the underlying assumption that carbon prices would rise through Phase III, creating the prospect of future value; companies held on to emission allowances they no longer needed and the carbon price proved surprisingly stable at around €15/tCO_2.

As a result, by the end of Phase II in 2012, the system had accumulated almost 2 billion tonnes of surplus allowances and credits overall, despite all the political blood which had been spilt to strengthen it.

This – against a backdrop of only weak and sporadic economic recovery – now looks sufficient to cover the entirety of expected shortfall right through to the end of Phase III in 2020. In principle, European industry has a binding emissions cap to 2020 – but in practice the surplus banked forward from Phase II makes that irrelevant.

The European Commission called for a unilateral move to a 30 per cent reduction by 2020 – creating expectations which helped to sustain the CO_2 price around €10–15/tCO_2 for most of 2010–11. But this proved politically untenable, as the EU economic crisis continued and it became clear there would be no major global deal before 2015 (and expectations for its content weakened). The expectation of future higher prices could no longer support the market. In autumn 2011 the price started to drop again, reaching €7 per tonne around the end of the year before bouncing back a little.

With the entire system increasingly confirmed to be in surplus right out to 2020, the main factor that prevented prices sinking further was the prospect of some other intervention to tighten the system again. However, the appeal to toughen Europe's target fell on deaf ears in a Europe grappling with the Eurozone crisis, and with Poland in particular becoming increasingly strident in its opposition. More modest measures were considered, and with each hope of action, the price nudged up, but with each setback signalling political difficulty, it sank further, hitting as low as €3/tCO_2 in February 2013.

As a way of establishing a carbon price, emissions trading in Europe had indeed proved easier to implement than its ill-fated carbon tax – though still not without a hard struggle. Getting it right, and adequate, has proved a lot harder.

7.4 The EU Emissions Trading System: ten lessons

Europe has been the world's laboratory for putting a price on carbon through cap-and-trade. Drawing on eight years' operating experience, this section summarises ten key lessons.

Fundamentals

(1) *Emissions trading works – incompletely*

This might seem a surprising statement to start with, given the roller-coaster ride of the EU ETS and its current problems. Yet it remains true, and central. The ETS captured boardroom attention like no other climate initiative, and its rapid introduction and impact contrasted with a decade of dispute over (failed) attempts to introduce a European carbon tax. Emissions trading enabled Europe to set a cap on the power sector and industrial emissions in a way which ensured compliance. A carbon price emerged, reflecting and incentivising the least-cost way of cutting those emissions across the continent.

Despite the history of (with hindsight) over-allocation, the EU ETS has cut emissions. Hard evidence comes from many directions. The periods in which the carbon price tracked the difference between coal and gas prices are testament to utilities using gas instead of coal to reduce their emissions. Cement companies increased their use of low-carbon fuels and some reduced their input of the carbon-intensive clinker in their operations. Numerous other changes can be traced to the door of carbon pricing. A survey of ETS participants found that 60 per cent of companies reported abatement or planned abatement in the early years of Phase II.[20]

Establishing *how much mitigation can be attributed to the EU ETS* in total is complex, since it requires estimating what would have happened otherwise. As noted, MIT estimated that the ETS cut the EU's CO_2 emissions in its first year by 90MtCO_2. A subsequent study estimated that, without a carbon price signal, emissions in the power sector would have been 90MtCO_2 higher in 2005 and 60MtCO_2 higher in 2006.[21] Obviously, the periods of low price have lowered the incentive. A survey of six different sources concluded that the average abatement has been around 30–70MtCO_2 per year over the periods studied – about 2–5 per cent of the total capped emissions.[22] In all, MIT's estimate was 120–300MtCO_2 total over Phase I, with similar rates estimated over Phase II, and with even more uncertainty as the financial crisis unfolded. In total to date, this implies that the ETS has cut emissions in Europe by at least 200–500 million tonnes of CO_2 – not bad for a system sometimes derided as hopelessly ineffective!

However, the success of the ETS is only *partial*, because over-allocation and periodic price instability reduced its effectiveness, implies periods with low incentive to cut emissions, and – most importantly – has rendered it ineffective as a means of stimulating low-carbon *investment*. The attributed savings have almost all been operational changes using existing stock. Industry has increasingly stressed that it cannot invest serious capital in low-carbon solutions based on a system in which prices have followed such a boom-and-bust pattern.

This contrasts sharply with the initial economic assumption that a carbon price would support low-carbon investment. This chapter later argues that the EU ETS requires a major overhaul for it to become an important tool for fostering investment. Hence our conclusion is that, as currently designed, emissions trading works, but incompletely.

(2) *Everyone will learn – and needs to*

Cutting carbon is a complex business and not all design choices will be right at the beginning – particularly given the considerable uncertainties and heavy lobbying involved. Learning about the system by government, but also industry and other participants can make the system more effective over time. Arguably, the smartest aspect of the EU ETS was its phased design. Each of the three phases has implemented improvements based on the outcome of the previous phase. Arguably its biggest mistake was assuming that after Phase I and a year or two of Phase II the EU had learned enough to lengthen Phase III into an eight-year period.

This sequential approach differs from the preceding cap-and-trade mechanisms implemented in the US for the pollutants SO_2 and NO_x. In these the trajectory and allocation could be agreed for a long time frame. However, given the scale of SO_2 and NO_x emissions and the comparatively lower effort required for pollution control, far less was at stake economically. Despite this, many of the challenges were subsequently mirrored in CO_2 trading schemes. For instance, the target level of stringency considered acceptable in the early years has been clearly far short of the stringency that ex-post would have been viable and necessary to avoid a price collapse.[23] Criticising the inadequacy of Phases II and III of the EU ETS is easy with

20:20 hindsight, yet both were stronger than would have been conceivable without the experience of Phase I delivering more emission reductions, at lower cost, than anyone had expected.

(3) *Prices can (and will) be impacted by numerous unforeseen factors which have generally reduced prices below expectations*

The record speaks for itself, but it is important to stress that the present difficulties are not due only to the recession. As noted, multiple factors – including global and internal developments – have led to emissions that are consistently below the expectations of the EU ETS and other quantified systems of emissions control. Phase I saw a price collapse before the credit crunch, despite rising gas prices that drove utilities back towards coal. The Phase II surplus occurred despite strong Commission intervention to cut allocations by 10 per cent to be consistent with Kyoto obligations. All other quantified cap systems to date have seen similar patterns of surplus and price collapse (e.g. the RGGI, the pilot UK ETS, UK Climate Change Agreements). We suggest that this systematic pattern can be traced back to three main underlying sources.

Firstly, it can be partly understood as a classic 'optimism bias', particularly (but not only) where allocation plans were built 'bottom up' from sector projections of industrial emissions. No industry plans for decline, or raises share capital, by proclaiming a difficult future. No government wants to project a dire economic outlook (thus 'undermining confidence'). Markets are a mechanism in which not everyone can meet their hopes and expectations for growth; some will be outcompeted. The overall growth of output will inherently be less than the sum of growth projected by each of the sectors or market participants. Indeed, the history of scenarios and forecasting shows evidence of a systematic upward bias in industry energy and emission forecasts in the order of 1 per cent per year.[24]

Secondly, this in-built tendency towards growth optimism is amplified if industries are negotiating with governments. The stronger the case they make for growth, the greater the value of the allowances they will get. Bottom-up allocation thus amplifies the in-built bias towards overestimating future emissions.

Thirdly, economic forecasting as an input to cap-setting is predominantly a *Second Domain* process, built on Second Domain assumptions. It thus tends to be blind to, or underestimates, the impact of First and even Third Domain processes. There may have been more abatement than expected in the initial years simply because the EU ETS seized the attention of European industry, even to options that were anyway cost-effective, and thus ignored in the 'abatement cost curves'. A crucial and related example of this is the huge volume of CDM credits that was created beyond all expectations and continued to grow even as the carbon price fell. This was partly because CDM became a mechanism for attracting *attention* to emission reduction projects globally – some, perhaps many, of which were in fact cost-effective anyway. These are largely First Domain effects, even if engendered by the creation of a new Second Domain instrument.

Finally, a key factor in the level of price sensitivity is that emissions trading has one feature different to most markets: supply is fixed while demand can be relatively 'inelastic' in the short term – meaning that emissions, which drive demand, may be very insensitive to the price. This implies a 'steep' demand curve which, combined with a fixed supply, creates an obvious potential for a highly uncertain price as illustrated in Figure 7.5.

With fixed supply and insensitive demand, it does not take much in the way of error, or unexpected shocks to emissions, to drive a major swing in the price. The collapse of Phase I

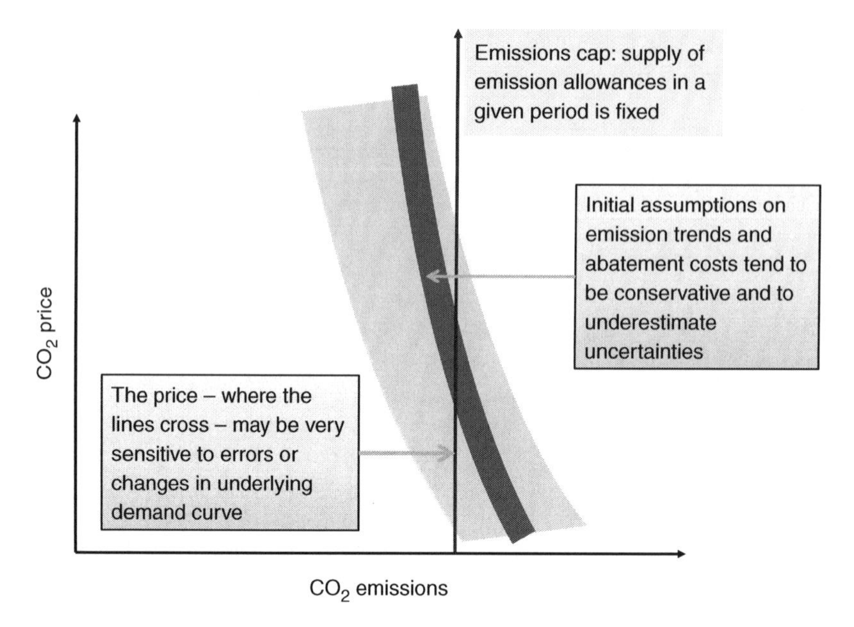

Figure 7.5 Source of price instability in an emissions trading system

prices occurred because the 'demand' line moved to the left until the lines no longer crossed at all, and the ban on 'banking' meant that the allowances had no value beyond 2007. In the next section, we show how and why 'emissions banking' can help – but only to a limit.

(4) *Robust regulation is required*

The EU ETS was established with strong systems for emissions monitoring, reporting and verification, and high penalties for non-compliance. This has underpinned its operational credibility. Despite this, as the value of the system attracted attention, there were problems with fraud and theft, which at one point required suspension of registries in some of the Member States. The key problems were:

- a 'carousel fraud' made possible by the different tax (VAT) treatment of allowances in different Member States and on international transfers – by setting up fictitious companies that bought allowances in one Member State, transferred them to another, charged VAT and then disappeared, governments were left liable for the tax;[25]
- a widespread 'phishing' attack on traders across Europe, with methods similar to those used in bank scams, in which traders were asked for their registry passwords. Indeed, one company which fell foul to the scam reported losses of $1.5 million;[26]
- the sale of CDM credits by the Hungarian government to a Japanese buyer, even though they had already been used for compliance within Europe – panic ensued when the credits were sold back into Europe and found to be invalid for compliance.[27]

The lesson learned from these fraudulent cases is that the valuable assets created from emissions trading will inevitably lead criminals – and sometimes governments – to find loopholes and imaginative ways to exploit the system. In this sense it is no different from financial markets except that emissions trading is newer, and political opponents of cap-and-trade seized upon the frauds to try and undermine its credibility. The reality is that fraud in finance has always existed and has never stopped us using money; the same is true of emissions trading. It does, however, underline the need for strong governance, with systems for security which must match those of finance itself.[28]

Cost impacts

Along with these 'four fundamental' lessons, another set of lessons concern the economic consequences. These are explored more fully in the previous and subsequent chapters, but can be summarised as follows.

(5) *Macroeconomic impacts are small*

The cost of the EU's 2020 goals was originally estimated to result in a one-off reduction of economic growth of about 0.5 per cent of GDP, or a few months of foregone economic growth, in the period up to 2020. As discussed in the previous chapter, if auction revenue is used effectively to offset negative impacts, to reduce distortionary taxes and to fund low-carbon investments, the cost impact can be lowered further (see Chapters 11 and 12 for a broader look at the macroeconomics of CO_2 abatement). In practice, the cost is now looking negligible, and the bigger problem is that programmes which had previously relied on revenues from the ETS (most notably CCS) are foundering. Ill-founded scare stories about macroeconomic costs should not be allowed to impede either the setting of goals or the use of economic instruments to meet them.

(6) *Industry can profit*

Emissions trading does *not* inevitably impose net costs on industry. Indeed, despite initially opposing the EU ETS, all participating industrial sectors in Europe have in aggregate profited from its operation to date – perhaps excessively. There are in fact three ways that an industry can profit from a trading system:

- The most obvious is if it receives emission allocations that turn out to be surplus to need, whether because it can cut emissions more cheaply than the market price or because other factors drive down emissions. Notably, with a fixed number of allowances, emissions trading is *counter-cyclical*: almost all industries in Europe have profited from the ETS because the recession reduced their emissions and left them with spare allowances to sell.
- If low and high emission sources are making the same product – such as electricity – the former will generally profit from any carbon price. The obvious example is that both renewable energy sources and nuclear power profited from the fact that coal and gas plants cost more to run due to the carbon price.
- The third way is more subtle: given free allowances, industries may gain a 'windfall profit' from the way carbon prices feed through to the price of their product in a competitive market. The mechanism is explained more fully in Chapter 8 but was the major cause of European power companies making many billions of euros of profit from the EU ETS in Phases I and II.

The mechanism with potential to generate 'windfall profits' in particular was not believed by most people – even some economists – before it happened. This illustrated again how most economists focused on the 'signalling' feature of prices, without thinking about the revenue implications. Only when data from power prices showed unambiguously that carbon prices were passed on to consumers even when allowances had been given out for free was this prediction widely accepted. This was nothing to do with market manipulation and indeed, ironically, it occurred mostly in *competitive* markets, reflecting the way they intrinsically operate (see Chapter 8). This helped to lay groundwork for the move to auction allowances in Phase III.[29]

(7) *International competitiveness impacts are limited to a small number of highly energy-intensive industry sectors*

For most manufacturing sectors, cost differentials due to labour and other input costs far outweigh any costs induced by the carbon price. Fluctuating exchange rates – and indeed raw materials more generally – are far more important. As a result, most sectors can accommodate carbon costs without a significant impact to their profits, sales or competitiveness (data are indicated in Chapter 8).

However, for a handful of carbon-intensive industrial activities, energy costs are an important factor and they face genuine competitiveness concerns, often specific to their sectoral characteristics. The specific situation of these sectors is discussed in the next chapter.

One of the most famous articles by the US economist Paul Krugman challenged the whole notion of 'national competitiveness' as a useful concept: competitiveness, he argued, is something that resides only in sectors trading with competing goods – not nations as a whole. The same might be said of emissions trading, and of carbon pricing more generally. There is scant evidence to suggest that a price on carbon creates a real problem for national competitiveness – whatever that means. It can, rather, create winners in unexpected places and potential losers in others. At least for the carbon prices historically experienced, or even the €40/tCO_2 once projected for the EU ETS, concerns about competitiveness impacts are a reason to look at options to help a few potentially exposed industries – not to duck the challenge itself.

Efficiency and effectiveness

The final three lessons concern the design of the system and the role of international processes.

(8) *Allocation, as well as the cap, matters*

The traditional economic mantra had been that only the overall cap on total emissions is relevant. The trading market, it was assumed, would settle at the least cost, and hence most efficient outcome, anyway. Whether the emission allowances were auctioned or given out for free, and how any free allowances were distributed between participants, was thus assumed to be irrelevant to the efficiency of the system.

Of course *politically* it was crucial, since allocation distributed the value. To gain industrial acceptance of the EU ETS, as noted previously, most governments started by handing out virtually all the allowances for free, comforted by the belief that this free allocation had no significant downside.

Unfortunately – and particularly in systems with sequential phases – free allowance allocation comes at a cost. Free allocation designed to cushion the impact on carbon-intensive sectors inevitably increases the burden on other sectors to achieve a given economy-wide emissions target. Moreover, it can easily undermine the incentive that the system is supposed to create. In the early days of the EU ETS, they were mostly *grandfathered* – based upon recent emission levels. Unfortunately, the expectation that higher emissions could lead to more free allowances in Phase II carried an obvious risk of weakening the incentive, if emitting more today might lead to more free allowances tomorrow.

This weakening of the incentive was amplified by the need to try and create a 'level playing field' between incumbent firms, closures and new entrants. New entrants to a market are nearly always considered a 'good thing', so they too were offered a special pool of free allowances to be issued to any new installation. Moreover, it was considered necessary to treat new entrants and closures symmetrically, and the value of all the allowances associated with a plant could lead a company to close it so as to sell the allowances. No government wanted the EU ETS to be blamed for politically unpopular plant closures, so most National Allocation Plans stipulated that allowances would be withdrawn if a plant was closed down. This of course created a perverse incentive to keep carbon-intensive plants running. Tying free allocation to production levels rather than emissions is little better.[30]

To summarise, in addition to the spectre of windfall profits, free allocation based on emissions or production levels creates a morass of potential inefficiencies across production, fuel and consumption choices. This is illustrated in Figure 7.6 (bottom row) – a diagram that became known as the 'pyramid of distortions'.

Some of these problems could be avoided by basing free allocation on 'benchmarks' – a standard level of allocation for a given technology, separated from how it actually performed (second and third rows in Figure 7.6). This prevents perverse incentives around production but does not preclude windfall profits, nor does it necessarily prevent international carbon leakage.

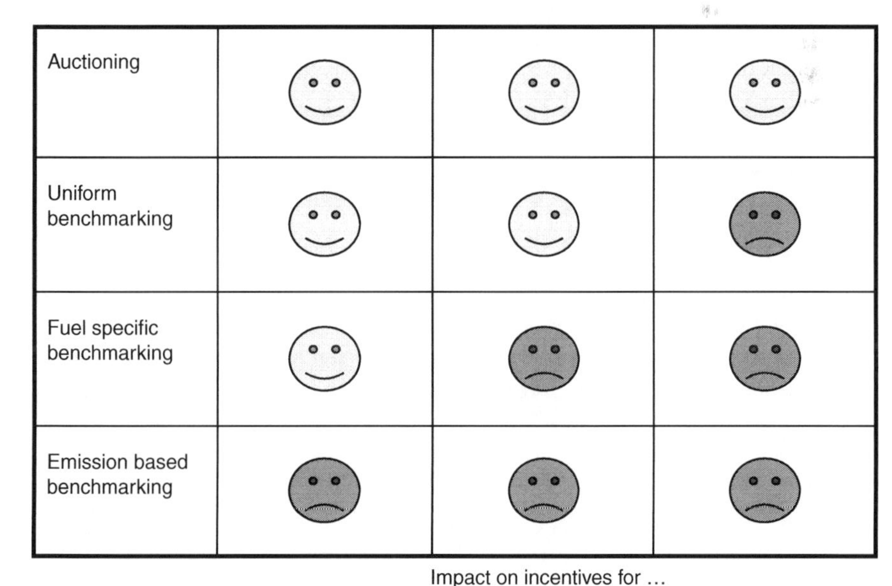

Figure 7.6 Allocation methodologies and efficiency

Source: Authors.

Benchmarking is not a panacea – it can still weaken the basic incentive, depending upon the exact method used, and is more complex than it sounds.[31] It was still rarely used in second-round National Allocation Plans. In its Second Allocation Plan, Germany used 'fuel-specific' benchmarks – i.e. bigger allocations for coal rather than gas plants – to reduce adverse impacts on coal. Far better, in terms of efficient incentives to cut carbon, would be uniform benchmarking – the same amount per kWh irrespective of the fuel source.

The simple insight is that *free allocation is not free*: it comes at a very real cost. Allocation methods should be designed to minimise net impacts on the aggregate profitability of incumbent sectors, whilst boosting the profitability of cleaner technologies and innovators. A balance of free allocation should strive to minimise economic distortions as well as windfall profits.

Unfortunately, all this complicates things further. The more fine-grained the effort, the more detailed the data requirements and the more extensive the technical assessments and negotiations need to be. Phase III of the EU ETS took the economic insights to heart, but policy-makers were still deeply concerned about the risk of carbon leakage abroad for energy-intensive industry. The result was almost two years of exhaustive negotiations before the 'benchmarks' for Phase III were finalised, with results that were still far from satisfactory to many.[32]

The simple lesson is an old one: there is much devil in the detail, and the detail in distributing free allowances will be driven by the lobbying pressures to resist change as much as possible.

(9) *There is a compelling economic rationale to increase auctioning over time*

This leads to our penultimate lesson: there are powerful reasons to maximise the degree of auctioning – selling allowances rather than giving them out for free. This removes the potential for enduring sector-wide windfall profits. Auctioning ensures that price signals remain intact to drive efficient corporate and private decisions on production, fuel choice and consumption, based purely on the carbon price – in principle including incentives for low-carbon investment and innovation. The tortuous complexities associated with handing out free allowances are thus avoided.

Of course, auctioning also provides revenues that could be used for public goals such as low-carbon technology development, to help compensate consumers as carbon costs feed through to product prices and/or for international programmes for technology transfers or economic assistance for adaptation.

By 2008, in Europe the intellectual battle had been largely won on this principle – a dramatic evolution from only three years earlier. It was considered that the principle could not be applied to any sectors that might be 'at risk of carbon leakage'. But power plants do not migrate, and the scope for importing electricity into Europe from outside the ETS was negligible.

Thus in its Third Phase, the ETS finally reversed the domination of free allocation. It established auctioning as the rational economic principle as Amendments to the original Directive, stopped free allocation to the power sector (except for politically essential derogations for eastern Europe) and set the objective to phase it out in other sectors over time.[33] In political terms, it was a remarkable turnaround.

(10) *A binding international treaty is politically, more than economically, important for effectiveness*

Finally, the EU experience points to important lessons about the wider international structure. The EU ETS would not have been launched without the international framework of the

Kyoto Protocol: the need for Europe to comply with its emission reduction obligations under the Kyoto Protocol drove the creation of the EU ETS. The EU ETS itself was designed and aligned with the Kyoto structure as the backdrop. Furthermore, access to the pool of 'offset' CDM credits – the fundamental architecture for engaging developing countries in emissions mitigation under Kyoto – was essential to help ease fears about the high costs of capping industrial emissions in Europe.

Most striking of all, though, vast lobbying pressures for excessive allocations in Phase II were only contained by the EU's Kyoto targets which, as described above, formed the fundamental legal basis on which the Commission won its battle to impose tougher allocation cutbacks in the Member States.

Finally, the EU's proposed tightening beyond the 20 per cent emission reduction target by 2020 was conditional upon an adequate and binding successor set of commitments. The fact that these remain far from materialising is, ultimately, the single most important reason for the current predicament. The EU is unlikely to be alone in finding out that binding international commitments help to bolster more effective domestic action.

There is another, quasi-international lesson. A former Polish government official noted that the eastern European countries had to enter a system they hadn't helped to design. This set the stage for a 'perpetually unhappy marriage' in a system subsequently seen to be unfair by countries like Poland which still bedevils attempts to rescue the ETS from its current predicament.[34] If a system spanning different countries is to work effectively, it needs to be seen as sufficiently fair by all participants; if not, it cannot be effective in the long run.

7.5 Investment, predictability and confidence

An old saying holds that the only certain things in life are death and taxes. If you want to invest a billion pounds, you need a bit better than that.

As noted, there is little evidence that the EU ETS, given the gyrations of its carbon price, has done much to support low-carbon investment in the way that was originally assumed. Low-carbon investors habitually call for 'certainty', and economists often consider that carbon taxes are better, based on the assumption that they give greater certainty and avoid the problems of free allocation.

That in itself is somewhat debateable, because tax levels in most countries are generally set as part of annual budget processes (and as noted in Chapter 6, carbon taxes too are often rife with exemptions). The 'certainty' of the UK excise fuel price escalator, as noted in the previous chapter – frozen after a popular revolt – is but one example. However, this misses the point. The core issue is identifying the needs of low-carbon investors and working out how best climate policy can help to meet these needs.

The multiple roles of carbon pricing

In an uncertain world, major investment involves more than prices. A study of investment decision-making in big companies highlighted that to impact on investment, a policy framework must:

- *Gain the attention* of the relevant decision-makers as something that will significantly impact their future operating environment. Particularly for big companies, it is the nature and credibility of long-term climate policy objectives that matter most for strategic decisions about how that company should deploy its resources. This is important too for innovation – interviews with 800 manufacturing companies showed that those companies which expect allowance allocation to be more stringent (including declining levels of free allowances) are more likely to pursue low-carbon innovation.[35]

- Be *clear enough* to help companies assess the new opportunities and challenges when making operational, investment and strategic decisions. The EU ETS Directive set a cap on ETS emissions that declines at a rate of 1.74 per cent per year from 2013, unless reset, until 2050. This does create a clear, long-term framework to inform company strategy on policy constraints. By creating a system to translate this cap into a carbon price, it also encompasses an enforcement mechanism to ensure that the trajectory is delivered. One of the core concerns about opening up the EU ETS to too many international offset credits has been that it undermines this strategic certainty, namely about what companies must ultimately deliver in Europe.[36]
- Help the *financial viability of specific low-carbon projects*, particularly relating to the balance of financial risks and returns. This is the classic role assumed for carbon pricing – increasing the profitability of low-carbon investments. However, the key word is *risk*. Seeking to decarbonise energy systems is asking companies to invest differently due to public policy. Any manager that promotes such changes wants to know that they will not be left high and dry by, for example, a project that proves uneconomic due to a collapse in the carbon price. Thus while the carbon price can contribute to the financial viability of a low-carbon investment, uncertainty about the future price complicates decision-making and may substantially drive up the perceived risks – and hence the financial hurdle.

Of course, the carbon price and climate policy are only two of several dimensions considered in any investment choice. Secure access to fuel and public perception are also important factors for power sector investors, for example, where investment decisions are further complicated by the vagaries of electricity markets.

Against this backdrop, how much does an uncertain carbon price matter?

- In terms of *attention*, some volatility of carbon prices will not impact the attention companies dedicate to climate policy. However, the persistent decline of the carbon price in Europe has been widely interpreted by companies as a declining dedication to climate policy. That matters.
- The *clarity* of the framework expresses itself in part through expectations of the future scale of markets for low-carbon technologies. In this case again, price uncertainties are of less concern as long as they do not undermine the credibility of the instrument in ensuring the quantitative targets will be delivered. Large and potentially persistent changes may, however, qualitatively change the outlook for certain technology areas. The most obvious example concerns CCS. With carbon prices projected to reach €40/tCO_2, and with dedicated financial support from the allowances set aside for the purpose, oil companies were able to make a strategic case for investing in CCS development. Yet they no longer can; it has to be bankrolled directly by governments and hinges one hundred per cent on subsidy, which makes it a far less compelling proposition for the energy companies. In this sense, the price is an indirect indication of lack of ambition to the pace and level of emission reductions that would demand CCS.
- A highly uncertain carbon price, however, is most obviously damaging to the *financial viability of projects* for which sufficiently stable and predictable revenue streams are essential. The risk of a low carbon price reduces the ability of companies to lower

> financing costs using debt; they become more dependent on high-cost equity. Carbon price uncertainty can then be highly detrimental and drive up costs for everyone. Confidence in the revenue stream is indeed why feed-in tariffs have been so effective in promoting renewables, while projects like CCS which depend on the carbon price have languished.

There is one other important dimension – the impact of uncertain carbon prices on investments in carbon-intensive assets. This does, however, require the firms involved to understand – and believe – carbon-related risks, which they tend to loath to do. Private sector initiatives can help with this.[37] One important example from 2009 was when the rating agency Standard & Poor downgraded the credit rating for the UK company which ran Drax – the UK's biggest coal power plant – citing 'Drax's rising business risk because of its focus on coal-based generation, which is subject to increasingly stringent regulatory and environmental requirements'.[38] Drax subsequently made big investments to adapt their station for the large-scale use of wood and it has now become probably the biggest biomass-coal 'co-fired' plant in Europe. A sustained low-carbon price, however, could lead others to hesitate before following their example.

Thus the *possibility* of significant carbon prices can deter high carbon investment and encourage relatively cheap conversions. Policy frameworks that ultimately can deliver carbon prices to match environmental goals can encourage business and the financial sector to explore lower-carbon strategies as a risk management strategy. Certainty is not required.

Confidence in the basic direction of energy and related climate policy, and its translation into significant financial impacts, however, is essential. To really turn major corporate investors towards low-carbon choices, moreover, a credible carbon pricing system needs to combine with suitable development of institutions, market structures and complementing infrastructure – as outlined in Chapter 10.

It is, in consequence, a matter of degree. Some price uncertainty in emissions trading systems may defer the time at which specific large low-carbon investments can hinge fully upon carbon pricing, but is not fatal to the broad task of 'turning the supertanker'. However, a major and sustained fall in the carbon price to levels that cause corporations to lose interest in the issue, as well as confidence in the framework and in the commitment of governments, may be devastating.

The role and limitations of 'emissions banking'

The key aim of allowing 'emissions banking' in trading schemes is to smooth the path of carbon prices – precluding the kind of cliff-edge collapse seen in the ETS first phase – by letting allowances be banked forward for future use, to give confidence in their continuing value even if they are not needed for the present.

As noted, that is indeed what happened for much of Phase II, which sustained a moderate price despite a radical economic recession. What then went wrong?

The answer seems to be a combination of two things. The first requires understanding *who* banks allowances and *why*. Many of the companies directly involved in the ETS like to hold allowances for a few years ahead; they can then be bundled with future power contracts, for example, without any risk. This means the system can readily accommodate some surplus without much problem. But these are not speculative players. The only people who hold on to allowances purely for the prospect of future profit are financial speculators (in the 'secondary' market). They demand higher rewards for taking bigger risks. Indeed, their aversion to risk has amplified since the credit crunch (and associated strengthened financial regulations), and their perception of the risk around future carbon prices and policy has escalated. They will only pay a few euros for something which they judge might (or might not) be worth much more, one day, on timescales they are not really used to handling.

By extending the relevant time horizon, 'emissions banking' has the effect of flattening the short-term 'demand curve' (see Figures 6.6(a) and 7.7), both by increasing the potential to cut emissions *and* allowing any short-term unexpected fluctuations to be absorbed in a longer (and hence larger) pool of allowed emissions. But this only works within limits. If the surplus grows further, it starts to fall over a cliff edge – the extant price reflects the short-term implications of surplus, but this time potentially encompasses not just the present, but the future period as well into which the surplus will be banked (Figure 7.7a).

Thus in Europe, the other factor was the growing scale of the surplus and the failure of attempts to toughen Europe's 2020 target unilaterally, which in late 2011 tipped the ETS over the edge, as the carbon price became detached from anyone's likely physical need for allowances, and into the hands of more speculative financial markets. Since short-term fixes and long-term (post-2020) prices both hinge fundamentally on politics, the result is that the carbon price is back in the trap from which it was designed to escape – in the hands of short-term fluctuations, politics and perceptions thereof. Banking can only solve the problem within limits. If pushed beyond that limit, it risks magnifying the consequences, by allowing the surplus from one period to swamp the next, and thereby undermine credibility along with the price (Figure 7.7(a)).

Steadying emissions trading systems

One response to the present malaise of the EU ETS, combined with the failure of the effort to get US cap-and-trade through Congress, are growing calls to revisit carbon taxation. Chapter 6 noted briefly that *one* of the key obstacles to carbon taxation – its revenue impact – can be overcome with the use of 'tax thresholds' which would be directly analogous to free allocations in trading systems. These would, however, introduce almost all the problems noted in this chapter with free allocation, including the tortuous negotiations, risks of perverse incentives, and in some cases the windfall profits. The big distributional and political challenges – as detailed in the next chapter – are fundamental to any way of pricing carbon.

Moreover, as noted in Chapter 6, the most recent economics thinking highlights that the classical 'tax *vs* trade' debate was grossly oversimplified, because what matters is the *type* of uncertainty.[39] The well-accepted conclusion is that long-term quantity goals make sense for defining the long-term objective (they also allow for consideration of the full mix of policy instruments, across all pillars, towards that objective). More recent research suggests, correspondingly, that setting a near-term price is more efficient if uncertainties around the cost/difficulty of cutting emissions are 'temporary', but setting the quantity is better if these uncertainties are 'permanent'. There is little doubt that the 'shocks' suffered by the EU ETS have enduring consequences – and the same would seem true of many plausible 'shocks' to expectations around energy and emissions.[40] To jettison a quantity-based approach thus does not make *strategic* sense. However. the repeated history of the EU ETS price underlines the standard warning of many economists, noted in Chapter 6, that relying on quantity alone for short-term goals risks huge instability. The solution developed for the Australian system was to have rolling, five-year commitments with a clear governance process around the setting of the sequential caps, that can respond to changing circumstances. From this perspective, the critical misfortune in the EU ETS was moving to an eight-year period, at precisely the moment when

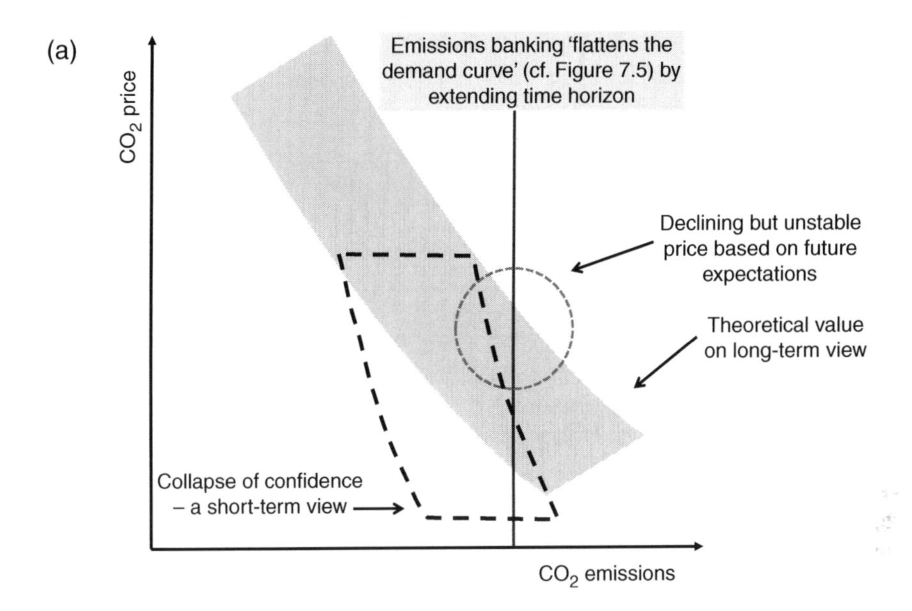

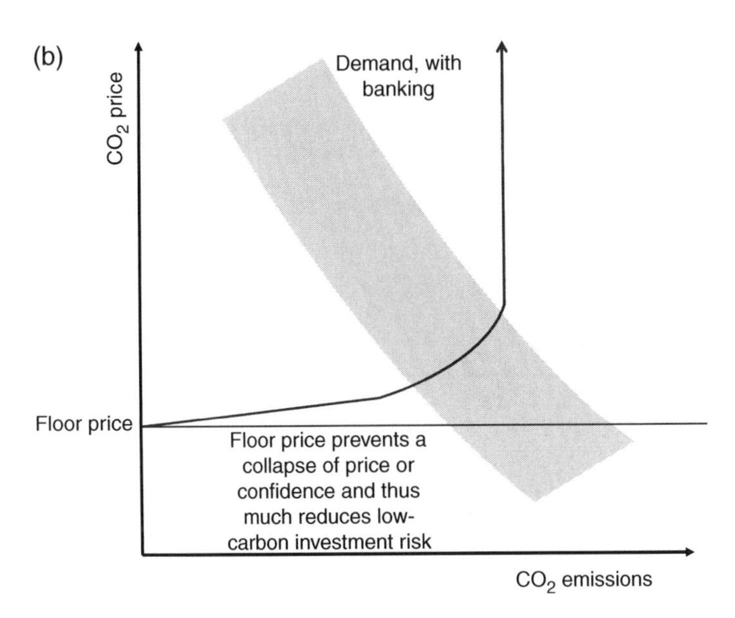

Figure 7.7 Impact of (a) banking and (b) floor on carbon price in ETS

the worst recession in at least two generations was about to descend adding to all of the other factors pushing the system beyond its limits – hence our reference to Murphy's Law ('if anything can go wrong, it will'). However, reality has a way of delivering surprises; what is needed is not only to fix the current problem (in the EU), but to make such systems more robust.

For the EU ETS, the case for 'doing something' rests on the basis that carbon pricing has an active, valuable and efficient role to play in fostering low-carbon investment, consistent with climate science and indeed the desire to reduce dependence upon international fossil fuel markets. It also has potentially a useful role in carbon finance. The current situation delivers none of these things. ETS targets were set with certain assumptions, but the combination of developments has been beyond the scope of any reasonable expectations and eight more years is too long to wait for adjustments.

Beyond possible short-term interventions, the real challenge is to map out more enduring solutions; the main ones are summarised in Box 7.1. They differ in their focus on either quantity or price-based adjustments, and on the balance between automatic rules and institutional mechanisms with discretion.

The most specific in terms of automatic price implications would be a minimum price, which would make explicit the implicit judgement signalled by all interventions about what constitutes a 'too low' price. Economics research has increasingly emphasised the value of 'hybrid' instruments which combine price and quantity features (see Chapter 6 and particularly the works by Pizer and Newell cited there).

A price floor would probably raise the converse issue, of mechanisms to avoid excessive prices. This could take different formats. A moderate level of surplus allowances in the system could provide a cushion for temporarily higher emissions. Also a fraction of the allowances could be held back for circumstances when either external triggers or price ceilings are exceeded. The pros and cons of a price ceiling, like a floor, depend significantly on the level.[41] In principle, it makes sense to design a system to cope with 'unacceptably' high prices, just as much as 'ineffectually low' ones. Planning for such a situation seems far better than the political panic that might ensue if circumstances – for example, a major gas crisis – drive both gas and carbon prices to politically intolerable levels. More moderately, directly addressing concerns about possible high carbon prices might also make it viable to set more stringent targets.

The Californian cap-and-trade system builds in a price corridor, starting with the range US\$10–40/tCO_2, and rising. Its impact and evolution will be interesting to monitor. In Europe the strategic choice is whether to treat the current malaise as a 'one-off' that can be addressed with a 'one-off' adjustment of quantities (whether by withdrawing a fixed quantity of allowances, renegotiating the 2020 target, or steepening the long-run trajectory), or whether to also add additional features to contain extreme price variations.

Mechanisms to achieve this may be either linked directly to the price (as in the Californian price corridor, with the quantity implicit), or implicitly through quantity triggers (e.g. determining when the cumulative surplus is too large) (see Box 7.1). Figure 7.7(b) illustrates the basic impact of a price floor.

An important question, then, is what happens to prices held back at the lower threshold (e.g. not auctioned due to a reserve price). One intriguing option is to only return such allowances if the system moves back to price or reserve levels closer to those originally expected and intended. At the opposite end of the scale, a similar approach to a 'ceiling' could help to deal with risks of extremely high prices, without permanently injecting additional allowances.

This approach to steadying the price in an emissions trading system is illustrated in Figure 7.8. Such mechanisms may enhance stability and allow for a system with price responsiveness, yet with a tendency for prices to gravitate back towards the ranges projected in the initial political deal. Thus it remains possible to maintain a reasonable balance of price and quantity, reflecting these dual goals that are anyway implicit when any emissions cap is negotiated. Making the bounds directly related to price (floors and ceilings) would provide more clarity for low-carbon investors and for any financial planning associated with the use of carbon revenues, but may have political or legal drawbacks. Different regions might thus

Box 7.1 Options for 'structural reform' of the EU ETS

'Structural reform' has become the term for finding enduring solutions to the collapse of carbon prices in the EU ETS. Many approaches have been suggested. Conceptually, the simplest would be to *strengthen the cap* for Phase III. This could be done in many ways: 'setting aside' or cancelling some allowances; formally tightening the 2020 target (and associated trajectory); and/or more stringent constraints on the use of offset credits (CDM). The current 2020 goals fall well short of a trajectory consistent with the goal for deep mid-century reductions. Toughening the targets would align better both short-term carbon prices and mid-term emission targets with the longer-term goal to provide a consistent and credible investment framework for low-carbon investors. Technically, any of these options would shunt the 'demand curve' in Figure 7.7 (a), which has drifted far to the left, back to the right.

The obvious problem is that this creates a risky precedent for intervening simply because policy-makers decide the price has gone 'too low' and they want to 'correct' the market. This amplifies political uncertainty; if there is one such intervention, there might be others. Moreover, it does not increase the *robustness* of the system. As noted, the history of the EU ETS is the biggest, but by no means the first, demonstration of the depth of uncertainties around energy and emission projections. With eight years of Phase III to run and the EU still in the midst of a fundamental structural economic transition, removing any given number of allowances may not have the price impact intended or resolve the periodic price instability implicit in having 'inelastic' demand set against fixed supply over a limited duration.

The EU ETS already embodies a *long-term* trajectory. Tightening this and bringing it into line with environmental aspiration could be effective but raises particular political problems.[a] An alternative balance may be struck if a long-term 'in principle' trajectory is accompanied by annual negotiation of binding caps five years ahead; this is broadly the Australian design but is not now an option for the EU.

With eight-year or longer periods, the challenge remains to design for intrinsically greater robustness. One approach focuses on the fact that emissions banking is desirable but only within limits; one could limit supply if *cumulative banking* exceeds a threshold. Specific ways of doing this have been proposed, though the full pros and cons have yet to be mapped out.[b] This has in common a focus on restricting excessive volume with uncertain implications for price but a degree of incremental and ongoing self-correction. Closely related are proposals *for volume-based adjustment rules* to moderate large supply/demand imbalances.

A far more ambitious suggestion is for a *carbon 'central' bank*, to buy allowances when they are cheap and sell them when prices are high. The two obvious problems with this are both, broadly, political. It is unclear why a government that has formulated an emission cap that results in a very low carbon price and that is unwilling to tighten the cap would nevertheless mandate a carbon bank to intervene. Moreover, setting up such an institution, with its rules and governance would clearly be a huge (and hugely sensitive) institutional undertaking. A carbon bank might eventually have a role, but it would probably be more credibly to do with smoothing problems of market operation rather than correcting long-term structural imbalances.[c]

A final approach is to directly build in a *price floor*. The move to large-scale auctioning in EU ETS Phase III introduces a (technically) easy way of doing this, by setting a reserve price for auctions. Allowances would not enter the market unless market participants were willing to pay at least the reserve price. This would guarantee that other trades would rapidly converge towards that level, or above. Unlike measures targeted mainly at trying to increase the price, a floor price would rather aim mainly to reduce the downside risk for investors[d] and secure a minimum level of government revenues available for multiple purposes. This would automatically adjust the volume according to how close the price was to the floor, as indicated in Figure 7.7(b) – if and as the demand reduces (demand curve shifts to the left), the price no longer collapses, but begins to converge more gradually towards the floor price level. An agreed *rising* floor price – with strategic advantages noted in Chapter 11 – would also create an incentive to purchase from auctions *earlier*, and thus bring forward the revenue streams associated with auctions.

Notes

a Tightening the overall trajectory and extending it towards the 2050 goals could restore confidence and value in 'banking' for Phase IV (post-2020) and beyond; this would be particularly effective if combined with removing the present surplus. However, extending the time horizon at the same time as trying to fix the Phase III problem amplifies political challenges. The EU has yet to embark seriously on a package for '2030' and is unlikely to finalise this before the global climate conference scheduled for 2015 to reach a new post-2020 deal. The EU is in no mood to commit to an even longer-term trajectory in the absence of a global deal.

b One expert, Erik Haites, suggests to 'Build automatic, small adjustments into the rules – if the accumulated bank exceeds X% of the annual emissions, the number of allowances issued for the next year is reduced by the excess … Adjustments, when they occur, are likely to be small, so they will have little effect on the market. And making them automatic means that political decisions, and the associated uncertainty, are avoided' (as cited in Grubb 2012, note 22). A similar approach has been suggested by the Carbon Market Investor Association, proposing that if a certain number of surplus permits have not been used after a period of three years, an equal number of permits should be removed from later supply (CMIA 2011, as cited in Sartor 2011; see also Grubb 2012, note 22). There would remain the difficulty not only of setting the levels, but of quantifying the actual surplus sufficiently clearly for a 'cap on banking' to bind without loopholes, given for example the complications of forward contracts for offsets. As noted, some surplus is necessary, e.g. to provide power companies an opportunity to hedge future power contracts. The mechanics thus remain to be detailed; also, to the extent that a cumulative surplus might well indicate that the region is on an accelerated emission reduction trajectory, limiting the degree of surplus might not adequately reflect the new situation and a shift of the cap trajectory instead would seem more appropriate.

c If the overall carbon price level is seen to be too low, then the 'central' bank could buy large amounts of allowances so as to sell them in the future at higher prices. However, this raises concerns about regulatory credibility and consistency. For investors, a high carbon price is only of value if they are confident that the carbon price level will be maintained. How can such 'central' banks be sufficiently independent to maintain this objective? Survey results suggest that investors in low carbon assets look at both the EU ETS and the long-term emission targets in their assessment of investment options. If a carbon price is only high due to the intervention of the 'central bank' but not consistent with the long-term emissions target, this might limit the value for low-carbon investments.

d For a study on the merits of hybrid instruments see, for example, Pizer (2002) and Weber and Neuhoff (2010).

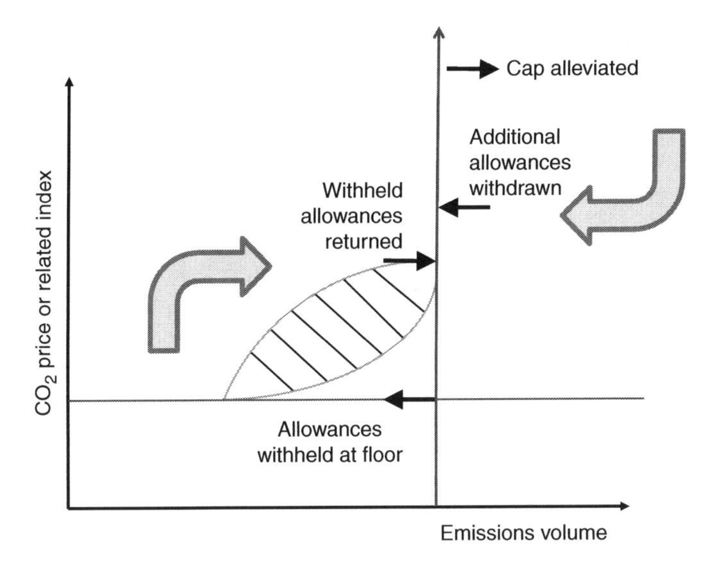

Figure 7.8 Steadying mechanisms for emissions trading systems

Note: The figure illustrates mechanisms to help emissions cap-and-trade systems deal with deep uncertainties, so as to maintain a reasonable balance of price and quantity objectives. The mechanisms are most simply illustrated with respect to price floors and ceilings, in which case the shaded area indicates the likely region of price and quantity for a system with substantial surplus allowances. However, the same principle could apply to other 'threshold' triggers, for example based on the level of cumulative surplus.

choose different approaches. The central fact is that mechanisms to make carbon markets more robust exist and should be used.

Any price steadying mechanism necessarily introduces another axis in managing the system. Yet all environmental policy in reality is a balance of economic and environmental objectives, embedded in uncertainty. Steadying mechanisms such as those outlined could and should not substitute for adequate targets, but should instead reflect explicitly these dual considerations, and so provide automatic insurance in the face of other economic and policy developments, and hence reinforce the system by removing – by design – extremes of risks.

The contextual fact is that the world is still in the very early days of pricing carbon; the level of environmental ambition in all these systems falls way short of what science suggests to be necessary; and there is no robust, global framework to lead convincingly in that direction. If and when that can be achieved, it might be possible to dispense with price steadying mechanisms. At present, however, they seem to be a key missing ingredient.

7.6 The Kyoto Mechanisms

There is one other major dimension that falls squarely in the realm of learning from experience with carbon pricing – that of the Kyoto mechanisms, to which we now turn.

Aims, expectations and experience

Although the EU ETS is the most visible cap-and-trade system in operation today, it is embedded in the larger trading system defined by the Kyoto Protocol. This allows for

International Emissions Trading between those countries that were covered by a mandatory cap (Annex I countries[42]). In addition, the Protocol established two other project-based flexibility mechanisms:

- The Clean Development Mechanism (CDM) through which Annex I countries can acquire emissions credits from emission reduction projects in countries without a cap (non-Annex I), which count against their own emissions; and
- Joint Implementation (JI) which allows Annex I countries to buy emission credits from emission reduction projects in other Annex I countries. The allowable emissions in the host country are reduced accordingly, thus maintaining the overall cap. The mechanism was primarily designed to fund projects in the post-communist Eastern European states.

The initial expectations of many authors (including of this book) were that the Protocol had too many mechanisms, and that Kyoto emissions trading – the exchange of national 'Assigned Amount Units' – would be the simplest and most dominant mechanism, with the CDM (and to a lesser extent JI) impeded by administrative difficulties.[43] The experience has been almost the opposite – with CDM dominating, JI being a sideshow and Kyoto trading being hampered partly by concerns over legitimacy, and the lack of much need (Figure 7.9).

Since this chapter is focused on lessons from pricing carbon, we needn't go into much detail about the Protocol's strengths and weaknesses. As with the EU ETS, a major lesson is the relative poverty of prediction. Russia, along with its east European neighbours, had seen its emissions collapse with its economic transition since 1990. It expected them to rebound as it recovered and negotiated emission targets accordingly. As noted in Chapters 1 and 11, the eastern European economies recovered instead (and necessarily) in large part by increasing efficiency, leaving them with a huge surplus of allowances.

Moreover, the withdrawal of the US (and then Canada) from the Protocol removed the major countries that might have needed to buy emission allowances. The Russian surplus alone proved to be several times larger than the conceivable needs of all other Parties. The remaining countries mostly baulked at purchasing what became dubbed as 'hot air' – an unearned surplus.

A partial solution emerged, however, in which some of these countries offered to earmark any revenues from the sale of allowances for energy efficiency and carbon abatement policies – Green Investment Schemes (GIS). The first such trade occurred in 2008 with Hungary selling 8MtCO_2 of Kyoto emission units to Belgium and Spain under contract to invest the money primarily to upgrade the efficiency of the Hungarian building stock. The scheme quite quickly ran into trouble, illustrating the risks of 'upfront payment' to another country in a time of crisis – but also the consequence: subsequent buyers can vote with their feet.[44]

In 2009 the sales of Kyoto emission allowances shot up to account for over a third of total sales from the mechanisms, mainly of Japanese purchases of credits from the Czech Republic and Ukraine. The former are backed up by a GIS but for Ukraine the situation is unclear.[45] The new Hungarian government elected in 2010 tried to re-establish its credentials and credibility as a trustworthy recipient of GIS funds but found that the ground was being cut away by the astonishing progress of what had already been dubbed the 'Kyoto Surprise' – the CDM.

The history of the Clean Development Mechanism

The idea is simple. Since climate change is a global problem, it was important to find a way of encouraging emission reductions in developing countries, which participate in, but do not

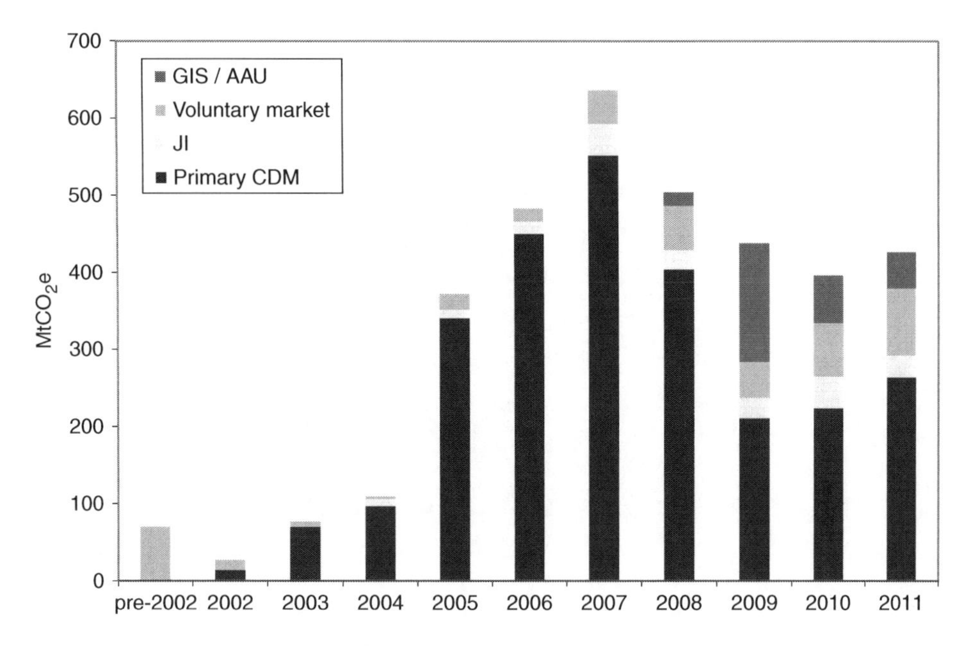

Figure 7.9 Evolution of the International Carbon Market

Note: 'Primary CDM' data are intended to record the original generation of CDM credits, unlike the 'secondary market' in which credits may be traded multiple times. However, the real extent of CDM projects is bigger than indicated because some projects developed 'unilaterally' within developing countries – without any foreign participants – and were sold directly into the secondary market and thus not included in this data.

Source: World Bank (2012).

have caps under, the Kyoto Protocol. Since industrialised countries were scared of the possible difficulty and costs of meeting their emission caps, they (and in particular the US) wanted as much flexibility as possible. The solution was to let emission-reducing projects in developing countries generate emission reduction 'credits', and to let industrialised countries buy these credits as a contribution to meeting their emission targets.

The reality is complex – in its design, development and outcome. To ensure emission savings, CDM projects are required to pass through a complex project pipeline, from project design, validation and approval to registration.[46] The development of the institutional machinery seemed glacial – negotiating the 'rulebook' (Marrakech Accords) took three years. A 'prompt start' provision enabled the institutional development to start long before the Kyoto Protocol entered into force (indeed the first projects started to appear in 2003), but it only really took off with the entry into force of the Protocol in 2005, together with the start of the EU ETS which generated demand for CDM credits through its Linking Directive. The outcome – in terms of 'good and bad' – remains contested: the CDM is a mixed bag of success stories, unexpected results, missed opportunities and regulatory limitations.

With the elements all in place, the first 'certified emissions reduction' was issued in October 2005; seven years later, in September 2012, the total emissions reduction passed a billion tonnes. Project validation, which confirms the expected number of emission credits to be generated over defined periods, was far bigger – reflecting the huge growth in the number of validated projects, which exceeded 2 billion tonnes by the end of the 2012 (see Figure 7.10).

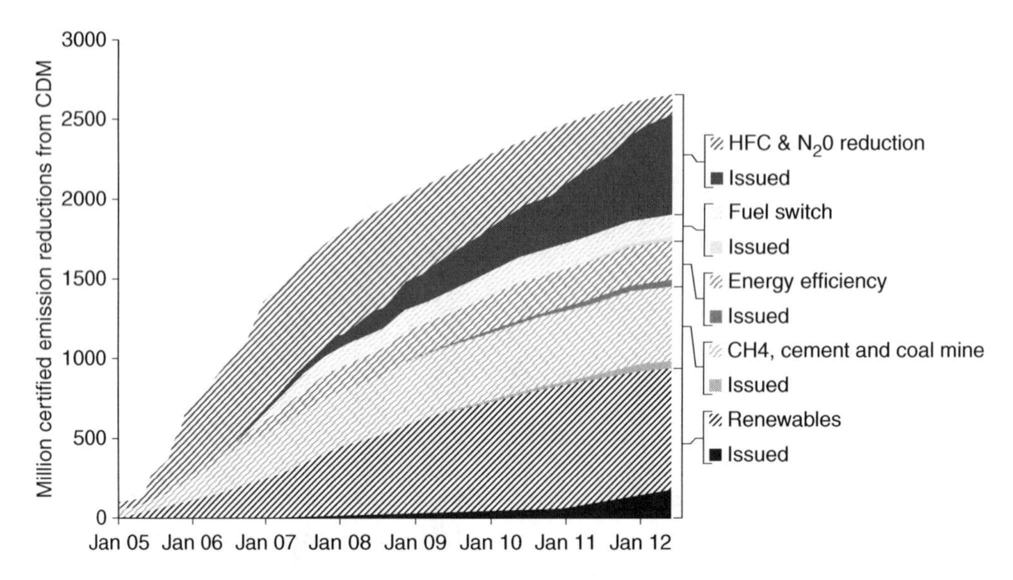

Figure 7.10 Project volumes – from validation to issuance of Certified Emission Reductions

Source: Based on data from UNEP Risø CDM/JI Pipeline Analysis and Database, 1 July 2012.

The CDM development thus paralleled (and was linked to) the EU ETS, but the volume of credited emission reductions has grown to between four and ten times the emission savings estimated from the EU ETS in Europe itself. Companies found it cheaper to source emission reductions from projects in developing countries than pay the EU carbon price. In total it is estimated that the CDM attracted private investment totalling around €150bn.

In this respect, the system behaved exactly as intended, allowing the cheapest means of compliance with Kyoto targets – so cheap, as it turned out, that the system ultimately was flooded, the price collapsed and the flow of new CDM projects has dried up. In common with other market mechanisms, completely unanticipated options emerged, as described shortly.

After an initial rush to industrial gas projects (discussed shortly), led by South Korea and Brazil, there followed a surge of renewable energy projects predominantly in China, but to a lesser extent in India as well, with wind becoming increasingly popular.[47] By 2012 Chinese projects accounted for over half of all projected emission savings.[48] There is a notable absence of projects in sub-Saharan Africa, though its share remains commensurate with its share in overall foreign direct investment.

The CDM was founded squarely on Second Domain principles: reward emission-reducing projects with a carbon price. One intriguing question is whether it is an adequate explanation either of its development or its consequences. Numerous economic analyses questioned whether the scale of financial rewards was sufficient to support many projects beyond industrial gases. The explosive growth of response suggests an attention effect – the mere existence of the CDM stimulated businesses across the developing world to look for potential emission reduction projects. The CDM rules allowed projects which also benefited from domestic policy support, and the volume of these projects grew rapidly.[49] Having established the institutions and policy supports to renewable energy, much of the growth in renewable projects continues irrespective of CDM credits. The CDM thus contributed to the transformation of institutions and attitudes to allow for the growth of renewable energy across Asia – an enduring legacy.

Thus, although fundamentally a Second Domain instrument, the fact that it was focused on something that previously hardly existed – rewarding clean energy projects in the developing world – meant that its impact was amplified by First Domain attention effects. Indeed, in changing attitudes and fostering institutions and policies associated with reducing greenhouse gas emissions, it also contributed to transformations more typical of the Third Domain.

The development of Joint Implementation

The progress of Joint Implementation – the system for crediting project-based emission reductions within and between industrialised countries – lagged behind that of the CDM, not least since credits were associated with national emission caps and so could not be redeemed until the start of the Protocol's commitment period in 2008.

The original concept of JI was that there would be minimal oversight of projects in countries that had met the Protocol's provisions for reporting, since this would ensure their overall emission caps were met. These provisions are extensive and many transition economies felt they could not meet the requirements. Given this fear, a second 'track' of JI was established which allows projects to be established in countries that did not meet these national eligibility requirements, with a JI Supervisory Committee established to verify projects and the issue of credits in ways analogous to the CDM. Project developers may also favour this track even in countries meeting the national eligibility requirements in order to gain the assurances offered by the independent Supervisory Committee.

The largest amount of emission savings from JI projects has originated from reducing fugitive emissions, initially from gas pipelines in Russia. In September 2007 these accounted for 60 per cent of the volume.[50] By 2012 the picture had changed, with 70 per cent of expected credits from industrial energy efficiency, capturing fugitive emissions including from landfill sites, industrial and chemical gases (HFC-23, N_2O) and coal-bed methane. Increased activity was particularly prominent in Ukraine and – somewhat surprisingly – in the Western part of the EU.[51] In total, by June 2012 JI credits corresponding to 169 million tonnes of CO_2 emission reductions had been issued.

The culmination was not a happy one. Regulatory dithering in the EU about whether JI credits would continue to be accepted after 2012 – along with the declining carbon price – contributed to a collapse in investor confidence. The UNFCCC Doha conference in December 2012 established that the overall surplus emission allowances accumulated in the transition economies could not be 'banked forward' for later use. The result was a last-minute stampede in which hundreds of millions of tonnes of emission credits, most notably from Ukraine, were funnelled through the largely unsupervised 'Track 1' process, contributing further to the global surplus and undermining further the reputation of Joint Implementation.[52]

Criticisms of the Kyoto mechanisms

The flexibility mechanisms of the Kyoto Protocol were designed to both allow the overall emission reduction targets to be met as cheaply as possible and to engage developing countries. The CDM was explicitly designed to contribute to sustainable development and deliver 'real, measurable and long-term benefits related to the mitigation of climate change' and emission reductions that were, crucially, 'additional to any that would occur in the absence of the certified project activity'.[53] How much the CDM has achieved toward either sustainable development or additional emission reductions has been hotly debated.

Assessing a project's contribution to sustainable development is left in the hands of host country governments. They have taken varied attitudes, but have been most interested in attracting investment – much literature questions the nature or extent of any other benefits to sustainable development.[54]

An inevitably thorny issue was defining and implementing the principle of 'additionality' (see Box 7.2). The difficulties were inevitable, were predicted more than a decade ago and include the paradox that the most cost-effective projects – the ones that require fewest CDM credits to be viable – are also the ones for which it is hardest to prove additionality, since by definition these are closest to being cost-effective anyway.[55] The approach, appropriately, has evolved to try and strike the best balance between rigour, consistency, predictability and administrative complexity (see Box 7.2), with results still the subject of extensive debate.[56]

Box 7.2 Additionality in the CDM

Additionality is a central concept in the CDM and refers to the principle that emission credits should reflect *additional* emissions savings associated with a CDM project, compared to what would have taken place without it. The principle was introduced in the Kyoto Protocol in 1997, with the CDM Executive Board mandated to define and operationalise it. The resulting 'additionality requirements' differ for small-scale and large-scale projects. For small-scale projects they take the form of eligibility criteria, i.e. the methodology specifies exactly which conditions must be met for the project to qualify as a CDM project.

The CDM Executive Board's method of reviewing projects and ensuring additionality has evolved. At first, it was up to each project proponent to find a way to show additionality, but in 2004 the CDM Executive Board published the first 'additionality tool' which set the standard for how additionality is proven. Over time the additionality tool has been enhanced with added flowcharts, questions to be answered and more steps required in order to specify and clarify different requirements.

The 'Validation and Verification Manual', first released in 2008, has further helped to ensure a transparent and consistent treatment of projects and has enabled non-additional projects to be rejected earlier in the process. In 2008, benchmarks for additionality were introduced, and applied first for energy-efficient refrigerators: the 'benchmark' represents a certain percentage of the most efficient refrigerators available on the market, and if a refrigerator meets this it is considered additional and its emission reductions can be credited in a CDM project. This procedure means that as technology advances and more efficient refrigerators come onto the market, higher efficiency is required to be considered additional – akin to the Japanese 'top runner' programme outlined in Chapter 5.

A possible further development would be to extend the use of technical criteria that are objective and easier to verify than project-specific conditions – perhaps including adaptations for projects in certain regions – to help clarify in advance what would be regarded as additional.

Source: Adapted from Raab (2012); for fuller analysis see reference in note 47.

Project coverage

One debate raged around the group of projects involved with reducing emissions from industrial gases. Most controversial is the greenhouse gas HFC-23. It has 11,700 times higher global warming potential than CO_2 and its emissions could be reduced very cheaply. This created the opportunity to generate lots of emission credits and big profits under the CDM.[57] Some factories might have remained open based upon the revenues from selling CDM credits alone, and it was feared that new factories would be set up just to generate CDM credits. In 2005 the UNFCCC ruled that no new HFC-23 producing facilities would be eligible for CDM.

Despite doubts about the true additionality of *some* of the emission reductions from these projects, there is little doubt that without some incentive they would not have been undertaken – with continued venting of an extremely potent GHG. Also, despite the existence of funds such as the Global Environment Facility (GEF) for many years, these emission reduction possibilities were not tapped until a market mechanism was created. The Chinese government taxed the proceeds from CDM projects (with HFC-23 projects having the highest tax rate at 65 per cent) and used the revenue stream in part to increase mitigation beyond offsets.[58]

In 2010 criticisms arose over coal plants in India and China. Twelve companies applied to the CDM for credits based upon constructing super-critical coal power plants rather than conventional coal plants, reducing carbon emissions by 30 per cent. These projects would be eligible for millions of credits, reflecting reduced emissions compared to conventional coal (and potentially advancing more efficient technology more generally), but there was strong opposition to the CDM funding what is still highly carbon-intensive investment with a life span of decades.

The CDM delivered projects in energy, industry and to a lesser extent waste; it has delivered next to nothing in buildings, transport and forestry, where projects cover a wide range of very small emission sources and many actors. Identifying individual interventions, determining additionality, and monitoring emissions reductions in these cases is difficult and costly.

Indeed, the *transaction costs* were substantial. Creating a market for something which otherwise would have no value requires strong regulatory oversight. Early experience points to transaction costs of several hundred thousand euros per project in the CDM.[59] Yet this does not appear to have limited the mechanism too severely, in particular as simplified rules were later introduced for smaller projects. 'Programmes of Activity' were also introduced more recently to cover suites of projects (such as an energy efficiency programme) under a single administrative umbrella, 37 having been registered by the autumn of 2010.

Despite these efforts, projects remain dominated by the energy and industry sectors, and there is a simpler way of explaining this fact. As noted in Chapter 3, these are the sectors which are dominated by large-scale decision-makers (big firms) focused on the Second Domain ideal of rational, financially based calculation. The CDM may have raised attention, but as noted in Chapter 4, the obstacles in buildings particularly are far more extensive than this. Their negligible role in the CDM, despite massive cost-effective potential, reflects the fact that a Second Domain instrument is not going to solve First Domain barriers. Transport too is a field in which the major determinants tend to reside in other domains, and the CDM has correspondingly been largely irrelevant. Forestry was deterred also as the EU refused to accept forestry-related credits in its EU ETS.

Distribution

The distribution of CDM projects already noted above focused on those emerging economies which already had the institutional capacity and stability to attract investment. Virtually none went to sub-Saharan Africa. Despite strong capacity building efforts, the CDM has yet to reach the poorest countries. This is due not only to the inherent institutional difficulties (that contribute to the poverty of these countries), but also the particular challenges in applying the methodologies in such contexts, given inadequate data and the lack of stable conditions against which to assess the viability and additionality of projects.

The experience from the Kyoto mechanisms teaches us to expect the unexpected when dealing with market mechanisms. Initial perceptions of the mechanisms were that trading between countries would dominate, with the project-based mechanisms playing much smaller roles. In practice, the CDM has grown to dwarf the use of international emissions trading or JI. The surge of early industrial gas projects in the CDM was also highly unexpected and delivered emission reductions where governments had struggled to negotiate the implementation of regulations before. It also brought large profits for developers and criticisms over inefficiency.

Yet, despite their problems, the Kyoto mechanisms have delivered a lot. They have drawn attention to abatement opportunities in many parts of the world and have mobilised a large amount of private finance to help achieve these aims. The value of CDM/JI credits has amounted to tens of billions of US dollars, leveraging far greater amounts of private investment into cleaner investment.[60] Overall the UNFCCC cites the gains from the CDM by 2012 as having:[61]

- attracted $150 billion of private investment;
- reduced over 1 billion tonnes of GHG emissions in 79 countries;
- uncovered opportunities and cost information in unexpected sectors, such as landfill and industrial gas sectors;
- vastly increased greenhouse gas accounting, monitoring and reporting expertise;
- positively influenced awareness and understanding of clean technologies, emissions trading and future action for climate change in both private and public sectors;
- enabled developing countries to gain first-hand experience and enhance local human capacity and institutions;
- built significant carbon market infrastructures for project development, verification and financing services;
- attracted finance for clean technology transfer which has improved the livelihoods of millions, reduced local air pollution and biodiversity loss, improved gender equality and increased access to electricity.

The CDM experience itself is unlikely to be replicated. The EU ETS, in setting binding caps on its industry that could also be met by purchasing credits, was the main economic driver; North America absented itself from the system. The flood of CDM credits into Europe has been a major factor in the collapse of ETS prices, and the value of CDM credits has fallen to practically zero. The CDM is thus a victim of its own success and in Europe it is widely seen as undermining the domestic effort.

Amid all the intricate complexities, the stark, basic lesson is that a market mechanism can only be sustained if there is both supply and demand for emissions reductions. In the aftermath of the credit crunch, with the changing global economic balance, the purely 'north–south' axis of the demand–supply balance implied in the Kyoto Protocol is unsustainable. If international market mechanisms are to remain relevant, something fundamental has to change.

7.7 The emerging global landscape

If 'a week is a long time in politics', then three years was an age in carbon pricing. The year 2010 started with Europe still promoting its vision of a largely harmonised, OECD-wide carbon market by 2015, linking the EU ETS with an expected US federal scheme to drive a new round of offsets. Beyond that, the Grand Vision would then evolve onwards to a unified system covering most of the world by 2020. In stark contrast, after the collapse of the US proposals in mid-2010, many pundits (and much of US industry) believed that carbon pricing was dead for at least a decade, except for a rump in Europe. The real world is looking far more interesting than either.

Following the demise of the US federal efforts, the US mid-term elections in November 2010 were widely seen as putting a nail in the coffin of US efforts to price carbon. Yet on the same day, California voters delivered the opposite verdict, voting down a proposition to suspend the Californian climate programme including its cap-and-trade bill. The Californian system duly came into effect in 2012. It initially covers electricity generation and large industrial facilities totalling 37 per cent of California's GHG emissions, but is due to expand drastically to 85 per cent of emissions in 2015 with the inclusion of transportation and other fuels, and natural gas (IETA 2012). The emissions cap is then scheduled to reduce at 3 per cent per year. Like previous emission trading schemes, the Californian scheme allocates many allowances for free.[62]

California was originally the core of a more ambitious 'Western Climate Initiative', at one time spanning eleven states. The evolution of the WCI in many ways provides a counterpoint to the EU ETS. Both were emboldened by a higher-level initiative (respectively the Kyoto Protocol, and the expectation of a US federal system). The WCI, however, could not draw on any pre-established political and legal structures (in contrast to the EU). Thus the WCI was left very vulnerable when the federal effort collapsed – the less committed states, often under heavy political pressure, withdrew. This left just the confident, powerful core (and less carbon-intensive) states of California and Quebec to proceed.[63]

However, the US experience also demonstrates that once a system is in place, it tends to survive: the Regional Greenhouse Gas Initiative (RGGI), which caps CO_2 emissions from power generators in seven North-East US states, has to date survived all the furious opposition thrown at it, in part because it generates revenues used to support crucial state programmes.

Several other rich-world regional governments, stymied at the national level, have also pressed ahead. As noted, Quebec remains linked with the Californian system despite the trenchant opposition of the Canadian federal government. Elsewhere in Canada, British Columbia introduced in 2008 the first major new carbon tax since the Scandinavian countries, almost twenty years earlier, with the price rising to \$30/tCO_2 in 2012 (see Chapter 6, Table 6.1 and associated reference).

Japanese industry successfully used the US federal developments to block a national cap-and-trade system in Japan, but in 2011 Tokyo became the first city to operate a municipal cap-and-trade system for carbon – an example swiftly emulated by Rio de Janeiro ahead of the 2012 'Rio+20' Summit.[64]

By 2010 New Zealand had already implemented its emissions trading system, including agriculture and a full set of linkages to the international system, having expected to have Australia alongside.[65] The Australian system became embroiled in vicious political battles. The government rescued its plans, which then came into force in 2012 with a three-year fixed price which is then due to morph into a trading scheme linked with the EU ETS. The system covers the 500 biggest polluters constituting 60 per cent of total emissions.[66] The Australian government then raised the stakes further by joining the EU in signing on for a second period of the Kyoto Protocol. Renewed uncertainty was then injected by the election of a climate-sceptic government in September 2013, committed to dismantling the system.

In practice, the tortured progress of US and Australian developments makes those in Asia more important and impressive. Of greatest consequence, continental Asian economies have started to fill the void left by US political gridlock and Japanese retrenchment. The pattern of emerging systems illustrated in Figure 7.11 shows a tentative but significant, global shift.

In May 2012, South Korean legislation for emissions trading finally passed its Parliament; it will cap 60 per cent of Korean emissions when it starts operating in 2015.[67] Singapore is also developing detailed legislation, and many other Asian countries have programmes at various stages of debate.

In China, the National Development and Reform Commission (NDRC) in July 2010 launched 'Low-carbon Pilot Development Zones' in five provinces and eight cities. Under its Five Year Plan (2011–15) five cities and two provinces will establish Pilot Emissions Trading Schemes.[68] Details on their design are emerging as these systems move into operation, and they will form an important input to debates on a China-wide carbon pricing mechanism which is under consideration for 2015 to help deliver the Chinese carbon intensity target formulated for 2020. This seems implausibly ambitious but with the pace of Chinese developments it is hard to be sure.

The Indian 'Perform, Achieve and Trade' (PAT) scheme further illustrates the sheer range of potential approaches. Established under the terms of the Indian Energy Conservation Act, PAT sets mandatory energy-intensity targets for large installations across eight energy-intensive sectors – broadly similar sectoral coverage to the EU ETS. Operators are given energy-saving certificates (ECerts) that form the basis of a trading market: installations that fall short of their targets will have to buy certificates from those that can over-comply. It operates in sequential phases, with three three-year periods spanning 2012–20 allowing successive improvements. The targets are intended to drive rapid improvements: for example, they would require the Indian steel sector to reach world best practice by 2020, something that on recent trends would otherwise take three decades. With a projected population by

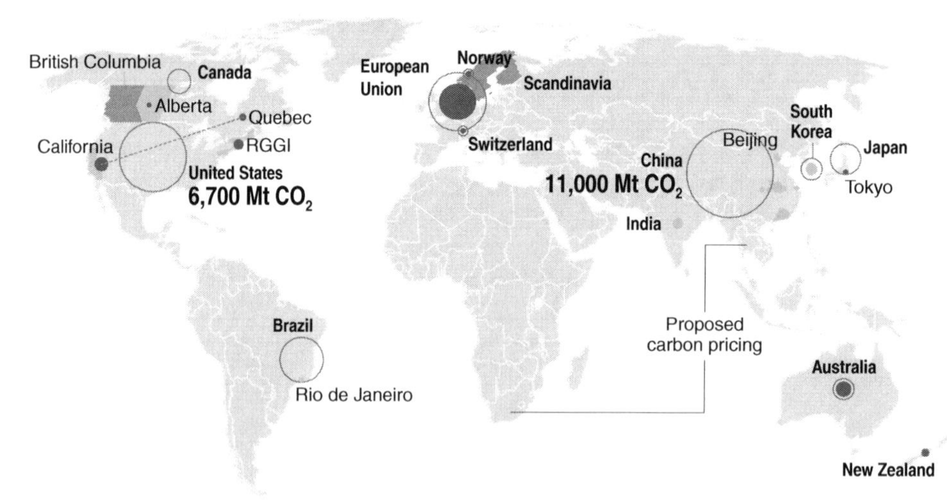

Figure 7.11 Existing and emerging carbon pricing systems

Source: Grubb (2012) 'Emissions trading: cap and trade finds new energy', first published in Nature, 491: 666–7, Nature Publishing Group, a division of Macmillan Publishers Limited.

then of 1.3 billion people and one of the fastest industrial growth rates in the world, the PAT scheme by 2020 could well cover energy on a scale comparable with the whole EU ETS.[69]

Two fundamental differences distinguish the Indian PAT scheme from cap-and-trade schemes for emissions: firstly, its focus on domestic energy rather than carbon; and secondly, its focus on the *efficiency* of energy use rather than total energy consumption.[70] It is not supposed to be a climate initiative – even though it will contribute to reducing carbon intensity – but reflects fear that Indian economic growth risks being hampered by its dependence on unstable oil markets and on poor-quality coal resources that are inflicting increasing stress on the Indian environment and transport systems.

India's PAT scheme tries to be consistent both with the reality of the development imperative and with the official Indian international position that climate change is a rich-country problem. Yet, alongside this potentially huge scheme to drive industrial energy efficiency, India has been among the world's pioneers in renewable energy and, after Brazil, has some of the most ambitious renewable energy targets as well as nuclear programmes. Thus the reality is that India, with a fifth of the world's entire population and a burgeoning industrial sector, is putting in place serious binding structures to limit its energy consumption and promote low-carbon energies. Consistent with the reality of India's huge vulnerability to climate change, it now includes the world's second-biggest system to put a direct economic value on energy efficiency and associated emission reductions.

All this underlines two basic facts. Neither the failure of US federal legislation nor the travails of the EU ETS have halted global progress, but they have handed the crucial initiative to Asia. About 10 per cent of global emissions are now covered by a carbon cap or tax (not accounting for India's PAT scheme); by 2015, Korea together with the Chinese pilot schemes will take this closer to 15 per cent – and more are emerging.

However, the effort will look nothing like the Kyoto vision, or the EU's imagining of an OECD-wide market. The systems emerging will not be copies of the EU ETS (or the Californian system): each will be adapted to national priorities and circumstances, and may look very different indeed. It is possible that systems will converge over time but that is clearly a task mostly beyond 2020. For at least a decade, the world will comprise a veritable forest of different circumstances and systems.

7.8 Conclusions

As the old Chinese proverb says: be careful what you ask for – you may get it. Common sense, as well the economic principles and data laid out in Chapter 6, dictates that a price should be placed on damaging emissions in a market economy. With the Kyoto Protocol, the world asked for market mechanisms to be applied to tackling climate change globally, and the EU built its domestic approach from this platform. In the event, Kyoto seemed to prove that international law matters – the countries which remained part of the agreement complied with its rules and commitments, thus powering the Kyoto mechanisms as discussed – but the non-participation of the US and Canada's subsequent withdrawal also emphasised the voluntary nature of participation. Overall the response to the world's grand experiments with economic instruments has been quite unlike anything predicted. There are innumerable detailed lessons shown throughout this chapter, including the specific 'ten lessons from the EU ETS', but the big themes are really quite simple.

First, carbon markets largely do what they are designed for: they hunt out the cheapest ways of cutting emissions within the specified rules and prevailing conditions. Thus the advantage to this is their ability to tap into unsuspected opportunities. The downside is the corollary: if people really wanted the international system to deliver a host of things other than cheap carbon emissions reductions, the CDM was the wrong instrument.

Second, it was widely assumed that the biggest challenge under Kyoto and its market mechanisms would be meeting the emission caps. The reality has proved to be precisely the opposite: the biggest structural challenge has been an over-achievement of emission targets – for multiple reasons – with correspondingly huge surpluses (EU ETS, Eastern Europe, the CDM). This leaves the question: how to prevent the surplus swamping the efforts of others, or of the future. Total flexibility combined with the combined biases noted in the chapter create this fundamental risk. This is an additional reason to consider hybrid instruments which combine price and quantity goals or other steadying mechanisms such as those outlined in this chapter.

Third, though they were conceived mainly from relatively static principles of equilibrium economics, in practice all economic instruments evolve and can learn. Aside from the eight-year Phase III of the EU ETS, which is proving ineffective without steadying mechanisms, each phase of the EU ETS has been an improvement on the former. CDM governance has had to evolve to keep pace with experience and debates on additionally and the risks of perverse incentives. Once there is a system to give value to emission reductions, someone somewhere will try to abuse it; the legislation, governance, monitoring and enforcement must keep pace. Economic instruments, in that sense, are not so different from financial systems. Problems with financial systems and markets have not stopped us using money or banks; the flaws revealed in carbon markets should not stop us putting a price on damaging emissions.

Fourth, a market instrument based on Second Domain assumptions can have some impacts through other Domains – the attention (First Domain) effects noted in both the ETS and CDM, and potentially the strategic influence of the ETS as a mechanism capable of giving economic force to long-term targets (Third Domain). Underestimating the impacts of First (and maybe Third) Domain effects has indeed contributed to the persistent over-achievement of targets and resulting over-supply of emission allowances. Moreover, revenues can also be harnessed explicitly for such purposes, as with the East European Green Investment Schemes, and the funding for CCS from selling EU emission allowances. Yet market instruments cannot *in themselves* overcome barriers arising in other domains – it was a fantasy to suppose that the CDM would deliver 'negative cost' reductions in buildings, for example, because the barriers are little to do with cost.

Fifth, the role of 'offset' mechanisms (generating credits from individual emission-reducing projects not covered by (binding) caps) may be most important as a transitional instrument. Offsets are inherently problematic both because of intrinsic complexities and uncertainties in estimating 'additionality' (see Box 7.2), and because they are chasing a moving target as technologies, costs and policies evolve. Offset systems are also economically unstable because success breeds both an excess of supply and resentment at the scale of economic transfers from those operating under emission caps which drive the system. However, the CDM in particular, and JI for eastern Europe to a lesser extent, have not only led to major emission reductions and associated investment flows; they have also had a transformational impact on these regions' involvement, capacity and interest. Such instruments, like everything else, must evolve or die.

Finally, a statement of the overlooked obvious: a market can only work if someone pays. People, generally, do not like paying – especially for something that was previously free or is seen to benefit mainly others. The US spearheaded the creation of market mechanisms, citing their obvious advantage in global efficiency. Yet in practice, the US then proved unwilling to pay and withdrew. Canada subsequently followed the US example. That left the EU ETS as the centrepiece, and it could not bear the weight of the global response through the CDM. The question of 'who pays' is, more fundamentally, at the heart of most political opposition to market instruments and is the focus of the next chapter.

Despite all the difficulties and often glacial progress, the spread of efforts noted in the foregoing sections does suggest an emergent coalition that recognises carbon pricing (and market instruments more broadly) as integral to a credible strategy for sustainable energy, innovation and growth.[71] Whether and how these may link – in their own countries, or looking internationally, for example, to include poorer regions through CDM-type mechanisms – remains to be seen. But the multiple arguments for a carbon price in some form are so compelling that it now looks like a river that ultimately finds a way through, no matter how many obstacles are flung in its path and however powerful the opposition.

Notes

1. For a compilation of many of the best-informed and most influential academic writings by one of the leading researchers involved in US debates, see Stavins (2013). In applying such ideas, however, Waxman-Markey became deeply politicised and overly complex, but also fell in the wave of concern about cost impacts on the general public combined at a time of unprecedented energy prices, along with general post-credit-crunch antipathy to complex market-based systems amenable to manipulation.
2. Personal carbon trading is a useful broad term used by Fawcett and Parag (2010) to cover a range of more specific proposals, which had been advanced under the names cap and share, tradable energy quotas, tradable consumption quotas, personal carbon allowances, household carbon trading and tradable transport carbon permits. The Special Issue of *Climate Policy* on the topic (Parag and Fawcett 2010) explores many dimensions of these proposals, including the complex issues of equity, administration and enforcement.
3. The first discussions of the possibility of an EU-level trading system began in the wake of the Kyoto conference in 1997. After the UNFCCC meetings in Buenos Aires in 1998 and Marrakech in 2001 it became apparent that the UNFCCC would be unable to create strong enforcement mechanisms, so the EU started to consider an EU-level arrangement that would fall under the purview of the European Court of Justice. Emissions trading creates valuable assets: strong governance, monitoring and enforcement is crucial. A discussion ('Green') paper by the European Commission on emissions trading in 2000 moved the tone of the debate from 'should there be?' to 'how could there be?' such a trading system. Progress was possible also due to the support of industry bodies such as the Union of Industrial and Employers' Confederations of Europe (now renamed Business Europe), which fiercely opposed the EU carbon tax, for an emission trading scheme. The proposal was drafted in 2001 and then submitted to the EU Council of Ministers and the European Parliament before final adoption late in 2003 (European Parliament and Council 2003).
4. In practice, three types of trading have emerged in the market: bilateral trading between large players, over-the-counter trades that now represent about 70% of all EUA transactions, and exchange-based trading through dedicated new exchanges such as the European Climate Exchange and Bluenext, and existing energy exchanges like the European Energy Exchange and Nordpool.
5. European Parliament and Council (2004).
6. Kettner *et al.* (2008).
7. By the start of the trading scheme at the beginning of 2005 four major emitters, Poland, Italy, the Czech Republic and Greece, still had not yet had their plans approved by the European Commission. Indeed the last of these, Greece, was not approved until June 2005, six months after the start of the scheme.
8. In the event only Ireland, Hungary and Lithuania actually conducted auctions, while Denmark sold the reserve into the market through financial intermediaries.
9. Neuhoff (2008). With higher gas prices a higher carbon price was necessary to create the economic incentive to burn gas instead of coal, which was seen as an essential element necessary to comply with emission caps. Hence the carbon price increased with the gas price.

10. Ellerman and Buchner (2008).
11. A major source of criticism was the way in which information came out – environment ministries perhaps not being conversant with the subtleties in handling market-sensitive data. The first leaks came from the Netherlands and the Czech Republic, which both announced significantly lower emission levels than their allocations. This was followed by announcements and leaks from a number of other countries, including a technical fault which leaked the data for 20 countries on 12 May 2006, precipitating the price crash.
12. The EU ETS Directive explicitly rules out borrowing of allowances between different phases of the system, which it was feared would undermine the integrity of the system. In practice the rules did make it possible for Member States to bank allowances forward and France and Poland sought to do so, but this decision was reversed once the Commission announced that any banked allowances would be removed from the Phase II cap.
13. With the exception of the new entrants of Romania and Bulgaria, which only entered the scheme in 2007.
14. In evaluating the first eleven National Allocation Plans (reduced to ten after the French government withdrew its plan a few days before), in November 2006 the European Commission rejected all but the UK's as inadequate. In fact the Commission went further than this. It clarified its interpretation of the Directive in terms of specific allowable allocations – setting an overall cap on national allowances that would be deemed compatible with the rules given expected economic growth, energy intensity and Kyoto targets. Specifically, total allocations could not exceed 2005 levels multiplied by projected economic growth corrected for trends in energy intensity (energy per unit of economic output), with the economic growth projections and energy intensity corrections taken from international (EU) sources, not Member States' own offerings. The Kyoto obligation was translated by a straight line from 1990 emission levels to the Kyoto target, after taking account of other aspects of Member State implementation plans, including provisions for the purchase of international Kyoto credits. Of the 15 Member States at the time, ten were determined to need to cut emissions below 'business-as-usual' levels and those were capped by this condition, with the remaining five allowed to produce NAPs up to 'business-as-usual' levels.

    The Kyoto targets were those as determined under the EU's 'Burden-Sharing Agreement' subsequent to the Kyoto Protocol. The Protocol itself simply stipulates the EU target as keeping emissions to 8% below 1990 levels, with the same indicated for each Member State. Article 4 in the Protocol, however (the 'bubbling' clause), was designed to allow countries – notably EU Member States – to club together to redistribute their targets within their overall cap, prior to ratifying the Treaty. It is the subsequent Burden-Sharing Agreement which thus defined the responsibilities of individual EU Member States.

    The implication of the Commission proposal was to slash something like 10% off the emission caps that countries had proposed, based on their extensive negotiations with their industries. The Commission's announcement seemed to come as a shock, and by presenting decisions on such a big group of countries simultaneously it raised the stakes enormously. A legal challenge could have brought the system crashing down or locked it up in legal disputes for months or even years – and certainly was unlikely to be resolved in time to be of much use to investors wanting to know the rules for Phase II. A furious German Economics Minister announced that Germany would challenge the Commission's decision, but a few weeks later he was overruled by the German Chancellor Angela Merkel. Germany was, after all, hosting the G8 Presidency in 2007, and she had put climate change high on the agenda. She could hardly claim to be leading global action (and berating the US Bush Presidency) while suing the European Commission to let German industry emit more. As Germany backed off, so did the other western European countries.
15. Neuhoff (2008).
16. Most of the east European countries made legal challenges to the European Court of First Instance, based on interpretation of the economic growth rules and projections – they were clearly going to meet their Kyoto targets anyway – and there were some rulings in their favour. These disputes dragged on, and the Polish situation was only finally resolved in April 2010 with the level of cuts not amended, only the internal allocation (EurActiv 2010).
17. The special treatment included derogations allowing (a declining level of) free allocations to power generators in eastern Europe, and 'fuel-specific benchmarks' which enabled them to give more free allowances to coal plants than to gas, for example.
18. Latest numbers can be obtained from http://unfccc.int/Statistics/index.html

19. Verified emission data from 2008 were released in April 2009 and showed that the national caps were appropriately set below emissions in major emitter countries such as the UK, Italy, Spain and Germany. However, the generous quotas for the use of international offsets (CDM and JI) avoided the need for ambitious domestic emission reductions (Consilence Energy Advisory Group Limited 2009). In 2009 it was a different story. Already by April 2009 data showed emissions dropping 11% compared to 2008, giving a surplus of around 80 million tonnes of CO_2 under the cap (Point Carbon 2010).
20. Point Carbon (2009a).
21. Delarue *et al.* (2008).
22. Climate Strategies: Grubb *et al.* (2012).
23. For a thoughtful analysis of lessons learned from the US trading experience, and in particular the US SO_2 scheme, see Schmalensee and Stavins (2013).
24. Grubb and Ferrario (2006).
25. Several people were prosecuted for such carousel fraud, but to avoid the problem the Commission subsequently passed a Directive allowing Member States to implement a reverse charge mechanism for VAT on EUAs, levying the tax on the purchaser rather than the seller.
26. Phillips (2009).
27. The EU has subsequently passed legislation banning such recycling of credits.
28. Problems were not confined to the EU ETS itself, and market manipulation is of course another kind of fraud. With intergovernmental trading of Kyoto emission allowances, in Slovakia four (!) environmental ministers had to resign due to a large sale of Kyoto allowances to an obscure US company at half of the market value.
29. Indeed, in 2005 I was berated by the Chief Economist at the European Bank for Reconstruction and Development for a presentation in which I predicted the scale of windfall profits that would accrue to European utilities – he echoed a common view in stating that it wasn't possible for companies to profit from environmental constraints. Along with explaining the basic mechanisms, the hard evidence was first published in a Special Issue of *Climate Policy* (Sijm *et al.* 2006). Within two years it became accepted fact, paving the way for the move to auctioning in Phase III.
30. Tying allocation to production levels – known as intensity-based (emissions per unit of production) or output-based allocation – can tackle windfall profits and leakage. However, this reintroduces or even amplifies the inefficiencies of emissions-based allocation. EU policy-makers consistently rejected industry pressures for output-based allocation or rebates on a mix of practical, environmental and efficiency grounds.
31. If the benchmarks applied are uniform and industry-wide, the inefficiencies are limited to a reduced increase in the price of the final carbon-intensive product, with a reduced incentive to switch to less carbon-intensive substitutes. If the benchmarks are fuel-specific, however, as was the case in much of the free allocation to the power sector in the EU ETS, the incentives to switch to low-carbon fuels are reduced. For example, if coal-based producers of electricity receive more allowances than gas-based producers, who in turn receive more than hydropower producers, the incentives to switch to low-carbon fuels are reduced, but the incentives to improve production efficiency are still maintained. For a more detailed analysis of the perverse incentives that might arise see Neuhoff *et al.* (2005) and Sterner and Muller (2008).
32. For example, there was a bruising battle in the cement industry over whether the benchmarks should be related to *cement* or to *clinker* – the raw input material. In the end, the company that had done most to reduce its emissions by reducing the use of clinker lost out to rivals who successfully lobbied for benchmarks based on clinker use.
33. European Parliament and Council (2009). Derogations were driven in particular by Polish opposition to the transitional shock of auctioning allowances to its large fleet of coal power stations. The exceptions for sectors that might be at risk of carbon leakage are discussed in Chapter 8.
34. Personal communication.
35. Based on interviews with 800 manufacturing companies: in Karoline Rogge (ISI Fraunhofer), Tobias Schmidt (ETH Zürich) and Malte Schneider (ETH Zürich), *Relative Importance of Different Climate Policy Elements for Corporate Climate Innovation Activities: Findings for the Power Sector*; study available at: http://www.climatestrategies.org
36. The allowed volume of CDM credits in Phase II could allow emissions from EU ETS installations to exceed the cap by an average of 5.4% in the period 2008–12. This potential was used to a degree that clearly also had a major impact on the EU ETS price.

37. Private sector initiatives like the Carbon Disclosure Project and financial products like a low-carbon index tracker, the Low Carbon 100 Europe Index or the Standard & Poor's/IFCI Carbon Efficient indices are starting to provide corresponding information. This is essential, because pension funds and other investors seeking fixed income from 15–20 year bonds require transparency about the risks they are facing.
38. Standard & Poor's (2009) 'On the Drax Power Ltd issuer rating cut to "BB+" on weak UK power prices and increasing business risk; Outlook Negative' (Ratings Direct).
39. Parsons and Taschini (2013).
40. The industrialised economies cannot conceivably return to the level of economic output that was being projected before the 2008–9 credit crunch; indeed, after recessions economies hardly ever recover to the previously expected trajectories (Bowen and Parker 2010). The unexpectedly rapid growth of renewable energy is unlikely to go into reverse. Even the unexpectedly large contribution from CDM emission credits would also have endured and expanded into Phase III had the price not collapsed and the rules been changed to prevent this. Temporary shocks are mostly a figment of imagination in economic models; in the real world, most have enduring consequences.
41. The combination of a floor and ceiling price has been derided as 'price management'. Yet some 'price management' is implicit in all intervention options, and fundamentally appropriate given the recognition that prices too low or too high are inconsistent with strategic objectives. The degree of 'price management' would be defined by the width of the collar and the economic and political principles on which the floor and ceiling are based.
42. The Kyoto Protocol defines two main groups of countries: Annex I are those covered by a mandatory cap, predominantly the developed countries, while Non-Annex I are those which do not have caps and are mainly the developing countries.
43. For a diverging view, giving the CDM and JI more relative importance than IET, see, for instance, Woerdman and van der Gaast (2001).
44. When the gathering economic crises pushed Hungary to the brink of bankruptcy the government deferred the GIS expenditure to forestall the need for an emergency IMF loan. The final outcome remains in negotiation.
45. Kossoy and Ambrosi (2010).
46. Validation by Designated Operating Entities (DOEs: an accredited third-party auditor who assesses the project against given criteria), approval by national Designated National Authorities in the host country (who check whether the project contributes to sustainable development) and then through to registration with the CDM Executive Board (an independent oversight body under the aegis of the UNFCCC).
47. For a more detailed analysis of CDM performance see Climate Strategies: Michaelowa and Castro (2008), and Grubb *et al.* (2010).
48. UNEP Risø Centre (2012).
49. This became known as the 'E+/E–' rules. The risk was that the need to prove projects as 'additional' could deter countries from improving domestic policies. To negate this potential perverse incentive either to adopt bad policies (to inflate baseline emissions) or to eschew good ones (which could make low-carbon projects happen in the absence of the CDM), in 2004 the Executive Board adopted the following principles: (a) Policy changes that drive up baseline emissions ('E+') could not be counted in the baseline if they were implemented after adoption of the Kyoto Protocol. (b) Policy changes that give positive comparative advantages to less emissions-intensive technologies ('E–') and that have been implemented since the adoption of the Marrakech Accords in 2001 (that defined the rules for implementing the Kyoto Protocol) need not be taken into account. Countries thus cannot generate more credits by making bad policy, and emission-reduction policies do not reduce CDM credits because 'the baseline scenario should refer to a hypothetical situation without the national and/or sectoral policies or regulations being in place' (Carbon Trust 2009: 56, online at: http://www.carbontrust.co.uk).
50. Korppoo (2007).
51. For a more in-depth analysis of JI project development see Korppoo and Gassan-Zade (2011) and Carbon Trust (2009). The 2012 numbers are from Risø JI Pipeline.
52. Zhenchuk (2012). In the aftermath of the Doha decisions and the 'Track 1' surge at the end of 2012, the Joint Implementation Action Group – the NGO which had been monitoring Joint

Implementation – wrote a reflective final letter to the UN Joint Implementation Supervisory Committee before disbanding (available at: http://www.jiactiongroup.com).

53. UNFCCC (1998).
54. Olsen (2007) and Olsen and Fenhann (2008).
55. Early studies included Michaelowa (1998), Grubb *et al.* (1999) and Jackson *et al.* (2001). For a detailed account of the regulatory history of additionality determination in the CDM see Michaelowa (2009).
56. See Cames *et al.* (2007), Michaelowa and Purohit (2007), Schneider (2007) and Grubb *et al.* (2010). For an extensive overview of the evolution and future possibilities see Michaelowa (2009: 248–71).
57. Wara (2008), Green (2008).
58. Liu (2010).
59. Fichtner *et al.* (2002) and Michaelowa *et al.* (2003).
60. Carbon Trust (2009).
61. UNFCCC (2011) *Benefits of the CDM*, available online at: https://cdm.unfccc.int/about/dev_ben/pg1.pdf, as cited in Rabb (2012: 8).
62. See IETA (2013) for an overview of the Californian emissions trading system. For the electricity sector and those industries deemed vulnerable to external competition, free allowances will comprise 95% of the total emissions cap, gradually declining to 42% in 2020 (EPRI 2011). For industrial emissions sources this initial figure is 90% and will reduce in proportion to annual product output. Learning from the European experience of windfall profits, the allocation in the power sector is not granted to generators but to power consumers.
63. 'Creating a carbon market is more than a technical or political challenge – it is a social process. The WCI experience highlights the importance of the logic of collective action, the need for jurisdictions to see individual benefits, the role of evidence from other policy contexts, and the need for broad agreement about the purpose of policy' (Klinsky 2013).
64. World Bank (2012).
65. New Zealand is another example of the political interplay between carbon tax proposals – ultimately killed politically – and the emergence of an emissions trading system in its place. The New Zealand scheme has been roundly criticised for its weaknesses in Bertram and Clover (2009).
66. Technically, for the initial years of the Australian system, a fixed price for carbon allowances was set at AU$23 in 2012 increasing by 2.5% per year up to 2015 (Meltzer 2012). Market participants can acquire unlimited amounts of allowances at this price in an auction and the government is committed to cover any demand beyond the domestic target with offset credits. After 1 July 2015 the supply of allowances would be capped, but participants will be able to comply by buying EU ETS allowances.
67. Republic of Korea (2012).
68. The participating Chinese cities are Beijing, Tianjin, Shanghai, Chongqing and Shenzhen; the two provinces are Guangdong and Hubei.
69. Specifically, the sectors and participation thresholds in the Indian PAT system are: 30,000 tonnes of oil equivalent (toe) for thermal power plants, cement, fertiliser, iron and steel, and pulp and paper; 7,500 toe/yr for aluminium; 12,000 toe for chlor-alkali production; and 3,000 toe for textiles. For a short summary of the PAT scheme see the Briefing Note by Joyashree Roy at http://www.climatestrategies.org. For fuller and updated analysis see Roy *et al.* (2013). By setting these thresholds, the Indian PAT scheme covers just under 500 major installations and avoids one other problem of the EU ETS which stemmed from its inclusion of many thousands of small installations – with large relative transaction costs – when actually the vast majority of emissions come from just the top 5–10% (Carbon Trust 2008).
70. The focus on energy efficiency (intensity) reflects the difficulty of determining a target for energy use in a country with high and historically uncertain economic growth rates. It is also politically less challenging, as 'only' efficiency potentials are captured, but no attempt is made to encourage a shift from energy-intensive to less energy-intensive commodities and services. The scheme will constrain the energy per unit of electricity, cement, steel and other basic goods, not the amount of their primary production.
71. Paterson (2012).

References

Bertram, G. and Clover, D. (2009) 'Kicking the fossil fuel habit: New Zealand's ninety percent target for renewable electricity', in Fereidoon P. Sioshansi (ed.), *Generating Electricity in a Carbon Constrained World*. Amsterdam: Elsevier.

Bowen, A. and Parker, C. (2010) *Economic Growth, the Recession and Greenhouse Gas Emissions*, Report for Climate Strategies. Online at: http://www.climatestrategies.org

Brack, D., Grubb, M., Windram, C. *et al.* (2000) *International Trade and Climate Change Policies*. London: RIIA/Earthscan.

Cames, M., Anger, N., Böhringer, C., Harthan, R. O. and Schneider, L. (2007) *Long-term prospects of CDM and JI*, Report by Öko-Institut and ZEW. Berlin: Umweltbundesamt.

Carbon Trust (2008) *Cutting Carbon in Europe: The 2020 Proposals and the Future of the EU ETS*. London: Carbon Trust. Online at: http://www.carbontrust.co.uk

Carbon Trust (2009) *Global Carbon Mechanisms: Emerging Lessons and Implications*. London: Carbon Trust.

Climate Strategies: Grubb, M., Laing, T., Sato, M., and Comberti, C. (2012) 'Analyses of the effectiveness of trading in EU-ETS', Climate Strategies Synthesis Report, London, www.climatestrategies.org

Climate Strategies: Michaelowa, A. and Castro, P. (2008) 'Empirical analysis of performance of CDM projects', Climate Strategies Synthesis Report, London, www.climatestrategies.org

Consilence Energy Advisory Group Limited (2009) *Accord. Agreed COP15 18/12/2009 by USA, China, India, Brazil and S Africa*, 'Noted' by UNFCCC. Copenhagen: Consilience Energy Advisory Group Ltd.

Delarue, E., Voorspools, K. and D'haeseleer, W. (2008) *Fuel Switching in the Electricity Sector Under the EU ETS: Review and Prospective*, TME WP EN2007-004. KU Leuven Energy Institute.

Ecofys, Fraunhofer Institute for Systems and Innovation Research and Oko-Institut (2009) *Methodology for the Free Allocation of Emission Allowances in the EU ETS Post 2012: Report on the Project Approach and General Issues*. Berlin: Ecofys.

Ellerman, A. D. and Buchner, B. (2008) 'Over-allocation or abatement? A preliminary analysis of the EU ETS based on the 2005–06 emissions data', *Environmental and Resource Economics*, 41: 267–87.

EPRI (2013) *Exploring the Interaction Between California's Greenhouse Gas Emissions Cap-and-Trade Program and Complimentary Emissions Reduction Policies*. Online at: http://eea.epri.com/pdf/ghg-offset-policy-dialogue/workshop14/EPRI-IETA-Joint-Symposium_Complementary-Policies_041613__Background-Paper_FinalPosted.pdf

EurActiv (2010) 'Czechs vie for top in Eastern European R&D league', *EurActiv*, 7 June.

European Environment Agency (2013) *EU Emissions Trading Systems (ETS) Data Viewer*. Online at: http://www.eea.europa.eu/data-and-maps/data-viewers/emissions-trading-viewer

European Parliament and Council (2003) Directive 2003/87/EC of the European Parliament and of the Council establishing a scheme for greenhouse gas emission allowance trading within the Community and amending Council Directive 96/61/EC, *Official Journal of the European Communities*, L275: 32–46.

European Parliament and Council (2004) Directive 2004/101/EC of the European Parliament and of the Council amending Directive 2003/87/EC establishing a scheme for greenhouse gas emission allowance trading within the Community, in respect of the Kyoto Protocol's project mechanisms, EU.

European Parliament and Council (2009) Council Directive 2009/29/EC to improve and extend the greenhouse gas emission allowance trading scheme, *Official Journal of the European Communities*, L140: 63–87.

Fankhauser, S. (2011) 'Carbon trading: a good idea is going through a bad patch', *European Financial Review*, April–May, pp. 32–5.

Fawcett, T. and Parag, Y. (2010) 'An introduction to personal carbon trading', *Climate Policy*, Special Issue, 10 (4): 329–38.

Fichtner, W., Graehl, S. and Rentz, O. (2002) 'International cooperation to support climate change mitigation and sustainable development', *International Journal of Environment and Pollution*, 18 (1): 33–55.

Fliechter, T., Hagemann, M., Hirzel, S., Eichhammer, W. and Wietschel, M. (2009) 'Costs and potential of energy savings in European industry – a critical assessment of the concept of conservation supply curves', *Eceee 2009 Summer Study – Act! Innovate! Deliver! Reducing Energy Demand Sustainably*, Paper No. 5376. Ecofys, Fraunhofer Institute for Systems and Innovation Research.

Green, R. (2008) 'Carbon tax or carbon permits: the impact on generators' risk', *Energy Journal*, 29 (3): 67–89.

Grubb, M. (2012) 'Emissions trading: cap and trading finds new energy', *Nature*, 491: 666–7.

Grubb, M. and Ferrario, F. (2006) 'False confidences: forecasting errors and emission caps in CO_2 trading systems', *Climate Strategies*, Report. London: Climate Strategies.

Grubb, M., Brack, D. and Vrolijk, C. (1999) *The Kyoto Protocol: A Guide and Assessment*. London: Earthscan.

Grubb, M., Brewer, T. L., Sato, M., Heilmayer, T. and Fazekas, D. (2009) *Climate Policy and Industrial Competitiveness: Ten Insights from Europe on the EU Emissions Trading System*. Washington, DC: Report for the German Marshall Fund Climate and Energy Paper Series No. 9.

Grubb, M., Laing, T., Counsell, T. *et al*. (2010) 'Global carbon mechanisms: lessons and implications', *Climatic Change*, 104 (3–4): 539–73.

IETA, EDF (2013) *California: The World's Carbon Markets: A Case Study Guide to Emissions Trading*. Online at: http://www.ieta.org/assets/Reports/EmissionsTradingAroundTheWorld/edf_ieta_california_case_study_may_2013.pdf

Jackson, T., Begg, K. G. and Parkinson, S. D. (eds) (2001) *Flexibility in Climate Policy: Making the Kyoto Mechanisms Work*. London: Earthscan.

JIAG (2012) *CDC Climate: Extensive Report on Joint Implementation*. Joint Implementation Action Group, available at: http://www.jiactiongroup.com

Kettner, C., Köppl, A., Schleicher, S. and Thenius, G. (2008) 'Stringency and distribution in the EU Emissions Trading Scheme: first evidence', *Climate Policy*, 8: 41–61.

Klinsky, S. (2013) 'Bottom-up policy lessons emerging from the Western Climate Initiative's development challenges', *Climate Policy*, 13 (2).

Korppoo, A. (2007) *Joint Implementation in Russia and Ukraine: Review of Projects Submitted to JISC*, Climate Strategies Briefing Paper, October. Online at: http://www.climatestrategies.org/research/our-reports/category/19/12.html

Korppoo, A. and Gassan-Zade, O. (2011) *Dangers of the Endgame: Engaging Russia and Ukraine during the Gap*, FNI Climate Policy Perspectives 2. Lysaker: FNI.

Korppoo, A. and Moe, A. (2008) 'Joint Implementation in Ukraine: national benefits and implications for further climate pacts', *Climate Policy*, 8 (3): 305–16.

Kossoy, A. and Ambrosi, P. (2010) *State and Trends of the Carbon Market 2010*. World Bank, May.

Liu, X. (2010) 'Extracting the resource rent from the CDM projects: can the Chinese government do better?', *Energy Policy*, 38: 1004–9.

Meltzer, J. (2012) *Carbon Pricing in Australia: Lessons for the United States*, Brookings Institute Working Papers. Online at: http://www.brookings.edu/research/opinions/2012/07/02-carbon-australia-meltzer

Michaelowa, A. (1998) 'Joint Implementation – the baseline issue', *Global Environmental Change*, 8 (1): 81–92

Michaelowa, A. (2009) 'Interpreting the additionality of CDM projects: changes in additionality definitions and regulatory practices over time', in D. Freestone and C. Streck (eds), *Legal Aspects of Carbon Trading*. Oxford: Oxford University Press, pp. 248–71.

Michaelowa, A. and Purohit, P. (2007) 'CDM potential of bagasse cogeneration in India', *Energy Policy*, 35: 4779–98.

Michaelowa, A., Stronzik, M., Eckermann, F. and Hunt, A. (2003) 'Transaction costs of the Kyoto Mechanisms', *Climate Policy*, 3: 261–78.

Neuhoff, K. (2008) 'Learning by doing with constrained growth rates: an application to energy technology policy', *Energy Journal*, 29 (2): 165–82.

Neuhoff, K., Grubb, M. and Keats, K. (2005) *Impact of the Allowance Allocation on Prices and Efficiency*, Cambridge Working Papers in Economics 0552. Faculty of Economics, University of Cambridge.

New Carbon Finance (2009) 'Emissions from the EU ETS down 3% in 2008', *New Energy Finance*, 16 February.

OECD (2013) *Climate and carban: aligning prices and policies*. Environment Policy Paper no. 1. Paris: Organisation for Economic Cooperation and Development.

Olsen, K. H. (2007) 'The clean development mechanism's contribution to sustainable development: a review of the literature', *Climate Change*, 84: 59–73.

Olsen, K. H. and Fenhann, J. (2008) 'Sustainable development benefits of clean development mechanism projects: a new methodology for sustainability assessment based on text analysis of the project design documents submitted for validation', *Energy Policy*, 36 (8): 2773–84.

Parag, Y. and Fawcett, T. (2010) 'Personal carbon trading', *Climate Policy*, Special Issue, 10 (4): 329–38.

Parsons, J. E. and Taschini, L. (2013) The role of stocks and shocks concepts in the debate over price vs. quantity', *Environmental and Resource Economics*, 55: 71–86.

Paterson, M. (2012) 'Who and what are carbon markets for? Politics and the development of climate policy', *Climate Policy*, 12 (1): 82–97.

Phillips, L. (2009) 'EU emissions trading an "open door" for crime, Europol says', *EU Observer*, 10 December.

Point Carbon (2009a) *Carbon Market Europe*, 30 January.

Point Carbon (2009b) *CDM Host Country Rating*. Available at: http://www.pointcarbon.com/research/carbonmarketresearch/cdmhostcountryrating/cdm/

Point Carbon (2010) *Carbon Market Europe*, 30 January.

Raab, U. (2012) *Market Mechanisms – from CDM Towards a Global Carbon Market*, FORES Study No. 8. Stockholm: FORES.

Republic of Korea (2012) *Act on Allocation and Trading of GHG Emissions Allowances*, 2 May.

Rogge, K., Schimdt, T. and Schneider, M. (2011) *Relative Importance of Different Climate Policy Elements for Corporate Climate Innovation Activities: Findings for the Power Sector*, Working Paper, Berlin: Climate Policy Initiative and London: Climate Strategies, available at: http//:www.climatestrategies.org

Roy, J. (2010) *Iron and Steel Sectoral Approaches to the Mitigation of Climate Change*, Briefing Note, 9 December. London: Climate Strategies.

Roy, J., Dasgupta, S. and Chakravarty, D. (2013) 'Energy efficiency: technology, behavior, and development', in A. Goldthau (ed.), *The Handbook of Global Energy Policy*. London: Wiley-Blackwell.

Sartor, O. (2011) *Closing the Door to Fraud in the EU ETS*, Climate Brief No. 4. CDC Climate Research.

Schmalensee, R. and Stavins, R. N. (2013) 'The SO_2 allowance trading system: the ironic history of a grand policy experiment', *Journal of Economic Perspectives*, 27 (1): 103–22.

Schneider, L. (2007) *Is the CDM Fulfilling Its Environmental and Sustainable Development Objectives? An Evaluation of the CDM and Options for Improvement*. Berlin: Öko-Institut.

Sijm, J., Neuhoff, K. and Chen, Y. (2006) *CO_2 Cost Pass Through and Windfall Profits in the Power Sector*, Cambridge Working Papers in Economics 0639. Faculty of Economics, University of Cambridge.

Standard & Poor's (2009) *Europe – Standard & Poor's*. Standard & Poor's Financial Services LLC.

Sterner, T. and Muller, A. (2008) *Output and Abatement Effects of Allocation Readjustment in Permit Trade*, RFF DP 06-49. Washington, DC: Resources for the Future.

Tietenberg, T. H. (2006) *Emissions Trading – Principles and Practice*. Washington, DC: Resources for the Future.

UNEP Risø Centre (2012) *UNEP Risø CDM/JI Pipeline Analysis and Database*, UNEP.

UNEP/IUC/99/2, Geneva.

UNFCCC (1998) Kyoto Protocol to the United Nations Framework Convention on Climate Change.

UNFCCC (2011) *Benefits of the Clean Development Mechanism 2011*. UNFCCC.

Wara, M. (2008) *A Realistic Policy on International Carbon Offsets*, Program on Energy and Sustainable Development Working Paper No. 74. Stanford, CA: Stanford University Press.

Weber, T. A. and Neuhoff, K. (2010) 'Carbon markets and technological innovation', *Journal of Environmental Economics and Management*, 60 (2): 115–32.

Woerdman, E. and van der Gaast, W. (2001) 'Project-based emissions trading: the impact of institutional arrangements on cost-effectiveness', *Mitigation and Adaptation Strategies for Global Change*, 6 (2): 113–54.

World Bank (2012) *State and Trends of the Carbon Market Report 2012*. Washington, DC: World Bank.

Zhenchuk, M. (2012) *The Integrity of Joint Implementation Projects in Ukraine*. Kiev: National Ecological Centre of Ukraine.

8 Who's hit?

The distributional impacts of carbon pricing and how to handle them

Don't tax you, don't tax me –
Tax that fellow behind the tree ...
– Russell B. Long, US Democratic Senator. Cited in Mann (2003: 333)

8.1 Introduction: who pays?

If there is a price on carbon, someone has to pay it – indeed, that's the point. Market economies are efficient to the extent that people pay the real cost of something. But paying is never popular – particularly for something which, like CO_2 emissions, has previously been treated as free. Chapter 6 presented the evidence on why price matters and the basic economic case for pricing carbon. Long ago, the rich countries' club – the OECD – agreed the economic principle of 'polluter pays' (Chapter 1).

In the case of CO_2, we are all polluters. Moving from principle to practice is thus all the more difficult and raises the question of who really does, and should, pay: how much, and how they may respond.

In principle, those that make or use high-carbon products will face higher costs, and those that make or use low-carbon products will pay relatively less. The shift in prices helps to change behaviour but also can transfer wealth between different groups. Paying a carbon price at a level sufficient to make a difference – or remotely equivalent to most estimates of climate damages – can mean a lot of money changes hands. The value of all the emission allowances in the EU ETS, for example, amounts to tens of billions of euros annually.

Much of the early concern about carbon pricing was about the overall societal costs – the overall impact on GDP or welfare. In fact, as described in Chapter 6, these are relatively modest. Politically, the impact of carbon pricing on the 'pie' of national income, or national employment, is really a sideshow because it is likely to be dwarfed both by other factors – like the overall rate of economic growth – and by the scale of transfers between groups.

The most basic distributional fact is that *not* pricing carbon is itself a transfer, in this case avoiding costs to ourselves by imposing risks on our children. As noted in the previous chapters, however, that observation helps little when it comes to designing and implementing carbon pricing in practice. The more salient issues emerge about the impacts between industry and consumers, and between different industry sectors and groups of consumers. This chapter explores the distributional minefield of 'who pays' – which, as already noted, is the real driving force that determines what can and cannot be delivered – and explores how to chart a path through it.

8.2 Profit and loss with emissions trading

For most readers (and voters) the biggest interest lies in the final impacts of emission trading on consumers. But to get a full picture it is useful to start the journey 'upstream' – with the industrial emitters. They have been the focus of the most intense lobbying and economic worries (including those concerning 'carbon leakage'), and they affect how carbon costs may (or may not) flow through the economy to the final consumers.

As emphasised in Chapter 6, money does not disappear. With a carbon tax – or if governments auction emission allowances – the revenues go to the government. The cost incurred by industry will then be passed on to the products. We all then pay as consumers of energy (and energy-intensive products), while benefiting as citizens (e.g. through reduced other tax payments). These 'downstream' distributional implications are explored in section 8.5.

If emission allowances are given out for free to industry, the effects are (even) more complicated. As noted, giving out free emission allowances – mainly in a bid to avoid redistributing money – has been crucial to the initiation of all the trading schemes in existence. Chapter 7 explained how, despite the cutbacks imposed to comply with Kyoto targets, the evidence is that many companies, in all sectors in the EU ETS, have actually profited to date.

Profits can arise for two reasons. First, where a sector ends up with surplus allowances, it can sell these. Second, product prices may rise by more than the overall costs incurred. The EU ETS experience confirmed a general economic principle illustrated in Figure 8.1: the impact on profit or loss depends on the combination of free allocation (vertical axis) with how much carbon costs are passed through to the cost of the product (horizontal axis):

- If there is no free allocation and the company has to pay for all its emissions but does not pass any of the cost on to its products (for example due to intense foreign competition), then it sits in the bottom left-hand corner, with (in the example of steel illustrated) severely reduced profit margins.
- Conversely, if a company receives free allowances to cover almost all its emissions, but adds the full carbon *price* on to its products anyway, it will be in the top-right corner of the diagram, with hugely increased profits.

At first, the latter looks like blatant profiteering. The paradox is that it is *most likely to occur in competitive markets*, particularly in those with continuous competition based on short-run operating costs – like liberalised electricity markets. This is because in such markets, profit-maximising companies will tend to set their price in relation to short-run operating costs: the cost of producing an extra unit is balanced against the value of the additional sales. Anything less would imply operating at a loss. Capping emissions increases this *incremental* (marginal) cost, since companies either have to buy allowances, or forego the opportunity to sell allowances, to cover the extra emissions. Electricity economists have for decades extolled the virtue of such 'marginal cost pricing'. The result, providing that all directly competing companies face the same incentive, is that the final price will tend to rise to reflect this 'opportunity cost'.[1]

In the first two phases of the EU ETS, governments gave out free allowances to all sectors, expecting them in effect to sit at the upper left-hand point of Figure 8.1 – with minimal revenue transfers but still an incentive to cut emissions. In reality, competitive electricity markets in Europe moved to the top right, and generators profited by many billions of euros. It was this reality (far more than the potential efficiency losses associated with free allocation noted in Chapter 7), which underpinned a wholesale move in power generation from free to auctioned allowances in Phase III of the EU ETS (2013–20).[2]

The underlying principles, however, apply to any competitive market, subject to the constraints of competition from countries without carbon pricing. Given the numerous complications associated with free allocation, the economic ideal is to land at the bottom

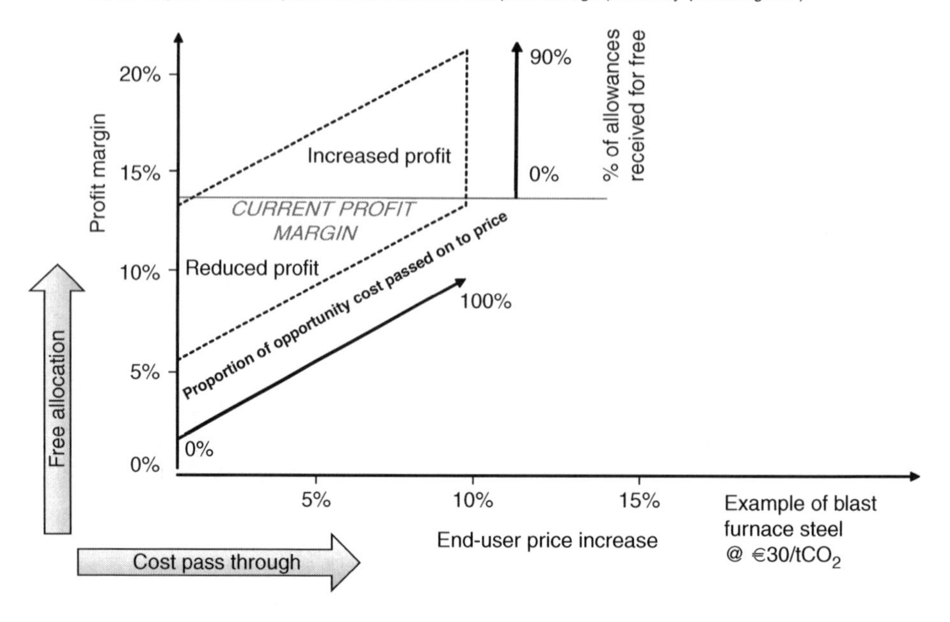

Figure 8.1 Profit and loss depend on the combination of free allocation and cost pass through to final products

Note: The numbers on the graph correspond to one of the most carbon-intensive products, namely steel, in Europe at a carbon price of €30/tCO2, before international effects.

right-hand corner of the graph. This is roughly where the EU ETS has arrived for power generation in Phase III. For other sectors, the story is altogether more complicated.

As noted, the impact of a carbon price depends not only on the degree of free allocation, but on whether (or how much) the price is passed through to products. Figure 8.2 illustrates three key processes.

On the *industry* ('supply') side, to the left, a carbon price creates incentives for manufacturers to cut carbon emissions (e.g. improve efficiency, find alternative fuels) and to innovate better ways of doing this.

If costs are passed through to the *consumer* ('demand') side, to the right, the carbon price creates incentives for consumers to use lower-carbon products and services and for companies serving those markets to create products to meet this demand.

With the aim of curtailing emissions, these are all desirable outcomes. But in the middle are the potential undesirable side effects if carbon prices differ across regions. If industries do not pass through their costs to product prices, profitability declines and investment is driven away. If industries do pass through costs, their products may be more expensive than those of their competitors producing in other regions and they lose ground to imports (or lose exports). Either raises the spectre of 'carbon leakage', of *investment* or *operation* respectively.

Numerous academic studies have sought to estimate the scale of potential carbon leakage. Most suggest the overall numbers to be small – certainly far less than the scale of expected emission savings.

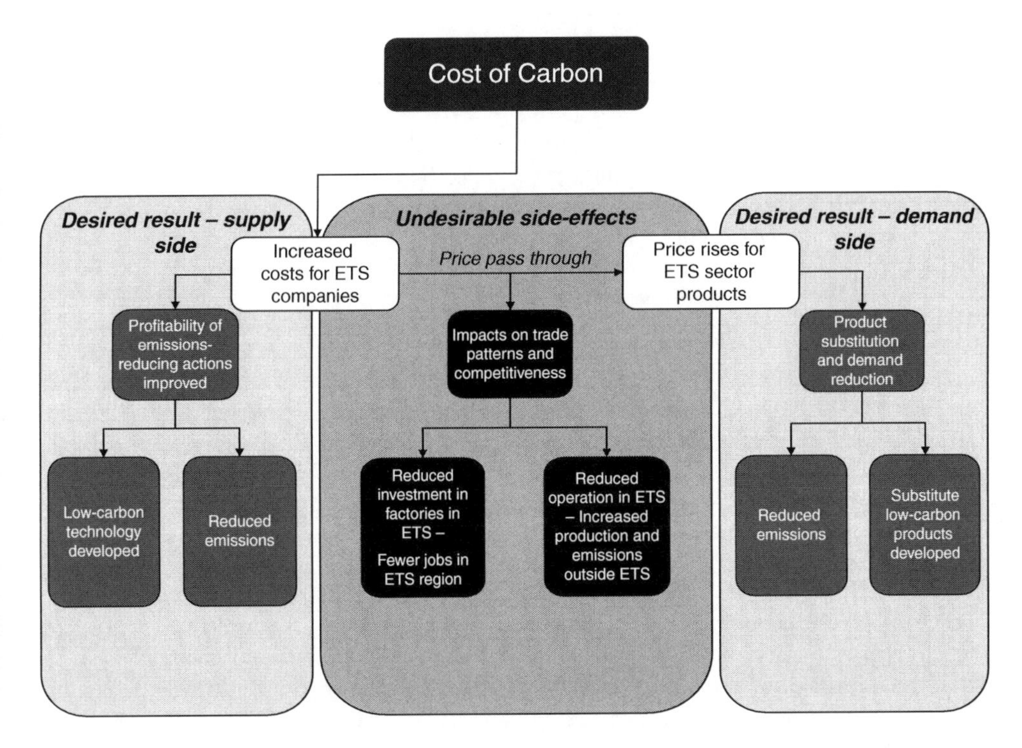

Figure 8.2 Desirable and undesirable effects of carbon prices

Source: Adapted from Carbon Trust (2010).

However, the limited scale of potential carbon leakage is not the issue. No politician, however committed, is likely to defend a system that may result in reduced exports or investment moved abroad – lost value and jobs, but without saving emissions. Politically, it is dynamite. No carbon pricing system in the real world has made progress by dismissing it as a small issue: indeed, it has been probably the single most important topic, and has led to special treatment of internationally competitive industries in almost all cases.

At the same time, risk of carbon leakage differs hugely between sectors. The key to a sensible policy lies in understanding these differences. The cost incurred by a sector is the obvious driver for such leakage concerns. But the cost alone can only serve as a first screen for identifying 'leakage risks'. This is obvious when considering the electricity sector. Carbon costs are unquestionably important here, but in general the industry cannot 'move' – UK consumers cannot buy electricity from Russia, the US is unlikely to dramatically increase electricity imports in response to a carbon price. The industry can switch to lower carbon sources and can pass the costs on to consumers without losing out to imports.

More generally, the trade world is not flat. Local production can often command some premium compared to imports, as they can garner customer trust, local market specificities and favourable regulation. Imported goods, in contrast, face additional costs such as transportation and handling costs, and exchange rate risks. Location-specific factors may mean that a sector could lose more money by relocating than it could save in reduced carbon costs. Transport infrastructure may be inadequate. Legal, political and other regulations in areas outside the carbon price may be unfavourable. All these factors can impede imports, inhibit the scope for relocation and increase the scope to for any added costs to be passed through in the market. However, when cost differentials are considerable – like the labour cost differences that have driven outsourcing of a lot of industry (and some services) to emerging economies – the costs may exceed these barriers.

Thus solid numbers and analyses are indispensable. We previously explained that companies have the *potential* to profit from emissions trading if they are given free allowances but still pass the carbon costs through. Whether or not they can actually do so essentially depends on how big the carbon costs are relative to other costs, differentials and the barriers as illustrated in Figure 8.3:

- If the goods they produce have no competitors and their customers do not respond much to price rises, then industry can (and does) pass on the costs to consumers. The industry will then tend to profit from free allocation, and may make even more money in this case if it can reduce its emissions.
- If there is stiff competition from other producers with lower carbon costs, or if consumers can easily move away, then carbon-intensive producers would not be able to pass on much of the carbon cost.

Thus for clearly differentiated products with few competitors, of modest carbon intensity or for products otherwise protected from foreign competition, it is likely that costs can be passed through. There may be a case for transitional free allocation or thresholds for a limited time, to partially compensate for 'stranded assets' from carbon-intensive investments made before the investors could reasonably have anticipated carbon pricing. Economically, there is no other, or enduring, reason why producers of such profits should be protected from paying full carbon costs.

However, companies producing homogenous goods with lots of competition that does not face a carbon cost may struggle to pass through any additional cost. In this case firms either have to absorb the extra costs themselves, reducing their profits, or pass on the costs and risk losing sales and market share. This is a more enduring problem.

Chapters 6 and 7 illustrated how industrial opposition to carbon pricing has been fierce and was indeed one of the key reasons why the European carbon tax and the US BTU tax floundered in the 1990s. Thus understanding what types of industry will be affected by carbon pricing – and how – is crucial.

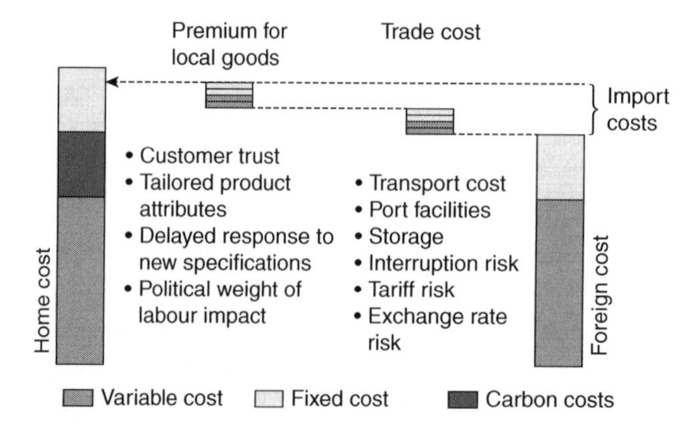

Figure 8.3 Carbon cost impact, local premium and trade cost

Source: Neuhoff (2008).

8.3 Industry impacts – winners, losers and movers

In analysing profit and loss under carbon pricing, the previous section pointed to two apparent paradoxes.

The first is that many academic studies of the risk of 'industrial carbon leakage' have suggested the scale of the risk to be small, and yet as noted (also in the previous chapters) fear of carbon leakage has been a huge obstacle in every attempt to implement carbon pricing.[3] The second paradox is that many sectors seem to have the potential to profit from carbon cap-and-trade systems, and yet most (though not all) industry has opposed them.

This section explains these paradoxes by looking more closely at the real structure and data around the 'carbon leakage' problem. Indeed, data are indispensable in making sense of 'carbon leakage'. The *potential* significance of a carbon cost is best measured by its scale relative to the sector's 'gross value-added' (GVA) – broadly, the difference between the money it gets through sales and the cost of its material inputs.[4]

Figure 8.4 lays this out for all the 27 Member States of the European Union, by ranking industries in terms of the costs if companies had to pay for all their carbon emissions – and the associated impacts on electricity costs – at a price of €30/tCO_2, which is the highest level reached in the EU ETS. The horizontal axis shows their GVA – how much these activities contribute directly to EU GDP.

Just a handful of industries really figure. Those for which a price of €30/tCO_2 increases costs by 5 per cent or more (relative to value added) represent about 2 per cent of the European economy – but these account for the lion's share of overall manufacturing emissions.

That is not just a European result: industrial emissions are very concentrated in the primary commodity industries that account for a very small part of 'value added' in industrial economies. There are some differences between countries in terms of scale and ranking, but the most cost-impacted activities are consistent worldwide. Together with electricity production itself, around half a dozen big material- and chemical-based activities stand out as being particularly impacted by carbon costs:

- iron and steel
- aluminium
- refining
- cement and lime
- basic inorganic chemicals (principally chlorine and alkalines)
- pulp and paper.

In terms of carbon intensity, fertiliser production also ranks among this top group, and there are about half a dozen other, smaller sectors – like glass – next in line, though these are small by comparison in both total emissions and 'value at stake'.

The 'Big Six' or – depending on how one chooses to draw the boundaries and labels – 'Dirty Dozen' commodity industries account for most of manufacturing emissions, and typically 20 per cent or more of national emissions. But they frequently account for only a couple of per cent of industrialised country economic output, and less than 1 per cent of jobs, because they are very resource- and capital-intensive sectors. The shares are bigger in the emerging economies – in China (one of the most energy- and manufacturing-intensive economies in the world), the corresponding share of GDP in such energy-intensive activities is closer to 10 per cent.

Of course, our economies depend on the products from these basic commodity industries. Moreover, there may be significant economic linkages between these and more 'downstream'

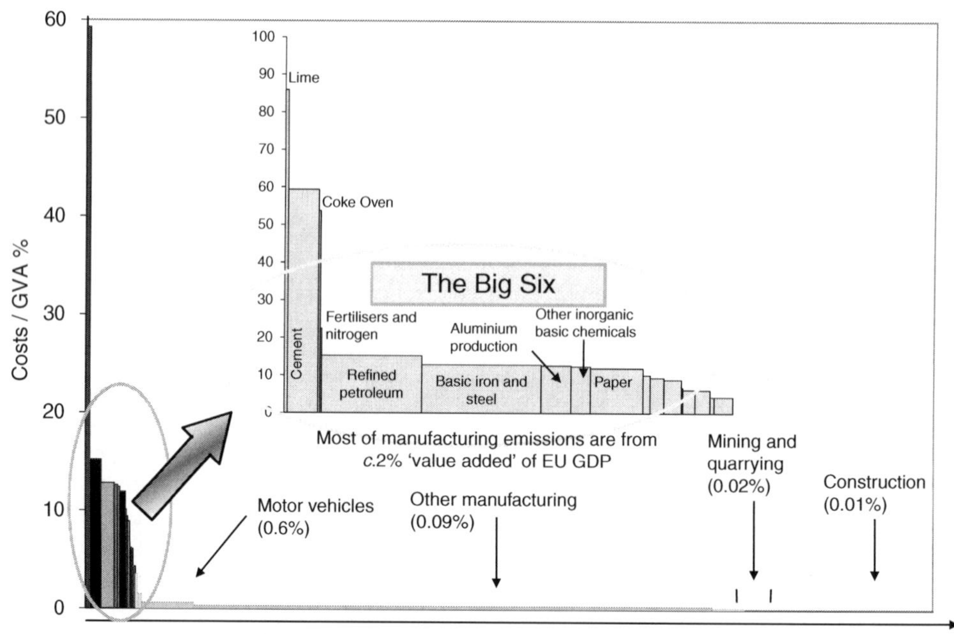

Figure 8.4 Impact of carbon pricing on EU industry sectors and their share of the EU economy

Note: The horizontal axis indicates the scale of the sector's Gross Value Added (GVA) – its contribution to GDP. The vertical axis shows the implied cost increase if sectors pay the full cost of CO2 at €30/tCO2 (around US$40/tCO2), as a percentage of the sector value added). The area of each column is thus proportional to total CO2 emissions. Note that electricity is shown as a separate sector in addition to its carbon costs being added to the other sectors.

Source: Generated from 'Results of the Quantitative Assessment of Sectors at NACE 4 Level'. Accessed from http://ec.europa.eu/environment/climat/emission/carbon_en.htm. The original data for the UK, including the contribution of electricity, are in Climate Strategies: Hourcade *et al.* (2007) and Carbon Trust (2008). Parallel US and Japanese data are in Grubb *et al.* (2009). For similar US findings see also Aldy and Pizer (2011).

sectors. Vehicle manufacturing, for example, depends on steel and aluminium inputs. If there is tight local linkage between these activities, the combined 'value at stake' will be bigger, though by the same token it will be harder for the primary activity to move.

This helps to explain the first apparent paradox. Most academic studies of 'carbon leakage' over the past 20 years have either looked for historical evidence of 'pollution havens' or have used global economic models originally developed for other purposes.

Examples of industrial migration to escape pollution controls to date are extremely limited, even for heavy industry – a long history of economic research into fears of 'pollution havens' has found precious little evidence.[5] But that may be partly because the fear of job and emission leakage is so powerful, that governments offer exemptions or other measures that prevent industries seeing the costs. And as Figure 8.4 suggests, carbon emissions are exceptionally concentrated in a few bulk commodities; evidence from other pollution control may thus be of limited relevance.

The economic models employed to *project* leakage, in turn, mostly aggregated sectors on the basis of economic value. From this perspective, something like cement is simply far too small to bother with. Indeed, often such models simply lump all manufacturing industries together. Not surprisingly, they then conclude that the cost differential – and hence the scale of industrial leakage – would be modest (see notes 17–19 concerning other modes of leakage).

Some of the more perceptive studies noted that although their share of total production value is limited, these carbon-intensive activities are nevertheless an important part of emissions – and hence abatement opportunity – and hence that leakage can be a significant problem.[6] That, combined with the underlying politics, is at the heart of the challenge. The data point to half a dozen energy-intensive activities which dominate the overall potential to significantly win, lose or move in response to carbon pricing.

Figure 8.4 itself does not give an insight into either the potential barriers to moving or the potential to cut emissions. If there are lots of cheap abatement options, the sector could merely cut its carbon emissions and reduce its exposure to carbon prices. In practice, as we have discussed in Chapter 3, the abatement potential in emission-intensive sectors is limited, at least in the short run. In the long run, new technologies and options can offer larger abatement potential. This illustrates the importance of the time dimension – both for corporate decision-making and for the assessment of the positive impacts and potential side effects of policy instruments.

'Moving' involves carbon leakage – carbon emissions that effectively escape the carbon price. As noted, losing out on trade or investment – *without even saving emissions* – is political dynamite, particularly for politicians whose constituencies include one of the heavy industries with most at stake.

As noted, the profit, loss or leakage risk of a sector depends on several combined factors:[7]

- the impact of carbon prices upon the cost structure;
- the level of free allowances (or exemption); and
- the ability to pass on the carbon costs to consumers in the light of competitors that are not exposed to a similar carbon cost;
- the abatement potential;
- the regulatory and legal framework.

As a proxy for all these factors, the EU ETS classifies sectors 'at risk of leakage' based on two simple metrics, namely carbon cost intensity and trade intensity.[8] The latter is intended to represent the extent to which they compete in the same market as firms without carbon costs. The EU classes a sector as 'at risk', and thus qualified for free allowances, if either its costs or trade exposure are high, or a combination of both. The sector then qualifies for leakage protection – notably through free allowance allocation.

With a carbon price in Europe since 2005, does the evidence suggest that fears of carbon leakage may be justified? The problem may have been exaggerated by industry, and the criteria for assessing sectors used by the EU may be too lax. Indeed, the rules have already been subject to strong criticism, as the small impact of carbon costs typical of most of manufacturing (as shown on Figure 8.4) is unlikely to have much of an impact on trade. This is supported by survey evidence that the problem has been exaggerated in the EU industrial debate, where far more sectors have been classified as 'exposed' than can be justified.[9] In fact, all sectors in the ETS have in aggregate profited, as noted in Chapter 7.

Why then has industry often opposed the system? This question has many possible answers. No producer likes to see its input costs increase, even if they can pass the costs on (customers will still blame them). Also, few sectors really understood or believed the mechanics that could lead to profits – and as noted, they were in good company (Chapter 7, note 29). Most were also of course bullish about their growth prospects, and thus did not expect to end up with the large volume of surplus allowances that have led to their becoming dubbed the 'fat cats' of the EU ETS.[10]

Yet there is an additional factor, which perhaps more than any other explains the paradox of industries opposing something from which they have so far profited. These industries are all long-term and capital-intensive. They want to invest. In the European system, though, they do not really know the future rules. Despite being classified as 'at risk of carbon leakage' – or, arguably, because of it – the industries have little clue about what the rules might be in the next phase of the EU ETS after 2020. Witnessing the public anger at the profits they have made to date, they are not sure free allocation will last. Hence, despite short-term profits they do not know or struggle to judge the longer-term future. Other regions – growing faster, with more demand and without a carbon pricing system yet in place – seem less risky.

Hence, despite all the giveaways, carbon leakage may occur not by a conscious decision to relocate, but by a slow cumulative process of mutual dithering on the part of industry and governments combined. Dispensing free allowances has serious drawbacks; uncertainty about whether they will continue in future periods just amplifies this further. At present, therefore, the approach to carbon leakage risks being the worst of all worlds.

8.4 Tackling 'carbon leakage' – an evolutionary approach to production, consumption and trade

So carbon leakage may be a problem in a world of unequal carbon prices, but it is likely only to affect a few sectors. It does not justify failing to price carbon, but in the big, trade-exposed sector, we need to think about systematic solutions.

The options fall logically into three main categories: adjusting the carbon costs downwards for those firms within the pricing regime, adjusting the costs upwards for those firms outside the scheme, or adjusting the costs at the border for imports and exports (Figure 8.5).

As noted, the main mechanism to address leakage concerns in the EU ETS used to date is free allocation. With these allocations now in most cases fixed out to 2020, the implications are surprisingly complex.[11] Firms could have an incentive merely to sell the free allocation and move abroad anyway. To stop this, 'closure rules' prevent an installation which closes receiving further allowances – which creates an incentive to keep old plants in operation.

Using free allocation to help solve the leakage problem is far from ideal, as we saw in Chapter 7. If costs can in fact be passed through by firms and sectors receiving free allowances then they make windfall profits. And if costs cannot be passed through by firms, the incentives to shift the consumption towards less carbonised products is lost. The effects (and effectiveness) of free allocation will vary between sectors.

More generally, Figure 8.2 reminds us why it is ultimately good to pay for pollution. On the left-hand side of the diagram, making producers pay creates the incentives to cut their emissions and to innovate cleaner production processes. On the right-hand side, passing these costs through to consumers sends signals to consumers to use less of such polluting goods and makes less polluting substitutes more profitable. But in the middle of the diagram, leakage is an unintended and undesirable side effect. The key to good policy is to deter leakage, without losing the environmental benefits on either the supply (left) or demand (right) side.

There are other possible options for reducing the carbon costs for firms in the system. Direct subsidies could be given to those who invest in new low-carbon technologies in exposed sectors; other taxes (like corporation tax) could be reduced for those sectors; or costs could be compensated directly from governmental budgets. As noted in Chapter 6,

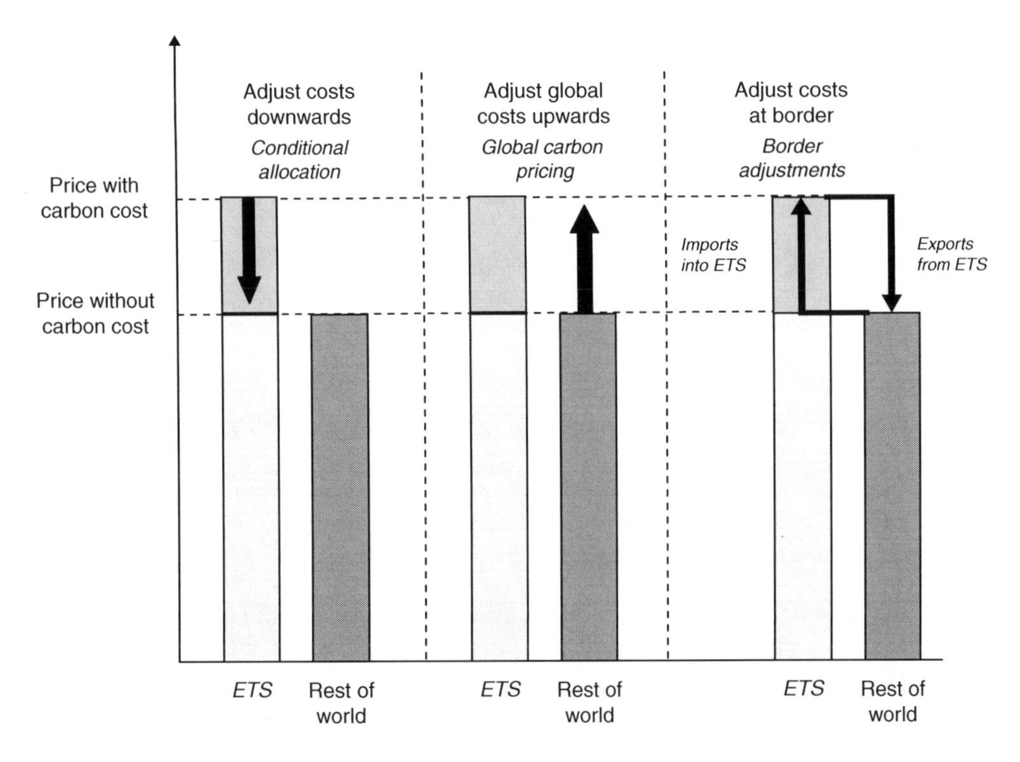

Figure 8.5 Options for tackling leakage

Source: Adapted from Neuhoff (2008).

indeed, smart use of revenues may help to alleviate a number of the potential adverse economic consequences of carbon pricing.

The second class of options illustrated in Figure 8.5 is to adjust upward the carbon price of producing the same goods in the rest of the world. Unfortunately, the discussion of Chapter 7 suggests how fantastically difficulty this would be. In the first decade after adopting the Kyoto Protocol, there was much talk of 'sectoral agreements' to try and solve the problem of competitiveness between regions taking action and those not doing so. It turned out that the one thing such 'global sectoral agreements' were most unlikely to do was to equalise the carbon costs faced by industry – at least at any level greater than zero.

As noted, pricing carbon is highly political and relies on complex domestic legislation. Even the system so far adopted – the various carbon taxes and the existing and emerging carbon trading schemes – are all different. They all developed in different ways, according to the progress of domestic politics. The idea of coordinating a carbon price globally, applied at similar levels to precisely those industries with most at stake, has proven to be a complete mirage. Indeed even its theoretical foundations are shaky, as outlined in the concluding section of this chapter.

The third broad option is to ensure that all *consumption* of carbon-intensive goods in a region face a carbon cost, while exports to regions which refuse to put a price on CO_2 emissions are reimbursed for CO_2 costs incurred in production. One way or another this implies levelling costs 'at the border', so that exports out of the carbon price regime have their costs reduced, while imports into the area incur the carbon price. Instruments to implement such adjustments could take a number of different forms, at least theoretically.[12] Firms that import into the scheme could be required to purchase allowances up to the carbon emissions associated with their product, while firms that export could be allowed to discount the emissions from their exports from the allowances they need to surrender. Or the adjustments could be made in the form of a tax, with importers paying a tax based upon their carbon emissions while exporters would receive a rebate. Or producers could be given free allowances while consumers pay a carbon-related charge at the point of consumption, irrespective of where a good was produced.

All the options raise some legal, political and administrative questions. The main legal questions that arise surround the rules of the World Trade Organization (WTO). There has been a wide discussion on whether border adjustments could be made compatible with WTO rules and the conclusion is, 'it depends' on the detail of implementation and the justification.[13]

A move toward levelling carbon costs at the border means that those in the carbon pricing region start to pay a price for the carbon emitted in producing the goods they consume, wherever they are made. Moves in this direction ultimately mean that consumers pay for the carbon 'embodied' in the products they buy and use, irrespective of their origin. Anything else, in practice, requires consumers in a given region to discriminate against their own producers – who pay for carbon in ways that foreign competitors do not. Given that trade flows are increasing around the world, with increasingly more products crossing borders at least once during their production, the embodied carbon consumed by a country (its 'footprint') is increasingly divergent from what it produces. Levelling carbon costs at the border ultimately embodies a principle of 'consumer responsibility'.

As emphasised at the start of Chapter 3, global energy and environmental problems are ultimately driven by consumption – particularly for richer consumers. A shift towards 'consumer responsibility' and associated pricing structures thus makes impeccable logic in principle and would resolve major practical problems. Unfortunately, there is considerable devil in the detail.

Where to start and how?

Despite the theoretical attractions, this shift is likely to be a long, complex and arduous journey, given the legal, political and administrative minefields of doing anything at the border. One potential approach might be to base a border adjustment (whether a tax or requiring importers to purchase permits) on a 'benchmark' – a standardised rate based, for example, on the best available technology for producing the product. This would imply that no domestic product faces lower obligations than imports (satisfying the WTO principle of 'National Treatment'), but still much reduces incentives for relocation to avoid carbon prices.[14]

A variant on this theme would be to set the 'default' adjustment at some average level but with a 'right-to-refute' whereby importers prove that their emissions are lower than the default level.[15] This could create incentives also for producers outside the scheme to reduce their emissions and produce the associated evidence of lower-than-average emissions. However, to the extent that sufficient new and usually efficient plants for energy-intensive commodities are located outside of the scheme, the differentiated import adjustments might encourage the new and efficient plants to dedicate their output to exports with other plants serving local demand. In this case transport volumes might increase with little additional incentives for emission reductions.

The challenge for simple, non-discriminating border adjustments is probably more political than legal in nature, and imbued with mutual suspicion and fear about the potential for measures to be abused or distorted into protectionist or discriminatory means. More complex and more far-reaching objectives associated with border levelling, proposed to create incentives for foreign producers to reduce emissions (or to punish countries not complying with international expectations), play into these fears. Limiting the use of border levelling to narrowly defined objectives, and implementing a full carbon price for domestic consumers and manufacturers without discriminating against foreign producers could facilitate acceptance, communication and understanding.

As noted, the most appropriate option to address leakages may vary from sector to sector. Free allocation may reduce some of the incentives to move to low-carbon production and may lead to windfall profits. Coordination across all countries to implement a similar carbon price for industry is an ideal to pursue in at least some sectors, but achieving (and implementing) such a global agreement seems improbably hard and certainly slow. In the absence of this, border levelling may come in many different forms, some of them compatible with WTO rules at least for – from an emissions perspective – homogenous goods, such as clinker or steel.[16] This might offer solutions for some of the major 'at risk' sectors.

Finally, our account would be incomplete without noting two other channels of 'carbon leakage' in a world of unequal efforts, one with predominantly negative environmental effects and one with potentially positive effects.[17] Neither, in fact, is driven by carbon pricing itself; rather they concern more general side effects of almost any way of tackling emissions.

The first additional leakage channel concerns fossil fuel prices. If actions in one region reduce its fossil fuel consumption, this reduces the pressure on supply and hence global fossil fuel prices – which may boost demand for fossil fuels (and emissions) in the rest of the world. This is, in effect, a generalised, global version of the 'energy efficiency rebound' discussed in Chapter 5. It became known as the 'green paradox' and generated considerable academic interest.[18]

Many of the claims around the 'green paradox' seemed poorly grounded in essential facts; it is unclear whether the simple story holds true in a more detailed analysis. To the extent that policy initially encourages a shift from coal to gas consumption, it could also increase gas prices. For coal, the global resources base is huge, hence it is unclear how much a reduction in demand would in the longer run reduce extraction costs. In the mid-term, uncertainties about potential coal demand could reduce investments in coal mining, and thus even increase scarcity of production capacity and coal prices. Thus concerns about leakage through the 'fossil fuel channel' have focused particularly on oil – the dominant source of energy for road, aviation and maritime transport. We briefly return to the role of oil in Chapter 10.

The second additional leakage channel relates to technology innovation and diffusion. Regions tackling CO_2 emissions are likely to accelerate innovation in low-carbon technologies and processes, as discussed in Chapter 9.[19] Other regions may adopt these new technologies as their costs decline. This 'technology' leakage yields positive environmental benefits to the world as a whole.

Essentially, if industrial relocation can be prevented without wrecking the incentives to get cleaner, the net result will be determined by the balance between these two other channels. It is then a question of whether enough of the world can move onto a low-carbon path quickly enough, and with sufficient innovation, for the low-carbon economy to dominate – or whether their reduction in fossil fuel consumption and resulting emissions is picked up by other regions. The innovation dimension is the subject of Pillar III (Chapters 9–11). The answer will also depend in part upon how we treat the rest of our economies – the 95 per cent or more of economic activity that is not, for the present, plausibly exposed to industrial carbon leakage.

8.5 Consumer price impacts

Carbon policy is likely to affect households by raising the price of energy and other carbon-intensive goods and services. Chapter 6 noted that countries already have hugely differing levels of taxes on gasoline. Any carbon price being actively considered at present, in comparison, would be pretty trivial. Also as noted, most of the actual 'cap-and-trade' systems in existence focus on power generation and industry, and to date the latter have largely been sheltered by free allowances. Hence, the main focus here is on electricity and gas prices and their distributional impacts.[20]

Three main factors determine the impact of carbon policy on electricity bills. One is the interaction of any free allocation or threshold with the regulatory structure. (As explained in section 8.2, a competitive electricity market will tend to pass through carbon costs irrespective of free allocation, which is unlikely to be allowed where prices are regulated.) Second is the mix of power sources. Countries which generate most of their electricity from coal – like India, China and Australia – will face a bigger price impact than those based on hydro-electricity or nuclear, like Brazil, France or Sweden. Third is the potential impact of *other policies*, which are often charged through to – but also may benefit – the final consumer.

Many countries still have highly regulated electricity markets and retail prices are often subsidised (think back to the discussion in Chapter 6). Handing out free emission allowances may in these cases shield the consumer. The same is true where utilities use a 'shadow' – hypothetical carbon price – to help guide investments, as trialled by South Africa's state monopoly, ESKOM. However, a real carbon price, operating in a competitive electricity market, will be passed to the consumer, along with the cost of other programmes.

The UK has a relatively diverse mix of power generation sources, a wide range of environmental programmes and a mixed building stock. The impact of these factors on bills has been extensively scrutinised and makes for a useful case study. At the time of highest EU ETS prices in 2009, the average impact on household bills was estimated at £24 – £31 per average household, 5–10 per cent of electricity bills and 2–4 per cent of overall energy bills.[21]

Energy bills also reflect other environmental policies, including programmes to support the efficiency improvement of buildings (Chapter 4) and deployment of renewable technologies (Chapter 9). Such programmes have ramped up in the UK and the most recent and detailed data are shown in Figure 8.6. The price impact of programmes related to domestic gas (which is not affected by the EU ETS) was about 4 per cent. Paying for the energy efficiency and renewables programmes, outlined in Chapters 5 and 9 respectively, each added about 6.5 per cent to the price of electricity, with the EU ETS having shrunk to a negligible portion of wholesale electricity costs given the collapse of carbon prices (and rising gas prices).

Of course these other programmes also bring benefits. Efficiency improvements reduce the amount of energy used by domestic consumers (thus also their future electricity bill), and renewable energy can reduce the *wholesale* power price. The extent to which higher *prices* can be offset by more efficient usage is crucial in the assessment of the overall impact on bills, as illustrated by two contrasting UK studies. One claimed that climate-related policies in 2020 could raise annual bills by £268 – £435 (with the EU ETS accounting for just £40 to £60 of this increase).[22] An alternative governmental study estimated the increase at just £13 (or 1 per cent of total bills).[23] The results of the latter study are not due to lower prices[24] but projections that bills are reduced through lower energy use – partly as a result of price effects and partly due to energy efficiency policies.

This result points to the importance of a holistic view, as underlined by the German situation. Germany is known for its high domestic electricity prices – which include the funding

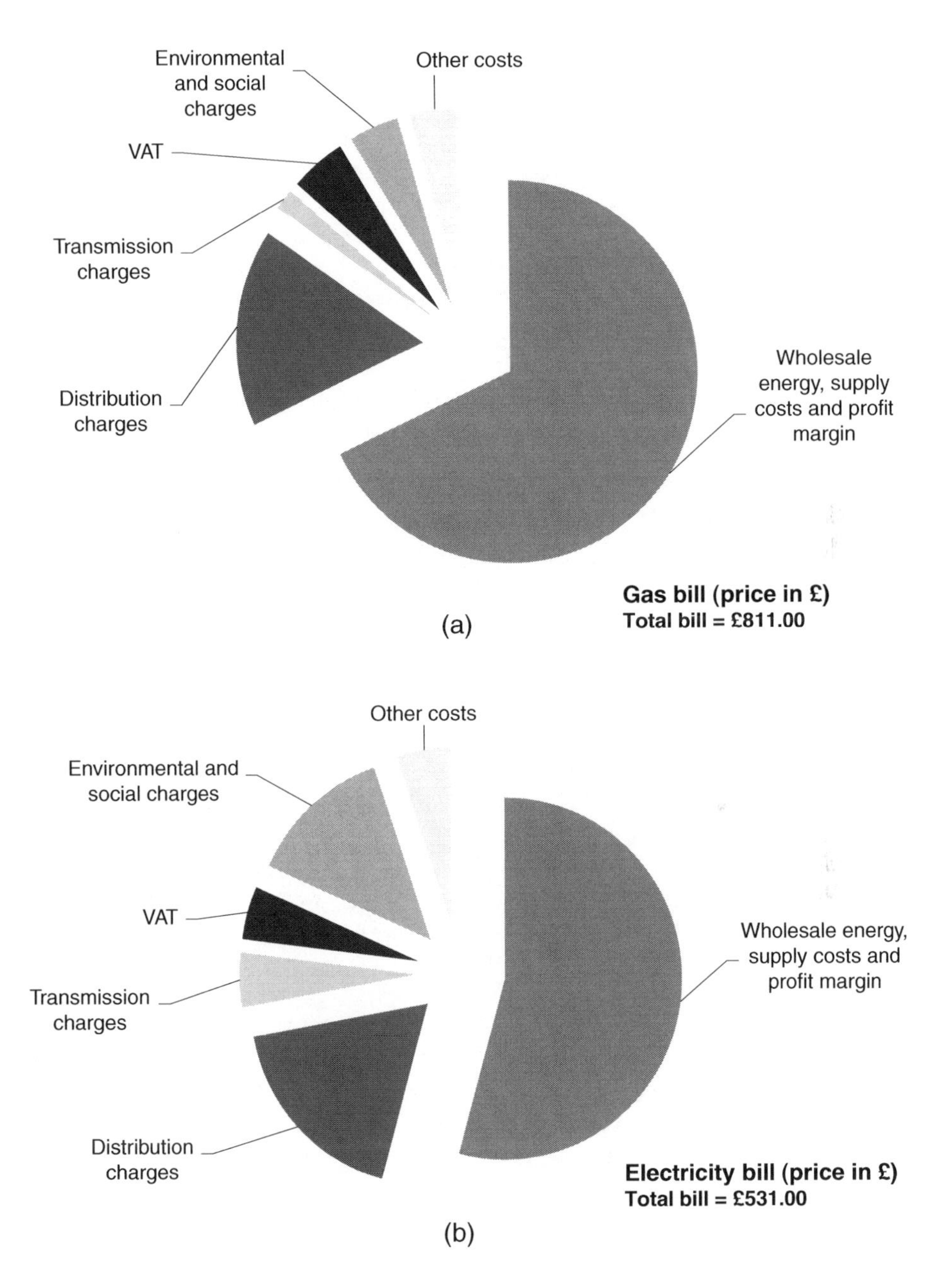

Figure 8.6 Breakdown of average domestic (a) gas and (b) electricity bills in the UK, 2013

Source: Data supplied by Ofgem, 2013.

of strong environmental policies – and yet the health of the German economy combined with the impact of strong energy efficiency programmes means that the average household bill in 2010 was 2.5 per cent of one's income, compared to 2.3 per cent twenty years earlier. All this is consistent with 'Bashmakov's constant' (Chapter 1) and further underlines the broad message of 'the most important diagram in energy economics' (Figure 6.1): higher prices do not necessarily translate into higher bills in the long run.

In the US, the Energy Information Administration estimated that the proposed Waxman Markey bill could increase the consumer price index for energy by between 8 per cent and 62 per cent in 2030 compared to reference scenarios.[25] The nature of the allocation proposed in the system, with energy distributors receiving free allowances in early years of the scheme, implies that price increases for consumers would be smaller in the early stages.

8.6 Who's hurt (and what to do about it)?

Although carbon-related policies to date have only increased energy prices by a few per cent, the impact will increase as policies tighten. As indicated, the overall impact will depend on a number of factors, such as how much energy you use, the percentage of income spent on energy, which options are available to reduce your energy demand, and whether you can access low-carbon energy without paying such charges.

The distributional challenge is that energy price rises are generally 'regressive' – they hit the poorest in society relatively more than the rich.[26] Energy serves many basic needs – keeping us warm, cooking our food and lighting our houses. This implies that we all have a basic level of the energy services that we require. As we get richer, we might purchase more electrical appliances or desire more heat (or cooling) and use more energy, but this generally accounts for a smaller *share* of income.[27] Hence the poor generally spend more of their income on energy than the rich. This is illustrated by the shares of household expenditure on energy in the UK, where the poorest 20 per cent of the population spend half as much on energy as the richest 20 per cent – but that accounts for almost three times as much as a proportion of their overall expenditure on (Figure 8.7).

In some countries, gasoline prices differ from this trend in that often the poorest people do not own a car and are thus less affected.[28]

The amount of income spent on energy does not just depend on your income level, however. The energy efficiency and size of houses plays a crucial role. Chapter 3 noted that the UK has a particularly inefficient building stock and this, combined with rising prices, has fuelled a big debate on 'fuel poverty'. In the UK, a household was defined as being in fuel poverty 'if it needs to spend more than 10 per cent of its income on fuel to maintain a satisfactory heating regime'. In 2008 4.5 million households in the UK fell into this category (about 15 per cent of the total); 80 per cent of them were among the 20 per cent of households with the lowest income.[29]

The distributional impacts were made starker by the types of houses inhabited by the 'fuel poor'. Almost 90 per cent of the 'fuel poor' live in the worst three categories (E–G) of the UK's Energy Performance Certificate ratings (Figure 8.8). Thus any measures which increase electricity and gas prices are likely to increase the absolute numbers of the fuel poor.[30]

'Fuel poverty' is an arbitrary definition but a telling condition. It arises as a result of actual poverty and/or poor housing conditions, which are often interlinked. The poorest members of society may lack the income and access to capital to upgrade housing stock and install energy-saving measures like insulation without strong governmental support. The poor also often use the least efficient appliances.[31]

The problems are particularly stark in Central and Eastern Europe, where the removal of subsidies involved a big increase in fuel prices in countries which had inherent poor building stock and also suffer severe winters. In countries like Hungary, nominal gas prices more than doubled between 2000 and 2007 and electricity prices rose by 75 per cent.[32]

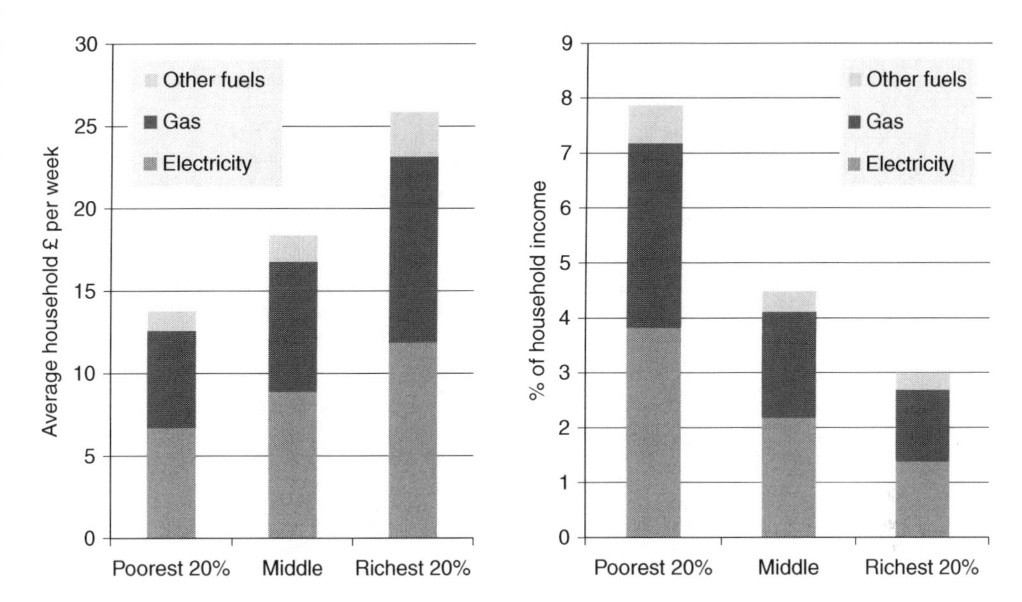

Figure 8.7 Household energy expenditure in the UK in different income groups, 2008

Source: Office of National Statistics (2009).

The inefficient infrastructure that had grown up around cheap energy prices also meant that huge amounts of energy were, and still are, wasted. This led to increasing numbers of people who were unable to afford the rising energy bills needed to heat their houses adequately. By 2005 almost 15 per cent of the Hungarian population could not afford to heat their home adequately, and almost 17 per cent were in arrears with their utility bills.[33] These trends are common across many Central and Eastern European states, with the poorest 10 per cent of households spending more than 10 per cent of their income on electricity alone in Croatia, Georgia and the Slovak Republic.[34]

Critics also point out that the same approach could be applied to many other issues (e.g. 'food poverty'). For both good and bad, it does serve to focus attention on the distributional impacts and has helped to make it a political issue – and not just in the UK. In France, for example, an estimated 10 per cent of the population (3 million households) suffer from fuel poverty.[35] It is a matter of debate whether the index provides a political rod used by opponents in an attempt to stop efforts to clean up energy, or as a useful indicator for governments when the pain of a policy may be greater than some parts of society should bear. It does both, and raises the question of what to do about it.

Subsidise? The first and most obvious solution would be to subsidise energy prices for the most vulnerable. The drawbacks have, however, been noted in the discussion of subsidies in Chapter 6. Consumption subsidies create incentives to use more energy less efficiently. Clear evidence can be seen from Germany, where welfare recipients' heating expenditures are completely covered by the government and the heating costs are 7–8 per cent higher than average households.[36]

Change the tariff structure? A 'rising block tariff' would charge less for the first chunk of electricity – enough to meet basic needs – with a higher price as consumption increases. Thus the incentive to save energy is maintained once consumption goes beyond basic needs.

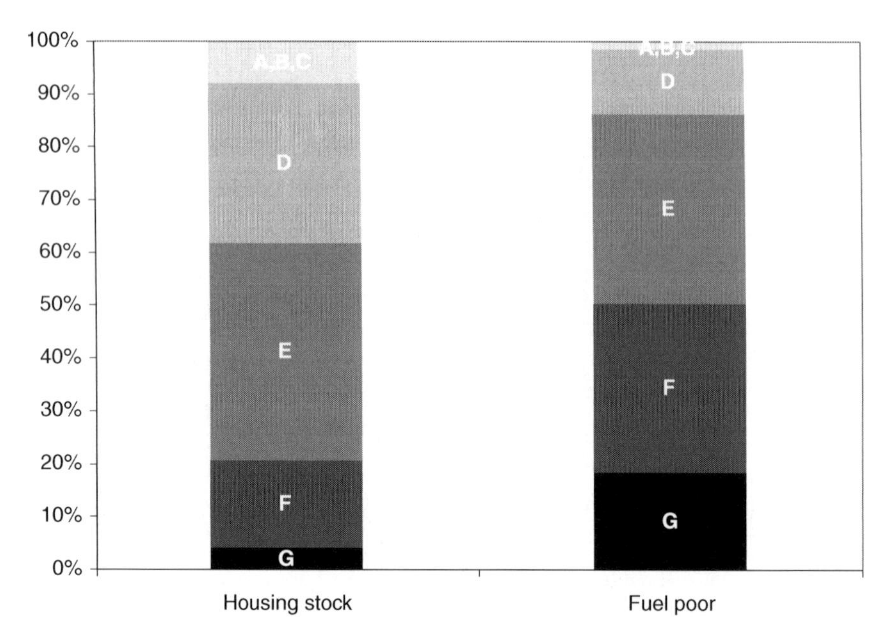

Figure 8.8 The energy rating of the housing stock and the fuel poor in the UK, 2006

Source: Boardman (2010).

Unfortunately, the message in Figure 8.8 implies that this can end up also hitting poor people, who inhabit the least efficient buildings.

Compensation – giving the money back. Given that most of the vulnerable households are among the poorest in society, 'compensation' is plausible, but targeting support is tricky. The fuel poor are not homogenous: they are pensioners (though by no means are all pensioners fuel poor), unemployed, single parents, single-person households, the disabled and so on. The UK has a Winter Fuel payment which pays out between £125 and £400 per person to people of pensionable age, but this means that pensioners with high incomes in highly efficient houses may receive support with their fuel bills, while an unemployed single mother living in a poorly insulated home receives no such direct support.

In the proposed US cap-and-trade schemes such income transfers were the main method of support to low-income consumers, allocating at least 50 per cent of allowance value to either low-energy income assistance or a consumer rebate fund. In addition, the early years of the schemes were marked by a proposal that allowances would be given to local distribution companies to mitigate energy price rises.

Amelioration – increasing efficiency. Instead of focusing on the 'poverty' dimension, an alternative is to reduce the amount of 'fuel' energy. Since the 'fuel poor' live in worse housing conditions than the general population, targeting energy-efficiency improvements for these buildings not only benefits the individuals in the form of lower bills, but can also represent some of the most cost-effective opportunities for energy saving and CO_2 reductions, as noted across Pillar 1 of this book. In the UK, the least efficient 10 per cent of homes would emit 12.7 tonnes of CO_2 on average if adequately heated, more than four times as much as the most efficient 10 per cent.[37] Fitting basic insulation, replacing old boilers and purchasing more efficient appliances could lift 400,000 households out of fuel poverty in the UK by 2022.[38]

A variety of schemes have been implemented around the world to support households by insulating existing buildings. In Germany, the state-owned bank KfW offers grants and reduced-interest loans to fund thermal retrofit measures. In the UK, electricity and gas suppliers are required to deliver a certain amount of energy-efficiency improvement measures for households. To address fuel poverty, 40 per cent of the measures have to be implemented among poor households (Chapter 5, note 43).[39] Such programmes, despite their complexities, offer the most enduring solutions and emphasise the need to think holistically about both carbon prices and other policies and how they bear overall on the final bills.

8.7 Energy access and carbon pricing in developing countries

If carbon pricing is tricky in terms of impacts on the poor in industrialised countries, how can it fare in developing countries? Curiously, the answer is more subtle than it might seem.

As noted in Chapter 1, in the least developed parts of the world (predominantly sub-Saharan Africa and rural India), over 2.5 billion people still depend on biomass for heating and cooking, and about half of these are not connected to electricity.[40] Even outside these regions, where there is almost universal access to electricity, many people have limited access to natural gas. This restricts a household's ability to heat their homes cheaply and effectively and raises their vulnerability to increases in electricity (or coal or heating oil) prices.

Creating access to electricity and natural gas is a key priority in many developing countries. It is useful not only for industry and business but reliable lighting also helps children study for longer, fridges help keep medicines cool and electric irrigation helps improve crop yields.[41] Switching to electricity for lighting and cooking can help cut down on dangerous fumes from using traditional biomass for cooking that causes huge health problems across the developing world.[42] Providing this access supersedes any other energy concerns.

How does this affect the prospects for carbon pricing? One paradox is that inadequate access to commercial energy implies that carbon prices in such countries would be more 'progressive' than in developed countries.[43] It is the rich in these societies who spend greater shares of their income on the energy they need for their automobiles and air conditioners. Of course, as emerging economies develop, their middle classes grow, and more people increase their energy use and would thus be hit by carbon prices. Higher energy prices also risk raising the ladder for those trying to get on to it. The determining factors thus include the politics of raising prices to the middle classes and whether carbon pricing might actually help to facilitate energy access for the rest.

This is not impossible. In the first place, a key reason why so many people remain without access to electricity is partly because governments control prices below costs – to aid access – leaving the industries unable to invest adequately in expanding energy infrastructure. That lesson has not been lost. The other lesson is that cheap energy has led to widespread inefficiencies in both industry and private use.

This creates a variety of opportunities. An Indian study showed how the macroeconomic costs of carbon pricing can be reduced significantly by combining global carbon prices with domestic policies to reduce the many inefficiencies in the power sector.[44] Inefficiencies also amplify the vulnerability of those embarking on the ladder of energy access.

Thus in developing countries, solutions may be different because the problems are different – creating access and providing the benefits of electricity, while providing the right signals to both reduce current emissions and create low-carbon electricity systems for tomorrow.

However, the fundamental objective will be the same as in developed countries and this similarity points to the potential synergies between economic and environmental goals:

highly efficient, and preferably low-carbon, energy developments that leave individuals less exposed to energy and carbon prices, and the country less dependent on the vagaries of international fossil fuel markets. The key role for carbon pricing in relation to developing countries, in other words, is to help them avoid getting trapped in the high levels of fossil fuel dependence that characterise most industrialised countries – and particularly the poorer people in them.

8.8 The false god of a global carbon price

Faced with the global problem of climate change, the instinct of most economists is to reach for one overarching solution: a global carbon price set at the level of the global damage it would prevent. The basic logic was set out in Chapter 6. It seemingly offers an efficient and effective solution: efficient because it would create the same level of incentive to cut emissions everywhere rather than a patchwork of higher and lower cost efforts thus maximising 'global bang for the global buck'; and effective because it would prevent carbon leakage (since industry would face the same price everywhere). In academic writings, this could be envisaged as either a global tax, an agreement on national taxes, a global cap-and-trade or a set of interlinked national cap-and-trade schemes – and all these approaches would deliver the same ultimate vision.

It is, however, a vision that has been challenged even at the first base of principles. It was already pointed out in the mid-1990s that establishing the same carbon price globally is unlikely to maximise welfare. Because a given level of energy price hurts poor people more than it hurts the rich, imposing the same price on all does not optimise welfare, unless there are compensating financial transfers.[45] There is a stark distinction between 'efficiency' – defined in terms of least cost and a 'global optimum' level of control – and maximising welfare. Of course, the same underlying idea is also true within individual societies, but in most societies there are all sorts of corrective mechanisms – progressive income taxes, public services, social security mechanisms, as well as finer-tuned measures like the UK's Winter Fuel payments. These help to ameliorate adverse distributional impacts while realising the benefits (and practicality) of a single price across a single jurisdiction.

Internationally, such corrective mechanisms are very weak. The economists Chichilnisky and Heal produced the maths to underline the extent to which, without such distributional corrections, global welfare would be 'optimised' if richer countries were to have higher carbon prices. A global carbon price is only 'optimal' if it is accompanied by international financial transfers to offset the differing relative impacts.

The world initially circumnavigated this point through the crude north–south division embodied in the Kyoto Protocol – in which a global 'price' is envisaged as a possibility, but would emanate from industrialised country commitments and would extend only to specific projects in developing countries where foreign investment can yield carbon credits. As explained in Chapter 7, the CDM has proved a partial solution but it cannot be the mainstay of future global efforts.

Moreover, the realities documented in Chapter 7 demonstrated that key rich countries baulked at paying other countries to cut emissions, especially through 'offsets'. It is not only the practicality of markets which depend on some participants being willing to pay to others: such willingness underlies the entire idea of a single market price bringing mutually beneficial outcomes.

In practice, it is difficult to project a route to one global carbon price: no country is going to yield such huge direct tax revenues to the UN or to a centralised global body, or let a UN forum dictate the design and coverage of a cap-and-trade system to be applied domestically. Nor are they likely to let a global deal dictate the level of carbon tax they should impose domestically – given all the difficulties traced in this Pillar (Chapters 6–8). Even if they did, it would be hard to define a meaningful comparable price, against the varied backdrop of different subsidy and fuel tax levels sketched in Chapter 6 (not to mention exchange rates).

The least incredible approach towards widespread international carbon pricing would be for governments to pursue domestic systems and then link them. This tempting vision received a severe knock with the set-backs to legislation in the US, and hence the focus has shifted to the patchwork of carbon pricing systems emerging elsewhere, as outlined in Chapter 7. Looking at the real difficulties caused by potential distributional impacts sketched in this chapter, this really should come as no surprise: it is a reality that has to be faced. The question is whether and under what conditions domestic systems could realistically be linked, and on what price terms.

As Chapter 6 has already observed, carbon taxes at least have only succeeded where they have been presented and accepted as an intrinsic part of competent economic and environmental management. Carbon pricing more generally should be recognised as appropriate when countries reach a basic level of industrialisation and institutional development. Establishing an expected *base level* for such countries makes sense. There would be a case for exempting the poorest countries for a multitude of reasons, including their limited institutional capacity. It does not follow, however, that all other countries should have the same price. Internationally a far more credible, and justifiable, approach would be to establish a 'base' price, expected to rise over time. This – and establishing the corresponding institutions – would send a powerful signal, particularly to domestic and multinational corporations. As suggested at points in these Pillar II chapters and elaborated in the rest of this book, there are multiple reasons why individual countries might realise higher carbon prices – not unlike the way that gasoline taxation, as noted in Chapter 6, has become almost universal, yet at levels which vary widely according to domestic circumstances.

These difficulties do not undermine the value of the growing efforts in many parts of the world. On the contrary, after several millennia of developing financial systems, we still do not have a single global currency – nor would it necessarily be a good idea. But we all have currencies. Moreover, they interact, against a general principle that prices should reflect costs. Paying at different levels for the carbon at the heart of our energy systems and the risks that creates – and placing different bets on the low-carbon economy – is still far better than not paying for it at all. What it does mean is that carbon pricing needs to be embedded in domestic economic and political circumstances, the challenges of 'carbon leakage' need to be addressed as an enduring, not transitory, problem, and the potential multiple national benefits associated with carbon prices and revenues needs full attention.

8.9 Conclusion

An old economics joke tells of a man searching for something under a lamp post at night, who explains to a passer-by that he has lost his keys. When the stranger asks the man where he dropped his keys, he replies: 'Over there, where it's dark – but the light is better here.'

There is another old joke – that all the world's economists laid end to end still could not reach a conclusion. It is not true. If there is one point of almost universal consensus it is that without carbon pricing we have dropped the keys. To tackle our dependence on carbon and its impacts without putting a price on it makes no sense in a market economy: everything else we might try would then be pushing uphill, against the grain of economic incentives.

The list of all the good reasons to price carbon, outlined in Chapter 6, is entirely valid: pricing carbon should be at the heart of an economic policy response. But the economics profession has focused its attention under the lamp post: how to design the 'optimal' instrument and how to quantify the macroeconomic cost.

In reality, these issues have proven almost irrelevant. Earlier chapters explained how the choice between the two possible ways of delivering a carbon price – taxation or cap-and-trade – has been driven by the basic politics: taxes generally transfer a lot of money from business (and/or consumers) to government, whereas cap-and-trade can exempt the most powerful opposition – or most vulnerable participants – with free allowances. The macroeconomic impacts of either, in turn, are quite trivial in relation to the size of our economies and projected growth.

What really matters are the differential impacts of a carbon price on different groups in society, and particularly the transfers involved. Carbon pricing creates winners and losers. The old adage that 'all politics is local' acts with vengeance: the challenge is how to deal with it without losing the incentive to switch to lower-carbon inputs, production processes and products. The science that is needed is no longer optimal efficiency but the political economy of pricing – to spread light to the dark and dirty recesses that prevent effective policy.

As the need for action has become more widely accepted, objections tend to be formulated in terms of effectiveness and distributional impacts. Concerns that carbon pricing could drive industry abroad certainly offer a powerful rhetoric. This chapter has shown how these concerns are in fact only warranted for about a half a dozen major industrial activities, like basic steel or cement. The approach to date – providing free allowances – limits the risk but also undermines the intended incentives and involves enduring uncertainty. Effective CO_2 mitigation across industrial activities depends on the ability of policy-makers to evolve towards full carbon pricing. This is probably only possible with the parallel introduction of border levelling. Adjusting for the carbon content of a few basic commodities like steel based on the carbon intensity of the best available technology could limit complexity and address WTO concerns. The sensitivities, however, remain high and the complexities real.

The other concern is about the impact on the poorest members of society. In developed countries they are largely the group with the highest share of income spent on energy and are therefore relatively more impacted by a carbon price increase. Generically subsidising their energy is a bad option (as we saw in Chapter 6), although 'rising block' tariffs *may* play a role. Taxes or cap-and-trade systems can provide a large source of revenue that can be recycled to compensate affected households, although targeting support is often tricky. Improving the energy efficiency of homes and appliances – exploiting to the full the potential synergies between the First and Second Domains – offers the opportunity to reduce energy bills while also reducing emissions.

Despite all its theoretical attractions, there are various reasons why a single global carbon price is unlikely and indeed not necessarily desirable. The fact that carbon prices cause different sorts of impacts in different sectors and in different parts of the world adds to the difficulties in creating a single global carbon price. We live in a world full of imperfections and political difficulties and with widely differing income levels.

Given all this, the only sensible way to approach the challenge of implementing carbon pricing is to see it as a journey. If countries start the journey just with a carbon tax, it is likely either to be at a very low level or to be riddled with exemptions and derogations, for example to heavy industry – or both. The challenge will be ramping up the level of tax and trying to narrow the exemptions – a tricky combination. If countries start the journey with cap-and-trade, the challenge will be to reduce the level of free allocation over time and to tighten the cap.

As such systems strengthen, two other aspects must evolve:

- More industries would face a growing price difference compared to some foreign competitors. Countries will not and should not be expected to prefer foreign production over their own, so the system must evolve towards pricing consumption and not just towards production – which implies some kind of border levelling. This can initially be very confined, but if and as carbon prices rise, it will need to extend to more commodities.
- The money raised from consumers will become more substantial, and without other measures the impacts and inequalities will become more pronounced. The poorest may justifiably need protection. And all will demand a proper democratic say in how such sums should be recycled back into their economy and to solving the global challenge. The concluding chapter offers theoretical reasons for some 'earmarking' of carbon revenues – but politics is likely to drive it anyway.

Completing the circle

Thus carbon pricing is necessary but not sufficient. The irony is that, when properly understood, the 'key policy' advocated by most economists turns out to be intimately nested in the complementary triad laid out in this book. Without strong policies to improve energy efficiency – particularly in the poorer parts of society and in poorer countries – the impact will make it even harder to get carbon pricing off the ground. Without innovation and appropriate infrastructure to facilitate low-carbon solutions, carbon prices will fail to wean societies away from fossil fuels and will face more and more opposition – because they will be seen as punitive rather than effective, with voters struggling to reduce the impact on their bills.

In turn, carbon pricing can provide both the incentives and the financial resources for such programmes of efficiency, innovation and investment. It heightens consumer interest and awareness and provides resources and a wider political context for major programmes of energy efficiency (Chapters 4 and 5). And as explored in Chapters 9 and 10, it can help both to fund R&D programmes and necessary infrastructure and makes low-carbon innovation more profitable.

Wind back to 'the most important diagram in energy economics', at the beginning of this Pillar (Chapter 6). Carbon pricing, with all these accompaniments, can extend the chart: towards societies in which energy and carbon prices reflect their real costs, while efficiency and innovation make the journey to higher prices politically tenable – by helping us, as individuals and overall societies, to manage the total bill.

Notes

1. It is much as if energy prices rose and were fed through the economy, but governments then compensated companies with cash transfers. A foundational report on this topic, which included the modelling of five industrial sectors, first predicted that most sectors would profit from the EU ETS (Carbon Trust 2004); the basic mechanism has been borne out by experience. Capping carbon means that emitting it is no longer free. The impact on firm profits and competitiveness will depend upon who gets the resulting 'economic rents'; free allocation enables business to benefit from this. Note that irrespective of free allocation, companies in principle should still strive to reduce their emissions as long as the costs of doing so are lower than the market price of an emissions allowance, bearing in mind the various potential distortions documented in the previous chapter. For more detailed discussion of the mechanisms and first evidence from the EU ETS, see papers in Grubb and Neuhoff (2006a).

2. Sijm *et al.* (2006). For a wider analysis of the impact of carbon prices on electricity prices and evidence from both econometric and model studies see Sijm *et al.* (2008). The scale of such windfall profits across Europe has now been estimated in many studies, reviewed in Climate Strategies: Grubb *et al.* (2012). They span a wide range because of different periods, methodologies and whether or not the study focuses on price pass-through effects or whether surplus allowances are included. However, all are in the region of billions of euros a year.
3. A wide number of studies have attempted to estimate how much leakage might occur. For a full review see IPCC reports, and Climate Strategies: Droege and Cooper (2010). Most of these are from a European perspective, but some look at future carbon pricing regimes considered in the US, Japan and Australia. Note that, as explained in the Climate Strategies: Droege *et al.* (2009) studies, there are other two other channels of leakage. One, associated with fossil fuel price impacts, is adverse (reducing fossil fuel use in one region tends to lower global prices and hence promote more consumption elsewhere). The other is positive (global diffusion of low-carbon technologies associated with low-carbon innovation in one region).
4. Gross value-added is a measure of the production of an industry, calculated by taking output and subtracting intermediate inputs.
5. The production channel of leakage is based on the concept of the Pollution Haven Hypothesis, whereby firms relocate to areas with weaker environmental regulations to avoid costs; for an early discussion see Copeland and Taylor (1994).
6. The leakage ratio is defined as the amount of emissions 'leaking' to the rest of the world, divided by the reduction in emissions in the carbon price regime. Assuming full auctioning of allowances and long enough periods for adjustments rates of up to 39% for steel, 21% for aluminium and 16% for clinker were simulated. As many other sectors are less impacted, leakage – the increase of emissions outside of Europe – corresponds to about 10% of emission reductions delivered in Europe. (See Monjon and Quirion 2009).
7. For a wider discussion of these criteria for assessing sectors at risk of leakage see Climate Strategies: Droege and Cooper (2010).
8. This is defined as the ratio of exports to outside the EU plus the value of imports from outside the EU to the market size in the EU.
9. A study of managers found that the average firm in each sector studied was at no risk of closure and only a few had a small risk of downsizing. For the full study see Anderson *et al.* (2011).
10. Sandbag (2012).
11. The EU scheme differs from many proposals in insisting on a principal that the volume of allowances is fixed for a given period, independent of the realised industrial output. Others have proposed 'output-based' allocation – with allowances in proportion to the levels of industrial output. In practice, the EU ETS provisions on closure and new entrants represent special forms of output-based allocation, and recognition that there could still be carbon leakage – with plants running at low levels to cash in free allowances and import instead – led to Phase III rules that allowances would be curtailed for plants operating below certain thresholds. Emerging evidence suggests that, as expected, this in turn leads to the alternative perverse incentive of encouraging plants to maintain output just above the threshold – in some cases, leading to 'reversed leakage', with cement produced under these conditions going to export. There is, quite simply, no ideal way of allocating free allowances, and which way is 'least bad' may vary by sector (Quirion 2009).
12. For a discussion of the options see Monjon and Quirion (2010).
13. WTO rules at heart aim to ensure that domestic goods cannot be treated more favourably than identical imported goods, and that all WTO members must be treated equally. This could be addressed by ensuring that any carbon-related adjustment level does not exceed the emissions that would result from production with best available technology. Alternatively, the requirements could be motivated by the protection of a 'global resource', a legal basis for allowing exemptions from some of the traditional trade requirements and thus allowing for more flexibility in the design of the adjustment. For a broad introduction to the interactions between climate change and trade, see the two-part survey by Brewer (2003) and Brewer (2004). See also WTO/UNEP (2009).
14. Ismer and Neuhoff (2007).
15. Indeed consultation with exporters and importers is likely to be a key element in securing WTO compliance and this is most likely to have to occur prior to the setting of the tariff.
16. For a fuller discussion of sector-specific problems and solutions see Carbon Trust (2010) and Climate Strategies: Droege *et al.* (2009), with more detailed development in Quirion (2009).

17. For a more extensive discussion of these other channels see Climate Strategies: Droege *et al.* (2009) and Carbon Trust (2011).
18. The economic research network CES-Info convened a major conference on the 'Green Paradox' in 2009. For a useful overview of issues and an entry point into the literature, see one of the resulting survey papers, van der Ploeg and Withagen (2012).
19. This is the basis of the Porter Hypothesis, proposed by Michael E. Porter in 1995 in Porter and van der Linde (1995); see Chapter 9, section 9.8 'Porter's Kick'
20. The data in figure 8.4 give a first indication of the potential significance of cost pass-through in manufacturing, *if* these sectors were subject to effective carbon pricing. All the sectors with a high exposure are primary commodities. Cost pass-through might thus have significant incentive effects in the early stages of the production chain, but any price impact on final consumers would be very highly diluted compared to the impact of electricity and gas prices. Final consumers buy electricity and gas, but they buy cars and houses instead of primary flat steel or concrete slabs.
21. Boardman (2010).
22. Owen (2008).
23. Department of Energy and Climate Change (2010a).
24. They predict that electricity prices will rise by 33% while gas prices will rise by 18% as a result of climate policies.
25. Energy Information Administration (2009).
26. For results from the Danish experience with energy and carbon taxes see Wier *et al.* (2005). For evidence as to the distributional impacts of the Kyoto Protocol see Speck (1999) and Zhang and Baranzini (2004).
27. In fact there is evidence that the relationship between energy expenditure and income follows an S-shaped Engel curve: as we get richer we initially spend more on energy as we have more income to meet our needs; as we get richer still, although the absolute level increases, still it increases at a lower rate, leading to a plateau at high levels of income. This is similar to the consumption of food, for which Engel curves were first derived. For evidence from the UK see Jamasb and Meier (2010).
28. Sterner (2011).
29. Usually 21 degrees for the main living area and 18 degrees for other occupied rooms (DECC 2010b). For a more detailed description of the fuel poor and future prospects see Roberts (2008).
30. The UK Committee on Climate Change estimates that climate policy support (carbon prices and additional support for renewable generation) would lead to an additional 600,000 households entering fuel poverty by 2022 (Committee on Climate Change 2008).
31. These are often second-hand and are the most inefficient for a number of reasons, including the simple fact that they are old and thus behind the technological frontier; they may have been subject to older, less rigorous efficiency standards; and they may be faulty and poorly maintained.
32. Tirado-Herrero and Ürge-Vorsatz (2010).
33. Ibid.
34. Fankhauser and Tepic (2005).
35. ADEME (2008).
36. When controlling for other factors (Rehdanz and Stöwhase 2008).
37. Boardman (2010).
38. Committee on Climate Change (2008).
39. This forms part of the CERT scheme indicated in the bills in Figure 8.6.
40. Saghir (2005).
41. Ibid.
42. For a full discussion see Bruce *et al.* (2000).
43. For a discussion in a Chinese context see, Brenner *et al.* (2007). For South African context see van Heerden *et al.* (2006).
44. Guivarch and Mathy (2012).
45. Chichilnisky and Heal (1995) underlined the conflict between 'welfare' and 'efficiency' in the debate about a global carbon price. The two counter-arguments to this both turn out to be problematic. For efficiency maximisation to turn into welfare maximisation requires a degree of international compensation that seems implausible. And the idea that a single carbon price improves welfare because climate change is a global problem expressible in a 'social cost of carbon damages' also overlooks the fact, for example, that the poor tend to have higher discount rates, for

obvious reasons – and discounting is one of the biggest determinants of a 'social cost of carbon'. Ultimately, such issues are hard to separate from more fundamental debates about international ethics, equity and finance.

References

ADEME (2008) 'The weight of energy expenditures in household budgets in France, ADEME and you', *Strategy and Studies*, no. 11, 3 April.

Aldy, J. E. and Pizer, W. A. (2011) *The Competitiveness Impacts of Climate Change Mitigation Policies*, NBER Working Papers 17705. National Bureau of Economic Research.

Anderson, B., Lieb, J., Martin, R., McGuigan, M., Muuls, M., de Preux, L. and Wagner, U. (2011) *Climate Change Policy and Business in Europe: Evidence from Interviewing Managers*, Occasional Paper No. 27. Centre for Economic Performance.

Boardman, B. (2010) *Fixing Fuel Poverty: Challenges and Solutions*. London: Earthscan.

Brenner, M., Riddle, M. and Boyce, J. K. (2007) A Chinese Sky Trust? Distributional impacts of carbon charges and revenue recycling in China', *Energy Policy*, 35 (2): 1771–84.

Brewer, T. (2003) 'The trade regime and the climate regime: institutional evolution and adaptation', *Climate Policy*, 3 (4): 329–41.

Brewer, T. (2004) 'The WTO and the Kyoto Protocol: interaction issues', *Climate Policy*, 4 (1): 3–12.

Bruce, N., Perez-Padilla, R. and Albalak, R. (2000) 'Indoor air pollution in developing countries: a major environmental and public health challenge', *Bulletin of the World Health Organisation*, 78: 1078–92.

Carbon Trust (2004) *EU Emissions Trading Scheme*. London: Carbon Trust.

Carbon Trust (2008) *EU ETS Impacts on Profitability and Trade: A Sector by Sector Analysis*, Carbon Trust Report CTC278. London: Carbon Trust.

Carbon Trust (2010) *Tackling Carbon Leakage – Sector-Specific Solutions for a World of Unequal Prices*. London: Carbon Trust.

Chichilnisky, G. and Heal, G. (1995) *Markets for Tradeable CO_2 Emission Quotas: Principles and Practice*, OECD Economics Department Working Paper 153. Paris: OECD Publishing.

Climate Strategies: Droege, S. (2009) 'Carbon Leakage and the EU ETS', Synthesis Report, London: www.climatestrategies.org

Climate Strategies: Droege, S. and Cooper, S. (2010) 'Tackling Leakage in a world of unequal carbon prices', Synthesis Report, London: www.climatestrategies.org

Climate Strategies: Grubb, M., Laing, T., Sato, M., and Comberti, C. (2012) 'Analyses of the effectiveness of trading in EU-ETS', Climate Strategies Synthesis Report, London: www.climatestrategies.org

Climate Strategies: Hourcade, J. C., Neuhoff, K., Demailly, D. and Sato, M. (2007) 'Differentiation and dynamics of EU ETS industrial competitiveness impacts', Climate Strategies Synthesis Report, London: www.climatestrategies.org

Committee on Climate Change (2008) *Building a Low-Carbon Economy – The UK's Contribution to Tackling Climate Change*, Report by the Committee on Climate Change. UK Government.

Copeland, B. R. and Taylor, M. S. (1994) 'North–south trade and the environment', *Quarterly Journal of Economics*, 109 (3): 755–87.

Daly, J. G. and Riedy, C. J. (2008) *Cogeneration Stakeholder Workshop: Summary Report*, prepared for NSW Department of Environment and Climate Change, Institute for Sustainable Futures, Sydney.

DECC (2010a) *Estimated Impacts of Energy and Climate Change Policies on Energy Prices and Bills*. Department of Energy and Climate Change, UK Government.

DECC (2010b) *Average Temperature of Homes*. Department of Energy and Climate Change, UK Government, online at http://2050-calculator-tool.decc.gov.uk/assets/onepage/29.pdf

EIA (2009) *Energy Market and Economic Impacts of H.R. 2454, the American Clean Energy and Security Act of 2009*. Energy Information Administration, Department of Energy, US Government.

European Commission (2009) Results of the quantitative assessment of sectors at NACE 4 level, online at http://ec.europa.eu/clima/events/0038/20090429results_quantitative_assess_sectors_nace4_en.pdf

Fankhauser, S. and Tepic, S. (2005) *Can Poor Consumers Pay for Energy and Water? An Affordability Analysis for Transition Countries*, EBRD Working Paper No. 92. European Bank for Reconstruction and Development.

Grubb, M. (2008) *Carbon Prices in Phase III of the EU ETS*, Report. London: Climate Strategies.

Grubb, M. and Neuhoff, K. (eds) (2006a) 'Allocations, incentives and industrial competitiveness under the EU Emissions Trading Scheme', *Climate Policy*, Special Issue.

Grubb, M. and Neuhoff, K. (2006b) 'Allocation and competitiveness in the EU Emissions Trading Scheme: policy overview', in Grubb and Neuhoff (2006a), pp. 7–30.

Grubb, M., Brewer, T. L., Sato, M., Heilmayr, R. and Fazekas, D. (2009) *Climate Policy and Industrial Competitiveness: Ten Insights from Europe on the EU Emissions Trading System*, Report for the German Marshall Fund, Climate and Energy Paper Series. Washington, DC.

Guivarch, C. and Mathy, S. (2012) 'Energy-GDP decoupling in a second best world – a case study on India', *Climatic Change*, 113 (2): 339–56.

IPCC (1996) *Second Assessment Report*. Cambridge: Cambridge University Press.

IPCC (2004) *Third Assessment Report – WG3*. Cambridge: Cambridge University Press.

IPCC (2007) *Fourth Assessment Report – WG3*. Cambridge: Cambridge University Press.

Ismer, R. and Neuhoff, K. (2004) *Border Tax Adjustments: A Feasible Way to Address Nonparticipation in Emission Trading*, Cambridge Working Papers in Economics 0409. Faculty of Economics, University of Cambridge.

Jamasb, T. and Meier, H. (2010) *Household Energy Expenditure and Income Groups: Evidence from Great Britain*, Cambridge Working Papers in Economics 1011. Faculty of Economics, University of Cambridge.

Mann, R.T. (2003) *Legacy to Power: Senator Russell Long of Louisiana*. Lincoln, NE: iUniverse.

Monjon, S. and Quirion, P. (2009) *Addressing Leakage in the EU ETS: Results from the CASE II Model – Final Workshop Presentation*. London: Climate Strategies.

Monjon, S. and Quirion, P. (2010) 'How to design a border adjustment for the European Union Emissions Trading System', *Energy Policy*, 38 (9): 5199–207.

Neuhoff, K. (2008) 'Learning by doing with constrained growth rates: an application to energy technology policy', *Energy Journal*, Special Issue, 29 (2): 165–82.

Office for National Statistics (2009) *Family Spending: A Report on the 2008 Living Costs and Food Survey*. London: Palgrave Macmillan.

Owen, G. (2008) *Towards an Equitable Climate Change Policy for the UK: The Costs and Benefits for Low Income Households of UK Climate Change Policy*. EAGA, online at http://www.carillionenergy.com/downloads/pdf/3867%20eaga%20Equity%20&%20Climate%2036ppg%20PROOF.pdf

Porter, M. E. and van der Linde, C. (1995) *Green and Competitive: Ending the Stalemate*, Reprint 95507. Cambridge, MA: Harvard Business Review.

Quirion, P. (2009) 'Historic versus output-based allocation of GHG tradeable allowances: a survey', *Climate Policy*, 9: 575–92.

Rehdanz, K. and Stöwhase, S. (2008) 'Cost liability and residential space heating expenditures of welfare recipients in Germany', *Fiscal Studies*, 29 (3): 329–45.

Roberts, S. (2008) 'Energy, equity and the future of the fuel poor', *Energy Policy*, 36 (12): 4471–4.

Saghir, J. (2005) *Energy and Poverty: Myths, Links and Policy Issues*, Energy Working Notes No. 4. Washington, DC: World Bank.

Sandbag (2012) *Carbon Fat Cats Could Make 5.6 Billion Euros from ETS*. London: Sandbag, online at http://www.sandbag.org.uk

Sijm, J., Hers, I., Lise, W. and Wetzelaer, W. (2008) *The Impact of the EU ETS on Electricity Prices*, Final Report to DG Environment of the European Commission. Energy Research Centre of the Netherlands.

Sijm, J., Neuhoff, K. and Chen, Y. (2006) *CO_2 Cost Pass Through and Windfall Profits in the Power Sector*, Cambridge Working Papers in Economics 0639. Faculty of Economics, University of Cambridge.

Speck, S. (1999) 'Energy and carbon taxes and their distributional implications', *Energy Policy*, 27 (11): 659–67.

Sterner, T. (2011) *Fuel Taxes and the Poor – The Distributional Effects of Gasoline Taxation and Their Implications for Climate Policy*. Washington, DC: Resources for the Future Press.

Tirado-Herrero, S. and Ürge-Vorsatz, D. (2010) 'Trapped in the heat: the post-communist genre of fuel poverty, working paper', *Energy Policy*, 49: 60–8.

van der Ploeg, F. and Withagen, C. (2012) 'Is there really a green paradox?', *Journal of Environmental Economics and Management*, 64 (3): 342–63.

van Heerden, J., Gerlagh, R., Blignaut, J. *et al.* (2006) 'Searching for triple dividends in South Africa: fighting CO_2 pollution and poverty while promoting growth', *Energy Journal*, 27: 113–41.

Wier, M., Birr-Pedersen, K., Jacobsen, H. K. and Klok, J. (2005) 'Are CO_2 taxes regressive? Evidence from the Danish experience', *Ecological Economics*, 52 (2): 239–51.

WTO/UNEP (2009) *Trade and Climate Change*. Geneva: World Trade Organisation and UN Environment Programme.

Zhang, Z.-X. and Baranzini, A. (2004) 'What do we know about carbon taxes? An inquiry into their impacts on competitiveness and distribution of income', *Energy Policy*, 32 (4): 507–18.

Pillar III

Strategic investment for innovation and infrastructure

Overview

Providing energy for the twenty-first century will cost trillions of dollars every year. Key to determining the long-run costs and impacts will be innovation, and investment in infrastructure, broadly defined as 'long-lived capital stock'.

Innovation is not just about scientific breakthroughs and government-funded R&D. The huge diversity of energy uses, systems and resources (Chapter 3) preclude a 'magic bullet', while useful innovation in practice involves extensive development, commercialisation and cost reductions associated with scale and learning-by-doing, as both industry and markets expand and develop.

However, the main energy-related sectors – particularly those related to electricity production and the use of heat – spend only a tiny fraction of their income on innovation compared to more dynamic sectors in the global economy. Well-mapped structural barriers in these energy sectors create a big gap between the results of publicly funded R&D and the widespread diffusion of better technologies. This largely severs the market feedback and financing loops which drive innovation in some other sectors (like pharmaceuticals and information technology – see Chapter 9).

Markets generally under-invest in innovation, compared to the overall economic benefits, and industries invest principally in technologies to extend their established positions. This is particularly true for such energy-related sectors – thus the deep pockets of fossil fuel industries invest far more in oil-related technologies than in clean energy solutions. So accelerating innovation in low-carbon technologies is likely to be economically beneficial, as well as holding promise to tackle public challenges, but requires extensive public intervention. 'Technology push' must meet 'demand pull' – for which policies have to span the entire innovation chain (where small and medium sized enterprises are crucial), or otherwise support the degree of confidence and financing required for large corporations to make long-term bets that coincide with the public interest. Either option provides obvious opportunities for governments to accelerate and/or align innovation towards public goals of energy security and low-carbon development (see Chapter 9).

The key policies for clean energy innovation group broadly into three stages:

- technology R&D and demonstration which depends principally upon direct public funding;
- commercialisation involving a mix of 'incubators' (the creation of small businesses based around new technologies) and 'market engagement' (field trials etc. to provide commercial demonstration, feedback, user confidence and adaptation to local environments);
- strategic deployment with either capital subsidies or policies which create a 'premium price' for new and desirable technologies to benefit from scale economies and learning, and to overcome the barriers arising from incumbent interests and existing infrastructure (see Chapters 9, 10).

Historical experience emphasises the fact that technological transitions also involve transformation of infrastructure and institutions, in a process of co-evolution of socio-technical systems. This makes them highly 'path dependent' – what is possible at any given time is heavily influenced by previous investments and decisions. Such systems are thus also prone to 'lock-in' – interrelated sets of technology, infrastructure and institutions which have huge inertia and tend to perpetuate the established interests. Present systems of fossil fuel dependence display all these characteristics. Currently, the dominant trajectory is towards increasing innovation and investment in fossil fuels: the world is investing trillions of dollars in systems and technologies aimed at accessing ever more difficult and remote carbon resources, and thus also in ways to pollute the atmosphere more cheaply. Low-carbon investment has grown faster in percentage terms but remains modest in comparison.

Turning the supertanker away from fossil fuel dependence undoubtedly requires several trillion dollars more investment over the next decade or two, but global simulations suggest there are a wide range of possible future developments and costs. Moreover, as oil production extends to increasingly difficult and/or remote resources, and countries deal with the multiple environmental problems of coal extraction and combustion, rising investment will be required anyway (see Chapter 10).

Reorientation towards low-carbon development will need not just credible carbon pricing and innovation, but also smart strategies to enable new entrants. Key mechanisms include both niche accumulation – the expansion of low-carbon technologies from niche applications to wider markets – and 'hybridisation', in which low carbon technologies can ride on the back of existing infrastructure and slowly adapt or extend it to their needs. Solar cells form a classic example of the former; the use of biofuels and hybrid vehicles typify the latter. Against this background, transition pathways can be mapped out for key sectors: vehicles, electricity and buildings – particularly in the context of urbanisation. Specific examples from the Americas, Europe and Asia respectively underline the intertwined roles of innovation and infrastructure, combined with the other domains. In each case the transition offers clear potential strategic benefits, but realising them requires some associated upfront investment of both political and financial capital (see Chapter 10).

Though decarbonisation requires more upfront investment, there is no compelling evidence that low-carbon trajectories need ultimately be more expensive, even aside from the climatic and security implications. There are instead divergent paths between default (carbon-intensive) and lower-carbon responses. Thus economic systems can evolve in many different directions, with widely divergent patterns of energy consumption, production and emissions (see Chapters 10, 11). Given the obvious inherent risks associated with the current trajectory, its continuation thus reflects a condition of individual short-term rationality but collective failure reminiscent of the factors which led to the global financial crisis.

Breaking out of this trap requires effort and innovation. Theories of economic growth have long recognised that innovation is a key but very poorly understood force – the 'residual' identified in classical growth models. The simplest theories of economic growth assume either that such innovation is an external, unalterable force, or that markets provide an 'optimal' level and direction of innovation. Neither assumption stands up to scrutiny. Innovation responds to needs, investments and incentives, and is impeded by multiple failures in the innovation chain (see Chapter 9). It is further constrained and shaped by existing infrastructure and the interests of major incumbent interests, of which the energy sector provides clear examples (see Chapter 10). Economists throughout the ages have identified that motivation is a key but poorly understood factor in economic renewal, and cycles of economic growth have often been associated with waves of industrial transformation (including 'creative destruction').

This 'dark matter' of economic growth comprises innovation in both institutional and technological systems; these can be identified directly with the processes of the First and Third Domains. Consequently, there is no reason to believe that government-led policies to shape efficiency, innovation and infrastructure in ways compatible with energy and climate security need compromise economic development. They do, however, require smart policies, motivated actions by citizens and companies, and greater upfront investment (both political and financial) for the sake of longer-term benefits (see Chapter 11).

The scale of financial investment required is in the order of 1 trillion in Europe to 2020, and tens of trillions of dollars globally over the coming decades. Economies depressed in the aftermath of the financial crisis may, ironically, be best able to harness such investment as a vehicle for renewed growth. This is because they have both large unutilised resources and historically low interest rates: large amounts of labour sit unemployed, and huge amounts of private capital are currently earning trivial rates of return. The investment required is a modest fraction of the wealth held in institutional investment funds (e.g. pensions, insurance and sovereign wealth funds), which could be particularly relevant for long-term secure investments, and energy infrastructure is a leading candidate. Presently, uncertainty and a lack of confidence deter such investment and associated employment. The physical constraints of a finite atmosphere, however, imply a solid science-based value in saving carbon, which should rise over time. Policy to translate that fact into clear and robust incentives thus has the potential to spur renewed investment, innovation and clean growth (see Chapter 11).

9 Pushing further, pulling deeper

Bridging the technology valley of death

'Let it be remembered that the closet-philosopher is unfortunately too little acquainted with the admirable arrangements of the factory; and that no class of persons can supply so readily … the data on which all the reasoning of political economists are founded, as the merchants and manufacturer … Nor let it be feared that erroneous deductions may be made from such recorded facts: the errors which arise from the absence of facts are far more numerous and more durable …'

– Charles Babbage, *On the Economy of Machinery and Manufactures* (1832) (mathematician and analytical philosopher generally cited as the 'father of modern computing' – the originator of the idea of the programmable computer)

9.1 Introduction

In his masterful book on oil, *The Prize*, Daniel Yergin charts the story of 'Colonel' Edwin Drake, the first man to drill an oil well. It took an idea, technology, time, experimentation, determination and money. Arguably the most profitable invention in history, it nevertheless illustrated the numerous obstacles and risks. Drake was only saved by the tardy US postal service: he finally struck oil after Townsend – the brainchild behind the venture and its last remaining financial backer – had run out of money and posted the letter instructing Drake to quit.[1]

Creating and commercialising a new product has always been a risky business. History demonstrates a reality which is far from the caricature of a 'eureka!' moment, with a new invention falling like manna from heaven and transferring easily to large-scale use. It almost always involves a long series of potentially costly and time-consuming enquiries, developments and experimentations. As scales and costs rise, innovators are in frequent tension with their funders. Economic rewards may take years, maybe decades, to materialise.

In its survey of energy technologies, Chapter 3 showed that innumerable technologies could, in principle, help to deliver a secure and low-carbon energy system at reasonable cost. At the same time it underlined that the diversity of energy uses, systems, resources and national contexts means that there can be no 'magic bullet' solution. Contrary to popular imagination, the core challenge is not to discover a new, revolutionary technology that will solve all our energy problems; a century of disappointments has caused even enthusiasts to lower their expectations for salvation from a single technology, be it nuclear fission, fusion, solar power or hydrogen. The challenge is rather to understand, and harness, the broader forces of innovation.

Against this backdrop, the chapters of the Third Pillar (Chapters 9–11) review what we have learned about the *process* of innovation, the transformation of energy systems and how these may relate to economic development. The present chapter is about the 'microeconomics' of innovation, particularly relating to energy technologies: how they develop; what drives – and blocks – growth and cost reductions; why, left to itself, more effort goes into fossil fuels than alternatives; and what can be done to accelerate innovation to meet the challenges of the twenty-first century. The argument is not just that innovation matters, but that energy is different – posing problems yet also offering opportunities.

9.2 The glittering prizes

To an important degree, innovation remains the 'dark matter' of economics and a source of confusion in academic, policy and public debates. Indeed there are two alternative caricatures, each equally wrong and unhelpful. In one, governments just need to fund enough smart researchers to come up with solutions. In the opposite caricature, innovation is simply a particular feature of economic processes, driven by markets to provide an optimal level of investment in new ways of producing and selling profit-making goods. Neither view bears much resemblance to reality, but all agree that innovation in energy and environment is important.[2]

The economic analysis of innovation is often traced back to the work of Joseph Schumpeter in the 1930s, though as always there were antecedents. Schumpeter introduced the concept of 'creative destruction', used to describe the way that radically new technologies and systems could supplant established industries and provide new foundations for economic growth; we touch on this in the subsequent chapters. Within economics, he is also known for identifying three distinct steps associated with major technological changes: invention, innovation and diffusion.[3]

Townsend and Drake's essential breakthrough was invention – the radical idea of drilling for oil – but its practical application was almost stymied by the need for innovation – developing and applying the technology to actually do it. Invention takes brains and imagination. Innovation takes time and money. Studies of historical transitions in energy emphasise that diffusion itself takes far more still. Indeed, the subsequent 'Shumpeterian wave' that established the oil age, like other major energy system transitions, took around three-quarters of a century.[4]

Since then, the energy game has subtly shifted. The 75 years after Townsend and Drake's breakthrough were dominated by the discovery of new oil resources, but the pace of genuinely new resource discoveries has declined. The emphasis has moved towards the innovation of better and cheaper ways of mapping and accessing resources that were previously inaccessible or uneconomical to exploit. At the same time, our understanding of and capacity to tap into other energy resources has grown. This applies equally to coal, gas, uranium and the wide variety of renewable sources. The 'fossil fuel frontier' is characterised by trying to get more out of reservoirs thousands of feet beneath the ocean's surface, buried under the Arctic ice cap, stuck in oil sands or squeezed out of rocks kilometres underground at incredibly high pressures ('fracking'). There is no inherent reason why this is cheaper than just exploiting the natural energy in the sun, winds and water, nor indeed tapping more deeply into the huge 'resource' of energy efficiency, where as outlined in Chapter 3 the evidence is that we still waste around 90 per cent of the primary energy we extract.

Thus, as argued in the conclusion of Chapter 3, we are not short of energy options per se, nor is one resource clearly globally dominant over others. We have entered an age of competition between diverse energy technologies and systems, which may have radically different economic and environmental properties. One clear difference, however, is that since a fossil fuel deposit is finite and emits CO_2 to a limited atmosphere, it offers a time-limited fix, unlike renewables. The glittering prizes of the twenty-first century will ultimately go to the technologies, firms, industries – and countries – that win the race to provide at scale energy that is clean, secure, affordable and enduring.

Innovation should justify commensurate effort. Plentiful evidence, detailed shortly, shows that the cost of most technologies decline as they are developed and used at scale. In the context of innovating for a low-carbon future, for example, consider Figure 9.1, which sketches conceptually the evolution of costs and benefits (vertical axis) for a new, low-carbon technology for power generation. Initially, it may cost several times as much as established technologies which have benefited from more than a century of development. Any new, clean generating technology has the prospect for costs to decrease as learning and scale economies accumulate over time (the x-axis), and with expansion of the applications (the third dimension of the chart). If and when the technology reaches a point of competitiveness, it can start to make a profit against established, fuel-based generators. The additional cost of initial deployment applies only for a limited time and market – illustrated by the initial wedge. As the technology takes off at scale, the benefits – the volume on the right-hand side, with mass deployment – have the potential to vastly outweigh the initial investment costs.

The benefits of innovation in a clean technology will be even bigger if one factors in the costs of environmental damage which it may avoid. Figure 9.1 illustrates this with the upper wedge, in terms of the potential value of CO_2 avoided, which is likely to rise over time. In theory, this additional (environmental) value is there whether or not governments implement it directly with environmental pricing. In principle, governments could support the development and drive diffusion of the cleaner technology at scale with subsidies or standards. However, the figure also illustrates the many ways in which carbon pricing can support innovation in higher efficiency and low-carbon technologies. By raising the cost of a conventional carbon-intensive technology, it reduces the cost gap that a cleaner technology has to shoulder, either through direct public subsidy or private learning investment. Also it means that the new technology can compete earlier against incumbent technologies, thus reducing the policy risks associated with reliance on subsidies (where support is publicly provided) and/or financing costs (where firms bear the learning investment). A rising carbon

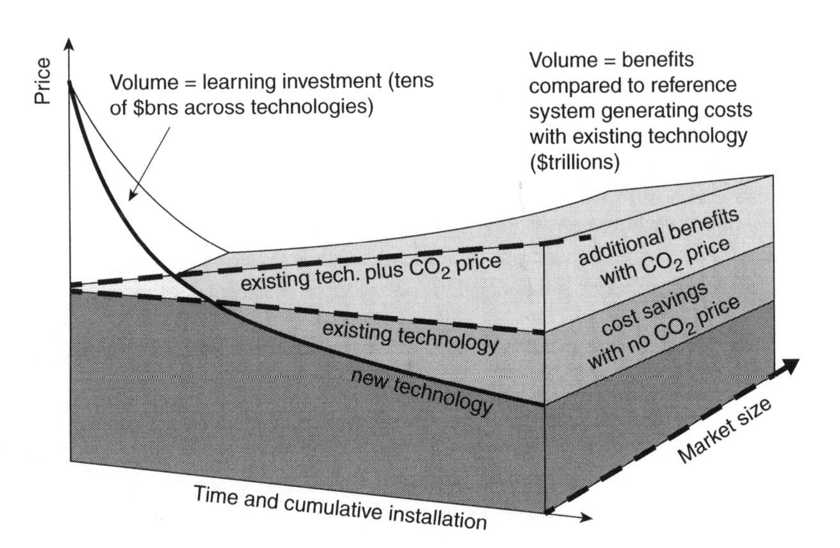

Figure 9.1 Immediate costs and future benefits of low-carbon technology innovation

price would also enhance the strategic viability of a new low-carbon technology and thus the wider case for innovation strategy and the commercial benefit for private investors.

As noted in Chapter 6, the idea that somehow research and development (R&D) offers an alternative to carbon pricing is thus one of the more bizarre propositions in the field. Someone has to bear the cost; without environmental policy the profits and benefits will only be realised much later – and if they remain unpriced, they would also be largely unvalued. As the old saying goes, one can lead a horse to water but not make it drink. Without an incentive to reward uptake of cleaner technologies, such innovation is likely to falter.

However, the converse belief – that environmental policy can supplant the need for technology policy – is equally misguided, for reasons replete throughout this chapter and the next. To reverse the usual analogy: offering an animal the prospect of distant water does not guarantee it will become a thoroughbred stallion. In a desert, it may instead die of thirst. Until recently, this was closer to the reality for many clean energy innovations, as detailed in this chapter.

The dominant approach to innovation in mainstream economics (as opposed to specialists who work in the field) is to treat innovation in terms of 'market failures' associated with 'spillovers'. The main problem is considered to be a failure of intellectual property rights: that an innovator may be unable to profit from all the knowledge they gain before key aspects of it 'spill over' to others. Drawing on this, the high-level energy and climate literature tends to talk of the 'dual market failures' of externalities (i.e. unpriced environmental impact) and property right failures, and to pose the answer as being to combine environmental pricing with strong property rights (patents).

While this usefully dispenses with the false dichotomy of posing environmental pricing and innovation policies as alternatives and correctly emphasises their complementarity, this 'double externality' approach remains a seriously incomplete diagnosis. Patent rights are essentially a legal protection for monopoly – not something that economists are usually so keen on. It reflects, to some degree, a peculiarly Anglo-Saxon debate, driven by the principle of clear separation between the state and the private sector, combined with recognition that a company is only going to innovate insofar as it can capture the benefits – which a monopoly ensures. In most other regions, the role of the state in supporting major innovation, including nurturing industrial development, is widely accepted. Even in North America and the UK, the idea that the state should keep out of innovation except to fund basic R&D and enforce intellectual property rights is widely 'honoured in the breach' (i.e. not observed in practice). Economic advisors have for decades pointed to a broad raft of other factors that get in the way of adequate innovation and justify a stronger governmental role.[5]

This chapter explains why this is particularly true for energy. The fact that the barriers to innovation are amplified further when we are seeking innovation to provide 'public goods' – like clean energy – merely adds an additional layer of complexity. The Third Pillar of policy is much richer, and much deeper, than just the funding of R&D and protecting intellectual property. But for a book grounded in facts, R&D funding is a useful place to start.

9.3 Energy research and development (R&D) – public and private

The idea of R&D policy is of course not new, least of all in the energy sector where it has a long history. In the 1950s, governments became active in nuclear research, in part because of the military connections and the desire to secure civilian spinoffs. In the 1970s, in the aftermath of the first oil shocks, governments embraced energy R&D more broadly, though still with a strong nuclear emphasis. After the 1980s, attention to energy matters faded, along with related government budgets for R&D. Rising concern about the environment, and more recently about energy prices, helped to drive a slight increase from the turn of the century, and R&D was a major beneficiary of major 'stimulus package' expenditures in 2009 that sought to stave off economic collapse in the aftermath of the credit crisis (Figure 9.2).

The results of the boom in public funding for energy R&D in the decade from the mid-1970s were mixed. Though there was some valuable progress, some big US programmes became notorious cases of 'pork-barrel' efforts – continuing heavy expenditure for largely political reasons, with little prospect of success.[6] The collapse of oil prices in 1986 seemed to remove a major justification for the big energy innovation programmes, leading to widespread retreat. Governments still funded some basic R&D, but most took the view that if a technology was worth developing further, the private sector should and would do so.

There were important exceptions. The commitment to nuclear R&D continued, though at less than half its 1970s peak of $10bn/yr, partly to also address concerns around nuclear waste as well as for the huge international collaboration on fusion research. Support for solar PV, considered one of the more successful 'high tech' programmes, continued, while some countries – notably Denmark – maintained R&D efforts in support of its emerging wind

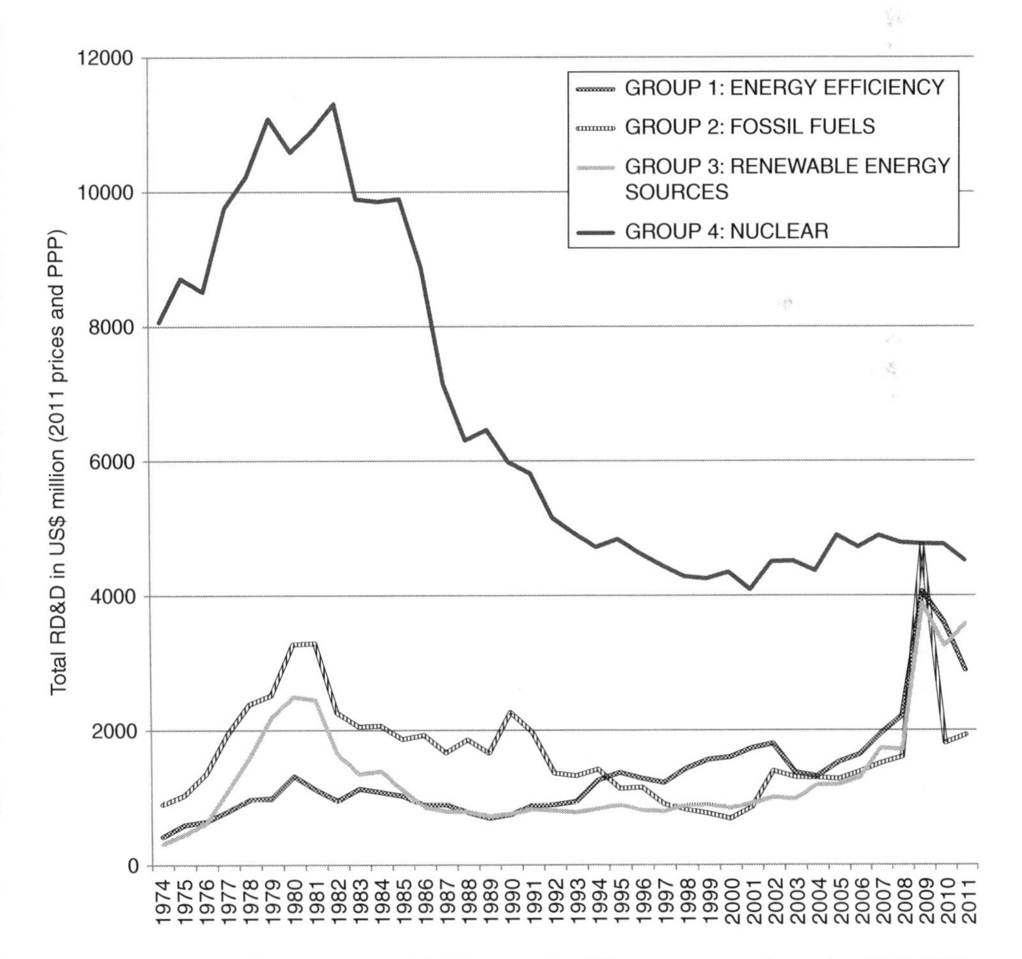

Figure 9.2 Total OECD government RD&D budget for different energy technologies, 1974–2011

Source: IEA Energy Statistics (2013), online at http://wds.iea.org/WDS/ReportFolders/reportFolders.aspx

energy industry. Overall public funding for renewables R&D maintained at below $1bn per year for 15 years, and the scale of overall energy efficiency R&D was similar, with both since rising as noted. All this compares with total global energy RD&D (Research Development and Demonstration) expenditures, including the private sector, estimated at around $50bn, and overall energy system investments globally measured in trillions of dollars (as summarised in Chapter 10).[7]

The idea, of course, was that public R&D would generate technology ideas that would be picked up and developed by industry, and that industry itself would do the 'heavy lifting' of demonstrating and commercialising new technologies. For these next stages can be difficult and more expensive, and commercial incentives – to manage risks while developing economically viable products – should play a bigger role.

In the 1980s, the oil sector became an increasingly internationally competitive business while most electricity systems were run by large, publicly owned entities. Some of these did indeed pursue sizeable technology programmes. The most prominent of these were the big nuclear programmes in France and Japan, both of which were essentially strategic collaborations between the government and the major electricity companies that they effectively governed. As noted in Chapter 3, the French programme from 1975 to 1990 cost an estimated €120 billion (in 2008 money), and transformed the French electricity system to one dominated by nuclear output.[8]

While the French and Japanese programmes – and the companies involved – became sources of national pride, the experience in most other countries was less encouraging. The UK, for example, between its government and the Central Electricity Generating Board, took disastrous decisions on nuclear power which permeated the system for decades after. Such national utilities could certainly risk money, but complaints grew about their inertia, commitment to big, established technologies and their lack of entrepreneurial culture. Combined with the US 'pork-barrel' programmes, the idea of governments and their public companies driving energy innovation fell widely out of favour, consistent with wider disdain for 'industrial policy' noted in the concluding part of this chapter.

Liberalising the electricity sector – pioneered in large measure by the UK – was supposed to fix this problem: the assumption was that the private sector would take on the R&D challenge and would do it better. In fact, liberalisation seemed to achieve the opposite. The new world of private ownership and competition wanted quicker returns and less risk; the long timescales and large uncertainties scared away potential investors, stultifying innovation. Such energy R&D as remained in the state monopolies largely collapsed after they were privatised. The history of energy technologies from the 1970s to the 1990s became littered with the corpses of those that never made it across the 'technology valley of death', which we examine in sections 9.5 to 9.7.[9]

While the tendency of society to under-invest in R&D is well known, less familiar is the fact that the degree of investment in R&D differs enormously between sectors. Figure 9.3 shows that top companies in the information technology and pharmaceuticals sectors, for example, spend typically 10–15 per cent of their gross turnover on developing new products. For electricity and construction, the figure is *less than a tenth of that.*

In many sectors, it seems, competition is a vibrant force for innovation – but not so in energy. A full understanding of the pattern is complicated by classification, but the discrepancy remains striking.[10] It implies that in tackling global energy and environmental challenges, we are seeking radical innovation in some of the least innovative sectors in our economies. Unless we understand what makes them that way – as set out in the sections on the 'technology valley of death' – policy is unlikely to succeed.

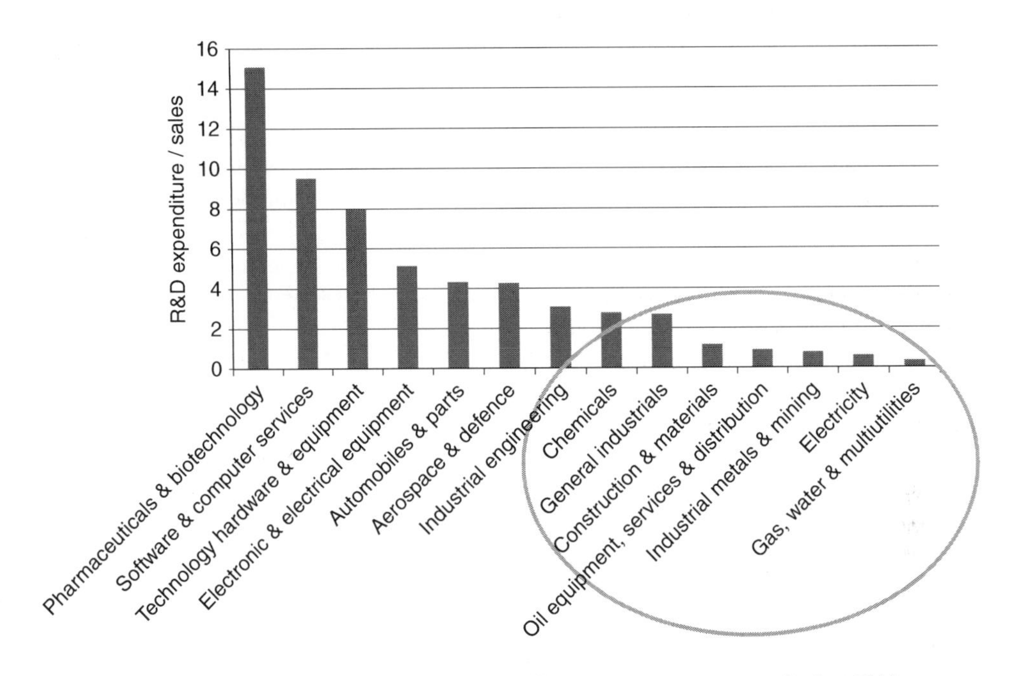

Figure 9.3 R&D expenditure by top companies in different sectors as per cent of sales, 2011

Note: The chart shows average sectoral R&D intensity from a database covering the world's top 1,500 companies. Service sectors are omitted.

Source: Data from EU Joint Research Centre on Industrial Investment and Innovation, R&D Scoreboard 2012, http://iri.jrc.europa.eu/scoreboard12.html

Moreover, the problems are further amplified if we seek innovation towards a global environmental goal, where the value to any innovator hinges upon the adequacy of – and confidence in – future environmental policy. The penultimate section of this chapter ('Porter's Kick') underlines more generally why controlling CO_2 emissions cannot be the *primary* lever for taking technologies beyond just R&D. We need to look more closely at all the interrelated factors which impede effective innovation – and what to do about it.

9.4 On learning

The public imagination tends to equate innovation with R&D. One key intellectual development of the last few decades is the understanding that it is in fact about far more. Indeed it is half a century since the Nobel economist Kenneth Arrow published his 1962 paper 'on the economics of learning by doing', which hypothesised – and argued – that much technology development and cost reduction does not in fact arrive like manna from heaven, or from R&D laboratories. In the real world, we often learn by doing. At the end of that same decade, the Boston Consulting Group, which had been contracted to analyse industrial cost reductions, introduced the concept of the 'experience curve' – a statistically observed reduction in costs which occurs as a technology grows in its scale of application.[11] These respectively offered 'top-down' and 'bottom-up' analytic breakthroughs to new ways of understanding innovation.

Like many crucial ideas, these concepts seemed to languish for a decade or two, slowly gestating, until an explosive growth of literature began in the 1980s, developing the concepts and evidence. Some of these related to the implications of assuming that technology costs would respond to scale underpinned by painstaking empirical studies.

The seemingly simple and unsurprising observation that costs per unit output generally decline as a technology or industry grows has huge implications. At the 'micro' level it suggests that numerous proposed or emergent technologies which currently appear dauntingly expensive may become much cheaper and could become competitive if and as they are developed at scale. Moreover, plugging such assumptions into models began to undermine common assumptions of a natural 'least cost' future determined by the resources and technologies available: such models could generate multiple different pathways for the development of systems, depending upon the direction of the initial effort, as shown in the rest of this book. Results like this formed one important input to the emerging ideas of evolutionary economics outlined in the next chapter, and indeed to broader 'chaos theory'.

The typical measure of such technology-specific learning effects became known as the 'learning rate' which charts by how much the cost declines for each doubled volume (of capacity or output). During the 1990s, researchers measured such learning rates for more and more technologies, and typically found a 10–20 per cent cost reduction associated with each doubling. If sustained, such improvements would have huge potential implications: if the production volume of a given technology increases tenfold, a learning rate of 20 per cent implies that costs would more than halve, while even a 10 per cent learning rate would imply that costs would fall by over a quarter. Plugging such numbers into models that project the costs of nascent technologies could turn them into something of a Wild West lottery, in which the prizes went to whichever technologies were the first to capture sufficient volume and keep others out – a far cry from the conventional view of a natural 'least cost' technology defined by some external fixed reality.

In practice, as the literature expanded inevitably the story became more complex. In the first place, causality works both ways: an expansion may cause cost reductions, but cost reductions should also fuel expansion. More careful analysis (of both data and causal factors) has gone a long way to disentangling these factors and continues to demonstrate a key role for causal (volume → cost) experience curve effects, but also underlines the mutual interdependences of the two processes.

However, learning rates were found to be quite uncertain and varied according to the technology and period of measurement (also prices are much easier to observe than actual costs). Figure 9.4 shows the ranges estimated for various (a) energy-consuming and (b) supply technologies. From hundreds – perhaps now thousands – of estimates, almost all are positive – there have only been very few observed examples of apparent 'negative learning' – but the ranges in almost all cases are wide.[12] The implication is that while learning curves confirm the likelihood that costs will decline with scale and may suggest potential magnitudes, they are far weaker as a robust predictor of the extent to which costs may come down.

Many strands of economic analysis and modelling still ignore such learning effects, because doing so simplifies analysis. Like the long battles over the evaluation of environmental damages, ignoring a known factor just because it is uncertain (or hard to model) is equivalent to assuming it to be zero – a value that is entirely contradicted by the evidence.

A closer look at the data has revealed some other important features and caveats – most crucially, that learning rates tend to change as a technology matures. High learning rates have been most associated with emerging and smaller-scale technologies (solar PV being the classic example), and lower learning rates are associated with more established technologies

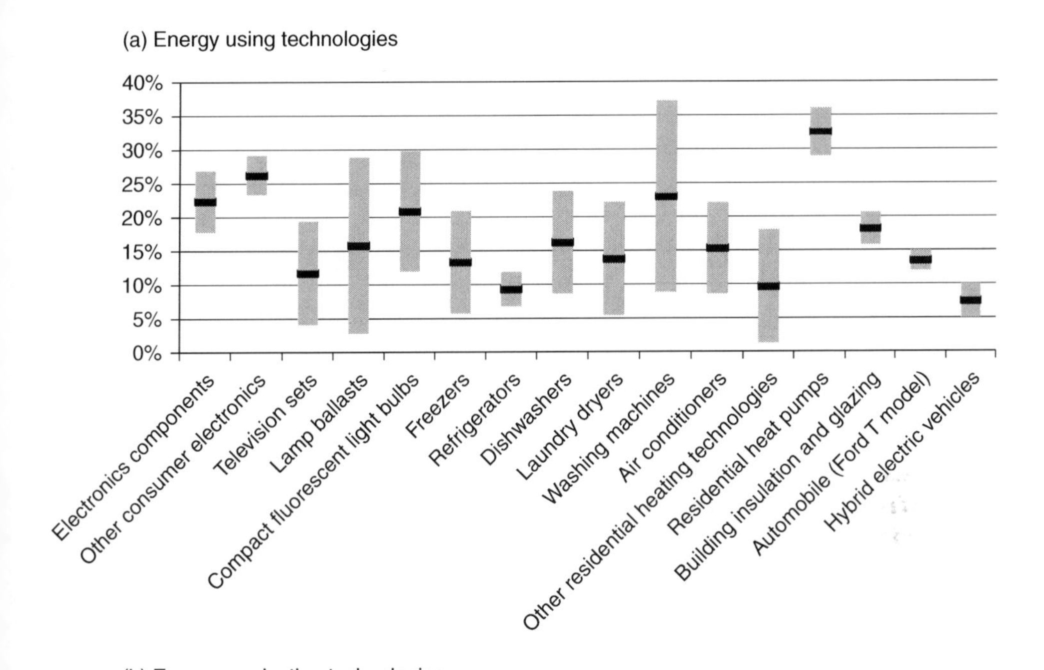

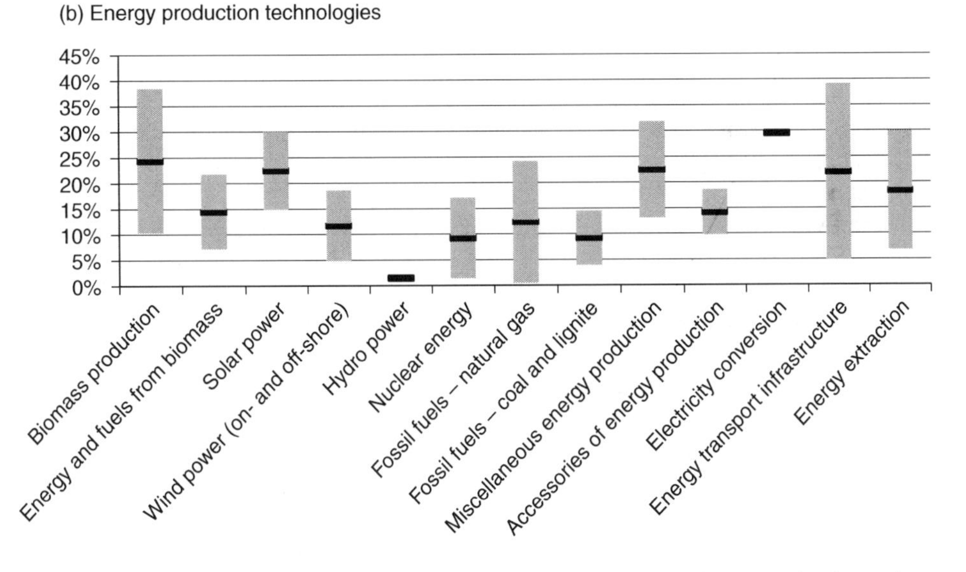

Figure 9.4 Cost reduction associated with a doubling of installed capacity or production volume ('learning rate') for various energy technologies

(a) Energy using technologies

(b) Energy production technologies

Source: Based on data from Weiss *et al.* (2010), Annex 1, supplemented with data on hybrid vehicles from Weiss *et al.* (2012).

and with larger industrial plants (such as coal stations). That is qualified, however, by the observation that in the very early stages, costs can go *up* as initial stages of deployment bring to light technological problems 'in the field' over full plant lifetimes. Thus, for example, one of the most extensive reviews suggests a general pattern in which learning rates may rise in the earliest stages of deployment (but with considerable uncertainty), plateau and then gradually decline as a technology becomes well established at scale.

As predictive tools for specific technologies, experience curves really need to be complemented by additional approaches, with due caution about the range of application.[13] As aggregate data, however, they underline not only the reasonable expectation that costs will fall as an industry grows, but the essential need to understand the fuller span of the innovation process.

9.5 The innovation chain

The previous two sections point to important issues at opposite ends of the innovation process: the research and development essential to the exploration of new technology ideas, and the extent to which cost reductions in a new technology may hinge on the scale of deployment. We can consider these 'push' and 'pull' forces in the innovation process. For innovation involves a complex chain of phases. Each involves different actors, barriers and policy influences. The *innovation chain* covers the whole path that starts with a technical idea and culminates in the diffusion of a new process or product.

As noted, following Schumpeter, economists became accustomed to thinking about three main stages: invention, innovation and diffusion.[14] In practice these divisions are still very coarse. If you have ever used a Mac computer you will be familiar with the idea of a line composed of many elements, any one of which can expand if you hover the mouse over it. Innovation is similar in that any one step can be expanded hugely when you take a closer look.

Figure 9.5 shows an illustration with the wide span between invention and diffusion broadened out into four distinct stages to make a total innovation chain comprising six stages.

- The process starts with an idea and *applied research*, that is the discovery of the core technology and its main functioning. This is far from a product, since the focus is on theory and principles more than applications. The solar photo-voltaic (PV) effect was observed for the first time in 1839. The first photovoltaic cells were built in 1941, but it took the space race to turn them into a product delivering power for useful applications.[15]
- *Technology-specific research and development* takes the basic discoveries and applies them to produce a prototype which is technically functional. The focus is on finding the design, materials and techniques to make it work: economic considerations are absent or secondary. The development may well include an element of demonstration, but at this stage mostly for internal testing purposes. This R&D stage may generate a discrete technology but it is not yet a product. What comes out of engineering labs can rarely be immediately commercialised. For example, people cannot use PV cells that just convert light into a voltage. They want panels which can be easily installed on the roofs of their buildings and which are trusted to produce a reasonable amount of usable electricity.
- Thus the next step adds a second D: *demonstration*. After testing that the basic technology works, it needs to be developed into a viable product that can be demonstrated to

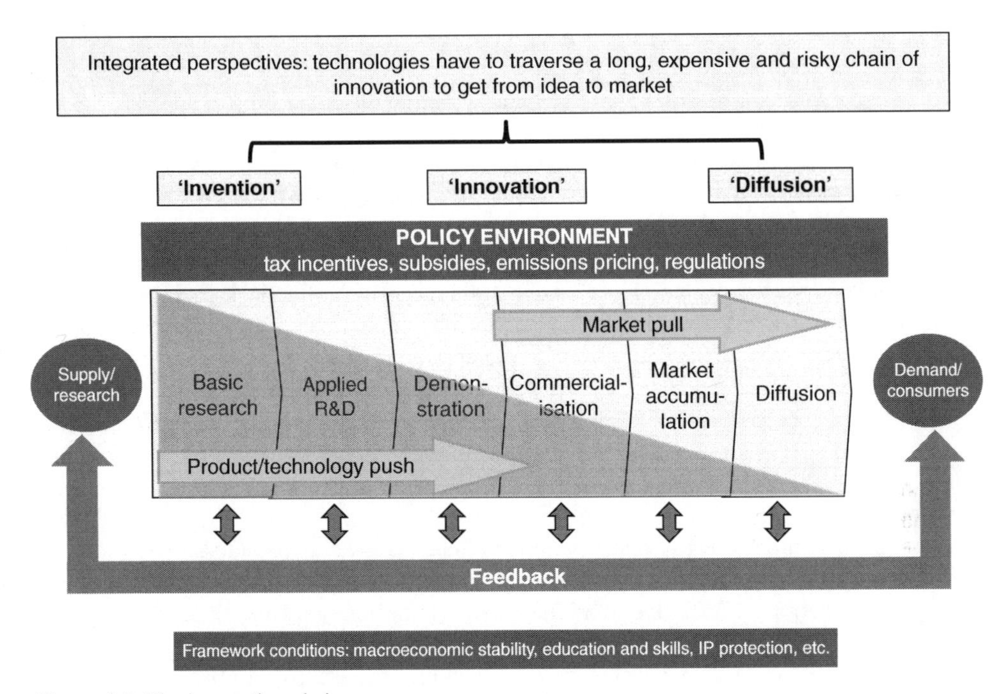

Figure 9.5 The innovation chain

Source: Authors, building on Grubb, Haj-Hassan and Newbery (2008).

potential investors – and users – to prove that the creation is something that has the potential to be used in practice. This process may be internally funded (for example, for R&D labs in major companies), or externally funded either from government, or more rarely from venture capital or innovators' private equity.

- In the *commercialisation* phase, a firm must manage the whole process of making the product and seeking customers, and all the financial and other considerations in between. For products which have emerged from independent sources (individuals, universities, etc.), this transition is one of the most difficult, because it involves a wholly different world – dependent on different skills such as management, business planning and marketing, and different funding sources, as discussed below.
- After this (precarious) transition from technology to commercial product, *market accumulation* is required in many dimensions: to feed learning-by-doing and economies of scale, develop supply chains, outlets and a first base of customers. Step by step, this may also involve infrastructure and regulatory changes. Revenues start to grow – but may still easily be outstripped by costs.
- The final phase illustrated in Figure 9.5 is the *diffusion* of the technology on a large scale: the product is moving to maturity and is no longer perceived as a novelty. Potential improvements do not affect its main functioning but, as illustrated by learning curves, contribute to lower costs and increased efficiency.

Needless to say, reality is not so simple nor linear. The deeper one looks, the more complex the overall process becomes. The fact remains that to be successful in application, any technology generally has to pass through all the stages indicated. If any are broken, it will fail.

Since the basic 'innovation chain' diagram was published in 2004, there have been dozens of variants, depending on the exact context, interests and expertise of the authors. Scholars of *invention* explore the many dimensions of that 'first' step (applied research); those who study *diffusion* also emphasise the extraordinary complexity of that 'final' step. The rest of this chapter explores the impediments in the central stages; the next chapter goes on to outline some of the obstacles in the way of diffusion and the transformation of systems.[16]

Note that the early stages are driven mainly by technology push efforts while the later stages are more dependent on the demand for the innovation. Several stages, however, involve some mix of both. This is partly for financial reasons, but more centrally the literature emphasises many points of feedback between the different stages. Scholars of innovation now recognise the innovation chain as too simplistic because of its visual implication of a linear path and have moved on to generate innovation maps of huge complexity to show linkages of different actors and feedback processes. Indeed, in practice most technologies traverse at least some parts of the chain repeatedly through successive product cycles (see Box 9.1). All the steps can also be influenced both by broad framework conditions which may either encourage or deter innovation, and by specific policies, as also indicated in Figure 9.5.

The 'learning' processes associated with experience curves documented in the previous section occur mostly in deploying a technology through the stages of commercialisation and market accumulation. This is the stage depicted by the learning (upper left) portion of Figure 9.1, and it poses a particular policy dilemma. Economic approaches to innovation, as noted, tend to see it in terms of property right protection with patents. Studies do indeed find that patent filings increase substantially during early deployment – which is another illustration of the various feedbacks between stages.[17]

However, it is not feasible to patent many dimensions of learning in these deployment stages. Moreover, patents essentially confer monopoly rights – yet it is particularly in these stages, as a technology reaches the market and demand forces start to dominate – that competition is most valuable for driving down costs. No one wants an emergent technology to be controlled by one company as it reaches maturity (except that company).

As noted, many technologies go through iterative stages, with improved models coming on to the market in regular stages (see Box 9.1) – consumer electronics are an obvious case. Only rarely do new technologies make a giant leap following either of the extremes illustrated in Figure 9.6. Most technologies follow some variant of the iterative path across the middle, with repeated rounds through various stages of the innovation chain as learning and markets expand.

9.6 When the chain is broken

Against this background, we return to the puzzle of section 9.3: why do some sectors spend 10–20 times more on R&D than others? Given the multiple reasons for interest in innovation, one would think this to be a central question, yet in all the research for this book we failed to find any systematic cross-sectoral analysis of this question – and indeed very few acknowledgements of its existence. There are, rather, separate literatures that analyse innovation in different sectors. In energy, a big topic is what has become dubbed the 'technology valley of death' – the observation that many promising technologies seem to get stuck in the early stages of the innovation chain.

Box 9.1 Iterating innovation

Innovation is an inherently dynamic process, not only because of the need to go through the stages of the innovation chain but because most products do so repeatedly as product improvements are brought to market. This is illustrated in Figure 9.6, where the upper-left corner represents a new technology with no market share (horizontal axis) and high initial cost (vertical axis). After its initial introduction, costs can decline and market share increases through a continuous process of R&D and learning, whereby small but continuous improvements in the technology foster its initial diffusion (technology push). Beyond a certain point, higher market shares are acquired through economies of scale and the corresponding cost reduction (market pull).

This iterative path – shown through the middle of the diagram by line (c) – typifies most technologies. There are other possibilities. Line (a) illustrates a large technological programme or 'breakthrough' innovation in which the three first phases of the innovation chain succeed in cutting the cost so much – or providing something so new and appealing – that once introduced to the market it can very quickly benefit from large economies of scale and rapidly move to its full market potential. Alternatively (b) it is possible to 'artificially' create a market for a technology through government subsidies, which combine incentives for strong efforts through the first three stages of the innovation chain alongside learning-by-doing in the field, with costs falling – and the subsidies being reduced – as these improvements come through.

France provides good examples of both extremes. Its nuclear power programme typifies path (a) (though even this drew on more extensive rounds of reactor development in the US military). Its Minitel services – which provided a terminal for IT-based information services in almost all French homes a decade before the Internet became widespread – followed a path more like (b). In doing so, each illustrates both the benefits and risks of such 'forcing' strategies.

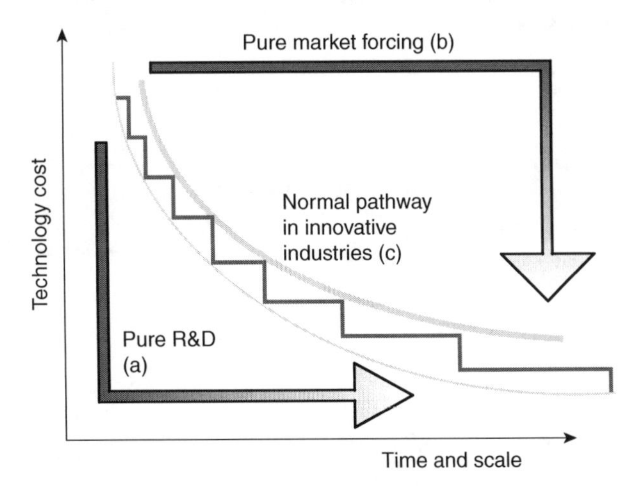

Figure 9.6 Technological diffusion pathways

Source: Authors, developed from CIRED, France.

We depict the basic issue in Figure 9.7. It is all about connectivity across the chain, and in particular between 'technology push' and 'market pull'. Determining factors include the *cost and timescales* of technology development in the former, and the *depth of consumer engagement* in the latter.

Consider the iPhone. It is a unique device and Apple had reason to believe that consumers would pay a high premium to get one; there was nothing else like it. It was expensive to manufacture and took a few years to develop – but that was trivial compared to the rewards from selling a product that hundreds of millions of people would want and no one else could produce. Apple continues to innovate to try and maintain the competitive edge by establishing new consumer products. Moreover, in the broader IT sector, the cost and timescale for developing new products – particularly software – is modest compared to that involved in big engineering. Indeed, innumerable small companies arise around software developments and the successful ones expand and may be soon bought out. Along with new products, the innovation process itself creates new and profitable markets. The result is a huge and rapidly evolving ecosystem of innovation, with close and continual feedback between consumers and innovators and small firms often leading the way with radical innovations.

In pharmaceuticals too, the companies know that a successful drug is a unique product. Consequently there is huge potential for demand – if the drug meets a need, health services and users will absolutely want to buy it. Moreover, since each drug is a specific unique and identifiable product, it can be readily protected by patents. The cost of development (and regulatory clearance) has grown over time, but there is a profitable and largely protected

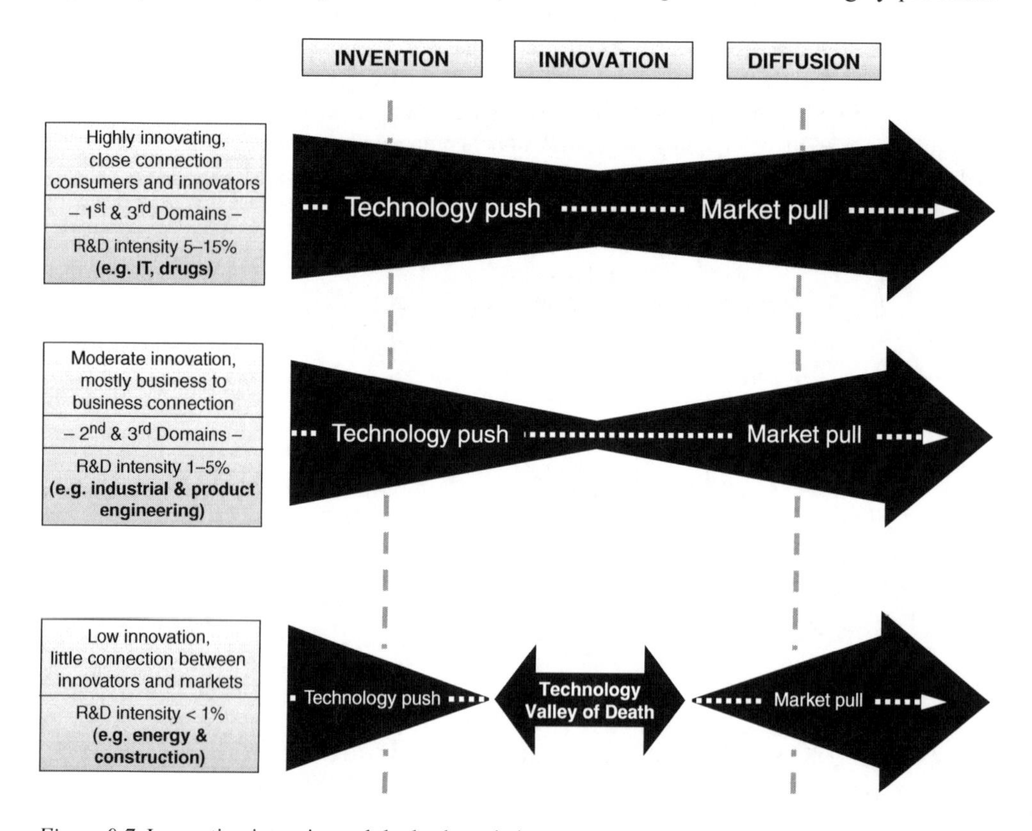

Figure 9.7 Innovation intensity and the broken chain

market in which companies can charge pretty much whatever it takes to recoup costs and make profit, providing the drug offers medical improvement or meets a new need.

In both these sectors, therefore, the innovation chain is well connected: technology push easily meets the pull of market profits.

Major classes of energy-consuming products, like vehicles and appliances, arise from sectors with lower but still substantial levels of R&D spend. Sectors like vehicles and electronic and electrical equipment manufacturers, which on average spend around 4–5 per cent of turnover on R&D (see Figure 9.3), share much in common with the high R&D sectors in that they have direct connections with the final consumers. Depending on the technology, development costs may be higher and the timescales longer, while the products meet well established needs rather than having the potential to create new markets. To the extent that energy is an 'embodied' property of the product and largely invisible to the final consumer, there is limited incentive for innovation to reduce energy use; however, if that changes as a result of energy or environmental policies, there may be a sizeable impact – see for example the impact of appliance efficiency labelling and standards noted in Chapter 5, and studies of patents in low-carbon vehicles noted later in this chapter (section 9.10). By connecting First Domain drivers to Third Domain innovation processes, environmental or energy efficiency policy can have a big impact on technology, and this helps to explain the remarkable and enduring response to some of the energy efficiency policies outlined in Chapter 5.

Contrast this with the energy and fuels businesses themselves. Consumers buy electricity or liquid fuels for their car. Any new production technology is still producing and selling the same basic product – energy. To get anywhere, it has to undercut the incumbents, yet consumers are wholly disengaged from the innovation process. Unless there is market support or protection, the technology must sell the same output (energy) at a lower price against technologies that have a century of development behind them. There is little prospect of large profit margins arising from a new patented product seizing the attention, enthusiasm – and pockets – of consumers in a new market.

This echoes a strand of mainstream economics literature which, in trying to understand some of the puzzles in economic growth, has suggested that product variety and quality play a significant role, not least because they promote innovation which in turn fuels growth.[18] These authors, however, never seemed to connect this generic macroeconomic reasoning to the unusually low levels of energy R&D.

This low level of innovation is, despite its huge potential *overall value*, the right-hand wedge depicted in Figure 9.1. It is very hard for any individual innovator to capture such value: some of it arises from slightly undercutting the technologies of incumbent equipment and energy suppliers (who will try to keep competitors out); some hinges upon future policy (which investors cannot guess); much of the overall value may be a public good (not reflected in private money); and most of the big benefits are long term – increasing both the risks and the likelihood of innovations spilling over to competitors.

This might matter less if the investments required were quick or small, but they are often not. Energy is fundamentally about *conversion* and the physics of converting large amounts of energy imply big engineering which takes time and money. 'Technology push' is thus also deterred. The innovation process for such technology is slow, not fast. The economic margins are small, not massive. The risks are high, the rewards are low, and the barriers, correspondingly high. In a nutshell, technology push does not directly meet market pull.

The 'technology valley of death' has been analysed and characterised in terms of institutional mismatches and financing gaps as well as straight economic incentives, but the result is the same: for these sectors, the bottom class in Figures 9.3 and 9.7, there is little or no direct link between the earlier stages of the innovation chain and final consumer markets.[19]

There is of course some continued innovation, but it is mostly conducted by large engineering firms and equipment providers, which carry the huge range of expertise and financial strength to develop complex new engineering systems. Clearly these do innovate, and tend to be in the lower ranks of the middle tier of R&D intensity. However, the relative expenditure is limited both by the (financial- and time-) scales of investment involved, and the fact that they are dealing with business-to-business relationships. These are, for example, the general industrials, industrial engineering and chemicals sectors that spend around 3 per cent or less of their turnover on R&D (Figures 9.3 and 9.7). Their technology development does touch their market, but they are not dealing in final consumer products. They are, generally, aiming to make incremental improvements in the performance and costs of what they sell to other businesses.

Moreover, in the energy sector, these engineering firms are selling to incumbent supply companies – the fossil fuel and electricity companies – who themselves are selling a wholly undifferentiated product to the final consumer. Competition is largely cost-based, without the potential for new and high profit margins associated with new consumer products. This dampened incentive for innovation (typified in the extraordinarily low innovation intensity of the energy suppliers themselves) feeds through the supply chain.

In Chapter 11 we note that innovation is accepted to be an important but poorly understood driver of economic growth, and that one of the factors suggested in mainstream economics is the contribution of product variety in driving incentives to innovate. This supports our suggestion that the lack of such diversity in energy is one explanation of energy's low R&D intensity – and again implies that accelerating innovation in such a sector has a high chance of being economically beneficial.

However, the observation has more problematic implications. While the major engineering companies are otherwise indifferent to the kinds of energy their equipment uses and produces, their industrial customers may not be. Incumbent industries will, obviously, be most inclined to spend their resources to further their own advantage in their own sector, largely to protect their existing interests and business models. Diversifying into wholly new technology areas where they have no comparative advantage – and which could undermine part of their existing asset base – is a pointless risk for them. They are certainly interested in technology improvement, but there is inevitably a strong bias in private R&D towards incremental improvements of established – mainly fossil fuel – technologies. The last thing they want is a wave of Schumpeterian 'creative destruction'.

This tendency is backed by painstaking research which demonstrates path-dependence in the innovation process itself – 'dirty' companies tend to focus on 'dirty' innovation.[20] Thus, while oil companies, with deep pockets and facing competition mainly from their rivals, spend vast amounts on exploration, this is matched by the amounts their service companies spend on new technologies for helping their clients find or extract oil from ever more remote and difficult resources. Such innovation is not helping to decarbonise the energy system – indeed, quite the reverse.

Major electricity companies, as noted earlier, tend to be either state-backed monopolies with little incentive to innovate or have been privatised to shareholders who have shown even less interest in supporting long-term innovation; several of the major utilities also have stakes in the fossil fuel business, though other energy multinationals have diversified into renewables.[21]

The weakness of the innovation chain in these sectors thus exacerbates the problems of path-dependency and lock-in described in the next chapter. As noted, radical innovations tend to come from new entrants – for obvious reasons – but in the conditions outlined, that path to innovation is blocked. The system instead tends to reinforce incremental improvements in existing technologies – and typically in consequently bringing down the cost of extracting and using fossil fuels, rather than fostering alternatives.

As a result, purely laissez-faire innovation in these sectors may be of little help (and indeed could make things worse). This is reflected in the results of some modelling work. For example, in response to vibrant economic debate around the importance of 'induced' technological change, Professor Nordhaus – noted in Chapter 1 as one of the first top economists to attempt global cost-benefit modelling of the climate problem – extended his model to 'endogenise' technological change. He duly incorporated empirical estimates of the role of R&D in the fossil fuel sectors. Not surprisingly, he found that the corresponding impact was small. When the problems lie in the innovation chain itself, environmental policy will not fix them.[22]

9.7 The technology valley of death

Consequently, to extend innovation in new energy supply technologies in particular, the innovation chain must be bridged in ways that support the entry and growth of new and cleaner technologies and associated industries.[23] As noted earlier in this chapter, however, governments have a mixed record in driving technological development, and as noted in Chapter 1, demonstrations of 'market failure' do not on their own make a case for intervention – the cure must be demonstrably better than the disease.

Addressing this challenge requires first looking more closely at the nature of the 'technology valley of death'. As outlined in the previous section, this concept describes the situation in which the nature of the industries and markets serves to disconnect the 'push and pull' forces in the innovation chain. This disconnect may be amplified by various barriers to entry and the very limited progress of governments in reflecting one of the key potential drivers – environmental benefits – in actual prices (as described in Pillar II of this book). Some of the clearest cases apply to low-carbon electricity sources which, as previously noted, must be a fundamental part of any low-carbon strategy.

The problem ultimately expresses itself in finance. The weakness of 'demand pull' forces blocks finance from customers while bank and equity finance is further deterred by all the classical limitations on private R&D funding (spillovers, etc.). As noted in section 9.2, R&D in clean energy sources consequently depends heavily on public finance. The initial stage of R&D funding, whether by academia, industry or their consortia, can be relatively cheap. However, the scale of costs starts to rise rapidly as a technology moves into demonstration or initial commercialisation.

At the opposite end of the chain, final diffusion of a product provides both increased finance and reduced uncertainty. But as outlined above, a technology may have to go through extensive stages of market-based learning and cost reduction to get this far – which itself requires a path to market.

Thus the wide span of intermediate phases are the most difficult – demonstration and commercialisation where public financing is increasingly stretched, and initial deployment where a technology has not yet benefited much from learning-by-doing and scale economies – are the most difficult.[24]

Figure 9.8 shows in graphical terms the essence of the problem. The valley of death occurs where the costs of further development are increasingly hard to support from direct public finance and there is a desire to inject private finance and competitive forces – but private capital may be either non-existent or only available at prohibitively high interest rates.

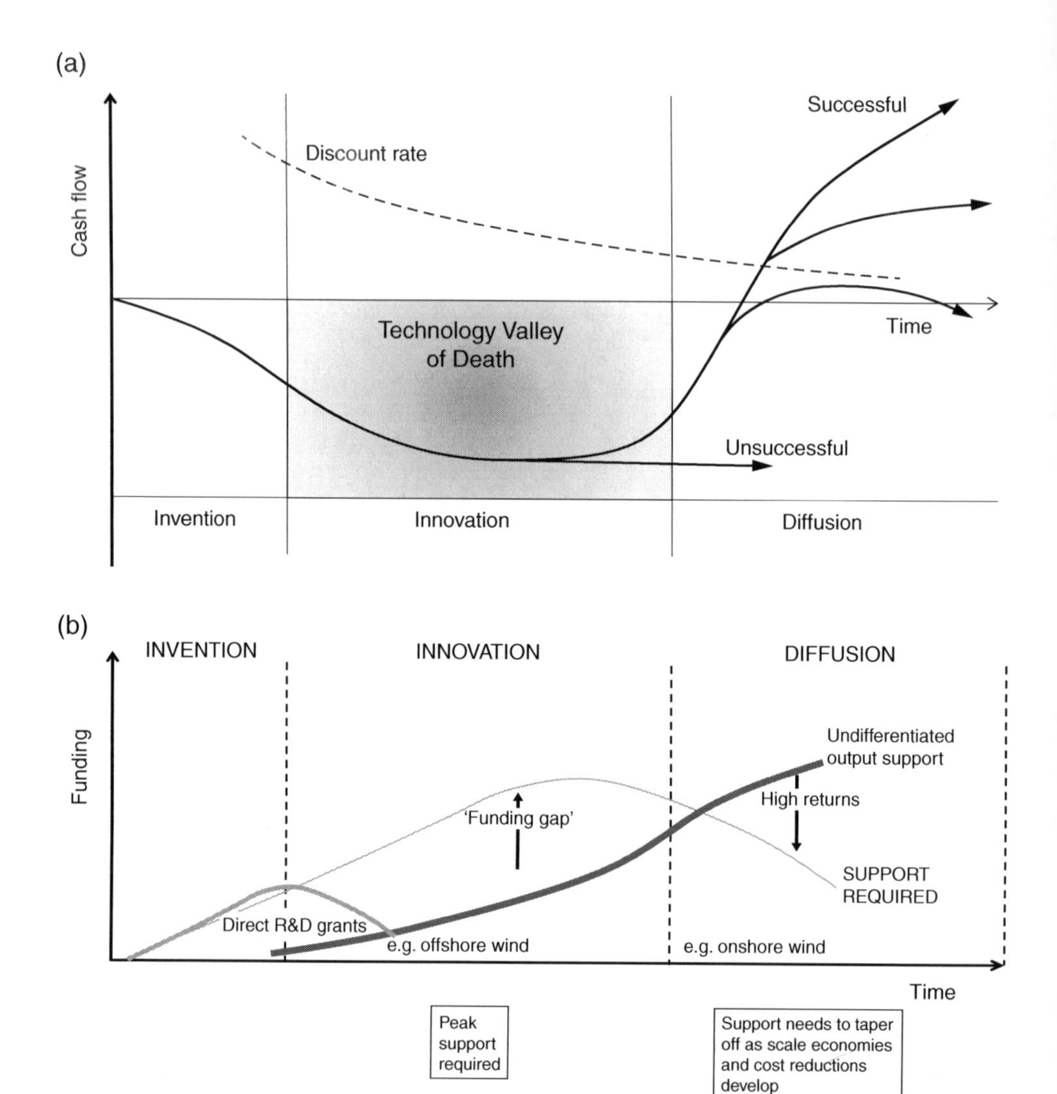

Figure 9.8 (a) The technology valley of death. (b) The funding gap under R&D plus undifferentiated demand-pull subsidies

Sources: (a) Adapted from Murphy and Edwards (2003). (b) Adapted from Carbon Trust (2006).

Different factors inhibit progress across the four central components of the innovation chain (Figure 9.5). Three factors in particular impede the 'technology push' into the commercial world:

- *Uncertainties and knowledge divisions around risks and rewards* may be of limited interest to the real inventor, but to investors – public or private – they are fundamental. Risk and uncertainty are endemic in many economic activities but 'R&D investment appears qualitatively different' for reasons akin to the fundamental characteristics of Third Domain security risks outlined in Chapter 2: rewards to invention are not only uncertain but may be dominated by low probability, high consequence outcomes.[25] As a technology moves through the innovation chain, the engineers may gain confidence about technical performance but have little understanding of the markets their product might sell into; the reverse is likely to be true of potential investors. For any individual investor there is therefore a strong temptation to wait and see what emerges. From a public policy standpoint, however, this is useless as it simply defers the process of innovation and diffusion. Even if there were no IP/spillover problems, a better pace of innovation thus takes public money.[26]

- *Transitions from technical to management and commercial skills* are particularly problematic in the earlier stages of innovation. A good technology development team will be dominated by researchers, but they may have little understanding of potential markets and business practice and may be unable to formulate a convincing business plan. Investors are more likely to back projects proposed by groups with greater business expertise. This also exacerbates the risk of uncertainty and information asymmetry that drives 'adverse selection', akin to phenomena in the First Domain (Chapter 4): an inability to discriminate between better or worse technologies may mean that investors' money becomes easily diverted to those with inferior technology but the best sales pitch.[27]
- *High capital costs and long timescales* are typical of renewable energy, since the resulting fuels are largely free. The more investors have to commit up front and the longer they have to wait to see a return on their investment, the greater the uncertainties (including around government policies) and risks, thus deterring investors.[28] Investors are in effect being asked to bear not only the technology and market risks, but those of public policy and politics over successive election cycles.

As a result, private investors prefer to support projects at more advanced phases, when they can better assess and manage the risks. Yet three additional factors serve to undermine finance for this potential 'demand pull', given the absence of the strong potential driver associated with producing new consumer products:

- *Economics of scale and experience.* As noted in section 9.4, innovation requires ongoing improvement of the product including learning by doing and scale effects. Yet initial plans have to focus on market niches and not on a large customer base. Learning-by-researching may be of declining value as a technology moves into production, but there is insufficient volume to fuel enough learning-by-doing to reach a sustainable market.[29] Potential financers are obviously aware of the learning processes, but bankrolling scale-up as the R&D funding declines (as illustrated in Figure 9.8) involves much higher funding and thus amplifies the risks. Again, the resulting hesitation can interrupt the growth – perhaps fatally. The 'solar surprise' (Chapter 3, Box 3.4) happened largely because special tariffs stimulated the demand while some countries – notably China – took the investment plunge to make the leap in the scale of production facilities.
- *Misalignment of private and public goals.* Government involvement for public policy purposes – such as to encourage environmental innovation – may not sit well with the motivations of investors, particularly if governments do not 'put their money where their mouths are' in terms of emissions pricing or other policy. To get R&D funding, the innovators themselves may have been focused on the public goal (e.g. low emission technology), but investors are likely to heavily discount public policy promises that are not backed by action, and may conclude that technologies backed by governments have been motivated for reasons not aligned with commercial interests and are thus more risky.

- *Incompatible public policies and understanding of the full innovation chain.* A key problem can be the failure of policy in some areas to reflect the intent in other policy areas, stemming either from lack of a common view on the objectives or on inadequate understanding of the innovation process. In-firm innovation requires companies to have some financial surplus and an eye to the strategic future, which may be hard to sustain in the face of potentially hostile takeovers based on maximising short-term return for shareholders. Yet, EC merger assessments tend to focus solely on maximising competition in the short run, and not consider impact on innovative capacity.[30] More generally, establishing new industries requires the coordination of many elements, including associated infrastructures and regulatory frameworks (see Chapter 10).

Examples of the latter effects have been rife within energy policy. For example, acknowledging the justification for renewable energy support combined with the desire to minimise government involvement in technology choice led the UK initially to implement a 'one-size-fits-all' level of support for renewable energy output (Box 9.3). The result was a first-past-the post effect, in which the most mature and lowest cost technologies (in this case onshore wind energy developed mostly in other countries) received excessive funding as their scale grew, while other crucial technologies that involved greater risks (but big long-term potential) languished.

If there is one thing we have learned it is that, at least in energy, maximising competition and minimising costs is not synonymous with maximising innovation – which may in part explain the economic success of countries like Korea and China, which have taken more pragmatic approaches to the challenges of innovation. Other regulatory, institutional and infrastructural barriers impede the large-scale diffusion of technology, as indicated in Chapter 10, and obviously these feed back into inhibiting the potential scale of 'demand pull'.

Rebuilding the innovation chain in these circumstances thus requires policy to both:

- *push further*, with public funding of R&D, demonstration of both technologies and their applications, and direct funding support for the transition from products to a business; and to
- *pull deeper*, with policies which create a price premium for newer technologies that generate the same product (e.g. electrons) but in better (notably, cleaner) ways.

A full survey of policies and experiences is way beyond the scope of this chapter – they are increasingly well charted in publications of the International Energy Agency, for example (IEA 2012) – but a few key aspects and examples are useful to note.

9.8 Pushing further

As noted, many governments have long financed both basic research and some energy technology development. European rules to contain 'state aid' are relaxed for R&D support. In addition most governments encourage corporate R&D, for example through tax breaks. The key point in bridging the technology valley of death is that simply financing R&D is not enough: if the chain is broken, the process needs to extend a technology to become a viable product with enough credibility to attract a sale or investment.

Indiscriminate financing of technologies without clear criteria, assessment and monitoring is obviously risky (see note 6 on the US experience of 'pork-barrel' projects) and ultimately unproductive: careless investment choices are not only a way to 'pick losers', but also a way to divert resources from better projects. Yet restricting government policy to

R&D is itself a policy choice – one which, in these sectors, is likely to leave the results of R&D stranded. There is a need for smart policy that helps to navigate through the barriers noted.

The most radical innovation tends to occur in start-ups and small companies, but these companies rarely have the resources or skills to turn their innovation into a viable product. Large companies can bring more weight to expand technologies through the later stages on their own – and also more lobbying power which could distort policy. Exceptions may make sense for inherently big challenges, particularly when linked with existing technologies like CCS. But in general, efforts in these earlier stages of 'pushing further' are helping new entrants and smaller companies.

Incubators

As noted, it is an illusion to believe that once an innovation is integrated into a product, it can be transferred to market. Entrepreneurs, financiers and potential private backers share neither common language nor motivations: no careful investor will fund an innovation that has no sense of market, management or at least the bare bones of a business plan. Governments – and universities – have become more accustomed to this, and it is now common practice to fund various *incubator* and *technology transfer* services. These offer start-up companies connection between funding sources on one side and research institutes like universities on the other. Although these networks are generally not yet able to make a profit and still need public funding, they can generate considerable social benefits (i.e. employment, new products and higher return from R&D investments).

Incubators can have different purposes: in some cases, their aim is to connect different geographical areas and fill the economic gap between them, like the Business Innovation Centres created by the EU. Others aim to foster conversion, for example from declining to emerging industries. Incubators which specialise in providing entrepreneurial skills to high-tech start-up business tend to have the strongest connections with university and research centres.

Public–private partnerships

Public–private partnerships tend to form a key role in innovation particularly for technologies with high social benefits but with low immediate market returns and/or some of the larger technologies. They need to follow an industry-driven strategy: more than setting technical priorities they help innovators understand what the market will buy. They then target support towards projects which are more likely to be successful once commercialised and engage the private sector.[31]One well-known energy example was the US PV:BONUS partnership.[32] Major examples in the UK include the Energy Technologies Institute and the Carbon Trust's 'Offshore Wind Accelerator' (see Box 9.2).[33]

Such cases illustrate ways to engage larger companies to handle bigger technologies on a more institutionalised basis, though managing the competing pressures and negotiating the intellectual property rights is complex.

Market engagement

The concept of market engagement is not new. An example often cited is the Risø test laboratory in Denmark, sometimes credited as the cradle of the wind industry. During the 1970s – and building on a much longer heritage of Danish wind energy enthusiasts – Risø began

Box 9.2 Building the missing link – the Carbon Trust's technology accelerators

Since its inception, the Carbon Trust aimed for all activity to be grounded firmly in commercial reality. With a recognition that private capital is key to mobilising the transition to a low-carbon economy, the end market played a key role in its innovation programmes. Almost by definition, low-carbon innovators operate in an area replete with market failure, but it proved possible to create activity engaging end users, thus encouraging the private sector to do as much lifting as possible. The Carbon Trust set up a range of interventions targeted at market failures, including applied research funding, incubation support, venture capital investment and sector-wide technology accelerators that focus on the specific challenges each market faces. All activity is focused on leveraging in as much private sector capital as possible.

Some sector-wide technology accelerators focused on demonstration projects, such as smart metering, micro-wind and micro combined heat and power (CHP) field trials, in which the critical challenge has been deemed proving (or not!) the benefits of the technology. Successful demonstrations in end-users' sites enabled the Carbon Trust to compile robust market-tested insights into their performance.

In other sectors the Carbon Trust has brought together key market players to reduce costs and lower risks. The 'Offshore Wind Accelerator' convened the majority of project developers to reduce the cost of offshore wind by 10%, focusing on a set of key problems from foundation designs to site access systems. Importantly, all are users of the technology and don't compete on the technology itself. This not only creates significant market pull for innovators, but also allows the group to take on and share more than they could stomach individually – every pound invested by a developer is leveraged 13 times due to contributions from others.

This approach to accelerating market pull by convening the ultimate customers of innovation is also being used in the areas of marine energy and developed for aviation bio fuels. Such approaches not only allow market-relevant innovation to be developed at a faster pace than would otherwise happen, it also leverages significant private investment from participants. The process is analogous to the crowd-funding models often popular in consumer product development.

subjecting the faltering technology of wind turbines to rigorous, independent and publicly accessible test regimes. It laid the basis for what has become a global industry.

A rather different model emerged with the UK Carbon Trust's programmes which operated across a wide range of technologies and came across the barriers in the innovation chain in very concrete ways. Numerous would-be vendors claimed to offer technologies that would cut energy consumption, costs and emissions. Potential buyers remained stubbornly uninterested or sceptical. It became clear that the 'natural' process of selection would be painfully slow. To overcome this, the Carbon Trust launched two pilot 'technology accelerators' to field test technologies, one on micro generation and the other on 'smart meters'. The contrasting results of these showed the value of an independent entity carrying out practical field tests to validate technology claims. Where the technology could not deliver what the customer wanted, the results fed back to the developers and spurred them to adapt or improve their product – not their marketing. Where the technology delivered, the independent assurance and data provided along with it proved invaluable in boosting market confidence, giving clarity around performance and the most effective situations for deployment. On the basis of the Carbon Trust's first two Accelerators, one technology (micro-wind generation) was in effect sent back to its proponents with the message 'needs to improve', the other gained such a level of credibility that the government felt confident enough to accelerate its adoption by developing a national deployment programme, now with substantial business support on the basis of the field trials.[34]

Institutions for accelerating industrial innovation are not unusual. A well-known example is the German system of Fraunhofer Institutes. These host creative researchers that effectively allow companies to outsource some of the innovative activities (and pursue them with expertise and scale that might be difficult to pursue internally). For researchers these institutes provide a stimulating environment to explore challenging tasks, while not being exposed to the job risk of start-ups. A key feature is thus that such industrial research institutes are broadly engaged with the established manufacturing industrial base in Germany. Consequently, there are already industrial 'deep pockets' able to pick up the ideas and solutions emerging and run with them through the rest of the innovation chain.

The challenge for low-carbon innovation is that the technologies needed are not neutral, and may have the potential to disrupt existing business structures. This may require the creation of new companies, not just feeding existing ones with new ideas. Established companies are not neutral with respect to this kind of innovation. Existing industries may more easily span the innovation chain, but as noted in this chapter and the next will have some bias towards technologies that help them build on their existing comparative advantages – not to move in wholly different directions. Change requires the symbiotic development of new technologies either with new companies or changing strategies in existing companies as a response to the developing policy landscape.

The Carbon Trust was unusual in focusing on commercially-oriented innovation with a *public* purpose – decarbonisation – and operating more widely across the supply chain and across more diverse and smaller technologies rather than the fashionable 'big ticket' items. It turned its attention to end-use and interface technologies. Based on its initial experience, the Carbon Trust went on to launch several other such technology accelerators, ranging from low-carbon buildings to marine energy, with each adapted to the particular needs of the developer–market interface. In some cases, convening the major market players helped to share risk and create stronger market pull, as in offshore wind. In other cases, including biomass heat and marine energy, new skills were injected into established industries to disrupt the status quo and bring fresh, objective thinking to opaque industries.

The level of flexibility and business buy-in enjoyed by the Carbon Trust is unusual. The underlying message, that industrial markets can be very slow to trust an emerging technology is not a new one. Independent testing, evaluation and commercially-oriented demonstration can help. Specialist and government-funded institutions can have a particularly important combined role to play if they have sufficient resources and freedom. State-backed institutions operating in this area always seek to leverage private finance; its unusual characteristics may have enabled the Carbon Trust to gain a higher level of leverage (seven-to-one) than more traditional state-backed institutions.[35]

Such initiatives particularly help to overcome the first two barriers to 'pushing further'. By helping to reduce the perceived risks, they also help to reduce the cost of financing and hence the third barrier of capital cost and timescale risks, particularly for smaller-scale technologies. Reducing the risks for bigger-ticket and longer-term investments – more specifically, sharing the risks between the private and public sectors given the potential public benefits of a clean technology – requires higher levels of capital support. These may be best mediated through more traditional channels of direct government grants, though public-private partnerships may also involve direct capital finance to buy down the risks of pushing larger technologies more widely.

9.9 Pulling deeper

A bridge needs two ends. For spanning the technology valley of death, the other end comes from the market side. Because of the very modest incentive associated with simply producing the same consumer product (fuels or electricity) in different ways, and particularly given the hesitant progress towards valuing environmental benefits, how best can we support cleaner technologies to expand and progress along the learning curve? These policies usually require strong government involvement, and the rising scale means costs also rise, to close the gap in profitability between incumbent, carbon-intensive technologies and emergent alternatives.

Some form of transitional subsidy is implicit in accelerating the public advantages of getting such technologies along the learning curve. The second barrier noted – the cultural gap between public and private finance – implies that support needs to be channelled through the market. Over the past two decades, a great deal has been learned about ways of doing this.

Niche markets and 'strategic deployment'

The logic of 'pulling deeper' points to a crucial role for early market experience. One approach is to identify and support initial niche markets, where the technology has a particular advantage. A whole literature has emerged around the concept of 'strategic niche management' as a bridge to larger-scale diffusion.[36] Obvious examples include the application of solar PV starting in remote areas where energy is exceptionally expensive, and for moving into grid-connected applications in regions with solar-driven peak heating loads, where PV is also particularly valuable. In helping to build up an industry, niches can also help to overcome infrastructural and institutional impediments, as discussed in the next chapter.

In essence, niches for energy technologies provide a route to expansion equivalent to the classical process of sequential models for consumer products. Either approach enables the technology to traverse the central, iterative expansion-and-learning route depicted in Figure 9.6. In the absence of natural niches, the policy option is to drive the process through strategic deployment – subsidising the use of technologies before the market would otherwise do so, for strategic reasons: to reduce the costs through market learning, industry building and economy-of-scale effects. This in essence means providing publicly driven funding for the technology learning portion – the top left quadrant – of Figure 9.1.

Two broad categories of support are available, one linked *directly* to the investment decisions, the other *indirectly* through support linked to the output – power production, in the case of renewable electricity sources.[37]

Investment can be supported *directly* through grants, preferential tax treatment or the provision of concessionary loans. Grants typically involve limited transaction costs and thus allow for the support of smaller-scale installations. Tax benefits have some attraction for policymakers, but also drawbacks.[38] Concessionary loans reduce the financing costs and can address potential constraints on the access to debt for project finance for emerging technologies.

Such upfront support is particularly helpful for demonstration projects and those at the early stages of commercialisation, where the multiple uncertainties reduce project developers' ability to use debt finance.[39] As technologies start to climb up the other side of the valley and be applied on a larger scale, the nature of support needs to change. If the upfront payment is a significant share of the total project value, developers may focus on capital subsidies rather than lifetime performance; early deployment programmes for wind power in both California and India linked support to capacity installed, resulting in poor-quality wind turbines which yielded low output and frequently broke down. Simple capital support also increases the burden on today's users, while later users gain the benefits of the cheap subsequent output.

For these reasons most support for renewable energy is linked to the production of power over the operational period of the projects. This helps both to incentivise good performance, and spreads the cost better across the beneficiaries. Three main approaches are used to provide this support:

- *Renewable obligations* (known in the US as 'portfolio standards') define a certain percentage of total electricity sold by each supplier that must be produced by renewable energies, usually implemented through a system of tradable certificates. These have supported the growth of renewable energy, but at the same time have raised concerns that the approach transfers a regulatory risk to the private sector, contributing to the increased cost linked with insurance premiums and eventually causing leakages of funds from the beneficiaries.[40] The UK experience offers salutary lessons (see Box 9.3).
- *Feed-in tariffs* fix a specific price per unit of electricity (kWh) generated with renewable energies. Feed-in tariffs reduce uncertainties while maintaining competitive pressures on the industry, and have been fundamental for the large-scale adoption of wind energy in Germany, Denmark and Spain. The relative certainty made them most appreciated by investors and the sheer success in terms of volume has led to its widespread use. However, the more recent experience has also underlined the need for more sophisticated adjustment mechanisms. This particularly applies to solar PV, where high costs early in the 2000s necessitated high levels of feed-in tariffs. The policy became a victim of its own success: the 'solar surprise' (see Chapter 3, Box 3.4) combined unimagined rates of growth with rapid cost reductions. Countries that had implemented such tariffs, like Germany and Spain, were not able to adjust the rates for new installations quickly enough, generating sizeable profits for those who moved fast and larger-than-expected bills for everyone else. Spain and the UK made hasty and retrospective attempts to reduce the tariffs, undermining investor confidence. Increasingly, feed-in tariffs now incorporate automatic adjustment mechanisms to alter the tariff level in response to deployment volumes.
- *Auctions* for a specified amount of renewable generation capacity offer another way to support new energy sources. The UK's non-fossil-fuel obligation was an early and problematic attempt (Box 9.3). California recently implemented a Renewable Auction Mechanism (RAM) based on the principle of reverse auction, in which competitive bids are used to set the support level.[41] Despite the theoretical attraction, the approach has struggled to deliver sufficient predictability on timing, volume and clearing price of auctions, and subsequent delivery was associated with the well-known problems of 'winner's curse'.[42]

A single approach is unlikely to suit all circumstances. For example, some developing countries in particular may suffer from conditions in which market and competition forces do not support industrial development, innovation and widespread diffusion. Impediments may include inadequate infrastructures (especially in rural areas), government tariffs to protect local industries, insufficient market demand and transparency, severe imperfection in financial

Box 9.3 How not to do market pull – the history of UK renewables policy

The UK government first embraced renewable energy in 1990: it did so in the context of bold policy experiments to open up UK energy to competitive forces, reflecting the liberalisation philosophy of the Thatcher government. It turned out that the markets could not be persuaded to buy nuclear power and an enterprising civil servant slipped in renewable energy under the banner of support for 'non-fossil energy'. Nuclear power was subsequently withdrawn from the Non-Fossil Fuel Obligation, leaving the NFFO as a mechanism to ensure premium payments for electricity generated from renewable energy.

The NFFO reflected a belief that the private sector should find the cheapest price to deliver a certain amount of renewable energy – and invited them to tender bids in different technology categories. Of course, the cheapest bids in a given category won, running the risk of 'winner's curse': projects that used unrealistically optimistic assumptions won bids, but then had to face the reality of risking hard money on construction (see note 42). Without a penalty for companies that didn't implement their 'bargain basement' proposals, a significant number of winners never proceeded to completion. With cost minimisation as the driving concern, UK renewable capacity slipped behind continental counterparts.

In 2002, the government switched policy to undifferentiated Renewable Obligation Certificates (ROCs). Similar to US 'Portfolio Standards', this mandated a fixed degree of renewable capacity. Fears about the costs of the policy were assuaged by a price cap (3p/kWh). With an underlying view that governments should not seek to 'pick winners', all renewables competed equally. Most of the support ended up going to the least risky, best-established technologies – mainly onshore wind projects and co-firing biomass in existing power stations. The fact that the UK domestic renewables manufacturing had lost out in the 1990s meant that foreign manufacturers were big beneficiaries. This contained costs, but also undermined any UK innovation benefits; this weakened public support, which manifest as growing public opposition to planning consents. By 2008, UK renewable capacity ranked almost bottom among European countries, despite having some of the best resources.

Faced with overwhelming evidence of these problems, in 2006 the government introduced 'banding' in which the less developed renewables received multiple credits to foster innovation and completed the price cap with a 'ski-slope' that ensured ROC values would be maintained should targets be overachieved, to give investors the price confidence they had been saying all along was needed. The UK had in effect been dragged into the messiest and most complicated way of delivering feed-in tariffs yet conceived.

markets and cultural barriers. Policies have to build on strengths and domestic interests, and address the weaknesses of each country, that manifest themselves in a far wider set of issues than just financial support. Nevertheless, recent evidence appears to confirm the logical extension of the 'demand pull' effect to international policy: because anticipated diffusion drives innovation, policies that encourage international diffusion of clean technologies can have a major impact on innovation activities.[43]

9.10 Porter's kick

More generally, the scale of finance required to pull technologies through the innovation chain tends to rise as their scale grows and the obvious question arises: does the use of a 'stick' for environmental policy also help innovation? Can environmental policy itself – controlling emissions – supplant much of the complexity discussed?

The role of the 'stick' came most controversially to the fore with the publication of a short paper by Harvard's business guru Michael Porter (1991), in which he suggested that environmental regulation might improve industrial competitiveness by stimulating innovation. It amounted to the proposition that kicking companies with regulatory tools – in this case to address environmental problems – created a demand for environmental improvement which could best be met by innovation, and that this innovation ultimately could improve firm competitiveness. The 'Porter hypothesis' provoked howls of incredulity from those who considered it ridiculous to assume that a constraint could improve performance. Yet Porter presented convincing case studies and it reflected his wider analysis of business competitiveness, which many in business and management studies found compelling. He followed up his contention with a far more extensive article in the *Journal of Economics Perspectives*, setting out five key arguments in addition to further evidence.[44]

Like many ideas, it slowly gestated, and attracted a small but growing band of academics to probe further. A sizeable body of research has now accumulated and has started to add some light to the heat. One useful clarification was to distinguish between a *weak* and a *strong* form. The weak form simply asserts that environmental regulation can stimulate innovation. This is now clearly established by studies of the impact of environmental regulation on patent filing in several sectors – including clear evidence that gasoline taxation stimulated innovation in more efficient vehicles for example.[45] This is an environmentally specific form of the induced innovation hypothesis – that innovation will be affected by market conditions – which is now widely accepted and, indeed, is common sense.

The strong form of the Porter hypothesis adds a second step: that the resulting innovation can improve firm competitiveness. This is far more contentious. To the extent that firms are assumed already to optimize their level of innovation, focusing it in a particular direction may detract from innovation efforts elsewhere (the 'crowding out' hypothesis). At a general level, theory increasingly challenges this presumption; the conclusion from empirical research is 'not proven' and varies according to the case.[46] In Chapter 11 we suggest reasons why it may be logically impossible to resolve the debate about the strong form of the Porter hypothesis *in general*.

However, interesting patterns are emerging. For example, instead of focusing on patents – which cannot themselves measure competitiveness – a recent study looked at how different forms of environmental regulation correlate with the profitability of German firms. It finds that environmental regulations which have required improvements in resource efficiency have statistically been associated with improved economic performance, whereas those requiring 'end-of-pipe' solutions have done the opposite.[47]

The research notes that carbon pricing straddles these two characteristics. Empirically, there is evidence that despite the low-carbon price, the EU ETS has encouraged low-carbon innovation as measured by patents.[48]

All this brings us to the heart of the relationship between Second and Third Domain processes and associated policies. The fossil fuel industry is massive. Our dependence on fossil fuels is deeply rooted and consumers themselves do not have easy access to alternative sources. So the distributional impact of carbon pricing, including concerns about fuel poverty and carbon leakage, heavily constrain how hard that lever can be pulled (Chapters 7 and 8).

Because carbon is so pervasive, the economic consequences of carbon control spread wide, impeding the scope for stronger 'pull' as discussed in Pillar II. Carbon pricing – or regulatory equivalents like carbon standards for power plants – are examples of 'sticks', but in terms of innovation, they are blunt ones, and no government has yet dared to try to use them at the scale required to drive pull-incentives deep into the innovation chain.[49] Thus a carbon price cannot supplant effective policies to span the innovation chain, but it can reinforce them, amplify the benefits of successful innovation policies, and rise over time as such innovations accumulate.

Yet there is indeed more work to be done to establish the industries required to respond adequately. For the success of a technology, as it climbs out of the 'valley of death' towards widespread diffusion, hinges upon many factors other than price alone. These include coordination between various actors including the manufacturer, the end user and the many intermediaries required to install and maintain the technology. Appropriate skills to span the resulting supply chain have to be developed and mobilised at each step of diffusion.[50] Thus beyond direct financial supports, and particularly as technologies move into market expansion and wider diffusion, a range of more structural and institutional barriers can impede progress. Some of these barriers are considered further in Chapter 10, in looking at the broader challenge of transforming systems. Before we turn to this, however, the wider considerations point to one inescapable fact: success in bridging the 'technology valley of death' is ultimately linked to wider debates about industrial strategy.

9.11 Industrial strategy revisited – risks, rewards and principles

In many OECD countries, industrial strategy used to be a dirty word, linked with the historical costs of 'supporting losers' in declining (but politically powerful) industries and historical failures in 'picking winners'.[51] No longer. Many factors, including the evident success of countries with clear industrial strategies like Korea and China, combined with the shock of the credit crunch and subsequent stagnation, particularly in Europe and Japan, have brought home a reality that whatever the complexities and risks of industrial strategies, having none is worse.[52]

The major debates are increasingly around the 'what' and 'how', and this sheds much light particularly on the later stages of the innovation and diffusion process. It turns out that much of the literature on wider industrial strategy complements well the insights from the innovation chain, and offers important additional guidance to the challenges of energy sector innovation and transformation. It helps to provide a bridge not across the 'technology valley of death' per se, but a bridge to the processes of sectoral transformation considered in the next chapter.

A recent major review by the OECD surveys the historical context and more recent developments in thinking about industrial strategy. It starts by expanding the 'final stages' of the innovation chain depicted in Figure 9.5, by noting a general classification of industrial phases as emerging, growth, maturity and decline. The corresponding challenges are the identification of emerging sectors and aiding their subsequent growth, so that countries capture more of the economic benefits of production in the mature phase – while also managing the displacement and decline of old industries and hopefully helping to transfer and transform some of their skills and capabilities towards rising sectors.

The broad phases of industrial *policy* can be simply characterised as widespread active support for manufacturing in the immediate postwar decades, followed by retreat at least in most OECD countries in the 1970s to 1990s, and subsequent resurgence with far greater attention to the question of 'how' to conduct industrial policy to maximise the benefits and minimise the risks. The OECD review notes that this has drawn on the evolution of five different rationales:

- *Laissez-faire arguments* based not so much on the belief that markets are perfect but rather that governments are worse: that government failures eclipse market failures and the best *practical* approach is to let market forces draw out emergent industries and weed out declining ones with minimal government involvement.
- *Traditional ownership-based arguments*, which involved attempts to stimulate certain sectors through production subsidies, state aid and trade-related policies to favour domestic industries. In some countries this took extreme forms of nationalisation or forced mergers, or the crude use of taxes and subsidies to support sectors that had developed strong lobbying powers and political connections. It was this de facto corruption of 'statism' and the corresponding failure of many such policies that caused strong reactions against industrial strategy.
- *Classical market-failure correcting and Pigouvian tax/subsidy-based analysis* consequently placed the burden of proof on demonstrating clearly the exact nature of market failures and how government policy could sensibly address them: classic examples concerned externalities (like IP spillover and environmental damage), market power and capital market failures, generally 'reflecting a mismatch between the structure of private and social benefits in a particular economic activity'.[53]
- *New growth, technological capabilities-based approaches* 'represent a dynamic extension of the neo-classical approach. Firms benefit not only from static economies of scale and scope but also, over time, from the cumulative learning embodied in building up and maintaining a production process …' The result is a strong emphasis upon cumulative technological capacity – implying broad-based support across the innovation chain above and beyond the specific 'valley of death' arguments – to help gain national benefit from the accumulation of knowledge and spillover benefits to the wider economy.[54]
- Finally, *systems approaches* go beyond the formal frameworks of neoclassical growth theories to build on insights derived from wider studies of technological transformation and institutional economics, including both the more radical ideas of Schumpeterian 'creative destruction' and the evolutionary economic perspectives outlined in the next chapter. The essential features include recognising the centrality of uncertainty and imperfect information, and the role of institutions and systems in facilitating experimentation, knowledge transfer and the coordination of the complex interlinked developments implied in technological transformations.[55]

Corresponding to these progressive levels of understanding behind the evolution of industrial policy, in other words, there is a parallel evolution towards the fundamental issues that help to define the Third Domain of human decision-making and economic development: the domain characterised by the evolution of complex systems over extended time periods. This implies a correspondingly sophisticated suite of policies that are based neither in philosophies of central planning nor in a belief in the supremacy of free markets. Rather the focus is on the capacity of government to aid, accelerate and shape the evolutionary processes of long-term economic development, which combine many elements of both private market and public goods.

A selection of associated policies in the arenas of capital, technology and systems/institutions are summarised in Table 9.1. In common with much of the literature on industrial strategy, this divides policies between 'horizontal' policies – ones broadly generic across the

Table 9.1 OECD typology of industrial policy instruments by policy arena

Policy arena	*Horizontal policies*	*Selective policies*
Capital	Loan guarantees; Corporate tax/capital allowances; Macro/financial stability; Financial market regulation	Strategic investment fund; Emergency loans; State investment bank; Inward investment promotion
Technology	R&D tax credit; Science budget IPR regime	Green technology; Lead markets; Public procurement for innovation; Patent box; Selective technology funding; Centres of expertise
Systems/institutions	Entrepreneurship policy; Scenario planning; Distribution of information; Overall competitiveness strategy	Indicative planning; Foresight initiatives; Identifying strategic sectors; Sectoral competitiveness strategy; Clusters policy

Source: Derived from Table 4 in Warwick (2013). Note that the original table also contains policies in arenas of product markets, labour and skills, and land (such as land planning regulations, enterprise zones and place-based clusters).

economy – and 'selective' policies that focus on particular sectors or technology areas. These of course correspond to the framework conditions and the policy environment around the innovation chain diagram (see Figure 9.5) – and help to underline the view that success in energy innovation is likely to align with wider economic success.

Horizontal policies are safer, because they avoid the connotation that governments are trying to second-guess 'winners', but ultimately are inadequate. Even policies that appear to be horizontal – cutting across many sectors – may in fact have radically different implications in different sectors. This is because – as underlined by the data in section 9.3 and subsequent discussions – sectors differ enormously. The nature of barriers and opportunities differs by sectors and the best results flow from policies which recognise this reality. The key prescription to minimise the risks historically experienced, rather, is to maintain broad-based policies to support innovation in key sectors judged to be of strategic economic importance. In many countries the world over, energy has been identified as one such key sector.[56]

9.12 Conclusion and synthesis: the multiple journeys

Innovation is complex. Multiple strands of evidence and analysis underline this central fact. Yet it is also true that we have learned a great deal about innovation processes in general, and in particular the obstacles to effective innovation in energy and environmental technologies. A useful way to gain some order and insight into the complexity is to think of the processes as involving not only multiple stages of the innovation chain, but the multiple journeys required to traverse it.

As with the innovation chain itself, the different journeys can be expanded or contracted according to the focus and level of detail required. In its most extensive form, the Carbon Trust produced a six-by-six matrix, which in Figure 9.9 we condense into a three-by-three summary in which the three classical stages of invention, innovation and diffusion have to be traversed in three concurrent and interrelated journeys.[57]

The entwined journey of the *technology and organisational unit* has been an important focus of this chapter. Many technologies start with the ideas of individuals operating in a research-based organisation (or corporate R&D lab). As the technology moves into the

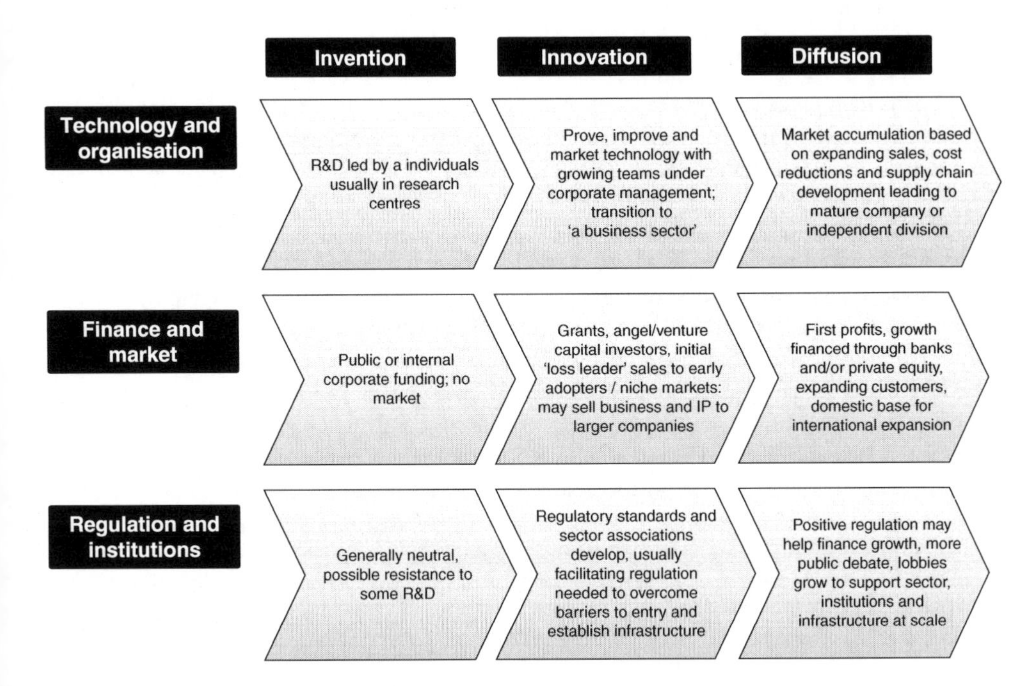

Figure 9.9 The multiple journeys of the innovation chain

Source: Adapted and condensed from Carbon Trust, Memorandum 64 to House of Commons Select Committee on Innovation, Universities, Science and Skills (2008), online at http://www.publications.parliament.uk/pa/cm200708/cmselect/cmdius/216/216we80.htm.

innovation phase, however, the scale grows and the emphasis shifts towards producing some kind of workable product that can be sold. Hence managerial skills and interfaces with the private sector become important, as the technology starts to become a product and a business – the classic point at which the transition falters and technologies languish for years or decades. If and as the technology navigates its way across this chasm, and the organisational unit successfully transfers or transforms with it and starts to extend its links with others in the required supply chain, these may start to climb the market accumulation foothills of the diffusion process. In doing so, the technology enters a race, seeking to gain enough in terms of learning experience and scale economies to enable the emerging sector to become an established industry complete with supply chain to support its growth.

The journey of *finance and the market* starts with public or (sometimes) internal corporate funding: there is no market, there is merely faith that the idea may prove to yield something of value. However, hope and intellectual curiosity cannot sustain a technology as it grows through the innovation process: that requires money, which sooner or later will start to outstrip the capacity of public funding and may test the patience of parent corporate funders. New entrants, not shielded by a parent organisation, may be particularly vulnerable as they seek to move beyond pure public grant finance to persuade a bewildering array of sceptical private sector 'angel' and 'venture' capital fund managers that their idea is worth backing. Or the venture may quit that effort and find a way to sell out their assets and IP to larger companies that have the financial muscle to sustain the journey, providing it doesn't threaten to undermine their own incumbent core interests. Specialised niche markets – or publicly funded strategic deployment – are likely to be key in either case.

If the technology makes that transition, the associated venture should start to make its first profits. This then gives it access to more extensive and less costly finance, including debt, to provide a larger platform for an expanding base of customers and sales networks, potentially moving from domestic to international scales. While this chapter has touched upon the initial stages of the financial journey, notably in terms of R&D expenditures (section 9.3) and the funding gap that underpins the technology valley of death (section 9.7), the full interplay between financing, innovation and growth is played out in Chapter 11.

The third set of journeys are those associated with regulation, institutions and infrastructure. *Regulation* can initially be neutral, but is more likely to be negative towards new applications (for instance, where it has been designed around an incumbent product and is poorly suited for the characteristics of a new technology). Standards are important in industry, but can also take a long time to establish. As the technology moves towards market, regulatory barriers may become more pressing, though globalisation may help as industries look around for the most favourable regulatory environments. Institutional structures must also evolve: a growing sector needs to develop its own associations that are eventually able to stand alongside incumbents in the struggles for influence over key regulatory, financing and governance decisions. Yet as the scale grows, the alignment – or not – of a technology with established infrastructure may become more and more important. Thus the motor car only finally came to realise its potential as horse tracks gave way to tarmac roads. The most obvious modern energy analogy is the mismatch between power systems designed for one-way flows from huge centralised stations generating baseload power, compared to the characteristics of decentralised renewable energy sources and smart meters which may embed the generation and dynamic use of electricity throughout the system. The challenge of transforming key systems is the focus of the next chapter.

Truly indeed do we need to 'find ways to treat technical and organisational innovation in a combined manner' (Chapter 2, p. 59). The presence of so many interconnected and different factors, combined with the characteristics of energy systems, explains why energy innovation is such a slow, ponderous and drawn-out process. The huge capital commitment required can inhibit technology push; the lack of consumer product differentiation undermines market pull; and the inadequacy of policies to price emissions and value technology spillovers means that most of the benefits from such innovation are public and would not accrue to innovators – who may anyway be trapped by the multiple financial and institutional obstacles noted. By the same token, successful intervention to accelerate low-carbon innovation may yield big benefits, not just for the environment but for the wider economic gains that may be expected from accelerating innovation in such relatively sclerotic sectors. Yet it is over-simplistic to think that one technology or one measure operating at one specific 'point' of the innovation chain could fill the gaps: that is one of the main messages of this chapter. All this explains why, in trying to develop a dynamic low-carbon future, clean technologies have had to be supported at different stages of the multiple journeys (see Figure 9.9) – and why there is still a long way to go.

Notes

1. 'The progress was slow, and the investors in Newhaven were becoming more and more restive at the lack of progress ... eventually Townsend was the only one of the promoters who still believed in the project, and when the venture ran out of money, he began paying the bills out of his own

pocket. In despair, he at last sent Drake a money order as final remittance and instructed him to pay his bills, close up the operation, and return to Newhaven. Drake had not yet received the letter when, on August 27th 1859, at sixty-nine feet, the drill dropped into a crevice … Uncle Billy came out to see the well … and saw a dark fluid floating to the top …' (Yergin 1991: 7). The innovation of drilling for oil was itself built upon multiple other innovations and experimentations, including several years of investigation into the possible uses and refinement of the 'rock oil' which seeped to the surface in some regions. With hindsight the idea of drilling for it seems obvious – like many inventions – but it was considered highly dubious and extremely risky at the time.

2. A superb overview of economic debates around technological change and the environment, with an emphasis on different theoretical approaches within economics, is found in Jaffe *et al.* (2005).
3. Schumpeter (1939).
4. There are many different definitions. The Global Energy Assessment (GEA 2012: Ch. 24) defines *invention* as the 'origination of an idea as a technological solution to a perceived problem or need', *innovation* as 'putting ideas into practice through an (iterative) process of design, testing, and improvements including smaller-scale demonstration or commercial pilot projects, and culminating in the establishment of an industrial capability to manufacture a given technological innovation', and *diffusion* as 'widespread uptake of a technological innovation throughout the market of potential adopters'. Historical energy transitions are reviewed briefly in Chapter 1 of the Global Energy Assessment.
5. Against the backdrop of widespread privatisation, leading UK economists Dasgupta and Stoneman (1987) charted seven major barriers in the way of adequate private-sector innovation. In contrast to the economists' typical focus on 'where are the market failures', a whole discipline has emerged which aims more programmatically to understand *innovation systems*. A review in *Research Policy* by Bergek *et al.* (2007) usefully organises this often disjointed literature into seven specific functions that an effective innovation system needs to deliver. The Global Energy Assessment (GEA 2012: Ch. 24) gives a good overview of the main themes emerging from the innovation systems literature applied to energy. This chapter uses the innovation chain as a structured way to help understand the key insights that emerge from this broad body of research.
6. 'Synthetic fuels, the breeder reactor, fusion power, most renewable technologies, and the persistence of the fuel cell option testify to this tendency. For the most part, however, these programmes have been either expensive failures or only slightly less expensive technological successes that serve limited markets' (Fri 2003). Note in qualification that the level of funding involved in nuclear power dwarfed all funding to renewable technologies, and some renewables support (e.g. Danish wind power) did prove cost-effective (Cohen and Noll 1995).
7. The most comprehensive analysis of energy-related investment across the innovation chain is in the Global Energy Assessment (2012: Ch. 24). In an important advance on most of the literature, the GEA makes every effort to encompass data also on private sector investments – acknowledging a number of data uncertainties – in its estimate that R&D and demonstration investments totalled around \$50bn, market formation investments (relying on directed public support like feed-in tariffs) some \$150bn, and with an estimated \$1–5trn in mature energy supply and end-use technologies (technology diffusion).
8. Concerning the French nuclear programme, see Grübler (2010), as cited in Chapter 3, note 42. Interestingly, only a small fraction of the cost is attributed explicitly to R&D, and despite low construction after 1990 the overall cumulative costs almost doubled again to 2000, due to cumulating operations and maintenance (O&M) and fuel costs.
9. For analysis of these 'unintended consequences' – the collapse of R&D in the electricity industries following privatisation – see Dooley (1998). The fact that almost all major innovations have originated with publicly funded research is rigorously document by Mazzucato in her book *The Entrepreneurial State* (2013), which goes on to argue for mechanisms which enable governments to reap directly more of the benefits associated with their investment in innovation, rather than relying purely on taxpayer funding but with the profits made from successful research accruing mostly to the private sector. The rest of this chapter is thus not about the generation of ideas for clean energy technologies, which abound, but the complexities of translating them into commercial products used at scale.
10. For example, oil and gas *producers* spend very little on R&D but there is huge expenditure (though still modest in percentage terms) by their suppliers in oil equipment, services and distribution. Similarly, some of the innovation expenditure associated with electricity is upstream

in general and engineering industrials (e.g. turbine manufacturers). Nevertheless the figures are relatively small, and the discrepancy with IT and pharmaceuticals is striking. In addition, the electricity R&D intensity is skewed by the continuing French effort and inclusion of AREVA, the French nuclear manufacturer, in the electricity classification; the R&D intensity in the rest of the global electricity sector is below 0.5%.

11. Arrow (1962); Boston Consulting Group (BCG 1974).
12. The most commonly cited example of 'negative learning' is that of nuclear power – for a discussion of both the data and the arguments as to why unit costs appear to have increased as the programme expanded, see the Grübler 2010 paper cited in note 8.
13. The argument that learning rates rise and then fall with the stage in the technology life cycle is most clearly spelled out in a substantial review paper by Kahouli-Brahmi (2009), who classified technologies as emerging, evolving and mature, and presented learning data on each. One of the leading analysts goes further and suggests that as a predictive tool, experience curves – at least with fixed learning rates – are most suited 'for the analysis of established technologies and forecasts of mid-time ranges'; however, her analysis does then go on to match experience curve data against two other approaches to assessing cost evolution, and concludes that within these ranges the experience curve estimates are likely to be robust as a tool for predicting the cost evolution of energy production technologies (Neij 2008).

    In recent years, an additional complication for the literature has been the fact that the cost of *most* energy technologies rose with the global increase in commodity prices in the second half of the last decade, as most rely extensively on inputs of both energy-intensive commodities – like steel and cement – as well as energy itself – and as the cost of financing also rose. The 'solar surprise' (Chapter 3, Box 3.4) and US shale gas were the biggest exceptions. To an important degree, learning effects are most robust as an indication of relative cost evolution among energy technologies, more than absolute. See also Kahouli-Brahmi (2009).
14. Schumpeter (1939).
15. Mumtaz and Amaratunga (2006).
16. Different people tend to want to expand different parts of the chain, depending on their profession and experience and the problem in hand. In reality all are important. For example, an IEA information paper on RD&D (Chiavari and Tam 2011) compacts the chain to omit commercialisation as a discrete step. One of the best summaries of US research and experience relating to innovation and energy is that by Alic *et al.* (2003), who similarly divide the process into R&D, development and commercialisation, and learning and diffusion, and trace the multiple instruments required at each step to accelerate the innovation process. My own experience with the Carbon Trust taught me the huge complexity, importance and difficulty of the commercialisation process – and diffusion may in practice only occur at scale years or even decades after initial commercialisation attempts.
17. See IEA (2011) and in particular the background paper cited by Mizuno (2010).
18. The key authors were Romer (1990) on variety, and Aghion and Howitt (1996) on product quality: see Chapter 11, note 22.
19. Moore (2006) focuses mainly on the institutional gaps between different kinds of producers and adopters; Murphy and Edwards (2003) produced the classic analysis of the technology valley of death in electricity production technologies in terms of the financing gap, and particularly the gap between public and private finance. The special characteristics of venture capitalists can play an important role (Gompers and Lerner 2001) but still do not remotely fill the gap given its multiple causes and special characteristics in the energy sector.
20. Aghion *et al.* (2012) demonstrate that there is path-dependence in the direction of technical change. The analysis shows that firms in economies which have innovated a lot in dirty technologies in the past find it more profitable to innovate in dirty technologies in the future. This path-dependency, when combined with the environmental externality, induces a laissez-faire economy to produce and innovate too much in dirty technologies compared to the social optimum. The approach is developed in Acemoglu *et al.* (2012), which concludes that climate change may be tackled relatively cheaply by a combination of carbon pricing with R&D, shifting the whole structure of innovation from dirty to clean technologies. The simplicity of the assumptions and recommendations are, however, strongly criticised by Pottier *et al.* (2014), who underline the need for a more realistic model of both the innovation process and the inertia and resistance of

the energy industry to rapid changes; their analysis concludes that applying empirical data from the energy sector implies that a much broader and more sustained effort would be required to achieve the optimistic results suggested by Acemoglu *et al.* In effect, the Pottier *et al.* analysis points to the intellectual gap between the theoretical economics literature and the innovation systems literature (see this chapter, note 5) and argues that while both underline the potential for innovation to solve the climate change problem at moderate cost (or indeed with net economic benefit), the policy prescriptions arising are substantially different. See also the section on the modelling of low-carbon transitions in Chapter 11.

21. In Europe, the major utilities Eon-Rhurgas and RWE are heavily into coal, ENI has a major interest in hydrocarbon production, while CEZ is involved in coal mining and GDF-Suez with other hydrocarbons. Several do also have some renewable energy holdings, including Centrica's (which owns British Gas) significant commitment to offshore wind, while Dong and Satoil are examples of oil and gas firms getting heavily into renewables.
22. For the fullest account of these studies see Nordhaus (2008). Continuing updates of the DICE model and the version with induced technological change (RICE) are available from Professor Nordhaus' website http://nordhaus.econ.yale.edu/. The review by Jaffe *et al.* (2005) also notes that 'to calibrate the model, he [Professor Nordhaus] needed parametric estimates of the private and social returns to fossil-fuel related R&D … using the existing R&D intensity of the fossil fuel sector to derive these parameters, he found that the impact of induced innovation is modest … he concluded that induced innovation has less effect than factor substitution on optimal emission levels.'
23. Energy efficiency in some products is less prone to the problems because of closer connection to end-users.
24. Freeman (1974); Murphy and Edwards (2003).
25. Quote from Jaffe *et al.* (2005), who elaborate: 'not only is the variance of the distribution of expected returns much larger than for other investments, but much or even most of the value may be associated with very low-probability but very high value outcomes', with reference to Scherer *et al.* (2000).
26. A better pace of innovation implies a rate that is closer to the social optimum – most obviously in this case with respect to improving environmental performance. Analysts concur that one of the most important consequences of the lack of perfect information flows is the delay of backing for a technology – whether R&D investment or later in the stages right through to its diffusion, even in those cases in which the investment presents a profitable net present value. See, for example, Carruth *et al.* (2000). The consequences in the energy sector are again a tendency to back mature, 'dirty' technologies, waiting for someone else to develop other solutions. In addition to perceived technology risks, another element difficult to predict are the so-called *soft assets*: human resources, trade secrets, skills and patents for which the innovator is much more aware of their intrinsic value and potential, but may be reluctant to share such knowledge with the investor. For a wider discussion of the role of information asymmetries in economic theory, see Stiglitz (2002).
27. Murphy and Edwards (2003).
28. Nemet (2008).
29. For a wider discussion on learning curves and energy technologies, see Jamasb and Köhler (2008).
30. Liz Hooper, personal communication.
31. For a wider discussion about the role of public-private partnership in the innovation process, see Stiglitz and Wallsten (2000).
32. One example of public-private partnership is PV:BONUS, a US Department of Energy (DOE) programme started in 1993 with the specific aim to develop photo-voltaic and hybrid technologies suitable for the residential and commercial sectors; the final aim was to develop a product that could be easily installed without specialised background. One of the most interesting aspects of the programme was the aim to foster diversity along with early use of competitive pressures: out of 16 partnerships started, only seven were confirmed for funding, leading to five commercially available products to which demonstration support was granted. From the beginning of the programme, most of the attention was focused on important requirements with possible high impact on future market profitability of photo-voltaic panels, such as design, flexibility, compatibility with building materials and safety. For a detailed description of the programme, see Thomas and Pierce (2001).

33. The Energy Technologies Institute is a public-private partnership between global energy and engineering companies and the UK government. It funds projects across a portfolio of nine technologies, projects which are intended to bridge the gap between laboratory R&D and commercial deployment (Energy Technologies Institute 2012: 3).
34. Based on the results of the smart meters field trial, the Minister responsible (Ruth Kelly) announced in 2009 '... We think the use of smart meters could be vital in helping businesses measure and manage energy use. Government wants advanced electricity metering for large business customers to begin to be deployed from next month, with completion by 2014. This is essential, and we would like to see this extended to all businesses.'
35. Its independence and level of government backing enabled the Carbon Trust both to build extensive technical expertise and to put initial equity investment into new ventures. The combination led naturally to a role as a lead investor; once the Carbon Trust committed to a project, it could attract many private sector venture capital investors.
36. This is the argument made by the Strategic Niche Management scholars – see Kemp *et al.* (1998).
37. See Grubb *et al.* (2008). The chapter draws on Carbon Trust (2006) led by Nadine Haj-Hasan which proved highly influential in persuading the UK government to move away from a 'one size fits all' approach to its renewable energy policy, i.e. to start introducing banding for different technologies to receive less or more than one 'renewable obligation certificate' according to their stage of development.
38. Tax incentives mean the costs are not fully incorporated into national budgets – the money doesn't flow through Treasuries – but they are more difficult to tailor to the specific circumstances of investment projects. Also their value depends on corporate circumstances – they are of most value to companies making a healthy (and otherwise taxable) profit, rather than new entrants that may still be struggling.
39. Bürer and Wüstenhagen (2009).
40. Grubb *et al.* (2008). For a more detailed comparison of UK and German renewable support systems with reference to risks, see Mitchell *et al.* (2006).
41. Public Utilities Commission of California (2009). For a report on the initial auctions, see http://blogs.reuters.com/environment/2010/12/16/california-approves-reverse-auction-renewable-energy-market/
42. The 'winner's curse' refers to the problem that a competitive auction, by awarding contracts to the cheapest bidders, inevitably creates pressures to be over-optimistic in bidding. As a result, there is a risk of the auction selecting projects that struggle to be viable, and indeed there has been a problem of winning bidders subsequently failing to proceed with the submitted project, presumably because they subsequently concluded the remuneration was insufficient to cover the full cost and risks. In an uncertain world, creating artificial pressures to select the most optimistic bidders is thus a risky strategy – yet another example of the risks of 'adverse selection' noted first in the context of behavioural economics and information asymmetries (Chapter 4). In response, the Californian auctions require participants to deposit significant collaterals, which the government keeps if the winning projects do not proceed. This does, however, increase the financing and transaction costs for participants. Auction design is itself a complex field, with many important lessons to be learned.
43. Dechezleprêtre *et al.* (2011).
44. As summarised most usefully in a 20-years-on review workshop on the Porter Hypothesis and published in journal form as Ambec *et al.* (2013): 'Porter and van der Linde (1995) explain at least five reasons why properly crafted regulations may lead to these outcomes', namely that regulation:

    - signals companies about likely resource inefficiencies and potential technological improvements;
    - focused on information gathering can achieve major benefits by raising corporate awareness;
    - reduces the uncertainty that investments to address the environment will be valuable;
    - creates pressure that motivates innovation and progress;
    - levels the transitional playing field. During the transition period to innovation-based solutions, regulation ensures that one company cannot opportunistically gain position by avoiding environmental investments.

Finally, they note: 'We readily admit that innovation cannot always completely offset the cost of compliance, especially in the short term before learning can reduce the cost of innovation-based solutions.'

45. Aghion *et al.* (2012) show that fuel taxes have redirected innovation away from dirty technologies (i.e. internal combustion engine-based vehicles) and towards clean technologies (i.e. electric and hybrid vehicles) in the car industry. Crabb and Johnson (2010) also show that higher oil prices lead to increased innovation in energy-efficient automotive technology – but do not find such an effect from the US CAFE standards, presumably because those standards were not strong at the outset and remained static for thirty years (Chapter 5).
46. The review by Ambec *et al.* (2013) concludes:

    The theoretical arguments for the Porter Hypothesis appear to be more solid now than when they were first discussed as part of the heated debate in the *Journal of Economic Perspectives* in 1995 ... On the empirical side, the evidence for the 'weak' version of the Porter Hypothesis (that stricter environmental regulation leads to more innovation) is fairly clear and well established. However, the empirical evidence on the strong version of the Porter Hypothesis (that stricter regulation enhances business performance) is mixed, but with more recent studies providing clearer support.

47. Wagner (2006).
48. Calel and Dechezleprêtre (2012).
49. Alongside EU carbon pricing and a floor price of CO_2, the UK has introduced a 'CO_2 emissions performance standard' for power plants as part of its Energy Market Reforms, while in June 2013 President Obama instructed the US Environmental Protection Agency to move forward with CO_2 standards for power plants. In neither case, however, is there any expectation that the standard will be set at a level to prohibit new gas plants, and without this, the standards will do little to motivate innovation in low-carbon power production.
50. For detailed insights about major barriers in Asian developing countries, see Koh *et al.* (2004).
51. Crafts (2012) reviews examples of failed UK 'industry policies' from this era which include a strong bias towards subsidising ailing industries such as shipbuilding (Wren and Simpson 1996) and subsidising high-tech national champions in civil aircraft (Gardner 1976), computers (Hendry 1989) and nuclear power (Cowan 1990). There were of course some successes (such as the rescue of Rolls-Royce automobile company (Lazonick and Prencipe 2005) and in other countries (all citations from Crafts 2012).
52. In no particular order, the OECD review (Warwick 2013) cites the development of industrial policy initiatives in France, Japan, Korea, The Netherlands, Turkey, the UK, the US, Brazil, China and India. The review goes on to suggests five reasons for the rapid re-emergence of interest in industrial strategy in the OECD countries: the responses to the financial crisis and subsequent stagnation; the extent to which the crisis undermined faith in market mechanisms, particularly relating to the flow of finance; the need to maximise the benefits from the deployment of bailout finance; the extent to which the crisis has underlined structural imbalances in economies that had become unhealthily dependent on finance and services at the expense of manufacturing; and the dramatic success of the emerging Asian economies in particular which had pursued clear industrial strategies. Other parts of the review, however, note that the re-emergence of interest in industrial strategy substantially pre-dated the financial crises, having been spurred by concern about structural imbalances and declining international competitiveness, contrasted with the astonishing success in particular of South Korea's industrial strategy followed by China.
53. As cited in Warwick (2013). Note that Crafts (2010) highlights three specific types of market failure related to infant-industry-related capital market failures, agglomeration externalities and rent-switching via strategic trade policy. The first two have clear parallels with the analysis respectively of the technology valley of death and learning-by-doing effects analysed in this chapter.
54. Citation from Sharp (2001).
55. 'The common factor is a focus on the processes for the generation, absorption and commercial exploitation of knowledge. In contrast to neoclassical approaches, which view knowledge as homogeneous and believe that it can disseminate instantly, systems approaches suggest that knowledge is heterogeneous, context-specific, tacit and "sticky". Actors face genuine uncertainty, to which

they have to adapt, rather than dealing with the more tractable risk distributions of neoclassical economics. In the Schumpeterian view of the world, the emphasis is on the cumulative nature of technological progress and the importance of learning from accumulated knowledge passed on by word of mouth within institutions … Networks of formal and informal connections between relevant institutions make up the national system of production and innovation. In this view, the role of industrial policy, hand-in-hand with innovation policy, is to create and develop institutions to promote networking and collaboration and to devise strategies to make best use of these institutions. Current policy thinking also increasingly adopts a "systems failure" approach to industrial policy and innovation. This approach is somewhat wider than – but does not displace – the "market failure" rationale …' (Warwick 2013: 21).

56. Most of the countries listed in note 52 include energy-related sectors, often with an emphasis on low-carbon technologies, as among priority sectors. The UK, traditionally among the most sceptical about any industrial strategy, has now moved decisively in this direction, with the Secretary of State for Business, Innovation and Skills declaring (7 March 2013) that 'that case has been won – and I think is now broadly accepted on Right and Left. Work is now well underway with industry to develop long-range strategies for 10 important sectors.' The ten sectors included four 'enabling' sectors, of which three were energy production: nuclear, oil and gas, and offshore wind, together with construction. Vehicles were also included as one of the advanced manufacturing sectors.
57. The initial version started with four journeys: technology, company, market and regulation. By the time the Carbon Trust presented to a House of Commons Committee enquiry, it had grown to include financial and institutional journeys; infrastructure adds a seventh. It seems more manageable to combine these multiple (and potentially expanding) lists into the more limited and compact set shown in Figure 9.9, recognising that any individual line can, like the innovation chain itself, be further disaggregated into components.

References

Acemoglu, D., Aghion, P., Burzstyn, L. and Hemous, D. (2012) 'The environment and directed technical change', *American Economic Review*, 102 (1): 131–66.

Aghion, P. and Howitt, P. (1996) 'Research and development in the growth process', *Journal of Economic Growth*, 1 (1): 49–73.

Aghion, P., Dechezleprêtre, A., Hemous, D., Martin, R. and Van Reenen, J. (2012) *Carbon Taxes, Path Dependency and Directed Technical Change: Evidence from the Auto Industry*, No. w18596. National Bureau of Economic Research.

Alic, J., Mowery, D. and Rubin, E. S. (2013) *US Technology and Innovation Policies: Lessons for Climate Change*. Washington, DC: Pew Centre on Global Climate Change.

Ambec, S., Coheny, M. A., Elgiez, S. and Lanoie, P. (2013) 'The Porter Hypothesis at 20: can environmental regulation enhance innovation and competitiveness?', *Review of Environmental Economics and Policy*, 7 (1): 2–22.

Arrow, K. (1962) 'The economic implications of learning by doing', *Review of Economic Studies*, 29: 155–73.

Babbage, C. (1832) *On the Economy of Machinery and Manufactures*, 3rd edn. London: Charles Knight, Chapter 17 'Of Price as Measured by Money'.

Bergek, A., Jacobsson, S., Carlsson, C., Lindmark, S. and Rickne, A. (2008) 'Analyzing the functional dynamics of technological innovation systems: a scheme of analysis', *Research Policy*, 37: 407–29.

BCG (1974) *The Experience Curve Reviewed*, Perspectives Reprint No. 124, 125, 128, 149. Boston: Boston Consulting Group.

Broadberry, S. and Crafts, N. (2001) 'Competition and innovation in 1950s Britain', *Business History*, 43 (1): 97–118.

Bürer, M. J. and Wüstenhagen, R. (2009) 'Which renewable energy policy is a venture capitalist's best friend? Empirical evidence from a survey of international cleantech investors', *Energy Policy*, 37: 4997–5006.

Calel, R. and Dechezleprêtre, A. (2012) *Environmental Policy and Directed Technological Change: Evidence from the European Carbon Market*. Grantham Research Institute on Climate and the Environment Working Paper No. 75.

Carbon Trust (2006) *Policy Frameworks for Renewables – Analysis on Policy Frameworks to Drive Future Investment in Near and Long-term Renewable Power in the UK*. London: Carbon Trust.

Carruth, A., Dickerson, A. and Henley, A. (2000) 'What do we know about investment under uncertainty?', *Journal of Economic Surveys*, 14 (2): 119–53.

Chiavari, J. and Tam, C. (2011) *Good Practice Policy Framework for Energy Technology, Research, Development and Demonstration (RD&D)*, IEA Information Paper. Paris: International Energy Agency.

Cohen, L. and Noll, R. (1991) *The Technology Pork Barrel*. Washington, DC: Brookings Institution.

Cohen, L. and Noll, R. (1995) 'Feasibility of effective public-private R&D collaboration: the case of cooperative R&D agreements', *International Journal of the Economics of Business*, 2 (2): 223–40.

Cowan, R. (1990) 'Nuclear power reactors: a study in technological lock-in', *Journal of Economic History*, 50 (3): 541–67.

Crabb, J. M. and Johnson, D. K. N. (2010) 'Fueling innovation: the impact of oil prices and CAFE standards on energy-efficient automotive technology', *Energy Journal*, 31 (1): 199–216.

Crafts, N. (2010) *The Contribution of New Technology to Economic Growth: Lessons from Economic History*, CAGE Online Working Paper Series 01. Competitive Advantage in the Global Economy (CAGE).

Crafts, N. (2012) *Creating Competitive Advantage: Policy Lessons from History*, CAGE Online Working Paper Series 90. Competitive Advantage in the Global Economy (CAGE).

Dasgupta, P. and Stoneman, P. (1987) *Economic Policy and Technological Performance*. Cambridge: Cambridge University Press.

Dechezleprêtre, A., Glachant, M., Hascic, I., Johnstone, N. and Ménière, Y. (2011) 'Invention and transfer of climate change mitigation technologies: a global analysis', *Review of Environmental Economics and Policy*, 5 (1): 109–30.

Dooley, J. J. (1998) 'Unintended consequences: energy R&D in a deregulated energy market', *Energy Policy*, 26 (7): 547–55.

ETI (2012) *Annual Review 2012*. Energy Technologies Institute.

Freeman, C. (1974) *The Economics of Industrial Innovation*. Harmondsworth: Penguin.

Fri, R. W. (2003) 'The role of knowledge: technological innovation in the energy system', *Energy Journal*, 24 (4): 51–74.

Gardner, B. L. (1976) 'The effects of recession on the rural-farm economy', *Southern Journal of Agricultural Economics*, 8 (1).

GEA (2012) *Global Energy Assessment – Towards a Sustainable Future*. Cambridge and New York: Cambridge University Press and Laxenburg, Austria: IIASA, Chapter 24.

Gompers, P. A. and Lerner, A. (2001) *The Really Long-Run Performance of Initial Public Offerings: The Pre-NASDAQ Evidence*, NBER Working Papers 8505. National Bureau of Economic Research.

Grubb, M., Haj-Hasan, N. and Newberry, D. (2008) 'Accelerating innovation and strategic deployment in UK electricity: applications to renewable energy', in M. Grubb, T. Jamasb and M. Pollitt (eds), *Delivering a Low-Carbon Electricity System: Technologies, Economics and Policy*. Cambridge: Cambridge University Press.

Grübler, A. (2010) 'The costs of the French nuclear scale-up: a case of negative learning-by-doing', *Energy Policy*, 28: 5174–88.

Hendry, J. (1989) *Innovation for Failure: Government Policy and the Early British Computer Industry*. Cambridge, MA: MIT Press.

IEA (2011) *Deploying Renewables: Best and Future Policy Practice*. Paris: International Energy Agency.

Jaffe, A. B., Newell, R. G. and Stavins, R. N. (2005) 'A tale of two market failures: technology and environmental policy', *Ecological Economics*, 54 (2–3): 164–74.

Jamasb, T. and Köhler, J. (2008) 'Learning curves for energy technology and policy analysis: a critical assessment, in M. Grubb, T. Jamasb and M. G. Pollitt (eds), *Delivering a Low-Carbon Electricity System: Technologies, Economics, and Policy*. Cambridge: Cambridge University Press, pp. 314–32.

Kahouli-Brahmi, S. (2009) 'Testing for the presence of some features of increasing returns to adoption factors in energy system dynamics: an analysis via the learning curve approach', *Ecological Economics*, 68 (4): 1195–212.

Kemp, R., Schot, J. and Hoogma, R. (1998) 'Regime shifts to sustainability through processes of niche formation: the approach of strategic niche management', *Technology Analysis and Strategic Management*, 10 (2): 175–95.

Koh, L. P. *et al.* (2004) 'Species extinction and the biodiversity crisis', *Science*, 305: 1632.

Lazonick, W. and Prencipe, A. (2005) 'Dynamic capabilities and sustained innovation: strategic control and financial commitment at Rolls-Royce plc', *Industrial and Corporate Change*, 14 (3): 1–42.

Mazzucato, M. (2013) *The Entrepreneurial State: Debunking Public vs Private Sector Myths*. London: Anthem Press.

Mitchell, C. *et al.* (2006) 'Quotas versus subsidies – risk reduction, efficiency and effectiveness – a comparison of the renewable obligation and the German feed-in law', *Energy Policy*, 34 (3): 297–305.

Mizuno, E. (2010) *Renewable Energy Technology Innovation and Commercialisation Analysis*, Report prepared for the IEA Renewable Energy Division, Cambridge Centre for Energy Studies (CCES), Judge Business School, University of Cambridge.

Moore, G. (2006) 'Managing ethics in higher education: implementing a code or embedding virtue?', *Business Ethics: A European Review*, 15 (4): 407–18.

Mumtaz, A. and Amaratunga, G. (2006) 'Solar energy: photovoltaic electricity generation', in T. Jamasb, W. J. Nuttall and M. G. Pollitt (eds), *Future Electricity Technologies and Systems*. Cambridge: Cambridge University Press.

Murphy, L. M. and Edwards, P. L. (2003) *Bridging the Valley of Death – Transitioning from Public to Private Sector Financing*, NREL/MP-720-34036. US Department of Energy.

Neij, L. (2008) 'Cost development of future technologies for power generation – a study based on experience curves and complementary bottom-up assessments', *Energy Policy*, 36: 2200–11.

Nemet, G. F. (2008) 'Demand-pull energy technology policies, diffusion and improvements in California wind power', in T. Foxon, J. Köhler and C. Oughton (eds), *Innovation for a Low Carbon Economy: Economic, Institutional and Management Approaches*. Cheltenham: Edward Elgar.

Nemet, G. F. (2009) 'Demand pull, technology push, and government-led incentives for non-incremental technical change', *Research Policy*, 38 (5): 700–9.

Nordhaus, W. (2008) *A Question of Balance: Economic Modeling of Global Warming*. Yale Press continuing updates of the DICE model and the version with induced technological change (RICE) are available from Professor Nordhaus' website, http://nordhaus.econ.yale.edu/

Porter, M. (1991) 'America's Green Strategy', *Scientific American*, 264: 168.

Porter, M. and van der Linde, C. (1995) 'Toward a new conception of the environment-competitiveness relationship', *Journal of Economic Perspectives*, 9 (4): 97–118.

Pottier, A., Hourcade, J. C. and Espagne, E. (2014) 'Modelling the redirection of technical change: the pitfalls of incorporeal visions of the economy', *Energy Economics* (in press).

Public Utilities Commission of California (2009) *Approval of Reverse Auction Renewable Energy Market*. Online at: http://blogs.reuters.com/environment/2010/12/16/california-approves-reverse-auction-renewable-energy-market/

Romer, P. M. (1990) 'Endogenous technological change', *Journal of Political Economy*, S71–S102.

Scherer, F. M., Harhoff, D. and Kukies, J. (2000) 'Uncertainty and the size distribution of rewards from innovation', *Journal of Evolutionary Economics*, 10 (1): 175–200.

Schumpeter, J. (1939) *Business Cycles: A Theoretical, Historical, and Statistical Analysis of the Capitalist Process*. New York: McGraw-Hill, 2 vols.

Sharp, M. (2001) *Industrial Policy and European Integration: Lessons from Experience in Western Europe Over the Last 25 years*, Centre for the Study of Economic and Social Change in Europe Working Paper No. 30. London: University College.

Sondes, K.-B. (2009) 'Testing for the presence of some features of increasing returns to adoption factors in energy system dynamics: an analysis via the learning curve approach', *Ecological Economics*, 68 (4): 1195–212.

Stiglitz, J. E. (2002) *Globalization and Its Discontents*. New York and London: W. W. Norton.

Stiglitz, J. E. and Wallsten, S. J. (2000) 'Public-private partnerships: promises and pitfalls', *American Behavioral Scientist*, 43 (1): 52–7.

Thomas, H. P. and Pierce, L. K. (2001) *Building Integrated PV and PV/Hybrid Products – The PV:BONUS Experience*. Golden, CO: National Renewable Energy Laboratory.

Venturini, F. (2012) 'Product variety, product quality, and evidence of endogenous growth', Economics Letters, 117 (1): 74–7.

Wagner, M. (2006), 'Innovation towards energy-efficiency and Porter's Hypothesis', *Zeitschrift für Energiewirtschaft*, 30: 4.

Warwick, K. (2013) *Beyond Industrial Policy: Emerging Issues and New Trends*, OECD Science, Technology and Industry Policy Papers, No. 2. Paris: OECD.

Weiss, M., Junginger, M., Patel, M. K. and Blok, K. (2010) 'A review of experience curve analyses for energy demand technologies', *Technological Forecasting and Social Change*, 77: 411–28.

Weiss, M., Patel, M. K., Junginger, M., Perujo, A., Bonnel, P. and Grootveld, G. (2012) 'On the electrification of road transport – learning rates and price forecasts for hybrid-electric and battery-electric vehicles', *Energy Policy*, 48: 374–93.

Wren, B. M. and Simpson, J. T. (1996) 'A dyadic model of relationships in organizational buying: a synthesis of research results', *Journal of Business and Industrial Marketing*, 11 (3–4): 63–79.

Yergin, D. (1991) *The Prize: The Epic Quest for Oil, Money and Power*. New York: Simon & Schuster.

10 Transforming systems

Two roads diverged in a wood, and I—,
I took the one less traveled by,
And that has made all the difference.
–Robert Frost (1916), in the collection *Mountain Interval*, New York: H. Holt and Co

10.1 Introduction: over the bridge?

The evidence to this point: so far, so good. Chapter 3 illustrated a huge array of technological options for the provision of clean energy at varied additional costs – sometimes 'negative' and often modest – across the wide diversity of our energy systems. The remainder of Pillar I explained why our energy systems are so wasteful (Chapter 4) and the extensive progress made in policies to improve efficiency (Chapter 5). The chapters of Pillar II illustrated the large cumulative influence of energy pricing (Chapter 6) and noted that despite all its complexities, setbacks and imperfections, progress is being made towards proper pricing (Chapter 7) and that there are options for at least partly addressing the distributional impacts that have impeded this (Chapter 8). The previous chapter (Chapter 9) explained the blockages that have so far impeded low-carbon innovation and the progress made to overcome them.

The energy world has moved a long way in a few decades – intellectually in its understanding of innovation, and physically in terms of the experience and growth of new energy industries. All the combined efforts put in, by governments, industry and others, have borne fruit. Most striking, the impact in particular of renewable energy support mechanisms has begun to accumulate with impressive results. Global investment in 'clean energy' rose from under US$50bn in 2004 to over US$150bn by 2008, and the global recession had less impact than in most other sectors, with investment growth already returning to pre-recession levels during 2009 after a brief dip. Initially dominated by Europe and the US, during this period the balance shifted rapidly: spurred on in part by the initial impetus of the Clean Development Mechanism (Chapter 7), Asian clean energy investment in 2009 surged past the level in America and reached a level on a par with European investment. The growth has been further aided by the stimulus packages adopted to avert the financial meltdown in 2009. Dominated by the US, China and Korea, stimulus expenditure that targeted clean energy rose to an average US$50bn per year over the three years 2010–12.

The cost of wind energy fell as investment mushroomed; the cost of solar PV, after a spike in prices caused by excessive demand, delivered its 'solar surprise' (Box 3.4), the full consequences of which have yet to be played out; and Brazilian ethanol has become competitive with petrol.[1] On the demand side, the energy efficiency of much individual equipment – notably lighting, appliances and cars – has improved dramatically.

Yet this does not yet mean that cleaner technologies have successfully crossed the bridge, and progressed far enough along the 'learning curve' (see Chapter 9) to land firmly on the other side. There has been impressive progress but it remains modest compared with the overall scale of the challenge. Despite efficiency gains, global energy demand continues to soar. The progress in a few renewable technologies for power generation remains focused in a few regions and is not matched more widely across other technologies; it has yet to make much of a mark on the carbon intensity of the world's energy supply (Chapter 1).

Moreover, most of the clean energy investment still depends either on public money, or on technology-specific regulatory supports like feed-in tariffs. Technologies have come down the curve depicted in Figure 9.1 – but most have yet to converge with the market price of conventional energy. Many countries (particularly developing countries) have yet to fully factor in local environmental damages, and carbon prices remain patchy and inadequate to reflect the full scale of planetary risks (Chapters 6 and 7). In only a few places are the resulting 'internalised' market signals sufficient to provide a clear point of unaided convergence for the new industries.

Moreover, as the scale grows, so too do the costs and the potential difficulties in maintaining high levels of support. The short-run positive side of the 'stimulus packages' has been offset by the impact of the economic recession, which led to hasty and sometimes retrospective cuts in support levels in some countries (like Spain), undermining investor confidence. Moreover, as capacities expand – particularly for variable renewables sources – they begin to bump up against the limits to which current electricity systems can efficiently absorb them. Similarly, with the dominant low-carbon fuels, bioethanol remains constrained by the 'blending wall' (Chapter 3).

Thus specific policies to bridge the 'technology valley of death' have helped to connect technology push with demand pull and thus enable some key technologies to cross the valley, but they are still climbing the other side (and more technologies are required to meet different niches and needs). In doing so, they are encountering a wider set of challenges, which arise from the fact that progress in discrete technologies is only a part of the story. Changing energy systems is not only a matter of funding and cost reduction. Fostering distributed and/or intermittent energy sources requires restructuring our transmission and distribution system. Supporting low-carbon fuels may involve different vehicles and/or networks of supply. All new industrial systems require a reliable supply chain, and the challenges go beyond the capabilities of a single technology or company. Our efforts to date have dug into the surface at many points – efficiency policies, pricing and innovation. But in doing so, they reveal beneath the awesome scale of the task – and in particular that the changes required involve transformation of entire systems. This chapter steps back to outline the challenges and opportunities in fostering larger-scale transformations of whole sectors.

10.2 No modest task

To see the scale of the task, consider the response to the extraordinary rise in fossil fuel prices since 2005, sketched in Chapter 1 (Figures 1.2 and 1.3). The backdrop, remember, was three decades of policy development with associated institutions, energy efficiency (and some pricing) policies and innovation, as illustrated throughout this book. By 2005 there was also a broad consensus on the risks of continued accumulation of CO_2 in the atmosphere. Many analysts suggested that the most effective single impetus towards global decarbonisation would be a major rise in fossil fuel prices. That, we got. Yet after a very brief pause, it has been accompanied by a sharp *increase* in global CO_2 emissions at a faster pace than ever, with a number of the major oil companies now retreating from their modest forays into alternative energies.

Macroeconomic analyses as sketched in Chapter 11 point to a related 'puzzle'. Many (though not all) global computer models estimate that the cost of a low-carbon future amounts at most to a few per cent of foregone global GDP, preserving economic growth at a level entirely consistent with the rapid development required to connect everyone to modern energy systems.[2] Indeed it equates to only a year or two of deferred economic growth. We could provide energy access for all, ensure energy security, protect the atmosphere and still be as rich in GDP terms just a year or two later. To buy stability and security in our energy and climatic systems for such a price seems an extraordinarily good deal – the core argument of the *Stern Review*. What then has impeded progress, and what explains the surge in global emissions despite almost unprecedented energy prices?

A more detailed look at the numbers begins to reveal the problem, for the 'net GDP' numbers do not spell out the real implications. The additional investment needed to achieve full access to electricity by 2030 is estimated at US$750bn by the IEA. To set this (and other numbers) in context, it is comparable with expenditure under the US 2009 Stimulus Package over this decade.[3] In this case, however, since the customers for 'energy access' would be the poorest people on the planet, they would have limited ability to pay for this as consumers, pointing already to some of the complexities. The estimated *net additional investment costs* associated with *low-carbon* development to 2030 (consistent with the 2°C trajectory indicated in Chapter 1) span from US$400bn to $1,200bn per year.[4] The investment costs would obviously be offset by the lower energy demand, lower fossil fuel prices (due to lower fossil fuel demand), and consequently radically reduced fossil fuel expenditure, which characterise low-carbon scenarios; the net effect explains the overall impact on GDP in the order of 1 per cent.

However, the 'additional' cost implied by changing the course of the global energy system towards decarbonisation is the difference between two even bigger numbers – the overall investment required to meet energy needs over the coming decades, one way or another. The money flow that is *redirected* in the low-carbon scenarios is considerably more substantial. The IPCC Special Report on Renewable Energy estimates the global *cumulative* investment in renewable electricity sources alone by 2030 to vary between about $3 trillion (in a 'business as usual' case) to up to $12 trillion (for a 2°C case). The IEA's World Energy Outlook (WEO) in 2010 estimated an overall investment cost to 2030 of a 2°C scenario at US$18 trillion. The New Policies scenario of the subsequent WEO 2012 estimated US$17 trillion global investment for the *power sector* to 2035.[5]

If compared against global GDP – currently around $70 trillion a year and expected to rise to, for example, $100 trillion during the 2020s – this implies redirecting close to 1 per cent of global GDP to transform the power sector alone – which is usually considered to be the easiest sector to decarbonise. This is a fantastic redirection of capital flows, with huge potential implications for the winners and losers.

We return to some scenario studies of price and cost implications at the end of this chapter and consider their macroeconomic implications in Chapter 11. For the present, the core point is that the 'net GDP' implication of changing course plays only a minor part. Financially, it requires *redirecting* investment worth up to tens of trillions of US$ over the period to 2030, with wholesale change in the energy sector: 1–2 per cent of global GDP, representing a far greater share of energy investment, on an ongoing and enduring basis.[6] All this money must be invested in different ways – in clean energy rather than dirty fuels, more in energy efficiency and less in energy supply. Aside from the money, it implies reshaping patterns of energy production, carrier systems and consumption. We need to reorient our entire energy system – the biggest industrial infrastructure on the planet. But beyond even that, low-carbon development means changing the course not just of the energy business, but of key manufacturing and construction sectors as well.

Since a growing share of those investments is – and will be – in the emerging economies, sustainable energy is not just (or even mainly) 'a Western issue'. Infrastructures built in the coming decades may induce carbon intensive patterns that are hard to reverse. The studies cited above generally estimate that at least half of the investment implications lie outside the current industrialised countries. It is a global problem and solving it involves changing the structure and financing of economic progress globally.[7]

A low-carbon transformation would be against the grain of dominant and established technologies, systems and interests. Starting from a *tabula rasa*, it would be so different. As noted in Chapters 3 and 9, there is no inherent reason why exploiting the natural energy in the sun and wind should be more expensive than extracting more fossil fuels out of reservoirs thousands of feet beneath the ocean's surface, buried under the Arctic ice cap, stuck in oil sands or squeezed out of rocks kilometres underground at incredibly high pressures. Yet the fossil fuel route benefits from a century of established development and infrastructure, and it matches the vast expertise and economic interests of some of the biggest companies on the planet.

Hence the current response to the energy price rises. Big oil companies know 'how to do fossil fuels' better than anyone else. Massively higher global energy prices boost their profitability enormously and also make it economic for them to spend billions of dollars developing the new frontiers to maintain the supply of fuels to feed through existing systems. By contrast, they have no comparative advantage in, and there were no global infrastructures developed around, non-fossil energies. As noted in Chapter 9 this also feeds back into the default direction of innovation. The idea that a higher global fossil fuel price would in itself move the world away from high carbon investment was thus a fantasy – indeed an inversion of reality – born of a complete failure to understand the nature of the systems and the interests involved.

This chapter is thus about how *systems* of technology, infrastructure and institutions play out in the energy sector and some of the essential elements for moving such complex systems in a different direction. It is about fostering investment and energy use along different paths. The core analysis focuses on three key examples: the transport challenge in the Americas; the electricity infrastructure in Europe; and expanding human settlements in Asia. First, however, we take a closer look at some of the general factors that keep much of the world on its current risky course.

10.3 Systems: evolving, locked and opened

Changing paths towards a low-carbon economy involves a pervasive transformation. Energy production and use results from the interplay between production and carrier systems, consumption patterns and technologies, with all also influenced by both institutions and the geographical distribution of activities. The move from stagecoaches to trains and then to cars did not only revolutionise the transport system. Electricity did not just provide a more convenient or cheaper fuel for the same activities. The distribution of human settlements and their urban form does much to determine energy needs for transport, heating and cooling. Moving to a low-carbon economy would not mean just doing the same thing with different supply technologies. To understand how to change course, we need to better understand these systems, what keeps us stuck on the current tracks and what can help to foster transitions.[8]

In setting out the third of the Three Domains, Chapter 2 noted the comparatively recent intellectual flowering of *evolutionary economics* – the study of how economic systems have evolved. This represents a coming-of-age of the seeds laid by a number of leading economists over the previous century, starting most notably with Thorstein Veblen who is widely credited with launching such enquiries in the late nineteenth century. The central point, now backed up by a vast array of evidence, is that our societies comprise evolving structures based on patterns of historical development. In particular, modern economies embody complex 'technological systems' – defined as 'interrelated components connected in a network or infrastructure that includes physical, social and informational elements.'[9]

It is no accident that theories of evolutionary economics, after lying almost dormant for decades, finally began flowering in the 1980s along with two other developments outlined below.

One was the rise of the field of behavioural economics. As long as it was assumed that economic systems could be reasonably represented in terms of 'economically rational representative agents' making uniquely defined optimal cost/benefit decisions, it remained possible at least in theory to argue that economic systems would develop defined and predictable ways based on the foresight of all these rational agents. Behavioural economics – like quantum theory in physics – introduces a qualitative element of inherent randomness arising from 'bounded rationality'. The observation that the behaviour of corporations and other organisations, as well as individuals, was often determined by 'tacit knowledge', in the form of rules and routines, myopia and imitation, underpinned seminal work by Nelson and Winter (1982) which is perhaps the most foundational single contribution in evolutionary economics.[10]

The other development was the increase of computing power, which enabled researchers to start simulating complex evolutionary processes for the first time, including learning-by-doing and selection. This was the basis for another key contribution in which Arthur (1988, 1994) demonstrated unequivocally the potential of complex systems to generate multiple different pathways.[11] Chaos theory became the popular embodiment of the fact that how such systems develop can be incredibly sensitive to the starting conditions or minor adjustments.[12] The accumulating evidence surrounding learning-by-doing and other endogenous learning processes outlined in the previous chapter (section 9.4) provides unshakeable empirical evidence of such factors in energy systems.

Thus the stable, least-cost equilibrium that is implicit in most Second Domain economic theories – the classical matching of 'supply' and 'demand' in a natural, all-encompassing market – in reality is at best a theoretical snapshot devoid of time. It can never actually be achieved because it is defined by technological and system boundaries that are themselves constantly evolving.[13] The different behavioural choices of different participants further deflects reality from any sense of optimum or equilibrium – but also helps to maintain some diversity of options. Individual markets may tend to balance supply and demand, but only as determined by larger systems which reflect the combined influence of infrastructure, institutions and interests. These systems express history as much as the present. We can influence their future evolution but cannot replace them overnight.[14]

The result is a formidable body of evidence-based theory that explains our economies in terms of inherently dynamic evolutionary processes. Our heritage constrains, while innovation generates options. Selection processes, in which different participants make different choices, winnow these down, but not to a single, sterile, 'optimum'. We inherit the combined historical evolution of technologies and infrastructures, with the social and institutional structures built around them. This makes these systems highly *path dependent*.

The evidence laid out in Chapter 1 indicates powerfully the extent to which this applies in energy systems, as illustrated even through very broad indicators: industrialised countries at the same level of wealth differ by a factor of at least two in key indices like per capita emissions and show no sign of convergence.

Such systems *do*, however, have an in-built bias towards a certain course – a path of least resistance – defined largely by existing technologies, infrastructure, the vested interests of incumbents and the habits and preferences of companies and consumers. The greatest single limitation of most global economic modelling studies is to equate this path of least resistance – usually termed 'business-as-usual' – with 'least cost'. Economic forces certainly generate some pressures to reduce costs, but only within the constraints of systems and structures, which themselves resist more fundamental change for multiple reasons.

Position rents

Chapter 9 outlined the evidence around 'learning-by-doing', which implies that incumbent technologies have benefited from decades of accumulated investment and experience. The previous section noted that existing fossil fuel industries inevitably have a core interest in further improving existing technologies, and accessing resources, in which they have comparative advantage. Moreover, since the system has grown around fossil fuels, they benefit from what economists call a 'position rent' – derived partly from the sheer fact of being the dominant technology:

- Gasoline cars, for instance, can find fuel anywhere quite easily from a multitude of filling stations. Alternative fuel vehicles like fuel cell or electric cars may face a daunting challenge of accessing convenient charging points, while other modes of transport, from bikes to trains, struggle against the dominance of road infrastructure.
- Our electricity systems have developed and optimised around large, concentrated power stations, transmitting the power one-way to passive consumers. Transmission systems may be ill-adapted to the location of major renewable resources. Also, along with fossil fuels comes the automatic storage of our primary energy. Thus current electricity systems were never designed in ways to make best use of variable, dispersed natural energy flows – for which we need smarter systems, as outlined at the end of Chapter 3.
- Buildings in many industrialised countries in particular reflect ages when local wood or coal was abundant and cheap, and urban layout particularly in North America reflects a century of cheap oil: these are some of the longest-lasting systems in our economies.

These are totemic examples of path dependency. Systems for utilising fossil fuels have become highly developed: they define our transport, dominate the structure of power systems and underpin our sprawling towns and buildings.

Note that the broad concept of path dependence applies across all three domains. First Domain characteristics may underlie the extraordinary ubiquity of inefficient energy use in buildings, but it is the historical legacy that makes it so much harder to change. The inadequacy of information or options for final consumers, which exacerbate First Domain inefficiencies, may frequently also be traced to systemic factors like the resistance of less efficient manufacturers to the efficiency labels and standards charted in Chapters 4 and 5. Similarly, the price structures that determine the Second Domain also carry a powerful historical legacy, as evidenced not least by the glacial pace of removing fossil fuel subsidies: price reform, as evidenced throughout Pillar II (Chapters 6–8), is generally the most politically contentious and difficult of all the steps needed. And as illustrated in Chapter 9, the mere existence of a potentially superior technology is not sufficient to ensure its diffusion.

The wastage inherent in First Domain characteristics and the political resistance to full-cost pricing in the Second Domain, along with the bias inherent in established technologies, 'position rents' and incumbent interests, lead to an unnecessarily inflated use of fossil fuels in developed economies. 'Business as usual' is certainly not efficient or optimal. For this reason Figure 2.3 is drawn unconventionally, with the curve sloping backwards at high levels of resource use: it is entirely possible for systems to generate an unhealthy over-reliance on key resources, like fossil fuels.

The wrong trousers

That does not, however, mean that change is easy, for all the reasons indicated. Academia has developed a rapidly expanding literature on the problem, with a simple term to describe it: lock-in. The idea of 'lock-in' has begun to permeate the language, but it is often equated with the simple lifetime of capital stock such as a power station. As indicated above, it is about much more. Lock-in also comprises established economic advantages (accumulated learning-by-doing and position rents), physical and institutional infrastructures, and the interests (including regulatory and lobbying power) of incumbent technologies and industries.

The incumbent advantages can in principle be overcome by sufficient public backing for innovation, including measures required to help new technologies bridge the 'valley of death' to become viable options (see Chapter 9). Most lobbying powers can ultimately be overcome or at least moderated with sufficient public concern and political commitment.

Yet the problem is all the more pervasive when considering not a single component, but a whole system. Even simple acts like filling a car or turning on a lamp involve innumerable technologies: from the oil well or power plant, through the refining or transformation systems, to the filling station or socket in the wall. These are not isolated components: the connections between the components are all needed to drive the car or cast the light. Many pieces of a puzzle only make sense as a part of the whole. The challenge is to transform the 'technological system'.

Of course, it is sometimes possible to change one component without affecting the rest (for instance, switching from incandescent to fluorescent lights). However, the limits to simple substitutions mean it is not long before they accumulate to scales that require a wider change of properties, involving other components, infrastructure and ultimately the whole system.

For example, even if alternative fuels could compete with gasoline for vehicles, or wind energy compete with conventional power sources, some of their intrinsic characteristics may set limits within the present systems. Small amounts can be easily accommodated; larger amounts raise other challenges. For alternative fuelled vehicles to become more widespread, or for wind to become a mainstay of power generation, requires radical change. The system-level investments involved are hard to justify without widespread use – but widespread use may be impossible without system-level investments. Inappropriate infrastructure can be a major deterrent to rapid diffusion of cleaner technologies, regardless of the benefits they may bring. Existing structures may block an ambition to change.

A crude analogy is when an overweight person commits to exercise. At first the effort may be rewarded by better health, but then their favourite clothes no longer fit. Trousers, shirts, jackets are all too loose and frustrate the ambition to look and feel better. Improvements need to accumulate – and, ultimately, change the system itself.

Growing niches

It is rare that an innovation is immediately embraced by the majority. The process is gradual. For many advanced technologies, initially only a segment of 'geeks' or 'tech-savvy' people wait for hours in front of stores to be the first to buy the new version of a mobile phone or a games console. Such phenomena have been noted long ago; one theory divides the adopters into five categories – *innovators* and *early adopters* who are the first to welcome a new application, the *early majority* and the *late majority* who constitute the largest mass of users, and finally the *laggards* who may actively resist innovations.[15]

Technologies thus need to grow through *market niches*. This fosters learning economies which help adapt the initial prototype to the needs of the market, attracting new investments and stimulating the development and the diffusion of the product.[16] Some niches arise for applications in which innovation is deemed essential. Thus solar panels were first used by the space industry; gas turbines were developed for military aircraft; and the earliest development of the Internet, 'spun out' from the scientific community of particle physicists at CERN, was pursued with ARPANET, a research network implemented by the US Department of Defense.[17]

The challenge then becomes whether (and why, how and when) a technology nurtured in one specialised environment may transition to a wider mass market. Predicting such markets is difficult. How many, twenty years ago, would have considered essential items to include a mobile phone, an Internet connection, a personal computer and many other objects that nowadays are perceived almost as essential as cars or food? The distinguishing feature in these cases, as emphasised in Chapter 9, is that they offered a fundamentally different service – an innovation attractive to final consumers. In contrast, innovation in clean energy generally provides much the same consumer product (fuel or electrons), just in a different way.

However, there are some exceptions. In terms of electricity supply, the most obvious is solar cells which enable new applications – mobile fridges, lighting in remote villages, power for boats, and so forth – because they produce electricity from a resource on the spot in scalable ways. Such PV is really competing with a battery that needs recharging. Its entry into grid-based systems has also been aided by niches where bright sunlight coincides with peak demand, notably for air conditioning. Public policy has vastly accelerated scale-based cost reductions, leading to the 'solar surprise' (see Chapter 3, Box 3.4), but PV does have a certain self-sustaining momentum through these expanding niche markets.

A niche environment and/or financial support may be a necessary condition for many new energy technologies to complete the transition to market, but not sufficient. A natural consequence of the fact that dedicated markets do not exist in early stages of the innovation chain is the need for innovation policies to test and develop the market environment along with the application itself, as explored in Chapter 9.[18] The failure of early efforts on battery vehicles, contrasted against the success of hybrid technologies, outlined in the next section, is a telling example which points to several generic factors.[19]

Infrastructure and piggybacking

As anyone intent on slimming must know, one of the best motivators may be to start buying a new wardrobe. However, if the infrastructure is too disconnected from market-readiness – if the clothes are too small – the effort may simply be wasted, costly and ultimately depressing.

Examples of premature investment in infrastructure are not common but do exist, with some cases of ambitious public transport or battery charging infrastructure that remained largely unused. Such failures again underline the interaction between developers, users and other third parties (like governments). Institutional change occurs in response to a social change, therefore trying to build a market *ex-novo* without considering the desires of consumers doesn't work; it can survive only in a highly subsidised environment and may not transfer to external reality. The same risks apply to *forcing* infrastructure at scale: we can build as many hydrogen fuel stations as we want, but people may not necessarily start to use fuel cell cars.[20]

There may be another way to manage the challenge of infrastructure transitions. Consider one of the most successful and pervasive infrastructure transitions of our time: the Internet. Wi-fi and optical fibre networks have experienced dramatic growth, allowing more and more users to exchange ever more data in less time. Just fifteen years ago, things were very different: the most common connection to the Internet was through a dial-up modem connecting to a telephone line, providing a connection speed that would be unacceptable to most users now – it became dubbed the 'World Wide Wait'. Those connections had an enormous initial advantage: they used a pre-existent, widely diffused network – the telephone line. A transition technology (the modem) enabled the innovation to exploit a pre-existing infrastructure to gain swift diffusion. The success behind the Internet (but also many other technologies) was to navigate a way through the infrastructure challenge – by 'piggybacking' on the existing phone infrastructure until the scale of use was sufficient to support new infrastructural investment in dedicated IT networks.

Thus moving new energy technologies to mass market requires at least four components: a technological pathway; a 'critical mass' of adopters; infrastructure suitable to support the new technological state; and markets capable of supporting expansion.

The rest of this chapter outlines lessons and potential transition pathways in some of the world's major energy systems: transport in the Americas, the European electricity system and human settlements in Asia.

10.4 Cool cars in the Americas

Cars are a good example to start with, for many reasons. Transport is vital for economic and human development. It involves long-lived infrastructure and is strongly influenced by many factors beyond energy or environmental considerations – numerous other determinants of collective and individual behaviours, like the location choices of both firms and households. Reducing transport's energy intensity and emissions requires progress on adopted technology (to decrease the energy and carbon intensity of given transportation modes), infrastructure (which defines the transport modes available) and behaviour (which influences the overall demand for mobility and choice of modes).[21]

Transport accounts for a large share of the world's oil consumption and about a quarter of global CO_2 emissions (see Chapter 3, Figure 3.1). Global oil price rises since the middle of the last decade hit people hard and exacerbated trade imbalances in many regions of the world. Sustained high international oil prices seem to reflect a new era as the cheap and easily accessible reserves cannot keep pace with demand, driving increasingly difficult forays into deep water or other frontier oil developments (see Chapter 3, section 3.6). Emerging megacities struggle with choking traffic. The nexus of global transport and oil faces some kind of inevitable transition. As mapped out in Daniel Yergin's book *The Prize* (1991), the 'black gold' of oil shaped the geopolitics of the twentieth century; it could do the same in this century too.

The Americas offer a useful focus for many reasons. The low population density and pattern of historical development drives an unavoidably high demand for transport; with less than 20 per cent of the global population, they account for over a third of global oil consumption.[22] The overwhelming need for oil constrains US military and foreign policy. Yet the same factors endow the Americas with plentiful resources: biofuels (as developed extensively in both the US and Brazil) and other renewables, and oil resources extending to offshore and unconventional deposits (like Venezuelan tars and Canadian oil sands). The latter in particular involve vast sunk capital and higher emissions. Moving deeper into higher carbon resources would amplify this further – a big gamble in the face of scientific uncertainty, let alone the actual preponderance of scientific concern – yet is currently forming the path of least resistance. However the continent also has some of the world's leading capacity for innovation, including in the car industry, which is now striving to catch up the lead it lost in recent decades.

Enhancing efficiency

As noted in Chapters 3 and 5, after the 1970s oil shocks the US established modest standards for vehicle energy efficiency which then became essentially stuck at the same level for three decades, steadily falling behind the rest of the world. It is probably no accident that the US automobile industry was also in steady decline during this period. The two sprang from the same root cause: political gridlock driven by the incumbent interests of US oil producers and refiners (the latter account for around 1 per cent of US GDP on their own) and auto manufacturers oriented to big cars with big engines and low efficiency.

It took more than just energy prices or environmental concerns to change this: it was in fact the combination of these factors with the unfolding crisis in the auto industry and the shattering impact of the credit crunch. Some of the most established vehicle manufacturers, facing bankruptcy, had to appeal to the US government to survive. President Obama bailed out the industry, but exacted the price of shaking up the industry and severing the cosy relationship of auto manufacturers and refineries. His re-election slogan – 'Osama bin Laden is dead and General Motors is alive' – came along with massive policy change, as Congress accepted a doubling of fuel efficiency standards (Chapter 5) and the industry capitulated in its opposition, accepting that it had no option other than to innovate.

It would be hard to find a starker illustration of the power of established systems and incumbent interests and the fallacy of assuming that modern economic systems naturally tend to an optimum. They can have, rather, periods of relative stasis and sclerosis, punctuated by periods of upheaval when multiple forces align for reform. In this case, the energy inefficiency of the US vehicle fleet was patently not 'optimal' in any conceivable meaning of the word, but rather reflected the overall economic inefficiency of the auto sector. Pressures arising from energy and environmental concerns alone were not enough to change this, but when aligned with other forces they helped to justify and reinforce much needed reforms.

Nevertheless, the expected growth in vehicle-miles travelled could offset much of the efficiency improvements in the US transportation sector to 2030, including rebound effects as described in Chapter 5.[23] Greater energy efficiency can alleviate but on its own cannot solve the energy and environmental problems.

Passenger cars and light commercial vehicles like pickups and vans are of particular interest for several reasons. Many low-carbon options are in principle available (see Chapter 3, section 3.5), but most involve different fuels and therefore different storage and/or distribution systems. Also, the number of final adopters is enormous (in comparison, for example, with power production), while the market is very competitive, with more than fifty manufacturer groups in the world, among which 17 produce at least one million vehicles each per year.[24]

At the moment, potential alternatives to conventional petroleum fuels are natural gas, biofuels, electricity, hydrogen, and coal to liquids. Among the industrialised countries, Brazil and the US have made some of the most strenuous efforts to increase the use of alternative fuels.

Brazilian breakthroughs

The Brazilian National Alcohol Programme was launched in the aftermath of the first oil price shock in the early 1970s as a way to reduce Brazil's dependence on oil imports while supporting the sugar sector. Government policies to support the ethanol industry were directly responsible for the success of the programme.[25] Although public support paused after the collapse of world oil prices in the mid-1980s, the technologies continued to evolve, in part driven by the continued interest of the sugar industry in marketing ethanol. Flexible-fuel vehicles (FFVs), which can run on varied mixes of traditional fossil fuels and ethanol, improved, and in 2003 were made eligible for the same tax breaks previously introduced for biofuel cars. Taxes on gasoline were also raised above those on ethanol.[26]

More recently, the government made compulsory the addition of a percentage of biodiesel in diesel consumed in Brazil. Starting with an optional 2 per cent mix in 2005, between 2008 and 2010 the addition of biodiesel became compulsory and rose to 5 per cent in January 2010.[27]

The results of these policies are presented in Figure 10.1. From 2004 on, the sales of flexible-fuel vehicles exploded and accounted for the 88 per cent of new light-duty vehicle registrations in 2009.

The Brazilian experience and the rapid switch from gasoline to mixed fuel contain at least two valuable lessons. First, sustained policies were needed over decades to support the intertwined innovations in fuels and vehicles along with all the associated standards and institutions, as noted in Chapter 9 (section 9.10). Second, such drastic change was facilitated through mostly existing infrastructure; ethanol piggybacked on much of the existing fuel distribution networks.

While the boom in ethanol use was initially triggered by the first oil price shock, i.e. by Second Domain mechanisms, the industry only became viable in the long run through public investments and associated institutional developments (Pillar III policies). A key political element of the government support for ethanol has been the co-opting interest of sugar manufacturers. Beyond the tax incentives introduced by the Brazilian government to support the industry, its long-term strategy to massively invest in infrastructure and R&D ensured the viability of the market in the long-run.

Fuelling the US – transport and beyond

The path in the US has had some common features with the Brazilian story but major differences.[28] Many different 'alternative fuel' technologies were pursued in the aftermath of the oil shocks, and a wide range of alternatives were deployed at modest scales in the 1990s. However, from a starting point that has been (and remains) totally dominated by conventional petroleum, the first decade of the century witnessed a constant increase in ethanol, mainly based on heavily subsidised corn, while other alternative fuelled vehicles declined.[29] The combined contribution to energy remains modest, but the contribution to farming incomes is sizeable.

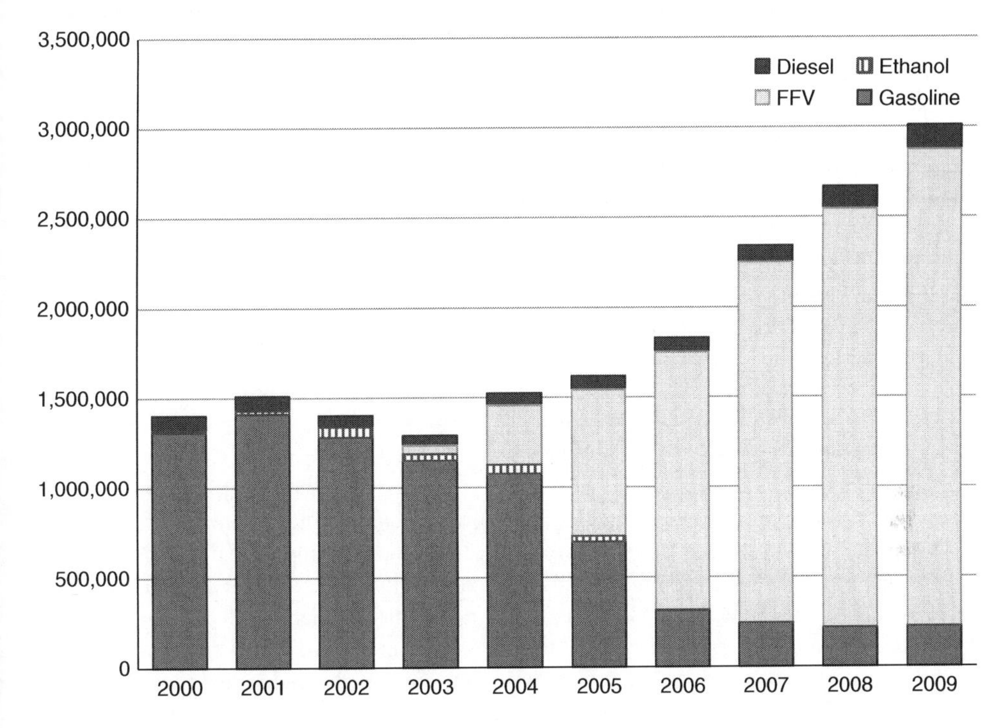

Figure 10.1 Registration of new light duty vehicles in Brazil per fuel type, 2000–2010

Source: ANFAVEA (2011).

Key instruments have included tax credits and production mandates.[30] While first presented as a way to reduce greenhouse gas emissions and improve US energy security, corn-based ethanol has little environmental benefit; as noted in Chapter 3, it is far less favourable than Brazilian sugarcane ethanol and may not even reduce greenhouse gas emissions at all.[31] However, compared with other alternative fuels, it had two major advantages: it can piggyback on existing infrastructure and it can support US farmers who are a formidable political force.[32]

Despite the boom of both bioethanol and biodiesel, the US biofuels transformation has little prospect of supplanting oil at scale, given the relative inefficiency and environmental drawbacks of biofuels, including the limits on the amount that can be produced without driving up global food prices.[33] It fuels farmers, but is a technology that risks getting stuck in the same interest-led trap that bedevilled US auto manufacturers. Enduring solutions lie elsewhere.

False starts

As noted in Chapter 3, one historic hope has been fuel cell vehicles run on hydrogen. Exaggerated claims about the miracles of hydrogen technology may in fact have done more harm than good, as governments seem now to be deeply uncertain about the merits of funding hydrogen projects.[34] Hydrogen also lacks any established infrastructure and thus faces the classic 'chicken and egg problem' of being locked out rather than in; as a representative of the fuel-cell industry noted, 'the real challenge as we move toward retail is that we need to change the nature of the fuelling.'[35] Without any obvious lobby to drive through these difficulties, the prospects look dubious.

Natural gas vehicles have a foothold in some cities, in part for their contribution to meeting local air quality standards.[36] With an established network to transport natural gas and the impetus of the US shale gas revolution set against continued high petroleum prices, some resurgence of natural gas vehicles would not be surprising. But, as with ethanol, their advantage in terms of greenhouse gases is probably marginal.

Electric vehicles have suffered a number of false starts in the US and elsewhere, but the technology has continued to improve enormously.[37] Vehicle manufacturers are thus placing their biggest bets on electric vehicles.[38] These are not only clean at the point of end-use, but electricity benefits from the flexibility derived from the multiple power production technologies and the existing ubiquitous transmission infrastructure – particularly in less densely populated regions like North America, most homes have space to park cars within easy reach of home electricity. Plug-in hybrid electric vehicles (PHEV) and range-extended battery electric vehicles (BEV) can thereby use electricity from the grid to power private mobility. The challenge of a sustainable transition for transport is eased if the final target is the widespread use of electric vehicles and the obstacle is making the transition for the vehicles themselves.

Obviously the fuelling system is only one of the many components of the transport system. A network of qualified mechanics and availability of spare parts are also essential but these again may benefit from the existing infrastructures – pieces will be distributed by the same means and, after the necessary training, mechanics should be able to continue to work in existing garages. Such factors do, however, amplify the underlying message that a transport transition is a complex and lengthy undertaking in which it makes sense to seek facilitating shortcuts.

From fuels to electricity: piggybacking hybrids and alternate infrastructure

Hybrid electric vehicles (HEV) – cars with both an internal combustion engine and an electric motor – appear to offer such a path. At low speed the electric motor is used instead of the conventional engine, allowing lower fuel consumption than conventional cars, especially in urban areas. Such range-extended hybrid vehicles, which drive on both electricity and petrol or diesel (or even LPG in Korea), give far greater range and flexibility than pure battery vehicles and may even be cheaper. The first commercialised HEV (the Toyota Prius) cut fuel consumption and emissions by about 30 per cent.[39] Price was a major obstacle, but learning effects have, as expected, rapidly closed the gap.[40]

Hybrid cars have formed a key plank in Japanese efforts to reduce oil dependence.[41] Looking back at the failure of Japan's efforts to deploy BEVs in the 1990s, perhaps this technology push was a premature effort at Pillar III transformational policies, lacking the sufficient groundwork targeted at early market support through other pillars. In fact, the sustained effort towards R&D, infrastructure and market support for BEVs[42] largely contributed to the later successful deployment of HEVs in Japan, which benefited from basic technologies developed for BEVs, from the existence of a well-known market and from the existing subsidy system.

Instead of integrating the old and new fuelling system into one vehicle, one option is to create a cheap alternative infrastructure. The major limitation of electric cars stems from the battery's long charging time, limited lifetime and small storage capacity. One solution is to set up a battery leasing system, as pioneered by Better Place in Israel, which provides a network of charging points where the battery can be exchanged.[43] This also can take advantage of the mismatch between consumption and production peaks for renewable energy by recharging batteries during sunny hours. Starting from the other end, Chapter 3 noted the rise of electric bikes in Asia, with the lower energy needs enabling greater distances within the constraints of battery technologies. The potential for different vehicles to better match travel needs with technological possibilities is clear.

These examples show that there are several ways to overcome the curse of infrastructural path dependency in transportation systems, allowing technologies to grow, respond and mature. The first plug-in points for vehicles may be simple electric plugs, but smarter control systems are bound to offer a better way of optimising charging cycles according to user demand, battery technology and power system development. How long the age of hybrid vehicles or battery switching stations will last is uncertain. In the case of the Internet, dial-up connections were rapidly substituted by broadband in response to user demand and large-scale government programmes. It might take longer to refine electric vehicles and their charging infrastructure.

The direction of this journey seems clear even while the end-point is not, for a wealth of innovation could emerge to better exploit the opportunities in matching different vehicle and transport needs to the fundamental assets of smart grids, better batteries and charging options. However, the present hybrid and related technologies will play a fundamental role in making the transition possible. Plausible scenarios have been developed for the evolution of transport systems away from fossil fuel dependency, through a series of technology waves. Figure 10.2 shows vehicle use for a scenario evolving low-carbon transport globally. Many decades of transition will require multiple technologies: a great diversity of fuels and

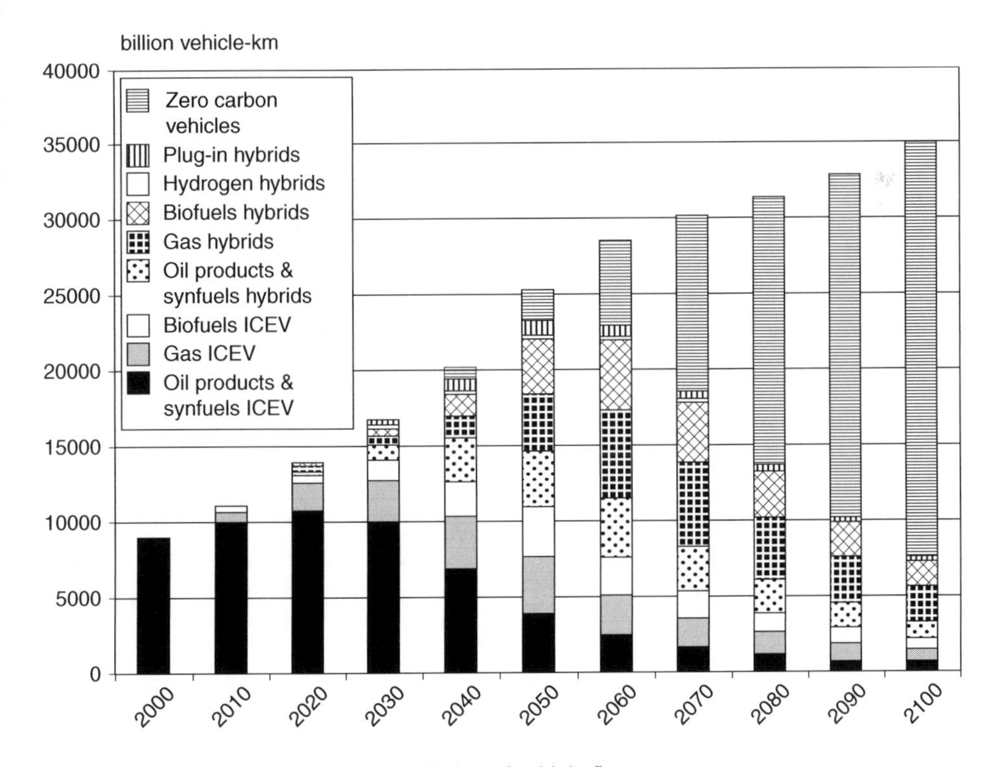

Figure 10.2 Global scenarios for the evolution of vehicle fleets

Source: Gül et al. (2009).

systems are implied by mid-century, with a mix of petroleum, biofuels and electricity. In the same way as electric trams and bicycles acted as stepping stones for the transition from horse carriages to automobiles in the US,[44] hybrid vehicles and infrastructures could introduce new mobility habits and preferences, and could lead the way to low-carbon mobility.

Conclusion

Changing the way we travel will require new technologies and fuels which are developing through specific niches: the first automobiles in the US were not designed to compete with horses for practical transportation purposes but were originally developed for entertainment.[45] However, given the subsequent dominance of the gasoline car, deploying alternative fuels and technologies beyond niche applications hinges upon policy. The Brazilian story demonstrates the need for policies that combine both Second and Third Domain drivers over an extended period: using guaranteed prices and sound innovation policy including regulation to drive the diffusion of flexi-fuel vehicles and public investment in infrastructure. In this way Brazilian technology has worked its way down the iterative steps of Figure 9.6. Still, although first-generation biofuels are now well established, their actual environmental benefits remain contested. CO_2 pricing, accounting for the full life-cycle emissions of biofuels and consumers' sustained concerns about their environmental, social and economic impact, could continue the push to develop still better liquid fuels.

Hybrids have successfully gone an important step further, by engaging First Domain drivers through their appeal to the wealthy clientele of hybrid cars, to overcome the inevitable cost penalty of dual-engine technology. Overcoming the limitations of electric-only vehicles will involve supporting specific niches with a focus on the potential users less affected by those disadvantages – such as fleet owners rather than single consumers.[46] The deployment and adoption of electric vehicles is another great illustration of the need to harness all three domains. Efforts must simultaneously create the right price incentives (Second Pillar policies) and support innovation along with adequate infrastructure and institutions (Third Pillar policies), while creating the conditions to change existing, carbon-intensive habits through public engagement with attractive options (First Pillar policies).

Such options would help to better meet diverse transport requirements. The need in many cities to drag a ton of metal with us on a local shopping expedition and waste 80–90 per cent of the fuel burned in doing so is a testament to the unspoken limitations of present motorised transport vehicles. As transport technologies evolve and diversify, so too will grow the possibilities for vehicles to match more closely our needs. For electric mobility, this means providing affordable electric vehicles along with attractive charging options. As noted, this may also require fundamental institutional changes. Following the example of PSA Peugeot Citroën in France, firms could offer full mobility services, for instance by providing customers who purchase 'city-size' electric vehicles with a choice of more traditional family cars for weekend use. Achieving low-carbon mobility will require close attention to patterns of infrastructures and institutions, which eventually both shape and respond to people's mobility habits and preferences. Yet the transition also has obvious potential for multiple other benefits.

So despite the pessimism around transport, much can be done to decrease the energy intensity of existing transportation modes (e.g. more energy-efficient vehicles) and to decrease the carbon intensity of fuels (whether through biofuels or decarbonised electricity). This illustrates a final point. Transforming transport requires going beyond that sector alone, to considering the wider systems. Figure 10.3 illustrates a scenario of transforming European transport, which

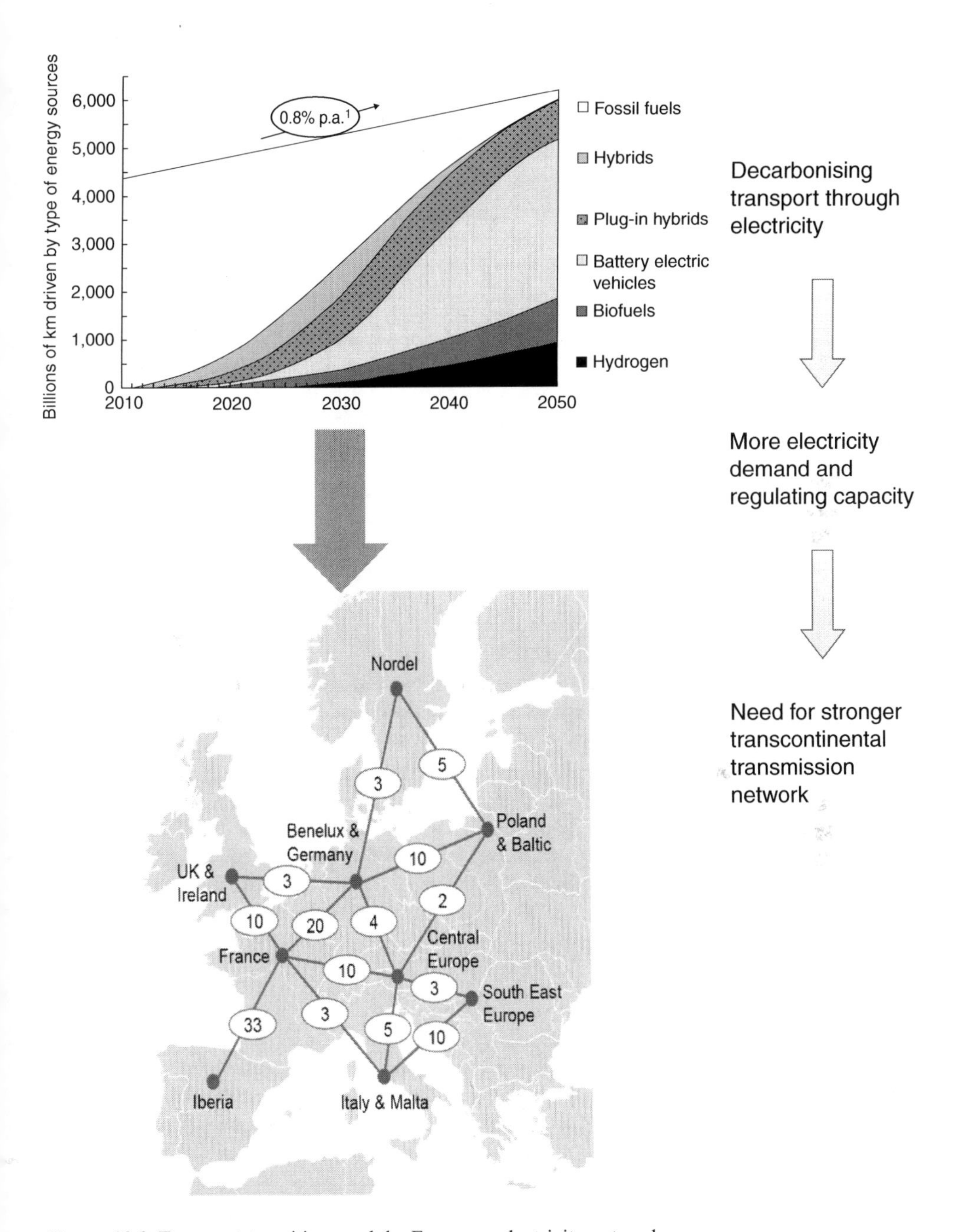

Figure 10.3 Transport transitions and the European electricity network

Note: The upper graph shows a scenario for decarbonising European transport, along with growing demand for mobility which relies heavily on hybrids and electric vehicles. The lower panel illustrates the scale of electricity interconnection (in GW) between different European countries emerging in such a scenario (which implies growing electricity use from transport) in 2050, met largely with a very high use of renewable energy which in turn requires far greater transmission capacity. The figures in ovals represent additional gigawatts of transmission capacity.

Source: McKinsey & Co. *et al*. (2010).

would have profound implications for both the scale and the nature of electricity demand in Europe. Thus, to make the transition clean as well as secure, the electricity system itself needs transforming.

10.5 Transforming European electricity

Chapter 3 has already indicated why electricity is becoming more and more central to energy futures globally. It is the flexible carrier that can accommodate and connect a vast array of valuable uses – a wide range of established and emerging demand-side technologies in homes, business and now in transport – to a broadening range of sustainable energy sources. Electricity production, however, is useless without the infrastructure to transform, transport and deliver it. The resulting combined systems form some of the most capital-intensive, long-lived and complex interlinked network infrastructures on the planet.

The extent to which the systems themselves need transforming is defined by how far the future vision may be from the current starting point. Most countries have designed, developed and regulated these systems to transport electricity from huge centralised power stations via a high voltage national transmission system, down through successive levels of distribution. In many countries, the generation investments are also conducted through centralised regulated companies but, as noted, from the late 1980s some countries began to inject competition into electricity generation, 'designing markets' where none existed before because everything has to flow through a single regulated transmission system.

Europe has taken forward such competition most extensively, establishing the basic rules for a Single Energy Market in recent legislation. Europe has also been the most ambitious in the pursuit of decarbonising its electricity system, which is seen as pivotal to the wider goal of deep emission reductions by mid-century. Standard economic assessments suggest that the electricity sector is the cheapest to tackle, which has been generally interpreted to also mean the easiest. While all European countries have committed to delivering an increasing share of renewables by 2020 and envisage using large shares of renewables to achieve the European emission reduction target of 80–95 per cent by 2050, some EU countries, including the UK and Germany, have additionally formulated a clear strategy to decarbonise the power sector.[47]

The simplest Second Domain theories would suggest that the central instrument to achieve this decarbonisation would be an adequate carbon price (Pillar II), backed up by R&D and technology-specific supports like feed-in tariffs to bring technologies down the learning curve as explained in Chapter 9. This section outlines some of the other realities of transforming power systems in Europe.

Technology, systems and markets

Of the three main classes of low-carbon technology, nuclear power sits most easily with the existing physical infrastructure: a big single source that feeds power into a national grid for widespread distribution. However, its scale, timescale and risks make nuclear investment inimical to the competitive market that Europe has established. France, the nuclear power leader in Europe, fought the effort to introduce competition, and as noted previously (see Box 9.2) it proved impossible to include even existing nuclear assets when selling off the UK generation business. The bid to secure new nuclear investment is a major driving factor behind a wholesale overhaul of UK electricity regulation, which critics see as largely ending the competitive market (on which more below). The physical infrastructure fits; the modern regulatory one does not.

Carbon capture and storage (CCS) from coal or gas power stations can also fit broadly with the existing physical infrastructure, but it faces other hurdles (in addition to technological uncertainties). Opposition to onshore storage has now precluded it from Germany's options. Offshore storage in North Sea reservoirs is possible, but more costly. The economics of fitting expensive additional equipment to fossil fuel power stations can only make sense when there is a high carbon price – a basis which, as noted in Chapter 7, is now in disarray. And although the electricity from CCS could use existing infrastructure, it would need to be complemented by a whole new infrastructure of CO_2 pipelines and disposal facilities, and the associated legal and regulatory structures. These current and prospective hurdles further impede the impetus for progress on CCS, which has since slowed to a snail's pace.

Renewable energies comprise a diverse range of options as also surveyed in Chapter 3, but they differ most from the classical model for three basic reasons. First, many renewables tend to be smaller in scale and may connect at the level of local distribution systems. Second, unlike nuclear or fossil fuels, production has to move to the location of the resources – not to convenient sites for centralised connection. Third, the biggest renewable resources are intermittent, varying with the cycles of sun and wind in particular. Though some other renewables (such as biomass) do not share these latter characteristics, the resource statistics backed up by the experience of the German *Energiwende* underline that these two resources offer the biggest renewable energy contributions – yet they are the most intrinsically different from the traditional ways of generating electricity.

Integrating renewables: from micro to macro (and back again)

To compact the resulting transformational challenges into a short space, it is useful simply to note the key issues at the successive scales of power systems – and then complete the circle back to end-use technologies.

Local generation

Distributed renewables, at the extreme of household PV systems or micro wind turbines, offer the unprecedented prospect of people generating their own electricity. At somewhat larger scales, wind turbines or solar arrays may benefit from some economies of scale and plug into local distribution networks. In practice, the variability of these sources means that people remain dependent on a connection to the wider system, but this does introduce the possibility of feeding into the grid by using smart meters to monitor and account for the power flows. This entails safety standards and regulatory changes governing the terms. But who then pays for the local network and the services it provides? Making such changes has not proved too difficult and opens up a social dimension previously absent: local engagement with power production, in the form of community- and farmer-owned wind farms, has been clearly charted as an important contribution to the social support for wind energy in Scandinavia and Germany. These are almost the only points at which power system transformation to date has been able to engage First Domain effects, moving public engagement from the traditional 'not in my back yard' (NIMBY) opposition (the typical resistance to change) to positive drivers based on the desire to contribute and help gain some control over the clean energy transformation.

Connection and distribution

However, the path of connection at local levels has not always been smooth. Distribution companies are regulated monopolies: they may have little incentive to connect assets which may undermine their own revenue base. In some countries (including the UK) delays in connection proved a major deterrent to renewable energy development by independent generators. There was a related, classic coordination problem: should connection agreement be established before or after a firm investment decision? The UK was probably not alone in encountering major struggles over connection, which resulted in the government stepping in to direct the regulator to adopt a 'connect and manage' regime to remove this blockage. This then of course placed an onus on the rest of the system to manage the resulting inflows before more wholesale reform had established the incentive structures for wider system reinforcement to cope with new types and directions of power flows. The ensuing lack of coordination has led to growing periods in which wind farms are paid not to generate (to avoid overloading distribution and transmission networks) – further feeding public resentment. Above all else, transformational change in a network requires some coordination and planning.

Transformational change also requires innovation, and distribution companies have tended to be at the bottom of the low scale of energy company R&D expenditure noted in Chapter 9 (Figure 9.3). In the absence of (network) competition, this too is almost entirely in the hands of regulators, and the UK regulator responded by establishing a £500m Low Carbon Network Innovation Fund, funded by network charges and governed by intellectual property (IP) rules to ensure that the resulting innovations are shared across the industry. The experience also testifies to the impossibility of separating 'carbon benefits' from the wider gains of accelerating inovation.[48]

Transmission and system balancing

The next step up takes us to the level of national transmission. This is usually the integrating system, and consequently the arena in which the need to continually balance national electricity supply and demand is played out. Two particular issues arise. One is the big brother of the local connection debates: the need to connect remote resources (e.g. major renewables), and potentially to reinforce the national network to cope with new power flows. In both Germany and the UK, for example, the bulk of wind energy resources are in the north (and offshore in the North Sea), but much of the demand is in the south. As wind capacity has risen in Germany, its neighbouring systems have been transporting its power south on windy days. UK wind energy growth has been slower and the UK is instead starting to construct 'bootstrap' power lines – with a key link being the UK's first offshore DC cable. The result is more investment cost, but it results in a stronger and more integrated power system – the weakness of north–south interconnection having been a long-standing issue.

The other challenge at this level is balancing supply and demand. The system anyway has to cope with sizeable swings in electricity demand and the risk of unplanned plant outages, but the challenge gets bigger and more complex if and as the inputs from variable renewable sources rise to comparable or greater scale. System balancing is a complex area beyond the scope of this book – like other challenges, it can be handled, but with some cost depending in large part on the sophistication of the response and the regulatory structures around it. The best options are bound up with another level of challenge – that of investment to maintain adequate capacity.

Enhancing interconnection

Beyond internal reinforcements of transmission is interconnection. The European power system evolved in fragments, very partially connected. A modelling study explored the prospects for accommodating exceptionally high levels of renewable energy sources. The principal analyst, Goran Strbac – a leading expert in modelling the costs of accommodating high levels of renewables – was used to the fact that this starts to get difficult as levels approach 40 per cent contributions in most individual EU countries. He expressed his surprise at the Europe-wide potential that emerged. With a more interconnected grid, the resulting Roadmap 2050 study described an engineering assessment for a power system that could operate with 80 per cent renewables, primarily based on solar and wind energy. With a focus on the transmission infrastructure, the result was a threefold expansion from present levels of interconnection (the lower panel in Figure 10.3) as the primary approach to smooth out short-term fluctuations and address seasonal differences.

Specifically with stronger north–south connection, southern Europe exports solar energy in the summer while wind energy flows south in the winter. This, however, would require massive transformational investments in the European grid infrastructure – with huge implications for public acceptability, coordination, regulation and cross-border cooperation, in addition to the investment costs. This also triggered discussions and further work to explore the possible trade-offs between local storage capacity, the potential spilling of some renewable energy when it could not be used and large-scale transmission expansion.

As with other major transformations, there would be potential co-benefits. Poland, historically nervous of its dependence upon Russian gas, might welcome the stronger connections into the EU system. The North Sea could re-emerge as an important European energy province, replacing the declining oil and gas outputs as its wind energy capacity grows. Moreover, an integrated system offers the scope to increase resilience by spreading resources across different technologies, in order not to be affected by, say, a gas crisis or a series of cloudy or windless days.

Another advantage of a broad and interconnected system is that each area can adopt some specialisation in its most productive technology – much like other forms of international trade. An isolated country has to produce all it needs, with obvious inefficiencies and limitations. In an integrated European system, countries with higher temperatures could specialise in solar technologies, Atlantic coast countries could exploit marine technologies, while others focus more on wind (or even, if Iceland were connected, geothermal energy). Each region could take better advantage of its own specific endowments, as well as benefiting from interconnection with higher diversification.

Of course, there would be no reason to stop at Europe's borders. Some of the more far-flung visions extend south to generate electricity from solar power in the Sahara deserts.[49] This is also seen as a component of stronger Mediterranean political integration, to help North African development for multiple reasons – including a possible reduction of immigration pressures (which could be exacerbated by the changing climate). As with all such grand visions costing hundreds of billions of euros, the key would be a realistic evolutionary path, starting from the domestic interests of the host countries which have begun to develop stronger interests in solar energy.[50]

Engaging storage and demand

Ultimately, a high level of variable electricity sources requires energy storage. Various *technologies* may provide this in bulk. Long-distance transmission may increase access to conventional hydro dams, but there is steady progress in a range of dedicated storage technologies ranging from compressed air to cryogenic or molten salt storage.

The other – potentially complementary – route brings us full circle, back to the role of electricity demand. As noted in Chapter 3, trends in all end-use sectors point towards growing electrification. Many of these uses also bring potential flexibility, mediated by smarter meters to facilitate two-way flows of information as well as electricity.

Thus electrification of vehicles would bring with it implicit storage. This could be connected to the electricity system when vehicles were plugged in to charge. With current battery technologies, there are concerns that repeated discharging – as implied in storage services provided to the grid – could erode the high battery performance needed for vehicles. Conversely, however, those high performance requirements mean that vehicle batteries would be replaced regularly – creating a stream of batteries still perfectly serviceable for grid balancing and backup purposes.[51]

In addition, almost all applications of electricity for providing cooling and heating demands (e.g. through heat pumps), and many industrial processes including water treatment, come with implicit cheap storage, through the ability to either reschedule the load or directly store the heat.

Investment, security and returns

Curiously, perhaps the biggest question around the future of Europe's electricity systems is not one driven purely, or even mainly, by the effort to decarbonise. It concerns the fundamental relationship between competition and investment in such big, long-lived systems. Over the previous two decades, injecting competition into generation was seen as a major success, bringing a new vigour into the sector and substantially driving down the costs to consumers. The EU 'Third Energy Package', extending competition in electricity and gas markets across Europe, was the natural culmination of Second Domain principles applied to the European power sector.

Yet, this is not an end point. Some of the cost reduction was due to greater economic efficiency, but much was due to reduced investment during a period of sufficient generation capacity. The newly competitive companies not only slashed R&D expenditure as noted in Chapter 9 (see section 9.3), but also eschewed much other investment. Where new power plants were built under the new rules – and excluding the supports for renewable energy – it was mostly gas-fired plants, which are cheap to build but subject to the vagaries of gas markets. As gas prices rocketed, so too did power prices – and where companies have been unable to pass the costs on through electricity prices, a number of such plants have been closed or 'mothballed' (taken out of service). With volatile fossil fuel markets, companies are reluctant to commit capital on plant that will last for decades when they have little certainty about whether or on what terms those plants may operate.

Second Domain theories suggest that the system should self-correct, as the prospect of power shortages with consequent high electricity prices attracts new investment. Given the multiple uncertainties, real companies and finance seem reluctant to take such bets. As with the EU emissions trading system, there is thus an emerging confusion about the real purposes and expectations: are competitive energy markets intended just to minimise short-run costs, or are they expected also to support adequate investment? And what should be done when evidence accumulates that (particularly with the uncertainties linked to a transition period) the two do not simply go together in the way that was assumed?

More fundamentally linked with this is the fact that markets which depend on a common monopoly asset (the grid) have to be constructed through regulatory rules. How they are designed has huge implications. In the current European approach, gas power stations tend to be the least risky investment, because they are relatively cheap to build. Brazil, however, has taken a wholly different approach, holding competitive auctions for long-term contracts to supply electricity over 20 years at a fixed bid price. With no subsidy – but with the backup of Brazilian hydro capacity for storage – wind energy beat gas, winning almost 80 per cent of the market.[52] Such long-term contracts reverse the location of risk on to plants like gas, with low construction costs but highly uncertain fuel prices. A study from INSEAD underlines how the economics of renewables *vs* alternatives may hinge more on market, contractual and system structures than on relative prices or carbon taxes. Cost is a relative concept which depends on how markets and systems allocate the risks and rewards.[53]

These realities point to a deeper question: to what extent is adequate investment, generally assumed to be a product of Second Domain forces of market-based price incentives, actually more closely related to Third Domain processes? When it comes to long-lived investment in assets whose main benefits are for security or environment, the boundary becomes increasingly blurred.

In this respect, there are interesting insights to glean from comparing and contrasting the worlds of oil and electricity. In the European context, North Sea oil was a massive development: after the oil shocks, UK investments in it rose to over £20bn annually. With hindsight it is viewed as one of the great bonanzas, yet it occurred only because of one massive forecasting error (oil price forecasts after the oil shocks), and proved beneficial only because of another (on technology). Having originally estimated that North Sea oil production could only be profitable above oil prices of $50/barrel, as innovation followed the decline of oil prices, North Sea rigs ended up being profitable at well under $20/barrel.[54]

Today, there is fierce political opposition to the investment of less than half as much to develop the equally huge (but enduring) resources of offshore wind. Yet, the proportionate value and scope for innovation are similar. The main difference is regulatory and cultural: the sense that governments, ultimately, control the price of electricity, and should always regulate to follow the path of least cost, which is a conservative and incremental one. Many have argued that the oil companies, given their huge offshore engineering expertise, should move into offshore wind. Some have to a limited degree, but a BP executive who led that company's evaluation said they reached one simple reason for not doing so: BP is not an electricity company and does not want to be one. By which he meant, quite simply, they concluded that huge risks and huge rewards are intrinsic to the oil sector, but that the UK government would never condone the scale of risks and rewards an oil company would expect in order to lead such a transformation in electricity.[55]

In pursuit of its overall *Energiewende*, Germany is perhaps showing such boldness; it recently reached the milestone of securing 25 per cent of its electricity from renewable sources, and is encountering the next set of issues. Marking the achievement, the director of *Agore Energiewende* observed that the first 25 per cent were politically very difficult, but technically very easy. The next 25 per cent, he predicted, would be the other way round – because it would involve changing the system itself.

That is the path to which Germany has committed, but as noted there are several routes to decarbonising electricity. Different regions would doubtless follow different roads and mixes of nuclear, CCS and various renewable energies, as well as varied emphases on end-use efficiency and smart electric services. The differences concern not just technology adequacy and cost. They involve societal choices and economic structures. Nuclear utilises an established but controversial technology and requires the greatest degree of centrally planned direction and investment. CCS can best preserve the incumbent interests of coal (or gas) industries and utilise associated national resources but requires big engineering innovation and new infrastructures. Renewables require the greatest change in the structure of electricity systems and their governance. The common currency is that price and cost are important but are far from the sole determinants: a major part concerns regulatory and strategic policy, for these will be key instruments of our choices and do much to determine the systems we get.

10.6 Urbanisation and Asia

In 2007, the world crossed an important threshold: for the first time in history, more than half of humanity lived in urban areas. The push towards urbanisation – especially in developing countries – is becoming stronger (see Figure 10.4): in 2050 more than 6 billion people will live in a city; most of these (probably 5 billion) will be in developing countries, which accounted for only 2 billion in 2007. Whatever progress is made in advanced vehicles and fuels, and the decarbonisation of electricity, a third key determinant of energy and environmental impacts will be the nature of these cities, their buildings and their transport systems.[56]

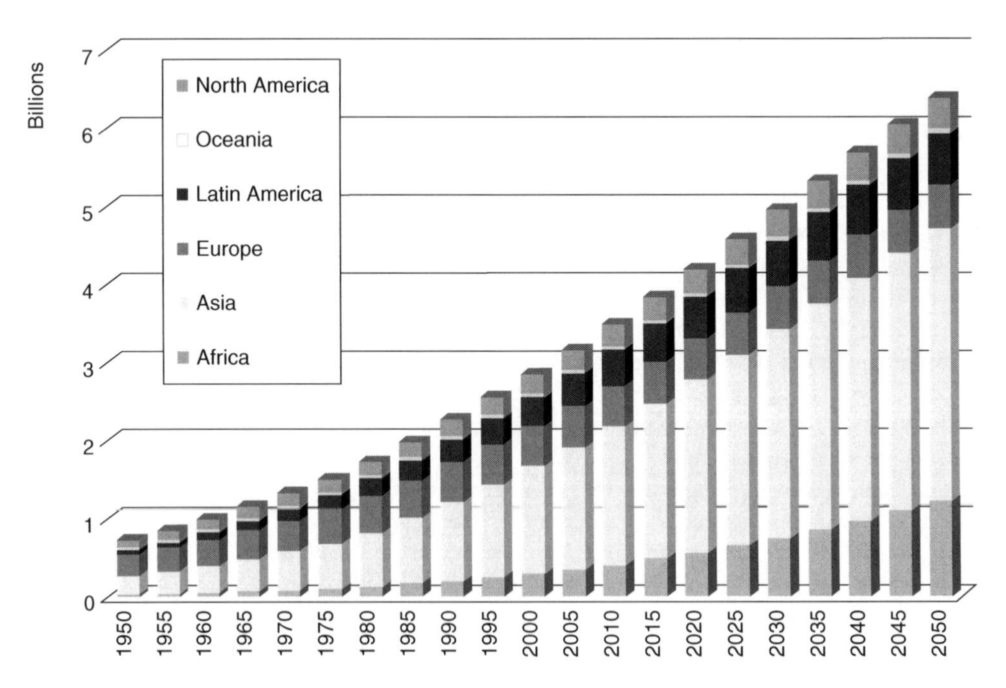

Figure 10.4 Urban population between 1950 and 2050

Source: UNDESA (2011).

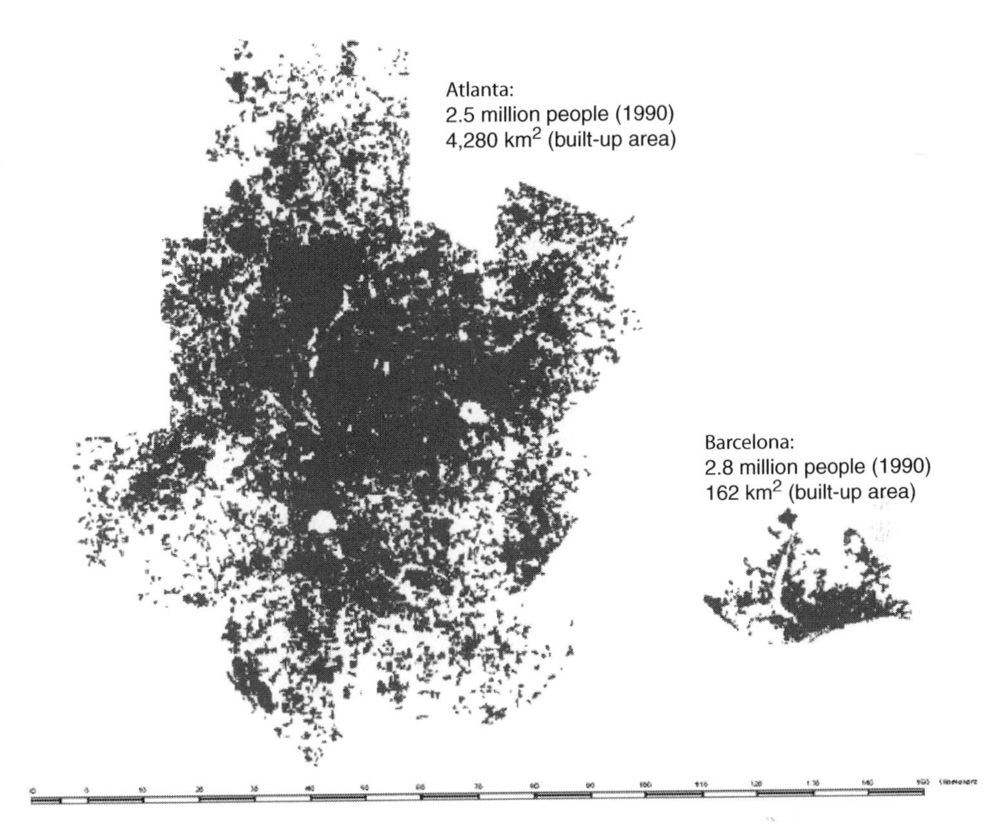

Figure 10.5 A comparison of the built-up areas of Atlanta and Barcelona to the same scale

Source: Bertaud and Richardson (2004).

To a large degree, for both Europe and the US, it is too late. One of the few things more enduring than a building is the structure of cities and their systems. As noted in Chapters 3 to 5, retrofitting buildings to improve their energy efficiency is far more expensive than getting it right first time; retrofitting a public transport system (particularly a metro) is likely to be much more costly, and less adequate, than designing it in as a city grows.

The diversity of existing cities does point to the huge differences that are possible. Figure 10.5 compares two cities of almost equal population: Barcelona on the Mediterranean coast and Atlanta, Georgia in the US. In 1990 the built-up area of Barcelona was 162 square kilometres; the urban sprawl of Atlanta covered almost 30 times that area. No one could reasonably assert that one is a much more desirable place to live than the other. Residents in Atlanta have the pleasure of large houses and gardens, but travel to work and social life depends unavoidably on the car. Barcelona is widely considered one of the most vibrant and desirable cities in Europe, but many residents do not need to own a car. The cities have evolved in very different ways for many reasons, including land and energy availability and prices, magnified by cultural choices that these factors help to create. Neither is likely to change: people, systems, technology and culture adapt.

It would be hard to find a more stark illustration of the evolutionary and path-dependent nature of our systems. The choices made in emerging cities the world over, over the next few decades, will have similarly major and enduring consequences. As indicated and illustrated in Figure 10.4, many of them will be in Asia, with growing numbers also expected in Africa.

Influences

The development of cities not only provides an excellent example of Third Domain processes – incredibly long-lived and fundamentally evolutionary – it also illustrates the impossibility of separating energy and climate considerations from other factors, as some kind of add-on and separately-costed choice. Cities, rather, evolve in many different ways as a product of numerous different forces.

Households and firms choose where to locate, or relocate, based on time, cost, employment possibilities and social preference. This translates partly into trade-offs between accessibility, overall cost (housing and transport) and the attractiveness of city centres.[57] Factors influencing such trade-offs include land and property prices, and transport times and prices. These in turn are affected by urbanisation rules, fiscal incentives, real-estate market regulations, public investments, transport infrastructure and the attractiveness of locations, along with the location and quality of jobs – which are also affected by the local environment and connections.[58]

Moving away from city centres seems to coincide with the aspiration of many to live in larger houses (with a garden) while benefiting from the amenities of open sceneries and less air and noise pollution. While this wish to 'escape the city' is not universally shared – it depends noticeably on age – and central cultural amenities, for example, continue to be desirable, observations of intra-city relocation movements suggest a common though not universal trend away from density.

Yet low density (and thus extended) cities involve far greater energy and emissions for multiple reasons: fuel consumption is directly proportional to distances; a greater number of areas to connect means a more complex and inefficient transport system which increases the share of private transport; more infrastructure is required at higher costs; and lower density of housing translates into higher heating/cooling requirements.[59]

An important question then is whether, when and how energy and climate concerns might influence the wider choices. In some cases they may be inconsequential; in others, however, they may have a catalytic role. For example, the decision to give planning approval for a major development may be linked to provision of an adequate public transport for multiple reasons, with energy and environmental concerns being just some of the factors. But in addition to any direct impact, the decision may have enduring implications for future locational choices and the overall development of transport networks and technologies as the city grows.

Emergent Asia

About half of all urban areas with more than 10 million people are already located in Asia. While all cities will continue to develop, the challenges in the emerging economies focus on providing infrastructure for an enormous number of emerging small and medium-sized cities in addition to big megalopolises. China is building urban housing at a rate of around one billion square metres a year – a rate equivalent to building every 15–20 years the floor area of the entire European building stock.[60] Obviously, such strong economic and demographic growth may swamp the gains of individual energy efficient technologies; in addition to the building stock itself, urban transport infrastructure in India and China will crucially affect growth in energy (mainly oil) and associated emissions.[61]

While energy consumption in developing countries used to be dominated by industrial activities, the importance of energy use in urban areas is increasing because of the emerging middle-class and lifestyle changes (and a significant part of industrial energy use itself is to provide the concrete and steel for urban expansion). In contrast to industrial production itself, these processes are again more subject to First and Third Domain processes – personal preferences, evolutionary forces and how their interplay is shaped by public investment and policy – and with consequently far greater scope for variation in outcomes irrespective of wealth levels.[62]

New challenges come with a growing urban population and associated lifestyles – congestion and local pollution in addition to wider energy and GHG emissions. All this clearly takes us beyond national policies focused on classical tools like pricing. Though some tools may be similar – for example, the Tokyo and Chinese pilot programmes on emissions-trading at city level noted in Chapter 7 – attention to urbanisation shows the need to engage also the more specific issues and opportunities at the urban or municipal level.[63]

Policies and urban governance

Sustainable urbanisation in developing countries depends prominently on the far-sightedness of urban policies, the degree of coordination between road, building and public transport planning, and the capacity to fund the necessary long-term investments – classic Pillar III policies. It also requires some interest in sustainable lifestyles for their citizens, to gain support for these efforts and to ensure that the wider range of efficiency and low-carbon solutions (like the use of public transport, CHP and energy saving in buildings) are effectively adopted by the mass of energy users – measures which are at the core of Pillar I.

These issues and options can be considered at three levels: (1) buildings; (2) energy system issues; and (3) urban planning (public transportation, limits to urban sprawl).[64]

Buildings

Poor building standards bring multiple problems – and safety is paramount. A lack of standards allows builders to cut corners to satisfy market demand and maximise profit. Improving standards can also readily embody better energy performance as part of a better construction method – an integrated approach to quality. The switch between construction techniques will take place gradually, and only if there is a sufficient commitment by local government to regulate and enforce building standards.[65]

The drivers thus include improvements in safety, health and thermal comfort as well as national energy and environmental concerns. The associated policies at the level of individual buildings are those reviewed in Chapter 5. Measurement and labelling is a bedrock because it needs (and creates) essential data and skills, which may be particularly lacking in developing countries and without which other policies cannot work in practice.[66]

Urban energy systems

In the urban context, these possibilities may be supplemented by other opportunities. Heat (or cooling) networks built into new urban developments enable greater systems efficiency – perhaps including the utilisation of industrial waste heat, co-generation and distributed renewable energy. Thus, for example, the Chinese Ministry of Construction has already started to implement urban district heating reform in northern cities with cooperation from the Ministry of Finance.[67] And, as noted in Chapter 3, growing numbers of Chinese cities are encouraging greater use of electric vehicles.

Urban planning

At the level of urban planning, a key challenge is to design and develop an efficient public transport system. The potential benefits include improved urban mobility, social cohesion, reduced congestion and reduced land pressures, as well as potentially cheaper and more stable travel costs due to reduced oil dependence. The barriers include high capital cost, lack of technological and financial capacities, and the fact that relevant actors are dispersed and scattered. Solutions include the mix of push and pull forces – pricing and tax instruments including congestion charging, integrated urban planning and the translation of low-carbon infrastructures policies into local levels; (again, examples include provincial or municipal government in China).

Bus Rapid Transit (BRT) systems combine improved road infrastructure (bus lanes and stations), vehicles and scheduling to allow urban mobility to reach the service quality of a light rail transit while enjoying the flexibility, cost and simplicity of a bus system.[68] By around 2010 more than 40 cities in Asia, including 15 Chinese, 11 Indonesian and 7 Indian, had already implemented BRT systems.[69] Capturing the increase in land value induced by new public transport infrastructure has also been suggested as a way of contributing to funding.[70]

Beyond transport patterns alone, there is a broader concept of improving the land use mix in cities – 'locating different types of land uses (residential, commercial, institutional, recreational, etc.) close together' can reduce travel distances, in particular if it allows affordable housing to be located near job-rich areas.[71]

These are no longer just academic concepts. In 2010, China designated eight cities (Tianjin, Chongqing, Shenzen, Xiamen, Hangzhou, Nanchang, Guiyang and Baoding) as low-carbon pilots. The Rockefeller Foundation Asian Cities Climate Change Resilience network has been testing approaches to build climate resilience in ten pilot cities in Indonesia, Thailand and Vietnam.[72]

As is often the case with complex evolutionary processes, key issues lie in planning and coordination. Lack of coordination between urban planning, public transport and road construction can greatly exacerbate problems. Deficient public transport planning and funding hinders appropriate land use and timely station building, leaving public transport networks often insufficient and slower, increasing the preference for private vehicles. The high upfront costs of urban infrastructures exacerbate this tendency. Moreover, the supply chains in buildings and urban transport tend to be very fragmented, requiring a diverse portfolio of policies to coordinate changes.[73]

Co-benefits and cooperation

Thus urban development faces multiple challenges. However, lack of appropriate urban planning implies a diffusion of informal settlements (e.g. slums) where many buildings and infrastructures (not only in informal settlements) do not meet fundamental safety standards.[74] This, combined with the poor services and infrastructure, make them particularly exposed to the risk associated with extreme weather events (such as waves, droughts, floods), heat or other climate-related risks.[75] Better buildings, in contrast, are likely to be more resilient as well as more energy efficient and pleasant to live in.

Addressing energy and climate change at the urban level can thus lead to a wide range of co-benefits. These may include social welfare benefits for low-income households, increased access to housing and transport services, improved indoor and outdoor air quality and employment creation.[76] One of the most obvious collective benefits may be improvement in urban air quality which, as noted in Chapter 1, is a huge and growing problem in many Asian cities.[77]

Greater recognition of these interlinkages has helped to drive expanding efforts, including linking national, transnational and international policy developments. In recent years, municipal involvement in transnational initiatives has broadened to include global and mega cities, especially those in Asia (of the C40 Cities Climate Leadership group with 40 members, 12 are located in Asia).[78]

The extension of the Clean Development Mechanism to include 'Programmes of Action' (Chapter 7) has increased its relevance to city schemes, and the evolution of international negotiations to emphasise and support the role of 'Nationally Appropriate Mitigation Actions' (NAMAs) has similarly increased the scope for urban-related policies to be recognised and encouraged in the international arena.

10.7 Integrating transitions

In each of the three cases considered – transport, electricity systems and urbanisation – there are evident strategies through which these sectors can evolve away from the present patterns of high fossil fuel dependence. We have chosen a broad regional focus to each case to give some specificity. In practice different regions will have different opportunities and will face different obstacles, according to their domestic resources, existing heritage and political and economic culture. Nevertheless there are common underlying themes.

At the most generic level, transitions are *possible*, but none are *easy*. All involve substantial changes in at least two of the three domains. Collectively, developments across all three domains – efforts on all three pillars of policy – are clearly vital to change the present collective course. The concluding chapter of this book looks more closely at such policy integration.

Moreover, to achieve the global goals outlined in Chapter 1, the transitions are interrelated. Thus even successful transport transitions to electric or fuel cell-based vehicles will be of little help if fossil fuels remain as the main source of the electricity (or hydrogen). Slow progress on either – vehicles or power production – could imply a greater need for biofuels, which are probably anyway the most realistic way to decarbonise fuels for heavy transport duties and flying. Higher bioenergy use would probably only be possible if urban forms become more compact, to leave more land available so as to minimise conflict with agricultural production. Low-carbon urbanisation may make it easier to use biomass for heating but is still likely to be more electric-intensive, not only for private vehicles but for public transport, local delivery services and household appliances. To be effective, the transitions are thus ultimately entwined.

Thus purely micro- or sector-focused approaches are limited if deep reductions are required. Most strikingly, curtailing transport sector emissions will involve transformations from all sides. Electric-only vehicles are only feasible if supported by convenient infrastructure; they are only low-carbon if powered by decarbonised electricity systems, and only likely to be attractive as the primary vehicle if nested in forms of urbanisation and institutional developments which make it easy to use other vehicles or modes for long trips. Oil is the most volatile component of the global energy system, but it is also the hardest to dislodge.

The previous chapter illustrated approaches to foster technology development. This chapter has switched from the 'micro' point of view of the single innovation to the sectoral framework of the infrastructure that embeds our energy systems and helps to define feasible technology paths. In each of the major systems that are important in our daily lives – transport, electricity and buildings – we have sketched the opportunities for their development in low-carbon directions.

... Or not. Third Domain processes involve strategic investment, and while major private companies – particularly multinationals – can undertake such investment on massive scales, they have no incentive to align such investment with public strategic goods like global energy security and climatic stability. On the contrary: as illustrated emphatically by the response to global fossil fuel price rises indicated in the introduction to this chapter, their incentive is to build upon their own comparative advantage in fossil fuels rather than trying to change course.

10.8 Conclusions: the carbon divide

This chapter therefore ends with the observation that the future opens options: the breadth of possibilities, and the emissions associated with them, inevitably broaden over time. There is a choice of avenues and they differ radically. Now we can better understand the big picture facing our energy systems.

More than ten years, ago, the International Institute for Applied Systems Analysis (IIASA) conducted a study of global energy futures that remains unsurpassed for the strategic import of its findings. It developed a computer model that included a wide range of technologies, with the potential for cost-reducing investments in either direction. The study remained largely unrecognised, because as so often with complex academic studies, the researchers couldn't communicate it clearly. What they produced, in the upper panel of Figure 10.6, showed the probability of different 'least cost energy futures', plotted against global CO_2 emissions at the end of the century. Its curious feature was the 'double hump': their projections did not cluster around a 'best guess', but rather split into two clusters. The lower panel is a translation of what it really means. As cheap oil becomes increasingly difficult to access, our investment in technologies and systems can cluster around lower-carbon solutions, or we can learn how to dig ourselves a high-carbon road more cheaply. Indeed, each has its own synergies:

- A high-carbon road with greater energy consumption is better able to fund the major, centralised investments in developing far-flung fossil fuel resources, scale economies in the conversion infrastructure and the potential protection of supply lines.
- A low-carbon road that has more emphasis on energy efficiency will be better able to meet its needs from renewable resources, which are more diffuse but also more widely spread – every country has some, even if international transmission is needed to make the best use of complementary resources.

A muddled mix – the central axis in the diagram – may be more costly than either, because the innovation effort is dispersed and systems are mismatched. The further we go down the

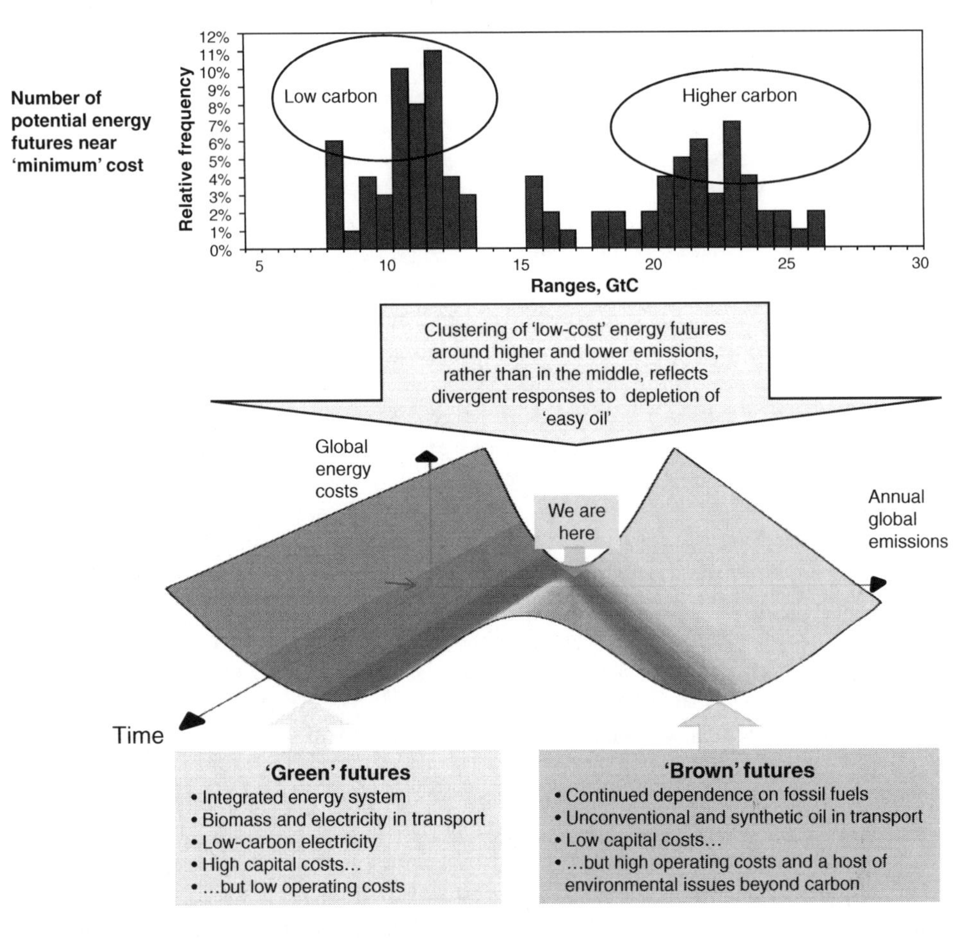

Figure 10.6 Two kinds of energy future – the carbon divide

Source: Upper panel: Gritsevskyi and Nakićenović (2000); lower panel: authors.

high-carbon road, the more costly it will be to cross the carbon divide, if and as accumulating climate damage forces us to do so.

The high-carbon paths do not look cheap, but we can learn to make them cheaper. We could invest in bigger and better coal mining and transport, improve our deep-water drilling technology, break open the Arctic reserves, extend our military investment in the protection of energy supply lines. To get really serious about it, we could gouge out large parts of Venezuelan tar and Albertan oil sands and escalate our investment to improve the way we get liquid fuels from coal. That indeed is the default direction. It is the future that incumbent companies know best how to build and that best fits the clothes – the infrastructure – we have created. It is the future that is easier for us to create, even if it turns out to be more expensive in the long run.

Note the paths are qualitatively different. The lower-carbon path requires more capital investment with less dependence on fossil fuel markets, more integrated energy systems with stronger grids, probably some electrification of transport and maybe heat, and a faster rate of innovation towards low-carbon technologies. This does not make it easy. It would require almost total decarbonisation of an expanded electricity sector by 2050, requiring a switch of investments towards more capital-intensive generation (renewables, nuclear, CCS) and transmission and a sustained commitment to energy efficiency. It would also change the way we use energy in buildings and on the road. The low-carbon road would require a transition towards energy-efficient cities.

Yet companies, technologies and societies have every potential to adapt to a world that is less dependent on fossil resources and is more environmentally sustainable. We will be able to do the same things using less energy, with lower emissions and with greater security.

In contrast, the 'business-as-usual' scenario of sustained investments in carbon-intensive technologies and their development is convenient, but may not be optimal; quite apart from atmospheric implications, over-dependence on fossil fuels does leave the world more vulnerable to the kind of instabilities already seen and discussed in Chapter 1.

Numerous energy modelling studies have looked at these transitions. One thing they show is that early signals will be required to trigger the transition towards a low-carbon energy system and meet stabilisation targets at acceptable costs. These signals include a sufficiently high carbon price (Pillar II measures), but also include specific policies to narrow the efficiency gap including behavioural and organisational changes (Pillar I) and to shape technical systems (Pillar III). The sooner these actions are taken, the lower the overall costs will be, as delayed climate action would eventually require capital to be replaced at a faster pace later in the century, incurring higher long-term costs than in the low-carbon scenario. Early action may also trigger technical change, allowing the economy to be more efficient than the default high-carbon scenario. However, the co-benefits of climate policies may not arise immediately, but may only become apparent once the transition is achieved. The timing of policies can thus relate to the inescapable trade-off between short- and medium-term costs.[79]

Changing course requires effort and involves risks (including policy failures). Industrial transformation – for this is the heart of the challenge – is indeed a complex process. Some of the lessons learned from innovation and industrial policy outlined in the previous chapter can help to mitigate the inherent risks.[80] Transformation also takes money – a lot of money. Electric and hybrid cars, as indicated, benefited from huge government investment – funded in part by high gasoline taxes in Japan that were used to help pay for the technologies to lessen Japan's oil dependence. Constructing charging infrastructure in North America for the potential successor of all-electric vehicles will not come cheap, nor will building a pan-European transmission infrastructure on the scale mapped out in the Roadmap 2050 studies. Particularly in the aftermath of the debt crisis, funding is one key challenge; the options are considered further in the next chapter.

Moreover, as emphasised above, creating options does not ensure that they are used. Developing a technology and its supporting infrastructure is only useful if investors and consumers want to buy it. The failure to acknowledge this difference has underlain some of the most expensive failures in the history of technology development: experience has demonstrated the dangers of governments driving or subsidising projects which are not going to diffuse outside the shell of government protection. Technology investment alone cannot solve the global energy problem, as our three examples illustrate:

- Even with massive support for hybrid cars, people could still purchase these cars but choose to run them mostly on the fuel component. Without using the plug-in potential, there would be no incentive to move on towards electric vehicles and America's gasoline dependence would remain undimmed. Even if drivers did move more towards electrification, it would still not help the global environment if the power system remained dominated by mid-west coal.

- A European commitment to strengthen the electric grid infrastructure would not ensure a low-carbon system. It could, indeed, open up the possibility of the rest of Europe using more of some of the dirtiest coal on the continent – lignite deposits in Germany and Poland – if that were the cheapest option.
- In Asia too, rising energy demand is unavoidable, however much governments try to avoid some of the mistakes the West has already made in its urbanisation. China and India have huge reserves of both coal and renewable energies. Which resources are used could amplify – or undo – all the efforts they might make to create more efficient urban systems.

To emphasise the point: large technology and infrastructure investment is *necessary* to enable a transition; it is not necessarily *sufficient*. The other part of the challenge is to ensure that businesses and people have incentives to invest in, and buy, low-carbon solutions. The challenge is not to centralise choices, but to combine strategic investment in low-carbon innovation and infrastructure with incentives for business and consumers that lead them to quench their thirst for energy in more sustainable ways. This points to the role of incentives and money across all three Domains, and the need to dig deeper into longer-term implications for economic development – the final part of the jigsaw to which we now turn in our penultimate chapter.

Notes

1. IEA (2012). Continued government support for the ethanol industry has allowed for the decline in ethanol production costs in Brazil. This decline has been largely driven by rising yields of sugar cane production and increasing scales of ethanol plants (van den Wall Bake *et al.* 2009).
2. IPCC (2007); Edenhofer *et al.* (2010).
3. US Congressional Budget Office (2009).
4. These are respectively UNFCCC ($380bn in 2030), IEA ($808bn per year average during the 2020s) and McKinsey ($1,215bn per year, average over 2026–30).
5. The IPCC SSRN report estimates cumulative investment in renewable energy for power generation at between US$2,850bn (reference scenario IEA) and US$12,280bn (450ppm target, IEA 2009) to 2030. The IEA's World Energy Outlook (2010) projected total system investment costs to 2030 at $18 trillion for a scenario consistent with a 2°C world; the 2012 analysis warned that 2°C was slipping out of reach and instead focused on a New Policies scenario, in which power sector investments alone totalled US$16.9 trillion to 2035.
6. For simplicity, US$15 trillion over 15 years (2015–30) equates to US$1 trillion per year; if global GDP over the period grows so that it averages US$100 trillion per year (an ambitious goal) it means redirecting 1% of total global GDP, and is of course a far bigger fraction of total investment, let alone energy sector investment.
7. World Bank World Development Report (2010).
8. Hourcade (1984).
9. The citation is from the classic reference on systems lock-in (Unruh 2000: 819).
10. Nelson and Winter (1982).
11. Arthur (1988, 1994).
12. Gleick (1988).
13. Robson (2001).
14. Thus: 'The result of these opposing forces of innovation and selection is a process of continuous change, without this necessarily leading to a state of equilibrium. Only with limited innovation or no innovation at all can a system converge under the influence of selection processes to a state of balance' (van den Bergh *et al.* 2007: 4).
15. Rogers (2003).

16. Unruh (2002).
17. Raven (2008). As observed previously and charted by Mazzucato (see Chapter 9, note 9) most major innovations have arisen from such government-funded programmes. However, the scope to transfer technologies to mainstream markets varies enormously. For instance, while electric cars have been used for a long time by golf players, they did not stimulate the demand of electric vehicles in other contexts (Cowan and Hultén 1996), since the requirements for a golf cart are very specific: it is not expected to cover long distances or reach high speed. Products for completely different applications may just remain in their niche.
18. Geels and Schot (2007).
19. Hoogma *et al.* (2002). Such 'other factors' include the need for adequate performance for mass markets, and for appropriate infrastructure.
20. Unruh (2002).
21. Jaccard *et al.* (1997); Bristow *et al.* (2008); McCollum and Yang (2009), Schäfer *et al.* (2009); Hickman *et al.* (2010).
22. EIA (2011), online at: http://www.eia.gov/cfapps/ipdbproject/iedindex3.cfm?tid=5&pid=5&aid=2&cid=ww,r1,r2,&syid=2008&eyid=2012&unit=TBPD
23. Morrow *et al.* (2010).
24. OICA (2008).
25. Government policies included investments in infrastructure and research and development, minimum blending requirements and price floors. These policies were backed by foreign investment, including US$1 billion loans from the World Bank (Hira and de Oliveira 2009).
26. Hira and de Oliveira (2009).
27. Maroun *et al.* (2011).
28. EIA (2011).
29. In 2009, gasoline formed 78% of total US fuel consumption, while diesel and biodiesel together accounted for almost all the remaining 22%; all other alternative fuels (notably ethanol, liquefied and compressed gas, electricity) were a tiny 0.1% of the whole consumption.
30. 2005 Energy Policy Act and 2007 Energy Independence and Security Act (Hochman *et al.* 2008).
31. Oliveira (2008); Solomon *et al.* (2007). Searchinger *et al.* (2008) find that US corn ethanol emits twice as much greenhouse gases as gasoline when accounting for indirect land use change-related emissions over a 30-year production period.
32. US support for bioethanol production and use has included exemption from federal gasoline excise taxes, exemption from road use taxes, a federal production tax credit and a federal blender's credit (Solomon *et al.* 2007).
33. On the spatial requirements of renewable energies, see Ashraf-Ball *et al.* (2009).
34. An example is the curious attitude of the US government towards hydrogen technology funding; see Wald (2009) and Whoriskey (2009a).
35. Catherine Dunwoody, executive director of the California Fuel-Cell Partnership (CaFCP). Reported by the Economist (2008). See also Struben and Sterman (2008) and Romm (2006).
36. See Flynn (2002) for a similar case (natural gas vehicles) in which high expectations were eventually followed by deleterious disappointment.
37. Since 1971, the Japanese Ministry of International Trade and Industry (MITI) has sponsored an ambitious programme to stimulate the diffusion of battery-powered electric vehicles (BEVs); it failed to take off, partly due to the lack of credibility of MITI's programme within the industry in the 1970s and 1980s, and partly due to the stabilisation of the oil market in the 1980s (Åhman 2006). Fifteen years later, California made a similar push with mandates for 'zero emission vehicles' hoping to jump-start the electric car industry. Their reputation in California suffered as the programme's requirements proved to be ahead of both compelling technology and recharging infrastructure, but this did help to drive developments in both. Providing niche support does not guarantee the success of an application: trying to 'skip' some steps by stimulating the immediate creation of a whole infrastructural system is a risky strategy.
38. Reisser and Roquelle (2009) and Whoriskey (2009b).
39. Vermeyen and Belmans (2006).
40. Weiss *et al.* (2012) makes the interesting point that the observed learning rate of 7% (± 2%) seen in hybrid cars since 1997 equates to a far higher learning rate in terms of closing the gap with conventional car costs – of 23% ± 5% – so that by 2010, HEV vehicles were only €10–50/kW more expensive than conventional cars (which average €150–200/kW). Their analysis suggests a

much longer road for battery vehicles, projected to reach breakeven in the latter part of the 2020s, and requiring around 100–150 billion euros investment to do so.

41. In 1997, the Ministry of International Trade and Industry launched the Advanced Clean Energy programme (ACE), which focused on the development of hybrid vehicles from the R&D to the early deployment phase. More than 500,000 HEVs were in use by 2001 (Åhman 2004).
42. Including leasing programmes, subsidies and standardisation (Åhman 2004).
43. Andersen *et al.* (2009).
44. In the US, the bicycle and the electric tram eased the transition towards automobiles by introducing new mobility habits, changing the perception of the function of streets, introducing higher-speed transportation and empowering public authorities for the administration of streets (Geels 2005).
45. Geels (2005).
46. Kemp *et al.* (2001).
47. The UK Climate Change Committee projected a reduction of the carbon intensity of power generation from current levels of 500g CO_2/kWh to 50gCO_2/kWh by 2030 as part of its review of the UK Fourth Carbon Budget (2023–27) which was adopted by the government. Germany's *Energiewende* aims at major industrial transformation of the country's electricity system to one based on renewable energy.
48. The governing panel of Ofgem's Low Carbon Network's Innovation Fund concluded that carbon saving, while a key motivator, could not be used as a metric to evaluate projects. It proved impossible to distinguish low-carbon innovation from broader innovation oriented to create more flexible and responsive distribution systems: innovation carries multiple benefits, and they frequently cannot be reasonably separated.
49. One of the most ambitious proposals is DESERTEC, with a declared aim to provide Europe with up to 17% of its energy needs in 2050 from Middle East and North African deserts. A parallel project (Medgrid, formerly known as Transgreen) aims to develop a transmission network under the Mediterranean Sea (Kanter 2011).
50. Egypt is planning to increase the share of renewable energy (mainly wind and concentrated solar power plants) to 20% by 2020 (ESMAP 2010). The biggest individual project in the North African region is being pursued by the Moroccan Agency for Solar Energy, which aims to install 500MW in Ouarzazate by 2015, reaching an installation capacity of 2,000MW by 2020 in five areas; up to now, the agency has selected four candidate companies for the Ouarzazate areas among nineteen bidding groups (MASEN 2013).
51. Taylor *et al.* (2012).
52. Brazil has established sophisticated systems for procuring electricity generation through competitive auctions of long-term (20-year) fixed-price contracts. Cunha *et al.* (2012) summarise the results as follows:

> In 2010, wind power was allowed to compete on equal grounds with small hydro and bioelectricity projects in two energy auctions for energy delivery in 2013... Wind power outclassed its competitors in both auctions, being responsible for nearly 80% of all energy contracted and reaching average prices of 80 US$/MWh (regular auction) and 73 US$/MWh (reserve auction) ... 1,500 MW of wind capacity was contracted under regular contracts, and 500 MW under reserve contracts. In August 2011, ... two energy auctions (one regular and one reserve) for delivery in 2014 were organized ... in the new energy auction wind power was allowed to compete directly with natural gas-fired thermal plants... An energy mix including 1,000 MW of wind capacity was contracted in this auction... wind and gas projects competed in the 2011 auction using a purely economic criterion [Brazil does not yet have a carbon price].... [T]he average wind energy price in these auctions was 60 US$/MWh, lower than the average natural gas energy price (62 US$/MWh).

Obviously this reflects the favourable characteristics of good winds, cheap land, efficient planning procedures with little opposition, and the automatic backup embodied in Brazilian hydro capacity, so it is unlikely that costs would be as low in Europe. Cunha *et al.* (2012) also suggest that wind costs may have been lowered by the post-recession slowdown in European demand for wind turbines. Nevertheless it underlines the extent to which the relative cost of wind- and gas-powered generation depends to an important degree on market and contractual structures.

53. Aflaki and Netessine (2011).
54. For UK government data on North Sea oil capital investment see: https://www.gov.uk/oil-and-gas-uk-field-data. UK capital investment in North Sea oil averaged over £5bn per year for more than twenty years; for additional information on the capital investment see: https://www.gov.uk/government/uploads/system/uploads/attachment_data/file/16102/ukcs-i-and-e-chart.pdf
55. Chris Mottershead, Special Advisor to BP Chief Executive, personal communication, *c*.2005. Chris emphasised to the author that this was the result of a two-year internal BP evaluation, which concluded that there were no engineering obstacles to offshore wind energy and plenty of scope to engineer down the costs. It was, he emphasised, a judgement about corporate comparative advantage and the regulated nature of the electricity business: 'It's quite simple. In oil we may risk £1bn on each of five developments, knowing that if one strikes well we'll make £10bn. If we made equivalent breakthroughs in offshore wind energy, we concluded that the government would simply change the rules.'
56. UNDESA (2011) and UN-HABITAT (2008a).
57. Schafer and Victor (2000). Energy prices and transportation policies therefore affect urban sprawl and suburbanisation. In practice, households and firms trade-off between accessibility, distance to attractive centres and real-estate price or rent (or equivalently affordable housing lot sizes) (Alonso 1964).
58. Transport times, urbanisation rules, fiscal incentives, real-estate market regulations, public investments, jobs distribution and particularly the development of transport infrastructures can be much more influential than the evolution of transport prices per se and they could be used as tools to help mitigate the urban sprawl induced by the combination of long-term decrease in transportation prices, increase in revenues and growing population (Brueckner 2000). Employment is also a major factor (Grazi and van den Bergh 2008). The relationship between the evolution of transportation prices (linked to energy prices, technological change and public policies) and urban area transformations is the foundation for several prospective models of city evolution – so-called land-use transport models (de la Barra *et al.* 1984; Lefèvre 2009; ITF 2010; Waddell 2002; Wegener 2004; Viguié and Hallegatte 2012).
59. Density is inversely correlated with the share of private transport: a highly extended city makes the creation of an efficient public transport system more difficult and a higher share of private motorised transport modes requires greater space. The total building floor area of a city is one of the major determinants of energy consumption, since for long distances the use of private cars appears more appealing than public transport in terms of convenience and costs.
60. Global urban population is expected to rise to 4.9 billion people (i.e. 60% of total population) by 2030 (Schroeder *et al.* 2012). China and India are expected to reach urbanisation levels of 62% and 41% by 2030, respectively (UN-HABITAT 2008b). In China, between 15 and 20 billion m^2 of urban housing were projected to be built between 2005 and 2020 to accommodate new urban dwellers, equivalent to the existing building stock in the EU-15 (Colombier and Li 2012). On urban growth in Asia see also Dhakal (2009).
61. Between 1991 and 2005, energy consumption from urban transport in 23 major metropolitan areas in India more than doubled, coinciding with a doubling of CO_2 emissions (Li 2011). Motorised traffic in India is expected to reach 13,000 billion passenger-km per year (of which 92% will be road transport), which would result in a fivefold increase in energy use and CO_2 emissions in transport by 2020 relative to 2000, assuming current energy and carbon intensities (Singh 2006). (See Colombier and Li 2012.)
62. On the shift from industrial to urban energy see Dhakal *et al.* (2003). Also, CCICED (2009) underlines the extent to which this shift allows greater variation and choices over levels of energy and emissions.
63. Muzones *et al.* (2010).
64. Appropriate urban planning can be achieved both by operating on the structure of the buildings (insulation, centralised or district heating, CHP), and through a reasoned planning of the city itself which, for instance, could ease the installation of an extended public transport system and avoid any excessive sprawl (Colombier and Li 2012).
65. Richerzhagen *et al.* (2008) and Schroeder *et al.* (2012).
66. Thus standards will play an insignificant role if builders themselves have not the necessary background to implement energy-efficiency solutions and when decisions at each stage of design, construction and operation involve multiple stakeholders (Jefferson 2000) or are confronted by

misplaced incentives and agent – principal barriers (agents responsible for investment decisions are different from those benefiting from the energy savings) (Scott 1997; Schleich and Gruber 2008; Levine *et al.* 2007).

67. See Li and Yao (2009).
68. A key feature of New Urbanism is to favour the development of public transport while discouraging the use of private vehicles in densely populated areas.
69. Schroeder *et al.* (2012).
70. The prospect of capturing the increase in land value induced by the implementation of public transport infrastructure is viewed as an ex-post means of funding or at least as a way of recouping losses. It is the subject of intense research activities, noticeably in the World Bank.
71. Litman and Steele (2012). The Land Use Mix concept is another key feature of New Urbanism. It includes mixing uses within buildings (e.g. retail and upper floor residential) and within neighbourhoods, and mixing housing types and prices to accommodate different income classes (Litman and Steele 2012). This concept has been used in industrialised countries and could be adapted to Asian cities (Schroeder *et al.* 2012). By reducing travel distances the improved mix also allows for the use of alternative travel modes (e.g. walking and cycling) (Kuzmyak and Pratt 2003; Litman and Steele 2012).
72. Schroeder *et al.* (2012).
73. Schroeder *et al.* (2012).
74. In 2005, 'informal settlements' comprised around 36% of urban population, with a peak of 62% in sub-Saharan Africa (UN-HABITAT 2008a).
75. Slums currently contribute little to global warming, but are likely to be affected more heavily than normal by its effects, being less protected against connected natural phenomena like floods, droughts and heatwaves (Ramin 2009 and Kjellstrom *et al.* 2007). The cause of such exposure to extreme climate events is mainly given by the fact that a large percentage of buildings and infrastructures – and not only in informal settlements – don't match some fundamental safety standards (Revi 2008).
76. IPCC (2007).
77. IGES (2008); Yedla *et al.* (2005).
78. Schroeder *et al.* (2012).
79. See Luderer, Bosetti *et al.* (2012) for an overview of a substantial model comparison project. The role of infrastructure investment and other long-term factors is explored in Waisman *et al.* (2012). The macroeconomic effects of timing are explored particularly in Bosetti *et al.* (2012).
80. See, for example, Olsthoorn and Wieczorek (2006) for a compilation of different disciplinary approaches to industrial transformation.

References

Aflaki, S. and Netessine, S. (2011) *Strategic Investment in Renewable Energy Sources*, INSEAD Working Paper 2012/59/TOM.

Åhman, M. (2004) *Government Policy and Environmental Innovation in the Automobile Sector in Japan*, Report No. 53. Lund University, Department of Environmental and Energy Systems Studies.

Alonso, W. (1964) *Location and Land Use*. Cambridge, MA: Harvard University Press.

Andersen, P. H., Mathews, J. A. and Rask, M. (2009) 'Integrating private transport into renewable energy policy: the strategy of creating intelligent recharging grids for electric vehicles', *Energy Policy*, 37 (7): 2481–6.

ANFAVEA (2011) *Anuário da Indústria Automobilística Brasileira – Brazilian Automotive Industry Yearbook*. São Paulo, Brazil: Associação Nacional dos Fabricantes de Veículos Automotores (ANFAVEA) – Brazilian Automotive Industry Association. Available from: http://www.virapagina.com.br/anfavea2011/

Arthur, W. B. (1988) 'Self-reinforcing mechanisms in economics', in P. W. Anderson, K. J. Arrow and D. Pines (eds), *The Economy as an Evolving Complex System*. Redwood City, CA: Addison-Wesley.

Arthur, W. B. (1994) 'Inductive reasoning and bounded rationality', *American Economic Review*, 84 (2): 406–11.

Ashraf-Ball, H., Oswald, A. J. and Oswald, J. I. (2009) *Hydrogen Transport and the Spatial Requirements of Renewable Energy*, The Warwick Economics Research Paper Series (TWERPS) 903. University of Warwick, Department of Economics.

Beinhocker, E. D. (2006) *The Origin of Wealth: Evolution, Complexity, and the Radical Remaking of Economics*, Cambridge MA: Harvard Business Press.

Bertaud, A. and Richardson, H. W. (2004) 'Transit and density: Atlanta, the United States and Western Europe', in H. W. Richardson and C.-H. C. Bae (eds), *Urban Sprawl in Western Europe and the United States*. Aldershot: Ashgate.

Bosetti, V., Carraro, C. and Tavoni, M. (2012) 'Timing of mitigation and technology availability in achieving a low-carbon world', *Environment and Resource Economics*, 51 (3): 353–69.

Bristow, A. L., Tight, M., Pridmore, A. and May, A. D. (2008) 'Developing pathways to low carbon land-based passenger transport in Great Britain by 2050', *Energy Policy*, 36 (9): 3427–35.

Brueckner, J. (2000) 'Urban sprawl : diagnosis and remedies', *International Regional Science Review*, 23 (2): 160–71.

CCICED (2009) *Annual Policy Report – Energy, Environment and Development*. China Council for International Cooperation on Environment and Development Secretariat, Beijing, China.

CLCF (2012) *Pathways for Energy Storage in the UK*. Centre for Low Carbon Futures.

Colombier, M. and Li, J. (2012) 'Shaping climate policy in the housing sector in northern Chinese cities', *Climate Policy*, 12 (4): 453–73.

Cowan, R. and Hultén, S. (1996) 'Escaping lock-in: the case of the electric vehicle', *Technological Forecasting and Social Change*, 53: 61–79.

Cunha, G., Barroso, L. A., Porrua, F. and Bezerra, B. (2012) *Fostering Wind Power Through Auctions: The Brazilian Experience*, International Association for Energy Economics Newsletter, 2nd Quarter. Available from: http://www.iaee.org

la Barra, Tomás, Pérez, B. and Vera, N. (1984) 'TRANUS-J: putting large models into small computers', *Environment and Planning B: Planning and Design*, 11: 87–101.

Dhakal, S. (2009) 'Urban energy use and carbon emissions from cities in China and policy implications', *Energy Policy*, 37: 4208–19.

Dhakal, S., Hanaki, K. and Hiramatsu, A. (2003) 'Estimation of heat discharges by residential buildings in Tokyo', *Energy Conversion and Management*, 44 (9): 1487–99.

Economist, The (2008) 'The car of the perpetual future', *The Economist*, 4 September.

Edenhofer, O., Knopf, B. and Luderer, G. (2010) 'From utopia to common sense: the climate mitigation challenge', in E. Cerdá and X. Labandeira (eds), *Climate Change Policies. Global Challenges and Future Prospects*. Cheltenham: Edward Elgar.

EIA (2011) *International Energy Statistics Online Database*. Energy Information Administration, US Government. Online at: http://www.eia.gov/cfapps/ipdbproject/iedindex3.cfm?tid=5&pid=5&aid=2&cid=ww,r1,r2,&syid=2008&eyid=2012&unit=TBPD

ESMAP (2010) *Annual Report*. World Bank, Energy Sector Management Assistance Program.

Fernández, J. L., Red Electra de España (Spanish transmission system operator) (2012) Presentation to European Climate Foundation 'Roadmaps to Reality' meeting, Brussels.

Finon, D. (2013) 'The transition of the electricity system towards decarbonization: the need for change in the market regime', *Climate Policy*, 13, Supplement 01: 130–45.

Flynn, P. C. (2002) 'Commercializing an alternate vehicle fuel: lessons learned from natural gas for vehicles', *Energy Policy*, 30: 613–19.

Geels, F. W. (2005) *Technological Transitions and System Innovations*. Cheltenham: Edward Elgar.

Geels, F. W. and Schot, J. (2007) 'Typology of sociotechnical pathways', *Research Policy*, 36: 399–417.

Gleick, J. (1988) *Chaos: Making a New Science*. East Rutherford, NJ: Penguin USA.

Grazi, F. and van den Bergh, J. C. J. M. (2008) 'Spatial organization, urban transport and climate policy: comparing instruments of spatial planning and policy', *Ecological Economics*, 67: 630–9.

Gritsevskyi, A. and Nakićenović, N. (2000) 'Modeling uncertainty of induced technological change', *Energy Policy*, 28 (13): 907–21.

Gül, T., Kypreos, S., Turton, H. and Barreto, L. (2009) 'An energy-economic scenario analysis of alternative fuels for personal transport using the Global Multi-regional MARKAL Model (GMM)', *Energy*, 34: 1423–37.

Hickman, R., Ashiru, O. and Banister, D. (2010) 'Transport and climate change: simulating the options for carbon reduction in London', *Transport Policy*, 17 (2).

Hira, A. and de Oliveira, L.-G. (2009) 'No substitute for oil? How Brazil developed its ethanol industry', *Energy Policy*, 37 (6): 2450–6.

Hochman, G., Sexton, S. E. and Zilberman, D. D. (2008) 'The economics of biofuel policy and biotechnology', *Journal of Agricultural and Food Industrial Organisation*, Special Issue, vol. 6, article 8.

Hoogma, R., Kemp, R., Schot, J. and Truffer, B. (2002) *Experimenting for Sustainable Transport: The Approach of Strategic Niche Management*. London and New York : Spon Press.

Hourcade, J.-C. (1984) 'Prospective Energy and Development Strategies in the Third World'. PhD in Economics, University of Paris VIII Vincennes in Saint-Denis, October.

IEA (2010) *World Energy Outlook*. Paris: OECD/IEA.

IEA (2012a) *Energy Prices and Taxes*, online data available at OECD elibrary.

IEA (2012b) *World Energy Outlook*. Paris: OECD/IEA.

IGES (2008) *Climate Change Policies in the Asia-Pacific: Re-uniting Climate Change and Sustainable Development*. IGES White Paper, Institute for Global Environmental Strategies (IGES). Available from: http://pub.iges.or.jp/modules/envirolib/view.php%3Fdocid=1565

IPCC (2007a) *Climate Change 2007: Mitigation of Climate Change: Working Group III Contribution to the Fourth Assessment Report*, Intergovernmental Panel on Climate Change. Cambridge: Cambridge University Press.

IPCC (2007b) 'Issues related to mitigation in the long-term context', Chapter 3, *Climate Change 2007: Mitigation of Climate Change: Working Group III Contribution to the Fourth Assessment Report*, Intergovernmental Panel on Climate Change. Cambridge: Cambridge University Press.

ITF (2010) *Transport and Innovation*. Berlin: International Transport Forum 2010.

Jaccard, M., Failing, L. and Berry, T. (1997) 'From equipment to infrastructure: community energy management and greenhouse gas emission reduction', *Energy Policy*, 25: 1065–74.

Jefferson, M. (2000) 'Energy policies for sustainable development', Chapter 12 in *World Energy Assessment: Energy and the Challenge of Sustainability*. New York: United Nations Development Programme.

Kanter, J. (2011) 'Energy security in uncertain times', *New York Times*, 10 October.

Kemp, R., Rip, A. and Schot, J. (2001) 'Constructing transition paths through the management of niches', in R. Garud and P. Karnøe (eds), *Path Dependence and Creation*. Mahwah, NJ and London: Lawrence Erlbaum, pp. 269–99.

Kjellstrom, T., Friel, S., Dixon, J., Corvalan, C., Rehfuss, E., Campbell-Lendrum, D., Gore, F. and Bartram, J. (2007) 'Urban environmental health hazards and health equity', *Journal of Urban Health*, 84 (Suppl. 1): 86–97.

Kuzmyak, R. J. and Pratt, R. H. (2003) *Land Use and Site Design: Traveler Response to Transport System Changes*, Transit Cooperative Research Program Report 95. Transportation Research Board, Chapter 15.

Lefèvre, B. (2009) 'Assessment of integrated "Transport – Land Use" policies potential to reduce long-term energy consumption of urban transportation: a prospective simulation in Bangalore, India', in Proceedings of the ECEEE Summer Study, Sweden, European Council for an Energy-Efficient Economy, p. 10.

Levine, M., Ürge-Vorsatz, D., Blok, K., Geng, L., Harvey, D., Lang, S. *et al.* (2007) 'Residential and commercial buildings', in B. Metz, O. R. Davidson, P. R. Bosch, R. Dave and L. A. Meyer (eds), *Climate Change 2007: Mitigation. Contribution of Working Group III to the Fourth Assessment Report of the Intergovernmental Panel on Climate Change*. Cambridge and New York: Cambridge University Press.

Li, B. and Yao, R. (2009) 'Urbanisation and its impact on building energy consumption and efficiency in China', *Renewable Energy*, 34 (9): 1994–8.

Li, J. (2011) 'Decoupling urban transport from GHG emissions in Indian cities – a critical review and perspectives', *Energy Policy*, 39 (6): 3503–14.

Litman, T. and Steele, R. (2012) 'Land Use Impacts on Transport: How Land Use Factors Affect Travel Behavior', online at: http://www.vtpi.org/landtravel.pdf

Luderer, G., Bosetti, V., Jakob, M., Steckel, J., Waisman, H. and Edenhofer, O. (2012) 'The economics of GHG emissions reductions – results and insights from the RECIPE model intercomparison', *Climatic Change*, 114 (1): 9–37.

McCollum, D. and Yang, C. (2009) Achieving deep reductions in US transport greenhouse gas emissions: scenario analysis and policy implications', *Energy Policy*, 37 (12): 5580–96.

McKinsey & Co. *et al.* (2010) *ECF Roadmap 2050: A Practical Guide to a Prosperous, Low-Carbon Europe. Volume 1 – Technical and Economic Analysis*. London: McKinsey & Co.

Maroun, C., Rathmann, R. and Schaeffer, R. (2011) *Brazilian Biofuels Programmes from the WEL-Nexus Perspective*. European Report on Development.

MASEN (Moroccan Agency for Solar Energy) (2010). Available from: http://www.masen.org.ma/

Morrow, W. R., Gallagher, K. S., Collantes, G. and Lee, H. (2010) 'Analysis of policies to reduce oil consumption and greenhouse-gas emissions from the U.S. transportation sector', *Energy Policy*, 38 (3): 1305–20.

Muzones, M. D. (2010) 'Bogor workshop: sustainable low carbon development in Indonesia and Asia', in K. Tamura Hayama (ed.), *Is Indonesia in a Good Position to Achieve Sustainable Low Carbon Development? Opportunities, Potentials and Limitations*. Japan: IGES.

Nelson, R. R. and Winter, S. G. (1982) *An Evolutionary Theory of Economic Change*. Cambridge, MA: Belknap Press.

OICA (2008) *Production Statistics Online Database*. International Organization of Motor Vehicle Manufacturers.

Oliveira, M. D. D. (2008) 'Sugarcane and ethanol production and carbon dioxide balances', in D. Pimentel (ed.), *Biofuels, Solar and Wind as Renewable Energy Systems*. Springer Netherlands, pp. 215–30. Available from: http://link.springer.com/chapter/10.1007/978-1-4020-8654-0_9

Oliveira-Martins, J., Gonand, F., Antolin, P., de la Maisonneuve, C. and Yoo, K. Y. (2005) *The Impact of Ageing on Demand, Factor Markets and Growth*, OECD Economics Department Working Papers 42. Paris: OECD.

Olsthoorn, X. and Wieczorek, A. (eds) (2006) *Understanding Industrial Transformation – Views from Different Disciplines*. Dordrecht: Springer.

Ramin, B. (2009) *Slums, Climate Change and Human Health in Sub-Saharan Africa*. World Health Organization.

Raven, R. (2008) *Strategic Niche Management for Biomass: A Comparative Study on the Experimental Introduction of Bioenergy Technologies in the Netherlands and Denmark*. Eindhoven Centre for Innovation Studies, VDM Verlag.

Reisser, S. and Roquelle, S. (2009) 'Carlos Ghosn : Priorité à la voiture électrique', *Le Figaro*. Online at: http://www.lefigaro.fr/lefigaromagazine/2009/06/27/01006-20090627ARTFIG00128--priorite-a-la-voiture-electrique-.php

Revi, A. (2008) 'Climate change risk: an adaptation and mitigation agenda for Indian cities', *Environment and Urbanization*, 20 (1): 207–29.

Richerzhagen, C., von Freiling, T., Hansen, N., Minnaert, A., Netzer, N. and Rußbild, J. (2008) *Energy Efficiency in Buildings in China*, German Development Institute Working Paper No. 41. Bonn.

Robson, A. J. (2001) 'The biological basis of economic behavior', *Journal of Economic Literature*, 39 (1): 11–33.

Rogers, E. M. (2003) *Diffusion of Innovations*, 5th edn. Riverside, NJ: Free Press.

Schäfer, A. and Victor, D. (2000) 'The future mobility of the world population', *Transport Research A*, 34 (3): 171–205.

Schäfer, A., Heywood, J. B., Jacoby, H. D. and Waitz, I. A. (2009) *Transportation in a Climate Constrained World*. Cambridge, MA: MIT Press.

Schleich, J. and Gruber, E. (2008) 'Beyond case studies: barriers to energy efficiency in commerce and the services sector', *Energy Economics*, 30 (2): 449–64.

Schroeder, H., Li, J., Bulkeley, H. *et al.* (2012) 'Enabling the transition to climate smart development in Asian cities', in A. Srinivasan, F. Ling and H. Mori (eds), *Climate Smart Development in Asia: Transition to Low Carbon and Climate Resilient Economies*. London: Routledge.

Scott, S. (1997) 'Household energy efficiency in Ireland: a replication study of ownership of energy saving items', *Energy Economics*, 19 (2): 187–208.

Searchinger, T., Heimlich, R., Houghton, R. A., Dong, F., Elobeid, A., Tokgoz, S. *et al.* (2008) 'Use of croplands for biofuels increases greenhouse gases through emissions from land use change', *Science*, 319: 1238–40.

Singh, S. K. (2006) 'Future mobility in India: implications for energy demand and CO_2 emission', *Transport Policy*, 13 (5): 398–412.

Solomon, S., Qin, D. and Manning, M. (eds) (2007) *Climate Change 2007: The Physical Science Basis: Working Group I Contribution to the Fourth Assessment Report*, Intergovernmental Panel on Climate Change. Cambridge: Cambridge University Press.

Struben, J. and Sterman, J. (2008) 'Transition challenges for alternative fuel vehicle and transportation systems', *Environment and Planning B: Planning and Design*, 35 (6): 1070–97.

Taylor, P. *et al.* (2012) *Pathways for Energy Storage in the UK*. Leeds: Centre for Low Carbon Futures.

UNDESA (2011) *World Urbanization Prospects, the 2011 Revision: Data on Urban and Rural Populations*. Available from: http://esa.un.org/unup/CD-ROM/Urban-Rural-Population.htm

UN-HABITAT (2008a) *State of the World's Cities*. New York: UN.

UN-HABITAT (2008b) *Harmonious Cities: Focus on China and India*, Policy Brief No. 1. UN-HABITAT.

Unruh, G. C. (2000) 'Understanding carbon lock in', *Energy Policy*, 28: 817–30.

Unruh, G. C. (2002) 'Escaping carbon lock in', *Energy Policy*, 30: 317–25.

US Congress (2005) *Energy Policy Act of 2005*, 109th Congress, USA.

US Congress (2007) *Energy Independence and Security Act of 2007*, 110th Congress, USA.

van den Bergh, J. C. M., Faber, A., Idenburg, A. I. and Oosterhuis, F. H. (2007) *Evolutionary Economics and Environmental Policy*. Cheltenham: Edward Elgar.

van den Wall Bake, J. D., Junginger, M., Faaij, A., Poot, T. and Walter, A. (2009) 'Explaining the experience curve: cost reductions of Brazilian ethanol from sugarcane', *Biomass and Bioenergy*, 33 (4): 644–58.

Vermeyen, P. and Belmans, R. (2006) 'Transport', in T. Jamasb, W. J. Nuttall and M. G. Pollitt (eds), *Future Electricity Technologies and Systems*. Cambridge: Cambridge University Press.

Viguié, V. and Hallegatte, S. (2012) 'Synergies and trade-off in urban climate policies', *Nature Climate Change*, 2: 334–37.

Waddell, P. (2002) 'UrbanSim: modeling urban development for land use, transportation and environmental planning', *Journal of the American Planning Association*, 68 (3): 297–314.

Waisman, H., Guivarch, C., Grazi, F. and Hourcade, J.-C. (2012) 'The Imaclim-R model: infrastructures, technical inertia and the costs of low carbon futures under imperfect foresight', *Climatic Change*, 114 (1): 101–20.

Wald, M. L. (2009) 'U.S. drops research into fuel cells for cars', *New York Times*, 5 July.

Wegener, M. (2004) 'Overview of land-use transport models', in Hensher, D. A., Button, K. J. (eds), *Transport Geography and Spatial Systems*, Handbook 5 of Handbooks in Transport. Kidlington: Pergamon/Elsevier Science, pp. 127–46.

Weiss, M., Patel, M. K., Junginger, M. *et al.* (2012) 'On the electrification of road transport – learning rates and price forecasts for hybrid-electric and battery-electric vehicles', *Energy Policy*, 48: 374–93.

Whoriskey, P. (2009a) 'The hydrogen car gets its fuel back', *Washington Post*, 17 October.

Whoriskey, P. (2009b) 'The deadly silence of the electric car', *Washington Post*, 23 September.

World Bank (2010) *World Development Report*. Washington, DC: World Bank.

Yedla, S., Shrestha, R. M. and Anandrajah, G. (2005) 'Environmentally sustainable urban transportation – comparative analysis of local emission mitigation strategies vis-à-vis GHG mitigation strategies', *Transport Policy*, 12 (3): 245–54.

11 The dark matter of economic growth[1]

'The best way to predict your future is to create it.'

–Abraham Lincoln

11.1 Introduction: on transformation and economic prescriptions

One of the most profound transitions witnessed in modern times was the fall of the Soviet Union. The connections to this book may not seem obvious, yet they are numerous and profound.

The most obvious connection concerns underlying economic philosophy. Central planning proved a disastrous way to run an economy. Its cumulative problems could for some decades be masked by the enthusiasm of the post-Second World War reconstruction imperative and by resource wealth – above all coal, oil and related gas. The previous chapter explained in general how it is possible for systems to get stuck on the wrong track; our first chapter underlined that dealing inadequately with environmental damage impairs both lives and economic development. Soviet central planning fell into both these traps. Massive resource-based and centralised industrialisation and agricultural 'modernisation' led to perverse outcomes. Fantastically inefficient use of energy undermined the value of the region's huge fossil fuel resources and exports. Accumulating environmental damage fuelled discontent and declining productivity. The collapse of oil prices in the 1980s exposed the fragile financial basis of foreign and domestic security, and weakened the state's ability to provide a minimum standard of welfare. All this paved the way for the subsequent political revolutions.

A second lesson learnt from the post-Soviet era is that simplistic economic advice in the wrong context is dangerous. As Russia emerged from its initial 'perestroika', Western economic advisers flocked to Moscow to recommend classical prescriptions of privatisation and competition. The result reinforced the breakdown of law and the rise of the oligarchs, laying the groundwork for the population to demand the return of a strong state. If the need is for structural change, pricing and markets are necessary but not sufficient. The classical tools of economics depend upon the right institutional context – an old adage sometimes forgotten but one that the World Bank and development economists now repeatedly emphasise.

But a third lesson – equally close to the heart of this book – lay at the geographical interface of east and west: in the subsequent reunification of Germany. In the aftermath of the Second World War, almost a third of Germany had been carved out into the sphere of Soviet influence. By 1990 its people were poor, by Western standards, and much of its infrastructure was old and decaying. Transforming East Germany into a modern economy was a daunting prospect. Indeed many economists, warned that the cost of reunification would be intolerable, that West Germany could not cope with the bill and that its own economy would be crippled by the effort.

Just two decades later, the reunified Germany was the strongest economy in Europe and the linchpin of the credibility of the euro currency. It also topped a global poll of the most admired countries on the planet.[2] Nothing could testify better to the fact that the common assumptions of Second Domain economics (that economic performance is driven by minimising costs and optimising the use of resources) tell us little about the Third Domain processes of transformation, innovation, renewal and growth. These are the core topics of this chapter.

After the 1980s collapse of global oil prices and subsequent demise of the Soviet Union, it seemed that Western economics was an all-encompassing philosophy and globalisation helped to spread the recipe for success. Yet following the 2008 credit crunch – preceded by another round of oil price rises as outlined in Chapter 1 – the 'growth engine' seems again to have ground to a halt, not even keeping pace with the population growth in many regions. Energy itself is not the primary cause of this stagnation but it is part of the process and energy policy reflects wider economic mindsets. Hence, this penultimate chapter seeks to elevate the debate about energy and environment to encompass the macroeconomics of economic growth itself.

This chapter will take the reader on a journey into the world of economic growth theories. This journey resembles the search for a philosopher's stone – the mythical source of eternal life. The modern quest of many economists has been to maximize economic growth; the quest of environmentalists is either to reconcile economic growth with a planet of finite resources and finite atmosphere or to conclude that economic growth itself is unsustainable. We can tell the outcome immediately: innovation matters, and technology can help to reconcile seemingly conflicting objectives, but they are not the philosopher's stones. What matters most is making the right strategic choices, around which a society, or sector, can focus and motivate its efforts.

This is the book's most technical chapter because it has to delve deeper into economic theory to prove its point and discuss the results of some economic modelling. The punch line, however, is quite simple. Economics, like other areas of life and indeed physics, is at some levels governed by irreducible uncertainties. In particular, theories of economic growth have always had ultimately to include a 'residual' factor attributable to unexplained dimensions of institutional and technological innovation. These correspond to processes in the First and Third Domains. There are limits to how well we *can* understand these factors, so there are limits to how well we *can* understand the impact of some kinds of constraints – and associated opportunities – particularly those which involve actions in these Domains.

The journey is complex but worth the travel. In the end, understanding why no philosopher's stone exists may help the reader to avoid becoming trapped in unresolvable debates about the exact costs of constraints at the expense of more fundamental issues around the role of energy and environment in economic development. With many economies around the world struggling, there is an essential economic need to understand why the historic model of economic growth is faltering. From this platform, the chapter explains the fundamental fallacy of the view that post-recession economies are 'too poor' to afford environmental concerns. Rather, it points to ways in which decarbonisation – intelligently pursued – can offer a lever to help find a sustainable way out of the enduring financial and economic malaise.

11.2 Economic growth and the search for the Philosopher's Stone

Economic growth, as it is conventionally understood, is typically expressed in terms of Gross Domestic Product (GDP). A lot of the confusion surrounding GDP is attributable to a common misunderstanding that it represents the sum of the market value of all the goods – shoes, cars, houses, etc. produced and sold in a year. In fact, GDP is the sum of the 'value added' of all products and services, i.e. the difference between their selling price and the cost of intermediary products used in their production (for example, the metal and the tyres required to manufacture a car). The value added is thus composed of wages, profits and a few taxes on production.

GDP is a monetary measure of the activity level. It is not a measure of welfare, nor is it a measure of economic prosperity. These are more complete aspirations for progress in society and a large body of literature exists which pushes beyond GDP as a metric, capturing important but more subjective concepts such as happiness, freedom and the depletion of 'natural capital'. Neither GDP nor growth are synonymous with welfare.[3] Instead of pursuing GDP growth, societies could, for example, choose to increase the amount of time people spend on leisure activities as opposed to working: this would lead to relatively lower 'value added', since labour is an input of products and services. Yet for all of its weaknesses, GDP remains an important economic indicator, because it represents a monetary social surplus which can be shared between today's consumption and investment to allow for increased future consumption – as long as growth is sustainable.

While acknowledging the importance of the debates about values and metrics, therefore, this chapter does nevertheless concentrate primarily upon the more conventional economic question of how energy and climate policy may affect GDP.

The 'rhetorical jousting' around energy, environment and growth

As shown in Chapter 3, at least half of all fossil fuel consumption globally is for intermediary purposes in economic systems: for producing material goods and for generating electricity. The remainder is split between transportation – which serves both production and consumers directly – and for basic welfare, such as keeping our homes warm. The way such inputs are treated in economic theory is thus critically important.

The consumption of primary energy resources to produce a car, like any other intermediary consumption, is not accounted as part of the 'value-added' of the automobile industry. If it is possible to produce cars with less energy input and, if the selling prices of cars are kept constant, this will increase the value added of the car industry; but the recorded value added of the energy sector which transforms primary energies into the energy needed for the production process will decline. Implicitly, in measuring economic growth as an aggregated measure of all of the value-added of all sectors in an economy at a particular time period, economic growth is defined as a 'value-creating machine', drawing on people (labour), machinery and infrastructure (productive capital), energy, land and raw materials.

On the one hand, there are fears that carbon constraints will hinder economic growth. Indeed, in the aftermath of the financial crisis and subsequent economic slowdown this is *the* dominant political concern held by most countries. The logic is straightforward. Just as the Industrial Revolution was powered by coal, so too have fossil fuels more generally been a key driver of economic growth. Stabilising the atmosphere must inevitably constrain their use or otherwise drive up energy costs (for example, through carbon capture and storage) and thus hamper economic recovery.

On the other hand, an opposing argument is that environmental action may be associated with reducing the need for resource inputs, improving the physical efficiency of use, and stimulating innovation and the emergence of new industries which produce more value with less inputs. At the micro (firm) level, this argument was put forward as the Porter hypothesis discussed in Chapter 9 – the idea that imposing environmental standards on industry might stimulate companies to be more efficient and innovative, and might also trigger the learning-by-doing and economies of scale described in that chapter.[4] A translation of similar arguments to the macro level would imply, for example, that carbon constraints may lead to cost-effective improvements in energy efficiency, accelerate innovation and bring down the cost of alternative energies, stimulating the growth of whole industries.

The counter argument to this in turn is that if companies and economies can become more efficient and innovative, they will do this irrespective of a carbon constraint. There is no doubt that environmental regulation can stimulate innovation. But debate over the existence of the 'strong' form of the Porter hypothesis – that this can lead to net economic benefits – has raged for almost two decades without any consensus emerging, but with some indications that it depends on whether the innovation concerns end-of-pipe fixes or deeper process innovations (see Chapter 9, notes 46 and 47). The macroeconomics of innovation have also been hotly debated, not least by arguing that innovation itself is a finite resource, dependent on a limited pool of research talent: that accelerating innovation in energy, for example, can only come at the expense of slowing down innovation in other areas (e.g. for medicine, food production, high-speed transport). This phenomenon is known as 'crowding out'.

The counter-counter arguments are that such 'crowding out' theories fail the test of reality: that innovation is an exploratory, not optimising, process, that societies consistently under-invest in innovation for multiple reasons (see Chapter 9), and that innovation in one sector may spill over into other sectors with net benefits. Prominent early examples include the development of nuclear power and gas turbines for power generation, both of which drew heavily on the foundations laid for military purposes. A more recent example is the way in which the improved design of electric batteries – triggered in part by the explosion of pocket calculators, micro-computers and cell phones – are now being applied to electric cars.

The difficulty in reaching any consensus on such issues goes to the heart of debates in modern theories of economic growth and difficulties in matching these theories to environmental and energy issues. Modern growth theory aims to support economic planning and policies, with an important focus on the choice between current consumption and current investment, and therefore between current consumption and future consumption. One important obstacle is that the 'long term' in growth models typically refers to a period of between 10 and 15 years, whereas the 'long term' of energy development trajectories encompasses 20–50 years – and for climate change even longer.

Another obstacle to understanding the relationship between energy and economy is an enduring disconnection between 'top-down' models of whole economies and 'bottom-up' models which emphasise technological and sectoral detail. The former 'top-down' approach favours very compact models in which energy is just another factor of production without any explicit consideration of the energy content of economic growth. The 'bottom-up' models have detail, but no representation of economic growth processes.

When analysts first tried to bridge the gap between engineers' and economists' views of technologies – the so-called bottom-up/top-down debates – for pragmatic reasons, the first macroeconomic efforts were to base the models on Solow's growth theories, presented in the next section.[5] These assumed that the energy sector is itself an optimising sector, defined and linked to the rest of the economy only in terms of resource costs.

This practice is nicely justified by the metaphor of the elephant and rabbit stew.[6] If GDP is like a stew made of elephant then this stew has the taste of elephant, so you can neglect the rabbit. At that time – the low energy prices in the early 1990s – the non-energy sectors represented 98 per cent of value-added and the 'rabbit' (the energy sector) only represented 2 per cent of value-added. Today the energy component is maybe twice that, and even more so in emerging economies. But the big question is whether this metaphor is really relevant given the pervasive input of energy across the economy – particularly when considering structural changes in energy demand, supply and growth of the nature considered in Chapter 10. Such transformations might imply growth pathways with less energy but a higher value-added in many of the large energy-consuming sectors. The metaphor also does not hold in the case of large shocks to energy prices, as illustrated by the impact of oil shocks on economic growth (see Chapter 1, notes 7 and 8).[7]

The rest of this chapter will examine the deeper economics underlying this rhetorical jousting. It will show why arguments which seem obviously legitimate from one perspective may in fact be hasty, partial and sometimes biased interpretations of what economics can really say about the links between energy, environment, innovation and economic growth. In so doing, the chapter also aims to shed light on the apparent wider disconnection between 'physical' and 'economic' views of the systems involved.[8]

Solow's residual – the discovery of dark matter

To explore something as big as economic growth, economists need to aggregate many things into a few basic categories like aggregate labour and capital, production, consumption and investment. Nobel economist Robert Solow first presented his model of economic growth in 1956 and it has since been elaborated and expanded on to create the foundations of neoclassical economic growth models. The model is important to understand as it underpins a lot of the discussions about the links between energy and growth modelling. It focuses on the choice between present consumption and investment, which helps to set the timescales implied.[9] It can best be explained using simple algebra.

Solow's theory sets out to explain economic growth through a 'growth engine' that draws on labour (L) and productive capital (K) to produce overall economic 'value added' – traditionally expressed in terms of GDP. Total economic output GDP is thus generated by a production function that draws on K and L: $GDP = f(K, L)$. The production can be either consumed (C), or invested (I) to increase income levels in the future: thus $GDP = C + I$. The stock of productive capital that can be used is determined by the level of equipment accumulated over years, built by saving part of the economic surplus instead of consuming it.

It is common practice to accept this equation as obvious, and after Solow's paper it became the underlying intellectual framework for analysing economic growth. Yet there is something implicit in it which should not be taken for granted. Namely, there are innumerable other essential inputs other than L, K and I. Capital, for example, cannot be wielded without transportation, and transport is thus both an 'intermediate good' (input to production) and a 'final good' (moving passengers). Similarly, it takes energy, transportation and water (and other factors) to produce food. In theory, all this intermediate production and consumption appears on both sides of the equation – it is both produced and consumed – and hence each cancels the other out if they are aggregated in terms of money, which is the common practice. However, there are limits to this reasoning – there are terms missing on each side of the equation as they are assumed to cancel out, but which in reality may not.[10] Ultimately, GDP is the value-added produced by the economy, not the sum of the value of products sold. For understanding the economics of energy and environment, this matters, as we will see later.

Investment is spent on accumulating more capital. Therefore K (capital in the broad sense of the word) depends on the level of investment and will determine future levels of GDP (unlike labour).[11] However, capital also needs maintaining as it depreciates in value each year through general wear and tear. In Solow's model, economic growth per capita falls to zero when an incremental investment increases production only just enough to compensate for this depreciation. As the level of equipment of a society increases – associated with its cumulative capital investment, so does the share of its product devoted to refurbish or renew equipment. This leads to a condition known as *zero-growth steady state*.[12] Note that this is *not* related to environmental concerns about the 'limits to growth' (as first popularised by the Club of Rome) since energy and resources play no direct role in this mechanism.[13]

The only way out of this zero-growth prediction is through innovation. More specifically, technical change has to improve the productivity of capital (its output for a given level of inputs) quickly enough to compensate for capital depreciation. Positive economic growth thus continues only with an additional injection of overall productivity, generally associated with innovation. Solow's model now becomes GDP $=A(t) \times f(K, L)$, where innovation drives an increase of A(t) over time, in this case independently of changes in capital and labour. Note that A(t) does not give any direction to growth. It acts as a fuel, similar to the way in which the quantity of gasoline determines the number of kilometres that can be covered by a car. It says nothing about the direction – or composition – of economic growth. The direction, the wheel, is the choice of techniques made at each point in time depending on the options and evolution of wages and capital costs.

Solow first tested his growth model against real data for US economic growth, capital accumulation and labour in 1957; this suggested that labour productivity had doubled between 1909 and 1949, with '87.5 per cent of the increase attributable to technical change and the remaining 12.5 per cent to increased use of capital'. The attribution to 'technical change'– innovation – which is neither labour nor capital availability, represented 40 per cent of economic growth over 1929–57 in the US. Similar studies in France suggest that this factor accounted for more than 50 per cent of French economic growth in the decades immediately after the Second World War.[14]

In 'growth accounting', this factor A(t) became known as 'Solow's residual', because it was the residual factor left after accounting for the impact of increased labour and capital, and the impact of capital accumulation on labour productivity. In this sense it represents the 'dark matter' of economic growth: unaccounted increases in the productivity of both labour and capital which thus enable more wealth to be generated from each input.

The striking result is that a significant part of economic growth – around half in many measurements – appears to come from this residual 'dark matter', sometimes historically referred to as the economic 'manna from heaven' of productivity growth. In many ways this is a weakness of the underlying model, but it also contains some intuitive wisdom. It shows that economic growth per capita depends not only on the classical mechanisms of economic theory – notably optimal use of resources and capital accumulation driven by relative prices and competitive markets – but also depends on many other factors. Collectively, these other dimensions of productivity – and hence economic – growth were broadly equated to 'innovation'. As outlined below, in practice they encompass a very wide range of factors, including institutional change.

One issue investigated early in the context of growth theories concerned the nature and direction of this 'innovation'.[15] The subsequent debate is important because it concerns whether or not there is a 'natural' pace and/or direction in technical change (an energy-saving one for example), and whether there is a 'wheel' to redirect growth towards a desired path.

Essentially there were two ways of trying to retain the primacy of purely economic forces in the growth story, which broadly represent two extremes. One was to assert that A(t) could not be influenced by purely economic decisions and was entirely an exogenous concept – 'manna from heaven' – so that the only viable aspiration for economic policy was markets that would make optimal use of resources and being grateful for the mysterious 'residual' addition of productivity gains. The polar opposite was not only to assert that productivity growth (A(t)) was itself driven by economic decisions but that it was intrinsically optimising – a view facilitated by the intellectual attraction in economics of optimising growth theories pioneered by Ramsey (1928). Part of this attraction comes from its mathematical elegance and consistency. These 'beautiful theories' of optimal growth had to assume a global 'representative agent' with optimal behaviour and perfect foresight, and neglect fluctuations and shocks. Not surprisingly, attempts to reconcile this with the observed data proved fruitless.[16]

This apparently abstract discussion has multiple implications for energy and climate change policy because it frames particular ways of thinking. If the pace and direction of innovation is somehow 'optimal' (or beyond reach), it implies that trying to change direction comes at a cost. Conversely, if innovation is neither inherently optimising, nor beyond the reach of other influences, the switch from carbon-intensive to carbon-saving technical change is not necessarily costly, nor will it rely only on prices. It involves changing what is behind the mysterious 'residual'.

Probably no modern economist believes that productivity growth A(t) is really at either extreme – either inherently optimised through markets or beyond influence. Nevertheless, the focus of growth models on resource accumulation has left a popular legacy, a mental map in which these factors have received the lion's share of popular attention. To underline the point more pragmatically, in the language of Chapter 9: if the energy innovation chain is broken, just public R&D + pricing will not fix it.

In the mainstream economic literature, analysis by Paul Romer (1986) is credited with first raising the discussion about barriers to innovation (Chapter 9) to the macroeconomic level. He underlined various general reasons why the value that accrues to innovators is likely to be much less than the full value of an innovation to society, and estimated a correspondingly large gap between the 'private' and 'social' returns to innovation. From a purely economic perspective it provides the conceptual justification for 'innovation policy' – but beyond justifying government R&D expenditure it says little about the wider span of what shapes productivity growth.

Where is the wheel?

Economists attempted many different ways of probing the 'dark matter' represented by A(t), and constructed numerous different models of 'endogenous technical change' in which innovation (translated as productivity improvement) is driven by economic factors. Some economists proposed that innovation and overall technical change are spurred mainly by the relative prices of labour and capital, that innovation will seek to improve the productivity of the factor which is relatively more expensive. Others seek lessons from economic historians and emphasise the importance of institutions, human capital, culture and research trends, which are hard to link to pure economic mechanisms.

Even if these efforts do not provide comprehensive explanations, they do contain important insights: they open the door to the search for 'the wheel' which can turn the structure of economic growth – and hence its environmental impact – in different directions.

One stimulating approach was to imagine a 'historic innovation possibility curve', which represents the *potential* improvements which could be generated, in the future, with a given R&D budget and a given human capacity.[17] This is the set of conceivable options, similar to the best-practice frontier referred to in Chapter 2, but the actual set of technologies available at a given point in time is determined by the historic path of relative prices of labour and capital. This makes technical change *path dependent*, as developed in the theories of evolutionary economics outlined in Chapter 10.[18]

Yet in this approach the *effectiveness* of relative prices is itself determined by non-economic factors and hence still does not tell us much about how growth can be steered in particular technological directions. Is innovation spurred by demand, which itself is directed by relative prices? Or is it pushed by accumulated knowledge and the internal dynamics of technological research? There is no winner in this controversy – not least because, as emphasised in Chapter 9, the full chain of innovation is a complex mix of supply and demand, or 'push' and 'pull' forces, the balance of which is bound to differ for different technologies at different stages of development.[19]

The question of what steers innovation, and at what cost, is central to addressing questions about the structural implications of innovation on the scale implied by global decarbonisation.

If economic development is thought of as the physical journey of a car leaving London, for example, most economic approaches imply that there is a unique 'best' journey for a given distance – say, with the destination of Oxford – to maximise the welfare of the travellers. The town of Cambridge (UK's second oldest university town) is a similar distance. But in the economics analogy, this destination may be less valuable, or the road more difficult (costly) to travel.

Treating technology as *exogenous* in economic development is equivalent to saying that the underlying factors – the preference for Oxford or the state of the roads – cannot be changed by economic choices. Endogenous growth theories equate to saying that if an increasing number of cars turn their wheel to take the road to Cambridge instead of Oxford, it is then more likely that road works will be made to improve the road to Cambridge. In this case, after a while, the cars will drive to Cambridge at the same speed. But it might be the case that investing in the road to Cambridge will crowd out other investments – such as those on the Oxford road perhaps, or on the quality of the cars – in which case the travellers to Cambridge may benefit (and so may the town), but there may still be an aggregate cost overall. The forces behind the 'wheel' of technical change matter.

Since Solow, there have been numerous efforts to endogenise drivers of innovation and technological change in growth models. Many of them sought to explain divergence in growth rates between countries. These models were mostly developed between the 1960s and 1990s, with varying degrees of success and accuracy. However, they suggest that there may be many possible endogenous drivers of productivity gains.

- *Learning by doing*. In 1962, Kenneth Arrow developed an economic model which included 'learning-by-doing'. He captured this concept by making knowledge and productivity (the 'A(t)') dependent on cumulative investment in growth models, rather than looking at investment levels in an isolated time period. In this model, knowledge and productivity increase in line with the growth in total investment and not with a specific investment. The implication is that redirecting technical change is costless, at least in the long run.[20]
- *Improving labour productivity*. A few years later, Hirofumi Uzawa (1965) showed several ways in which the quality of the labour stock could also be influenced by

investment – leading to the notion of accumulated 'human capital'. The effort was renewed after the observation that the original Solow model could not explain why capital in the USA appeared to be almost 60 times more productive than in India which leads to the paradox that capital should flow massively from India to the US. The solution was attributed to different levels of human capital arising from higher US investments in education and training.[21]

- *Product quality and product variety.* Other models explored the roles of product variety and product quality as major drivers of technical change. If consumers are willing to pay more for new or better goods and services, then there is an incentive for companies to invest more in R&D to provide them, and profit by being the first with a new product. Similarly, consumers may be willing to pay more for increased variety in food or clothes, for example. This echoes a core argument in Chapter 9, which noted that the energy sector (particularly utilities) has an unusually low level of R&D, and attributed this in part to the fact that energy is a homogenous good. For most consumers, there is no such thing as a 'better electron' – an observation which powerfully links the macroeconomic debate with energy implications.[22] The economic rents (and hence R&D) have been larger in oil for other reasons, some of which are geopolitical.

These so-called endogenous growth models were principally used to examine differences of productivity and income between countries, and to examine whether and how poor countries could grow faster thanks to a 'catch-up' of productivity levels. However, overall the growth theories proved inconclusive – ultimately because of the inherent difficulty involved in measuring some of the key processes (Sala-i-Martin 1996).[23]

Finally, we know that the dark matter of economic growth comprises numerous forces, often but loosely termed 'innovation', which in fact encompass factors as diverse as efficient regulation, institutional and technical change, education and infrastructure – as well as potentially more nebulous factors still, associated with culture, for example. These form the content of 'Solow's residual' A(t) required to fill the gap between the classical explanations and observed growth. We also know that empirical studies suggest that this residual represents up to half of the driving force behind GDP. We can hypothesise on the characteristics, drivers and pathways for innovation, but we cannot quantify their component parts with any great certainty.

Our opening examples at the beginning of this chapter help to underline the points. The structural problems which existed in the Soviet Union's economy could not be compensated by its almost unlimited abundance of cheap energy. In the aftermath of its dramatic political transition, the mental map – the underlying assumption – remained that cheap energy was essential to growth and a key part of the Russian endowment. But Russia's over-consumption of energy was part of its problem, not part of the solution.[24] Russia's economic recovery (particularly after 1997) was not matched by rising energy use. Instead, improved energy efficiency, flowing in part from a more liberalised economic system plus belated attention to environmental concerns, was a valuable economic contribution (which also left Russia with a huge surplus of emission allowances under the Kyoto Protocol).[25] Considering CO_2 caps as a constraint on economic growth wholly overlooked the fact that energy has to be understood as part of an overall economic strategy. In contrast, Germany built upon the dynamism and power injected by its reunification, and its *Energiewende* transition to decarbonise the energy system is widely seen as the next invigorating challenge. In neither example can energy and CO_2 be usefully divorced from the wider economic and social context.

The problem of aggregation

Almost half a century after introducing his growth model, Solow (2001) made a pessimistic diagnosis about the efforts to explain productivity gains A(t) in terms of purely economic forces or to endogenise technical change in growth models.[26] In the terminology of this book, we interpret this disappointing conclusion as indicating the role of First and Third Domain processes in the 'residual' – and hence in economic growth.

One major methodological problem in trying to formalise (in any quantified way) the links between the Three Domains stems from the extent of aggregation in economic growth models. We have seen that Solow's growth engine is like a jelly factory. This jelly has no structure: there is one sector, one form of consumption, which is to be traded off against aggregate investment. Such simplifications do not matter when studying the trade-off between savings and consumption in the whole economy, but it means such models cannot analyse structural changes in the share of different goods produced in the economy, or the contribution of different sectors to economic activity.[27]

Three particular problems arise from this which are of relevance to energy and environmental issues. One is that such models have no economies of scale (doubling the level of production only doubles the inputs in all production factors), whereas economies of scale are obvious in most industries, as demonstrated by learning curves (Chapter 9) and numerous other industrial examples. The assumption that economies of scale vanish *in aggregate* was made in part for technical reasons by economic modellers.[28] This makes it impossible for such models to represent the links between the existence of large economies of scale in some sectors and the overall dynamics of growth (e.g. the economies of scale in the steel, petrochemicals and car industries after the Second World War) – a topic we will revisit later in the chapter.

The second consequence of this aggregation is that 'technical change' in such growth models combines both the choice of techniques (the inputs used to produce the many products represented by the 'composite good') and structural changes (the set of products in the 'composite good'). This provides no grip to capture the interplay between changes in production technologies, consumption patterns and structural change.[29]

In reality innovation is triggered by the dreams and inventive capacity of humans, by the strategic choices of publicly funded research institutions (including military research), and by corporate investments.[30] Demand is also needed to help fuel the trial and error process of learning-by-doing and economies-of-scale. One typical example is the almost one hundred year process of incremental innovations in the automobile industry initiated by the 'five dollar workday' decreed by Henry Ford in 1914 – to which we will return later, after first looking explicitly at the energy dimension of economic growth.

11.3 Probing the past: energy in economic development

Perhaps surprisingly, most economists since the Second World War did not consider energy issues as an important topic for understanding economic growth until recently.[31] Consequently, as a result there has been remarkably little connection between works in mainstream economic growth theory and those focused on the link between growth and energy.

It is undoubtedly the case that energy has historically played an important part in major phases of economic development, such as the role of coal in the Industrial Revolution and of oil in US expansion and global development in the twentieth century. It was indeed perhaps because of this very abundance of cheap energy in the postwar years that growth models neglected it.

In fact, energy permeates and underpins the entire economic system. As shown in Chapter 3, globally about half of commercial energy consumption directly delivers essential welfare (warmth, light, transport) to the 'final consumer'; the other half is an essential input to production across both industry and commerce.

After the 1970s oil shocks highlighted this all too painfully, the classical growth models came under fire for not including any consideration of the limits of energy and natural resources. One option would have been a 'bottom-up' approach, to build sectoral models to track the input and outputs of energy flows within an economy (the intermediary consumption) and their relationship with the environment.[32]

But this was not the route followed by the mainstream of the economic profession, perhaps because of the lack of elegance and technical difficulties of models with large matrices, and the absence of a formal theory of structural change to project these flows over long time horizons. Rather, in response to the Club of Rome (Meadows *et al.* 1974) report on the (environmental and resource) *Limits to Growth*, Solow (1974 and 1986) expanded his model by adding energy (E) as a new factor in the basic equation of primary production, alongside capital and labour.[33]

In this revised model, the stock of the 'energy production factor' was determined by the amount of primary energy accessible for a given investment in technologies; limited energy and natural resources do not necessarily limit growth as they can be substituted by more labour and capital, depending on the relative prices of these inputs. Critics of the model question its 'technological optimism' regarding the ease with which these different inputs can be substituted and how they affect the rate of economic growth.[34]

Not all economic analysis stayed at this level of aggregation. The rise of energy and environmental concerns generated a completely new generation of modelling and other strands of literature. This literature is very diverse and published in specialised journals. Yet despite its diversity, it pointed to two major issues in understanding the role of energy in economic development: trends in energy efficiency and the economics of energy innovation. Thus at last we begin to investigate how economic modelling has approached the phenomena characterised in this book in terms of the First and Third Domains.

'Autonomous' energy efficiency – the dark matter of energy

When economists began constructing more detailed energy/economy models, their first instinct was that energy was an input into economic growth, so energy consumption should be proportional to GDP. But real-world evidence consistently indicated that the global energy to GDP ratio actually fell every year – energy efficiency seems to have been consistently improving and decoupling energy from economic growth, as shown in Chapters 1 and 5.

This observation presented a similar situation to Solow's model and his residual 'A(t)' – the 'dark matter' of economic growth. We know energy efficiency improves over time and so should therefore be included in energy models, but we do not fully understand what drives it. However, the question is extremely important, given the long timelines involved. If the global economy grows at 2 per cent per year, annual decoupling of either 0 or 2 per cent is the difference between a global energy demand which doubles or remains constant over 35 years.

In fact, evidence since the nineteenth century suggests that the energy content of growth follows a bell-shaped curve which tracks the trajectory of a country's industrialisation.[35] Initially, the energy content of growth increases, for instance due to the need for energy-intensive products like cement and steel to build infrastructure, and for the heavy machinery used in manufacturing; the energy intensity then falls as the country further develops. The long-term trends demonstrated in Figure 11.1 show that countries which have developed later have lower emissions intensity (energy per unit GDP) during the industrialisation phase compared to first movers (such as the US and the UK), as they have access to relatively more efficient technologies than those available in the earlier periods.

The rate of 'decoupling' does undoubtedly depend on the price of energy (see, for example, Chapter 5, Figure 5.7). Rising energy prices create an incentive to use energy more efficiently. But energy prices cannot explain all of the observed energy efficiency improvements. In the early 1990s, modellers introduced a coefficient of 'autonomous energy efficiency improvement' (AEEI) to represent the underlying trend, estimated at around 1 per cent per year for OECD countries.[36]

Thus the decoupling between energy and growth cannot be explained by prices alone. The question remains as to whether or not this so-called 'AEEI', in many modelling exercises, is stable and constant, or whether (and how much) policies and behavioural changes can affect it. Pillar I of this book (notably Chapters 4 and 5) presents the theory and evidence that energy efficiency is rife with First Domain effects, and documents the extent of corresponding impacts from energy-efficiency policies. When raised to the level of global modelling, drawbacks to the simple AEEI arise similar to the autonomous productivity gains (A(t)) in Solow-type growth models. If all sectors are aggregated into general 'energy efficiency',

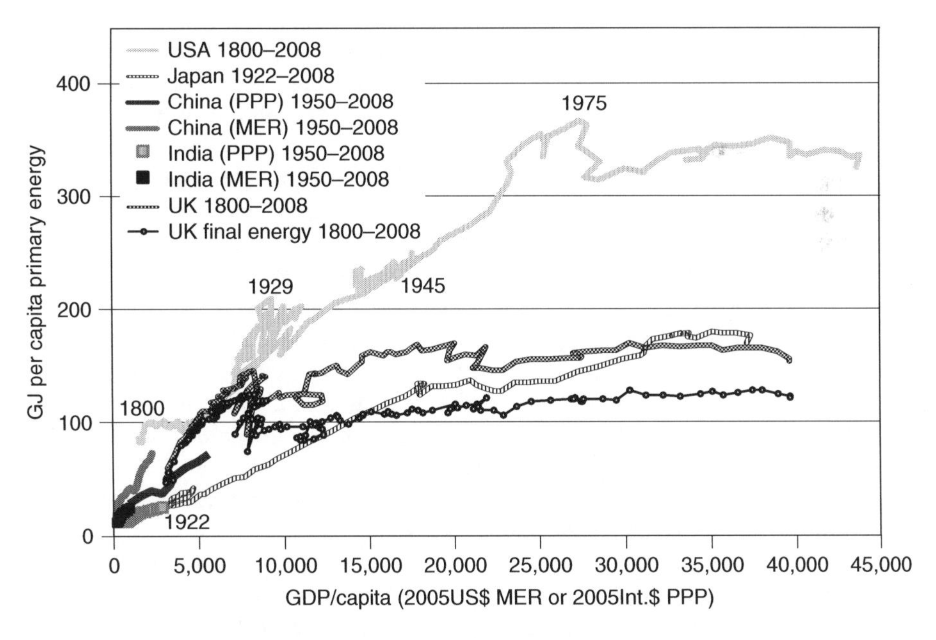

Figure 11.1 Long-run historical trends in energy consumption per capita for a few major economies

Note: This graph provides, for a few major economies, similar data to that in Chapter 1, Figure 1.7, but for energy (rather than carbon) per capita and over much longer periods, to enable comparison of India and China with the energy consumption levels of US, Japan and the UK when they had comparable levels of economic output per person. It also shows the influence of different exchange rate treatments for conversion of developing country GDPs. (MER = market exchanges rates; PPP = purchasing power parities. Note that GDP is expressed in US$ 2005.)

Source: Grubler *et al.* (2012).

the AEEI compounds both technical improvements and broader structural changes in the economy.

A strand of economic literature tried to overcome the limits of the AEEI by 'endogenising' the energy intensity in empirical energy and economic models to better understand its drivers and impacts. One early and influential approach to doing so estimated a small tendency for technical change to have *increased* energy intensity in the US.[37] However, like many econometric conclusions drawn from specific countries, periods and specifications, it was unclear whether this was really a systematic result or just a snapshot of a particular country and period which may moreover confuse correlation with causation.[38]

The lack of agreement – and indeed the inconclusive nature of subsequent literature about the links between energy and long-term productivity trends – should not mask the undoubted negative economic impacts associated with rapid and unexpected increases in energy prices.[39] Timescales are the key, an issue to which we will return later because an important 'spillover' of energy-saving technical change is to increase the resilience of an economy to such energy shocks.

Innovation and 'crowding out': a matter of judgement about spillovers

The other main area of debate was the extent to which energy-related innovation is triggered by R&D spending and/or learning-by-doing mechanisms.[40] The top-down models faced a conundrum related to that examined 'bottom up' in Chapter 9 of this book in terms of push versus pull forces: whether innovation should be represented in terms of low-carbon R&D or as a market response to carbon prices. In addition they argued over whether strengthening energy innovation by either route would lead to a 'crowding out' of other innovation, or conversely to a 'spilling over' of R&D into other sectors – with opposite macroeconomic implications.[41]

Goulder and Schneider (1999) highlighted that the net effect depends on cumulated R&D investments and learning experience in 'clean' energy compared with fossil energy and non-energy production. Specifically, they identified that it depends on whether: (a) the potential for cutting down production costs is high or low in low-carbon technical options; (b) past R&D investments and learning-by-doing have realised a significant share of this potential; and (c) there are spillover effects of innovation from any sector to other sectors. The last factor can potentially counterbalance the crowding out effect. In this case, knowledge spillover and overall productivity depend on total accumulated investments in R&D and knowledge accrued from this, irrespective of whether investments are directed to low- or high-carbon technologies.[42] We face a challenge to interpret these results, again due to the level of aggregation. Although these offer insights into economy-wide productivity, such insights are qualitative and cannot be applied at a smaller scale.[43]

Perhaps the large investments in military research provide an *ab absurdo* demonstration of the point. If there were no spillover from military R&D into other sectors then countries like the US, the UK and France would have achieved a remarkable feat by maintaining rapid economic growth despite very high 'unproductive' R&D expenditures. The truth is that they have reaped spillovers from military research to productive sectors of the economy, including the energy examples noted earlier.

Again, the crowding out versus spillover debate seems inconclusive and risks a pointless rhetorical stalemate. It diverts attention from a more important generic issue, which is the deficit of societal investment in knowledge, be it in the form of R&D or the learning-by-doing incorporated in new industries. This was the basic warning of Paul Romer in 1986 whose work as noted highlighted the gap between the social and private returns on investments. Private investments may benefit society as a whole, but private decision-makers may not be rewarded for the positive knowledge spillover of their investments. This can be thought of as a positive 'externality': the true value of R&D is undervalued in the market, so there is insufficient investment in knowledge.

This gap between private and social value of investment is not only due to spillover. It may be amplified by the regulatory regimes in energy sectors, investment risks related to the business cycles and risk-aversion related to uncertainties in final demand and the performance of new technologies. The systematic underinvestment in innovation may also be amplified by the overall transformation of the business context in the past twenty years, from a managerial regime that allowed firms to focus on maximising their long-term value to one dominated by the maximisation of shareholder value at each point in time.[44] The question of investment risk and finance is thus embedded in the fundamental evolution of the governance of industrial systems. We come back to this point in the last section 11.7.

These two broad issues, namely the macroeconomics of energy efficiency and innovation in the energy sector, can be broadly related to the mechanisms of the First and Third Domains, respectively. The bulk of the book has focused on the microeconomics, namely the practical realities uncovered by engineers, firms and policy-makers. The inability of macroeconomic analysis to reach stable conclusions about either 'autonomous' energy efficiency or the economics of energy innovation underlines our core message and elevates it to the level of macroeconomics and growth: optimal progress in these other arenas cannot be adequately understood only through the assumptions and theories of Second Domain economics.

The essential theoretical problem in understanding the role of energy in economic development thus comes down to the fact that it has two quite distinct roles. One role is that of energy as a raw resource: the physical power that can transform matter, move us around and keep us warm. The other factor is the role of innovation and change in energy-related activities, most notable in the huge swathe of energy-consuming industries, which can not only reduce dependence on energy (and typically other physical resources too) but also contributes to productivity growth more widely in the economic system.

Both factors can contribute to growth, but there is a tension between them. Cheap energy can fuel growth, but also encourages stagnation in energy-consuming activities. At the level of overall energy consumption, the tension between these forces helps to explain Bashmakov's constant (Chapter 1). To understand the point, one only needs to consider countries at the polar opposite ends of the spectrum: contrast Russia and Japan through the 1970s and 1980s, and consider which emerged more successful. Japan, whose dependence on oil imports meant it faced the full force of the oil price shocks, gained technological advantage which moreover contributed to its economic success in other markets – particularly compared to competitors who had kept energy cheap and not made the same efficiency adjustments, like US vehicles and steel production.

11.4 Framing the future: insights and blind spots from integrated assessment models[45]

Too good to be true?

The previous sections might lead the reader to the seemingly disappointing diagnosis that even decades of research have delivered no firm conclusions about the links between growth and energy beyond the obvious. However, this diagnosis contains useful information.

As noted, it points to the potential importance of forces outside Second Domain assumptions. More broadly, it indicates that economic history results from too many factors (including geopolitical accidents like the two oil shocks or the collapse of the communist system) to derive robust econometrical evidence about the underlying historical relationship of energy and growth.

An alternative approach is to start with structural assumptions and project their implications forward using 'integrated assessment' models to explore future scenarios. This can leave aside the non-energy-related factors – like geopolitical upheavals as well as commodity or financial shocks – which cloud the lessons from the past statistics.

There is now a database of hundreds of such scenarios which differ in their assumptions of the costs of low-carbon technologies and policy tools like taxes, carbon trading, technological transfers and R&D.[46] In spite of these differences, the overwhelming majority of scenarios deliver a few consistent messages:

- Deep cuts in greenhouse gas emissions are possible.
- The cost of these cuts will be a few percentage points of GDP over the century, and this cost is higher for more ambitious temperature targets (e.g. a 2°C rather than a 4°C rise), not taking into account the benefits of reduced climate change.
- The time profile of emission cuts matters: the implementation of large emissions reductions now would increase short-run costs but reduce long-run costs.[47]

The long-term macroeconomic implications of such results can be demonstrated by using simple back-of-the-envelope calculations. If economic growth was sustained at 2–4 per cent a year from 2015, GDP would be between two and four times its starting level by mid-century. If stabilising atmospheric concentrations cost 2 per cent of GDP by mid-century the corresponding GDP would be 1.96 to 3.87 times the starting level – a small fraction off the rate of global growth. If the impact were as big as 10 per cent by the end of the century (at the upper extreme of most results), it would be hardly discernible in comparison with the overall scale of growth – equating to cutting growth rates by around 0.1 per cent per year. The impact is equivalent to waiting another one or two years to restore the same level of GDP as without climate stabilisation.

As noted in Chapter 1, this seemingly small magnitude raises the question of why countries hesitate and refrain from taking actions to aim for a 2°C target – and why global emissions are indeed surging. The modelling results seem too good to be true. Part of the explanation is that in some senses the structural assumptions behind models are implausibly optimistic; in other respects, though, the converse is true. The underlying paradox remains to be explained: if smart global solutions are quite cheap and involve urgent action, why then does this seem so difficult?

The struggle for realism: when transitions matter

The answer to that paradox is to be found in the gulf between most global economic models and the realities charted in the previous chapters of this book. Most such models assume a 'global planner' with 'perfect foresight' between now and the end of the century, so that it is possible to find the least expensive pathway.[48] They also assume frictionless market mechanisms (each abatement opportunity is introduced when it is cheapest), which underestimate the complex interactions between the energy sector and the rest of the economy: in the event of any exogenous shocks, including a policy shock, the economy quickly returns to a state of equilibrium.[49] They actually describe an economy as though it were governed by Second

Domain mechanisms only, with a perfect global planner mimicking the operation of a perfect global market aided by autonomous gains in energy efficiency and innovation.

These are optimistic assumptions – yet such models also fail to represent the opportunities. By assuming that the system is perfect, both before and after introducing climate change, any constraint inevitably incurs a cost.[50] Most assume that innovation proceeds without effort, reducing the future costs to some degree, but in ways that cannot be influenced by the responses. As a consequence, they paint a very incomplete picture which limits their applicability to real policy.

One of the major efforts of the energy-economic modelling community over the past decade has been to get better at representing innovation, in part by harnessing some of the insights from mainstream endogenous growth modelling.[51] Among these efforts, the 'RECIPE' project sought to probe and compare in depth the results from three major models with endogenous technical change. As outlined in Box 11.1, the three models have much in common, and all conclude that a comprehensive global effort which would start immediately might enable the mid-century goal of halving global emissions to be achieved very broadly at a cost of around 1 per cent of discounted global GDP.[52] And yet, as shown in Figure 11.2, the actual time path of carbon prices and economic impacts associated with achieving a 2°C target still vary enormously.

However, the RECIPE results themselves already hint at some of the difficulties of achieving any common view. There is little empirical, robust, quantifiable basis around which to gain consensus about the prospects and mechanisms for innovation, or indeed energy efficiency, for the fundamental reasons observed earlier. As a result, most models largely ignore these issues or make exogenous and somewhat arbitrary assumptions on which differing teams and models express differing judgements.

All models find that results are sensitive to assumptions about underlying energy and economic growth and technology costs. There are many other generic observations in common, like the conclusion that actions restricted to smaller groups of countries raise the global cost of climate mitigation and makes achieving given goals more difficult (or impossible). Inaction closes options.

There are interesting insights to be gleaned from the few models which do attempt to probe First and Third Domain effects. Model 1 in the RECIPE studies (WITCH) underlines that neither efficiency nor innovation *on their own* are panaceas, if price is the only tool to drive them: this somewhat echoes the observation about efforts to include induced innovation in the archetypal model of Professor Nordhaus (see Chapter 9, note 22): if First Domain characteristics continue to limit consumer responsiveness or if the innovation chain is weak (and/or if the reach of low-carbon supply technologies is sectorally confined), then price is a blunt tool which has to be wielded very heavily to be effective.

Model 2 in the RECIPE studies (REMIND-R) shows the potential impact of more optimistic assumptions around the technology restrictions in particular. Induced innovation in renewable energy and CCS, combined with more sectoral detail, allows these technologies to disperse throughout the energy system – from which point carbon prices and costs stabilise at modest levels. While not explicit in the model, a moment's thought demonstrates that this scenario would require a wide range of actions beyond simply pricing carbon. In particular, it would require establishing huge scales of biomass production and supply chains as well as the physical and legal infrastructure needed to create a global industry in transporting and disposing of CO_2.

A different story again emerges from the third model in the RECIPE studies (IMACLIM), as illustrated in Figure 11.3. Its default run results in some of the highest costs (greatest

Box 11.1 Global energy economy modelling with induced innovation: the RECIPE results

Some useful insights about the consequences of expanding analysis to include at least some elements of Third Domain effects can be derived from the RECIPE project, in which a group of modellers systematically explored and compared their results around the implications of stabilising atmospheric CO_2 concentrations at 450ppm.

These three models generate a lot of common findings. They suggest that halving emissions from recent levels by 2050 (consistent with international goals outlined in Chapter 1) would cost around 1% of global GDP, averaged (with discounting) to 2050. They all showed a faster decarbonisation in electricity production compared to other sectors thanks to the expansion of electricity production from renewables. All three models concurred that the economic costs of electricity transformation would be a modest part of the total, because deep emission reductions in other sectors are more challenging and costly, particularly in transport.[a]

Despite these many common elements, the carbon price pathways differ radically across the three models (see Figure 11.2).[b] The two models that assume perfect foresight yield low carbon prices for the first decades since investments are driven by foresight of the rising cost, which by 2030 is in the range circa \$25–60/tCO_2 and rises sharply in the following decade. The carbon price then starts to diverge widely between the two models. In Model 1 ('WITCH') the carbon price rises to over \$1,000/tCO_2 in the second half of the century while in Model 2 ('ReMIND-R') it never rises significantly above \$100/tCO_2. This difference relates mainly to their relative degree of optimism regarding the scale and pervasiveness of low-carbon energy sources.

Model 1 ('WITCH') follows a path of sustained improvements in energy efficiency, with limited decarbonisation outside the power sector. The efficiency improvements *initially* help to contain energy costs and GDP losses. But the carbon price becomes very high in the second half of the century, both to contain rebound effects of increased energy consumption from efficiency improvements and to drive reductions in the non-electric category.[c]

In Model 2 ('REMIND') the carbon price remains low even at the end of the century because it predicts that there will be a huge scope for renewable energy, including large-scale biomass energy, at a modest cost; this echoes the rapid turnaround of the lower line in Figure 1.5 in Chapter 1. In particular, in later decades low-carbon biomass energy provides not only a 'renewable liquid fuel', but 'negative emissions' by combining biomass with carbon capture and storage. This enables the more 'difficult' sectors to continue without deeper structural changes, being either simply fuelled with biomass and/or with their emissions sucked back out of the atmosphere and then deposited underground.[d]

Model 3 ('IMACLIM') is completely different in its behaviour, not because of its balance between energy efficiency and decarbonisation (in which it lies between the other two), but because it does *not* assume perfect foresight: it assumes that investors take the existing carbon price at any point in time as the basis for the best guess of future prices. This lack of foresight is one reason why the carbon price has to rise sharply from the outset: 'strong lights are necessary to guide the partially sighted', and to persuade industry to invest away from fossil fuels. The model also has a direct representation of infrastructure, in particular for transport. In the absence of complementary policies, the impact of high carbon prices on an industry that is stuck in previous patterns of dependence on high carbon infrastructure initially results in economic losses of several percent of GDP.

Notes

a Decarbonising transport is relatively costly because of the absence of cheap low-carbon alternatives to current technologies and the existence of long-lived infrastructure. In ReMIND-R the energy demand for residential use is similar in climate policy and business-as-usual scenarios but this is due to an increasing share of biomass + CCS for producing liquid fuels and hydrogen. In all three models, residual emissions in climate mitigation scenarios are dominated by emissions from transport and other non-electric energy demand.

b Technically, these models contain a shadow cost of the carbon constraint, as the principal indicator of the effort required at any point in time.

c The version of the WITCH model used in the RECIPE studies illustrated in the text encompassed just two sectors, 'electricity' and 'others'. WITCH has since been expanded to model other sectors explicitly, but the key point remains: 'We are now running a version with a careful sectoral detail (agriculture forestry and land use, transportation, building in addition to the power sector). All sectors are carefully modelled including innovation. Results remain very similar and are explained by the small role of biomasses and negative emissions beyond 2050 ... We also find a limit in the development of solar technologies in the absence of an efficient and diffuse technology to store energy' (C. Carraro, personal communication, 16 September 2013).

d Such a large-scale use of 'biomass energy with carbon capture and storage' (BECCS) thus provides a kind of get-out-of-jail card as the atmospheric constraint tightens. The implications of this in terms of land use are not spelled out in the RECIPE studies (Luderer *et al.* 2011), but note brief discussion on resources in Chapter 3 (Popp *et al.* 2011) model the economic potential of bioenergy for climate change mitigation with special attention given to implications for the land system, one result of which is that 'forest protection combined with large-scale cultivation of dedicated bioenergy is likely to affect bioenergy potentials, but also to increase global food prices and increase water scarcity.' Bibas and Méjean (2013) argue for early climate action as a strategy to limit the risk of unsustainable biomass use.

economic losses) among the various models. However, the model allows strategic investment in transport and urban infrastructures to reduce mobility needs and urban sprawl and increase transport efficiency. This requires a wave of investment but these costs are recovered over subsequent decades because it reduces dependence on cars and oil (see Figure 11.3, line (b)). The other set of 'complementary policies' recycle the revenues of carbon prices into lower taxes on production (to offset the propagation of higher energy costs throughout industry) and to enhance the capacity of the labour force to move into new activities. The combined sets of measures then drastically reduce – and within two to three decades offset fully – the costs of transition to a low-carbon economy (line (c)).

Interpreted through the lens of the Three Domains of human decision-making, this confirms that achieving deep CO_2 reductions *purely* through carbon prices (Pillar II) would result in costs which almost certainly exceed Bashmakov's constant (Chapter 1), and not give enough scope or time for the adjustment processes that might help to contain energy bills (Chapters 4–6). The resulting price impacts would seem politically untenable according

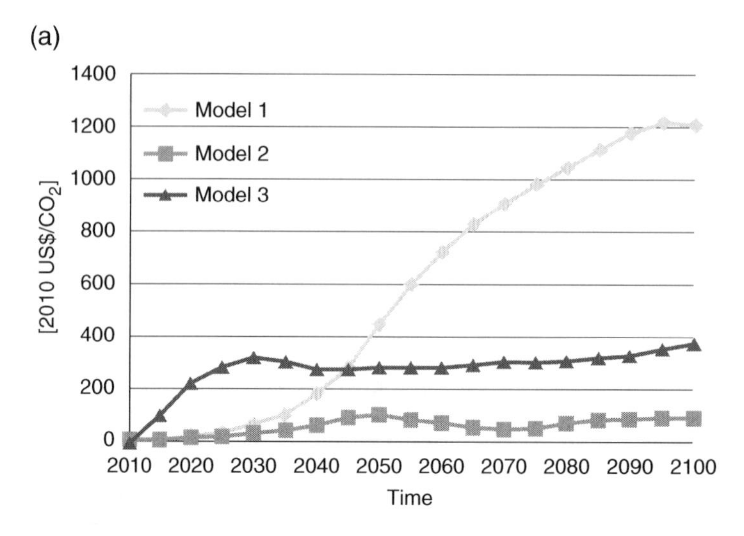

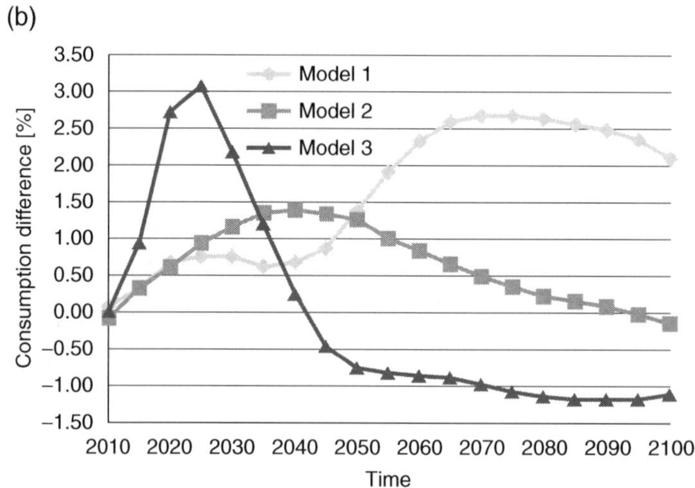

Figure 11.2 Global mitigation cost trends over the century in three different economic models: the RECIPE comparison

The graphs show the projected trend over the century of (a) the global carbon price and (b) the loss of global consumption relative to the 'baseline' projection, for scenarios in which the atmosphere is stabilised at a concentration of 450ppm. The three models all include induced innovation and the cost results do not take into account the (reduced) climate damages associated with the low emission paths.

Note: Model 1: WITCH; Model 2: REMIND-R; Model 3: IMACLIM-R.

Data source: RECIPE project (Edenhofer *et al.* 2009).

to our experience to date (Chapter 7) because of socially unacceptable distributional impacts (Chapter 8). These effects concentrate on a transitory phase, even if they are recoupled in the long term. Here lies the basic obstacle to ambitious carbon pricing: that the costs may be low in 2080 does not solve governments' current problems.

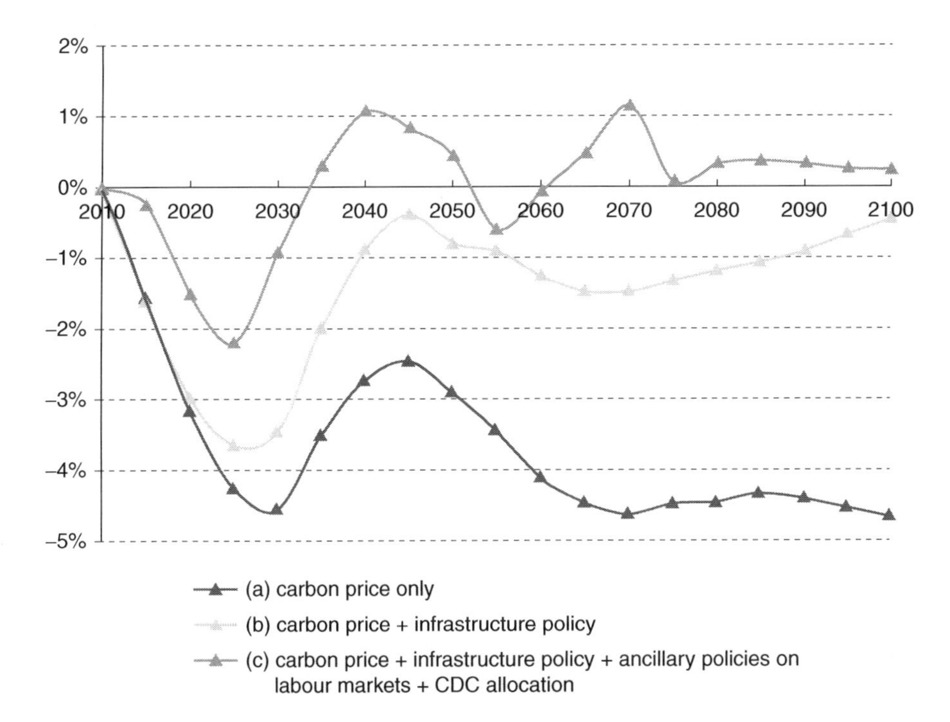

Figure 11.3 Global mitigation cost trends* over the century: dependence on infrastructure investment and other complementary policies in one model (IMACLIM)

* The graph shows the projected loss in GDP for scenarios in which atmospheric CO_2 is stabilised at a concentration of 450ppm, not taking account of (reduced) climate damages.

Note: Infrastructure policy refers to early action on spatial planning-related policies and transportation infrastructure that are governed by policy signals and price signals (real estate) other than energy prices. As explained in (Waisman *et al.* 2012 and 2013), these actions are represented through three main sets of assumptions: (i) a shift in the structure of investment in transportation infrastructure; (ii) the reallocation of transport infrastructure investment to low-carbon transportation modes (rail and water for freight transport, rail and non-motorised modes for passenger transport) and implicit urban forms (e.g. households' constrained mobility (commuting) declines from 50 per cent to 40 per cent of total mobility); and (iii) the organisation of logistics to decrease the transport intensity of production/distribution processes (a 1 per cent yearly decrease of the input–output coefficient between transport and production). Ancillary policies upgrade technical skills and labour markets to ease the sectoral shifts implied by decarbonisation), expressed through the level of the 'wage elasticity' (Guivarch *et al.* 2011). Under Common but Differentiated Convergence (CDC) allocation, the per capita emission allowances of Annex I countries converge to a low level within a convergence period, while the allowances of non-Annex I countries converge to the same level within the same period, but only starting when their per capita emissions are a certain percentage above global average (Höhne *et al.* 2006).

Source: Derived by the authors from source data in IMACLIM model runs. For assumptions and other results see Hourcade and Shukla (2013).

In contrast, the results portrayed in Figure 11.3 suggest that a more integrated strategy across the three pillars can contain the costs and potentially yield net economic benefits in the longer term. A carbon price remaining above \$200/tCO_2 suffices to deter a return to coal and the world is more insulated from rising oil prices; the *leverage* of the carbon price is amplified enormously by the other policies. In particular, the costs of the transport sector and its energy supply chain are reduced because of the infrastructural investments and technological evolution and by revenue recycling which sharpens the focus on carbon while containing overall industry cost impacts. The transition costs are far lower, the initial investment is recouped within a couple of decades and there is no net economic cost over the rest of the century; the world has transformed to the lower carbon trajectory sketched conceptually in the conclusion of the previous chapter (Figure 10.6). Contrary to what happens in stylised optimal growth models, carbon prices do not 'do the job alone' and prices and mitigation costs do not necessarily have to rise inexorably: efficiency, prices and strategic investment need to reach levels *sufficient* to transform the global energy system on to a new growth trajectory that would be intrinsically more compatible with global atmospheric constraints.

The RECIPE studies represent a small snapshot of modelling efforts around the world; far more will be available and reviewed through the IPCC Fifth Assessment in 2014. A notable trend is towards modelling combinations of different policy instruments, albeit with different approaches and for different reasons.[53] But some key messages seem robust. Solutions at moderate aggregate costs (to the point of insignificance) are possible, if there is good progress on energy efficiency and innovation and there is 'perfect foresight' of and confidence in rising carbon prices over decades. There are other ways too of reaching low carbon futures without significant long-run economic cost if governments act to shape infrastructure in appropriate ways. Translated into the investment language of Chapter 10, they all by implication involve redirecting trillions of dollars into the multiple sectoral transformations sketched in that chapter. And to achieve this, they all hinge upon putting enough effort in, early enough, to accelerate efficiency and change the course of investment and innovation.

There is of course infinite room for debate about the detailed nature of the transformations and the appropriate mix of options and efforts required to overcome the myopia of established investment and consumption patterns and the inertia of key infrastructure sectors. In the real world, these form questions that are far more interesting and relevant than simply debating how high the carbon price needs to go to squeeze carbon out of a system that is otherwise assumed to conform to the Second Domain ideal of being intrinsically and globally optimising.

11.5 *Changer de conversation*

Virtually all the global economic modelling of low-carbon trajectories has sought to quantify the costs of cutting emissions. To an economist – and indeed many others – it is *the* obvious question that must be answered. Yet, rather like the tortuous efforts to quantify the 'social cost of climate change' discussed in Chapter 1, after more than two decades of debate, the range of answers provided by economic analysts hardly seems to have narrowed: assuming optimal 'reference' systems and optimal responses to CO_2 constraints, the answer suggests that deep reductions in global CO_2 emissions would make GDP a few per cent lower than it otherwise would be, plus or minus a few percentage points.

While the cost of constraints seems like a natural economic focus, its oddity is suggested by the wide span of GDP 'results' displayed in Figure 11.4, drawn from the mainstream comparative studies of the Global Energy Modelling Forum. The figure plots the projected global GDP by mid-century across a wide range of economic models (convened in a global comparison by the Stanford Energy Modeling Forum) which seek to include some feedback between energy, GHG emissions and growth. This makes it plain that the main focus of attention – the cost of cutting emissions – is swamped by whatever drives economic growth: whether or not the world succeeds in cutting emissions in half by mid-century seems to have no discernible correlation with global GDP. The 'few per cent' cost of halving global CO_2

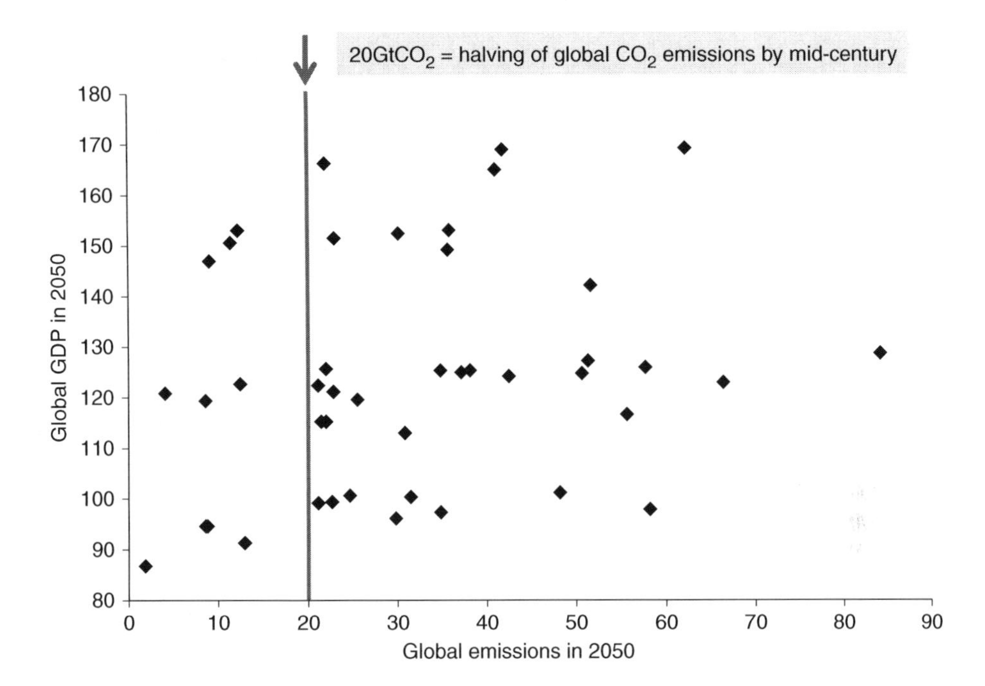

Figure 11.4 Gross world product in 2050 (excluding environmental benefits) from a wide range of models

Source: Energy Modeling Forum (EMF 2009).

emissions by mid-century is dwarfed in a context of economic growth which is also projected to yield GDP anywhere between US$100 and $170 trillion within the same timeframe.

The defence is to argue that of course a few per cent of GDP is still a relatively big number. This is true, but only in the context of another key assumption: that the energy transformation has no other implications for economic progress. In a world of purely Second Domain economics, everything can indeed be separately optimised: the economic policy need have nothing to do with energy policy, and CO_2 simply becomes an externality cost that has to be imposed on the system – this is the beauty of the theory. This explains the almost total separation of this huge literature on global energy modelling from wider economic debates. Correspondingly, the parallel assumption in mainstream economics underpins the strange absence of energy policy in most macroeconomic policy discussions, which until recently tended to focus on the monetary system more than any 'real' concepts such as infrastructure investments.

Yet what Figure 11.4 shows is that a new question has to be considered, namely what are the *interactions* between energy/climate policy and other dimensions of economic performance? Are there *spillovers* from energy/climate policy to macroeconomic policy and vice versa? If so, then the 'few per cent' of cost that even conventional models project, may be swamped by these interactions with macroeconomic factors, which determine whether we end up in the US$100 trillion or $170 trillion world.

Even without delving into the debates surrounding crowding out versus technology spillovers and theories of endogenous growth, this recalls the observation of Chapter 10 on why climate policy is difficult: it is not about incurring a modest cost, but about redirecting huge flows of capital investment from one activity to another. The estimated net cost is the difference between two much bigger numbers. Low-carbon scenarios involve redirecting a sizeable share of gross capital investment – the engine of economic growth. It is simply implausible to imagine that this can be done without any interaction with 'mainstream' economic performance.

The obvious antecedent is the discussion of CO_2 controls in Russia. Russian analysts ran models which showed the possible cost of CO_2 controls and the potential risk that the Kyoto targets would end up constraining the Russian economic recovery. From the outside, the proposition that one of the least efficient countries in the world would be impeded by a requirement to become more efficient seemed bizarre. But it was a natural consequence of separating the macroeconomic question – the inefficiency of the post-Soviet economic – from climate policy. The idea of energy efficiency as a key *plank* of Russian economic recovery took more than a decade to sink in and is still resisted in some quarters.

The equivalent implicit assumption built into the vast majority of global energy-economic models is that the rich countries of the OECD are much smarter – and that the economic ideas spread through globalisation mean that our economies are optimised and energy policy has nothing substantial to contribute. Looking at the world around us, external observers might be forgiven for doubting this bedrock assumption.

The discussion of models in the previous section underlined that if the core conventional (Second Domain) assumptions are modified by recognising their imperfect foresight and the inherent non-optimality of other processes, the result can be to widen the range of results further, to encompass either higher costs, or net benefits (before taking account of climate damages). Ultimately, this confirms our earlier observation about classical economic (Solow-related) growth models: there is no reason to assume that redirecting technical change must reduce long-term growth rates, because productivity growth is not driven by purely economic optimising mechanisms. Hence the potential for innovation to generate diverse future economic structures with similar costs and vastly different carbon contents (Chapter 10). We can construct detailed models in which low-carbon technologies become cheaper and consumption patterns dematerialise, and the economic contribution of these activities means there is no net cost difference.

However, most models are of little help in assessing the main *drivers* of the direction, the means of changing course to either improve economic progress (which in most models arises through exogenous assumptions) or its decarbonisation (which in most models is determined purely in terms of a carbon price) and to face the costs of any transition.

The underlying reason is quite simple: the challenge involves transformational changes, which implies substantial use of mechanisms and interactions across all three domains, which, for the reasons explained, we have no reliable way of quantifying. The changes inexorably entwine the 'dark matter' of economic progress. They depend on wider social choice, their projection is a matter of assumption and the actual economic impact will hinge on conjoined economic and energy policy.

Ignoring how the Second Domain is embedded in the First and Third Domains of human activity has thus stuck us in the wrong conversation. First and Third Domain processes are neither inherently optimising nor inherently separable – new technologies do not change one thing at a time (a low-carbon technology is not the same as high-carbon technology but with lower emissions); major new industry sectors do arise leaving everything else unchanged. Similarly, wider economic policy unquestionably has implications for energy and emissions – as illustrated, for example, by the impact of electricity liberalisation on long-term R&D noted in Chapters 9 and 10.

The interesting issue is thus not how much carbon controls may cost within an energy sector divorced from everything else. The bigger question is how energy transformation can be designed to fit within and contribute to policies for continued economic progress, and why and under what circumstances low-carbon development becomes an attractive and enduring proposition.

To answer that question, we need to step back further and look at the larger historical patterns of economic development.

11.6 Back to the future: industrial renaissance

The need for a 'new conversation' in part explains the appeal of 'Green Growth' concepts, which are increasingly called for, not least by many international economic organisations, but build on a much wider and deeper heritage.[54]

In a purely Second Domain world, resting on the principles that markets make best use of all available resources and extended by the idea that this paradigm should also produce optimal levels of innovation, no such construction is possible. From such sceptical perspectives, merely invoking the words Green Growth can raise fears that efforts to induce low-carbon innovation may raise energy costs and crowd out potential gains from other 'non-green' investments. Some developing countries have even portrayed it as a cloak for 'de-growth' as a way to manage finite planetary resources, thus locking in global inequalities. The term has thus already acquired some unfortunate connotations in the international political sphere, which obscure the real issues.

The first base is thus to reiterate the implications of all that has gone before in this book: the nature of economic forces across all three domains, their interactions and particularly how they shape the forces of innovation that imply a wide range of future energy scenarios to be possible at similar costs. Given the conjoined threats from instability in global fossil fuel markets and climate change, it makes sense to orient economic progress towards lower-carbon trajectories. However, as already noted, this requires effort and may only be realisable with greater investment by the present generation to force a change in course. This does raise distributional issues – between present generations and the next and within present generations. This makes it a lot harder, with continued potential to fuel the rancid global politics of global environmental action. It thus remains important to dig deeper.

Innovation waves and growth cycles

To dig deeper, it is useful to start by looking further back, to the time before the long cycle of postwar economic growth. Economics was in fact not always tantalised by the idea of stabilised, optimal growth. Prior to the 1960s, much of the interest was to explain the evident cycles and periodic crises in economic history.[55] Economists like Ricardo, Marx and Schumpeter differed in their interpretations and political ideologies but explained this history through the interplays between institutions and technical change. Schumpeter, for example, launched the notion of *creative destruction* to explain the dynamics of capitalism through the role of entrepreneurs whose innovations disrupt the existing order.

A stylised vision of the links between technical paradigms and growth cycles in history is sketched in Figure 11.5. After the first phase of mechanisation in the late eighteenth century, the Industrial Revolution was based upon the mastery of steam power and the deployment of railways; these two factors opened new economic frontiers with access to cheap coal and iron mines and a first phase of globalisation in manufactured (textiles) and agricultural goods. Most importantly, these goods could all now be transported over long distances.

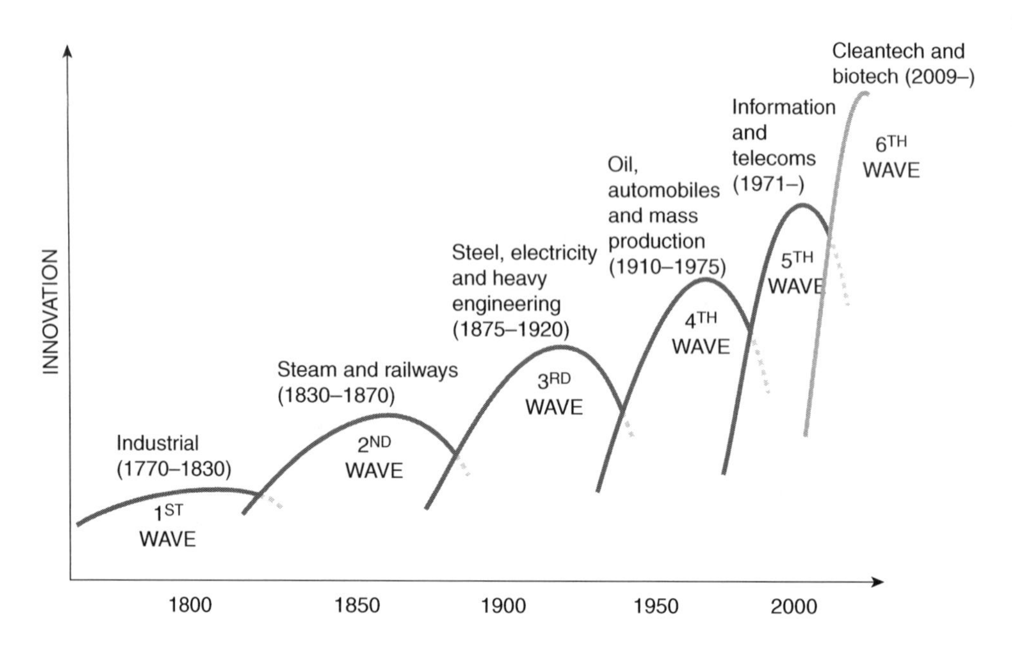

Figure 11.5 Waves of innovation

Adapted by the authors from Stern (2012) and Perez (2002).

The mastery of electricity in the late nineteenth century launched a new wave of growth thanks to unprecedented access to power for heavy industry and better access to education thanks in part to electric lighting: on this was based the belle époque, before its brutal termination by the First World War.[56]

The core economic engine of the twentieth century, which became the dominant economic force after the First World War up to the middle of the 1970s, drew on mass production with economies of scale, typically in the automobile industry and electric appliances, with oil as the rising energy resource. After inhibition by the World Wars, this powered an unprecedented wave of growth, the 'Thirty Glorious Years' leading up to the oil shocks of the 1970s.[57]

It is probably no accident that the mathematical theories of stabilised economic growth took hold during this period: with steady growth and the gradual extension of the underlying economic model beyond the rich countries of the OECD, they were products of their time. The fact that the fundamental model seemed to survive the oil shocks and reassert itself during the 1980s only reinforced this belief. The collapse of the Soviet Union, whose economic paradigm was still to a greater extent based on the earlier industrial model of electrification and heavy engineering (which still dominated at the time of the Russian revolution in 1919), further amplified confidence in the economic model of fossil-fuel based mass production led by consumer markets.

The model was also aided for the next 20 years by the additional wave of growth spurred by developments in information technologies and telecoms, though the overall economic contribution of the IT revolution as a fuel for growth seems already to be waning.[58]

Yet as noted, the dominant growth theory that embodied the post-Second World War age itself contained the observation that around half of economic growth could only be explained with reference to 'innovation' and in and of itself carried no structural analysis of what drives such innovation. The notion of successive waves driven by different socio-technical ages is thus perfectly consistent with the basic formulation, if one accepts that innovation has various causes associated with the dynamics of such transformations.

This abbreviated economic history thus emphasises the importance of the co-evolution of technical and institutional changes which together involve large-scale transformations in industries, lifestyles and the spatial distribution of activities. The economic frontier in the US would not have moved so quickly to the West without the railway revolution; education levels would not have risen at the same pace without access to light and the radio; urban sprawl or mass tourism would not have taken the same form without the generalised use of motor cars.

It follows from this view that enduring recessions in part are the 'manifestation of a "mismatch" between the socio-institutional frameworks and the technico-economic sphere' (Perez 1985). New technical paradigms cannot be deployed without appropriate institutions (property rights, political structures, social and business networks) and infrastructure. This is one of the key reasons why China missed the same Industrial Revolution launched in Europe in the eighteenth century (although both regions had comparable technical capabilities and income levels in the seventeenth century). It is also the reason behind China's impressive economic rise after Deng Xiaoping's reforms in the 1980s.[59]

This raises an obvious question: whether, in addition to the financial origins of the debt crisis itself, the enduring recession – and apparent global spread of the current economic malaise – may itself be a symptom of the exhaustion of the prevailing paradigm and a resulting mismatch, at the world scale, between the emerging technico-economic potentials and the existing institutions of economic governance.

To do more than speculate, we thus first need to correctly understand the nature of the economic engine that drove the productivity gains of the age.

Why the end of the Thirty (plus twenty) Glorious Years?

The wave of technological transformation dominated by oil, automobiles and mass production reached its zenith in the 'Thirty Glorious Years' after the Second World War. Centralised and more automated production techniques enabled huge economies of scale, both in intermediary industries and in the mass production of final goods. Particularly after the war, access to the vast pools of cheap Middle East oil also cut transportation costs and enabled industrialising economies to reach distant raw materials (energy, steel, non-ferrous goods) elsewhere. Institutional changes accompanied these advances to create virtuous economic cycles. They were deployed in different forms and at different speeds but shared these common attributes.[60]

The iconic example of such institutional changes is Henry Ford's $5 a day wage, introduced in his factories in 1914. Raising the minimum daily pay from $2.34 to $5 added to production costs but also fuelled the demand for cars. This enabled industry to expand production, reap huge economies of scale and thus reduce production costs so that, in combination, the elite's toy at the beginning of the century started to become a mass-appeal product.

This is why the growth regime is sometimes known as Fordism.[61] It provided the technical basis for the economic triumph of Roosevelt's New Deal in the 1930s, which dug the US – and subsequently the world – out of the massive depression that followed the 1929 financial crash. The New Deal included a huge public investment in infrastructures, creating both the physical and financial basis for the expansion of automobile, domestic appliance and material industries, drawing on strong US coal and oil production. The economic model came to command a social and political consensus, albeit in different forms in different countries, typically involving negotiation between trade unions and employers with a focus on the maximisation of consumption.[62]

This era saw the reign of Keynesian economics. The implicit deal was that wages increased along with overall productivity in a virtuous economic cycle, and that public spending could boost expenditure to compensate for economic downturns. But instead, the globalisation of markets progressively undermined this virtuous cycle, imposing competitiveness constraints on wages and on social security expenditures. In the late 1960s there was the first alert about the exhaustion of cheap oil in the US and growing dependence on the Middle East. The 1970s oil shocks then revealed energy as the Achilles heel of this growth regime.

Economic globalisation was both the product of Fordism and its other major threat. Its technological pattern was based on economies of scale which, increasingly, could only be realised when planning for international markets. For instance, middle-sized countries are large enough for the steel industry to invest in plants of 1–3 Mt per year, but a steel plant of 15–20 Mt per year makes sense only for international markets. It also became more efficient to standardise the components of a car in order to use them in many models while adapting the end product to the country-specific consumers' preferences. An increasing share of industrial sectors became exposed to international competition and these sectors responded by outsourcing parts of their production or creating global companies (e.g. Renault-Nissan or EADS-Airbus) to dilute investment risks in several markets.

The policies that had supported the 'Thirty Glorious Years' thus increasingly bumped into an invisible wall, because part of the increase of final demand generated by higher wages, public spending or monetary easing went into higher imports instead of higher domestic production.[63]

Globally, after the impact of the oil shocks abated, growth resumed apace through two main channels: adoption of the model by the emerging economies playing catch-up in terms of industrial production and drawing initially upon the returns sought by Western investors, and the quest of Western consumers for cheap goods. But the famous aphorism, 'what is good for General Motors is good for the US', no longer held true.[64]

In the 1990s it was explained that the difficulties associated with globalisation were without consequence and that, indeed, further liberalisation and relaxation of banking regulation was necessary to free a new growth cycle within the OECD based on information technology (IT) in a context of trade liberalisation. Along with globalisation, the IT revolution thus contributed to this extension of the Fordist era but has not fundamentally changed the dynamics. Indeed, as observed (see note 58) and displayed in Figure 11.5, its economic impact appears to have been quite short-lived. Instead, the rise of debt in the West fuelled an illusion of continued real growth.

Many accounts of the enduring economic malaise since 2008 have focused on such distributional effects, along with the misplaced financial wizardry which for two more decades enabled the West to continue the fantasy of ever-rising consumption divorced from material content.[65] The increase of real estate values, which occurred first and foremost in the USA and was subsequently imitated in several European countries, owed nothing to chance; it was allowed for, if not encouraged, by explicit policies to fuel consumption by an illusion of continued wealth.[66]

We return to the problems of debt and recession and their relevance to energy and climate policy in the final section of this chapter. For the primary need is to understand another, more fundamental reason for economic stagnation which is becoming a more global phenomenon, namely the fact that the socio-technical basis behind the virtuous cycle of Fordism itself began to run out of steam.

This is for three main reasons:

- *Saturating economies of scale.* The easiest economies of scale in intermediary industries like steel, in 'semi-durable' products (cars, electric appliances) and in the food industry have been exhausted. Remaining economies of scale are increasingly offset by the management costs of ever more complex systems and the investment risks due to the increasing time-lag between investment and its payback. Moreover the middle classes, once equipped with cars and semi-durable goods by mass production, redirected much of their demand to high-quality and diversified products and services less likely to bring economies of scale, like food and tourism. Productivity built on economies of scale cannot continue in perpetuity.
- *Delayed human and environmental consequences of consumption patterns.* The unintended environmental and social externalities of Fordist growth patterns had already started to become apparent in the 1960s.[67] As outlined in Chapter 1 (in 'What comes out' under section 1.2), more prosperous populations met with a declining environment. Environmental improvement was essential and beneficial but was not free, particularly where it involved 'end-of-pipe' solutions like desulphurisation. Mass food production is a typical example whereby, after the dramatic improvements of the 'Green revolution', the cumulative impact of pesticides, chemical fertilisers and energy use have had an increasingly counterproductive impact on agricultural productivity, and now generate instead a demand for higher food quality and security.[68]
- *Energy vulnerability.* The most obvious examples of this are the two oil shocks in the 1970s and the resulting economic crises (see Chapter 1, notes 7 and 8). During the 'Thirty Glorious Years', economies became more dependent on oil due to urban sprawl and just-in-time production processes which increased transportation needs. In the US, low and medium income households spent on average 26–29 per cent of their budget on transportation and an equivalent amount on housing. A further average of between 55 and 62 per cent of their income was spent for the new 'basic needs' – a personal vulnerability which also reduced their purchasing power for other products and hence further weakened the Fordist cycle.[69]

The fact that in the classical theories of economic growth, both environment and energy (along with other natural resources), are 'intermediate products' that cancel out in the growth equations has helped to blind economics to their central role. The complexity surrounding the nature and causes of productivity growth, with its attribution to the dark matter of 'innovation' as a proxy, may also have helped to obscure the central role of the ultimately limited engine of Fordist economies of scale, completing the disjuncture between the theories and impending realities. The expectation of unlimited economic growth is meeting the physical realities not just of a finite environment and resources, but of one of the basic 'fuels' that has sustained the last half century – the waning economies of scale in the dominant fossil-fuel based industries of the twentieth century.[70]

Green Growth, Part I: the technico-economic characteristics of low-carbon development

From this perspective, we can see a more alluring case for 'Green Growth', particularly associated with energy and decarbonisation. Growth demands innovation; Green Growth gives it direction and purpose. Decarbonisation helps to reduce the exposure of economies to the instabilities of fossil fuel markets. The combination requires us to accelerate innovation in what historically have been some of the least innovative sectors of the global economy. It also implicitly requires the differentiation of products – which as recognised in the mainstream literature above can be an important stimulation to innovation and growth – on the basis of their environmental footprint (Chapter 9). It offers a way to move beyond the industrial paradigm that seems to be reaching its point of exhaustion.[71]

Indeed at a pure technico-economic level (to use Perez's term), there are some clear economic advantages to a low-carbon transition. The main sectors concerned are infrastructures (transportation, construction, energy), material transformation, electric equipment and agro-industry, which collectively represent the majority share of investments. In terms of the 'elephant and rabbit stew' metaphor used to justify the assumption that the energy and climate responses would be peripheral to wider economic activity, the above-mentioned sectors are pieces of rabbit. The metaphor ignored the critical role of the energy-consuming sectors in capital formation.

From this perspective, the relative lack of interest in climate change displayed by much of the mainstream economics community reflects the abstraction of finance and GDP from physical realities; climate change offers an overarching context that could also help the field of economics itself reconnect with the physical basis of economic systems.

Because these sectors encompass both consumption and production goods, they hold the potential to generate a virtuous circle between demand (efficient buildings, electric vehicles, smart use of electricity, food diets) and supply (smart grids, new materials and infrastructures). Contrary to a naive vision of dematerialisation, information technologies may contribute to a new growth wave primarily through their capacity to facilitate the deployment of the 'low-carbon paradigm' in all these infrastructure sectors.

In the light of global imbalances, a related advantage of the low-carbon transition is that a good part of the concerned sectors are oriented towards domestic markets (buildings, transportation, domestic renewables).[72] It would thus lead to less export-oriented development strategies and might help to calm down wage competition between export-oriented industries and reduce tensions on currency manipulation. Chapter 3 (e.g. Figure 3.6) underlined the potential for reuse and recycling of materials which gain economic value from waste streams, reflecting emerging thinking in the business community about the 'Circular Economy'.[73]

Investing in sectors such as buildings, transportation and energy also contributes to a more inclusive growth. Not surprisingly, the trade unions that have accepted the need for change (as opposed to resisting it through the defence of high carbon industries) have now focused on articulating a 'Just Transition'.[74]

Note that this is more subtle than it seems and has to be based on understanding both the current malaise and the full span of the Three Domains. For there are two facts we know for sure. Firstly, from the world around us, policy-makers are struggling to find effective economic policies. And secondly, based on everything mentioned so far in this book, simply hitting societies with a high carbon price alone – divorced from the wider economic and social context – is neither a politically credible nor necessarily even an effective recipe. It is not politically credible for all the reasons articulated in Chapters 1 and 2 concerning the psychological distance of climate change and the impossibility of gaining consensus about 'the social cost of carbon'. And *on its own* it is not effective either, because it raises the distributional quagmire examined in Chapter 8 without addressing the key mechanisms that could help to tackle both distributional and aggregate cost impacts through the First and Third Domains.

To stress the point further: innovation involves change, and change can arise from a combination of sticks and carrots. If price is the stick for carbon-intensive activities – and we reiterate its central importance in the next section – the gains from the transformation will require carrots as well across the wide span of sectors involved.[75] Thus the argument starts to bring us round full circle, to emphasise the importance of harnessing all Three Domains, utilising all Three Pillars of policy. Attention needs to focus on the suite of economic and institutional changes which could trigger a new wave of long-term economic progress based on a transition towards high efficiency and low-carbon societies.[76]

As in the past, this virtuous circle will not be launched without reducing the 'mismatch' between institutions and the key characteristics of the new technological paradigm. This demands political choices which will not be made unless low-carbon policies can simultaneously respond to the immediate economic and financial tensions. This, in turn, brings us to our final set of linkages.

11.7 Lifting depression: finance and investment

The culmination of this argument leads, not surprisingly, to questions of finance, investment and value – this time in the global macroeconomy. Yet it shares a common root with our starting point in the First Domain on behavioural economics: the need to understand the realities of human and institutional behaviours rather than caricatures of perfectly optimising systems.

It would indeed take a brave person to argue that the global economy is in a state of perfection – that it is operating at the 'best practice frontier' (Chapter 2). Hence we return to the old dispute about 'negative costs options' and 'no-regret policies', but from a wider perspective. In the energy world, historically these debates focused on purely technical choices – 'bottom-up' assessments describe technical potentials but did not explain why cost-saving options are not sufficiently used already or what would be required to change this. Yet the opposing assumptions in macroeconomic modelling are no more helpful – we need to understand what to do in a world which is not governed by Second Domain assumptions but rather run by decision-makers with imperfect foresight in a turbulent age under numerous constraints and pressures. Most of the energy-climate models neglect much that matters – not only First Domain realities, but the roles of infrastructure, the dynamics of urban forms, the missing links in the innovation chain, tax distortions and the real behaviour of labour and capital markets.

Hence part of the discussion about 'negative costs' must concern macroeconomic policy. Just as choices of techniques and energy markets are shaped by micro behavioural and organisational realities, so too are labour markets and investment behaviours constrained by existing structures, social resistance to change and numerous distortions in domestic and international markets. These frictions have macroeconomic implications. Economies are not in a stabilised pathway; recessions and unemployment are symptoms of adjustment failures. These failures occur when industry lacks the skills, infrastructure or supply chains to adapt or, for example, when the innovation chain fails (see Chapter 9), in the case of lock-in effects (see Chapter 10) and crucially when savers refrain from investing.

Such undoubted imperfections can of course cut both ways. They can impede a low-carbon transition or mean its costs might be higher than suggested in models of perfection.

For example, one response to the credit crunch was huge publicly funded stimulus programmes, and some countries – most notably Korea – particularly focused their programmes on green technologies and industries. This made eminent sense, but many countries could not now repeat it due to their sheer level of debt: if the West is too debt-ridden to afford continued stimulus, then adding a climate label to that stimulus does not help. This is the limit of pure Keynesian stimulation in the modern world.

As indicated, the decarbonisation does have a number of positive macroeconomic features and models closer to real-world conditions can shed some light on these features, with due caution. One example in Europe has been a major study commissioned by the EC Energy Directorate to examine the potential impacts of the EU energy decarbonisation roadmap (to deep CO_2 reductions in 2050) on GDP and employment. A wide diversity of participating institutions, and two radically different economic models, did manage to agree that the GDP impacts would be small but disagreed on whether they would be positive or negative. They did agree, however, that the additional domestic investment involved would lead to a net increase in jobs and reduced unemployment.[77]

There is one last stage in the understanding of the roots of the potential for decarbonisation to simultaneously address the current macroeconomic malaise, namely understanding finance.

Buridan's donkey

We live in an age in which unprecedented debt levels coexist with very high rates of savings (45 per cent of GDP in China, pension and insurance funds in the 'advanced' economies and sovereign wealth funds mainly in major oil exporters). In reality of course these are related: government debt is owed to financiers who hold huge pools of capital. The savings glut results in part from the fact that, during the last twenty years, the business environment led more and more investors to prioritise short-term shareholder value rather than the longer-term value of the firm.[78] They indulged in practices (like leveraged buy-out) utilising borrowing facilities at the same time as innovative market finance instruments which had the attractive property – to the markets – of undervaluing risks in ways which were opaque to all. The result is that capital markets seeking risk-adjusted returns as high as 10 to 15 per cent stopped investing in industry and used savings to pursue capital gains, for example through real-estate bubbles. With the bubbles burst and financial controls necessarily tightening, the capital is unsure where to go.

The situation is akin to the legend of *Buridan's donkey* which died hesitating between oats on one side and the pail of water on the other.[79] This was a deliberate caricature of a theologian who argued that decisions should be delayed until all the facts are known and weighed. Savers do not know where to really invest in industry. Here lies the operational link between debt policy, growth policy and climate policy: the need to awaken Buridan's donkey out of its hypnosis. Reducing investment risk associated with low-carbon projects would attract investment, contribute to clearing the savings glut and address one structural cause of the financial crisis.

These are of course 'easier' options, notably to ease both monetary and environmental policy. This would essentially favour a revival of a Keynesian compact bankrolled by the hope of a new era of energy abundance based on shale gas and oil. This amounts to trying again to revive and extend the old paradigm of a growth model based on unsustainable consumption, fuelled by financial and natural debt, with numerous other potential problems to accumulate over time.[80]

For the reasons articulated in Chapter 10, this is the easy short-term path: to ease further the flow of capital into the traditional fossil fuel-related businesses to extend the fossil fuel frontier. But it is a zero sum game: competition to attract heavy industry with low energy prices, and competition between the short-run interests of present structures against the long-run global interest in a low-carbon economy and stable climate. The shale gas revolution in this context is a double-edged sword: in the right context and properly governed, it can help to displace coal and bring down national emissions and energy prices, but its own emissions and geological characteristics set limits on this contribution and suggest it to be partly a palliative or transitional option.

It is already apparent that shale gas does not mean the end of instability (or of high prices) in global fossil fuel markets, and the concluding chapter outlines why Europe, for example, cannot plausibly outcompete the US in this game. Without strong governance it risks being a distraction from the core need to move to a new growth paradigm based on sustainable finance for low-carbon innovation across multiple sectors. And ultimately, governments cannot plausibly protect fossil fuel investments (and investors) against the risks implied by climate change.

Green Growth, Part II: finance for low-carbon development

The key to achieving economic goals in ways compatible with environmental needs lies in attracting the huge pools of institutional capital – such as pension and insurance funds and sovereign wealth funds – to low-carbon investment. The volumes are huge – tens of trillions of dollars. At present, such finance is earning paltry rates of return, typically 2 per cent or less. This indicates its extreme aversion to risk – but also its huge waste of resources. The amount of additional investment required to decarbonise the development of energy systems globally (c. 0.5-1€trillion/yr; see Chapter 10, and Appendix) is a lot of money, yet represents just a small fraction of this surplus capital.

At a time of such unprecedentedly low interest rates, infrastructure investment can be an extremely attractive proposition. As noted in Chapter 10, the low-carbon transition heavily involves such infrastructure, much of it in the form of capital-intensive investments that yield a long-term stream of value – as with enduring reduced energy consumption from highly insulated buildings, high-speed railways or very low operating cost energy from renewable sources. As illustrated in Chapter 3, the net cost of renewables at low interest rates can be highly attractive. The deterrent is risk – partly technological, but much of it also political.[81]

Conceptually, the fundamental point is that a constraint can reduce risk and enhance value. Gold is more valuable than water because it is so scarce, but the desire for gold is a purely human construct. Energy is an essential need and the physical constraints on fossil fuel resources and on the atmosphere mean that the value of low-carbon energy must rise over time. Science can thus form the bedrock for a new source of value in the global economy – providing it is translated into a new certainty that attracts long-term finance into real investment.

There are many approaches to doing this. One strong element could be for a sufficiently large group of countries to agree a common minimum 'social cost of carbon', set to rise over time until emissions decline sufficiently. Obvious mechanisms for implementing this would include either carbon taxes (Chapter 6) or a floor price in emission trading systems (Chapter 7).

This would have a direct value, but the effectiveness of low-carbon value for the purposes of attracting investment could be enhanced through various other mechanisms. It is beyond the scope of this book to discuss the technical options in sufficient detail, which span from 'green investment banks' to project-specific low-carbon investment bonds and many other related proposals. Banks and institutional funds could in parallel issue 'carbon' financial products, aimed at attracting domestic savers and institutional investors, with guaranteed return on investment slightly above that of usual safe deposits. Carbon-based drawing rights from the central banks of the IMF, transformed into carbon certificates for helping development and investment banks to provide low-risk loans, could even culminate in the gradual emergence of international financial products based on long-term carbon value, helping to align the financial system with the real physical needs for long-term, sustainable investment.[82]

We conclude with one final paradox. We have emphasised that the transition is not easy and not costless, precisely because it requires substantial effort and investment to pull us away from the default course: the modelling results presented in the middle of this chapter suggest the required resources could easily amount to 1 per cent of GDP or more for a couple of decades. The natural assumption would be that it is harder to secure this in the economies that are in greatest economic difficulty, such as Europe. The paradox is that it is precisely those countries gripped by continuing recession, unemployment and financial uncertainty which have most unused resources of labour and finance. It is also Europe that has succeeded in establishing a legal instrument to price carbon – which is itself one of the most difficult steps. If that can be turned into a credible platform for low-carbon investment, complemented with enhanced policies on the First and Third Pillars, Europe's wasted resources could at last be mobilised in the conjoined pursuit of economic recovery, financial stabilisation and decarbonisation; (we return to consider some specific policy options in the concluding chapter). It may be precisely those regions most deeply mired in the recessionary consequences of the previous age which are best placed to launch the next.[83]

11.8 Conclusion

Much of the economic effort directed at analysing low-carbon futures has focused on estimating the global costs of meeting the atmospheric 'carbon constraint'. This, it is assumed, can then inform a 'cost-benefit' discourse about whether (or at what point) the costs exceed the benefits. In Chapter 1 we demonstrated and explained that, in actual fact, the economics community has been wholly unable to provide a useful, monetised estimate of the value of reducing climate change, and Chapter 2 hypothesised that the 'cost-benefit' framing might actually be asking the wrong question.

We are now, at the end of Pillar III, able to complete the picture by demonstrating why economics cannot provide a meaningful estimate of the long-term costs of cutting emissions either, and why understanding this fact enables us to start a different and more useful conversation.

Again, multiple lines of evidence have informed this conclusion. It is in fact implicit in some of the technological discussion of innovation and learning curves in Chapter 9 and the implications for possible sectoral and energy system transformations in Chapter 10. This culminated in Figure 10.6, which summarised the fact that global models of energy systems which include induced innovation and fossil fuel resource profiles are capable of generating a huge range of future emissions at similar costs.

This was paralleled in efforts by the IPCC to generate global emission scenarios, which resulted in sets of 'without policy' scenarios which themselves differed enormously, based on varied assumptions about development choices and technological evolution.

Such results have been resisted by many economists on the assumption that the economy should be designed as an optimising system and that adding a constraint (like CO_2) must inevitably therefore incur economic costs. This chapter has shown the error in this assumption.

Nothing in the archetypal model of economic growth theory says that an economic system will deliver 'optimal' growth: indeed, aligning the theory with observation suggests that around half of the forces driving economic growth have to be explained in other ways. Much of this 'residual' corresponds to processes in the First and Third Domains, which we have shown to have no inherent tendency to optimality. Indeed, the wider state of the world economy makes it hard to argue that even Second Domain economics is delivering close to what classical theory says it should.

Imperfection is reality. The question is whether and how smart policy on climate change, embedded in this context, can aid better outcomes. The chapters in Pillar I demonstrated how this could be true for energy efficiency, and Pillar II outlined conditions in which it could be true even within Second Domain conditions arising, for example, from the prevalence of perverse subsidies and the instabilities in commodity cycles and financial systems. Pillar III has underlined the contribution of innovation in technologies and systems, working from the characteristics of energy innovation (Chapter 9), through sectoral prospects of transformation (Chapter 10) and culminating in the macroeconomic implications, through both theories of growth and the modern realities of depression and instability (Chapter 11). All this underlines decarbonisation as simultaneously a cost and an opportunity.

Finally, we reiterate the cautionary note from the opening of this chapter: it has focused on growth and GDP, not welfare per se. The man usually credited with inventing the GDP indicator himself observed that 'the welfare of a nation can scarcely be inferred from a measurement of national income' (Kuznets 1934). The wider literature on global development and environment encompasses both 'economic' and 'social' narratives, the latter emphasising more directly the relationships between social, energy and environmental systems, and the quality of life they imply.[84] This chapter has confined itself to traditional economic metrics; a fuller analysis would encompass broader measures of welfare.[85]

Yet the underlying conclusion remains: the future is open, not predestined. Our economic wealth, and wider welfare, will be constrained not by our finite atmosphere, but by the choices we make in adding that reality to the innumerable others which shape the evolution of our economies and societies.

Notes

1 Valuable contributions from Aurélie Méjean (Cired) and Simone Cooper (Cambridge University) throughout this chapter are gratefully acknowledged.
2 BBC Poll, July 2013.
3 The most high-level assessment of the inadequacies of GDP as a guide for policy is probably that for the Sarkozy Commission, which brought together a number of the world's leading economists to study the issue. Their final report (Stiglitz *et al.* 2009) noted 'it is of particular concern [...] when narrow measures of market performance are confused with broader measures of welfare [...] what we measure affects what we do; and if our measurements are flawed, decisions may be distorted. Policies should be aimed at increasing societal welfare, not GDP.' In *Prosperity without Growth*, Jackson (2011) shows evidence that increasing consumption and material growth does not necessarily contribute to, and may even impede, human well-being in advanced economies and calls for the redefinition of prosperity. These issues of changes in metrics and human aspirations are important and acknowledged by many leading economists, among others, but are beyond the scope of this chapter. In relation to this book, one of the central limitations of GDP is its assumption that markets reflect valid expressions of value. But clearly they cannot, for example, reflect the value that future generations would place on our actions – including the costs of depleting finite natural assets or of leaving a legacy of pollution for them to deal with. This also, fundamentally, explains why most economists do not accept observed market discount rates as a legitimate measure of the social rate of time preference, as discussed in Chapter 1.

4 See Chapter 9, section 9.10, 'Porter's kick'.
5 More information on the so-called bottom-up/top-down debates is given in the Third and Fourth Reports of the Intergovernmental Panel on Climate Change (IPCC 2001 and IPCC 2007 – Working Group III Reports). Major papers which sought to bridge, or at least explain, the gap included Grubb *et al.* (1993) and Hourcade *et al.* (2006). The intellectual journey was endearingly captured in the title of a paper by one analyst, 'Been top-down so long it looks like bottom-up to me' (Huntingdon 1994).
6 The elephant and rabbit stew metaphor was introduced by Hogan and Manne (1977).
7 See also Ghersi and Hourcade (2006).
8 See, for example, the observation by Professor Stavins (Chapter 1, note 44) about the historical gulf between scientific and economic views of the climate problem. Though this gulf has narrowed over time, a parallel disconnect remains with regard to views on the nature of, potential for and likely cost of solutions to the problems, which this chapter seeks to explain.
9 Solow later described growth models as 'a framework within which one can seriously discuss macroeconomic policies that not only achieve and maintain full employment but also make a deliberate choice between current consumption and current investment, and therefore between current consumption and future consumption' (Solow 1987). In the context of debates about the non-optimality of economic systems and Keynesian policies, he later agreed: 'At short-term scales, I think something sort of "Keynesian" is a good approximation, and surely better than anything straight "neoclassical". At very long time scales, the interesting questions are best studied in a neoclassical framework and attention to the Keynesian side of things would be a minor distraction [...]. At a five to ten year time scale, we have to piece things together as best as we can, and look for a hybrid model that will do the job' (Solow 2000). The fact that 'five to ten years' is intermediate is suggestive that the 'very long time scales' of growth models in fact refer to maybe 15–30-year processes – long in terms of any formal economic theory but short in terms of the issues of energy and the global environment.
10 Formally therefore, the real equilibrium equation of economic growth could be written as GDP + IC = C + I + IC, where IC is all the intermediate consumption (and production). The sum of the products sold in an economy is GDP + IC: when I buy a tomato, I pay the cost of the energy, water and transportation to produce it and not only the wage and profits of the peasant and of the merchants. In an aggregated model, intermediate goods are a homogenous product like a jelly (Robinson and Naqvi 1967), which consequently can be cut from both sides of the equation. However, it may no longer be possible in a multi-sectoral model with many different discrete products which cannot all be aggregated and substituted. The classical treatment is arithmetically right only insofar as all the intermediate products can be aggregated in purely monetary terms, and thus tends to provide a somewhat incorporeal vision of the growth engine in which the main physical flows (such as energy and materials) throughout the economic system tend to become invisible.
11 Labour depends on the size of the population which cannot be explained by consumption or investment levels. In economic terms, K is endogenous – affected by economic decisions – while L is mainly exogenous – determined outside the sphere of the economic model, except insofar as wage rates may affect employment. The quicker productive equipment can be accumulated, the faster the economic growth.
12 That is to say an equilibrium in which the economy grows at a constant rate, here equal to zero.
13 Meadows *et al.* (1974)
14 See Solow (1957), the original model and analysis. For review of US data for the postwar decades see, for example, Denison (1985). The initial French study (Berthet *et al.* 1965) examined data for 1949 to 1963. Innumerable other studies have been conducted and almost all confirm the need for the 'Solow residual' to explain observed economic growth.
15 In the original formulation, the residual A(t) is entirely neutral, applying equally to all the factors of production (labour, capital, etc.) (known as the 'Hicks neutral' assumption). Alternatively it might be that technology could enhance the ratio of labour to capital productivity ('Harrod neutral') or the converse ('Solow neutral'). Formally: *Hicks neutral*: ratio of marginal product remains unchanged for a given capital–labour ratio ($Y = AK^{a}L^{1-a}$); *Harrod neutral*: relative input shares remain unchanged for a given capital–output ratio ($Y = K^{a}(AL)^{1-a}$); *Solow neutral*: relative input shares remain unchanged for a given labour–output ratio ($Y = (AK)^{a}L^{1-a}$).
16 Ramsey (1928). The effort to transform growth models into optimising systems was later described by Solow (1987) as: 'The end result is a construction in which the whole economy is assumed to

be solving a Ramsey optimal-growth problem through time, disturbed only by stationary stochastic shocks to tastes and technology. To these the economy adapts optimally. Inseparable from this habit of thought is the automatic presumption that observed paths are equilibrium paths. So we are asked to regard the construction I have just described as a model of the actual capitalist world. What we used to call business cycles – or at least booms and recessions – are now to be interpreted as optimal blips in optimal paths in response to random fluctuations in productivity and the desire for leisure.'

17 Ahmad (1966) was the first to formulate a simple theoretical representation of innovation in terms of a historical 'possibility frontier'.

18 The pioneer in broader mathematical explanations of economic evolution and path dependence was Arthur (1989); see Chapter 10, section 10.3, for fuller discussion of evolutionary economics and path dependence. In Ahmad's model, the succession of relative prices at each given point in time determines the amount of learning by doing accumulated in a given direction, which is clearly one subset of the overall evolutionary forces at play.

19 See Chapter 9, particularly sections 9.4 and 9.5, for a discussion of the mix of push and pull forces in the innovation chain. In the econometrics literature a classical demonstration of demand pull forces is given by Griliches (1957) for hybrid maize. Articles by Mowery and Rosenberg (1979) or the econometric study by Scherer (1982) tend to emphasise 'supply push' forces, in which innovation can be less readily explained by demand and relative factor prices.

20 The introduction to this chapter noted that the macroeconomic impact of investing in innovation in any given sector is implicitly a balance between 'crowding out' (using smart researchers or other resources that could have contributed to other innovation) and 'spillovers' (the benefits of innovation in one sector spilling out to others, as with the various spillovers noted from military research to nuclear power and gas turbines). Implicitly, Arrow's model translates into the view that these two forces balance in the long run – or at least, that there is no way of knowing which may dominate. In the case of clean technologies, this implies that the crowding out effect of investing more in a clean sector rather than in a dirty one (that is the slowing down of the learning-by-doing in the dirty sector) is, at the aggregate level, totally offset by knowledge spillovers from the clean sector to others.

21 Lucas (1990). The productivity gap has presumably substantially narrowed since then.

22 Romer (1990) identified the role of product variety, while Aghion and Howitt (1992) discussed product quality, both analyses concerned with their role in economic growth. Neither, however, seemed to draw the obvious conclusion that sectors with homogenous products, without such variety, would be likely to systematically under-invest in innovation compared to the *aggregate* macroeconomic value of innovations which reduced either their financial or their environmental costs. For the link with microeconomics of energy innovation see Chapter 9, particularly section 9.6.

23 Sala-i-Martin *et al.* (2004), reviewing the main determinants of long-term growth, found that 18 parameters have a significant influence among which non-economic factors like primary school enrolment were important. Rather than clear econometric relationships, instead there tended to be some convergence in clubs (countries with similar growth regimes); the challenge remained to understand how a country passes from one club to another. They were confronted with the difficulty of measuring, at an aggregated level and over time series long enough to support stable econometric relationships, parameters like R&D investments in the private sector (the number of patents is a poor measure of innovation in many sectors) or the level of education.

24 President Putin's economic advisor, for example, argued strenuously against Russia ratifying the Kyoto Protocol on the grounds that Russia would need to massively increase its CO_2 emissions as part of an economic recovery. In the event, these projections proved to be fantasy.

25 See Chapter 1 (Figures 1.6 and 1.7), and Chapter 7 (section 7.6). For historical background on the Russian projections of rising emissions, see http://www.climatestrategies.org.

26 'I have found that even these deeper and more circumstantial models of endogenous growth all rest at some key point on an essentially arbitrary linearity assumption, on the claim that the rate of growth of this is a function of the level of that, where 'that' is some fairly simple and accessible variable that can be manoeuvred by policy' (Solow 2001).

27 Such structural change is in part driven by changes in consumption patterns (e.g. private versus public transportation or the meat content of diets) and in part by the relative speed of increased productivity in each sector. This is why (Solow 1956) noted that simple growth theory can serve 'as a background on which to hang multi-sector models that probably try to do more than can be done,

and as a framework for simple, strong, loosely quantitative propositions about cause and effect in macroeconomics.' He also noted that 'the way remains open for a reasonable person to believe that the stimulation of investment will favour faster intermediate-run growth through its effect on the transfer of technology from laboratory to factory.'

28 With unbounded economies of scale the production tends to the infinite and there is no possibility to converge towards a stabilised growth pathway. Moreover, economies of scale are sources of rent and no rule as simple as the remuneration of production factors at their marginal level of productivity can be established for the distribution of this rent. One justification which was advanced was that the deployment of economies of scale in intermediary sectors (steel, non-ferrous, petrochemicals) and manufacturing (automobile, semi-durable goods) allows for an increase in final consumption but at a level not high enough to absorb the available workforce. This workforce is then redirected to other sectors (services, health, teaching, security, administration) which are more labour-intensive and benefit less from economies of scale and learning-by-doing.

29 This problem of aggregation is at the root of the enduring disconnect between 'top-down' and 'bottom-up' models and the unsatisfactory nature of most attempts to bridge the divide.

30 Thirtle and Ruttan (1987).

31 The major contribution of growth theorists to energy and environment issues was to incorporate these issues into existing models of exogenous growth, e.g. Solow (1974), Nordhaus (1974).

32 As proposed, for example, by Ayres and Kneese (1969).

33 The model specification thus becomes Y (GDP) = A(t) × F(K, L, E).

34 This criticism is in part misplaced; if energy-saving investments entail higher capital costs – without technological learning – then the capacity to purchase other productive capital (machines, infrastructures) will be lower for a given amount of savings, thus reducing economic growth. In terms of our analogy, the gasoline put in the car after the change of direction towards Cambridge is the same but the road is narrower and the speed of the car is reduced. However, the criticism is in part right since without an empirical basis for the aggregate substitution between energy and other production factors, this opens the possibility to argue that it will be always possible to 'enlarge the road' at low costs – to substitute something else in place of existing energy resources. The discussion between the so-called weak and strong sustainability ultimately relates to a judgement about the possibility of finding technical man-made substitutes for finite natural resources (Common and Perrings, 1992; Hartwick 1978; Pearce and Atkinson 1995; Toman *et al.* 1993) and/or to overcome entropy (Daly 1997; Ayres *et al.* 1998).

35 Following a similar pattern to an environmental Kuznets curve.

36 The AEEI was introduced by Manne and Richels (1992).

37 This was developed by Dale Jorgenson and Barbara Fraumeini in 1981, and underpinned the Dynamic General Equilibrium Model (DGEM) developed by Dale Jorgenson and Peter J. Wilcoxen (1993). Jorgenson and Wilcoxen's model includes thirty sectors. This model uses for each sector a F(K, L, E, M, t) production function with K, L, E, M standing for capital, labour, energy and other inputs respectively, and t standing for time. In their model, the energy content of each sector (and of the overall economy) depends on both the price of energy relative to the price of other inputs and an exogenous trend in energy intensity for each sector. If historical data and analysis show the energy intensity of a sector is increasing, it indicates that technical change uses more energy. From this and their US data they concluded that 'any increase in relative energy prices will lead to a slower growth in total factor productivity' – i.e. economic growth (Hogan and Jorgensen 1991). See, however, note 38.

38 Firstly, the employed methodology implicitly assumes that the optimal combination of technologies is chosen for a given energy price. Second, the time period of the study may have biased the results, emphasising caution not to confuse correlation with causation. This follows Pindyck and Rotemberg (1982) and Berndt (1983). An intrinsic limit of F(K, L, E) and F(K, L, E, M) functions is that they use data on the share of energy in the production costs of each sector to derive their implicit 'technical functions' thanks to the envelope theorem. The use of this theorem is only valid if the specific basket of technologies used is optimal for the prevailing relative prices. This is sometimes a heroic assumption. The meta-analysis conducted by Frondel and Schmidt (2002) over more than a thousand studies shows that this technique can be a deadlock because of drastic changes of prices over long periods. Moreover, the time period considered encompasses both the 'Thirty Glorious Years', a period of intense economic growth combined with cheap energy and lifestyles based on automobile use, and the twenty years of economic instability and slowdown after the first

oil shock. The interrelation between post-Second World War economic cycles and energy prices makes it difficult to disentangle their relative influence. For example, applying this methodology, the past thirty years in the US would attribute the recent economic slowdown to the energy price increases between 2002 and 2008 rather than the influence of the real-estate bubble in the US and its impact on the banking system. This is especially true when econometric correlations do not pass stability tests, e.g. when calibrating over 1960–2000 instead of 1955–95 substantially changes the results (Berndt and Wood 1984 and 1986).

39 As argued by Hamilton (1996 and 2012). Sudden energy price increases raise production costs and reduce purchasing power, and hence growth, particularly if they concern imports; the potentially positive effects of higher energy prices associated with revenues and investment (if domestic) and enhanced efficiency generally take longer. For more extensive discussion see Chapter 1, notes 8 and 9.

40 Hourcade *et al.* (1996). In most of these studies, energy prices continue to be a strong factor in accelerating the penetration of low-carbon technologies.

41 The argument goes as follows: 'The opportunity cost of a dollar of energy R&D is that one less dollar is available for any of three possible activities: consumption, physical investment, or investment in other R&D. [...] Since the social rate of return on R&D is four times higher that of other investment, [...] the cost of any research that crowds out other research is four dollars. [...] Removing the assumption of partial crowding out from energy R&D nearly doubles the gains from policy-induced energy R&D. [...] It suggests that models ignoring the costs of R&D necessary to develop new technologies, such as those relying solely on learning-by-doing, overstate the gains from policy-induced technological change' (Popp 2006).

42 This is the case in Arrow's model of learning-by-doing.

43 Firstly, there is the challenge of accurately modelling private research and development investments as they are accounted for differently among sectors and countries. Secondly, there is no obvious metric to capture innovation rates which in itself captures innovation at a micro level. The number of patents issued is often used as a proxy variable but it only captures – and only in some sectors – early phases in the innovation chain and not the entire process.

44 Hallegatte *et al.* (2008). The underlying point is that financial capital is much more mobile than physical infrastructure – and hence much more 'impatient' than the long timescales involved in engineering and innovation investment, particularly in the energy sector, as illustrated in Chapters 9 and 10.

45 The title of this section 'Framing the future' is a deliberate play on words. In English, 'framing' is often used to mean presenting a framework for thinking about issues in structured ways. But 'framing' can also mean creating an impression, or apparent evidence, which is false. The use of models based almost exclusively on Second Domain economic assumptions unquestionably has helped to frame ways of thinking; it may also have inadvertently 'framed' the result as implying unavoidable tradeoffs between environmental policy and economic growth.

46 These modelling results are periodically summarised in IPCC reports (Weyant *et al.* 1996; Hourcade *et al.* 2001; Barker *et al.* 2007).

47 The inertia of energy systems amplifies the costs of delayed action (Ha-Duong *et al.* 1997), a result since elaborated in numerous studies; for further references see Chapter 12.

48 A recent wave of comparative modelling studies have assessed various assumptions around technology and the implications of delayed participation for climate policies – confirming the intuition that delayed participation either results in increasing the costs of meeting a given climate target or having to abandon it. A useful overview is in Riahi *et al.* (2014), with a wider range of studies to be reviewed in the forthcoming IPCC Fifth Assessment Report, due for publication in April 2014. Other recent studies which focus on the macroeconomic dimensions of revenue recycling and innovation-related complementary policies include those in Barker and Crawford-Brown (2014).

49 This was a caveat listed in the 4th Assessment Report of the IPCC: 'Most models use a global least cost approach to mitigation portfolios and with universal emissions trading, assuming transparent markets, no transaction cost, and thus perfect implementation of mitigation measures throughout the 21st century' (IPCC AR4 WGIII SPM Box 3).

50 Because equilibrium-based models generally assume a baseline which is optimised, any constraint almost by definition implies a cost. To assume otherwise involves either forcing distortions (e.g. subsidies) into the baseline or adopting a fundamentally different approach. For example, in the

econometrically based E3MG model (Barker *et al.* 2006), the pace of transformation is constrained technically, so that decarbonising at a faster pace than technology turnover requires extremely high carbon prices, but the relationship between carbon price, investment and GDP is fundamentally different from other models. The econometric data suggest that societies are under-investing, so that high carbon prices which drive greater investment depress short-run consumption but boost long-run GDP. Critics argue that the potential GDP benefits of boosting investment should not be conflated with climate change policy – an issue considered in section 11.7.

51 Grubb *et al.* (2004). Two earlier review papers which helped to fuel widespread economic debate on the issue of induced innovation in relation to energy and climate change were Weyant and Olavson (1999) and Grubb *et al.* (2002). A useful subsequent paper (Ian Sue Wing 2006) compared and contrasted the two main approaches to modelling-induced innovation – learning-by-doing and knowledge-stock approaches – and in particular their implications for timing. He concluded:

> Though it may be tempting to interpret this dichotomy in terms of the debate over adoption of 'act now' versus 'wait and see' climate policies, it is important to realise that *both* of the approaches appear to indicate earlier action. The essential difference is qualitative [...] in the LBD approach, induced innovation is synonymous with faster micro-scale shifts to low carbon production [...] reflecting the guiding assumption of *complementarity* between abatement and the overall efficiency of production. By contrast [...] the stock of knowledge approach [...] incurs the opportunity costs of foregone 'economy growing' R&D – reflecting the guiding assumption of *substitutability* [...] All this [...] suggests the critical importance of investigating the empirical content of the assumptions which underlie the representation of induced innovation.

Again, this points to the importance of grounding economic analyses in facts, such as the analysis of energy-sector R&D intensities covered in Chapter 9.

52 Edenhofer *et al.* (2009). Of the three models involved in the RECIPE project, WITCH (Model 1) and ReMIND-R (Model 2) are inter-temporal optimisation models with 'perfect foresight' while IMACLIM-R (Model 3) is a recursive dynamic model under imperfect foresight. As a result, in ReMIND-R and WITCH, the long-run optimisation of the global energy-economy system results in carbon prices increasing exponentially for the first few decades. The accumulated impact of induced technological change can then lead to deviations from the exponential path. As a share of global total consumption over the century, the aggregated mitigation consumption losses in the RECIPE studies relative to the baseline, aggregated over the period to 2100 and discounted at 3%, amount to 0.1% (IMACLIM-R), 0.6% (ReMIND-R) and 1.4% (WITCH) (Luderer *et al.* 2011). Substantially wider ranges can be found in the studies of the Energy Modeling Forum (see, for example, Figure 11.4).

53 Modelling by Kalkuhl *et al.* (2011) concludes that 'small market imperfections may trigger several decades of lasting dominance by an incumbent energy technology over a dynamically more efficient competitor', and hence the value of complementary policies of which 'technology quotas and feed-in tariffs turn out to be only insignificantly less efficient than first-best subsidies and seem to be robust [...].' Later Kalkuhl *et al.* (2013) warn against over-reliance on such policies, i.e. that a rising carbon price remains crucial. Others are beginning to explore the value of a combination of instruments (e.g. see chapters in Barker and Crawford-Brown 2014). A common feature is that carbon pricing remains important in all. A number of papers on these themes of modelling multiple instruments have started to appear, particularly in the journal *Resource and Energy Economics*.

54 OECD (2011). For an overview of OECD analysis see http://www.oecd.org/greengrowth/key-documents.htm. The 'Green Growth Knowledge Platform' is now formed of an alliance between the World Bank, OECD, UNEP and the global Global Green Growth Institute based in Korea; the UN Development Programme is also active in considering the opportunities. The Green Growth Knowledge Platform is at: http://www.oecd.org/greengrowth/greengrowthknowledgeplatform.htm. To an important degree these concepts echo earlier calls for 'eco-development'. The expression of eco-development was formulated in 1972 by Maurice Strong, the first Chair of the UN Environment Programme (UNEP). It was systematised by Ignacy Sachs (1974, 1984) in which environmental issues are viewed as opportunities to reshape development patterns, as an alternative strategy to the 'zero growth' message of the Club of Rome.

55 The synthesis of the state of the art by Abramovich (1952: 176–7) shows clearly that before the dominance of mathematical growth models in the 1960s, discussions in growth theory were

primarily related to explaining historical facts like long-run or short-run cycles and the occurrence of economic crises (e.g. Kondratieff and Stolper 1935; Juglar and Wolowski 1862).

56 The 'belle époque' refers to the period of stable growth and deployment of both a rich wave of innovation in industry and a revolution in cultures and lifestyles between 1880 and 1910 in Europe and the USA.

57 This expression was invented by Jean Fourastié (1979) to describe the exceptional period of growth in OECD countries after the Second World War up to the oil shocks. It was used by many Anglophone scholars, including J. Bradford DeLong in his 'Thirty Glorious Years' (1997).

58 The apparent brevity of the *economic* boost from the IT revolution may in part be due to the lack of a virtuous cycle between technical change and the final demand. There is a major difference between the 'car' and the 'iPhone' as symbols of the 'old' and 'new' paradigms. During the post-Second World War growth cycle, wage increases were translated into higher demand for cars, and this higher demand for cars in return was translated directly into higher demand for steel, glass and pneumatics, generating large economies of scale. This propagation is not the same with the 'iPhone' and information technology because of their low material content. They contribute to reducing the production costs (management, planning, advertising) of the other sectors (impressively so in banking, insurance, commerce) and accelerate the transition towards just-in-time processes. But they do not change fundamentally the demand for the overwhelming majority of sectors, be it infrastructure sectors (energy, transportation, housing) or the production of food, clothes, cars and semi-durables. They triggered a revolution of services to consumers, but electronic games and easy access to thousands of movies does not have the same mechanical spillover impact as the development of automobile supply chains on materials throughout the economy, for example. Thus the Internet revolution, despite its deep impact on societies and on the organisation of many sectors, is not likely to be the equivalent of the 'steam engine', 'railways', 'electricity' and 'car' revolutions of the past in terms of growth dynamics. It could support a long wave of economic growth only by helping the extension of Fordism in other world regions where the demand for cars, semi-durables and food is not saturated or in helping the transformation of the prevailing technological patterns. It is that which is now slowing down as the emerging economies begin to mature.

59 For example, Pomeranz (2009) shows how the origin of the 'Great Divergence' between China and Europe before the modern time is rooted in the structure of the links between the central power of the Emperor and the structure of the peasant society; Aglietta and Bai (2012) show the recent rise of China after centuries of stagnation is typically due to the capacity of the communist regime after Deng Xiaoping to transform the destruction of the Chinese traditional peasant society in the Maoist period into a virtuous cycle where a regulated rural drift fuels the productivity increases of both agriculture and industry.

60 For example, the modern steel factory (the coke plus O_2 chain) produced 3Mt of steel per year after the Second World War and between 15 and 20Mt in the 1970s. The same figures for the production of naphtha were about 75,000t and 750,000t respectively.

61 This is the expression used by French analysts of modern capitalism (Boyer 1990).

62 The implicit political consensus encompassed Labour in the UK, the Christian Democrats and Social Democrats in continental Europe and the Democrats in the US. It was not really combated by Conservatives in the UK and Republicans in the US until the 1980s.

63 The globalisation of goods and capital markets was successful in helping a billion human beings out of the poverty trap in two decades. However (and this is the invisible wall), it accelerated outsourcing to locations with lower input costs (labour, energy) but without a quick enough rise of final demand in emerging economies because of the inertia in building a welfare state. Then if the wages in emerging economies increase less than productivity and if currencies are not re-evaluated (to prevent the accumulation of excess export surpluses), the extension of world markets does not allow for maintaining full employment in matured economies. These economies can respond by lower wages but, quite apart from the political minefield this involves, it risks further depressing their internal demand.

64 Theoretically, in a world with flexible exchange rates and a similar organisation of labour markets, higher imports would be compensated for by higher exports towards the countries in receipt of the outsourced segments of production. For example, the adverse impact on US industry of the huge US investments in Europe in the 1960s was to an extent compensated by higher growth in Europe and Japan and corresponding demand for US products. In practice, this did not work sufficiently well: the current account balance of the US was for the first time in deficit in 1966 and the trade balance

went negative in 1971; wages in the US started stagnating after 25 years of steady and fast increase. For the OECD, the market losses caused today by the competition of emerging economies should be compensated by new markets in these economies. This is not the case for reasons which are in part due to the lack of coordination of economic and social policies. There are now many accounts of the malaise and why the globalisation of Fordism turned out not necessarily to benefit all; one compelling account is by the Chief Economist of HSBC, Stephen King (King 2011).

65 The USA was the first victim of the new global context. Increased international imports from emerging economies, most notably in China, led to a large surplus of dollars abroad and a drop in the purchasing power of American households. The political necessity of hiding this explains Greenspan's laxity at the head of the Federal Reserve. Unprecedented debt facilities were given to private lenders, especially 'non-bank banks' (Krugman 2009) not subject to the prudential rules imposed on the banks. Both banks and non-bank banks were attracted by leveraged buy-outs and by property speculation. The increase of real-estate values gave the middle classes the illusion of a continued growth of their income (and their borrowing capacity). This mechanism, first and foremost in the USA but imitated in several European countries (e.g. Spain, Ireland) was allowed, if not encouraged, by explicit policies because the low and middle classes had the feeling of becoming richer because the value of their houses was steadily increasing. One of the best accounts of this phenomenon is by the former Chief Economist of the International Monetary Fund, one of the few to have warned in advance about the systemic risk posed by unrestrained credit; his book *Fault Lines* (Rajan 2010) warns that the problems have not been adequately addressed and that the global economy remains in peril.

66 As exemplified by the statement of G. Bush Junior, captured in the film *Inside Job*: 'You don't have to have a lousy home, the low-income home-buyer can have just as nice a house as anybody else!' Rajan (2010) emphasises the political imperative behind the resulting real-estate bubble.

67 These were pointed out by authors like Kapp (1950) in Germany or Mishan (1961) in the UK and by the warnings of the Club of Rome (Meadows *et al.* 1973) about the limits of non-renewable resources.

68 Alston *et al.* (2009), Binswanger-Mkhize *et al.* (2010).

69 Hickley *et al.* (2012); Lipman (2006). The range in part is determined by whether they are tenants or owners. Energy expenditures of low-income classes are lower in Europe but are still significant – in France, for example, they increased drastically to reach between 12 and 14% (Orfeuil 2013) which combined with a steady increase of housing costs.

70 Through the financial system, the accumulation of problems in the USA and some European countries spilled over throughout the world. Today the crisis does not only touch Europe and the OECD countries. Brazil experienced a strong decline in economic growth from 7.5% in 2010 to 2% in 2011 and 1% in 2012, leading to a social crisis in June 2013. India's growth rate passed from 8% to 5% in one year and another decline would have exacerbated interregional tensions. China's growth has slowed. Moreover, limits of the trickling down of the Fordist model in these economies can already be observed, with associated social exclusion and tensions. These include concerns about energy security; revolts against pollution in China and against the debased working conditions in Bangladesh; the agricultural modernisation pathway based on industrial intensification and urban sprawl disconnected from dense public infrastructures.

71 Stern and Rydge (2012) summarise the case bluntly: 'Low-carbon growth will be more dynamic, creative, cleaner and more bio-diverse; high-carbon growth will self-destruct.'

72 One example is the prevalence of economic studies which find that renewable energy in Europe has a relatively more positive impact on GDP than equivalent fossil fuels because of the higher share of domestic value-added. For a recent example see Ernst & Young *et al.* (2012).

73 The 'circular economy' would tend to increase value-added extracted from what are current waste streams, which again tend to be predominantly (though not exclusively) domestic activities. See in particular the Ellen McArthur Foundation (2012, 2013), available at: http://www.ellenmacarthurfoundation.org/business/reports.

74 This contrasts with the social tensions generated by an unregulated economic globalisation which has had a well-documented tendency to increase income inequalities and social dualism. An overview of 'Just Transition' ideas may be found in: http://en.wikipedia.org/wiki/Just_Transition. The Just Transition movement originated in the US and developed internationally; one publication outlining the principles and thinking is Trade Union Congress (2008).

75 This indeed was the central factor in the formation of the UK Carbon Trust which, as noted in Chapter 4, was established at the insistence of UK industry to help it manage the impact of the UK Climate Change Levy.

76 For one set of visions of low-carbon societies and analysis of some of their implications, see publications from the International Low Carbon Societies network.

77 European Commission (2013). The consortium comprised Warwick Institute of Economic Research, Ernst & Young consultants, Cambridge Econometrics, Exergia consultants, E3M lab and COWI, and used one economic model founded on general equilibrium principles (GEM-E3), the other an econometric model (E3ME). The 'Es' in each case denote models that have been developed with a reasonable level of energy, economic and environmental detail. As expected, the results differed but overall: 'The models predict that the [decarbonisation] scenarios will have a modest impact on GDP [...] E3ME predicts a slight increase (2–3% by 2050), whilst the GEM-E3 model suggests a GDP reduction of 1–2%, each relative to a baseline in which GDP increases by 85% from 2013–2050 [...] Both models predict an increase in employment levels in the scenarios, compared to the baseline [in the range] 0 to 1.5% depending on the scenario, with the results from the E3ME model roughly 1 percentage point higher than those from the GEM-E3 model.

78 Zenghelis (2011). The difference between the managerial regime in the 1970s described by Galbraith (1973) and the new shareholder regime is emphasised by Jensen (1986). Some implications for growth dynamics are demonstrated in Hallegatte *et al.* (2008).

79 This legend is a caricature of Jean Buridan, a theologian at the Sorbonne in the fourteenth century who argued that a wise conduct is to postpone decisions until all the necessary information is available.

80 Depending on the exact framework, regulations (or not) and national circumstances, these problems can include agricultural systems of mass production which marginalise remote rural areas, waste scarce water resources and destroy arable land, rent-seeking in real estates, land and raw materials, and contempt for environmental conservation.

81 For an extensive discussion see the UK House of Lords (2013) and associated testimonies (notably from Martin Wolf, Chief Economics Correspondent of the *Financial Times*, and from Dimitri Zhengelis, London School of Economics) available on the House of Lords website.

82 Illustrations of possible mechanisms are given in Hourcade *et al.* (2012), Neuhoff *et al.* (2010) and Bredenkamp and Pattillo (2010). The former suggests that carbon-based drawing rights from the central banks of the IMF, transformed into carbon certificates for helping development and investment banks to provide low-risk loans, could even culminate in the gradual emergence of international financial products based on long-term carbon value, helping to align the financial system with the real physical needs for long-term, sustainable investment.

83 For analysis of how climate finance might provide a lever for the Eurozone to combine a strict fiscal compact with a credible growth compact, see Agliatella and Hourcade (2012).

84 See note 3 for some entry points into the huge literature on the relationship between wealth and welfare. In the context of climate change, two major volumes from the international Low Carbon Societies project encompass a range of technological, economic and social dimensions of the relationship between low carbon societies, wealth and welfare (Strachan *et al.* 2008; Skea *et al.* 2013).

85 Jakob and Edenhofer (2013) underline this in arguing that:

> The popular concepts of Green Growth and Degrowth are both eventually misleading. As both concepts fail to make explicit which objectives are ultimately to be achieved, it remains unclear whether these objectives are better served by promoting or curtailing economic growth. That is, by focusing on economic growth instead of welfare both concepts ultimately confuse means and ends, i.e. they present influencing the rate of economic growth as an end rather than a means to achieve certain ends [...] the discourse on economic growth and the environment should be firmly based on the concept of social welfare instead of economic growth [...] [with] an approach of 'welfare diagnostics' that takes into account the broad spectrum of normative positions and the multi-dimensional nature of social welfare [...] and in particular what can be understood as 'basic needs' and 'minimal thresholds'.

Of more direct relevance to this chapter these authors also conclude that welfare diagnostics

> would aim at correcting over-use of natural capital as well as under-provision of public goods (such as public infrastructure). As both natural capital and public infrastructure have characteristics of

commons, managing a portfolio of different capital stocks of commons can be regarded as a central task of public policy. In particular, the possibility of appropriating natural resource rents to finance public investment creates a close relationship between managing natural capital and investing in public infrastructure [...]

This paper forms one of a wide collection of papers presented to a conference on the economics of 'Green Growth' (London, September 2013) held shortly before this book went to press, and several other papers from the conference, due to be published in the *Oxford Review of Economic Policy*, may be of interest to readers.

References

Abramowitz, M. (1952) 'Economics of growth', in B. F. Haley (eds), *A Survey of Contemporary Economics*, Vol. 2. Homewood, IL: Richard D. Irwin, pp. 132–82.

Aghion, P. and Howitt, P. (1992) 'A model of growth through creative destruction', *Econometrica*, 60 (2): 323–51.

Aglietta, M. and Bai, G. (2012) *La Voie Chinoise Capitalisme et Empire*. Paris: O. Jacob.

Aglietta, M. and Hourcade, J.-C. (2012) 'Endettement public et développement soutenable en Europe. Sortir des peurs?', in P. Weil (ed.), *80 Propositions qui ne Coûtent pas 80 Milliards*. Paris: Grasset.

Ahmad, S. (1966) 'On the theory of induced invention', *Economic Journal*, 76 (302): 344–57.

Alston, J. M., Beddow, J. M. and Pardey, P. G. (2009) 'Agricultural research, productivity, and food prices in the long run', *Science*, 325 (5945): 1209–10.

Arrow, K. J. (1962) 'The economic implications of learning by doing', *Review of Economic Studies*, 29 (3) : 155–73.

Arthur, W. B. (1989) 'Competing technologies, increasing returns, and lock-in by historical events', *Economic Journal*, 99 (394): 116–31.

Ayres, R. U. and Kneese, A. V. (1969) 'Production, consumption, and externalities', *American Economic Review*, 59 (3): 282—97.

Ayres, R. U., van den Bergh, J. C. J. M. and Gowdy, J. M. (1998) *Viewpoint: Weak Versus Strong Sustainability*. Tinbergen Institute.

Barker, T. and Crawford-Brown, D. (eds) (2014) *Decarbonising the World Economy*. London: Imperial College Press (in press).

Barker, T., Bashmakov, I., Alharthi, A. *et al.* (2007) 'Mitigation from a cross-sectoral perspective', in B. Metz, O. R. Davidson, P. R. Bosch *et al.* (eds), *Climate Change 2007: Mitigation*, Contribution of Working Group III to the Fourth Assessment Report of the Intergovernmental Panel on Climate Change. Cambridge: Cambridge University Press.

Barker, T., Pan, H., Kohler, J. *et al.* (2006) 'Decarbonizing the global economy with induced technological change: scenarios to 2100 using E3MG', *Energy Journal*, SI2006(01). Available online at: http://www.iaee.org/en/publications/ejarticle.aspx?id=2143

Berndt, E. R. (1983) 'Quality adjustment in empirical demand analysis'. Available online at: http://dspace.mit.edu/handle/1721.1/2032

Berndt, E. R. and Wood, D. O. (1984) 'Energy price changes and the induced revaluation of durable capital in U.S. manufacturing during the OPEC decade'. Available online at: http://dspace.mit.edu/handle/1721.1/60595

Berndt, E. R. and Wood, D. O. (1986) 'Energy price shocks and productivity growth in US and UK manufacturing', *Oxford Review of Economic Policy*, 2 (3): 1–31.

Berthet, J., Carré, J. J., Dubois, P. *et al.* (1965) *Sources et Origines de la Croissance Française au Milieu du XXe Siècle*. Paris: INSEE.

Bibas, R. and Méjean, A. (2013) 'Potential and limitations of bioenergy options for low carbon transitions', *Climatic Change*, forthcoming.

Binswanger-Mkhize, H. P., McCalla, A. F. and Patel, P. (2010) 'Structural transformation and African agriculture', *Global Journal of Emerging Market Economies*, 2 (2): 113–52.

Boyer, R. (1990) *The Regulation School: A Critical Introduction*. New York: Columbia University Press.

Bredenkamp, H. and Pattillo, C. A. (2010) *Financing Response to Climate Change*. International Monetary Fund.

Common, M. and Perrings, C. (1992) 'Towards an ecological economics of sustainability', *Ecological Economics*, 6 (1): 7–34.

Daly, H. E. (1997) 'Georgescu-Roegen versus Solow/Stiglitz', *Ecological Economics*, 22 (3): 261–6.

DeLong, J. B. (1997) 'The great Keynesian boom: 'Thirty Glorious Years', in *Slouching Towards Utopia: The Economic History of the Twentieth Century*. Profile Books. Available online at: http://www.j-bradford-delong.net/tceh/Slouch_Keynes20.html

Denison, E. F. (1985) *Trends in American Economic Growth: 1929–1982*. Brookings Institution Press.

Durlauf, P. S. N. and Blume, P. L. E. (2008) *The New Palgrave Dictionary of Economics*, 2nd edn. Basingstoke: Palgrave Macmillan.

Edenhofer, O., Carraro, C., Hourcade, J.-C. *et al.* (2009) *RECIPE: The Economics of Decarbonization. Report on Energy and Climate Policy in Europe*. Postdam: Potsdam-Institut für Klimafolgenforschung. Available online at: http://lpintrabp.parl.gc.ca/lopimages2/bibparlcat/23000/Ba445107.pdf

Ellen McArthur Foundation (2012/2013) *Towards the Circular Economy*, Vols 1 and 2.

EMF (2009) *EMF22 Database: Climate Change Control Scenarios*. Available online at: http://emf.stanford.edu/events/emf_briefing_on_climate_policy_scenarios_us_domestic_and_international_policy_architectures/

Ernst & Young, Acciona and EDF (2012) *Analysis of the Value Creation Potential of Wind Energy Policies – A Comparative Study of the Macroeconomic Benefits of Wind and CCGT Power Generation*.

European Commission (2013) *Employment Effects of Selected Scenarios from the Energy Roadmap 2050 – Final Report for the European Commission*. Brussels: European Commission DG Energy.

Fourastié, J. (1979) *Les Trente Glorieuses: Ou, la Révolution invisible de 1946 à 1975*. Paris: Fayard.

Frondel, M. and Schmidt, C. M. (2002) 'The capital-energy controversy: an artifact of cost shares?', *Energy Journal*, 23 (3): 53–80.

Galbraith, J. R. (1973) *Designing Complex Organizations*, 1st edn. Boston: Addison-Wesley Longman.

Ghersi, F. and Hourcade, J.-C. (2006) 'Macroeconomic consistency issues in E3 modeling: the continued fable of the elephant and the rabbit', *Energy Journal*, Special Issue No. 2: 39–62.

Goulder, L. H. and Schneider, S. H. (1999) 'Induced technological change and the attractiveness of CO_2 abatement policies', *Resource and Energy Economics*, 21 (3–4): 211–53.

Griliches, Z. (1957) 'Hybrid corn: an exploration in the economics of technological change', *Econometrica*, 25 (4): 501–22.

Grubb, M., Kohler, J. and Anderson, D. (2002) 'Induced technical change in energy and environmental modeling: analytic approaches and policy implications', *Annual Review of Energy and the Environment*, 27: 271–308.

Grubb, M., Edenhofer, O., Carraro, C. *et al.* (2004) 'Innovation modeling comparison project', *Energy Journal*, Special Issue.

Grubb, M., Edmonds, J., Ten Brink, P. *et al.* (1993) 'The costs of limiting fossil-fuel CO_2 emissions: a survey and analysis', *Annual Review of Energy and the Environment*, 18 (1): 397–478.

Grubler *et al.* (2012) 'Energy primer', in GEA, *Global Energy Assessment – Toward a Sustainable Future*. Cambridge and New York: Cambridge University Press and Laxenburg, Austria: IIASA, Chapter 1.

Guivarch, C., Crassous, R., Sassi, O. *et al.* (2011) 'The costs of climate policies in a second best world with labour market imperfections', post-Print, HAL. Available online at: http://ideas.repec.org/p/hal/journl/halshs-00724487.html

Ha-Duong, M., Grubb, M. J. and Hourcade, J.-C. (1997) 'Influence of socioeconomic inertia and uncertainty on optimal CO_2-emission abatement', *Nature*, 390 (6657): 270–3.

Hallegatte, S., Ghil, M., Dumas, P. *et al.* (2008) 'Business cycles, bifurcations and chaos in a neo-classical model with investment dynamics', *Journal of Economic Behavior and Organization*, 67 (1): 57–77.

Hamilton, J. D. (1996) 'This is what happened to the oil price-macroeconomy relationship', *Journal of Monetary Economics*, 38 (2): 215–20.

Hamilton, J. D. (2008) 'Oil and the macroeconomy', in S. N. Durlauf and L. E. Blume (eds), *The New Palgrave Dictionary of Economics*, 2nd edn. Palgrave Macmillan.

Hamilton, J. D. (2012) *Oil Prices, Exhaustible Resources, and Economic Growth*, Working Paper, National Bureau of Economic Research. Available online at: http://www.nber.org/papers/w17759

Hartwick, J. M. (1978) 'Substitution among exhaustible resources and intergenerational equity', *Review of Economic Studies*, 45 (2): 347.

Hickley, R., Lubell, J., Haas, P. *et al.* (2012) *The Struggle of Moderate-Income Households to Afford the Rising Costs of Housing and Transportation*. Center for Housing Policy.

Hogan, W. W. and Jorgenson, D. W. (1991) 'Productivity trends and the cost of reducing CO_2 emissions', *Energy Journal*, 12 (1): 67–86.

Hogan, W. W. and Manne, A. S. (1977) 'Energy economy interactions: the fable of the elephant and the rabbit?'

Höhne, N., den Elzen, M. and Weiss, M. (2006) 'Common but differentiated convergence (CDC): a new conceptual approach to long-term climate policy', *Climate Policy*, 6 (2): 181–99.

Hourcade, J.-C. (1996) 'Estimating the costs of mitigating greenhouse gases', in P. J. Bruce H. Lee and E. F. Haites, (eds), *Climate Change 1995. Economic and Social Dimensions of Climate Change.* Contribution of Working Group III to the Second Assessment Report of the Intergovernmental Panel on Climate Change. Cambridge: Cambridge University Press.

Hourcade, J.-C. and Shukla, P. (2013) 'Triggering the low-carbon transition in the aftermath of the global financial crisis', *Climate Policy*, 13 (Special Issue): 22–35.

Hourcade, J.-C., Perrissin Fabert, B. and Rozenberg, J. (2012) 'Venturing into uncharted financial waters: an essay on climate-friendly finance', *International Environmental Agreements: Politics, Law and Economics*, 12 (2): 165–86.

Hourcade, J.-C., Jaccard, M., Bataille, C. *et al.* (2006) 'Hybrid modeling: new answers to old challenges', *Energy Journal*, 2 (Special issue): 1–12.

Hourcade, J. C., Shukla, P. R., Cifuentes, L. *et al.* (2001) 'Global, regional, and national costs and ancillary benefits of mitigation', in B. Metz, O. Davidson, R. Swart *et al.* (eds), *Climate Change 2001*, Contribution of Working Group III to the Third Assessment Report of the Intergovernmental Panel on Climate Change. Cambridge: Cambridge University Press.

House of Lords (2013) *No Country Is an Energy Island: Security Investment for the EU's Future.* House of Lords European Union Committee 14th Report of Session 2012–13.

Huntington, H. G. (1994) 'Been top down so long it looks like bottom up to me', *Energy Policy*, 22 (10): 833–9.

IPCC (2001) *Climate Change 2001: Contribution of Working Group III to the Third Assessment Report of the Intergovernmental Panel on Climate Change*, eds B. Metz, O. Davidson, R. Swart *et al.* Cambridge: Cambridge University Press.

IPCC (2007) *Climate Change 2007: Contribution of Working Group III to the Fourth Assessment Report of the Intergovernmental Panel on Climate Change*, eds B. Metz, O. Davidson, P. R. Bosch *et al.* Cambridge: Cambridge University Press.

Jackson, T. (2011) *Prosperity Without Growth: Economics for a Finite Planet*, reprint. Abingdon and New York: Routledge.

Jakob, M. and Edenhofer, O. (2013) *Green Growth, Degrowth, and the Commons*, paper presented to Grantham Institute/GGGI conference on 'Green Growth', London, September 2013; forthcoming in *Oxford Review of Economic Policy*.

Jensen, M. C. (1986) 'Agency costs of free cash flow, corporate finance, and takeovers', *American Economic Review*, 76 (2): 323–9.

Jorgenson, D. W. and Fraumeni, B. M. (1981) *Relative Prices and Technical Change. Modeling and Measuring Natural Resources Substitution*. Cambridge, MA: MIT Press.

Jorgenson, D. W. and Wilcoxen, P. J. (1993) 'Reducing US carbon emissions: an econometric general equilibrium assessment', *Resource and Energy Economics*, 15 (1): 7–25.

Juglar, C. and Wolowski, L. (1862) 'Des crises commerciales et de leur retour périodique en France, en Angleterre et aux États-Unis/par le Dr Clément Juglar ... Guillaumin (Paris)'. Available online at: http://gallica.bnf.fr/ark:/12148/bpt6k1060720

Kalkuhl, M., Edenhofer, O. and Lessmann, K. (2011) 'Learning or lock-in: optimal technology policies to support mitigation', *Resource and Energy Economics*, 34: 1–23.

Kalkuhl, M., Edenhofer, O. and Lessmann, K. (2013) 'Renewable energy subsidies: second-best policy or fatal aberration for mitigation?', *Resource and Energy Economics*, 35: 217–34.

Kapp, K. W. (1950) *The Social Costs of Private Enterprise*. Cambridge, MA: Harvard University Press.

King, S. D. (2011) *Losing Control: The Emerging Threats to Western Prosperity*. New Haven, CT: Yale University Press.

Kondratieff, N. D. and Stolper, W. F. (1935) 'The long waves in economic life', *Review of Economics and Statistics*, 17 (6): 105–15.

Krugman, P. (2009) *The Return of Depression Economics and the Crisis of 2008*. Reprint. W. W. Norton.

Kuznets, S. (1934) 'National Income, 1929–1932'. 73rd US Congress, 2nd session, Senate document no. 124, page 7. Available online at: www.nber.org/chapters/c2258.pdf

Kuznets, S. (1955) 'Economic growth and income inequality', *American Economic Review*, 45 (1): 1–28.

Lipman, B. (2006) *A Heavy Load: The Combined Housing and Transportation Burdens of Working Families*. Center for Housing Policy.

Lucas, R. E. (1990) 'Why doesn't capital flow from rich to poor countries?', *American Economic Review*, 80 (2): 92–6.

Luderer, G., Bosetti, V., Jakob, M. *et al.* (2012) 'The economics of decarbonizing the energy system – results and insights from the RECIPE model intercomparison', *Climatic Change*, 114 (1): 9–37.

Manne, A. S. and Richels, R. G. (1992) 'Global CO_2 emission reductions: the impacts of rising energy costs', in J. C. White, W. Wagner and C. N. Beal (eds), *Global Climate Change*. Houten: Springer Netherlands, pp. 211–39.

Meadows, D. H., Meadows, D. L., Randers, J. *et al.* (1974) *The Limits to Growth: A Report for the Club of Rome's Project on the Predicament of Mankind*. New York: Universe Books.

Mishan, E. J. (1961) 'Welfare criteria for external effects', *American Economic Review*, 51 (4): 594–613.

Mowery, D. and Rosenberg, N. (1979) 'The influence of market demand upon innovation: a critical review of some recent empirical studies', *Research Policy*, 8 (2): 102–53.

Neuhoff, K., Fankhauser, S., Guerin, E. *et al.* (2010) *Structuring International Financial Support for Climate Change Mitigation in Developing Countries*, Discussion Paper, German Institute for Economic Research. Available online at: http://www.econstor.eu/handle/10419/36752

Nordhaus, W. D. (1974) 'Resources as a constraint on growth', *American Economic Review*, 64 (2): 22–6.

OECD (2011) *Towards Green Growth*. Paris: OECD. Available online at: http://www.oecd.org/greengrowth/towardsgreengrowth.htm

Orfeuil, J.-P. (2013) *Quand la Voiture Devient Contrainte*, Projet, No. 334(3), pp. 50–8.

Pearce, D. W. and Atkinson, G. (1995) 'Measuring sustainable development', in D. Bromley, (ed.), *The Handbook of Environmental Economics*. Oxford: Blackwell.

Perez, C. (1985) Microelectronics, long waves and world structural change: new perspectives for developing countries', *World Development*, 13 (3), 441–63.

Perez, C. (2002) *Technological Revolutions and Financial Capital: The Dynamics of Bubbles and Golden Ages*. Cheltenham: Edward Elgar.

Pindyck, R. S. and Rotemberg, J. J. (1982) *Dynamic Factor Demands, Energy Use, and the Effects of Energy Price Shocks*, Working Paper, Massachusetts Institute of Technology, Energy Laboratory. Cambridge, MA: Massachusetts Institute of Technology, Energy Laboratory.

Pomeranz, K. (2009) *The Great Divergence: China, Europe, and the Making of the Modern World Economy*. Princeton, NJ: Princeton University Press.

Popp, D. (2006) 'Comparison of climate policies in the ENTICE-BR model', *Energy Journal*, Endogenous Technological Change (Special Issue No. 1): 163–74.

Popp, A., Dietrich, J. P., Lotze-Campen, H. *et al.* (2011) 'The economic potential of bioenergy for climate change mitigation with special attention given to implications for the land system', *Environmental Research Letters*, 6 (3): 034017.

Raghuram, R. G. (2010) *Fault Lines*. Princeton, MA: Princeton University Press.

Ramsey, F. P. (1928) 'A mathematical theory of saving', *Economic Journal*, 38 (152): 543–59.

Riahi, K. *et al.* (2014) 'Locked into Copenhagen pledges – implications of short-term emission targets for the cost and feasibility of long-term climate goals', *Technological Forecasting and Social Change*, forthcoming.

Robinson, J. and Naqvi, K. A. (1967) 'The badly behaved production function', *Quarterly Journal of Economics*, 81 (4): 579.

Romer, P. M. (1986) 'Increasing returns and long-run growth', *Journal of Political Economy*, 94 (5): 1002–37.

Romer, P. M. (1990) 'Endogenous technological change', *Journal of Political Economy*, 98 (5): S71–S102.

Sachs, I. (1974) 'Environment and styles of development', *Economic and Political Weekly*, 9 (21): 828–37.

Sachs, I. (1984) 'The strategies of ecodevelopment … Ceres', *FAO Review on Agriculture and Development*, 17 (4) = no. 100, pp. 17–21. Available online at: http://agris.fao.org/agris-search/search/display.do?f=1985/XF/XF85001.xml;XF8546293

Sala-i-Martin, X. X. (1996) 'The classical approach to convergence analysis', *Economic Journal*, 106 (437): 1019–36.

Sala-i-Martin, X. X., Doppelhofer, G. and Miller, R. I. (2004) 'Determinants of long-term growth: a Bayesian Averaging of Classical Estimates (BACE) approach', *American Economic Review*, 94 (4): 813–35.

Scherer, F. M. (1982) 'Inter-industry technology flows and productivity growth', *Review of Economics and Statistics*, 64 (4): 627–34.

Solow, R. M. (1956) 'A contribution to the theory of economic growth', *Quarterly Journal of Economics*, 70 (1): 65–94.

Solow, R. M. (1957) 'Technical change and the aggregate production function', *Review of Economics and Statistics*, 39 (3): 312–20.

Solow, R. M. (1974) 'Intergenerational equity and exhaustible resources', *Review of Economic Studies*, 41: 29–45.

Solow, R. M. (1986) 'On the intergenerational allocation of natural resources', *Scandinavian Journal of Economics*, 88 (1): 141–9.

Solow, R. M. (1987) *Growth Theory and After*, Nobel Prize in Economics documents, Nobel Prize Committee. Available online at: http://ideas.repec.org/p/ris/nobelp/1987_001.html

Solow, R. M. (2000) 'Toward a macroeconomics of the medium run', *Journal of Economic Perspectives*, 14 (1): 151–8.

Solow, R. M. (2001) Addendum to Robert R. M. Solow's Nobel Prize Lecture 'Growth Theory and After'. Nobel Prize in Economics documents. Available online at: http://ideas.repec.org/p/ris/nobelp/1987_001.html

Stern, N. (2012) *How We Can Respond and Prosper: The New Energy-Industrial Revolution*, Lionel Robbins Memorial Lecture Series, Slide 5. London School of Economics.

Stern, N. and Rydge, J. (2012) 'The new industrial revolution and international agreement on climate change', *Economics of Energy and Environmental Policy*, 1: 1–19.

Stiglitz, J., Sen, A. and Fitoussi, J.-P. (2009) *The Measurement of Economic Performance and Social Progress Revisited*, Commission for the Measurement of Economic Performance and Social Progress.

Strachan, N., Foxon, T. and Fujino, J. (2008) 'Low-carbon society (LCS) modelling', *Climate Policy*, 8, Supplement 1.

Sue Wing, I. (2006) 'Representing induced technological change in models for climate policy analysis', *Energy Economics*, 28: 539–62.

Thirtle, C. G. and Ruttan, V. W. (1987) *The Role of Demand and Supply in the Generation and Diffusion of Technical Change. Fundamentals of Pure and Applied Economics*. New York: Harwood Academic. Available online at: http://catalogue.library.manchester.ac.uk/items/249194

Toman, M., Krautkrämer, J. and Pezzey, J. (1995) 'Neoclassical economic growth theory and sustainability', in D. Bromley (ed.), *Handbook of Environmental Economics*. Oxford: Blackwell, pp. 139–65.

Trade Union Congress (2008) *A Green and Fair Future: For a Just Transition to a Low Carbon Economy*. Available online at: http://www.tuc.org.uk/touchstone/Justtransition/greenfuture.pdf

Uzawa, H. (1965) 'Optimum technical change in an aggregative model of economic growth', *International Economic Review*, 6 (1): 18–31.

Waisman, H.-D., Guivarch, C. and Lecocq, F. (2013) 'The transportation sector and low-carbon growth pathways: modelling urban, infrastructure, and spatial determinants of mobility', *Climate Policy*, 13 (Supplement No. 1): 106–29.

Waisman, H.-D., Guivarch, C., Grazi, F. *et al.* (2012) 'The Imaclim-R model: infrastructures, technical inertia and the costs of low carbon futures under imperfect foresight', *Climatic Change*, 1–20.

Weyant, J. P. and Olavson, T. (1999) 'Issues in modelling induced technical change in energy, environmental and climate policy', *Environmental Modeling and Assessment*, 4: 67–85.

Weyant, J., Davidson, O., Dowlatabadi, H. *et al.* (1996) 'Integrated assessment of climate change: an overview and comparison of approaches and results', in J. P. Bruce, E. F. Haites and H. Lee (eds), *Climate Change 1995: Economic and Social Dimensions of Climate Change*, Contribution of Working Group III to the Second Assessment Report of the Intergovernmental Panel on Climate Change. Cambridge: Cambridge University Press.

Zenghelis, D. (2011) *A Macroeconomic Plan for a Green Recovery*, Policy Paper, Grantham Research Institute on Climate Change and the Environment. Available online at: http://www.cccep.ac.uk/Publications/Policy/docs/PP_macroeconomic-green-recovery.pdf

Integration and Implications

Each of the Three Domains has developed its own world: a way of thinking, language, base of evidence, academic community, ranked journals for research, policy recommendations, and associated constituencies. The key to Planetary Economics lies in integrating all three.

Analytically, no single domain offers a world-view that usefully corresponds with all the evidence. Behavioural and organisational theories (First Domain) are incomplete without an understanding of how price and technological innovation affect human behaviour and management structures. Neoclassical economics (Second Domain) cannot in its own terms adequately explain the efficiency gap (Chapter 4), the observed (ie. asymmetric and internationally divergent) responses to price (Chapter 6), or the enormous divergence between R&D intensity between different sectors (Chapter 9). Evolutionary and institutional economics (Third Domain) have their strongest theoretical foundations when building on the other two domains to explain how social and economic choices over time can combine to shape the evolution of complex technico-economic systems (Chapter 10). Efforts to explain the Solow Residual of economic growth are also in essence forays from classical economics into the other two domains (Chapter 11).

Practically, the most effective and efficient responses will involve all three associated Pillars of Policy, aligned with the different conceptions of risk (Chapter 2). Our long-run security depends upon translating an objective, strategic assessment of collective risks into practical policy in individual countries, companies and individuals (Chapter 12, section 12.3). This requires not only willingness to make the strategic investments required (Pillar III), but a social licence for governments to align economic signals in relevant markets and fiscal systems to reflect external costs (Pillar II), and to nourish and galvanise action by individuals and their organisations (Pillar I), all oriented around shared goals.

Combined responses across all Three Domains enable energy systems to adapt to constraints (section 12.4). First and Third domain processes in particular are largely progressive in the sense that they accumulate changes in behaviour and organisations, and infrastructure and innovation, which enable us to 'get more from less'. The resulting adaptability of energy systems has profound consequences, which can be demonstrated with simple numerical examples (Appendix).

Measures with an enduring impact on the pathway are likely to be several times more valuable, per tonne CO_2, than the assumed 'social cost of carbon emissions'. The optimal response may then evolve global energy-related systems towards zero emissions during the second half of the century. Such adaptability thus helps to bridge the gap between 'precautionary' and 'cost-benefit' approaches to climate change. Plausible assumptions consistent with the evidence in this book and other studies suggest that enhancing relevant investments by around €1 trillion/yr globally could secure this transformation, helping to avoid combined costs of climate impacts and abatement totalling potentially ten times this level later in the Century.

Securing such transformation would hinge upon integration at three levels.

Integration across three pillars of policy. The Three Domains embody complementary economic characteristics, including the relative scales of public and private returns to investment (section 12.5). No pillar on its own is either sufficient or stable. Pillar I policies (eg. efficiency standards) on their own would have to become increasingly stringent and intrusive to offset rebound effects. Environmental pricing (Pillar II) at the levels required if pursued on its own, is politically untenable due to the costs imposed particularly on vulnerable consumers and industries, whilst the beneficiaries are indirect and diffuse. The strength of incumbent interests and other lock-in effects raise insuperable barriers to new low-carbon technologies transforming systems (Pillar III) without the incentives arising from the other pillars (section 12.5).

In contrast: energy efficiency can support proper pricing by managing the impact on bills; carbon pricing enhances attention and incentives in relation to both energy efficiency and innovation, and generates revenues that can help to overcome structural barriers to adequate efficiency and innovation programmes; whilst innovation and infrastructure generate options for furthering energy efficiency and low carbon energy in response to associated policies (section 12.6).

Integration of climate policy with sectoral policies and co-benefits. Purely Second Domain processes imply a world in which decisions can be separately optimised, for example with energy policy focused on establishing markets whilst environment policy targets environmental pricing or caps emissions. Reality precludes such easy separation. Energy systems are by their nature heavily dependent upon regulation and infrastructure, which are essential to efficient market operation; the terms of these affect both the efficiency of energy use and investment choices in supply. Transport and urbanisation are similarly dependent on regulatory choices and infrastructure which all have long-run energy and environmental implications. These and other First and Third Domain aspects signify sectors which combine both public 'goods' (smarter choices, innovation and infrastructure), and 'bads' (eg. pollution). In addition to pricing the 'bads', delivering the 'goods' generally involves State-led programmes, but reliance on these can be reduced by designing pricing policies to maximise impact also on First and Third Domain decisionmaking (ie. private engagement and strategic investment). There is also a logical case for some of the revenues from pricing 'bads' to be earmarked to help finance the 'goods' (section 12.7).

In these circumstances there are demonstrable potentials for 'co-benefits' associated policies which combine climate policy with other objectives. Overall, co-benefits appear as a potentially important feature to the extent that climate policy can help to *motivate*, *stabilise*, *coordinate*, and *finance* policies with multiple benefits (section 12.8).

Integration of energy and climate policy with macroeconomic and development policy. The climate change problem in particular has been widely perceived as one of taking on an economic burden for future global environmental benefit. In reality the sectors most involved – energy, transport, heavy industry and urban structures – are crucial to our economic systems, and the need is to make strategic choices for conjoined long-term benefits of both economy and environment. Paradoxically there may be particular potential to secure such joint gains in regions struggling with recession, where a secure policy framework could help to galvanise underemployed human and private financial resources around low carbon investment (section 12.9).

Economic development itself involves a progressively shifting balance between the Three Domains. International frameworks which facilitate such development in low carbon directions, including in relation to enhancing energy access, could offer some of the biggest global joint gains of environment and welfare. Embedding commitments in an international framework could enhance their credibility for the purpose of attracting and focusing the investment required to change course. Any 'common platform' of commitments should also be designed to facilitate more ambitious countries, regions or industries to invest more in return for a larger economic stake in the transformations required.

One of the world's top economists described climate change as 'a hellish problem that is pushing the bounds of economics' (Chapter 1, note 29). Like many challenges, this is also an opportunity: to use the facts to extend the frontiers of economics itself, so as to more systematically help with our common planetary challenges.

12 Conclusions

Changing course

'Ignoranti quem portum petat nullus suus ventus est' [*No wind favours those who don't know where they are going*].

–Lucius Annaeus Seneca[1]

'Why not recognise these realities explicitly and take them as a starting point for theory, rather than trying to force theory in to the Procrustean bed of optimisation? Acknowledging the limits ... would be a fertile beginning for new discoveries and breakthroughs in all branches of economics. Plus, it would enable us to think more constructively ...'

–Stephen J. Decanio (1999)

'There is nothing so practical as a good theory.'

–Kurt Lewin (1951)

12.1 Introduction

Just because something is good does not make it easy. We all know about the benefits of being fitter but in societies that are sated with rich foods, sedentary lifestyles and automated transport, a healthy diet and exercise take effort that many cannot muster. Some reach a level of obesity that threatens their own health and happiness. We all know that bad habits can stick and be hard to break.

Just as humans are largely creatures of habit, so too are economic systems and the energy structures that underpin them. That is a core message of this book. The dominant models and theories of the Second Domain largely equate short-term ease and personal gain with long-run benefit. If it is easy and the market provides it, it is good for us; if something is hard, it is because it is costly.

The global financial crisis underlines the limitations of such thinking. Easy credit for mortgages was appealing for everyone. Myopia and herd behaviour led people down a dangerous path. Inadequate regulation of financial instruments helped to render long-term systemic risk invisible and untraceable. Second Domain thinking, without the disciplines of respecting real human behaviour on the one hand and systemic risk on the other, can be dangerous.

Despite the traumas resulting from financial debt, we continue to apply the same modes of thinking to natural debt: the uninhibited use of natural resources and the accumulation of the wastes in the atmosphere, in the hope that the future will take care of the consequences. The global challenges of energy and environment – notably climate change – deserve Lord Nicholas Stern's description as 'the greatest market failure in history'.

It is yet more than this, for the canvas stretches beyond 'market failure'. It is hard to think of any other arena in which the actions of every human being on earth eventually merge in ways that could ultimately harm all others and impact the planet itself over inter-generational timescales.

Obviously, issues at this scale involve basic ethical dimensions, as underlined by the discussion in the opening chapter on efforts to quantify climate impacts.[2] This book, however, has sought to focus on the ways in which ongoing economic research across several domains – interpreted as broadly as possible – can help to explain the issues and inform specific policy choices.

This chapter summarises key lessons and brings together some implications of integrated thinking across all three domains. It outlines how this shifts the economics from a static conception of sharing burdens for environmental gain to a dynamic appraisal of investment and returns along a different pathway. The policies required to deliver these benefits cannot be captured in a simple 'carbon price', but require intimate integration across the three pillars. The chapter summarises the evidence that such a strategy can bring multiple co-benefits and explains the mechanisms involved. The chapter then closes by outlining the integrated responses that could also aid European economic recovery, and offers thoughts on what Planetary Economics might imply for international approaches.

12.2 Three Domains: a short résumé

This book has argued that thinking about problems of both global and inter-generational reach – the great challenges of the twenty-first century – requires us to think across several domains of academic enquiry, broadening out from classical modes of economic thought.[3] We have probed, illustrated and applied these principles with reference to the challenges of energy and climate change, and entwined the development of theory, fact and experience. The challenges are without precedent in human affairs, and we have been struggling with them for decades, with distinctly limited success. Something needs to change in the way we think and act.

Chapter 1 outlined the extraordinary scale of global energy development, and the accelerating pace of global pressures on energy resources and environment. It also illustrated the huge complexity of academic efforts to quantify the severity of the challenges, but they tend towards one broad conclusion: the current trajectory is unsustainable. To move toward sustainable energy systems requires the dramatic acceleration of two distinct factors: the rate of energy efficiency improvement and the decarbonisation of energy supplies. There have been many efforts to respond to the apparent needs and opportunities, but they have lacked a coherent underlying framework, and have thus been insufficient to turn the trend of rising energy consumption and emissions. Moreover, the dominant economic theory carries the logical seeds of our apparent inability to solve the problem: the 'pessimism squared' with which Chapter 1 ended.

Yet there is huge technological potential to resolve our energy and environmental challenges. The potential for *smarter choices* is enormous, particularly – but not exclusively – in the ways we use energy in our homes, business and transport. Technologies for cleaner energy supplies are extensive, particularly for electricity production, but also encompass other energy carriers and are continually improving. The options span all sectors and are continually being broadened by innovation not just in individual technologies, but also in systems (see Chapter 3).

The book has revolved around the theme of how best to make our energy choices. Chapter 2 delineated three different domains of human decision-making in relation to economic and resource choices. The different domains can be broadly identified with different decision-making processes, operating over successively larger scales of space and time. This structure implies the need for three quite distinct pillars of policy, firmly grounded in an understanding of their respective domains.

The First Domain is defined by the ways in which practical behaviours differ from the theoretical ideal of rational foresight and optimisation: the scope for smarter choices by individuals and organisations within the confines of existing technologies and market structures. Alongside long-established analyses of structural barriers to 'economically efficient' choices – particularly concerning energy use – the underlying explanations have been greatly expanded by the evidence of experimental economics and the theories of behavioural economics that the experimental approach has spawned. All this has helped to overcome dogged scepticism about the prevalence and scale of potential for improvements, particularly relating to how individuals and organisations use (and waste) energy (see Chapter 4).

These characteristics point to Pillar I policies, which deliver smarter choices, principally through either *standards*, or by various approaches aimed at increasing public *engagement.* Improving energy efficiency is the main target, which frequently yields cost savings and improved energy services – irrespective of any environmental gain. The evidence unambiguously suggests that such policies can be (and have often been) beneficial – efficiency policies have reduced both energy use and energy costs as well as emissions. Yet experience indicates that the realised scale of savings is *not as big nor as cheap* as the pure technological potential suggests; moreover, the benefits of efficiency policies are *shared* between actual energy/emission savings and the 'rebound' increase in energy-using activities (see Chapter 5).

The Second Domain is that of rational foresight, in which people and organisations actively seek to optimise the trade-off of different inputs. Markets are generally (though not always) best at this and price is the defining signal – the central indicator around which rational economic choices can be made and aim to optimise resource use to the extent that prices reflect real costs.

These characteristics point to Pillar II policies, namely the use of markets, with 'proper pricing' to reflect external damages. In the energy sector, prices make important contributions to energy efficiency, yet they have an even greater potential leverage on decisions to choose cleaner and lower carbon technologies. In principle, policies to add the cost of environmental damage – including CO_2 – into energy markets could spearhead efforts to decarbonise energy supplies, providing energy markets function adequately.

Such pricing policies are the core recommendation of most economic analysis. Where implemented, carbon taxes have never made the 'sky fall in' and have had demonstrably positive results (see Chapter 6). The alternative approach to carbon pricing, through emissions cap and trade, has generally revealed that achieving the cap set (on emissions) turns out to be easier and cheaper than expected. Indeed, in contrast to fears of excessive costs, cap-and-trade systems have consistently faced the opposite problem: repeated price collapses undermine the incentives for cleaner investment. The best approach to pricing carbon is to combine elements relating to both price and quantity (see Chapter 7), which can also enhance the interaction with other pillars of policy.

Yet governments have encountered fierce opposition to 'proper pricing', ranging from removing energy subsidies to pricing carbon. Pricing affects everybody in transparent ways, and people object when they have to pay for something that previously seemed free. The benefits of proper pricing are indirect, widely spread and cumulative, while those most adversely and immediately affected – and even those who stand to benefit but do not (yet) perceive it – may strongly oppose it (see Chapter 8). Thus the instrument most recommended by economists is often the least favoured politically.

Nevertheless, there are carbon pricing systems either in place or under active development in an estimated 60 jurisdictions (of which about half comprise the 31 countries covered by the European emissions trading system). This points to the strength of the case for pricing carbon, despite the political obstacles. Yet the present and proposed systems do not yet look remotely sufficient compared to the scale of energy and climate change challenges. Real progress can only be incremental and differentiated between countries, according to different levels of economic and political development, and growing over extended periods (see Chapters 6–8).

The Third Domain is defined by the fact that technologies, systems and indeed ultimately human values and behaviour all evolve. The frontiers of technology and choice – which define the 'supply and demand curves' of neoclassical economics – are not fixed but are malleable over time, as innovation and infrastructure investments (along with social and institutional changes) open up new possibilities. However, our societies systematically under-invest in innovation – particularly in energy and some related sectors – while the default direction of investment favours the incremental improvement of existing and largely fossil-fuel based systems. There is thus enormous scope for Pillar III policies, using *strategic investment* to accelerate innovation and infrastructure towards cleaner technologies and systems.

To overcome the unusually low intensity of innovation in the energy sector which is reflected in the 'technology valley of death', a set of interlinked policies are required to push further and pull deeper across the innovation chain (see Chapter 9). Also, sectoral studies show that each of the major systems considered – vehicles and liquid fuel supplies, electricity, and urban development – are capable over the long term of enormous transformation. Yet such transformation is inhibited by the inertia of existing systems and the entrenched interests of incumbents (including, but not only, corporate interests) (see Chapter 10). Whether and how these three key systems (vehicles and fuels, electricity, and urban systems) are ultimately transformed will depend not only on strategic investment in the requisite innovation and infrastructure, but also on the demand for improvement generated across all three pillars of policy.

Finally, Chapter 11 located the Three Domains in relation to the historic debate surrounding economic growth and the cost of environmental constraints. It noted that traditional growth theories focused on the accumulation of capital, labour and other resources have themselves been unable to adequately explain economic growth without recourse to 'innovation' as the catch-all explanation for those factors which cannot be directly measured in the standard framework. This 'Solow residual' of economic growth comprises complex forces, including economies of scale and multiple forms of technological and institutional innovation; waves of combined innovation are associated with renewed phases of economic growth.

Energy, as a pervasive input to economic systems, can have an important role in such economic growth processes. In this chapter we show that First and Third Domain processes have much in common – they are both concerned with innovations in policy/regulation and technological systems respectively, which have the potential to create enduring benefits. Hence our conclusion that the First and Third Domains comprise the Solow residual – the 'dark matter' – of economic growth.

The neoclassical models of economic growth (see Chapter 11) converge in the long run towards an equilibrium (in which the accumulation of resources and capital offset depreciation and external costs such as environmental damage). Yet in reality, the framework conditions are themselves constantly evolving; the underlying assumptions of Second Domain economics thus dissolve over sufficient time spans into a much more complex world. The processes of the First and Third Domains are part of economic growth but are not inherently quantifiable or optimising. Behind them, moreover, lie questions about what motivates people to make an effort, improve, innovate, collaborate and construct for both themselves and the common good. Top economists throughout the ages have pointed to the fundamental question of motivation in economic development.[4]

It follows that the debate on whether environmental controls will hinder or contribute to economic growth cannot be logically or empirically resolved. Constraints generate both costs and value, just as the constraints on gold supply make it more valuable than water. If necessity is the mother of invention and institutional and technological innovation are central to economic growth, then the ultimate impact of environmental constraints on economic growth is unknowable, because it depends on our response. The possibility of positive impacts cannot be proven *in general*, but nor can it be excluded. What matters is credible and consistent policy across all three domains, designed to maximise the benefits associated with respecting environmental limits and utilising those limits to inspire and spur better performance. This chapter sheds more light on the policy implications.

Chapter 2 set out the nature of the Three Domains, including the alignment between risk and response perspectives, the evidence behind each and the respective theoretical bases. It also summarised evidence that for the energy sector, *all three domains matter*:

- the engineering (Chapter 3), experimental (Chapter 4) and policy (Chapter 5) evidence surrounding energy efficiency (First Domain);
- the measured responses to price changes (Second Domain) of both energy efficiency and technologies (including the divergence between domestic and international price elasticities and the asymmetry of responses between price rises and price falls) and the role of expectations and policy design (Chapters 6–8);
- the enormous physical and engineering potential for better technologies and systems (Chapter 3) combined with the observation that the R&D intensity of energy generation and major energy-using sectors (like construction) is exceptionally low for clearly traceable reasons (Chapter 9), while potential transformations in key energy-related systems are possible but impeded by huge inertia and path-dependence (Chapter 10).

Finally, the 'Solow residual' in growth theories (see Chapter 11) is itself evidence of the central importance of these multiple types of processes in economic growth, which reach far beyond the simple accumulation of factors of production.

No domain can be dismissed as irrelevant or as peripheral compared to the other domains in the context of transforming energy systems. Specialists in *individual components of the problem* may believe otherwise but, if so, they are simply revealing the limitations of their own horizons. Behavioural, neoclassical or evolutionary economics tend to disregard each other's insights.[5] Empirical specialists in buildings, electricity markets or future transport systems are likely to emphasise the corresponding theoretical frameworks and policies. Yet each are, in effect, the equivalent of the blind men connecting with just a part of the metaphorical elephant. With respect to the grand challenge of transforming the global energy system, all domains matter.

The framework of the Three Domains provides a simple way of understanding and organising many complex issues and raises some crucial questions. We start by briefly revisiting the relationship between optimisation and security, and the rest of the chapter then seeks to answer the following broad questions:

- What all this implies for the aggregate potential to 'change course', and for the appropriate nature and scale of efforts applied to reducing the risks of high carbon dependence.
- What the Three Domains add to debate concerning the relationship between public policy, competitive markets and energy prices.
- How the resulting three pillars of policy relate and can interact with each other, and what this might imply for practical, more enduring and effective policy responses.
- Whether (and how) responses, particularly to climate change, as a long-term collective challenge may bring about joint benefits with other aspects of improving overall human welfare.

The chapter then offers specific policy implications in the European context, focusing on the ways in which more coherent energy and climate policy could contribute to European economic recovery. It then briefly outlines possible implications for the broad shape of economic dimensions of international responses. Finally we conclude by reiterating the essential philosophy of the book: the need for our economic theories to be grounded in the facts that inform all the Three Domains.

12.3 On optimality, security and the evolution of energy systems

The message expressed in the opening line of this chapter has a mirror image: just because something is easy does not make it good. We need to understand where this 'something' may lead us. To continue the opening analogy, it is easy to eat the sweetest food, or drink the wine being offered to us, but if we are already overweight, or about to drive through thick fog, these may not be wise choices.

Two essential threads ran through Chapter 1: the risks that can arise in relation to complex systems we do not fully understand, and the huge inertia and path-dependency in energy systems.

The nature of the strategic risks associated with global energy trends have much in common with security challenges. Energy has long involved local inconveniences, typically solved through the reactive nature of traditional political processes to alleviate problems like cleaning up local pollution. Expansion of the global energy system over the past century has correspondingly elevated the associated problems to a level in which the dominant risks are profoundly uncertain in nature and timing, almost impossible to quantify, but potentially devastating, particularly when combined with the growing complexity and interdependency of our societies.

A strikingly clear example of underlying principles comes from an entirely different sector.[6] To change Chapter 11's metaphor of driving to the case of telecoms: Cambridge, the hub of the UK's 'silicon fen', generates huge data flows, as indeed do most cities. If the phone or data lines from Cambridge to London were the only route, communications quality would fall as the volume start to saturate the capacity of the lines: the phone would crackle and the data packets would be dropped. There is a local and immediate feedback of local inconvenience. However, data travels almost instantaneously, so if one line is congested, it is far more efficient to reroute the data via other lines which have spare capacity. That is the efficient local response, which if replicated adds up to an efficient global response – but only up to a point. That point, obviously, is the risk of a moment when every line becomes saturated – at which point the whole system could freeze. Note that the efficient response does not only save costs; no individual user experiences any inconvenience as their use of the system exceeds its local capacity, because others help out – as long as they can. However, the system itself – and everyone involved – becomes more fragile to a seizure which may affect all.

There is of course a way around this, but it requires monitoring the state of the system combined with responsible governance. The build-up towards a whole-system capacity crunch is much slower than for any individual line. Given adequate and objective monitoring and governance of the overall system, there is – in principle – time to take evasive action through agreed measures. The challenge is one of governance, to turn that 'in principle' into practice.

For this example echoes a growing concern about the fragility of highly optimised and complex systems. The Chatham House *Resources Futures* report highlights, for example, the vulnerability of global food and other supply chains given highly optimised just-in-time delivery systems.[7] The general conclusion is that *maximising efficiency in a system of expanding scales is only safe if it is accompanied by monitoring to provide continuous feedback concerning safe limits and a willingness to act in time to avoid systemic risk.* Combining economic efficiency with safety requires a deep connection and feedback between Second and Third Domain-type concerns. In the case of energy and the environment, scientists have expressed safe limits in terms of planetary boundaries (Chapter 1, section 1.5).

There are then two big challenges. One is that boundaries may be fuzzy and hence contested. The other concerns our collective willingness and ability to assess risks and constraints objectively, and to act in time. It is these characteristics which require us to combine foresight and caution.

Political debate is caricatured between 'optimists' and 'environmentalists'. The former can point to numerous past pessimisms, from predictions of food shortage to 'oil exhaustion', which turned out to be either false, exaggerated or more easily accommodated than expected.[8] Unfortunately such histories are a poor guide for one simple reason: they depend on humanity's remarkable ability to innovate and adapt *once problems occur.* We have been responsive, not anticipatory – as with the gush of energy efficiency after the oil price shocks. Similarly, well charted local environmental improvements in rich countries occurred mostly in response to legislation, adopted after the burden of pollution became politically unacceptable. We have had to suffer consequences before responding to the constraints. For the reasons detailed in Chapter 1 (in the section 'Driving blind', especially note 76), the timescales and nature of climate impacts in particular do not afford us that luxury. In particular, if the atmospheric constraint is to drive change and innovation in time, it needs to be translated through present policy.

Environmental writers frequently assert that we are 'at a crossroads'. If only that were so. If we were at a crossroads, then we would be forced to choose – and would be much more likely to make sensible choices. In practice, the global trend of energy consumption and emissions is more akin to drifting in a ship with the current. Yet we are the ones powering the boat; fuelling it are millions of companies, billions of people and trillions of dollars. Chapter 1 noted the analogy of driving through thick fog, but this does not capture the huge inertia explained in Chapter 10. Our habits, technologies and systems are designed for fossil fuels; resource concerns and scientific warnings seem puny forces in comparison.

Perhaps the best analogy is of an evolving, dynamic struggle to turn the course of a supertanker in the fog as the evidence accumulates of icebergs ahead while a rebellious and sceptical crew of incumbent interests fight to preserve the current course. Equating 'business as usual' with the optimum 'least cost' path is perhaps the most endemic mistake of much economic thinking. Our situation is more like being on board the *Titanic*, with hundreds of different skippers all arguing and none in control.

The inertia of energy systems (as well as the climate system), is thus the other essential thread running through the evidence outlined in Chapter 1 and elaborated particularly in Chapters 3 and 10. Our economies – together with energy systems, policies and the wider society – are complex, evolving systems. They may aspire to an optimum, but it is never achieved in any meaningful, static way. Instead, people, technologies, organisations, governance systems and even cultures all learn, change and adapt.

The global energy system has time and again demonstrated its inherent volatility (notably concerning prices) and confounded predictions. People respond to circumstances, experiences, available resources and what seem to be the best choice at a given time. With the latest rush of investment into frontier fossil fuel resources, spurred by historically exceptionally high energy prices, we could enter another period of surplus with a downward swing in fossil fuel prices. If so, we have no way of knowing how long it may continue: among other factors, the nexus of energy and geopolitics looks no less volatile in this century than the last.

What we do know is that the reactive choices of the time do not necessarily take us in sustainable directions. One manifestation of this is growing attention to the 'carbon bubble' – the fact that the volume of carbon-based fuels owned and valued by companies (and countries) is (without carbon capture and sequestration) inconsistent with the internationally agreed science-based goal for stabilising the atmosphere (see Chapter 1). This means that either that those limits will be breached or that a major re-evaluation of the value of fossil fuel deposits – and concomitant increase in the value of non-fossil energy resources – will occur. The longer the delay in effective action, the bigger will be the costs of the subsequent economic and/or environmental shocks.

12.4 Order from complexity: the economics of changing course

In the past few decades, we have learned a great deal about the nature of the supertanker that characterises our energy systems, and somewhat more about the icebergs ahead. Yet much of the quantitative economic analysis of climate change mitigation – the maths and the modelling applied to it – still struggles with the big issues. Stepping back from the detail and looking at some essential characteristics offers a route to pin down a key implication of this book's analysis for the global picture.

No history?

Innumerable global energy-economic modelling studies have examined the cost of CO_2 constraints and explored 'optimal' emission scenarios. Most have at their heart the classical Second Domain 'equilibrium' assumptions inherent in most mid- to long-term economic modelling.[9] For the purposes of energy economy modelling, at the core is a baseline projection, which is almost always assumed to be the optimal (least cost) path in the absence of the climate problem, along with equations which represent how the cost of cutting emissions is assumed to rise with the degree of cutback. The cost is thus defined purely in terms of how much emissions deviate from the baseline.

This has a curious feature: it involves no history. This is strikingly clear in probably the most widely used such model, Professor Nordhaus' DICE model: the equation that defines the cost of cutting emissions in 2100, for example, is wholly unaffected by anything that happens before. The world could do nothing until 2090, or it could undertake drastic emission reductions; neither would have any impact on the cost of a given level of cutbacks in 2100. That is what 'equilibrium' means: it is a timeless state, defined by externally imposed assumptions concerning the cost of and constraints on resource inputs to the economy in a given period. Many models may add to this some costs arising from, for example, premature replacement of capital, or constraints on how fast new technologies may expand. These enter in as constraints on the underlying equilibrium, but do not change the fundamentals.

In much of the early 'global cost-benefit' modelling, the underlying assumptions about the cost of cutting emissions compared to climate damages, particularly combined with the effects of discounting, meant that the computed optimal response involved only a modest deviation from the baseline. However, such results reflected an odd combination of assumptions. Modest climate damages reflect an assumption that we can adapt to climate change (plant different crops, move our coastal cities), even though the way in which climate change will manifest is fundamentally unpredictable and likely to express itself mostly in occasional extreme events (see Chapter 1, section 1.4 'Costing the climate'). Yet in contrast, the models assume that the energy system, in which policies and incentives *are* under our control, have no capacity to adapt.

This seems an intrinsically implausible combination, yet for two decades it informed the scepticism of many (though far from all) economists about the economic 'cost-benefit' case for substantial action towards stabilising the atmosphere. In practice, and recognising the fundamental uncertainties and controversies over valuing climate damages, many analysts moved away from 'cost-benefit' attempts to analyse just the costs of different trajectories to meet pre-set goals of atmospheric stabilisation. This is consistent with the precautionary and security-led approach outlined in Chapter 2, yet maintains a gulf between scientific and economic approaches which has not helped policy. Not till the Stern Review was there a serious attempt to bridge the gulf, and this – and the post-Stern debates – came to focus heavily on the debates about discounting and the evaluation of extreme planetary risks (Chapter 1, section 1.4 'Costing the climate').[10] The post-Stern debate was predominantly about the scale of the problem, not the structure of solutions or their interplay.

Now return to Chapter 2, Figure 2.3, which illustrated the different domain processes in relation to the 'best practice frontier' (a similar graphic, translated for policy implications, is displayed later in this chapter, as Figure 12.2). As noted, the frontier moves as technologies and systems evolve. The pure general equilibrium assumption (e.g. as reflected in the DICE model) is that what the frontier looks like – the technologies which define the trade-off of energy/emissions and costs – is defined at the outset for every future period.

In reality, as charted throughout this book, innovation responds to needs and costs typically fall with experience. This means that *how* the frontier moves will depend on three broad factors: how close 'current practice' is to the frontier; the extent to which relative prices change (driving movement along the frontier); and the impact of directed efforts to expand options, notably through innovation and infrastructure. Obviously these three factors correspond closely, though not uniquely, to the Three Pillars of policy. Since the frontier is evolving, we never actually get to it; it is more accurate to say that the closer we get to a given point on the curve (First Domain), the stronger the signal from relative prices (Second Domain), and the stronger our innovation policies (Third Domain), the faster that part of the curve is likely to move.

Thus if the crosses on Figure 2.3 move towards the right – economising on other inputs but not on energy and emissions – it will encourage innovation and evolution that pushes the curve to the right: we will simply increase our capacity to use energy and pollute the atmosphere more cheaply. However, if economies move down the axis, to save on energy and emissions, this will tend to push the curve itself downwards, and we will learn how to better improve our lives and the atmosphere with less use of fossil fuels.

If we map these factors on to the trajectories of energy and carbon intensities shown in Chapter 1, and the clear technological possibilities summarised in Chapter 3, we get something like Figure 12.1. By mid-century, fossil fuel consumption and emissions could be either much bigger or much smaller than the present levels, depending on whether we continue the current default path of rising emissions or turn it round. Where we get to depends on the path taken, *which in itself will have shaped the frontier of 'best practice' available in 2050*, and will then shape what is possible thereafter.

No going back

At the same time, we have also noted that energy investments – and of course broader socio-economic systems – are characterised by immense inertia. There is a default trajectory based on current systems, habits and interests, and hence a cost associated with changing course. Again, there is a parallel with individual psychology and the findings of behavioural economics: the tendency to 'anchor' preferences in the status quo. Even apparently 'good'

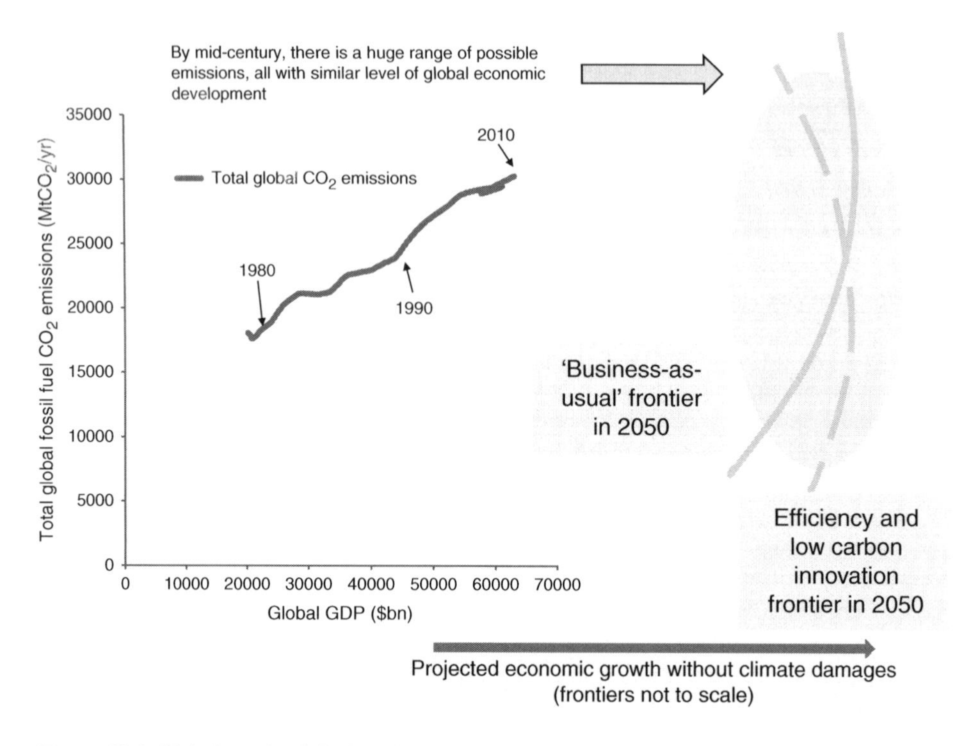

Figure 12.1 Global trends: default and potential trajectories

change involves a certain amount of discomfort and stress. Similarly, changes which may be good at the macroeconomic level, such as overdue structural reforms, can still involve transitional costs – like the personal cost for those that lose their jobs in one industry and have to find new opportunities in new sectors.

In reality we may not know how much of the cost is ongoing and how much is transitional related to the mismatch of technologies and current systems. Responses to a carbon price may include any mix of measures. Some of them may incur ongoing higher costs (for example, running gas plants instead of coal, when the latter would otherwise be cheaper), others may be enduring investments (such as a nuclear power station that lasts for many decades).

Now consider First Domain processes – the apparently 'free lunch' of smarter choices. One approach is to assume that the lunch is indeed free, and delivers a good meal of reduced energy consumption.[11] Yet it is ultimately more plausible to assume that the lunch is *not* free but has to be bought in other ways – notably by the effort of behavioural, policy and regulatory changes. As noted in Chapter 4, behavioural change often involves personal effort. All the energy efficiency programmes described in Chapter 5 also required someone to expend effort – to explore the possibilities, to correct the numerous market failures, to set the standards, to measure, to verify, to enforce and/or to nudge. Much the same is true, in fact, of other economically positive measures, like having to face down the politics of subsidy removal.

Yet *this still does not make such measures equivalent to the costs and benefits of Second Domain processes*, for one simple reason: the benefits are enduring. The standards and engagement (not to mention the other tools) of First Pillar policies end up saving money – hence our use of the term *smarter choices*. After achieving such an improvement, why would anyone go backwards and abandon such measures? Indeed, empirically, Chapter 5 noted a steady expansion of energy efficiency policies with few if any reversals. Who, after all, would want to expend more effort to undo a policy that saves both money and emissions? For example, as detailed in Chapter 5, inefficient fridges have simply disappeared from the European market in response to efficiency labelling and then standards (and the ones that replaced them are not noticeably more expensive). Pillar I reforms may not be free – though the cost may be measured in terms of effort to overcome personal or institutional resistance instead of money – *but the costs are transitional while the benefits endure.*

Now consider the Third Domain. Innovation generates options (see Chapter 9), as does infrastructure, whilst the wider processes of evolutionary economics embed the course of the system (see Chapter 10). Our summary of sectors (see Chapter 10) underlined the capacity of wider energy-related systems to transform – to become fundamentally different, not just higher cost versions of the same transport, electricity or urban systems.

Again, these changes do not come for free. Innovation requires investment. So does infrastructure. Major institutional reforms also require substantial efforts. Yet the resulting changes endure. Successful innovations diffuse, adding to our stock of knowledge, rather than disappear. Similarly a road or a railway, efficient buildings or new transmission lines – all require investment, but the results last decades or even centuries.

So the defining characteristic is that the costs of such strategic investment are transitional: not enduring. *Economically*, such measures thus share important characteristics with capital accumulation in the economic growth models explored in Chapter 11. *Environmentally*, they mean that subsequent generations inherit the combined benefits of this investment along with both less CO_2 from the past *and an intrinsically cleaner starting point for the future.*

The adaptability of energy systems

Indeed the combination of specific technology possibilities, sectoral assessments, observed differences between countries and all the literature covered under Pillar III all point to the same underlying feature: the *adaptability* of energy systems. Chapter 1 noted the degree to which different countries at similar levels of per capita wealth have substantially different patterns of energy consumption and emissions. Chapter 6 noted that countries with much higher energy prices seem to have offset this by higher efficiency, so as to end up spending a similar fraction of their wealth on energy. Each has adapted to its inheritance of energy resources and the decisions taken over decades. If technology choices and systems vary between countries, it is an obvious corollary – and it follows from all the analysis under Pillar III in particular – that they may also change over time.

In this respect, we finally see an important commonality between the First and Third Domain. While the Third Domain is broadly about innovation in technologies and systems, the First Domain is also about innovation – but in behaviour, regulations and policies that improve our efficiency. And since innovation is generally about finding ways of doing things better, it rarely goes backwards – whether the innovation is technological or takes other forms, such as in the evolution of policy on energy efficiency.

Moreover this characteristic, we argue, explains why the First and Third Domains are at the heart of productivity gains (and hence contribute to economic growth). If we had to repeat a given R&D investment every year in order to generate the same technologies again, obviously most of the benefits of innovation would disappear. The costs of energy efficiency include setting up measurement and enforcement systems and legislating the regulatory change, and all the costs of companies establishing their own monitoring, management and compliance systems. Again, if these costs were repeated every year, of course the gains would dissipate. It is the *fact* that the costs are transitional while the benefits of reduced resource use and waste are enduring that underpins their contribution to both economic and environmental improvement.

Whilst this characteristic of enduring benefit is dominant in First and Third domain processes, it is far from absent in Second Domain price responses. For example, most of the investments in renewables and industrial emission reductions under the CDM (Chapter 7, section 7.6) are also in practice largely irreversible: not only do the projects themselves continue to operate at a CDM price now close to zero, but a number of the industries and technologies established continue as 'best practice'. The reality is that technology, infrastructure and preferences are moulded by history and our choices. Given time, our systems have an enormous capacity to adapt.[12] In economic language, the classical Second Domain notion that supply and demand curves are fixed (or beyond influence), while useful as a short-term approximation within a specific country, is increasingly eclipsed over time by processes across all three domains through which systems can learn, innovate and adjust. The concept of adaptability can be quite unsettling to some economists, precisely because it does involve evolution in things that are often assumed to be externally fixed (notably supply and demand curves).[13] But it is a logical consequence of all we have shown about particularly (but not exclusively) First and Third Domain processes, and indeed is echoed in emerging strands of economic thinking more broadly.[14]

The economics of changing course

At first sight, one problem of all the ground covered in this book appears to be the sheer complexity of the issues, particularly when it comes to trying to aggregate implications at the level of global responses. The IIASA analysis of the 'double hump' in global energy futures (see Chapter 10, Figure 10.6) was the product of many years of technical research and involved groundbreaking mathematical advances.[15] Its conclusion that 'green' and 'brown' energy systems may have similar long-run costs (*excluding* environmental impacts) – and that a muddled mix may be more costly than either – remains little known outside the specialist arena, and like any projection over the century of course depends on numerous specific assumptions. A whole institute (the Santa Fe Institute) was established to analyse the mathematics of complexity and chaos which help to underpin the ideas of evolutionary economics.

Yet the basic insight of the discussion in this section suggests a way to boil all this complexity down to a simple proposition – that our economic systems and societies have substantial capacity to *adapt*. This indeed reflects a far wider strand of argument around policy towards complex problems: the need to learn and adapt.[16] The notion that energy systems adapt is itself a consequence of learning from the past century of evidence.

At the culmination of the journey, this enables us to explore some high-level implications in really simplified ways. The Appendix presents a simple numerical illustration in which the costs of changing course can be either *enduring* or *transitional*. The latter captures inertia, and the balance between the two represents the capacity of the system to adapt given time. The model is highly simplified: that is the point. It simply and transparently captures the essential insights of a far more sophisticated body of work extending back more than twenty years.[17]

Three important conclusions flow from this simple adjustment of including inertia and adaptability in modelling the optimal trajectory and associated costs of energy system responses to external costs or constraints (like climate damages).

First and most fundamental, *'optimal' responses can accumulate towards long-run solutions at moderate costs*. Adaptability in energy systems amplifies the benefits of action while constraining the long-run costs of deep emission reductions. With effort, solutions may accumulate, as responses across all three domains move the system towards stabilising the atmosphere during the course of this century. Moreover, this does *not* depend on motivating publics about the need for drastic action to avoid catastrophic risks, which as noted seems to have little political traction: it is the accumulation of incremental change which substantially also amplifies the benefits of action. If a low-carbon trajectory is not only safer but also represents an enduring change, then of course the case to get on that road is far clearer.

Second, *adaptability can narrow the apparent gap between 'cost-benefit' and 'precautionary' approaches*. As indicated in Chapter 1, the schism between scientifically based precaution and the economic calculation of identifiable climate damages has not narrowed to consensus. Through the extensive debates on risks, discounting and distribution, most economic evaluations have edged somewhat closer to the concerns of most scientists. Scientists for their part remain deeply concerned about the risks, but none of these seem sufficiently tangible, proximate and incontestable to motivate strong global action – particularly where the 'merchants of doubt', often funded by vested interests, set out to confuse publics.[18]

In the economics of changing course, however, the demonstrable adaptability of our energy systems narrows the gap: the costs of a more cautious approach (with lower energy/emissions) are reduced, and the benefits endure, whatever perspective on damages and risks one chooses.

Third, the optimal cost of action – to be more precise the effort worth exerting – rises when systems are adaptive because the benefits endure. Responses which help to adjust the long-run trajectory – through diverse possible mechanisms – may be much more valuable than just cutting emissions alone. This echoes the implications of Figure 9.1 (on declining technology costs meeting rising carbon prices), which illustrated the potentially huge benefits of learning investments. Given some pretty standard assumptions, the modelling in the Appendix suggests that the value of actions which adjust the trajectory can be several times the value of their direct emission savings.

Putting these three observations together completes the 'big picture'. To put specific numbers on it: the most detailed global studies tend to suggest that staying "within 2°C" will require 0.5–1€ trillion/yr of investment. The global 'cost/benefit' efforts outlined in Chapter 1 suggest that this is only 'economically optimal' if we place strong weight on the avoided climate damages, by assuming either potentially catastrophic risks, or a very low discount rate. Whilst there are good arguments for either (or both) of these assumptions, the analysis in the Appendix finds a similar level of investment to be optimal without assuming either – because the adaptability of the system multiplies the gains of early action.

Ultimately, it is this adaptive capacity of energy systems which has in the past kept our energy system broadly within the bounds of Bashmakov's constant (see Chapter 1). Action to ensure that accelerated efficiency and innovation can keep pace with rising energy prices similarly offers the prospect of containing the impact on costs and total energy bills.

To stress again: which processes matter depend on the scale of a problem. The approach does not dispute standard economic analysis applied to standard problems. The adaptive processes considered here may be quite insignificant for decisions relating to a single nation's economic performance over a few years, for example. But their importance rises with geographical and temporal scale.

Thus for Planetary Economics these learning, dynamic and adaptive processes are crucial. Planetary Economics at one end of the scale concerns the collective impact of individual choices by seven billion energy consumers. At the other end, it must combine the cumulative impact not only of these individual choices over time, but also of the full span of government policies around markets, pricing, innovation and infrastructure. The consequences of these combined choices will diffuse globally over decades through some of the most complex and networked infrastructures ever created. The simple modelling exercise set out in the Appendix, as informed by the evidence accumulated throughout this book, suggests that for such processes the combined forces of inertia and adaptability in energy systems is likely to dominate over the more traditional dimensions of calculated costs and benefits and thus fundamentally change the picture. Moreover the 'multiplier' benefits associated with the transformation of energy systems are more tangible and direct than the climatic benefits, and are more likely to accrue predominantly to those making the investment – which has potential to fundamentally change the politics.

The rest of this chapter explores the policy dimensions of taking such an integrated view across all three domains.

12.5 Not just nails

An English proverb observes that 'if all you have is a hammer, every problem looks like a nail.'[19] In its purest form, the abstracted concepts and modelling of Second Domain assumptions has traditionally led many economists to two fundamental propositions concerning climate change responses:

- rational responses involve a calculated trade-off of the aggregate costs of cutting emissions compared to the global benefit of reducing climate change;
- the optimal policy instrument is a global carbon price.

These two propositions indeed become conflated so that the optimal carbon price is a direct indicator of the costs that should be borne for the sake of the environmental benefit. In a wholly Second Domain world, price is *the* optimal instrument, and it has to work harder and harder to widen the gap from the default (rising) emissions trajectory. Many of the models, which are grounded in this assumption, predict that astronomical carbon prices will be required to stabilise the atmosphere – some rising to over a thousand dollars per tonne of CO_2 (unless they include a 'backstop' technology which then defines the long-run price).

It is an entirely reasonable question to ask whether such costs would be worth it – how much *should* we be willing to incur? Unfortunately, as noted in Chapter 1 (in the section 'Costing the climate'), it has proved impossible to answer this question if it is articulated as a global numerical trade-off: there are just too many imponderables, on multiple fronts of science, systemic risks and the ethics of aggregation over space and time. A debate framed in terms of security, respecting planetary boundaries, is essential. However, this still raises questions about the scale and nature of the insurance premium we should pay – and the willingness of individuals and voters, remote from the problem, to pay it.

Moreover, projections that the carbon price associated with staying within 'two degrees' of global warming could rise to many hundreds of $/CO_2$ contrasts sharply with the reality: even jurisdictions which have enacted a carbon price have struggled to get above $30/tCO_2$. The capacity of energy systems to adapt will in itself moderate the price levels needed to make the transition, as suggested by the previous section and some of the models in Chapter 11 (Figures 11.2 and 11.3; see also corresponding notes). Undoubtedly, carbon prices will need to rise substantially as part of any major decarbonising transition. Yet Chapter 8 (in section 8.8, 'The false god …') noted the fallacy of the idea of a unified 'global carbon price' as an optimal response in an unequal world.

Chapter 2 already suggested that such conundrums might best be tackled through the alignment of the different levels of risk conception with the different domains of responses (see Figure 2.6). Having established the evidence bases through Chapters 3–11, we can now finally understand this proposition more deeply in terms of the diverse underlying economic characteristics of the three pillars.

The cost of insurance: on models and mindsets

Models inform mindsets, and the dominant modelling approach in many economics ministries and much of academia is some variant of a 'general equilibrium' modelling grounded squarely in Second Domain assumptions.[20] As noted, if these are the only models used to evaluate policies, pricing looks like the universal answer and anything else is less efficient. Recognising the Three Domains which characterise real-world energy systems makes the problems both more interesting and more manageable.

The core point of the 'nail and the hammer' is not about modelling results but is instead about concepts and assumptions. The computer models that many academics use start to define the way they think. The models that underpin technical advice can start to become confused with reality. The aspects that *can* be quantified, even in a very loose, fragile fashion, capture more attention in policy than those which cannot.

This makes it all the more important to be clear about what such models assume, what they can say and what are their boundaries, and hence the kind of questions they can help to answer. General equilibrium models, developed with regional details, may, for example, generate valuable insights concerning the costs, emissions and distributional impacts of carbon pricing over a period of a couple of decades, for Second Domain processes are the primary focus of carbon pricing. The same model may be misleading if applied to studying the implications, for example, of the US vehicle efficiency standards – a policy predicated upon First Domain behaviours, with an objective to ultimately drive Third Domain transformation in the technology and competitiveness of more efficient vehicles.[21]

Fortunately, the representation of energy producers and consumers as perfect optimising entities operating in perfect markets is largely confined to an abstract academic world of global 'top-down' economic modelling. Such abstractions of theory have not greatly impeded progress on energy efficiency policies because most real-world policy analysis is not so stylised, particularly with regard to Pillar I policies. Balance is introduced into policy evaluation by the plethora of engineers and others pointing to continued technical potentials, by the legitimacy accorded to policy focused on removing 'market barriers' and by the enormous practical experience of energy efficiency policies. Government energy departments do tend to evaluate 'micro' policies with 'micro' tools and recognise that the gains from energy efficiency policies are real and enduring – hence their global spread. This approach may neglect or underestimate 'rebound' effects but, as noted above, these do not negate the welfare gains, since rebound instead means taking the benefits in other ways (like warmer homes – see Chapter 5, sections 5.3–5.6).

Far bigger problems arise with the modelling of Third Domain processes. Infrastructure, innovation and the global diffusion of better technologies and policies operate at much larger scales of space and time and make things much more complex. Consequently, it is far easier to ignore associated policy and technology learning-by-doing and to assume that 'someone else' will do the learning. Most models – and thus also much policy advice – reverts to standard assumptions when modelling national policy. Yet if everyone assumes that technology falls like manna from heaven, potential technologies stranded by the 'valley of death' (see Chapter 9) or otherwise locked out by incumbent systems (see Chapter 10) will stay that way – resulting in collective conservatism and an inability to change course.

Similar limitations make it much harder for governments to pursue the ideas of 'green growth' outlined in Chapter 11. Most of the economic models used by governments do not contain any Third Domain economic processes. In such models, environmental and climate policy inevitably appear as costs without (economic) benefits because the mechanisms that could generate benefits are absent from the models.

As noted above, not all abatement actions are comparable. Some may indeed help to foster low-carbon innovation. Others may have other enduring consequences, like building deeply insulated buildings, creating dense public transportation infrastructures (rail, waterways, electric buses) or investing in transmission systems that open up access to remote renewable resources. But some, like running gas plants instead of coal plants, may involve little or no enduring benefit beyond the actual, immediate, emission reductions. An economy-wide carbon price, which rises over time, is a crucial bedrock, but options which involve learning and other enduring consequences may justify a substantial 'premium' over this base. Thus a key implication of the Three Domains in general, and implicit in the results of the previous section, is that policy needs to be more discriminating. So do the corresponding tools of analysis and mental models.

For example, many economic critics of renewable energy policies in particular point to the fact that, per tonne of CO_2 saved, renewable energy sources look relatively expensive. The message of the analysis presented above is that it can be well worth paying a premium for measures which indeed do contribute *enduring* solutions, even if we cannot make robust estimates about the actual degree of learning or other benefits. Many of the benefits may be both enduring and public, and thus not captured in market prices or private incentives.[22]

More realistic modelling, capable of representing all three domains (and policy pillars), does *not* necessarily lower costs. A few such models have been developed for energy systems analysis and they frequently stress that making rapid *near-term* changes may be costly, difficult or even almost impossible; such models capture the inertia in the system better than classical models (hence the importance of inertia in the stylised model of our Appendix). But such analysis can better represent the cost reductions and propagation of new technologies and (if sophisticated enough) systems, which represent the heart of the transition process.[23]

Echoing the message of Chapter 6 (see Figure 6.6), therefore, there is indeed potential to align a strategic transformation goal with scientific constraints, while nearer-term pricing is moderated by concerns to manage costs. Ultimately, the challenge is to move from a mindset of making efficient marginal changes to one of understanding the transformation over time of complex systems, including the interrelationships between innovation, infrastructure, institutions and behaviour. In the real world, that is not something that can either credibly or efficiently be achieved with only a hammer, for many of the elements are not nails. As with any craftsman, the key is to use the right tools for the right tasks. This is the challenge for economic modelling – but even more fundamentally, a challenge for policy.

On investment and returns, public and private

One key step in moving from the broad concepts of the multiple domains to effective policy is to delve more deeply into the underlying economic structures, including the relationship between private and public benefits (as illustrated conceptually in Figure 12.2).

We have seen in Pillar I that the decisions in the First Domain typically have high monetary returns (10–25 per cent or even higher annual paybacks on investment); indeed the evidence

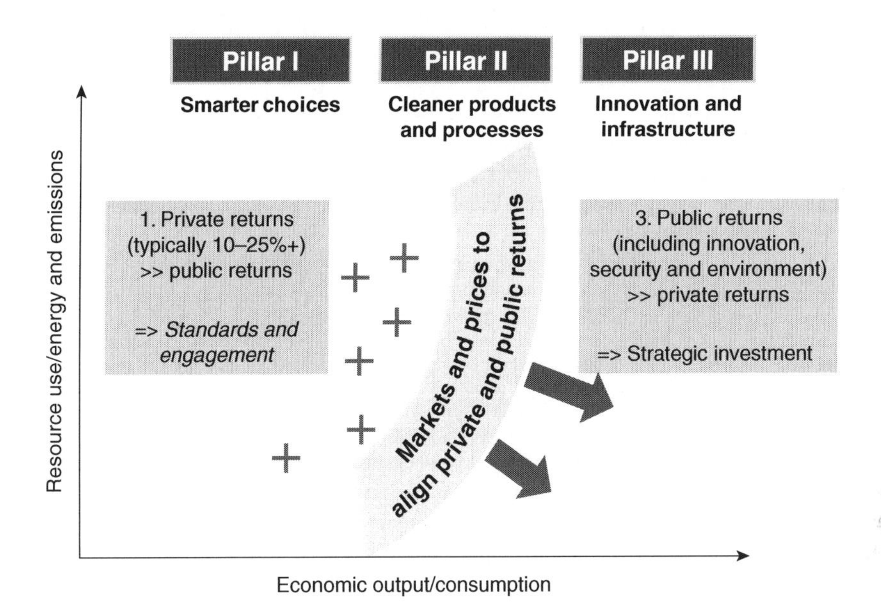

Figure 12.2 Public and private returns from the Three Pillars

in Chapter 5 carries examples of energy efficiency measures which pay back in a year or two, implying a rate of return exceeding 50 per cent. Chapter 4 explained why markets themselves do not deliver these choices, and Chapter 5 provided the evidence that public policy, such as efficiency standards, can do so with net benefit. The overall returns involved appear to exceed those typical of most government programmes; much of the benefit stays with the individual, alongside the public benefits of reduced fossil fuel dependence and associated emissions. Moreover, some of those public benefits may be local and tangible, such as reduced local air pollution: recall the observation in Chapter 2 that First Domain characteristics are most typical of smaller decision-making entities, focused on smaller scales of space and time. Where public policy can help to drive such actions, the gains can be substantial and, as noted, rarely reversed. We return to this theme in considering 'joint benefits' later in the chapter.

Pillar II policies can then be understood as aiming to better align private gains with the public gains. In these conditions, markets are almost always the most efficient system (notwithstanding distributional impacts) for all the reasons commonly advanced, notably the huge advantages of decentralised decision-making based on alignment of costs and benefits as mediated by prices. The 'beautiful theory' that underpins the idea of a market-based general equilibrium is, fundamentally, an expression of a world in which all prices reflect all costs and benefits, and hence one in which private gains equal public gains; the job of government is then simply to ensure that markets (including the cost of 'bads') exist to deliver these optimal decisions.

Pillar III policies can then immediately be understood as those for which public benefits exceed private gains. Infrastructure is the classical case: only in rare circumstances is it possible for a private investor to capture the benefits of building a road, for example. The 'spillovers' associated with innovation are almost universally recognised as justifying government funding of R&D. As Figure 9.1 indicated, the benefit-cost ratio of successful low-carbon innovation may be vast, but the rest of that chapter explained why only a tiny fraction of these benefits are likely to accrue to innovators through markets.

The link to the economics of changing course is obvious. First Domain characteristics are pervasive among the world's seven billion energy consumers (and their organisations); a good share of the gains from associated policies (mainly on energy efficiency) accrue to them in direct and tangible ways, enhancing the political appeal of these policies. In energy markets, charging for pollution to reflect the cost of environmental damages is the most efficient way of reducing the collective welfare burden of pollution. Securing the public benefits in the Third Domain, however, requires strategic investment, mostly funded or otherwise incentivised by government; the major returns are public, not private, and are likely to be slower and later.

Yet the overall gains from Pillar III policies, particularly as new technologies diffuse over space and time, may be huge. They justify commensurate investment, not only because of the global good but given the prospect that a substantial portion of the benefits may accrue either to the supporting government or its industries. Over time, benefits accumulate, for example, from investment in renewable energy sources that exploit domestic resources at low marginal costs or as domestic industries gain a key role in technology supply chains. Moreover, as environmental costs are progressively internalised, those benefits grow. Also, the greater the scale of such strategic investments, the less exposed a country will be to the risk of economic shocks likely as the world is forced to adjust to the limitations of cheap resources and/or the atmosphere – the bursting of the 'carbon bubble'.

It follows that the conception of climate mitigation as simply 'incurring costs for the global benefit' is woefully misplaced. The structure rather is that of investment and returns. In terms of government policies across the three pillars, the greatest benefits will go to those who best judge the appropriate level, mix and focus of investment in the low-carbon economy.

12.6 Only connect: integrating policy across the three pillars

Utilising these insights in practical policy will require becoming more sophisticated in our understanding of the three pillars and how they may interact.

The limitations of single pillar policies

Since the characteristics of energy systems and climate change span all three domains, changing course towards a low-carbon economy will involve working across all three pillars of policy simultaneously. Indeed, relying on a single domain or process is ultimately self-defeating:

- A focus purely on increased efficiency is clearly inadequate: success makes it cheaper to do things that consume energy or emit carbon, with the consequence of 'rebound' as people consume or do more in response (see Chapter 5). There are also likely to be diminishing benefits from efforts to bring us ever closer to the 'best practice frontier'. Only progress in the Second and Third Domains will then expand the scope for new cost-effective energy efficiency options.
- Relying on price alone is the favoured tool from the classical perspective, it maximises the efficiency of market transactions. Yet it assumes conditions which are not satisfied in the First or Third Domains, and political obstacles have hugely constrained the pace of introducing price measures. In the short run, the extent to which energy demand falls in response to price increases is limited (at least without complementary Pillar I policies) and the timescale of energy supply system responses to price changes can be

decadal. Moreover, the failures in the innovation chain blunt any innovation response; as noted in Chapter 9, if the innovation chain is broken, energy/carbon pricing will not fix it. The result is that rising prices hit consumers – who are voters as well – with rising costs, when they have very limited options to respond and cannot readily identify or relate to the potential benefits which are more distant and nebulous. Prices have the biggest distributional impact of almost any instrument, provoking strong opposition. Relying on price alone risks generating more resistance than positive action.

- Purely technology-driven approaches, applied without complementary measures, are also self-defeating. A key lesson explained in Chapter 9 is that successful innovation requires a mix of push and pull forces, and that both are weak in the energy sector. If there is no market-based pull, technology programmes will have to be entirely driven by the government. The products of such innovation will compete with well-established incumbents, which will be in a strong position to keep out new entrants. Without pricing, regulation or actively engaged consumers, it is all push and no pull. Without pull there is little market; without a market, innovation will either wither, remain confined to the laboratory or become totally dependent on subsidy. Successful Pillar I and II policies are required to generate demand for low-carbon products and processes.

A traditional idea of policy analysis has been that we can 'work along the abatement curve' shown in Chapter 2 (see Figure 2.3), starting with the 'negative cost' energy efficiency measures, subsequently introduce a carbon price, and then later move on to the higher cost options. We have shown this strategy is fundamentally flawed, for multiple reasons. Based on such reasoning around inertia and learning, a recent World Bank study pinpointed 'when starting with the most expensive options makes sense' – which is a far cry from the simple idea of starting at the cheap end of the 'abatement curve' and working up it to more expensive options over time.[24]

We need, in other words, to work with all three domains at the same time, using all the pillars in the smartest way possible. This will be all the more effective – and compelling – if we also understand more clearly the interactions between the Three Pillars, which have been sketched at the end of each of the corresponding parts of this book, and are summarised in Figure 12.3.

Interactions between the pillars

We can start at almost any point within Figure 12.3 and work from there. Starting with Pillar I, behavioural/organisational measures in the 'satisficing' domain will tend to improve energy efficiency and thereby reduce the adverse impacts of Pillar II action (pricing) on consumer bills. 'Nudging' consumers should also make them more responsive to rising energy prices. The international comparison data in Chapter 6 suggests that *long-run* responses may be sufficient to keep energy bills roughly constant in the face of rising prices.

If the scope of Pillar I policies also engages individual preferences from an environmental perspective, this may also be a powerful lever for innovation. The role of consumers and leading public figures in opening up the market for hybrid cars is an iconic example. Consumer interest in efficiency and environmental benefit – especially if aligned with style – may thus have an important potential to increase the effectiveness of Pillar III policies.

Moving to Pillar II – pricing – the interactions are even more striking. Rising prices will raise attention accorded to energy wastage (First Domain), thus reinforcing the effectiveness of related Pillar I policies. The revenue from economic instruments can also help to fund energy efficiency programmes, which for reasons elaborated under Pillar I can be highly cost-effective but still require public funding. Examples include rollout programmes for

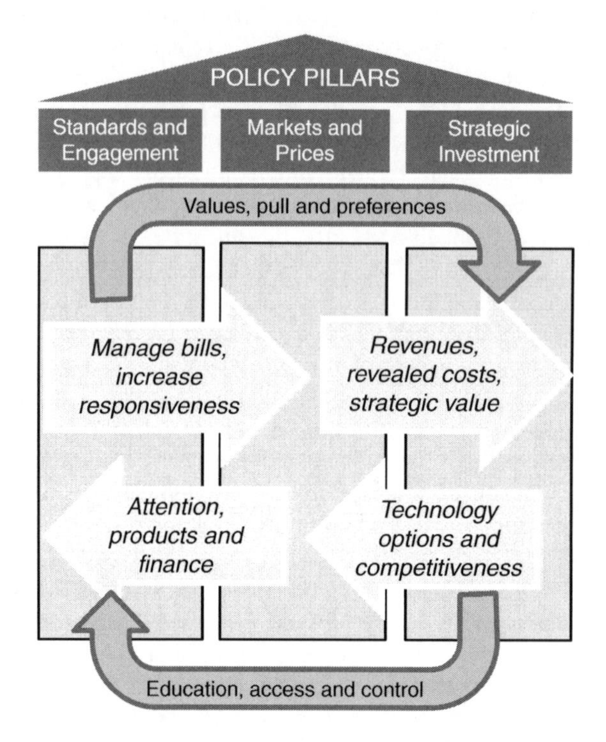

Figure 12.3 An integrated package: interactions between the Three Pillars

building insulation, for example. Such programmes typically bring various joint benefits as outlined in Chapter 5 and summarised in the next section.

Pricing will also interact with strategic Third Domain decisions and the corresponding Pillar III policies. Proper pricing will enhance the incentives for low-carbon innovation and the value of associated R&D and infrastructure. Rising carbon prices will increase the potential economic gain to be made by investors in low-carbon innovation. For the reasons discussed in Pillar III, this incentive is, however, seriously incomplete – especially in electricity, construction and some heavy industry sectors. Consequently, there is also an important role for funding of innovation, including the expensive processes of scale-up to accelerate the passage of big technologies through the innovation chain. There are additional reasons why policy may put more emphasis on such approaches, but again balance across instruments is the key: exclusive focus on innovation policies risks exacerbating lock-in.[25]

There is another, less often noted but strategically vital link from second to third pillar policies. The response to market prices reveals the real costs of technologies in ways which are far more credible than the cost estimates of the engineers and industries involved; after growth and learning, cleaner technologies have frequently turned out to be cheaper than predicted (Chapters 3 and 6). Price and its trend are a clock through which progress and difficulty can be tracked. A rising carbon price also provides a landing point for innovation – a clear point at which governments can get out of the business and leave low-carbon technologies more to broad-based competitive forces (see subsection below, 'First among equals?').

Finally, Pillar III policies feed back to other policy domains in numerous ways. Better, more efficient, cleaner energy technologies will help consumers respond more effectively to energy efficiency policies (Pillar I) and expand their range of options to respond to prices

(Pillar II). All these factors help to reduce the impact of rising carbon prices on bills and the wider economy while creating a clear potential for economic benefit as low-carbon industries grow in scale. Indeed, innovation is at the heart of managing the long-run costs of tackling a problem as profound as climate change.

The infrastructure dimension of Third Domain processes are also crucial in building a low-carbon economy. Infrastructure can extend our access to cleaner energy resources and expand the capacity to integrate renewables at large scales. Institutional innovations can increase investor confidence, thereby attracting long-run capital. It is also a realm of long-term thinking, including education, which can help people to become more informed about both the problems and responses, which in turn can feed back to individual preferences and social norms – i.e. to First Domain behaviour.

At the interface of Pillars II and III lies a crucial challenge, particularly in the realms of trade and state subsidies. Since competition is generally healthy and most valuable where there is a 'level playing field', huge efforts have been made in international trade and regional agreements to prevent or limit State subsidies. Yet for all the reasons articulated in Chapters 9 and 10, governments have to be inextricably involved in fostering the investments required to change the course of energy systems (and in practice, are deeply involved in energy anyway). There is no blanket solution to this tension, only the pragmatic need to strike a balance: to give room for effective transformational policies, to design them in ways to harness the benefits of competitive forces, and to design market instruments to contribute as much as possible to the transformations required (see "Expanding the market domain", in section 12.7).

To stress again, however: some Pillar III policies without an appropriate market context and interactions risk trying to solve the problems of energy and climate change with the tools of central planning. As Chapter 11 underlined, that doesn't work. Ultimately, constructive interactions with market forces are key to the success of Pillar III policies.

This lays the intellectual foundations for thinking more clearly about why and how different policy instruments should be combined. This is a complex topic and each jurisdiction needs to work through the implications based on specific circumstances and existing policy landscape; we return later to the European case.

Diversity and targeting in instruments and decision-makers

There is no global decision-maker; indeed there are not really unified national decision-makers. The universal 'representative agent' of most economic models, required to aggregate results based usually in response to price, is an approximation which can become deeply misleading when it comes to real policy. The Three Domains provide a natural framework also for a first level of disaggregation, as illustrated in Table 12.1, which depicts three 'archetypal' classes of decision-makers:

- *individuals and consumer-facing organisations* for which energy consumption is an incidental cost (over which they may have limited control), which hence may be easily eclipsed by other factors (many public organisations in the service sector also fall into this class);
- *investment and procurement decision-makers* in other companies, typically earlier in manufacturing supply chains, many of which have a higher energy intensity and are more detached from their final consumers; and
- *public authorities* charged with making decisions and policies for the public interest, and potentially board-level decision-making in multinational companies concerning strategic development of the firm's areas of interests.

Table 12.1 Different classes of decision-makers, investments and drivers

	Type of investment		
	Organisational or behavioural change	*Product/project*	*Strategic*
Decision-maker	*Principal decision factors*		
Individuals and consumer-facing organisations	Attention, habits, anchoring	Price and payback, values or brand	Expectations
Corporate	Outside drivers (consumer, competitor, regulator)	Price and payback time/ market discount rates	Strategic competition/ market trends and scale expectations
Public authority	Political priorities, targets or legal mandates		Strategic targets, public discounting or sustainability assessment

These three archetypes of decision-maker obviously broadly correlate with the Three Domains. Across the top of Table 12.1 are three different kinds of investment decision that can be taken by these entities:

- investing effort in organisational or behavioural change;
- investing in specific products (purchases) or projects; and
- strategic investments in a direction of travel or future options.

Similarly, these different types of decisions broadly correlate with the Three Pillars of policy. Moreover, in the way that different actors evaluate costs and benefits, there is a link to the fundamental economic debates about discounting noted in Chapter 1 (notes 55–60). Product purchases by consumers will reflect their (usually high) discount rates or rapid payback requirements. Companies will decide which projects to invest in guided by the market interest rates (and cost of equity) they can attract. Public authorities will use social discount rates for public investments, and for public-private partnerships may be best advised to evaluate specific project investments using hybrid approaches which combine the market cost of capital with social discount rates.[26]

Across the matrix of Table 12.1, price and quantity play different roles for different entities taking different types of decisions. This is another reason why the economic debate on 'price versus quantity' is misguided; both are required, because they perform different and often complementary roles for different decision-makers taking different decisions.

12.7 Expanding horizons

Thinking across the Three Domains not only leads us to consider the corresponding diversity of policies, their interactions and the different decision-makers involved, as sketched in the previous section. More fundamentally, it implies a broadening of horizons, particularly concerning the role and design of economic instruments and the institutional structures required to foster transitions.

Revenues and 'earmarking'

Economic instruments generally raise revenues, whether through taxes or through increased auctioning of emission cap-and-trade allowances. The previous section noted one potential interaction between pillars as being the use of revenues from economic (Pillar II) instruments to help fund activities on the other two pillars – programmes on efficiency and engagement, or financing innovation and infrastructure.

In principle, the most efficient way to spend money should be logically separate from how it is raised: for a central, unified decision-maker (notably a national government) they should be separate decision processes. The classical view is thus that there should be no 'earmarking' (also known as 'hypothecation') of revenue streams for particular expenditures.

In practice, reality does not conform to this ideal. Gasoline taxes are often used to fund roads or public transport infrastructure (see Chapter 6). Exceptions in the field of climate change include the deal that established the Carbon Trust (see Chapter 5 – though the UK Treasury carefully vetoed any *institutional structure* to the linkage), and the set-aside of 300 million tonnes of EU ETS CO_2 allowances, to be sold to support the development of CCS and other low-carbon technologies (see Chapter 9). The Japanese struggle over carbon tax or cap-and-trade has also largely been a shadow struggle over who gets the revenues, and the Waxman-Markey cap-and-trade scheme proposed by the US House of Representatives (but never taken to Senate) was suffused with linkage between revenues and expenditures. At the international level, revenues for the Adaptation Fund under the Kyoto Protocol are obtained in part from a transaction charge on the generation of emission reductions credits under the Clean Development Mechanism, subsequently extended to encompass all project credits under the Protocol (see Chapter 7).

So in reality, earmarking of revenues is rife. Much of this has simply been driven by the politics of trying to buy out various political constituencies, but there is more to it than this.

Where there are pure 'public goods' (like education or defence), there is a clear case for public expenditure funded by taxation. It makes no sense to raise public revenue from such activities and, regardless, these and other 'public good' activities generally and inevitably do not make much profit that can be taxed.

Conversely, where there are 'public bads', like pollution, the polluter pays principle applies as indicated, and these activities clearly should not be subsidised.

Energy and climate change seem almost unique in that policy has to deal with an intricate combination of 'goods' and 'bads': charging for pollution while supporting energy efficiency, innovation and infrastructure. Moreover, activities on these two pillars (I and III) not only contribute directly to the objective of reducing the 'bads' (e.g. emissions or energy insecurity), *they also* facilitate the use of economic instruments: as shown in Pillar II, the acceptability of carbon pricing is intimately connected with the extent of efficiency and innovation, for both consumers and industry. Revenue and expenditure are thus logically intertwined.

Given the importance of all three domains, the strength of policies across all three pillars should be directly linked to the perceived seriousness of the problems. Hence the level of expenditure under Pillars I and III should *in principle* rise along with the appropriate level of Pillar II carbon pricing.

This is the opposite of the classical assumption in which non-price policies are often seen as second-best alternatives to proper pricing (and hence are sometimes portrayed as inefficient substitutes or even as being in competition with pricing policies). The fundamental point is that policies across the three pillars are not competitors but are complementary. Hence there is a perfectly logical case to link revenues to expenditures in tackling a problem of this nature. Earmarking at least some proportion of carbon-based revenues to expenditures on energy efficiency, innovation and infrastructure is not only politically convenient but also reflects the underlying economic structure of a public good-and-bad issue spanning all three domains. It can make sense for pricing 'bads' to fund the public 'goods'.

There are potential distributional (and hence also political) advantages to using the 'bads' to fund the 'goods'. For example, if the revenues from pricing emissions from a given sector are in part used to fund innovation in that sector, this may help to alleviate competitiveness concerns. It may also of course ease the politics: people – and industries – will be less resistant to paying if they themselves receive tangible benefits.

There are, however, potential and important risks to earmarking, which justify the caution of most governments. Without good governance and clear principles, linkages between revenues and expenditures may compromise both. If *all* revenues from pricing bads were linked to specific expenditures, for example, this would preclude the wider 'double dividend' benefits discussed in Chapter 6 associated with a shift of taxation away from labour and/or investment taxes. The biggest risk is if the linkage is made too tight, so that emissions pricing is only accepted to the extent that it generates revenues for specific purposes. This appears to be part of the story in Japan, which recently agreed a small carbon tax reflecting principally a political deal on use of the revenues. Such an approach risks making emissions pricing subject to institutional capture (by the recipients of the revenues) and divorced from its *primary* objective – as an economic instrument to reflect the cost of emissions in decision-making.

First among equals? Expanding the market domain

Linkages do not only occur 'horizontally', between pillars. Figure 2.6 emphasised that each pillar of policy, while impacting most upon one domain, could have spillover impacts on others. In particular that figure suggests that markets and pricing can have impacts on both First and Third Domain behaviours. The extent to which they do may in fact depend heavily on how they are designed.

Given all the drawbacks of excessive reliance on government-directed policies and subsidies, there are benefits to maximising to the extent possible the impacts of carbon pricing across the other domains – trying to increase the 'indirect' impacts depicted in Figure 2.6. This could help to reduce the degree of reliance on purely government-driven programmes on energy efficiency and help governments to move beyond transitional technology-specific supports as soon as possible.

We noted in Chapter 7 that carbon markets appear to have had significant behavioural (First Domain) impacts, bringing to attention the risks and opportunities associated with tackling climate change. The cap set under the EU ETS also had significant strategic (Third Domain) influence on some of the sectors involved, in particularly clearly raising the risk of coal power investment. However, the relatively short-term and volatile nature of the carbon price has hugely undermined its relevance to specific low-carbon investments and indeed removed much of the attention impact.

The conclusions to Chapter 7 summarised a case for a 'hybrid' instrument, combining a quantity goal expressed through cap-and-trade with a price floor or corridor or similar adjustment mechanisms. The main argument there concerned increasing robustness in the face of external (and partly policy-driven) uncertainties to reduce risks for investors.

Stepping back to consider the role of prices and quantities across the three pillars shows multiple additional reasons for hybrid price-and-quantity instruments. The approach reduces price volatility and thus enhances revenue stability, and establishes a minimum base which is essential if governments are to budget and plan rationally to make the best use of revenues. The price element – if stabilised – is most relevant to product and project investments. But the quantity element is more relevant to strategic investments – how big the market may be for low-carbon innovations and whether to invest more in grid infrastructure to accommodate a projected scale of renewables, for example. However, targets on their own lack credibility without an implementation mechanism that can ultimately translate them into economic signals: a cap-and-trade system *gives market credibility* to targets. Markets and pricing also impact on First Domain behaviours where motivations are particularly important and varied between participants. As explained in Chapter 7, hybrid designs can also better harness the reality of both economic and environmental motivations in human decision-making, and have more scope to adapt to both technological and institutional learning.

In addition, considerations of uncertainty work in both directions and across all pillars. A pure quantity approach to economic instruments is even more exposed in the presence of the multiple pillars. As articulated clearly by the International Energy Agency, multiple instruments mean that the weight carried by carbon pricing (Pillar II) is itself intrinsically more volatile, since it carries the 'residual' after the impact of the other pillars.[27] With a hybrid design, the quantity goal does not only act as strategic guidance or to enhance confidence as compared to a pure carbon tax (which can be more easily revoked in budget cycles). It is also an insurance against delivery uncertainties in the other domains. If energy efficiency or technology-pull policies deliver more than expected, a price floor has set a bottom-line level of reassurance to other low-carbon investors, and if other policy pillars deliver less than expected, this will be compensated by a rise in the carbon price required to ensure that the economy remains on a pathway to the long-term goal.

A hybrid design of economic instruments within a broader policy package across all three pillars is thus not only the most effective in supporting investment across multiple actors, but also the most robust in the face of multiple uncertainties.

The impacts of pricing and markets on First and Third Domain behaviours will, however, be most amplified by enhancing their long-term clarity and credibility. This would have the effect of aligning private and public gains over a much wider sphere, by increasing confidence that more of the long-run Third Domain public benefits will be realisable by private investors in the transformations required. This is suggested conceptually by the broadening of the 'markets and pricing' band in Figure 12.2: a good design of economic instrument combining hybrid elements of quantity and price with a base price that rises strategically has potential to increase the breadth of alignment between private and public gains and thus start to reduce the reliance needed on direct government programmes to deliver the latter. Thus the importance not just of carbon pricing, but its effective and strategic design.

At first sight, this appears impossible with hybrid instruments. The logic spelled out in Chapters 2 and 6 (notably Figure 6.6) is that the nature of uncertainties implies a need for a long-term quantity emissions goal – in which case the price cannot and should not be fixed. Yet what could in principle be fixed is the underlying logic – that the carbon price should rise, until the long-run goal of securing a stable climate is achieved.

The ideal would be for this principle to be anchored in an international agreement. Even without that, translated at national level, countries can set a long-term national emissions goal, reflecting a sense of their share of the global effort, and commit to a price floor designed and reviewed periodically to rise until that goal is achieved. That in itself would enhance strategic certainty for investors about the direction of travel of a minimum carbon price as well as giving more force to the long-term quantity goal.

The carbon price becomes a base level across the economy. Countries wanting to use market forces more extensively can use cap-and-trade to allow the price to vary above this as needed to deliver the goals. Other more targeted and overtly transformational instruments can also build on the base to share the heavy lifting. The result is a combined strategy to accelerate efficiency, innovation and infrastructure investments – the other realms of opportunity that are equally essential to the transition because they enable the system to respond better to rising prices.

On institutional alignment

While this book centres around the economic principles of energy and climate change, not institutional design, it would be incomplete without acknowledging the enormous institutional challenges. Not for nothing does the literature on innovation, in particular, repeatedly come back to the centrality of 'man's greatest and earliest innovation' – social organisation – and the consequent 'need to find ways to treat technical and organisational innovation in a combined manner' (see Chapter 2, p. 59). The literature on induced technical innovation has also been extended to emphasise its corollary, namely the importance of induced institutional innovation.[28]

Climate change in particular is pushing the limits of human institutions' capacity to cope: social scientists have come to describe it as a 'wicked' problem.[29] Chapter 10 noted that institutional inertia is one of the major factors keeping us stuck on a destructive course. Changing course challenges both government organisation and the relationship of government to a range of other bodies, including regulators.

Politically, Pillar 1 policies are the easiest to introduce and manage. They can involve considerable technical complexity – a complexity which would be considerably increased if they were taken to the next level as outlined in the final section of Chapter 5. But, as noted, governments around the world have steadily improved and extended energy efficiency policies.

Establishing markets and systems for enforcing property rights ranks among the 'greatest of institutional innovations' and forms an essential step in economic development, along with the associated regulatory systems to help ensure competition and/or fair cost recovery.[30] The next step – ensuring that external costs (like pollution) are incorporated – is much harder still, as articulated throughout Pillar II, and is still developing.

Institutions which represent Third Domain concerns and opportunities are much more patchy and often quite rudimentary. Obviously, some are now taken for granted – education being the most pervasive, along with systems for public infrastructure investment. Many countries, predominantly the richer countries of the OECD, also have increasingly sophisticated systems for managing government R&D. The concluding section of Chapter 9 also noted the increasing sophistication of approaches to industrial policy in OECD countries, after it went out of fashion due to its earlier failures.

Concerns about the strategic implications of energy and environmental trends introduce a new dimension: the aim is no longer purely to support education, innovation and infrastructure, but to help steer these forces in directions compatible with our planetary boundaries. We need institutional structures not only to enhance development but also – in the language of Chapter 11 – to turn the wheel.

The development of institutions to represent the positive need to change course in energy systems – as distinct from institutions for implementing specific policies – is relatively new territory. One of the most ambitious efforts to do this is the UK Climate Change Act, which seeks to increase the strategic certainty for business investment in the low-carbon economy (see Box 12.1).

Box 12.1 Representing the future: UK legislation

To address the strategic, ethical and investment challenges of climate change, the central legislation in the UK emerged with the Climate Change Act, adopted by the UK Parliament in 2008, with a large cross-party consensus. The Act establishes an 80% greenhouse gas emissions reduction by mid-century in primary legislation and creates implementing architecture based upon five-year emission budget periods, to be adopted at least 15 years in advance. A statutory Climate Change Committee recommends appropriate budgets consistent with a cost-effective pathway to the 2050 goal and informs a powerful monitoring and incentive framework. The legislation took effect from 2008, when the first budget period started and the first three carbon budgets were adopted. In 2010 the new Conservative-Liberal coalition government accepted the Committee's recommendation for the Fourth Carbon Budget, 2023–7, which is based notably upon substantial decarbonisation of the UK power sector.

All this was possible in the UK because of a national consensus developed over two decades about the seriousness of climate change, including business desire for a consistent, predictable and evidence-based strategic framework. It helps to provide a more stable context for policy and industrial investment, but it does not itself resolve more detailed challenges of implementation.

The Energy Act of 2008 also clarified the primary duties of the UK Office of Gas and Electricity Markets (Ofgem, the energy regulator) as being to protect the interests of both existing and future consumers. As part of its response Ofgem established a low-carbon networks innovation fund.

The Climate Change Committee expressed concern about the ability of the existing electricity market to deliver investment, and in 2009 an extensive study by Ofgem ("Project Discovery") advised the government that the existing design of the electricity market would not on its own provide the investment required to provide either energy security or low-carbon investment. This led to a wholesale revision of the market structure, designed to reduce the cost of financing investment in the electricity transition, culminating with the UK Electricity Market Reform legislation of 2013.

Ofgem had to consider by what means and metrics a regulator can judge how to balance the interests of both existing and future consumers, and likewise concluded that conventional cost-benefit appraisal on its own could not provide useful answers to key questions raised by sustainability and transitions.

The UK government had also conducted a substantial review of the economics of sustainability, which likewise concluded that the traditional approach of monetised cost-benefit appraisal could not capture some of the key issues (GES 2010). After two years of investigation and consultation, in 2013 Ofgem changed its framework for Impact Assessment to include specific factors linked to the implications for future consumers, including optionality and diversity, security, and consistency with carbon budgets and the 2050 target of 80% national reduction (which implies around 90% reductions in energy sector emissions).

In the energy sector, crucial issues came to the fore in the context of investment in competitive markets. Chapter 9 noted that liberalisation, far from invigorating R&D as many expected, in fact decimated it. It has had a similar (though less extreme) impact on investment, as the short-term horizons of share ownership are in tension with the long timescales involved in energy system investments.

In some countries (notably Denmark and the UK), the institutional tensions within governments have been partly resolved by creating joint ministries of energy and climate change. In the UK, the resulting Department of Energy and Climate Change has then overseen the introduction of the Energy Market Reforms aimed at addressing the investment problem.[31] The regulator, in turn, saw its formal duties expanded to protect the interests of 'existing and future' consumers, with implications for its framework of decision assessment (see Box 12.1).[32]

As noted in Chapter 10, the entwined nature of electricity generation with the structure of electricity and gas networks makes these some of the most complex and path-dependent of all our socio-technical systems. Investment in energy systems thus forms a bridge between the Second and Third Domains. As outlined in Chapter 10, the nature and scale of the investment implied is generating tensions within European electricity systems. Europe is still trying to complete the task of fully establishing competition in electricity markets which, as indicated, necessitates independent regulation. Yet it is faced by the reality that such markets are unlikely to deliver the investment required; the amounts are enormous, but fleet-footed shareholder capital has limited appetite for long-run returns and little interest in energy security or the environment.

Future generations – as individuals and consumers – have an interest in inheriting strong electricity and gas grids, to facilitate the use and management of renewable energy for example. Expanding the regulator's duties so as to encompass those interests as well is another innovation; not least, distanced from electoral cycles, regulators may also be more able to enhance the investment stability required. Yet managing the nexus between energy markets (based on short-run marginal costs and shareholder interests) and the strategic concerns of Third Domain transformation remains work in progress.

One way or another, accepting the challenge is necessary for sustainable economic progress. No society takes its important decisions (or fails to) purely on the basis of economic calculation, because they involve competing interests and perspectives. The American War of Independence was launched with the slogan, 'No taxation without representation'. The same principle should apply to other forms of cost and risk we knowingly impose on others, including those we impose on our descendants. As this book has shown, it is entirely possible to change course; representing the voice of future generations at the table of our decision-making could help provide the overarching motivation to do so – and in time to reap the benefits.

12.8 Joint benefits

One vital fact emanating from the Three Domains may help to cut through the 'wicked' institutional and political complexities of tackling climate change in particular. For an intriguing feature of the evidence presented in the various chapters of this book is the prevalence, under each pillar of policy, of apparent 'joint benefits' – actions which deliver both traditional economic and/or social benefits while also tackling greenhouse gas emissions.

There are numerous 'joint benefits' examples uncovered within the book, and they help to illustrate in more practical terms a few distinct classes of co-benefits that can arise as a result of efforts to tackle climate change.

Joint gains: Pillar I

Pillar I has documented numerous examples of joint benefits associated with actions in the first domain, associated with low carbon and particularly enhanced energy efficiency. In the home, examples include the benefits of potentially better health (associated with reduced drafts or less extreme temperatures of either hot or cold) and reduced street noise (with double glazing). In business, more energy-efficient equipment is frequently 'embedded' along with more modern machinery and better control systems. In transport, more efficient cars carry a lesser fuel load, and/or achieve greater range between refuelling, for the same fuel, while better train services may free up commuter time to work, for example.

Sometimes the indirect benefits may be more substantial/considerable than the official objective itself. For example, the International Energy Agency, in urging countries to make stronger efforts to improve their energy efficiency, has estimated that the cost of energy efficiency improvements in buildings may pay for itself not only in terms of energy savings, but also in the reduced cost to health services struggling to cope with old people suffering from winter ailments in poorly insulated homes.

The more obvious benefits from energy efficiency are lower energy bills. Standards, engagement and other policies (including financing efficiency investments) that move decision-making closer to the expectations of neoclassical economics, cut bills for both individuals and organisations. This 'private' benefit aligns with higher-order benefits of reduced national fuel expenditure with potentially improved national energy security and lower environmental impact.

The interesting feature here is the way that higher-level concerns *motivate* policies that benefit also at smaller scales – even if few of those involved realise that is what they are doing. It is an individual's decision to waste their money; it is not generally of direct concern to government. But if governments are trying to tackle energy and environmental problems, the cost-effectiveness of the policy matters. If helping people to save energy also reduces their fuel bills, then so much the better. And if funding energy efficiency programmes is the best way of delivering the same – along with a reduced load on health services – then that is the best use of the taxpayers' money devoted to the environmental goal. The private gains (e.g. lower bills, warmer homes) are a major co-benefit of the government motivation to tackle larger energy and environmental problems. This reflects climate change as a *motivator*.

Joint gains: Pillar II

It is intuitively harder to conceive of joint gains under Pillar II (pricing policies), where neoclassical conditions imply a close existing alignment of personal and social benefits – and costs – if markets have been developed as close to the theoretical ideal as possible. Nevertheless, the reality appears to be somewhat more interesting.

The most obvious example of co-benefits associated with a pricing instrument is its potential role within a more efficient tax system, the 'double dividend' as discussed in Chapter 6 (section 6.5): governments need revenue, and all the ways of raising revenue imply some degree of economic distortion. There will be an overall gain if energy-related pricing instruments (tax or auctioned emission allowances) displace more distorting taxes and/or if this displacement is targeted towards strategic Pillar III objectives. The colloquial argument, namely that it is better to tax 'bads' than 'goods', is thereby reinforced and amplified.

Pillar II (notably Chapter 6) outlined the complex economic arguments around the potential extent of such gains. A more fundamental challenge is the view that this should not be considered as a benefit of energy or climate policy – the tax system should be optimised anyway. Yet this of course is easier said than done, because taxation is highly political. Large industries, for example, tend to be more powerful at lobbying than the individual taxpayer. If appealing to global environmental concerns can help to move societies towards a more economically 'optimal' way of raising revenue, then that is reasonably considered as a co-benefit.

The most widespread example of 'joint gains' in this vein is that of gasoline taxes, which have been used for at least three functions: reducing national oil dependence, funding public transport infrastructure and raising revenues for central government budgets.

Environmental co-benefits may also be particularly strong in relation to economic instruments which leverage investment away from coal and oil, and move them to gas and renewable sources. Chapter 1 (p. 9) noted that even in the US and Europe, the multiple costs associated with coal mining and burning seem to exceed its actual value in the economy. If that is true even after decades of environmental policy, it is even more so in the emerging economies, where local and regional air pollution are increasingly stark problems. Reflecting these factors, Chapter 6 outlined estimates of the potential value of health benefits associated with carbon pricing across a range of economies (see Figure 6.4). More generally, the academic literature on environmental policy is now replete with papers underlining the huge environmental co-benefits associated with curtailing coal and oil use.[33] Once more, therefore, 'higher domain' concerns (around energy security and climate change) may again *motivate* policies which may have multiple benefits.

An entirely different and often unspoken kind of co-benefit may arise if long-term goals and economic instruments associated with them can help *stabilise* investment expectations.[34] Chapter 1 noted the volatility of fossil fuel markets (not just oil). There is relatively little that any individual country (except possibly the US) can do about this, but national policies can make a difference to the domestic investment context. Chapter 7 noted how uncertainty deters investment and raises the cost of capital. Chapter 11 observed the large pool of institutional capital (pension, insurance and sovereign wealth funds) that is currently earning paltry economic returns. The chapter outlined why energy investment in principal could put such capital to good use – but this remains deterred by a lack of confidence in future directions.

This points to a second kind of benefit, of climate policy as a *stabiliser* of returns to investment. This rationale underpins the use of feed-in tariffs for renewable energy generators and the UK's energy market reform programme. On a larger plane, the hybrid designs for economic instruments discussed in the previous section have wider potential to increase financial confidence, particularly concerning low-carbon investments. If it is possible to ground investment incentives in a long-term, science-based perspective, this may offer a more stable long-term platform for energy sector investments. Thus Pillar II policies developed along the lines outlined in the previous section have the potential to reduce the cost of capital by providing a more stable and credible basis for low-carbon investment.

Increasing long-term stability for energy investments and improved international rules around carbon pricing systems are thus both examples of ways in which co-benefits may come from aligning Pillar II policies with the strategic goals of the Third Domain.

Joint gains: Pillar III

Perhaps the biggest potential for 'joint gains' – not surprisingly – may arise directly from Pillar III policies. The potential synergies between energy and climate security, particularly for fossil fuel-importing economies, are a classic example.

The economic opportunities flow from innovation. Chapter 9 summarised the empirical evidence that key sectors (notably electricity and construction) display very low rates of innovation and explained why. Chapter 10 observed that the direction of innovation is inevitably biased by incumbent interests. Chapter 11 charted the central role of innovation in economic growth. Climate change is obviously a potential *motivator* of innovation – the extraordinary wealth of ideas for low-carbon technologies seems to be motivated more by a desire to help save the planet than any realistic appraisal of the potential to make money.

Climate goals can also facilitate efforts to escape the traps of path dependence and lock-in discussed in Chapter 10. Associated measures can provide a framework for coordinating the numerous different actors along the value chain of a product. This can accelerate innovation and diffusion of a new product, process or service and give strategic confidence to the investments required. A national effort to enhance lower-carbon investment can serve to *coordinate* development of the associated regulatory frameworks and infrastructure development.

The potential gains from coordination can extend beyond national borders. Reflecting the natural gains from trade, Europe for example would benefit from having a more interconnected and integrated energy system – one estimate is that pooling generation resources through greater interconnection could forestall the need for 100GW extra generating capacity by 2030. Interconnection allows for a more efficient system: demand and intermittent renewable generation is more evenly spread over time, countries can pool their backup capacity and there is wider access to cheaper clean energy resources.

However, the process of developing a more integrated system has been very slow, and has often been resisted by Member States. It requires governance and motivation. The impetus for a more integrated system is stronger with a pan-European economic instrument (the EU ETS) and interconnection is not just a luxury but a necessity, with high levels of renewable energy. A higher-order challenge that motivates cooperation, in other words, can act as a coordinating influence – a common goal that carries an obvious potential for joint gains.

These observations point to three broad mechanisms through which responses to climate change may bring co-benefits: as a motivator, as a stabiliser and as a coordinator. These roles can most easily be identified within each of the Three Domains respectively, though in practice each may also be relevant across other domains.

On separability and finance

The classical economic response to claims of co-benefits has been that governments should seek such benefits independently: they should optimise fiscal policy, optimise health and air pollution policy and optimise climate policy, and should not mix up one with another. This rests on a fundamental assumption that different benefits can be *separated* from each other. A moment's thought demonstrates that this is a purely abstract assumption, with little basis in reality. We just referred to the politics of pricing, which frequently precludes governments from adopting 'economically optimal' policies.

Nor are technologies and investments separable and reversible at the whim of abstract policy decisions. China cannot choose to invest in desulphurising its coal power stations (which requires major investment) today, and then claim the money back in a few years if its climate policy later results in the plants closing down. The rational choice is to develop an overall strategy for the coal sector, taking account of multiple goals embodied in investment choices and taking account of the full range of external costs.

Perhaps most fundamentally though, the assumption of separability is incompatible with the norms of governance. Any major policy decision (and almost anything related to energy pricing is a major decision) hinges upon a coalition of sufficient interests to lead a decision through a bureaucratic and convoluted political system. Thus, for example, removing energy subsidies has been advocated as economically beneficial for decades but climate change is now a major additional rationale and force for reform and a source of international pressure to do the right thing. Co-benefits, consequently, are at the heart of real-world policy-making.

The need for realism applies also to finance. The previous discussion of *earmarking* pointed to a fourth potential role of climate policy, associated mainly with economic instruments. The most obvious dimension is as a *funder*, with revenues contributing both to reducing other taxes and to funding programmes across the three pillars. As noted earlier, the UK Carbon Trust was in part funded in relation to the UK climate change levy (as well as being a direct political response to it). Its programmes helped both to reduce the exposure of British business to the global energy price rises (see Chapter 5) and to launch new low-carbon technology companies (see Chapter 9). This obviously was a modest example of linkage across all three pillars.

At a larger scale, the use of EU ETS allowance auctions in Europe was intended to provide a key mechanism for funding the development of carbon capture and storage technologies. The collapse in carbon prices has undermined a key plank of European strategy for innovation; ETS reform could restore the associated joint benefits.

The use of revenues, however, may play a minor part relative to the use of economic instruments as a strategic tool to leverage huge pools of capital which currently sit in institutional investment funds: sovereign wealth funds, pensions and insurance. After the debt

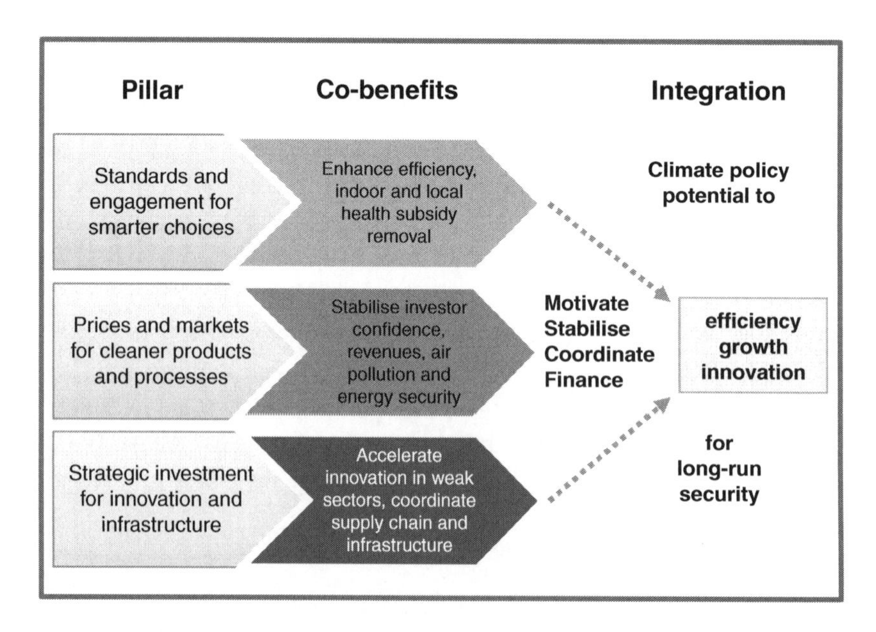

Figure 12.4 Potential joint benefits in energy and climate policy

crisis, such capital is typically earning less than 2 per cent per year interest rates. Bond markets in low-carbon finance are already developing, and two things of which we can be practically certain are that we will all still be consuming energy in the coming decades and that the value of decarbonising it should rise as long as we remain so far adrift from a prudent path. As outlined in Chapter 11, various tools can help to translate this into leverage for attracting such long-term capital: overall, climate change as a *financier*.[35]

Thus, as summarised in Figure 12.4, there appear to be four kinds of roles through which climate change can stimulate joint benefits. The central challenge is then the design and governance required to align such possibilities with actions. Realising such gains often requires cooperation; the larger ones require it internationally. There is a natural resistance against cooperation, based on the desire for sovereignty and autonomy. If such developments were easy they would have been taken further already. Yet the development of human civilisation itself has, to an important degree, been defined by improved organisational structures – and cultural developments – that facilitate greater cooperation. External threats help to unify and strengthen societies – providing that societies respond in time. We no longer need to invent or imagine such threats.

12.9 An application: energising European recovery

Of course, none of this is easy. The opening line of this book offered a rhetorical statement, which reflects the tone of many economics textbooks when they imply that problems would be simply solved if only policy-makers understood economics. The book has shown that tackling big challenges, like those posed by energy and climate change, is not easy, even in theory, because it requires integration across different domains of thought that are at best disconnected, and ignore and too frequently despise each other. Effective policy response is even harder. Even if joint gains are possible, the slow progress to date testifies to the political difficulties they face.

Although the principles and insights from mapping the Three Domains are universal, the way in which the corresponding Three Pillars play out in practical policy may differ greatly according to national circumstances. Domestic resources, economic systems, present circumstances and a wide range of cultural and institutional factors will all affect both the balance of emphasis and the choice of instruments. Countries struggling with weak institutions and subsidies may do best to focus particular efforts on Pillar I reforms; regions with good access to cheap resources may stress Pillar II; those with limited conventional resources and strong industrial capacity may make the biggest gains from Pillar III. Yet as emphasised above, the effort will only be ultimately sustainable if all three pillars are engaged.

The authors of this book are all European and hence we restrict our more specific policy conclusions to Europe. Our core recommendations flow from the combined need to stabilise the overall European economy and to emerge from the EU's crisis in a sustainable direction – fiscally and environmentally grounded in both economic and scientific realities.

The European crisis will not be solved with the same modes of thinking that landed us in it. It is widely recognised that the currency crisis, which still threatens to tear Europe apart, can be traced back to the launch of a common currency without more closely linked fiscal mechanisms. It is less recognised that this in turn reflected partly a problem of economic ideology. The European Community – originally in fact founded on agreements in energy-based sectors of coal, steel and nuclear power – grew into the European Union with its principal economic objectives as a free trade zone, along with lofty strategic political ambitions.

Thus Europe trades freely, but the capacity to shape investment for long-term growth resides with individual Member States. As long as it could be assumed that the two were largely synonymous – that free markets deliver optimal investment and growth – it did not matter. But as we have shown clearly, at least for the big infrastructure-based sectors like energy, Second Domain economics do not support strategic investment. The cost-cutting associated with competition is only part of these sectors' contribution to economic development. Yet, with the main exception of the European Investment Bank, European economic institutions and their legal mandates are mostly based on Second Domain thinking. The contributions of the First and Third Domains are central to sustainable growth strategies – economically as well as environmentally – but risk being seriously underplayed.

Both the decision to launch a single currency without more closely linked fiscal mechanisms and the fantasy that the EU ETS could – on its own and in its current form – provide a platform for hundreds of billions of euros of low-carbon investments are traceable to this fundamental intellectual mismatch. The irony is that in principle, climate policy and the EU ETS offer instruments that could aid European recovery, if they can harness the opportunity outlined in Chapter 11: principally, to create an attractor to the huge volume of surplus savings in strategic investment ('institutional') funds, so as to secure that investment into the European energy and related sectors in ways that could utilise the still-growing pool of underemployed resources currently languishing in Europe.

Europe currently risks a 'lost decade' of economic stagnation stemming in part from a lack of confidence about the future direction for the European economy and financial retrenchment in the aftermath of the debt crisis. Europe – and indeed other regions – are suffering from 'Buridan's donkey' syndrome (see Chapter 11) – an uncertainty and lack of confidence is deterring investment. As a result, huge pools of private capital sit in funds earning paltry rates of interest of around 2 per cent per year or less, while the real economy is desperate for investment. The European Commission estimates that institutional investment funds in Europe amount to around €14 trillion. As a recent report from the UK House of Lords noted, in principle the economic conditions are uniquely favourable for infrastructure investment – both to stimulate economic demand and to increase the future supply potential of the European economy.[36]

The energy sector accounts for the majority of infrastructure investment – buildings and transport are the other large opportunities for investment. All are relevant to climate change. And two things we can be certain of are that we will still be consuming energy in the coming decades and that the relative value of energy efficiency and low-carbon energy should rise for as long as we remain so far adrift from the scientific necessities. The European decarbonisation roadmap estimates that investments for a low-carbon economy could amount to €1 trillion over this decade. Compared to European GDP totalling over €100 trillion over the decade, €1 trillion may not be enough on its own to pull Europe out of its trap, but it would certainly be a big help.

It is no accident that Europe's strongest economies also tend to be the ones with the strongest climate policies (see Chapter 5, note 51). Sustainability, in both environmental and fiscal policy, requires a long-term view, with an appropriate mix of policies. Against this European economic backdrop, our analysis of the Three Domains points to a policy architecture based around the Three Pillars along the lines of the ideas set out in Table 12.2. While each involves many levels of potential detail, the broad elements can be interpreted as follows.

- *Pillar I: Enhanced energy efficiency policy for employment and cohesion.* The EU Energy Efficiency Directive (2012) and Ecodesign Directives (2009) created a European framework with the major implementation responsibilities resting with the Member

Table 12.2 High-level application of the Three Pillars in the European context

	Pillar I: Standards and engagement for smarter choices	*Pillar II: Markets and prices for cleaner products and processes*	*Pillar III: Strategic investment for innovation and infrastructure*
Foundations in economic theory	*Behavioural and organisational economics*	*Neoclassical and welfare economics*	*Evolutionary and institutional economics*
Archetypal policy instruments	Efficiency standards Information Reporting requirements	Fuel excise duties Emissions pricing through taxation or cap-and-trade	Innovation chain funding Infrastructure investments and network regulation
Principal potential co-benefits in Europe	Direct and transaction cost reductions Health benefits (e.g. buildings, urban air quality) Other benefits of improved energy efficiency – reduced domestic energy cost impacts particularly in less energy efficient/lower income parts of EU	Reduced exposure to fuel price volatility Enhanced investment due to stabilising price expectations at time of very low interest rates Demand stimulus in underemployed economies Health benefits (e.g. regional pollution) Revenues and linked dividends	Accelerated innovation in low-innovation sectors Efficiency gains from improved coordination and infrastructure Greater energy integration of central and east European economies Growth in economic supply potential from low marginal cost domestic energy sources
Key policy instruments with view towards '2030 package'	Implement commitments in *inter alia* European Energy Efficiency and Eco Design Directives More consistent use of behavioural insights beyond purchase decisions to energy use patterns Supply chain carbon accounting and labelling for embodied carbon	Structural reform of European Emissions Trading System, combining set-aside of accumulated surplus with mechanisms to stabilise price, e.g. price floor component Move towards consumption-based (levelling) framework for 'big six' energy-intensive sectors Progress EU Energy Tax directive to support energy and carbon taxation in other sectors	Codify strategic 2050 goal, and derive GHG target for 2030 with indicative framework of sectoral/Member State distributions Establish EU funding mechanisms for the Strategic Energy Technologies programme Clarify goals for key sectoral/industrial transitions, e.g. renewables target for power, low-carbon vehicles *Finance:* Pilot EU LTCFI* built on project bonds and/or basis for strategically rising floor price

*LTCFI = Long-term carbon-based financial instrument, e.g. carbon bonds, built on through European Investment Bank pilot programme of project investment bonds (see Chapter 11).

States. In addition to the energy and climate contributions, strong implementation has the potential to help stimulate the construction sector across Europe, and to particularly help the central European EU Member States to address the daunting social issues arising from the combination of their poor building stock, severe winters, lower incomes and high dependence on fossil fuels. The evidence accumulated around First Domain principles and policies (see Chapters 4 and 5 respectively) helps to inform the effective tools for delivering these gains. As well as drawing on the insights from behavioural economics, the policies could usefully be extended to include the supply chain, and embodied carbon and materials efficiency, which as indicated in the final part of Chapters 5 currently remain only weakly addressed.

- *Pillar II: Strategic reform to stabilise the EU ETS and turn it into an instrument that can support investment.* The EU Emissions Trading System now carries such a huge surplus of emission allowances that there is a strong case to 'set aside' some emission allowances and/or strengthen the targets to 2020, but this will not address the need for structural reform (see Chapter 7). The move to widespread auctioning from 2013 enables a simple mechanism for implementing a price floor through an agreed Europe-wide reserve price on the auctioning of emission allowances or related mechanisms. It has also created the only negotiated EU-wide instrument that could (and was originally expected to) raise hundreds of billions of euros for Member States out to 2020. To fulfil its potential, the ETS needs some such mechanism to reduce volatility, reassure investors and enable governments to budget for the effective use of revenues. Clarification of goals for 2030 is also an urgent need, though this may take longer to implement. In the context of a three-pillar approach for 2030, a reformed ETS with a target consistent with Europe's long-term climate goals would not only help reduce investment risks, but would also insure against under-delivery in other domains, enhancing market credibility for the overall long-term direction. Additionally, by helping to stabilise revenues, it will both stabilise the instrument politically and help to provide the financial links across to other domains along the lines discussed in the previous section.
- *Pillar III: Multi-sector strategies for innovation and low-carbon transformation.* First steps in this direction have been taken with the EU's Strategic Energy Technologies plan and sector-specific roadmaps. These provide the background to guide innovation and infrastructure policy. For example, in the power sector, R&D support over the past two decades combined with the targets for the deployment of renewables by 2020 provide the basis for corresponding grid developments, and inform government and regulatory choices on planning systems, market design and remuneration mechanisms.

  When looking to 2030, the energy framework would need to encompass more fully the role of 'smart and integrated' networks, to better engage consumers (establishing a link to First Domain processes) and to make the best use of Europe's renewable resources – both those that are remote and those that are embedded. The EU also needs to develop a transformative strategy for heavy industry, with more clarity on technology potentials and related supports to help EU industry innovate and enhance its competitiveness. For the 'big six' energy intensive sectors this might need to be linked with ETS reforms to shift the carbon price towards consumption-based principles, to avoid discriminating against domestic production and the risk of relocating in the face of rising carbon costs (see Chapter 8, sections 8.2 and 8.3). In transport, policies on e-mobility, fuel efficiency and network investments could accelerate innovation and transformation. Particular emphasis rests on the combination of innovation funding, e.g. through the SET-Plan and national policies, with deployment policies as reflected in e-mobility schemes and investments in transport infrastructure.

These broadly correlate to the potential roles of climate policy to help *motivate, stabilise* and *coordinate* actions that contribute to economic progress. The second element in particular also points to the potential role in relation to *finance*, both directly through auction revenues, and (indirectly) to attract private finance. The latter may be only partially addressed through structural reform of the EU ETS. The benefits of agreement on a floor price would be enhanced if accompanied by reaffirmation of the commitment to a mid-century goal to reduce greenhouse gas emissions by 80–95 per cent (with corresponding pathways for caps under the EU ETS), combined with a principle that the floor price would rise over time until the EU delivers the overall goal. As underlined, these complementary measures relate to different decisions taken by different decision-makers. This could also cement the value of the various proposals for carbon-related financial instruments outlined in Chapter 11, doing a lot more to unleash the scale of investment that Europe needs at low financing costs: climate policy as a *financier*.

None of these policies are easy. Nor is it obvious that the economic benefits associated with the 'joint gains' accrue at the same time and in the same 'pockets' as the transitional costs are incurred (though this is more likely in the European context of severe recession with available unutilised resources; see note 36, and Chapter 11, section 11.7). Changing course has to overcome established systems and interests, and thus requires effort and a sense of direction, which need to come from a concern for the future.

12.10 The Three Domains in economic development

One final question concerns how the Three Domains relate to different stages of economic development. It is indeed likely that the balance of emphasis would shift in the course of development. Least developed countries and emerging economies still struggle with the need to tackle extreme poverty. As stressed in Chapter 1, global economic progress in the past two decades has been remarkable, yet 2.5 billion people depend on biomass or other traditional heating fuels, and half of them are still without access to electricity – many of them in the rural areas of emerging economies like India. Development remains their top priority, yet their choices are not simple: development goals are also threatened by global instabilities in energy or climate systems.

It is tempting to think of the Three Domains and associated policies as sequential. The earlier sections of this chapter have underlined the fallacy in this, drawing on all the evidence in the book. Energy efficiency remains the most basic opportunity in developing countries for good reasons, but Chapter 2 suggests a continued large First Domain potential also in the rich countries of the OECD. Many reasons for the continued 'energy efficiency gap' in supposedly efficient economies are given in Chapter 4, but there are additional reasons linked to the legacies of the development process itself.[37] In principle, developing countries have potential to 'leapfrog' to superior technologies. In practice this is not so easy, but at least in industry it has become part of the pattern in some emerging economies, particularly (but not exclusively) as multinational companies extend leading-edge investments into countries with lower labour costs.

All this is occurring against the backdrop of the search for a new phase in international action on climate change. As already flagged, this book has not endeavoured to address the fantastic complexity of international coordination. Its starting point was that any successful international approach will need to be grounded in domestic economic and political realities; a key question then is what is 'nationally appropriate'? The analysis does suggest some general principles and trends.

There is bound to be an evolution of what is appropriate and a shift of balance between measures over time. The benefits of energy efficiency, as charted in Chapters 4 and 5, are universal even if the form they take (including rebound effects) may differ at different stages of development. Many developing countries still use consumer energy subsidies to help the poor (though as noted in Chapter 6, in practice more of the subsidies go to the emerging middle classes), despite the acknowledged economic drawbacks. Yet the major emerging economies, and many others particularly in South East Asia and Latin America, now have well established systems of governance and markets; their growth owes much to their adoption of key principles of Second Domain economics as applied to open markets and globalisation. Yet their astonishing rates of economic growth have also of course propelled emerging economies to become major consumers of energy and emitters of greenhouse gases.

The idea of a global carbon price as the universal answer to climate change was flawed in theory, as explained in Chapter 6. As then charted in Chapter 7, the 'common global architecture' approach to carbon pricing has also crumbled in practice, yet many of the emerging economies are considering or embarking upon carbon pricing, in myriad different forms. At the time of going to press, three of the regional pilot systems in China had 'gone live' and generated their first carbon prices, and the Indian PAT system is operational. Other regions have proposed some highly creative ideas on how carbon pricing could be designed to help them also with the other pillars of policy.[38]

Turning to Pillar III, analysis in Chapter 9 underlined the extent to which other and 'higher cost' measures can make eminent sense, if and as they form part of a strategy for innovation and industrial transformation. Chapter 10 charted the extent of potential lock-in, but also the scale of potential transformation in all the key sectors of transport, electricity and urbanisation. Chapter 11 then elevated all this to the macroeconomic level, stressing the relevance of energy choices to the pace and direction of economic growth.

The OECD countries and emerging economies share a common interest in the challenges and principles of Pillar III policies. Yet, again, the emphasis may differ between countries and regions. Much of the world's innovation capacity remains concentrated in the rich countries of the OECD (though that balance is also shifting). There would be a certain economic logic to a widely used (and rising) base price of carbon – or equivalent index of effort.

Emerging economies are in the process of building both their infrastructure and social systems to try and meet both basic human needs for their poor and to satisfy the aspirations of the rising middle classes. On both, they face choices about which path to follow – bifurcating choices which will have enduring consequences.

The pace of infrastructure development is breathtaking, as illustrated by the process of Chinese urbanisation and its energy implications (see Chapters 5 and 10). This is intrinsically energy-intensive, but there are enormous degrees of freedom about the legacy this leaves. Strategies of urban planning, along with industrialisation and rural development, need to embed the carbon implications of those choices, by mobilising policies across all three domains. Financial reforms including international finance, for example contributions to help align Energy Access for All programmes with strategic environmental goals, will be crucial to help decarbonise the process of infrastructure development.

As emphasised, however, technology and infrastructure also depend upon the wider economic context, and the principles of innovation and path dependence apply as much to institutional choices as they do to physical ones. This highlights the relevance of how emerging economies choose to develop their social and fiscal systems. Funding the needs of health care, social security and retirement is not cheap; the choices that emerging economies make in deciding whether they should be funded by taxes on 'goods' or 'bads' may ultimately be just as important as the specific investment choices in high or low carbon infrastructure.

Citing the example of California, Chapter 11 underlined the potential investment, financial and macroeconomic benefits that could flow in particular from such mechanisms to help to bridge processes of the Second and Third Domains. Countries can upon this build a raft of more overtly transformative Third Domain policies.

The next phase of the global effort cannot be based on a crude assumption that tackling climate change is a matter of incurring costs for the global benefit; we have underlined rather that it is a question of investment and returns. Ultimately, countries will self-select whether they want to be part of the 'transformers club' of countries aiming both to make the investment and reap the rewards of low-carbon investment and innovation. Such an approach would be all the more effective if it could be combined with international cooperation and finance to accelerate the diffusion of low-carbon technologies that also meet development needs, perhaps funded in part by revenues from carbon pricing and the other financial dimensions touched on in Chapter 11.

Thus the big choices are all entwined. As increasing numbers of leading economists stress, it is simply not helpful to offer prescriptions dominated purely by the principles of liberalisation and the globalisation of markets. The struggle for those principles of the Second Domain was won long ago, and the aggregate benefits are manifest, but their blindness to distributional and strategic consequences breeds discontent.[39] Those principles alone are not adequate to the challenges we face. The economics of sustainable development must embrace all Three Domains.

12.11 Conclusions

A few years ago, the Royal Society – the oldest establishment of what we recognise today as modern scientific knowledge – celebrated its 350th birthday. In its anniversary volume, one contributor noted that something we take for granted today was at the time a deliberate and controversial choice.[40]

The intellectual flowering of the Renaissance had led to divergent strands of enquiry. Some – the philosophers – were theorists, thinking through from first principles how the world should or must be, and from that, deriving general rules about how the world was assumed to work. Others – sometimes seen as the more lowly class – were experimenters, observing the world around them, and devising ways of testing how the world actually behaved. The founders of the Royal Society insisted that the two must be brought together under the same institution.

The *Oxford English Dictionary* once defined economics as 'the science of political economy' and defined political economy as 'the art of managing the resources of a people and its government'. If economics is to lay claim to being more of a science than an art, it too must be grounded in facts. The key issue is not whether economists use data – many, perhaps most, academic economic papers these days draw upon huge banks of data. The issue is whether data is used simply to populate the assumed structure of economic processes or to actually test and probe the structural assumptions being made.[41]

The difference lies at the heart of this book. We clarified Three Domains of decision-making processes in relation to economic systems, drawing on a huge span of research and experience, and explored their significance and implications in relation to energy and the global environment. The results show that all three domains matter. Moreover, though they appear to be disparate, the key to Planetary Economics is to understand their relative roles, their relationships and their unifying themes.

The domains reflect more fundamental characteristics of human nature and systems, and it is helpful to acknowledge this. Experimental psychology and neurology have established that the human mind encompasses at least two discrete decision-making systems: a short-term, reactive and impulsive component, along with a separate and distinct capacity to think ahead: to plan and calculate rationally for the future.[42] That higher-level ability is a key characteristic which we associate with mature humans; the extent to which even young children can resist immediate temptation is one of the most reliable indicators of success later in life.

In turn, the growth of civilisation has involved the development of social capacity to pursue the common good, as noted through all manner of academic lenses, including 'institutional economics' and the original studies of political economy. The capacity to develop institutional systems that extend beyond individuals, family, tribe, and which endure beyond a single generation on larger scales of both social groupings and time, is a fundamental characteristic of the development of civilisation.

To a large degree, we accept an implicit hierarchy between these levels, that the actions at one level have to be compatible with the higher levels. We teach children to moderate their immediate impulses, to think ahead and to cooperate. The very term 'adult behaviour' implies a degree of rational planning, accepting the rights of others, and a respect for the laws and institutions of our societies that have developed over decades or indeed centuries. Often, over time, these laws have become so ingrained in our societies as to become inseparable from a general social consensus – as with, from an economic perspective, the existence and enforcement of property rights, which most people (and certainly economists) tend to take for granted.[43]

These are big issues which academics have approached from many directions and from several/various/many disciplines of study. The potential connection to the three domains of economics that have framed this book is obvious and directly reflects patterns and classifications in branches of institutional research, for example. We have not delved into these wider, non-economic theories, but rather focused on what the accumulated evidence says about the nature of economic processes in relation to energy systems and the atmosphere, and the implications for policy.

The specific policy implications are readily summarised in terms of the Three Pillars of policy, which flow directly from the Three Domains. This concluding chapter has demonstrated five cross-cutting, generic implications:

- *Energy systems have a large capacity to adapt, and this increases the long-term value of action.* The first and third domains in particular involve largely adaptive processes, which result in enduring changes to energy systems that are unlikely to reverse. The benefits of associated actions in terms of changing course may be far bigger than the value of their directly attributed impact on energy and emissions.
- *We need to broaden the tools of analysis, particularly relating to Third Domain processes.* Theoretical and computer models can illuminate and inform us about many of the future implications of today's decisions – and hesitations. They can also constrain and mislead, where the assumptions are inappropriate, when they can narrow the intellectual horizons of decision-making. Notable is the fact that many of the Third Domain processes of induced innovation, systems evolution, lock-in and path dependence are rarely represented in the models used by most academic and government appraisals. This leads to an underestimation of some of the most important potential strategic gains

from policy appropriate to the long-term nature of the issues at hand, including industrial transformations.

- *Multiple instruments are required* to reflect the different processes in each domain, the complementary contributions of each policy, the complexity of motivations and markets, and the different structure of private and public returns to investment under each pillar. Improved design of economic instruments may extend the useful role for markets but cannot supplant other pillars. The detailed choices and balance will vary by country and stage of economic development.
- *Joint benefits are pervasive and inseparable*. There are many ways in which actions on energy and climate change may be associated with other benefits (and vice versa). These arise particularly from the intrinsic characteristics of First and Third Domain processes, but are likely to be biggest for integrated strategies across all three pillars of policy. In the real world it is not possible to 'separate' each benefit with different decisions: both technically and politically, most decisions involve multiple dimensions. Smart policy on climate change has the potential to help to *motivate*, to *stabilise*, to *coordinate* and to attract *finance* to actions which are economically and socially beneficial in other respects.
- *Only integrated strategies across all three policy pillars are effective or stable*. Measures focused purely on First Domain efficiencies will eventually run dry and their impact will be offset by rebound effects. Relying purely on Second Domain principles of 'internalising external impacts' in prices is politically unsustainable without First Domain efficiencies and engagement, and the generation of new options to respond in the Third Domain. However, options generated in the Third Domain will never realise their potential without the drivers generated by the other pillars of policy.

For the reasons explained in Chapter 11, there is no generic answer to the question of how tackling dependence on fossil fuels and associated CO_2 emissions, aside from the environmental benefits, will influence economic development, for the answer rests both on untestable assumptions and the strategies we choose. What we do know is that the economic and environmental crises spring from the same source of short-termism. The previous sections suggest ways in which, for example, effective energy and environmental policy consequently can be used as a lever to help Europe extricate itself from its current morass.

We end, however, on a more philosophical note, and an analogy. No single theory should be expected to apply to everything. When Newton uncovered the laws of classical mechanics, it seemed as if they were universal – a view which largely prevailed in science for two centuries. It was only in the dawn of the twentieth century that the accumulating evidence of observation, at the opposite extremes of physical and temporal scales, cracked the consensus of appealing simplicity. The emergence of quantum mechanics at one end and relativity at the other finally established that the scope of Newtonian mechanics was not unbounded. Physics still searches for a unifying theory but has learned to accept the dominance of different physical processes at different scales. Even in the absence of a unifying theory, it is understood that even the smallest and largest domains interact: that the great patterns of the universe, observed for example in the microwave background radiation, cannot be explained without the principles of quantum mechanisms.

In the twenty-first century, the same maturity is now possible in economics. Painstaking research has established the realities of behavioural and organisational economics, the sources and nature of instabilities in real markets, and the evolutionary and path-dependent characteristics of large-scale techno-economic systems. None on its own offers a universally useful description of human decision-making in relation to economic and environmental processes. Faced with issues that span from the actions of seven billion people to the future of the planet over century-long timescales, we need all three.

Indeed the three different fields of theoretical advances match particularly strongly the observed characteristics of energy systems: the structural asymmetries between energy supply and use; the roller coaster of international fossil fuel markets and their impact on economies; and the mind-bending scale and longevity of energy systems and networks. The intellectual challenge is to map better the boundaries and explore more deeply the interactions between them.

As indicated in this chapter, the insights arising also offer hope. But to paraphrase the old Chinese proverb: unless we make an effort to change course, intellectually as well as in policy, we are bound to end up where we are heading.

Notes

1. Numerous translations of this two-thousand-year-old saying are available on the web. The most literal translation of the Latin is 'To the ignorant of the harbour he aims, no wind will be his wind.'
2. See in particular Chapter 1, section 1.4 on 'Costing the climate'.
3. 'Classical modes of economic thought' is of course a dangerously sweeping term, but is here used as shorthand for those based on Second Domain assumptions, of which neoclassical models amenable to quantification are an important part. Economic debates themselves, of course, have always reached more widely. The conclusions of this book, for example, have some echoes with the insistence of Polanyi (1957) that any economic system has to be understood in relation to the wider institutional and social context within which it operates, and consequently that economic theories suitable to one context may not be universally applicable. This set Polanyi at odds with both the neoclassical schools and the Marxists, as a result of which his thinking was largely sidelined.
4. For example, Keynes emphasised the essential contribution of 'animal spirits' in motivating change, innovation and growth. More recently, George Akerlof's 2007 Presidential Address to the American Economic Association was published in the *American Economic Review* titled 'The Missing Motivation in Macroeconomics', arguing this question had to be at the heart of economic debate. See also the final footnote of this chapter.
5. Note, however, the discussion of evolutionary economics in Chapter 10, and associated references, some of which highlight intimate intellectual connections between behavioural and evolutionary economics. Also notable is that an excellent short review of the policy implication of evolutionary economics (Marechal and Lazaric 2010) stresses the connection with behavioural foundations as key to avoiding lock-in.
6. I am grateful to Frank Kelly, Master of Christ's College Cambridge and a top Cambridge mathematician, for pointing out this example.
7. Lee *et al.* (2012): see Chapter 1, note 72, which also cites the US Department of Agriculture as warning that global pre-harvest corn stocks in 2012 were falling to the lowest levels since 1974. The complexity of such systems is also at the root of concerns about 'X-events' (Chapter 2, section 2.1), and a burgeoning body of academic literature on risk and resilience.
8. A classic example of this line of argument is *The Rational Optimist* by Matt Ridley (2010). Lomborg (2001) presented much data on the improving environment, but similarly his subsequent assessment of climate change has rested largely on assertions that technology will solve the problem *without* the incentive of legislative constraints; see discussions in section on 'Driving blind' in Chapter 1 and associated notes.
9. Thus:

   The prevailing manner in which economic analysis is conducted … is firmly based on the idea that there is an equilibrium path around which the economy can travel which essentially pre-exists.

Also, that equilibrium path is the basis for the actual path followed by the economy in the sense that the economy will oscillate around the equilibrium path. Furthermore, part of the information set held by those individuals is knowledge of that equilibrium path … [which is] already 'out there' to which their decisions will essentially have to confirm. In most economic analysis, where the economy starts and which path it follows through time does not affect the final equilibrium position in these models. (Arestis and Sawyer 2009: 2)

10. One puzzle is why, after the Stern Review, the big name economists focused mainly on these dimensions of the climate system rather than the empirical characteristics of the energy system, its inertia and its capacity to adapt. Thus the main debate focused on the problem but not the solution, or the interplay of the two in terms of the costs and the pace of action and/or of damages, given the uncertainties they were debating (Hourcade *et al.* 2009).
11. Thus the International Energy Agency estimates that strengthened energy efficiency policies could halve the projected rate of growth in global energy demand, which makes everything else cheaper and more manageable. This has a crucial bearing upon the cost of and ability to meet climate goals (we include a simplified illustration in the Appendix).
12. In the language of economics this amounts to changing and reshaping the 'supply and demand' curves. But it has a far wider connotation, and indeed it can be argued that at its root there is an even deeper philosophical issue, harking right back to debates about determinism versus free will. The classical formulation – with exogenous technology and production functions – recognises that we can change the future, but insists that it is only at an indefinitely recurring cost as compared with the 'optimal' path. An adaptivist perspective asserts that we can choose our future and – through our ingenuity and adaptability, we can learn to love it.

    The argument holds equally well even when innovation policy is included as part of the classical formulation. As analysed in Chapter 11, many economic attempts to grapple with innovation assume that there is a limited 'innovation resource' and that societies will, with due attention, tend toward this optimal pace and direction of innovation based mainly on price signals, supplemented by public research and development. The idea of 'optimal' is fundamental in Pillar II but, as all the chapters in Pillar III have shown, is deeply problematic. Chapter 9 demonstrated that the energy sector under-invests in innovation, and that systems can innovate and evolve in many directions. It is not possible to know at the outset which will be more attractive and many mathematical constructions result in 'multiple pathways' which cannot be simply ranked in terms of economic preferences.
13. One recent economics graduate I worked with, shown the underlying paper on which this section rests, handed it back saying she simply couldn't relate to it at all. We seem to steep our economics students so deeply in the idea of 'fixed supply and demand functions' that the idea of technologies and preferences evolving and adapting over time, influenced by decisions we take today, is too hard to grasp and too threatening to the accumulated body of economic analysis.
14. See, for example, the papers collected in Arestis and Sawyer (2009).

    One line of economic research that may hold the promise of something like a unifying theory across the domains is that of *adaptive preferences*. In seeking ways to 'maximise human welfare', one basic assumption of welfare economics has been that human preferences are well defined and largely fixed. However, unassailable evidence has shown that preferences can change and adapt to circumstances: what we want is strongly influenced by what we have experienced in life and what we see around us. The idea of 'adaptive preferences' seems at first sight to strike at the very core of welfare economics, which in theory aims to maximise welfare by designing economic systems to give us what we want. How can policy do this if what we want changes? Yet a leading economist, Professor Dr Christian von Weizsäcker, has explored more carefully the economics of 'adaptive preferences' and has shown how it is still possible to maintain a real and important meaning to the goal of maximising welfare – but its characteristics then share much in common with both behavioural and evolutionary economics. Neoclassical economics is a special case, applicable to choices and timescales over which preferences are stable.

    Fundamentally, this perspective allows for a far greater degree of human choice over our destiny, in ways which cannot be readily reduced to conflating resources, wealth and welfare as one. It also sheds light on some of the dilemmas of 'collective action'. If preferences adapt to the present, people collectively are more resistant to change (echoing the 'risk aversion' findings of behavioural economics). This helps to underline the importance of individual freedoms and the benefits of markets – which allow individuals to experiment with new choices without the need for everyone to agree on them – while also explaining why it is so hard for societies overall to move in a better direction, however clear the overall rationale. This gives rational and 'leadership' policy-making a high importance while also explaining the political difficulties that improvements are bound to face.

The conceptual basis was already indicated in von Weizsäcker (2005). Building on this, the mathematical and philosophical implications of adaptive preferences have been worked through in an extraordinary series of subsequent studies, now brought together as *Freedom, Wealth and Adaptive Preferences* (Max Planck Institute for Research on Collective Goods, Bonn, Draft February 2013).

15. The conclusion of the IIASA work may be summarised as follows. As illustrated in Chapter 3, we are not globally constrained by a lack of energy resources: in addition to the potential for energy efficiency, solar and wind resources easily match those of fossil fuels. Which we use is a function of technology and our systems. Through its analysis of technology learning and system interactions, the IIASA modelling (see Chapter 10, Figure 10.6) translates these underlying facts into the following conclusion: that the belief that a 'green' energy system – apart from its environmental benefits – is more expensive than a 'brown' one is just that: a belief, which reflects in part the limits of current technologies and institutions – and of our thinking.
16. The popular economics writer Tim Harford (2010) neatly summarised the evidence and argument on this in his book *Adapt: Why Success Always Starts with Failure.*
17. Modelling studies of the impact of inertia and uncertainty on decision-making range from Ha-Duong *et al.* (1997) through to various reviewed in Hourcade *et al.* (2009) and subsequent debates focused on infrastructure (e.g. Guivarch and Hallegatte 2011). They all concur that the combination of inertia and uncertainty strengthens the need for early action – because the costs of delay are amplified for systems with high inertia, and particularly so in the event of having to act faster in the event of bad news, the probability of which of course increases as a function of the degree of acknowledged uncertainty. Other models include sophisticated models of induced innovation in energy systems, and the numerous 'guardrail' studies of the economics of achieving specific CO_2 stabilisation levels, increasing numbers of which include at least some degree of induced innovation (see, for example, Chapter 11, section 11.4). Related studies include analysis of the impact of capital stock lifetimes and inertia on decision-making in an uncertain world – for example, the World Bank studies cited in Chapter 2 that caution against interpreting the 'McKinsey curve' as a guide to just doing cheap things first.
18. Oreskes and Conway (2011), in *Merchants of Doubt*, draw a clear distinction between the healthy expression of doubt with scepticism expressed through intellectual debate, and the propagation of confusing or misleading information in popular media – often supported by vested interests – divorced from the norms of intellectual review and evidence. The book traces the extraordinary influence in the US of a small group of people repeating this practice from debates on tobacco and ozone depletion through to global climate change.
19. Attributed to Mark Twain.
20. This is somewhat paradoxical, since the Treasury looks at the short term whereas the Second Domain assumptions are most valid for mid-term equilibrium. This partly reflects intellectual reflexes which are rooted in the kind of economics taught in first- and second-year economics classes, and the fascination for a stabilized world in which the simple recipes of conventional wisdom work. It is also just much easier to build computer models in that way.
21. For example, the MIT-EPPA model has generated a number of useful insights into the likely mid-term impacts of carbon pricing systems. Application to modelling vehicle standards (e.g. Karplus *et al.* (2013)), where in practice cost evolution may depend on the regulatory environment and the degree and timing of deployment, is more problematic. The modelling effectively assumes that standards are more costly, yet another paper from the same research group (Saikawa, 2013) finds that 'there is indeed a race to the top of automobile emission regulations, including in developing countries' – an observation that surely puts into question the Second Domain-constrained assumptions of the economic modelling. Again this underlines the importance of grounding economic modelling in emperical observation. The likely behaviour (of both consumers and producers) is particularly complex. For a wholly different modelling approach see Mau *et al.* (2008).

22. Indeed, even if learning were to decline towards zero, there would still be some case for favouring renewable investments: it is only the conflation of investment and returns in a simple 'cost per unit energy', within a framework which assumes equivalence between depletable and renewable assets, which assigns a zero value to this. Note here a clear theoretical link to the debates on discounting, and in particular the enduring disagreement among economists between 'observed' and 'ethical' pure rates of time preference. A measure of 'cost per unit energy' of a given technology reflects a conflation of investment and returns aggregated at a market-oriented discount rate. That unit cost may then be used in modelling the overall system and evaluating public costs and benefits (which may use a lower discount rate). An extreme example would be tidal barrages, as structures likely to last centuries, which are in most cases assessed to be uneconomic in terms of cost per unit energy, but which when assessed explicitly in terms of investment and returns may be highly positive – along with low (economic) risk since the technology of dams is simple and well established – if the investment is evaluated at a Stern-type 'ethical' discount rate.
23. Mark Jaccard and his colleagues in Canada are to be credited with pioneering some of the most rigorous modelling work along these lines. Jaccard *et al.* (2003) set out fundamentals of their CIMS model and Rivers and Jaccard (2006) trace some key policy implications (the NEMS model developed by the US Department of Energy adopts a similar approach). The latter paper suggests that 'strategic deployment' of emerging technologies is frequently not the *most efficient* approach, but that the cost difference may be small compared to the more efficient (but unattainable) goal of more precisely targeted cost internalisation plus R&D. From the same school, Mau *et al.* (2008) explore the dynamics of the diffusion of new vehicles influenced by 'neighbor effects', thus extending their work more explicitly into the dimensions of First Domain behaviour.
24. Vogt-Schilb and Hallegatte (2011).
25. Thus Rozenberg *et al.* (2013) note:

    Policymakers have good reasons to favor capital-based policies – such as CAFE standards or feebates programs – over a carbon price. A carbon price minimizes the discounted cost of a climate policy, but may result in existing capital being under-utilized or scrapped before its scheduled lifetime, hurt the workers that depend on it, and inflict an immediate income drop. Capital-based policies avoid these obstacles, but can reach a given climate target only if implemented early enough. Delaying mitigation policies may thus create a political-economy lock-in (easier-to-implement policies become unavailable) in addition to the economic lock-in (the target becomes more expensive).

    See also Fuss *et al.* (2012).

26. This approach involves amortising investment costs using the market interest rates typical for a company or sector, but then discounting the overall cash flow over time using the governmental social discount rate. In the UK it is known as the 'Spackman approach' and is the approach recommended by the UK Joint Regulators Group.
27. As noted in Chapter 2, the International Energy Agency, among others, has already highlighted the relevance of 'complementary instruments' akin to the Three Pillars. In their key publication on this, Hood (2011) goes on to illustrate schematically how an equal contribution from each of the Three Pillars means that the carbon pricing pillar is potentially exposed to far greater volatility in the face of baseline uncertainties if it bears the 'residual' after the contribution of other pillars is fixed.
28. Ruttan (1997).
29. Indeed Levin *et al.* (2009) describe climate change as a 'super-wicked' problem because of the way it conflates numerous complex problems at both global and intergenerational levels.
30. Such regulation is of course particularly important where there are natural monopolies, as with electricity and gas systems.
31. For an overview of the economic arguments and conclusions around UK energy market reform, see Newbery (2012).
32. Recognising the importance of innovation for the regulated networks, Ofgem established a Low Carbon Networks Innovation Fund. See Ofgem (2012) for information on Ofgem's Strategic and Sustainability Assessment approach, which was adopted as part of a revised Impact Assessment framework in 2013.

33. Journals carrying a great deal of literature on co-benefits include *Environmental Science and Policy* and *Energy Policy*. Others include *Climate Policy, Climatic Change, Energy and Environmental Management* and numerous specialist journals on energy efficiency, including the new house journal of the International Energy Agency. A brief survey of just the first of these, *Environmental Science and Policy*, revealed dozens of papers on environmental co-benefits in different countries and sectors over the past few years. A major Special Issue on regional air pollution and climate change in Europe, for example, concluded that 'there are strong financial arguments for developing *joint* policies to reduce air pollution and greenhouse gases in Europe' (Alcamo 2002). On the links between regional air pollution and climate change in China see, for example, Mao *et al.* (2012).
34. It is curious how this dimension is almost never considered in the literature; this is the consequence of the fact that most models do not model uncertainty, have no explicit financial sector and assume already stabilised expectations based on 'perfect foresight'.
35. See the final section of Chapter 11. For information on the market development of low carbon bonds see HSBC (2012: 13).
36. House of Lords, European Union Committee,14th Report of Session 2012–13, *No Country Is an Energy Island: Securing Investment for the EU's Future*, 2 May 2013. HL Paper 161:

    The time is right for infrastructure investment, including in energy, because it can have a multiplier effect, it can provide secure energy at a stable cost and it can boost technological advance. Low carbon generation and system infrastructure in particular can provide domestic energy production for decades at low and stable operating costs but at a high capital cost. We conclude that such investment is particularly appropriate at a time of historically low interest rates and recession ... Institutional investors hold €13.8 trillion of assets but, in order to invest in energy projects, even at a time of historically low interest rates, they need confidence in policy. That is why agreement on a 2030 policy framework, by 2015, must be a priority for the EU. Without that clarity and investment, the EU will be uncompetitive and over-dependent on elsewhere to meet its energy needs, and it will fail to seize an opportunity to make a material and enduring contribution to European economic recovery.

37. Additional factors behind the 'efficiency gap include the interaction between inertia and innovation that continues to evolve the 'best practice frontier'. Thus buildings constructed in the last century could not incorporate modern technologies but have enduring lives: the legacy of investments that might have been close to the frontier when built inevitably start to lag as technology evolves. Also poverty can emphasise short-termism. As time horizons extend, other technologies – more capital intensive with returns generated over long periods – may become more attractive.
38. One notable example is the proposed South African carbon tax, within which one proposed feature is to allow 'offsets' restricted to targets of particular priority, like the funding of efficient motor drives; thus a modest tax on carbon pollution would be used, for limited periods, to help drive the establishment of the supply chains and 'norms' of more efficient drives, which in turn will help South Africa to manage its power system.
39. Stiglitz (2002).
40. Goldstein (2010).
41. For a clear technical illustration of this point – whether data are used to help probe the structure of assumptions or merely to populate the parameter values of assumed structures – see the clinical review of theories of the economics of innovation by Jaffe *et al.* (2003). See also Chapter 11, note 51, notably the paper by Ian Sue Wing which similarly emphasises the impossibility of making generalisations about the nature of innovation processes without empirical foundations.
42. As reviewed in Kahneman (2011).
43. Note that in practice, the literature tends to identify a fourth level, more fundamental still, commonly called 'embeddedness', associated with culture and beliefs: 'informal institutions, customs, traditions, norms and religions' (Williamson 2000). In the Jewish and Christian traditions they tend to be associated with the Ten Commandments, but they exist in similar forms in most religions, though societies differ in the manner and extent to which they are incorporated into formal law.

References

Akerlof, G. (2007) 'The missing motivation in macroeconomics', *American Economic Review*, March, pp. 5–36.

Alcamo, J. (2002) 'Introduction' to Special Issue on Regional Air Pollution and Climate Change in Europe, *Environmental Science and Policy*, 5: 255.

Arestis, P. and Sawyer, M. (eds) (2009) *Path Dependency and Macroeconomics*. Basingstoke: Palgrave Macmillan.

Bryson, B. (ed.) (2010) *Seeing Further*. New York and London: Harper Press.

Carbon Tracker and Grantham Institute (2013) *Unburnable Carbon 2013: Wasted Capital and Stranded Assets*. Online at: http://www.carbontracker.org/wastedcapital

Decanio, S. J. (1999) 'Estimating the non-environmental consequences of greenhouse gas reductions is harder than you think', *Contemporary Economic Policy*, 17 (3): 279–95.

Ericksen, P., Ingram, J. S. and Liverman, D. M. (2009) 'Food security and global environmental change: emerging challenges', *Environmental Science and Policy*, 12: 373–7.

Fuss, S., Szolgayová, J., Khabarov, N. and Obersteiner, M. (2012) 'Renewables and climate change mitigation: irreversible energy investment under uncertainty and portfolio effects', *Energy Policy*, 40: 59–68.

GES (2010) *Review of the Economics of Sustainable Development*. UK Government Economic Service, DEFRA. Online at: http://archive.defra.gov.uk/evidence/economics/susdev/documents/esd-review-report.pdf

Goldstein, R. N. (2010) 'What's in a name? Rivalries and the birth of modern science', in B. Bryson (ed.), *Seeing Further*. New York and London: Harper Press.

Grubb, M., Chapuis, T. and Ha-Duong, M. (1995) 'The economics of changing course: implications of adaptability and inertia for optimal climate policy', *Energy Policy*, 23 (4): 1–14.

Guivarch, C. and Hallegatte, S. (2011) 'Existing infrastructure and the 2°C target', *Climatic Change*, 109 (3–4): 801–5.

Ha-Duong, M., Grubb, M. J. and Hourcade, J. C. (1997) 'Influence of socioeconomic inertia and uncertainty on optimal CO_2-emission abatement', *Nature*, 390 (6657): 270–3.

Harford, T. (2012) *Adapt: Why Success Always Starts with Failure*. London: Abacus.

Hood, C. (2011) *Summing Up the Parts: Combining Policy Instruments for Least-Cost Mitigation Strategies*. Paris: International Energy Agency.

Hourcade, J.-C., Ambrosi, P. and Dumas, P. (2009) 'Beyond the Stern Review: lessons from a risky venture at the limits of the cost–benefit analysis', *Ecological Economics*, 68 (10): 2479–84.

House of Lords (2013) European Union Committee,14th Report of Session 2012–13, *No Country Is an Energy Island: Securing Investment for the EU's Future*, 2 May.

HSBC (2012) *Bonds and Climate Change: The State of the Market* (and update report 2013). Online at: http://www.climatebonds.net

Jaccard, M., Nyboer, J., Bataille, C. and Sadownik, B. (2003) 'Modeling the cost of climate policy: distinguishing between alternative cost definitions and long-run cost dynamics', *Energy Journal*, 24 (1): 49–73.

Jaffe, A., Newell, R. G. and Stavins, R. N. (2003) 'Technological change and the environment', in K. G. Maler and J. R. Vincent (eds), *Handbook of Environmental Economics*, Vol. I. Dordrecht: Elsevier Science BW.

Kahneman, D. (2011) *Thinking, Fast and Slow*. London: Allen Lane.

Karplus, V., Paltsev, S., Babiker, M. and Reilly, J. M. (2013) 'Should a vehicle fuel economy standard be combined with an economy wide greenhouse gas constraint? Implications for energy and climate policy in the United States', *Energy Economics*, 36: 322–33.

Lecocq, F., Hourcade, J.-C. and Ha-Duong, M. (1998) 'Decision making under uncertainty and inertia constraints: sectoral implications of the when flexibility', *Energy Economics*, 20: 539–55.

Lee, B., Preston, F., Kooroshy, J., Bailey, R. and Lahn, G. (2012) *Resources Futures – A Chatham House Report*. London: Chatham House.

Levin, K., Cashore, B., Bernstein, S. and Auld, G. (2009) 'Playing it forward: path dependency, progressive incrementalism, and the "super wicked" problem of global climate change', *IOP Conference Series: Earth and Environmental Science*, 6: 50.

Lewin, K. (1951) *Field Theory in Social Science: Selected Theoretical Papers*, ed. D. Cartwright. New York: Harper & Row, p. 169.

Lomborg, B. (2001) *The Skeptical Environmentalist: Measuring the Real State of the World*. Cambridge: Cambridge University Press.

Mao, X., Yang, S., Liu, Q., Tu, J. and Jaccard, M. (2012) 'Achieving CO_2 emission reduction and co-benefits of local air pollution abatement in the transportation sector of China', *Environmental Science and Policy*, 21: 1–13.

Marechal, K. and Lazaric, N. (2010) 'Overcoming inertia: insights from evolutionary economics into improved energy and climate policies', *Climate Policy*, 10 (1): 103–19.

Martenson, C. (2011) *The Crash Course: The Unsustainable Future of Our Economy, Energy and Environment*. Hoboken, NJ: John Wiley & Sons.

Mau, P., Eyzaguirre, J., Jaccard, M., Collins-Dodd, C. and Tiedemann, K. (2008) 'The "neighbor effect": simulating dynamics in consumer preferences for new vehicle technologies', *Ecological Economics*, 68 (1): 504–16.

Newbery, D. (2012) 'Reforming competitive electricity markets to meet environmental targets', *Economics of Energy and Environmental Policy*, 1 (1).

Nordhaus, W. (2008) *A Question of Balance: Economic Modeling of Global Warming*. New Haven, CT: Yale University Press.

Ofgem (2010) Project Discovery: options for delivering secure and sustainable energy supplies, UK Office of Gas and Electricity Markets. www.ofgem.gov.uk (published 3 February 2010).

Ofgem (2012) *Strengthening Strategic and Sustainability Considerations in Ofgem Decision Making*, 23 July; summary published also in *Chronicle of the International Council of Energy Regulators* (2013 forthcoming).

Oreskes, N. and Conway, E. (2011) *Merchants of Doubt: How a Handful of Scientists Obscured the Truth on Issues from Tobacco Smoke to Global Warming*. London: Bloomsbury.

Polanyi, K. (1957) *The Great Transformation*. Boston: Beacon Press.

Ridley, M. (2010) *The Rational Optimist: How Prosperity Evolves*. London: Fourth Estate.

Rivers, N. and Jaccard, M. (2006) 'Choice of environmental policy in the presence of learning by doing', *Energy Economics*, 28 (2): 223–42.

Rozenberg, J., Vogt-Schilb, A. and Hallegatte, S. (2013) 'Efficiency and acceptability of climate policies: race against the lock-ins', *Review of Environment, Energy and Economics*, forthcoming.

Ruttan, V. W. (1997) 'Induced innovation, evolutionary theory and path dependence: sources of technical change', *Economic Journal*, 107 (444): 1520–9.

Saikawa, E. (2013) 'Policy diffusion of emission standards: is there a race to the top?' *World Politics*, 65 (01): 1–33.

Shalizi, Z. and Lecocq, F. (2009) Climate Change and the Economics of Targeted Mitigation in Sectors with Long-Lived Capital Stock. World Bank. Online at: https://openknowledge.worldbank.org

Stallworthy, M. (2009) 'Legislating against climate change: a UK perspective on a Sisyphean challenge', *Modern Law Review*, 72 (3).

Stern, N. and Rydge, J. (2012) 'The new industrial revolution and international agreement on climate change', *Economics of Energy and Environmental Policy*, 1 (1).

Stern, Sir N. (2006) *The Economics of Climate Change: The Stern Review*. Cambridge: Cambridge University Press.

Stiglitz, J. E. (2002) *Globalisation and Its Discontents*. London: Allen Lane.

Vogt-Schilb, A. and Hallegatte, S. (2011) *When Starting with the Most Expensive Option Makes Sense: Use and Misuse of Marginal Abatement Cost Curves*. World Bank. Online at: https://openknowledge.worldbank.org/handle/10986/3567

Vogt-Schilb, A., Meunier, G. and Hallegatte, S. (2012) 'How inertia and limited potentials affect the timing of sectoral abatements in optimal climate policy', *World Bank Policy Research*.

von Weizsäcker, C. (2005) *The Welfare Economics of Adaptive Preferences*, Max Planck Insititute for Research on Collective Goods, 2005/11.

Williamson, O. E. (2000) 'The new institutional economics: taking stock, looking ahead', *Journal of Economic Literature*, 38 (3): 595–613.

Appendix: The importance of inertia and adaptability

A simple model[1]

The evidence for adaptability: a brief retrospective

An important generic theme for which evidence builds up throughout this book is the capacity of energy technologies and systems to evolve and hence adapt. This was already flagged in Chapter 1, which presented the sustained differences in energy consumption between countries, and noted 'Bashmakov's constant' – the relative constancy of long-term national energy expenditures (as a proportion of income) despite large price variations. Chapter 1 also underlined the tremendous inertia in energy systems.

Chapter 3 outlined the huge array of energy technologies available or under development and the extent of both fossil fuel and renewable energy resources. Chapter 4 identified the sources of behavioural and structural inertia which leave many economically and environmentally efficient technologies unused. Chapter 5 charted the remarkable responses to the 1970s oil price shocks and subsequent policies.

The chapters of Pillar II (6–8) presented economic data to underline how much energy systems have responded to price increases, given time; the basic data charted by Newbery (Figure 6.1) underline all the above points (perhaps the apparent long-term constancy of national energy expenditures should be known as the 'Bashmakov-Newbery constant').

Chapter 9 then introduced the evidence on learning effects and underlined the extent to which energy technology development depends on the extent and orientation of push and pull forces. Chapter 10 summarised the principles of evolutionary economics and the evidence around path dependence, and illustrated the extent to which the three dominant sectors of transport, electricity and urban systems all have the potential to be transformed over the next few decades. Chapter 11 raised the issues to the macroeconomic level, noting also that economic systems themselves have been characterised by innovation, transformation and path dependence.

Finally, Chapter 12 pinned down the fact that First and Third Domain processes in particular imply a capacity to adapt. Second Domain processes, indeed, do *not* exclude that possibility, though they have generally been interpreted otherwise in modelling.

On representing learning

Analytically, an apparent drawback of all this is that the individual processes and possibilities involved seem complex; together it seems they make for unmanageable complexity. Moreover, many factors are intrinsically hard to quantify. Chapter 9 (Figure 9.4), for example, noted the wide uncertainties in technology learning rates – the extent to which technology costs decline with volume – and emphasised the risks in using learning rates for specific projections of technology costs.

Indeed, shortly before this book went to press, an article by Professor Nordhaus warned against 'the perils of the learning model', based upon such uncertainties.[2] He pointed out that it is hard to disentangle induced learning from other forms of learning, and hence the risk that technology learning rates (i.e. cost reductions driven by scale) are exaggerated.

The difficulty is that *ignoring* such effects just because they are complex and hard to quantify amounts to 'throwing out the baby with the bathwater'. As noted in Chapter 12 (section 'No history?' p. 454), classical Second Domain assumptions, as implemented in most models, imply *no* capacity to adapt and learn in response to energy or environmental policy. This is clearly inconsistent with the evidence replete throughout this book.

Assuming something to be zero just because it is hard to measure is often the most misleading approximation of all. It can easily lead people to ignore or neglect important issues. The difficulty is that trying to represent learning processes directly, for example through learning rates or equations that directly link investment to cost reductions, involves making numerous assumptions on a wide diversity of technologies. The studies of the Innovation Modelling Comparison project underlined the problem that models with induced technical change tend to be sensitive to assumptions (the 'butterfly effect' of chaos theory) and that the empirical data is never good enough to provide robust estimates. This reflects the combination of the inherently unpredictable nature of innovation and the difficulty of measuring learning rates robustly. The greater differences between results of the three models of the RECIPE project outlined in Chapter 11, for example, are largely about such differences.

A simple model of inertia and adaptability

Consequently, this Appendix introduces a simple analytic way to cut through many of the complexities. Though some of the maths can get complex, the basic elements are not.

We start with the classical notion of a default 'baseline' projection, in this case of energy and related global CO_2 emissions. As noted in Chapter 12, the vast majority of global energy-economic studies then assume that the cost of cutting emissions is primarily related to the degree of cutback – the extent to which emissions are reduced below that baseline (see Chapter 12, section 'No history?' and note 9).

In contrast to this classical approach of defining a cost of lowering emissions related purely to the *degree* of reduction (abatement), we explore the implications of making this cost also dependent on the *rate* of deviation from the baseline.

Making the abatement cost depend on the *rate* of abatement obviously is one way of capturing inertia. Shifting the balance of costs from purely the *degree* of the effort to the *rate* also then enables a simplified way of representing the capacity of a system to adapt.

In a fully adaptive system, the costs would be *only* transitional. In this case the cost represents an *investment* in changing the pathway, incurred as long as we are trying to increase the degree of cutback from the reference case. After this investment, however, the baseline has shifted and stays on that new (lower emitting) pathway at no extra cost.

To keep things as simple as possible, the resulting model is denominated in terms of global CO_2 emissions, the costs associated with their reduction, and the costs of their accumulation in the atmosphere. A moment's thought suggests that the costs of emission reductions are likely to rise non-linearly both with the *degree* of abatement and its *rate*.[3] We assume these costs rise quadratically, so that:

$$\text{Abatement cost at time } t = cost_a \times (\textit{degree} \text{ of abatement})^2 + cost_b \times (\textit{rate} \text{ of abatement})^2$$

or more compactly:

$$\text{Abatement cost } C(t) = C_a \cdot \varepsilon(t)^2 + C_b \cdot (d\varepsilon/dt)^2$$

where $\varepsilon(t)$ is the degree to which emissions are below the baseline trend at time *t*, and $d\varepsilon/dt$ its rate of change. The degree and rate of abatement refer to the deviation at any time *t* from the 'default' or 'business-as-usual' trajectory, and the constants C_a and C_b (listed as $cost_a$ and $cost_b$ in the model equations below) represent the magnitudes of enduring *vs* transitional costs respectively. The first term represents the ongoing cost element, with reversible consequences; the second term reflects inertia, i.e. the cost of *changing* the level of abatement.

Lowering the ongoing component C_a and raising the transitional component C_b means that abatement cost (or effort) is increasingly dominated by the inertia of moving from one state to another, relative to the recurring costs of staying at any given distance from the 'default' path. The ratio of C_b to C_a thus represents the adaptive capacity of the system: raising C_b and lowering C_a increases the influence of transitional costs, i.e. efforts to overcome inertia which have an enduring impact in lowering the cost of subsequent cutbacks or altering the underlying pathway.

The task of applying this to studies of optimal global responses is greatly eased by the finding from the scientific community that global temperature change at a given time is closely related to cumulative emissions to that point. This enables a simple representation of climate impacts linked to temperature change; (the vast majority of 'integrated assessment' studies, which try to compare the cost of cutbacks with the cost of avoided damages, express climate damages in terms of global average temperature increase). A central estimate is that 500GtC cumulative emissions increases global temperature by about 1°C (there are some time lags but they are of secondary importance for most practical emission trajectories). In the results here, it is assumed that global damage increases in proportion to the square of temperature change:[4]

Annual damage from climate change at time t,
d(t) proportional to $(\text{temperature change})^2 = (E(t)/500)^2$

where E(t) is the cumulative CO_2 emissions in billion tonnes (gigatonnes) of carbon (GtC) at time t ($1GtC = 3.7GtCO_2$). Emissions since the late nineteenth century to 2010 amount to around 500GtC, and annual emissions in 2010 were around $40GtCO_2$ (10.8GtC) per year.[5]

For any given level of assumed damage associated with accumulating CO_2 in the atmosphere, we can explore how the path of theoretically optimal responses depends on the balance between enduring abatement costs and the transitional costs of overcoming inertia set against the long-run benefits associated with reduced atmospheric change. The full model specification, for those interested in tracing the details and verifying results, is given below.

Numerical assumptions

We adopt the following key assumptions. While the main interest is in the influence of shifting costs between enduring and transitional components, to illustrate the impact of adaptability in the energy system, a number of other assumptions are necessary and chosen in part to facilitate comparison of the conventional cost case (with non-adaptive abatement costs) with other modelling works in the field.

- *Real discount rate 2.5 per cent per year.* This is a compromise between the 'prescriptive' and the 'descriptive' rates (see Chapter 1, section 'A Stern warning about time' p. 25, and associated notes 56–64), though leaning more towards the latter. This leads to significant discounting of costs after a few decades; the results show that the economic case for substantial action is not dependent on very low (Stern-type) discounting assumptions.[6]
- *Climate change damage $3 trillion per year for an additional 500GtC emission.* As discussed in Chapter 1 the cost associated with any given degree of atmospheric change is extremely uncertain and needs to factor in corresponding issues of risk. If an additional 500GtC (from present) emissions increases global average temperatures to 2°C above pre-industrial levels, $3 trillion per year damages amount to around 2 per cent of projected global GDP in the corresponding decades (around or soon after 2050); this is well within 'standard' assumptions and modest compared to many risk-weighted estimates.
- *Reference emissions growth 800MtCO_2 per year.* As charted through the first few chapters of this book, global emissions growth has tended to be approximately linear over extended periods while fluctuating significantly. There are various reasons why emissions growth is not exponential and is even less likely to be so in the future. We thus use linear projections for the reference case, and emissions are reduced relative to this:

$$\text{Emissions } e(t) = e_0 + e_1 t - \varepsilon(t)$$

where e_1 is the linear growth rate. The historical growth rate has varied widely; over the past few decades the average increase of fossil fuels CO_2 has been about 1.5 per cent of 2010 emissions, but was substantially higher in the early 2000s (with the Asian boom, before the energy price rises and the credit crunch). Also, historical data already include the impact of extensive energy efficiency measures as charted in Chapter 5. We take as a reference (no action) case a pessimistic view in which global emissions rise at e_1 = 800MtCO_2 per year = 2 per cent of 2010 total CO_2 emissions, which corresponds closely to the reference projection of the IEA (2012).

Abatement costs parameters and relationship to the Three Domains

The abatement cost parameters are derived with reference to a 50 per cent cut in global emissions *relative to projection* by 2040 as follows:

- Purely enduring costs (C_b = 0): cutting global emissions to 50 per cent below reference projection in 30 years time (i.e. a cutback of 32GtCO_2 by 2040) costs $2trn/yr (e.g. 2 per cent of GDP@$100trn). This is towards the pessimistic end of literature.
- Purely transitional costs (C_a = 0): the same cutback, on a linear trajectory of abatement, results in the same total integrated cost over the 30-year period, but these are now attributed as transitional costs of reorienting the energy system over these decades.

As outlined below, a great deal of insight can be gained simply by comparing the results of these two opposing conceptual assumptions – contrasting the two extremes of zero adaptability, or total adaptability. These are the cases shown in bold lines (dashed and solid respective) in Figure A1.1 below.

However, it is of course quite likely that reality lies somewhere between. Going further requires some estimate of how much of the apparent costs of any given path may actually be ongoing, and how much are transitional (with enduring benefits). This book has offered indications that, for transitions like those implied in Chapter 1 – achieving big changes in the efficiency and energy mix of energy systems over a period of several decades – all three domains are of comparable importance in terms of the potential magnitude of their contributions (see Chapter 2, Box 2.1).

We have shown that two of the three domains (the First and Third) involve predominantly *transitional* costs (or efforts). The other (the Second Domain) is a mix of enduring and transitional costs which are hard to disentangle because of the complexity of induced innovation, energy resources depletion, 'double dividend' debates around carbon revenues and their spillover relationships with the other Domains.

As part of our exploration, we examined two different ways of representing the impact of energy efficiency:

- In one case ('external efficiency'), strengthened efficiency policies directly reduce the reference emissions, at no cost, to 1 per cent per year; this corresponds closely to the estimate of the IEA (2012) in their high-efficiency scenario.
- In the other case ('adaptive efficiency') efficiency is assumed to be part of the *adaptive capacity* of the energy system in response to Pillar 1 (energy efficiency) polices which are *not* free, but for which the cost is *transitional* while the benefits endure, as discussed in Chapters 5 and 12.

The fainter dotted lines in Figure A1.1 show the latter case – maintaining the (high growth) reference case, but showing the implications if two-thirds of the abatement costs are transitional, i.e. associated with the costs of moving to a different path rather than enduring costs of deviating from the 'default' trajectory. The absolute trajectories in fact are relatively similar in the two efficiency variants. The difference is that in the cases displayed, the efficiency gains are not free (no 'free lunch') but are represented as part of the investment of effort required to change course.

These intermediate cases allow for the possibility that we continue to develop fossil fuels in ways intrinsically cheaper than the alternatives. This contrasts with the fully adaptive assumption that systems have huge capacity to innovate, evolve and adapt over time, and that there is no inherent reason why a system based on fossil fuels should ultimately be cheaper than one that is based on more intensive energy efficiency and zero carbon energy resources, as suggested for example by the analysis in Chapter 10.

Results

The results are displayed in Figure A1.1, which shows the optimal trajectories of emissions, cumulative emissions, abatement costs and climate damage over the century.

For the classical, non-adaptive/low inertia case, there is a substantial jump of initial abatement which then increases slowly as climate damages accumulate. The effort is defined by distance from the default trajectory, and the abatement cost directly reflects the assumed 'social cost of carbon damages' as discounted. This is fundamentally the implication of a purely Second Domain world applied to the energy system, with little inertia and no adaptability of the energy system.

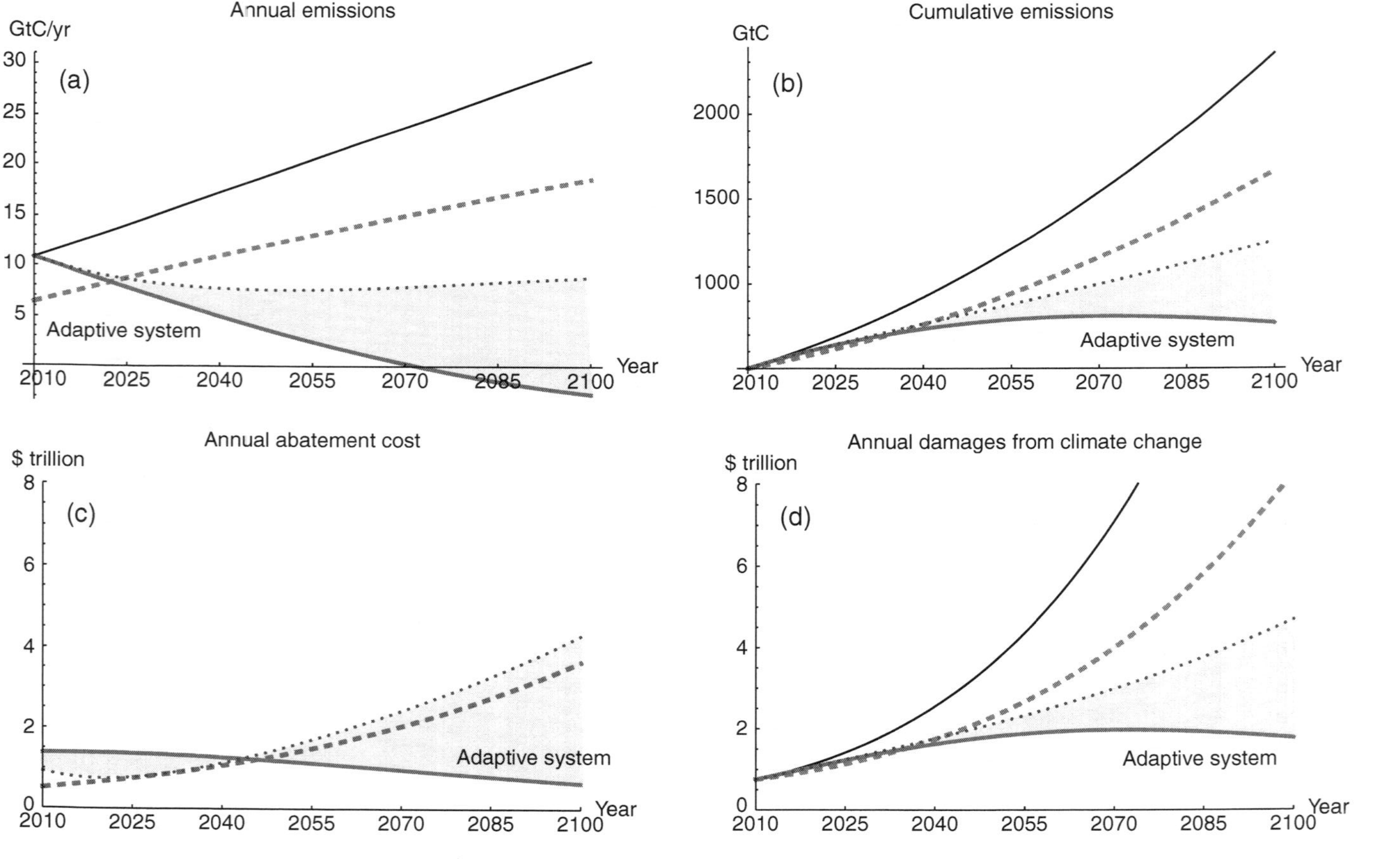

Figure A1.1 Implications of inertia and adaptability for optimal responses.

Notes: The figure shows the least-cost global response given different assumptions about the structure of energy systems costs. The top panels show trajectories of annual (left) and cumulative (right) CO_2 emissions (in MtC); the bottom panels, the corresponding costs of abatement (left) and climate damage (right), in trillions of dollars per year. The thick dotted lines reflect classical assumptions in which abatement costs relate purely to the degree of abatement, relative to baseline global emissions that are steeply rising (the top line in the emission figures). The lower solid lines reflect an adaptive system in which abatement costs (the degree of effort) relate to the rate of abatement but shift the trajectory thereafter. The dotted line in between reflects a mixed case, with costs divided three ways, of which two-thirds (notionally, First and Third Domain responses) are adaptive. The underlying assumptions include moderate damages and substantial discounting (2.5 per cent per year real), as outlined in the text. Despite this, the fully adaptive case results in steadily declining global emissions, reaching zero in the second half of the century (panel (a)) and limiting the additional cumulative emissions to about 350GtC (panel (b)). For the likely range as illustrated (hatched), it is worth investing up to $1 to 1.5trillion/yr (panel c) to mid century [illegible]

The combination of such assumptions means that after the initial prompt cutback, global emissions continue to rise, as shown in the top 'classical case' (thick dashed) lines of Figure A1.1; the abatement cannot keep pace with the rising emissions of the reference case. This is broadly the result that emerged from many of the modelling studies embodying classical assumptions (particularly prior to the debates provoked by the Stern Review, which have tended to increase the impact of damage estimates (see Chapter 1)).[7] Cumulative emissions from 2010 to the end of the century reach around 1,500GtC.

In contrast, if the energy system itself is highly adaptive in the long run (offset by high inertia as discussed), the pattern is quite different ('adaptive system' lines as indicated in Figure A1.1). The deviation from the default trajectory rises to exceed the 'steady-state' level of the classical case after 10–15 years and it carries on diverging at a rate which does not slacken. The optimal response in this case involves global emissions halving before 2050 and continuing to decline, reaching zero in the middle of the second half of the century. The corresponding cumulative emissions reach around 600GtC, after which atmospheric concentrations slowly start to decline.

Note that *the assumed damage associated with a given degree of climate change in the two cases is identical*. It is the dynamics of response that differs. At first glance, this appears to be somewhat paradoxical – one might suppose that the effort would be less when inertial/transitional costs increase. Yet this is not the case, because abatement in the adaptive case is associated with an enduring change in trajectories. The benefits are not only those of the immediate emission reduction, but they extend over time – initial efforts carry through to a pattern of more extensive abatement spanning decades.

The second row in Figure A1.1 illustrates the pattern of the costs involved. In the classical (non-adaptive) case, for the given assumptions, optimal expenditure on abatement starts at a little over $500bn per year and then rises steadily, exceeding $3 trillion per year by the last decade of the century.

The optimal cost of initial action – to be more precise, the effort worth exerting – is substantially bigger in the adaptive case (because the benefits are much larger), but then declines. For the given assumptions, the initial effort is around $1.4 trillion – almost three times the classical case – and declines towards $1 trillion per year by mid-century, continuing to decline thereafter.

This shows that responses which help to adjust the long-run trajectory through the various mechanisms analysed in this book may be much more valuable than the value of cutting emissions alone. In this example they are worth almost three times as much. This echoes the implications of Figure 9.1 (on declining technology costs meeting rising carbon prices), which in turn illustrates the potentially huge benefits of learning investments. The investment worth making in such actions, in other words, is almost three times whatever value is assumed for the 'social cost of carbon'.

The corresponding costs of climate damages are illustrated in the final panel of Figure A1.1. In the classical, non-adaptive case, the level of damages arising from the 'optimal' response increases to over $8 trillion per year by the end of the century, since emissions carry on rising. In the adaptive case, the damages stay below $2 trillion per year. For optimal responses, the total equivalent-cost by the end of the century is thus less than $3 trillion per year for an adaptive energy system, and over $10 trillion in the non-adaptive case.

The importance of this is not so much the absolute numbers but rather the more generic headline insights summarised in Chapter 12 (the absolute numbers are interesting, but more dependent on assumptions).[8]

One essential insight is that an adaptive energy system can greatly lower the overall costs associated with climate change impacts and responses, but only if requisite effort is put upfront into changing course. Given this, it also serves to narrow the gulf between the 'cost/benefit' and the 'security' approach to the problem. An adaptive energy system implies that the costs of remaining secure – or of avoiding many trillion dollars of climate damages – will end up much lower than classical approaches suggest. With an energy system that is largely adaptive, thanks to the mechanisms examined across all three Domains in this book, either approach to the costs and risks in interfering with the planetary heat balance then yields similar conclusions about the benefits of strong action.

Indeed, it is interesting to note that these numbers span quite closely the range of estimates of the investments required to get on course for a 2°C target (US$ trillion 400-1200/yr, cited in Chapter 10). The numbers in Figure A1.1 though are derived from a simple balancing of assumed (relatively high) abatement costs against monetised climate damages, in ways that vary depending on the degree of adaptability of the energy system (and energy efficiency: the optimal additional investment goes down to around $1trillion/yr with the alternate (free) representation of energy efficiency).

Of course, we do not need to decide today on the long-run trajectory itself: what matters now is how much effort we make over the next few years. A consistent rough translation could be that investing in the range £500bn – 1 trillion/yr can get the world on a course of substantially declining emissions over the next few decades. This would expand our subsequent options, and stand a good chance of leading to substantially lower combined energy and environmental costs. It would keep open the pathway to stabilise the atmosphere close to 2°C, should the unfolding science confirm common scientific judgement around climate sensitivity and risks. Eschewing such investment incurs global risks, and the inertia of the energy system – poorly represented in many economic models – would then make it extremely expensive (or impossible) to catch up.

As the results here show, the extent to which energy systems are indeed adaptive is a very important economic question. There is need for more considered debate and analysis about the extent of adaptability, including the evidence laid out in Pillar III (Chapters 9–11) which amount to a proposition that energy systems may be fully adaptive given sufficient time, particularly as the easiest fossil fuel deposits have been exploited. Unquestionably, the degree of adaptability deserves attention. The idea that energy systems have no capacity to learn and adapt to constraints and incentives is clearly inconsistent with the evidence and this appendix shows how much this matters.

Detailed mathematical formulation

The mathematical specification of the model outlined in this Appendix is as follows:

Marginal emissions $e(t)$

Cumulative emissions $E(T) = E_0 + \int_0^T e(t)dt$

Reference emissions $e_{ref} = e_0 + e_1 \cdot t$

Marginal damage (X = temperature change)[9] $d(t) = d_1 \cdot X(t) + \frac{d_2}{2} \cdot X(t)^2$

Net Present Value (NPV) of damages
(r = real discount rate) $D(T) = \int_0^T e^{-r \cdot t} \cdot d(t)\,dt$

Cost abatement type A $c_A(t) = \cdot\ cost_A \cdot \left(e_{ref}(t) - e(t)\right)^2$

NPV abatement cost type A $C_A(T) = \int_0^T e^{-r \cdot t} \cdot c_A(t)\,dt$

Cost abatement type B $C_B(t) = \frac{1}{2} \cdot cost_B \cdot \left(e_1 - \dot{e}(t)\right)^2$

NPV abatement cost type B $C_B(T) = \int_0^T e^{-r \cdot t} \cdot c_B(t)\,dt$

The economic objective is then to find the path of cumulative emissions E(t) which minimises the total of these different costs, $F(T) = D(T)+C_a(T)+C_b(T)$. Mathematical theory enables this to be calculated as:

$$\int_0^T F(t)\,dt = \int_0^T e^{-r \cdot t} \left\{ \begin{array}{l} d_1 \cdot E(t) + \frac{d_2}{2} \cdot E(t)^2 \\ + \cdot\ cost_A \cdot \left(e_{ref}(t) - \dot{E}(t)\right)^2 \\ + \cdot\ cost_B \cdot \left(e_1 - \ddot{E}(t)\right)^2 \end{array} \right\} dt$$

Euler Lagrange method to find
the optimal trajectory $\frac{\partial F}{\partial E} - \frac{d}{dt}\left(\frac{\partial F}{\partial \dot{E}}\right) + \frac{d^2}{dt^2}\left(\frac{\partial F}{\partial \ddot{E}}\right) = 0$

It proves possible to solve the resulting equations analytically to derive the cost-minimising path of cumulative (and hence annual) emissions for any combination of the different types of cost coefficients. This results in a small compact model which can be run easily on any PC with appropriate mathematical software (we implemented it in Mathematica) to explore the impact of different assumptions.

Notes

1. Michael Grubb with Rutger-Jan Lange and Pablo Salas, Cambridge University, respectively Faculty of Economics and the Centre for Climate Change Mitigation Research, at the Department of Land Economy.
2. Nordhaus (2014). Prof Nordhaus noted the potential importance of induced technical change in introducing his book *A Question of Balance*, and he went to significant effort to include elements of learning in some variations of his DICE model, but with limited impacts for reasons outlined in Chapter 9 (see section 9.6 'When the chain is broken', and note 22). This underlines the point that when complexity and the difficulties of measurement are obstacles, it can help to think laterally of different ways of characterising more simply key processes, which is what this Appendix offers.
3. Doubling the degree of emission cutbacks means moving to higher-cost measures, so will cost more than twice as much; doubling the rate of cutback, similarly, is far more likely to incur additional transitional costs or involve premature retirement of capital stock, for example.
4. A much earlier study by the author (Grubb *et al.* 1995) presented the basic idea of the mitigation analysis and showed results that emerged if the damages from climate change were assumed to be proportional to the atmospheric concentration of CO_2. At the time this seemed the analytically tractable approach and a useful approximation to illustrate the underlying themes. However, maths does what is specified and this treatment had the serious drawback that, especially in cases with a highly adaptable energy system, results could be driven by the long-run benefits of negative emissions which reduce concentrations, without limit. Particularly at low discount rates (or high damage coefficients) this could go to implausible extremes. The treatment in this Appendix, in which damage is related directly to the square of temperature change since pre-industrial levels, avoids this problem since the benefits of reducing concentrations decline non-linearly, and turn negative if global temperature drops below pre-industrial levels. I am grateful to Rutger-Jan Lange for pointing out that the case of quadratic temperature change could be solved by transforming the analysis to make cumulative emissions the control variable, and formulating the resulting code.
5. Perhaps unfortunately it has become standard in the field to cite most emission rates in terms of carbon dioxide – the gas emitted - whilst scientists tend to use absolute carbon (the element) for cumulative volumes, which generates the convenient amounts of c. 500GtC extracted to date and another 500GtC to reach 2 °C temperature change on central estimates. The two measures differ by the added weight of the two oxygen molecules: 1tC = ((12+2x16) /12) tCO_2 = 3.7tCO_2. 2010 global CO_2 emissions of around 40$GtCO_2$ comprise 33$GtCO_2$ from fossil fuel combustion, with the rest divided between cement, land use change and a few other much smaller sources
6. As with the Ramsey discounting formula (see Chapter 1, note 59). For example, if global average economic growth is 2.5% per year, the 2.5% real discount rate would correspond to: a 2.5% pure rate of time preference (PRTP) with inequality aversion 1; a 1.25% PRTP with inequality aversion 1.5; or a 0% PRTP combined with inequality aversion 2. In the authors' view this is a high level of discounting for a problem like climate change but is necessary to replicate results along the lines of many in the field. The resulting graphs (Figure A1.1) serve to underline the extent to which such discounting, combined with classical assumptions, essentially loads costs on to the future: in the conventional case, the initial abatement effort is less than $1 trillion, leading to combined abatement + damage costs exceeding $10 trillion per year before the end of the century. That is the consequence of such discounting in the classical case. Note how adaptability in the energy system serves to narrow the implied intergenerational inequality.
7. It always seemed to the authors that this classically common combination of economic assumptions embodies an inconsistency. Modest climate damages reflect an assumption that we can adapt to climate change (plant different crops, move or protect our coastal cities, increase resilience to droughts or floods), even though the way in which climate change will manifest is fundamentally unpredictable and likely to express itself mostly in occasional extreme events. Yet this is set alongside an assumption that energy systems, in which policies and incentives *are* under our direct control, have no capacity to *adapt* to emission constraints. This seems an intrinsically implausible combination, yet for two decades this combination informed the scepticism of many (though far from all) economists about the economic 'cost/benefit' case for acting to stabilise the atmosphere.

   In practice, and recognising the fundamental uncertainties and controversies over valuing climate damages, many analysts moved away from 'cost/benefit' attempts to analyse just the costs of different trajectories to meet pre-set goals of stabilisation. This was consistent with the precautionary and security-led approach outlined in Chapter 2, yet underlined a gulf between 'scientific thresholds' and 'economic cost/benefit' approaches, which has not helped policy. It was this gulf which the Stern Review attempted to bridge, and this – and the post-Stern debates – came to focus heavily on approaches to discounting, and the evaluation of extreme planetary risks (Chapter 1, section 1.4 'Costing the Climate'). Thus, these debates focused on the structure of the problem, far more than the structure of solutions at source (i.e. changing energy systems), which has been the focus of this book.

8. Technically minded readers may wish to email the author for a copy of the model (which is implemented in Mathematica) and experiment with different assumptions, including those on damage equivalence and discount rates which can have a strong bearing on the absolute results, and to compare energy-related assumptions with other studies with strong empirical content such as those by the International Energy Agency. A recent analysis of how results in such modelling depend on key assumptions – and the consequent implications of uncertainty for estimates of the 'social cost of carbon' – is given in Hope (2012). For a longer history of sensitivity studies and related debates on quantifying the monetary-equivalent value of climate damages see the many sources cited in Chapter 1 (from Ackerman, right through to Yohe and Tol).
9. To avoid confusion with the time horizon T in the model, X(t) is here used to denote temperature change; as explained in the text this is approximately proportional to cumulative emissions: X(t) = E(t) * 500. In all the modelling work presented here we set $d_1 = 0$, so that the focus is simply upon the quadratic damage function.

References

Grubb, M., Chapuis, T. and Ha-Duong, M. (1995) 'The economics of changing course: implications of adaptability and inertia for optimal climate policy', *Energy Policy*, 23 (4): 1–14.

Hope, C.W. (2012) 'Critical issues for the calculation of the social cost of CO_2: why the estimates from PAGE09 are higher than those from PAGE2002', *Climatic Change*, online, 12 December 2012.

International Energy Agency (2013) *World Energy Outlook Special Report 2013: Redrawing the Energy Climate Map*. Paris: IEA.

Nordhaus, W. D. (2014) 'The perils of the learning model for modelling exogenous technological change', *Energy Journal*, 35 (1): 1–13.

Index

Page numbers in italic refer to figures and tables and page numbers followed by 'n' refer to notes

abatement 248, 291, 462; cost curve *see* McKinsey curve, *see also* emission reduction(s)
acid rain 10, 215, 226
ad valorem tax 228–9
adaptability (of energy technologies and systems) 19, 457–8, 458–9, 486, 495–505
adaptation 21–2, 23, 31
Adaptation Fund 469
adaptive preferences 489n
additionality 268, 269, *see also* Clean Development Mechanism
ADEME 162, 163
adverse selection 140, 155n, 169, 333, 350n
affective processes 29, 48, 53
Africa 6, 25, 167, 380, *see also* North Africa; South Africa; sub-Saharan Africa
air pollution 9
algal fuels 109, 111, 115
alternative fuels 362, 366, *see also* renewable energy
aluminium 93, 96, 289, 290, 306n
American Economic Review 27
anaerobic digestion 89, 118n
apathy 53
Apple 328
appliances 84, 87, 88, 137, 140, 167, *168*, 169, 172, 184, 298, 304, 383, 420, 423
architecture 87
Arrow, K. 59, 321, 403
Asia 66; carbon dioxide emissions 17, 36n; clean energy 204, 356; emissions trading 272; energy consumption 2, 13; energy demand 387; energy efficiency 134; labelling 167; low carbon electricity 98, 100; transport use 105; urbanisation 380–1
Assigned Amount Units 241, 264, *see also* emission allowances
associative processes 29, 48, 53
asymmetric damages 233n
A(t) factor *see* Solow residual
Atlanta 379
atmospheric warming 10; smooth adjustment assumption 21
Atomic Energy Agency (AEA) 163
attention: to climate policy 255–6
attention effect: of CDM 266
auctioning (allowances) 242, *253*, 254, 262, 478
auctions: renewable generation capacity 339
audit subsidies 177–8
Australia 15, *16*, 167, 177, 226, 229, 238, 239, 258, 271, 272
auto industry 108, 169–72, 194n, 398, 421, 422, *see also* cars
autonomous energy efficiency improvement (AEEI) 130–1, 406–8, 409
'availability' effects 53
aviation 105, 107

bagasse 92
'balance of system' (BOS) costs 101, 102n
banding: UK renewables policy 340
Barcelona 379
Barrett, S. 32, 39n
baseline emission projections 227
baseline energy performance 189, 190
Bashmakov's Constant of Energy Expenditure 19, 63, 208, 297, 409, 413, 459
battery electric vehicles (BEVs) 110, 368–9, 376, 399
Baumol, W. 215
behavioural economics 29, 54, 55–6, 142–50, 156n, 193, 360, 451, 455, 487
Belgium 178, 264
benchmarks 246, 253–4, 268, 277n
best practice frontier 60, 61, 62, 69, 129, 192, 425, 455, 492n
bilateral trading 275n
biofuels 109, 112, 113, 123n, 366, 367

biomass 1, 83, 88, 89, 90, 92, 103, 109, 111, 301, 383
'Black Swan' events 18, 24, 51, 74n
blast furnaces 92
blend wall: ethanol use 109, 123n
'boil the rich' 186, 197n
bond markets 478
border levelling 293–4, 294–5
bounded rationality 50, 143, 360
BP 149
Brazil 15, *16*, 109, 266, 366, 377, 389n, 436n
British Columbia *228*, 229, 271
broad-based energy tax 228
BTU tax 228–9
buildings: codes and standards 179; energy consumption 84–8, 150, *151*, 191; energy demand 81–3; energy efficiency 134, 135, 139, 150–2; energy efficiency policy 178–82, 188–90; key challenges 88–9; urban policies 381–2, *see also* cooling; heating
Burden-Sharing Agreement 276n
Buridan's donkey 426–7, 480
Bus Rapid Transit (BRT) 382
business cycles 7, 56, 409
business-as-usual 62, 64, 244, 361, 362, 386, 453

California 180, 260, 271, 279n, 339
calories 85
Canada *16*, 112, *214*, *228*, 229, 264, 271, 275
cap-and-trade 203, 216, 230, 231, 237–79; Californian 260, 271, 279n; emerging global landscape 270–3; investment, predictability and confidence 255–63; Kyoto Mechanisms 263–70; opponents of 234n; scope and coverage 238–40; tax versus 223–6, *see also* European Union Emissions Trading Scheme
capabilities-based industrial policy 343
capital accumulation 401
capital constraints 137, 177
carbon bubble 454, 464
carbon capture and storage 65, 105, 113, 373; costs *95*, 121–2n; power generation 98; slowing down of innovation 119–20n; and utilisation 94
carbon 'central' bank 261, 262n
carbon constraints 220, 398, 399, 428, 454
carbon dioxide emissions 1; baseline projections 227; cost-benefit 21; industrial *91*, 204; labelling schemes, new cars 174; measurement 85; natural gas-fired generation 97; pollution and climate change 10, 11, 12; pre-industrial levels 10; pricing *see* carbon pricing; reduction *see* emission reduction(s); transport 107; trends 3, 14–17
Carbon Disclosure Project (UK) 278n
carbon divide 384–7
Carbon Emission Reduction Targets (CERT - UK policy instrument) 181, 195n
carbon financial products 427–8
carbon footprint 17, 80, 121n, 155n, 176, 190
carbon index tracer 278n
carbon intensity: GDP 4, 15, 36n; global energy system 13; infrastructure 359, *see also* carbon-intensive sectors
carbon leakage 253, 254, 341; carbon price and 289–92; estimation 286, 306n; reversed 306n; sectoral differences 287; tackling 204–5, 292–5
Carbon Market Investor Association 262n
carbon markets 204, 270–1, 273–4, 470
Carbon Pollution Reduction Scheme (Australia) 226, 239
carbon price(s): atmospheric stabilisation 460; collapse 246–7, 478; global 302–3, 460, 484; innovation 62; responses to 456; willingness to pay 70, *see also* Climate Change Levy
carbon pricing 67, 449; appropriateness 303; China 272; emission reduction 237; energy access 301–2; impacts 204–5, 284–308; low carbon innovation 317–18, 342; low carbon investment 259–60; multiple roles 255–7; R&D as an alternative to 318, *see also* pricing pollution
Carbon Reduction Commitment (CRC - UK policy instrument) 183, 230
carbon sinks 95
carbon taxes 203, 258, 303, 427, 449, 492n; concerns about 216; difficulty of implementation 231; economic recommendation 1; European Union 227–8; versus cap-and-trade 223–6
Carbon Trust 135, 137, 148, 161, 163–4, 166, 176, 178, 185, 335, 336–7, 344, 477–8
carbon-based drawing rights 428
carbon-intensive sectors 252–3; activities 289, 291; products 16, 88, 96, 293–4, *see also* carbon intensity
carousel fraud 250, 277n
cars 108, 174, 364–70, *see also* electric vehicles; fuel cell cars; hybrid cars
Carter, J. 131, 161
catastrophic: impacts/events 28, 29, 49, 70; risks 429
cellulosic butanol 109
cement/industry 88, 90, 93–6, 119n, 248, 277n, 289
Central Electricity Generating Board (CEGB) 163, 320
Central Europe 131, 298, 299
central planning 396
centrally planned economies 15, 131
chaos theory 360
chemical-based industries 289
Chernobyl Syndrome 98
Chichilnisky, G. 302
China 421; building construction 88; carbon dioxide emissions *16*, 17; CDM credits,

coal plants 269; economic growth 436n; emissions savings 266; energy consumption 13, 15; energy inefficiency 15, 131; energy subsidies 211; environmental degradation 9–10; Great Divergence 435n; innovation 334; Low-carbon Pilot Development Zones 272; nuclear power 98, 99; resource availability 111; taxation on proceeds from CDM projects 269; travel choice 106; urbanisation 380–1, 382
choice(s): energy-related 148; welfare maximization 489n, *see also* rational choice; smarter choices; travel choices
circular economy 424, 436n
clarity of policy: and investment 256
classical economics 46, 59–60, 153n, 193, 423, *see also* neoclassical economics
Clean Air Act (US) 9, 226
Clean Development Mechanism (CDM) 204, 241, 246, 264, 458; additionality 268, 269; credits 204, 249, 255, 265, 270, 469; criticisms 267–8; history of 264–7; Programmes of Action 269, 383; project coverage 269; project distribution 269–70; sustainable development 267–8
clean energy: benefits of innovation 129, 317; investment 356, 357; 'markets and prices' response 68; policies 311; production 90–6; research and development 331
climate change 8–12, 447, 472; addressing, urban areas 383; as an externality 30–1; attitudes to 20; collective action 32–3; costing the impact of 20–9; damage studies 34n; denial 32, 135–6; developing countries 189; and innovation 476; lack of interest in 424; as the perfect moral storm 30, 217; resilience 382; responses 31–2; risk conceptions *48*, 49, 50, 52, 53; risk to global 39n; as a security issue 47; tipping points 37n
Climate Change Act (UK) 472, 473
Climate Change Committee (UK) 307n, 389n, 473
Climate Change Levy (UK) 163, 183, 477
climate change policy 429; benefits estimates 29; co-benefits 478; energy bills 296; energy efficiency role 152–3n; fuel poverty 307n; and investment 255–7; as stabiliser of returns to investment 476
Climate Technologies Initiative 162
Clinton administration 228, 229
Club of Rome 401, 406
co-benefits 134, 217–18, 474–8, 486–7, 491n; enhancing interconnection 375; transforming urban development 382–3
co-firing 103
coal 7; -biomass 115; -to-liquid technology 71, 113; consumption *3*; demand 295; energy and emission flows *82*, 83; environmental costs 9; investment 470; prices 7, *8*; resource availability 111; subsidies *212*
cognitive bias 147, 156n
coke 92
collective action 32–3, 489n
collective responses 144
combined heat and power (CHP) 89, 91, 118n, 135, 160
'command and control' measures 215
commercial buildings 134–5, 139, 191
commercial energy system 81–3
commercialisation 325, 326, 331, 338
commodity price speculation 17–18
compact fluorescent lamps (CFLs) 87
'comparative static' studies 21
compensation 23, 24, 50, 70, 300
competition 222, 320, 330, 376, 479
competitive markets 285, 296
competitiveness: industrial 341; international 170, 252, 293, 320, 351n; technological 317
computing power 360
concentrating solar power (CSP) 100, 103
concessionary loans 338
concessions 227
condensing boilers 132, 153n
Confederation of the European Paper Industry 96
consumers: carbon pricing 286, 296–8; energy efficiency policy 165–73; energy subsidies 173, 211, 212; pollution pricing 225–6
contractual constraints 149–50
control systems (modern) 87
conversion technologies 111
cooling 87, 89–90, 381
cooling (atmospheric) 10
coordination 73, 192, 295, 374, 375, 382, 477
Copenhagen climate conference 33, 247
Corporate Average Fuel Economy (CAFE) 169–70, 172, 194n
corporate culture 177
cost curve 63, 66, 68, 75n, *see also* McKinsey curve
cost pass-through *286*, 307n
cost-benefit analyses 459; climate change 20–9, 50, 53; demand-side programmes 196n; energy efficiency policies 185; intellectual dominance 71; low carbon innovation 317, 463–4; modelling 454–5; utility maximisation 57
cost-effective efficiency 88, 129, 189, 193, 300
cost-sharing 23
costs: carbon capture and storage *95*, 121–2n; electric cars 110; emission control 223–5; energy efficiency 136, 457–8; environmental regulations 226–7; EU ETS 251–2; insurance 461–2; low carbon electricity 100, 101, 102,

103, 104–5; low carbon future 357; nuclear power 98; research and development 331; technology deployment 324; technology development 328, 329, *see also* hidden costs; negative costs; opportunity costs; transaction costs
counter-cyclical emissions trading 251
creative destruction 312, 316, 343, 419
crises, and reform 58, 396
crops (energy only) 109
cross-country measurements 63–4
crowding out 341, 399, 403, 408–9
crude oil prices 5, 214
crystalline silicon PV cells 101
cumulative banking 261
cumulative investment 358
cycling 106

'dark matter of economic growth' 396–438, *see also* innovation
Dasgupta, P. 27, 30
data assessing government programmes 184–5
deaths: emission-related 9
debt 17, 423, 426, 447; crisis 480
DeCanio, S. 147, 156n
decarbonisation 14, 403; positive macroeconomic features 426; roadmap 96, 426; trends 13, *see also* low carbon; transforming systems; zero carbon
decision-makers 467–8
decision-making: mitigation of rebound 192; organisational 147, 148–9; three domains *see* three domains
decoupling 63, 406, 407
default adjustment 294
default domain of habit 50
demand: for innovation 405, *see also* energy demand; supply and demand
demand elasticity 18
demand pull 331, *332*, 333, 340
demand-side programmes 196n
demonstration 324–5, 331, 338
Denmark: auctioning allowances 242; carbon taxation 228; energy efficiency policy 160, 176, 177, 178, 181; energy intensity 182; institutional tensions 474; R&D 319–20; renewable energy 89, 335–6, 339
derogations 245, 254, 276n, 277n
descriptive approach to discounting 26, 27, 33, 38n
Designated Operating Entities (DOEs) 278n
developing countries: carbon dioxide emissions 15; climate change 189; economic growth 1; emissions reduction 246, 264–5; energy access and carbon pricing 301–2; energy consumption 381; energy efficiency 153n, 182, 189; energy entitlements 117n; fossil fuel subsidies 211; impediments to innovation 339–40; pollution 9–10; rebound *186*, 187; sustainable urbanisation 381
Diamond, P. 55
DICE model 454
Dion, S. 229
direct reduced iron (DRI) 92
direct subsidies 172, 292
dirty energy 387
dirty innovation 330
discounting 25–6, 27, 33, 37n, 38n, 137, 141, 144, 155n, 454
disempowerment 49
Dismal Theorem (Weitzman's) 22, 27–9, 30, 46, 223–5
Display Energy Certificates 182
distribution/systems 114, 374
distributional choice: consumer subsidy programmes 173
district heating 89
double dividend 219–20, 233n
double glazing 87, 140
double hump 384, 458
Downing, T.E. 27, 28
downstream pricing 239–40
Drake, E. 315, 316, 347n
Drax 257
driving habits 108
Drowning in Oil 4
dual market failures 318
dung 9
Dynamic General Equilibrium Model (DGEM) 432n

E+/E- rules 278n
Eastern Europe 15, 89, 131, 134, 212–13, 244, 255, 276n, 298, 299
eco-labelling 139
ecological economics 52
economic: cycles 56; forecasting 16, 249; indicators 13, 26; instruments 215, 219, 231, 469, 470–1, 486; liberalisation 1, 404; prescriptions 396–7; processes 59–62; recession 6, 17, 204, 229, 249, 251, 257, 259, 356, 357, 421, 423, 425, 428; volatility 7
economic growth/development 4, 398–405; classical theories 423; 'dark matter of 312, 316; energy in 405–9; fossil fuels 1, 398; inflated projections 18; innovation 62, 64, 312, 401, 402, 403, 419–20; neoclassical models 450; oil price shocks 217; risk/management 29; Three Domains in 483–5
economic modelling 425, 467; aggregation 405; assumption of perfect foresight 17; autonomous energy efficiency 130–1, 406, 407–8; bottom-up 399, 406, 408; carbon leakage 290–1; cost-benefit 454–5; emission

reduction and GDP 416–17; equilibrium-based 433–4n, 461; global energy futures 384, *385*; inertia and uncertainty 490n; interconnection 375; learning 321–2; local carbon future 357–8; price response 210; pricing pollution 220, 233n; top-down 399, 408, 461; transition to low carbon systems 386, *see also* integrated assessment models
economics: of changing course 454–60; cost-benefit *see* cost-benefit analyses; of energy 14–20; of innovation *see* macroeconomics; microeconomics; as resource allocation 59; of sustainability 473; taxation of emissions 1; traditional 2; Western 397, *see also* behavioural economics; classical economics; evolutionary economics; Keynesian economics; mainstream economics; organisational economics; welfare economics
economies of scale 100, 103, 325, 333, 343, 405, 421, 423, 432n
El Niño 10
electric vehicles 106, 108, 110, 115, 368, 369, 370, 376, 382, 383, 386
electric-arc furnaces 92
electrical appliances 88, 298
electricity: access to 301; bills 296, *297*; consumption 84–5, 91; energy efficiency 134–5; global cumulative investment 358; innovation 330, 348–9n; low carbon 97–105; markets 181, 222, 285, 296, 320; oil-based generation 9; prices *170*, 204, 296–7, 298, 307n, 376; subsidies *212*; transforming 117, 372–8; unit measurement 85; zero carbon 113, *see also* renewable energy
electrolysis 92, 110
embodied energy 190, 191, 192, 329
embodied hardware 139
emerging economies 7, 77n, 152, 182, 269, 289, 301, 359, 380, 400, 422, 483–4, *see also* Brazil; China; India
emission allowances 215, 216; auctioning 242, *253*, 254, 262, 478; benchmarking 246, 253–4; borrowing 276n; bottom-up 249; free allocation 226, 230, 241, 251, 252–4, 285, 292, 298; intensity-based 277n; output-based 277n, 306n; over-allocation 248; surplus 204, 226, 241, 242, 247, 260, 262n, 267, 285, 404; value, in EU ETS 219, *see also* National Allocation Plans (NAPs)
emission reduction(s) 16–17, 18; 2 degree target 12, 53; certified 246, 265, 266; developing countries 264–5; in the EU 204; gas power generation 97; and GDP 416–17; industrial 92–6; industrialised countries 2; plausibility of 31–2; savings potential, buildings 154n; transport 107, *see also* abatement
emission-based benchmarking *253*
emissions banking 242, 250, 257–8, *259*, 261, 267
emissions trading: profit and loss 204–5, 230, 241, 245, 251–2, 253, 284–8; sulphur dioxide 226, *see also* cap-and-trade
Emissions Trading Directive (EU) 241
employee motivation 148
endogenous growth 403–4, 411
endowment effect 144, 156n
Energiewende 373, 377, 404
energy: agencies 162–4; audits 135, 137, 141, 174–5, 176; cheap 132, 185, 211, 299, 301, 404, 406, 409; economic development 1, 405–9; and economy 14–20; emission ratio and GDP 4, 13; expenditure 18–19, 298, *299*, 307n; imports 6; indirect costs 213; inefficiency/wastefulness 15, 88, 129–57, 190, 299; innovation *see* innovation; leakage 87; research and development 318–21; as a security issue 47, 52; trends 2–14; units and conversion factors 85; vulnerability 423
energy access 301–2
Energy Access for All 12, 115, 129–30, 484
energy bills 18, 296
Energy Company Obligations (UK policy instrument) 195–6n
energy consumption: behavioural economics 145–6; buildings 84–8, 150, *151*, 191; cutting per capita 13–14; delayed consequences of 423; developing countries 381; energy price 208; global 2–4, 13, 55, 85, *91*, *106*, *151*; industrial *91*, 191; patterns 192; personal decision-making 50; transport *106*, 191, 390n; United States 85, 131, 150, 388n; urban areas 381
energy demand: Asia 387; coal 295; economic boom and trebling of 1; electricity 376; global 33–4n, 81–3, 115; oil 4, 33–4n; per capita 111; reducing 130, 488n; transport 105; urban 112
energy efficiency 4, 129; autonomous 130–1, 406–8, 409; barriers to, and drivers of change 135–41; behavioural realities 142–50; buildings 134, 135, 139, 150–2; cost-effective 88, 129, 189, 193, 300; costs 457–8; energy prices 209, 407; gap 61, 134, 135, 138, 143, 492n; improvements 15, 192, 210, 296, 300, 301; industrial 273; macroeconomics 409; as a national priority 174; obligations 181; PAT scheme 273; physics of 86; quantitative targets 164–5, 174; rebound 295; resource standards *see* supplier obligations; role, climate change policy 152–3n; Russia 404; theory and practice 63; trends and potential 13, 61, 130–5, 160, *see also* vehicle efficiency
Energy Efficiency Commitment (EEC- UK policy instrument) 195n

Energy Efficiency Opportunities Programme (Australia) 177
energy efficiency policies 160–95, 329; buildings 178–82, 188–90; end-use tool box 165–78; energy demand 488n; global spread 174–5; initiatives and institutions 161–5; price and 210–11; rebound 186–8; scope 190–2; successfulness 182–6
Energy or Extinction 111
energy intensity 279n; and GDP 4, 13; national 131, 153n, 182; targets 272, *see also* energy-intensive industries
Energy Labelling Directive (EU) 167
energy ladder 4
energy management 137, 176–7, 195n
Energy Market Reform (UK) 105, 351n, 474
energy markets 203; regulation 179, 372, *see also* emissions trading
energy modelling *see* economic modelling
energy performance: buildings 87, 189, 190; certificates 182, 196n; large organisations 177; league table 183; standards 138, 179, 180, 351n; thermal 134, 150, 179, 189
energy price(s) 182, 298; energy consumption 208; energy efficiency 209, 407; forecasts 17; global escalation 246; political nature 203; responses 18–19, 63–4, 208–11; subsidisation 299; volatility 6–7, 17–18, *see also* coal; electricity; gas; gasoline; oil
energy production factor 406
energy rating: UK housing stock *300*
energy reporting 176–7, 195n
energy saving certificates (ECerts) 272
Energy Savings Trust (EST) 135, 163–4, 166, 185
energy security 130, 213–14, 218
energy services companies 140–1
Energy Star 166–7, 173, 196n
energy storage 114, 373, 375–6
energy subsidies 211–13, 299
energy system(s): adaptability 457–8, 486; adjustment to price changes 61; evolution of 452–4; importance of three domains in transforming the global 63–4; inertia 20, 61, 433n, 452, 453, 495–505; integrated 117, 477; urban 381–2; uses, channels, fuels 81–4
Energy Technologies Institute (UK) 335, 350n
Energy Technology Perspectives 115, 162
Energy Technology Support Unit (ETSU) 163
energy-intensive industries: concessions 227; emissions 91; emissions trading 252, 254; products 219, 406–7; supply chain efficiency 96, *97*, *see also* energy intensity; heavy industry
engagement 68, 192, 328, 335–7, 449
Enhanced Capital Allowance Scheme (UK) 178
The Entrepreneurial State 347n
Environmental Defense Fund 226
environmental economics 52, 72
environmental Kuznets curve 14
environmental policy(ies) 21, 263; development of 9; energy bills 296; lagging behind desired level of protection 12; as supplanting the need for technology policy 318; and technology development 329
environmental regulation 215, 220, 226–7; and economic performance 72; industrial competitiveness 341; lagging behind impacts 9
environmental stressors 31
epsilons 156n
equilibrium economics 61, 360, 433–4n, 454, 461, *see also* general equilibrium
ESKOM 296
ethanol 109, 123n, 366, 367
ethics 27, 30, 32, *72*, 73
Europe/European Union 479; biofuels 109; CDM credits 270; decarbonisation 426; development as Third Domain characteristic 58; economy 246; energy efficiency policy 16, 160, 161–5, 167–9, 180; energy taxation 214, 227–8, 229; recovery 479–83; Third Energy Package 376; travel choice 106; wind power 100
European Union Allowances (EUAs) 241
European Union Emissions Trading Scheme 227; addressing carbon leakage 292; allowances *see* emission allowances; amendments 225; carbon leakage 291, 306n; closure rules 292; coverage *238*; demand for CDM credits 265; emission reduction 204; evolution of 240–7; focus 238; lack of foresight 18; lessons 247–55; long-term trajectory 261; low carbon innovation 341; profit and loss 230, 241, 251–2, 285; robustness 261; steadying 258–60, 261–2; structural reform 261–2, 482; Third Domain influence 470; upstream application 239
evolutionary economics 54, 58–9, 359–60, 451, 457, 488n
exchange-based trading 275n
exergy efficiency 84, 86, 91, 118n
experience: curves 321, 322, 324, 326; and innovation 333; prior 48, *see also* learning
export-oriented development 424
exports: out of the carbon price regime 293, 294
externalities 9, 215, 318, 409

facilitating conditions 144, *146*
Factor 4 solution 14
feed-in tariffs 339, 357
Fifth Assessment (IPCC) 417
finance: energy efficiency 137, 141; and innovation 345–6; low carbon 427–8, 478; realism in 477–8, *see also* investment
financial incentives 175, 179

financial reform 484
financial viability low carbon projects 256–7
Finland 227, *228*
First Commitment Period (Kyoto Protocol) 241, 242, 245
First Domain 55, 61, 69, 418, 448–9; energy efficiency 152, 407; energy system 83; energy use and waste 129, 132, 133, 150, 191, 362; hybrid vehicles 370; innovation 312, 329, 457; investment and return 462–3, 464; markets and pricing 470, 471; path dependence 361; pricing pollution 217, 249, 267; rebound 187; risk perception 49, 50; smarter choices 449, 450, 456; timescales 73; urbanisation 381
first generation biofuels 109, 112
fiscal measures 172–3, 177–8, 184
Five Year Plan 2011–15 (China) 272
fleet fuel efficiency 169–70
flexifuel vehicles 109, 366
floating foundations (wind power) 100
food: energy and 7
Fordism 421, 422, 423, 435n
foresight: lack of 17–18, *see also* imperfect foresight; perfect foresight; rational foresight
fossil fuel(s): consumption 12, 13, 362, 398; conversion to coal-biomass 115; dependency 487; economic development/growth 1, 398; environmental problems 1, 9, 12; frontier 2, 7, 316, 427, 453; leakage channel 295; prices 6, 210, 295; subsidies 211, 212, 213, *see also individual types*
Fourth Assessment (IPCC) 134, 135, 154n
fourth generation reactors 99
fracking 7, 113, 316
France: carbon dioxide emissions *16*; economic growth 401; electric mobility 370; energy agencies 162; energy efficiency policy 160, 178, 180; fuel poverty 299; nuclear programmes 98–9, 120n, 320, 327, 347n; pricing pollution 213, *214*, 229, 233n
fraud 250–1, 277n
Frauenhofer Institutes 337
free allocation 226, 230, 241, 251, 252–4, 285, 292, 298
free-riding 32, 173, 177, 195n
fuel cell cars 110, 367
fuel economy 170, *171*
fuel poverty 173, 298, 299, *300*, 301, 307n, 341
fuel price escalator 214, 225, 255
fuel taxes 351n
fuel-specific benchmarking *253*, 277n
fuelling transport 109–10

'game-changing' technologies' 170–2
gas: -to-liquid technologies 113; bills *297*; consumption *3*; energy and emission flows *82*; industry efficiency 135; pipelines 267; power stations 377; prices 7, *8*, 298, 307n, 376; reserves 7, 17, *see also* natural gas; shale gas/oil
gasification 89
gasoline: prices 298; taxes 173, 213–14, 231, 366, 469
general equilibrium 56, 219, 461, 463
General Motors 170, 422
General Theory 57
geo-engineering 32, 39n
geothermal energies 103, 121n
Germany 244; carbon dioxide emissions *16*; consumption subsidies 299; electricity prices 296–7; energy efficiency 132, 301; energy efficiency policy 160, 180; energy intensity 182; halt on CCS projects 98; industrial innovation 337; pricing pollution 213, *214*, 254; renewable energy 339, 374, 377; retreat from nuclear power 98; reunification 396–7
glaciers 10
global carbon price 302–3, 460, 484
global economy 6, 406, 424, 425, 427
Global Energy Assessment (GEA) *107*, 115, *116*, 130, 150–2, 188, 347n
Global Energy Technology Strategy Program 124n
Global Environment Fund 269
global goals 12, 39n
global modelling 130–1, 233n, 357–8, 407
global oil market 4–6
global population 1, 12
globalisation 346, 397, 418, 419, 422, 435n
goal misalignment and innovation 333
goal-oriented approach 70
good practice 87, 88
Goulder, L.H. 408
government: data assessing programmes 184–5; policy and innovation 333, 334; R&D funding 463; role, collective information provision 141
grants 338
Green Deal 196n
Green Growth 72, 419, 424–5, 427–8, 434n, 437n, 462
Green Investment Schemes (GIS) 204, 264
green paradox 295, 307n
green revolution 1, 423
greenhouse gases/emissions 83; stabilisation 35n, 227, *see also* carbon dioxide emissions; HFC-23; methane; nitrogen oxides; water vapour
gross domestic product (GDP) 4, 6, 9, 13, 15, 36n, 217, 358, 398, 400–1, 416–17
gross value-added (GVA) 289
Gulf Wars 6
Guyana 6

habit(s) 50, 55, 69, 108, 132, 147, 152
Haites, E. 262n
Heal, G. 302
heat pumps 89–90
heating 83, 86, 87, 89–90, 132, 186
heavy industry 83, 96, 204, 240, 245, 427
heavy oils 112
HFC-23 269
hidden benefits 63, 66, 139
hidden costs 63, 66, 134, 136, 137–9, 140
hidden subsidies 215
high carbon investment 257
high carbon path 384, 385
high energy prices 208–9, 210–11, 301, 351n
high grade energy 86
high-voltage direct current (HVDC) cables 100
'historic innovation possibility curve' 403
holistic approach: costing climate change 28
Home Energy Efficiency Scheme (HEES UK policy instrument) 173
horizontal industrial policies 343–4
household energy efficiency 173
household energy expenditure 298, *299*
housing 132, *300*
Hungary 264, 298, 299
hybrid cars 108, 110, 368, 370, 386
hybrid instruments 225, 262n, 470–1
hybridisation 312
hydroelectricity *3*, 100, 111, 120n
hydrogen 96, 110, 367–8
hyperbolic discounting 144
hypothecation 469

ignoring risk 49–50
imperfect foresight 418, 425, 434n
imports: into the carbon price regime 293, 294
in-country measurements: price response 208
incidental costs 150
income: energy expenditure 18–19, 298, *299*, 307n
income transfers 300
incremental innovation 331, 405
incubators 335
India: building construction 88; carbon dioxide emissions 15, *16*; carbon-GDP intensity 15; CDM credits, coal plants 269; climate damage 25; deployment programmes, wind power 339; economic growth 436n; energy consumption 15; energy managers 176; energy subsidies 211, 212; eradication of distribution losses 114; macroeconomic costs, carbon pricing 301; Perform, Achieve and Trade (PAT) scheme 272–3, 484; resource availability 111
indifference to risk 49
indirect emissions 88–9
indirect subsidies 211
inductive reasoning 51
industrial ecology 96
industrial renaissance 419–24
Industrial Revolution 1, 10, 111, 398, 405, 419, 421
industrialised countries 1, 2, 4, 9, 15, 89, 107, 112, 152, 180, 185, 186, 211, 213, 214, 265, 302, 360–1
industry: carbon dioxide emissions *91*; carbon pricing 286, 288, 289–92; cleaner production 90–6; energy consumption *91*, 191; energy efficiency 273; energy and emission flows *82*, 83; migration 290, 304; policy/strategy 342–4, 351n, 352n; profit from emissions trading 251–2; research 337; transformation 386
inequality aversion 27
inertia 20, 31, 39n, 49, 55, 61, 69, 433n, 452, 453, 490n, 495–505
information: campaigns 165–6, 176, 179, 191; failure 140–1; flows 148, 349n; as a public good 194n; through price 208
information technology(ies) 88, 320, 420, 422, 424, 435n
infrastructure 458, 461; carbon intensity 359; cleaner and renewable energy 467; emission reduction 67; energy wastage 299; and innovation 346; investment 364, 415, 421–2, 427, 463, 480, 491–2n; pace of development 484; as a policy response 68; transforming systems 363–4, 367, 369, 382, 387
innovation 19, 405, 461; adoption 363; capacity 484; chain 324–40, 403; and change 425; climate change and 476; competition 320; complexity of 344–6; crowding out 408–9; economic growth 62, 64, 312, 401, 402, 403, 419–21; economic modelling 411–14; emission reduction 65, 66–7; energy-related 311; industrial strategy 342–4; iterative path 327; low carbon 295, 311, 317–18, 337, 341, 463–4, 466, 485; macroeconomics 399; microeconomics 315–52; as a policy response 68; price and 210; renewable energy 105; as resource neutral 62; under-investment 311, 409, *see also* institutional innovation; radical innovations; technological innovation
INSEAD 377
institutional alignment 472–4
institutional change 134, 364, 421
institutional economics 54, 57, 58–9
institutional failure 58, 61, 140–1
institutional innovation 59, 62, 312, 467
institutional investment 428, 478, 480, 492n
institutional reform 457
institutional systems 485–6
insulation 64, 87, 132, 135, 173, 301
insurance 51–2, 461–2
integrated assessment models 409–16

integrated energy systems 117, 477
integrated policy 192, 487
Intelligent Energy Executive Agency (IEEA) 164
interconnection 114, 374, 375, 477
Intergovernmental Panel on Climate Change (IPCC) 24, 25, 107n, 134, 135, 152, 154n, 358, 416
'internalising the externality' 9
International Energy Agency (IEA) 113, 115, 135, 152, 162, 184, 334, 358, 471, 474–5
International Institute for Applied Systems Analysis (IIASA) 384, *see also* Global Energy Assessment
International Standards Organisation (ISO) 177
invention 316, 324, 326
investment: capital accumulation 401; in clean energy 356, 357; efficiency 153n; in electricity systems 376–7; fossil fuels 453; infrastructure 364, 415, 421–2, 427, 480, 491–2n; in innovation 333, 338; low carbon 248, 255–7, 259, 292, 312, 313, 358, 428; North Sea Oil 377; nuclear power 372–3; private and social value of 409; renewable energy 358, 490n; research and development 320; and return 26, 50, 73, 149, 333, 334, 462–4, 476; risk reduction 426; risks, in CCS 98; subsidies 173, 177, 184; in technologies and systems 384, 386–7, *see also* strategic investment
iPhone 328
Iran 211
Ireland 166, 178
iron and steel 92–3, 289
irreversibility 20
Italy 173, 178, 213, *214*

Japan 409; cap-and trade 271; carbon dioxide emissions *16*; energy consumption 150; energy efficiency policy 16, 160, 172, 183–4; energy managers 176; gasoline tax *214*; hybrid cars 368; nuclear programmes 98, 320; price elasticity 208, *see also* Tokyo
Joint Implementation (JI) 264, 267
joule 84, 85
Just Transition 424, 436n

Kahneman, D. 48, 55, 143
Keynesian economics 57, 422, 426, 488n
Knightian uncertainty 51
knowledge 351–2n, 360
knowledge divisions 332–3
knowledge-stock approach 434n
Krugman, P. 252
Kuznets curve 14
Kyoto Protocol 16, 22, 33, 204, 227, 230, 237, 241, 255, 302, *see also* Clean Development Mechanism; Joint Implementation

labelling 139, 140, 166–9, 174, 183, 184, 194n
labour productivity 403–4
laissez-faire 331, 343
land use mix 382, 391n
Latin America 17, 25, 167
'league table': of energy performance 183
learning 321–4, 326; by doing 321, 325, 333, 360, 361, 403, 405, 408, 434n; rates 322–4, 348n; vicarious 144, 145
least cost 18, 57, 59, 322, 360, 361, 384, 453
least developed countries 6, 483
least resistance 361, 365
levelling price/cost 205, 293–4, 294–5
light duty vehicles 134–5, 169–70, *171*, 366, *367*
light emitting diode (LED) bulbs 87
lighting 87, 169, 186, 195n
Limits to Growth 406
Linking Directive (ETS) 241, 265
liquid fuels 107, 112, *113*, 114
liquid hydrocarbons 109
lobbying 255, 335, 362
local generation 373
lock in 88, 362, 476
Lomborg, B. 22
loss aversion 143, 147
Lovins, A. 160
low carbon: development 424–5, 427–8; electricity 97–105, 116; finance 478; generation 492n; innovation 295, 311, 317–18, 337, 341, 463–4, 466, 485; investment 248, 255–7, 259, 292, 312, 313, 358, 428; path 384, 385–6; transforming systems 356–91; urbanisation 383
Low Carbon Network Innovation Fund (UK) 374, 389n
Low Carbon Pilot Development Zones (China) 272
low grade heating and cooling 84, 86, 89–90

McKay, D. 112
McKinsey curve 64, 65–6, 133, 134, 148, 154n, 189, 220–2
macroeconomic costs: carbon pricing 219, 220, 301; EU ETS 251
macroeconomics: of energy efficiency 409; of innovation 399; of tax and cap-and-trade responses 226
mainstream economics 22, 50, 55, 57, 318, 329, 330, 402, 424
malleability 58
management studies 56
mandatory audits 141
mandatory labelling 140, 167, 168–9, 184
mandatory reporting 176
marginal cost pricing 285
market accumulation 325, 326
'market' discount rates 27

market engagement: in innovation 335–7
market failure(s) 61, 75n, 136, 139, 215, 318, 331, 343, 351n
market imperfections 339–40, 434n
market liberalisation 181, 320
market niches 312, 338–40, 363
market structures 68
'markets and prices' policy option 68, 203–5, *see also* cap-and-trade; carbon pricing; pricing pollution
materials efficiency 96
'merchants of doubt' 459, 490n
methane 9, 83, 92, 111, 267
Mexico 218
microeconomics 55; of energy efficiency 193; of innovation 315–52
military R&D 408
million tonnes of oil equivalent (Mtoe) 85
minimum price 260
'missing contract' problems 61–2
mitigation: attributable to EU ETS 248
moral hazard 140, 155n
motivation: in economics 312, 471; employee 148
motor systems: drives 118n, 139; electricity use 92
municipal emissions trading scheme (Tokyo) 239
Murphy's Law 246, 259

National Academy (US) 9
National Alcohol Programme 366
National Allocation Plans (NAPs) 241–2, 243–4, 245, 253, 254, 276n
national implementation agencies 135
Nationally Appropriate Mitigation Actions (NAMAs) 383
natural gas: access to 301; cleaner production 92; emissions 83, 97; resource availability 111; subsidies *212*; vehicles 109, 368, *see also* gas
natural systems 10, 12, 32
near-term price 258
negative costs 63, 66, 76n, 133, 134, 135, 154n, 222, 425
negative learning 322, 348n
Nelson, R.R. 360
neoclassical economics 54, 56–8, 59, 61, 156n, 186, 450, 451, *see also* Second Domain
Netherlands 132, 177, 178, 182, 227
New Deal 421–2
New Institutional Economics 58
New Zealand 238, 271, 279n
Newtonian mechanics 129, 487
niche markets 312, 338–40, 363
NIMBYism 103, 373
nitrogen oxides 107
Non-Fossil Fuel Obligation (NNFF) 339, 340
non-market impacts 24, 29, 30
non-optimising behaviours 55
Nordhaus, W.D. 21, 23, 25, 411, 454
North Africa 103, 375, 389n
North Sea Oil 17, 214, 377
Norway 228
nuclear fusion 99
nuclear power 98–9, 378; energy consumption *3*; investment 372–3; radioactive waste 98, 233n; research and development 318, 319, 320
Nudge 55

Obama administration 170, 229, 230, 351n, 365
ocean acidification 12
ocean surface temperatures 10
OECD countries 16, 66, 174, 182, 227, 342, 418, 420, 484
offset credits *see* Clean Development Mechanism
Offshore Wind Accelerator 335, 336
offshore wind power 100
oil: -GDP relationships 34n; consumption *3*, 364; demand 4, 33–4n; dependency 105, 368, 422; drilling 315, 316; energy and emission flows *82*, 83; environmental costs 9; international competitiveness 320; prices 1, 4–6, 52, 61, 99, 105, 130, 131–2, 210, 214, 217, 319, 351n, 364; resource availability 111, 112; as a security issue 52; subsidies *212*; waves of innovation *420*, *see also* North Sea oil; shale gas/oil
oil sands 112
onshore wind power 104, 123n
Opower 145, 146
opportunity: three realms of 62–7
opportunity costs 285, 433n, 434n
optimising *see* Second Domain
optimism bias 249
optimists 453
optimum equilibrium 56
Organisation for Economic Cooperation and Development (OECD) 9, 162, 284, 342, 343, 434n, *see also* OECD countries
Organisation of Petroleum Exporting Countries (OPEC) 4
organisational economics 54, 55–6, 148, 487
organisations: behavioural realities 147–50; energy efficiency 173–8; energy management 137; and technological innovation 344–5
overachievement of emission caps 204
overestimation of costs 226–7
overreaction 69
ownership-based industrial policy 343
oxy-fuel combustion 94

Pareto improving 61
particulates 9, 34n

'passive' heating and cooling 87, 89
patents 326, 341
path dependence 58, 312, 330, 348n, 361, 369, 379, 403, 452, 476
Pathways for 2030 and 2050 (GEA) *116*
payback thresholds 137
perfect foresight 17, 21, 402, 410, 412, 416, 434n, 491n
Perform, Achieve and Trade (PAT) scheme 272–3, 484
Performance of Buildings Directive (EPBD) 180, 182, 196n
personal carbon trading 239, 275n
personal utility 57
petroleum industry 135
pharmaceutical innovation 328–9
phased design: EU ETS 241–7, 248
photo-voltaic (PV) cells 100, 101, 102n, 105, 114, 319, 324, 339, 363
photosynthesis 109
piggybacking 363–4, 366, 367, 368–70
Pigou, A. 215
Pigouvian taxes 215, 226, 343
Pillar I 71, 79–80, 429, 456, 457, 461, 465, 469, 472; energising European recovery 480; energy efficiency policy 160–97; energy and emissions 81–124; energy wastefulness 129–57; First Domain 449; joint gains 474–5; transforming systems 386
Pillar II 71, 203–5, 429, 465–6, 469, 472; cap-and-trade and offsets 237–79; distributional impacts of carbon pricing 284–308; energising European recovery 480–1; investment and return 463; joint gains 475–6, 476–7; pricing pollution 207–34; Second Domain 449; transforming systems 386
Pillar III 71, 311–13, 466–7, 469; dark matter of economic growth 396–438; energising European recovery 482; investment and return 463–4; microeconomics of innovation 315–52; transforming systems 356–91
Pilot Emissions Trading Schemes 272
planetary boundaries 31, 39n
planning: central 396; economic 399; rational 486; strategic 70; urban 144, 382, 390n, 484
plug-in electric hybrids 110, 368
Poland *16*, 178, 247, 255, 375, 387
policy(ies): capital-based 491n; carbon-related 298; clean energy innovation 311; consistency 71; cost reduction 138; government-led 312–13; industrial 342–4, 351n, 352n; inefficiency 71; innovation and public 333, 334; integrated 192, 492; limitations of single pillar 464–5; packages/mix 179, *180*, 227; renewable energy 340; three pillars *see* three pillars of policy; urbanisation 381–2, *see also* climate change policy; energy efficiency policy; environmental policy(ies)
political economy 18, 29, 57, 485
'polluter pays' principle 9, 284, 469
pollution 8–12, 14, 106, 469; pricing *see* pricing pollution
pork-barrel programmes (US) 319, 320
Porter hypothesis 72, 321, 341–2, 350–1n, 399
portfolio standards 339, 340
Portugal 176, 178
position rents 361–2
poverty 1, 130, *see also* fuel poverty
power generation: United Kingdom 296; windfall profits, emissions trading 245
'Power of One' 166
precautionary principle 28, 455, 459
prediction 264
preferences: adaptive 489n; business/companies 230, 361; consumer 361, 422; economic 489n; individual 55, 57, 106, 361, 370, 381, 458, 465, 467; political 104; social 38n; stable 56, 57–8
preferential loans 178
prescriptive approach to discounting 26, 33, 38n
price(s): ceilings 225, 260, 278n; collapse 249, 319, 478; corridor 260; elasticity 208–11, 232n; energy systems adjustment to 61; floors 225, 260, 262, 278n, 427, 471; general equilibrium 56; management 278n; as a policy response 68; resource depletion and 1, *see also* carbon price(s); energy price(s)
pricing pollution 207–34, 464; 'co-benefits' and 'double dividend' 217–20, 475–6; emission reduction 18; flexibility 215–16; gasoline taxes 213–14; politics of 229–31; in practice 226–9; principles 215–17; reforming energy subsidies 211–13; response options in the Second Domain 220–3; tax versus cap-and-trade 223–6, *see also* cap-and-trade; carbon pricing
principal-agent 56, 61, 140, 148, 149, 155n, 176
private sector: funding 157n, 345; initiatives 278n; innovation 347n, *see also* pubic-private partnerships
The Prize 315, 364
process emissions 90
procrastination 144
product efficiency 165–73, 191
product quality/variety 329, 330, 404, 431n
production subsidies 211, 213
productivity 21, 401, 402, 403–4, 423, 457
profit and loss: emissions trading 204–5, 230, 241, 245, 251–2, 253, 284–8
Programmes of Action (CDM) 269, 383
propagation effect 233n
proper pricing 205, 210, 237, 449, 466
psychological distance 49, 74n
psychology 29, 32, 56, 61, 455, 485
public sector: capital constraints 177; contractual constraints 149–50; funding 319–20, 345, 426; policies, and innovation 333, 334

public-private partnerships 335, 349n, 350n
pulp and paper 96, 289
'pure rate of time preference' 26
'push and pull' forces: in innovation 311, 324, *325*, 326, 328, 329, 331, 332–3, 334–80, 403; urban planning 382
pyrolysis 89

quantity triggers 260
quantity-based approach: carbon pricing 256, 258
quantum physics 129, 487
Quebec 271
quotas 215

radical innovations 92, 95, 96, 320, 330–1, 335
radioactive decay 111
radioactive waste 98, 233n
radiosonde 10, 35n
rail transport 105, 106, 110, *420*
Ramsey, F.P. 402
Ramsey Rule 26, 27, 38n
'rare earth' materials 103
ratchet effect 209
rational agent assumption 56, 57
rational choice 143, 477
rational expectations 17, 56
rational foresight 56, 57–8, 449
'rational trade-off' approach 28
rationality 50, 55, 143, 156n, 360, 486
reactor designs 99
rebound 66, 108, 130, 134, 153n, 186–8, 191, 192, 193, 194n, 295, 461, *see also* respending effects
RECIPE studies 411–13, 416, 434n
Reduce, Reuse and Recycle 96
reductionist studies 22, 23
refrigeration 155n, 167, 169, *170*, 179, 268
refurbishment standards 180
Regional Development Agencies 166
Regional Greenhouse Gas Initiative (RGGI) 238, 271
regulation: energy markets 68, 179, 372; and innovation 346, *see also* environmental regulation; standards
relative significance 156n
remote risks: lack of attention towards 49
renewable energy: growth 115; high costs and timescales 333; India 273; induced innovation 411; investment 358, 490n; policies 340, 462; projects 266–7; reduction in wholesale power price 296; resources 99–103, 111, 373; subsidies 212; support 334, 339; technologies 356–7; transforming systems 373–8, *see also* solar power; wind power
renewable obligations 339, 340
representative agents 56, 70, 402, 467
research and development 177, 318–21, 324, 329, 331, 333, 408, 463
resource allocation 59
resource curse 6–7
resource depletion 1, 217
resource economics 57
resource neutral: innovation as 62
resource trade-offs 59–61
resources 110–13; renewable energy 99–103, 373
Resources Futures 452
respending effects 187, *see also* rebound
response: three fields of theory 53–9
retrofits 87, 88, 140, 189–90, 301
revenue(s): earmarking 264, 469–70, 477; leveraging of capital 478; recycling 183, 219, 220, 229, 413, 415
right to pollute 216
right to refute 294
'rising block' tariffs 299, 304
risk(s) 25–6, 256, 337–8, 386
risk aversion 55, 149, 152, 258, 409, 427, 489n
risk conceptions 29, 47–53, 55, 70, 74n, 461
risk matrix *23*, 26, 28, 31, 52
risk-free market interest rates 26–7, 38n
Riso test laboratory 335–6
road transport 105, 382
Romer, P. 402, 408–9
rotary kilns 93
Royal society 32, 485
rules of thumb 142, 147, 177, 294
'run-of-the-river' plants 100
Russia 409; carbon dioxide controls 418; carbon dioxide emissions 15, *16*; consequences of simplistic economic advice 396; emissions collapse 264; emissions savings 267; energy efficiency 130, 404; energy inefficiency 131; energy subsidies 211, 213; price elasticity 208

safety codes 179
safety thresholds 53
'Sahara forest' project 115
Sarkozy, N. 229
satisficing: theory 143, *see also* First Domain
Saudi Arabia 4, 211
SAVE programme 161
savings glut 426
Scandinavia 132, *see also* Denmark; Finland; Norway; Sweden
Schneider, S.H. 408
Schumpeter, J. 316, 419
Second Assessment (IPCC) 24
Second Domain 70, 134, 418, 447, 450; economic performance 397; energy efficiency 193; energy system 83; ethanol use 366; European electricity system 376; firms/organisations 147; least cost equilibrium 360; low carbon electricity 105; modelling 461; path dependence 361; pricing pollution 217,

220–3, 239, 249, 266, 274, 449, 460; rational foresight 449; risk conceptions 50; timescales 73, *see also* neoclassical economics
second generation biofuels 109, 113
sectoral agreements: emissions trading 293
sectoral R&D expenditure 320, *321*
secure/transform 70
security: risk perceptions 51–3, *see also* energy security
selection processes 360
selective industrial policies 344
self-interested individuals 61
separability 477
service efficiency 84, 86
shale gas/oil 7, 97, 112, 426, 427
shipping 105, 107
short-termism 7, 26, 137, 152, 487
Singapore 36n, 176, 272
smart building design 87–8
smart meters 114, 336, 350n
smarter choices 64, 67, 68, 448, 456
smarter systems 114–15
Smith, V. 143
'social cost of carbon' 21, *23*, 27, 28, 70, 71, 427
social norms 50, 142, 144, 146n, 245, 467
solar cells 101, 312, 363
solar power 89, 100–3, 121n, 375, *see also* photo-voltaic (PV) cells
solar radiation 10, 99, 103, 111, 120n
Solow residual 64, 131, 400–2, 404, 405, 451
Solow, R.M. 399, 405, 406
South Africa 296, 492n
South Korea 36n, 99, 167, 266, 272, 334, 426
Soviet Union 15, 212–13, 396, 404, 420
space cooling/heating 89, *188*
Spain 103, 173, *214*, 264, 339, 379
speculation 17–18
spillovers 68, 318, 399, 408–9, 463
split incentives 139–40, 141, 148
Sports Utility Vehicles (SUVs) 170
Standard & Poor 257
standards 138, 169–72, 174, 176, 177, 179, 180, 183–4, 190, 194n, 346, 381, 390n
stationary energy systems 83
status quo 69
Stavins, R. 21, 29, 230
steam energy 111
steel (iron and) 88, 90, 92–3, 289, 290, 306n, 422
Stern, N. 12, 25, 447
A Stern Warning about Time 12, 25–9
stickiness: of economic features 57
stimulus packages 164n, 170, 178, 228, 319, 356, 357, 358, 426
storage (energy) 114, 373, 375–6
strategic deployment 338–40
strategic investment 68, 384, 413, 450, 457, 479, *see also* Pillar III
strategic judgement 71
strategic planning 70
strategic risks 51, 70, 452
strategy 51–3
stratosphere 10
Strbac, G. 375
strike prices 105
structural adjustment/change 58, 131, 261–2, 482
sub-Saharan Africa 22, 266, 270, 301
substitution(s) 64–5, 67, 83, 220, 232n, 362
sulphur dioxide 9, 34n, 226, 240
supplier obligations 178, 181, 196n
supply chains 140, 325, 382; efficiency 96, *97*
supply and demand 18, 249–50, 270, 360, 403, 489n
sustainability 52, 473
sustainable development: CDM and 267–8
Sustainable Energy - Without the Hot Air 123n
Sweden 89, 177, 180, 182, 228
system balancing 374
System Benefit Charge (US) 181
System of Environmental and Economic Accounts (SEEA) 52
Système Internationale (SI) 85
systemic technologies 117
systems-based industrial policy 343

tacit knowledge 360
Taleb, N. 51
tariff structures 299–300, 304, 339, 357
tax: benefits 338; credits 172; deductions 173, 184; incentives 177, 178, 350n, 366; perception of 230; thresholds 230, 258, *see also individual types*
technical change 61, 401, 402, 403, 405, 408
technological innovation 59, 62, 185, 295, 312, *see also* low carbon, innovation
technology(ies) 458; accelerators 336–7; diffusion 19, 295, 324, 325, 326, 327, 331, 461; electricity storage 375–6; endogenous growth 403; energy using and energy production *323*; inferior 135; investment 384, 386–7; leakage 295; and organisational unit 344–5; price and energy-efficient 209; and rebound 187
technology: transfer 335, 432n
technology(ies): *see also* information technology(ies)
technology valley of death 320, 326, 329–30, 331–4, 357, 450
technology-specific learning 321–4, 326
technology-specific research and development 324

temperature trends 10–11
tenant-landlord (and other) divides 139–40, 141, 180, 182
theory: three fields of 53–9, *72*
thermal energy: solar 89
thermal performance 134, 150, 179, 189
thermal units 85
thin-film modules 101, 102n
Thinking, Fast and Slow 48, 55
Third Domain 52, 70, 418; energy system 83; EU development 58; industrial policy 343; infrastructure dimension 467; innovation 62, 312, 329, 457; institutions 472; investment 377, 384; low carbon electricity 105; markets and pricing 470, 471; models and mindsets 461; Pillar III policies 450; pricing pollution 217, 240, 249; timescales 73; urbanisation 380, 381
Third Energy Package 376
third generation: biofuels 113; liquid fuels 109; reactors 99
Thirty Glorious Years 420, 421, 422, 423
thorium 99, 111, 120n
three domains (decision-making) 46–74; alignment 68–72; broadening of horizons 468–74; in economic development 483–5, *see also* First Domain; Second Domain; Third Domain
three pillars of policy 59; domain alignments 68–72; implications 486–7; integrating 464–8; response approaches 67–8, *69*, *see also* Pillar I; Pillar II; Pillar III
tidal energy 103, 111, 121n
time: inconsistency in treatment of 144
time consistency 57–8
timescales 26, 32, 61, 73, 88, 217, 328, 329, 330, 333, 408
Tokyo 239, 271, 381
'Top Runner' programme 172, 183–4
Toyota Prius hybrid 108, 110, 368
trade-offs: costing climate change 28; investment costs and energy savings 139; resource 59–61; savings and consumption 405; in urbanisation 380
transaction costs 137, 141, 148, 155n, 165, 169, 173, 269, 338
transformation: economic prescriptions 396–7
transforming systems 114, 117, 356–91, *see also* Third Domain
transitions: infrastructure 364; integration 383–4; low carbon 428; technological 312, 325, 333, 345, 363
transmission networks/systems 114, 374
transport: electricity and 115; energy consumption *106*, 191, 364, 390n; energy and emission flows *82*, 83; transforming 105–10, 114, 117, 364–72, 382; vehicle efficiency 108–9, 134–5, 165–6, 170
travel choices 106–8
troposphere 10

Ukraine 264, 267
unacceptable outcomes 51–2
uncertainty 17–18, 28, 57; behavioural factors 144, 147; carbon price 98, 256–7; climate change 26; costs of emission cut-backs 223; modelling impact of 490n; potential coal demand 295; risk conception 51; technology push 332–3
unconventional oils 112
underlying energy demand trend (UEDT) 232n
Unfold the Future – 2050 Roadmap to a Low-Carbon Bioeconomy 96
uniform benchmarking *253*, 254
United Kingdom: carbon dioxide emissions *16*; climate change legislation 472, 473; downgrading of credit rating 257; electricity regulation 372; energy agencies 163–4; energy bills 297, *297*; energy efficiency 132, 137, 138, 140, 160, 301; energy efficiency policy 173, 178, 179, 181, 182, 183; energy expenditure 298, *299*; fuel poverty 298, 300; institutional tensions within government 474; North Sea Oil 17, 377; nuclear power 320; power generation 296; price volatility 6; pricing pollution *214*, 225, 255; renewable energy 105, 334, 339, 340, 374; resource availability 111; Winter Fuel Payment 300
United Nations Framework Conventions on Climate Change (UNFCCC) 267, 269, 270, 275n
United States 4; air pollution control 9; biofuels 109; carbon dioxide emissions *16*; energy consumption 85, 131, 150, 388n; energy efficiency 131, 139, 148, 160; energy efficiency policy 166–7, 169–72, 173, 180, 181, 182; ESCOs 141; income transfers 300; low carbon electricity 100, 103; oil dependency 105, 422; price elasticity 208; price volatility 6; pricing pollution *214*, 226, 228, 229, 230, 237, 238, 239, 260, 271, 274–5; R&D funding 319; transforming transport 364–70; withdrawal from Kyoto Protocol 264, *see also* Atlanta; California
upfront costs 99, 137, 139, 172, 173, 382
upstream pricing 239
uranium 111, 316
urban: air pollution 9; energy demand 112; heat islands 88; planning 144, 382, 390n, 484; population 378, 390n
urbanisation 378–83, 391n
utilitarianism 57
Uzawa, H. 403–4

Validation and Verification Manual (CDM) 268
value at stake 289, 290
value of statistical life 24, 25, 37n
value of time 25, 26
value-added 398, 400
Vartiainen, H. 55
vehicle efficiency 108–9, 134–5, 170, 365–6
Vehicle Excise Duty (VED) 173
Venezuela 112, 211, 365, 385
vicarious learning 144, 145
visual and non-visual properties 140
volatility: energy prices 6–7, 17–18; global energy system 453; market 218
volume-based adjustment rules 261
voluntary labelling 167
voluntary reporting 176–7

waste energy 90, 92
waste management 96
water: energy and 7
water vapour 10, 35n, 107, 226
wave energy 111
Waxman-Markey Bill 237, 239, 298
wealth: carbon dioxide emissions *16*
Weitzman's Dismal Theorem 22, 27–9, 30, 46, 223–5
welfare 12, 27, 129, 187, 307–8n
welfare economics 54, 56–8, 61, 437–8n, 489n
Western Climate Initiative 229, 271
When Starting with the Most Expensive Option Makes Sense 222
'whole buildings' efficiency 134
willingness to pay/accept 24–5, 70
wind power 100, 104, 111, 120n, 122n, 319–20, 335–6, 339, 374, 375
winner's curse 339, 340, 350n
Winter Fuel payment 300
Winter, S.G. 360
wood 1, 9, 92, 93, 103
World Bank 34n, 58, 182, 222
World Energy Council 174
World Energy Outlook 162, 358
World Trade Organization (WTO) 294, 295, 306n

X-events 51

Yergin, D. 315, 364

zero-carbon 89, 113, 121n, 179
zero-growth steady state 401
zero-interest loans facility 178